9th Edoardo Amaldi Conference on Gravitational Waves (Amaldi 9) and the 2011 Numerical Relativity – Data Analysis Meeting

(NRDA 2011)

Journal of Physics: Conference Series Volume 363

Cardiff, United Kingdom
10-15 July 2011

ISBN: 978-1-62276-780-9
ISSN: 1742-6588

Printed from e-media with permission by:

Curran Associates, Inc.
57 Morehouse Lane
Red Hook, NY 12571

Printed by Curran Associates, Inc. (2013)

For permission requests, please contact the Institute of Physics
at the address below.

Institute of Physics
Dirac House, Temple Back
Bristol BS1 6BE UK

Phone: 44 1 17 929 7481
Fax: 44 1 17 920 0979

techtracking@iop.org

Additional copies of this publication are available from:

Curran Associates, Inc.
57 Morehouse Lane
Red Hook, NY 12571 USA
Phone: 845-758-0400
Fax: 845-758-2634
Email: curran@proceedings.com
Web: www.proceedings.com

Papers

Preface

The Ninth Edoardo Amaldi conference on gravitational waves (Amaldi 9) and the 2011 Numerical Relativity – Data Analysis meeting (NRDA 2011) were held on July 10–15, 2011 in Cardiff, UK.

The summer of 2011 marked the beginning of a crucial time for the field of gravitational-wave astronomy. After the successful completion of long-duration data taking, the initial LIGO and Virgo detectors were shut down and the era of first-generation laser interferometric gravitational-wave detectors came to an end. With the analysis of the last set of first-generation data approaching completion, the field now faces the challenge of preparing for the era of second-generation detectors, and, hopefully, the regular detection of gravitational waves.

The first of the advanced detectors should begin operation in 2014, but the intervening years are far from a time of sitting and waiting. This will be a hectic time for experimenters, who have to fight the limits of current technology to make their theoretical noise curves a reality. During Amaldi 9 we were led through the upgrades to advanced detectors, and the challenges that the experimenters face. We also heard about proposals for going yet further with third-generation and space-based detectors.

This is also a crucial time for theoretical work. With the increased sensitivity of advanced detectors, we hope to not only make the first detections of gravitational waves, but to learn about their sources, and interpret what this means for astrophysics. We need more complete source modelling, more sophisticated and efficient search pipelines and parameter estimation tools, as well as a greater understanding of what we can learn about the universe. Amaldi 9 included a number of talks on the status of these efforts, and the open questions that will be the focus in the coming years.

Sunday July 10th was devoted entirely to NRDA 2011. In recent years the NRDA meetings have brought together numerical relativists who model compact-binary sources, and gravitational-wave data analysts, who want to use the numerical results to aid gravitational-wave detection and parameter estimation. This work has coalesced around the Numerical INJection Analysis (NINJA) project, which is now in its second incarnation, and will be used to test and refine data-analysis procedures on black-hole binary waveforms injected into real LIGO–Virgo data. This NRDA meeting focused on preparing the set of numerical waveforms, and paving the way for data analysis projects.

The rest of the week consisted of Amaldi and NRDA-specific sessions, which contained talks on the most pressing issues facing gravitational-wave science. The participants enjoyed a successful meeting that included plenty of time scheduled for informal discussions, and popular poster sessions. Most evenings featured public lectures by prominent Cardiff academics following wine and cheese receptions. The Wednesday afternoon social excursions saw participants exploring the fairy-tale Castle Coch, the imposing Caerphilly Castle, and downtown Cardiff itself. The conference banquet on Thursday evening was hosted in the National Gallery, which included viewing of works by Renoir, Monet, and Cezanne in the Gallery's Impressionist wing.

The organization of these meetings was overseen by the Gravitational Wave International Committee (GWIC), the NRDA Scientific Organizing Committee, and the Amaldi Local Organizing Committee. The organisers wish to extend special thanks for the generous support of Cardiff University, in particular to

Samantha Emmott, Cardiff & Co., and the students and staff of the Cardiff University School of Physics and Astronomy, who bore the brunt of the effort at ground level. We thank IUPAP for financial support. We also wish to thank Adam Day and Ben Sheard of IOP for their assistance in the preparation of this Special Issue. Finally, to all the participants, we say thank you for making these meetings a success.

These proceedings represent a small fraction of the science presented, discussed, and imagined at Amaldi 9 / NRDA 2011. Additional proceedings are published in the accompanying edition of *Classical and Quantum Gravity*. It is our hope that these articles will be a resource for the field for years to come.

The Editors

Amaldi 9 and NRDA 2011 IOP Publishing
Journal of Physics: Conference Series **363** (2012) 012001 doi:10.1088/1742-6596/363/1/012001

Low Frequency - High Sensitivity Horizontal Inertial Sensor based on Folded Pendulum

Fausto Acernese[1,2], Rosario De Rosa[3,2], Gerardo Giordano[1,2], Rocco Romano[1,2], Silvia Vilasi[1,2], Fabrizio Barone[1,2]

[1] Dipartimento di Scienze Farmaceutiche e Biomediche, Università degli Studi di Salerno, Via Ponte Don Melillo, I-84084, Fisciano (SA), Italia

[2] INFN sez. di Napoli, Complesso Universitario di Monte S.Angelo, Via Cintia, I-80126 Napoli, Italia

[3] Dipartimento di Scienze Fisiche, Università degli Studi di Napoli Federico II, Via Cintia, I-80126 Napoli, Italia

E-mail: `fbarone@unisa.it`

Abstract. This paper describes a new implementation of monolithic horizontal sensor, developed at the University of Salerno, based on the Folded Pendulum architecture, configurable both as seismometer and as accelerometer. The large low-frequency band ($10^{-6} \div 10\,Hz$), the high sensitivity ($10^{-12} m\sqrt{Hz}$ in the band $0.1 \div 10\,Hz$) and the high quality factor in air ($Q > 1500$) are largely better than all the previous Folded Pendulum implementations. Moreover its monolithic implementation of the whole mechanics, coupled with a full tunability of its resonance frequency ($70\,mHz \div 1.2\,Hz$) obtained with a specially designed calibration procedure and with an integrated laser optical readout, guarantees both compactness, robustness and immunity to environmental noises. This makes this sensor suitable for a large number of scientific applications, also in high vacuum and cryogeny. Applications of this sensor are already started in the field of geophysics, including the study of seismic and newtonian noise for characterization of suitable sites for future underground interferometric detectors of gravitational waves.

1. Introduction

The present generation of interferometric detectors of gravitational waves, GEO600 [1], LIGO [2], TAMA [3] and VIRGO [4], although very close to their design sensitivity, have not yet detected gravitational waves. For this reason a second generation of detectors, (*Advanced LIGO* [5] [6] and *Advanced Virgo* [7]) have been designed to improve the global sensitivity of at least one order of magnitude, so that, according to the current models of GW sources, they should be sensitive enough for the first detection of gravitational waves. Different is the already outlined scientific approach to the third generation of interferometric detectors, whose design is more oriented to the observational aspects than to the detection capability [8] [9]. Within this context, the reduction of the fundamental and technical noises limiting the first and second generation of interferometric detectors assumes great relevance.

In the low frequency band ($0.1\,Hz \div 10\,Hz$), this goal can be achieved only if seismic and newtonian noises are largely reduced. Seismic noise reduction requires the installation of detectors in carefully chosen underground sites and the improvement, at the same time, of the mechanical suspensions performances. Newtonian noise reduction, still an open theoretical and experimental problem [10], requires instead a suitable design of the detector geometry in connection with the structure of the hosting underground site.

Therefore, on the basis of what said above, the general scientific interest in the development of new seismometers and accelerometers, with architectures that guarantee very good sensitivities, especially in the low frequency band ($1\,mHz \div 10\,Hz$), coupled to large immunity to environmental noises, becomes clear. On the other hand, the development of low frequency large-band high-sensitivity seismic sensors has been quite slow in the past decades. Only few large bans instruments are available today (e.g. STS-2 [11], Trillium-240 [12]).

Many architectures exist to build seismic sensors (seismometers, velocimeters, accelerometers), but only few of them are suitable for the design and implementation of large-band (ideally $0 \div 100\,Hz$) high-sensitivity seismic sensors. Among all the possible architectures, available in literature and/or currently used in experiments of physics or in commercial instruments, we chose the Folded Pendulum (hereafter FP) one. The FP, called also *Watts-linkage*, is a classical suspension system developed in 1962 [13], but recently rediscovered for application in gravitational wave research as ultra-low frequency mechanical suspension for vibration isolation in interferometric detectors of gravitational waves [14]. More recently single-axis monolithic FP accelerometers have been developed as sensors for the control system of advanced seismic attenuators [18], demonstrating the feasibility of compact and scalable FP suitable for application as sensors in the control of the mechanical suspensions of interferometric detectors of gravitational waves.

Moving from these experimental results, some years ago we started a research line aimed to the development of low frequency large-band high-sensitivity sensors for geophysical applications (exploration of the low frequency band of the seismic spectrum), as a stand-alone sensors or as part of large and geographically distributed seismic networks, specializing some of them also for applications of interest in the field of gravitational waves detection [15] [16] [17].

In the following sections we will describe the completely new architecture of the horizontal monolithic FP sensor developed at the University of Salerno, that removes all the configuration asymmetries limiting its performances both as seismometer and as accelerometer. We, then, compare the theoretical performances, as predicted by the theoretical/numerical models, and the experimental measurements and sensitivities in connection with different optical configurations of the integrated readout. Finally, we show the first experimental results obtained in its application at the INFN Gran Sasso National Laboratory, where some prototypes are operational since December 2010, discussing further developments and improvements, including the scheduled first tests as sensor on the top stage of a TAMA like mechanical suspension.

2. FP Theoretical Model

The simplified Lagrangian model developed by J.Liu et al. [14], is very useful to describe the basic FP dynamics, main characteristics and expected performances. On the other hand, the design and implementations of optimized FP sensors require a more detailed analysis of the FP dynamics, that can be obtained only with simulations based on very accurate numerical models, like the one we developed for this specific task [15].

The FP mechanical scheme, shown in Figure 1, consists of two vertical beams of equal length, l, a pendulum of mass m_{p_1} and an inverted pendulum of mass m_{p_2}, concentrated in their centers of mass in P_1 and P_2, respectively, at $l_b = l/2$. The central mass, m_c, is modeled, instead, with two equivalent masses, m_{c_1} and m_{c_2} ($m_c = m_{c_1} + m_{c_2}$), concentrated in the pivot points C_1 and C_2, respectively at the same distance, l_p, measured with respect to the pivot points of the pendulum and of the inverted pendulum arms. The distance between the pivot points C_1 and C_2 is fixed and equal to l_d. All these hypotheses, are well satisfied in all our mechanical implementations of monolithic FP sensors. Then, for small deflections angles, θ, the

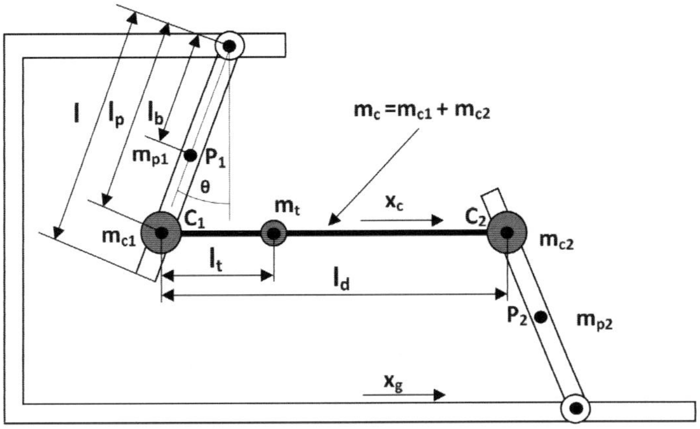

Figure 1. Folded Pendulum Mechanical Scheme

FP resonance frequency, f_o, is

$$f_o = \frac{\omega_o}{2\pi} = \frac{1}{2\pi}\sqrt{\frac{K_{g_{eq}} + K_{e_{eq}}}{M_{eq}}} = \frac{1}{2\pi}\sqrt{\frac{K_{eq}}{M_{eq}}} \tag{1}$$

where $K_{g_{eq}}$, the equivalent gravitational linear stiffness constant, and $K_{e_{eq}}$, the equivalent elastic constant, are defined as

$$K_{g_{eq}} = (m_{p_1} - m_{p_2})\frac{gl}{2l_p^2} + (m_{c_1} - m_{c_2})\frac{g}{l_p} \qquad\qquad K_{e_{eq}} = \frac{k_\theta}{l_p^2} \tag{2}$$

and M_{eq}, the equivalent mass, as

$$M_{eq} = (m_{p_1} + m_{p_2})\frac{l^2}{3l_p^2} + (m_{c_1} + m_{c_2}) \tag{3}$$

Equation 1 is the classic expression of the resonance frequency of a spring-mass oscillator with an equivalent elastic constant K_{eq} and mass M_{eq}. This equation well describes f_o also for the the tuning procedure, performed with a tuning mass, m_t, as described in [15] [17]. It is here only worth underlining the importance of the FP tuning sensitivity for a comfortable and stable tuning, obtained deriving Equation 1 with respect to the position of the tuning mass, that is [15]

$$S_{f_o} = \frac{df_o}{dl_t} = \frac{g}{2\pi l_p l_d}\frac{m_t}{\sqrt{M_{eq}(m_t)K_{eq}}} \tag{4}$$

As expected, Equation 4 demonstrates that the FP sensitivity is function of the value of the tuning mass, m_t, and that to obtain the same Δf_o, the heavier is the tuning mass, the smaller is its necessary displacement.

Defining, then, the coordinate of the FP frame (fixed to the ground) as x_g and the coordinate of the FP central mass (m_c) as x_c, then the mass displacement transfer function with respect to the ground displacement is [15]

$$\frac{x_c(\omega) - x_g(\omega)}{x_g(\omega)} = \frac{(1 - A_c)\omega^2}{\omega_o^2 - \omega^2} \tag{5}$$

where

$$A_c = \frac{\left(\frac{l}{3l_p} - \frac{1}{2}\right)(m_{p_1} - m_{p_2})}{M_{eq}} \tag{6}$$

is the parameter related to the centre of percussion effects [14]. This transfer function describes the dynamics of a FP seismometer, that is the ground displacement, $x_g(\omega)$, as function of the measured mass displacement, $x_c(\omega) - x_g(\omega)$.

Actually, this transfer function only qualitatively predicts the dynamical behavior of FP sensors, because in the Liu et al. model [14] the dissipation effects have been neglected, that are real limitations to the FP performances and are measurable. To globally solve this problem, we introduced a global dissipation term in the lagrangian model, that led us to rewrite Equation 5 in the Laplace domain as

$$\frac{x_c(s) - x_g(s)}{x_g(s)} = \frac{-(1 - A_c)s^2}{s^2 + \frac{\omega_o}{Q(\omega_o)}s + \omega_o^2} \tag{7}$$

where $Q(\omega_o)$ is the global Quality Factor. The dependence of Q on the FP resonance angular frequency, ω_o, that we introduced in Equation 7, has been experimentally demonstrated and will be discussed in the following section. This Equation can then be rewritten in the Fourier domain as

$$\frac{x_c(\omega) - x_g(\omega)}{x_g(\omega)} = \frac{(1 - A_c)\omega^2}{-\omega^2 + i\frac{\omega_o}{Q(\omega_o)}\omega + \omega_o^2} \tag{8}$$

3. The UNISA Horizontal Seismometer

The UNISA horizontal seismic sensor is a new FP implementation, developed starting from the FP original architecture [13] and from the monolithic design by Bertolini et al. [18], that allows very compact and robust FP implementations. The main limitation of the Bertolini implementation is the very small dynamics of the arms (few hundreds of microns), consequence of the use of circular hinges, that allow only small rotational angles of the joints. Therefore, the force feed-back configuration (accelerometer) was an obliged choice, with the consequent limitations in band and sensitivity due to the control electronic noise. We solved this problem enlarging the gaps among the arms, the central mass and the frame, increasing the sensor dynamics to few mm, but, as drawback, we are obliged to use elliptically shaped flexures, as suggested in [18] and experimentally tested in [15] [19], that have the advantage both of a stiffness scaling down as $1/e$ (e is the ratio of the major to minor axes) [20], of a reduced stress and of a strain distributed over a larger length [21].

It is important to underline that, although the instantaneous rotation centre is better defined with circular hinges (compared to the elliptical ones), a suitable choice of the hinge ellipticity factor in connection with the mechanical dynamics minimizes all the coupling effects. As result, we designed and implemented very compact broadband single-axis monolithic FP horizontal sensors of reasonable size with very low natural resonance frequencies (down to $\approx 70\,mHz$) to be used both as seismometers and accelerometers (with or without external force feed-back control systems). Furthermore the innovative application of laser optics techniques for the implementation of the FP monolithic readout (laser optical levers and laser interferometers), has improved its sensitivity, especially in the low frequency band, increasing, at the same time, its immunity to environmental noises [15] [19]. It is important to underline here that the sensitivity of the interferometric optical readout depends on many parameters, like the optical configuration, the quality of the laser, the open loop error signal extraction techniques used, etc.. Many configurations and techniques exist in literature suitable for this purpose (e.g. [22] [23] [24]), and applied to seismic sensors since long time (e.g. [25] [26] [27]), so that any improvement of the optical readout may be simply a matter of suitable choice, and does nor require special dedicated studies.

Actually, the main problem that prevented the use of monolithic FP as open loop sensors, is geometric. In fact, all the previous versions of FP were designed with the constraint that the four joints, necessary for the mechanical implementation, work in tension. This technical choice led to the implementation of FP with asymmetric arms. In particular, the moment of inertia of the inverted pendulum arm cannot be minimized, becoming a real problem for the quality of the FP dynamics. These asymmetries become very relevant in the open loop configuration FP sensor. In fact, being the rotation center of the elliptic hinges time and position dependent, the coupling of the inverted pendulum arm motion with the central mass motion largely increases the couplings of different degrees of freedom, reducing the quality factor and increasing the noise on the horizontal axis.

This problem was solved in the UNISA Horizontal Seismometer [28], characterized by a symmetric FP configuration, that allows the optimization of the moments of inertia of both the arms according to the specific application. In this new design, the two joints of the inverted pendulum work in compression. In order to compare the results obtained with the UNISA Horizontal Seismometer and the previous monolithic FP versions we implemented this new Seismometer using the same material (Aluminium Alloy 7075-T6), the same dimensions ($134\,mm \times 134\,mm \times 40\,mm$) and the same ellipticity (16/5) and thickness ($100\,\mu m$) of the hinges of the previous versions. In Figure 2 the UNISA Horizontal Seismometer is shown together with some particulars of the hinges, the box used for thermo-insulation and the tuning mass, necessary for changing its resonance frequency. It is clear from what said above that the

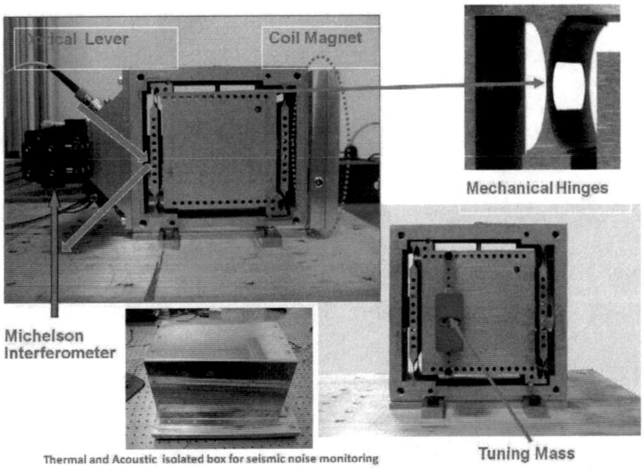

Figure 2. UNISA Horizontal Seismometer

application of the sensor as seismometer (no force feed-back control) has the great advantage that no limitations to the band and sensitivity are introduced by the control electronics, so that the quality of the instrument depends mainly on a careful and optimized mechanical design. The real limitations to the performances of a mechanical monolithic sensor become then the thermal noise, the sensitivity to the external temperature and acoustic noise (in air) and the readout sensitivity. Being the latter actually a laser optical readout (optical lever and laser interferometer) [15] [16] [17], its quality is a problem of cost and portability of the sensor. In particular the monolithic horizontal FP sensor can be used both as stand alone seismometer/accelerometer for seismic noise monitoring for geophysics environment and as seismometer/accelerometer for the automatic control of suspensions of the present and future generations of interferometric detectors of gravitational waves.

Of course, although the new FP configuration presented here solves many technical and sensitivity problems, the problem that the FP configuration couples horizontal forces and frame tilts is still an open problem, only partly solved, for example, if they are separate in band or if the FP is coupled with a tiltmeter of comparable sensivity [29] [30] [31] [32].

4. Tests of the UNISA Horizontal Seismometer

In order to understand performances of the UNISA Horizontal Seismometer we performed a series of tests aimed both to demonstrate that the prototype follows the predictions of the theoretical/numerical models developed for simulation and design. Their reliability was proved by comparing the experimental monolithic FP transfer function at its natural design resonance frequency [15], made using a standard measurement procedure used in control theory to obtain the transfer function of a linear system injecting white noise. The results of this first test are shown in Figure 3 where the theoretical and experimental transfer functions are shown. Another

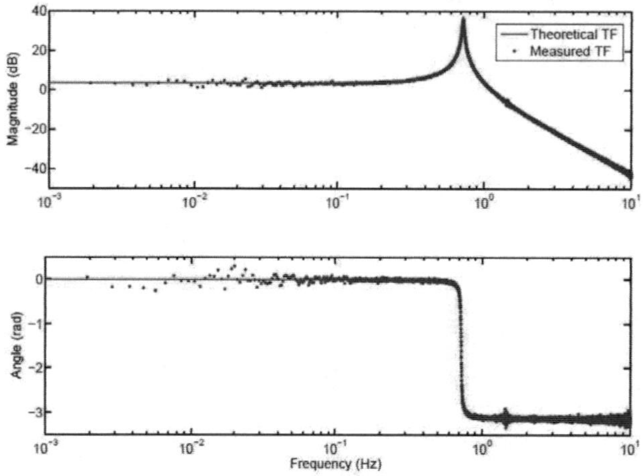

Figure 3. FP Horizontal Seismometer measured Transfer Function

important feature of the UNISA sensor is the very effective tuning procedure developed: a resonance frequency of $66\,mHz$ has been obtained. This result is still more relevant if the small dimensions of the monolithic FP are taken into account. It is anyway important to underline that tuning the FP at its lowest possible natural resonance frequency improves the sensor measurement band at low frequencies, but at the same time reduces the restoring force of the pendulum to external perturbations, increasing the probability for the test mass to touch the frame, saturating the sensor output. Although this may be again only a problem of dynamics for the UNISA Horizontal Seismometer (that can be partially solved enlarging the gaps among the central mass-arms and arms-frame), it is not at all a problem if it is configured as accelerometer, being the central mass always forced in its rest position by the force feed-back control.

The real problem is that the FP quality factor, Q, decreases together with its natural frequency, so that the FP performances decrease moving its resonance frequency towards the low frequencies region. This effect is fully taken into account in Equation 8, where the theoretical prediction and/or the experimental measurements of the function $Q = Q(\omega_o)$ of the mechanical system becomes relevant. We remind that the function $Q = Q(\omega_o)$ depends also on the value of the tuning mass, m_t. This dependence is function of the ratio m_t/m_c: the larger is this ratio, the larger is the increase of Q. In order to validate the performances of the UNISA sensor we performed a series of tests to experimentally evaluate the function $Q = Q(\omega_o)$. The tests were performed positioning the sensor in a vacuum chamber and measuring the function $Q = Q(\omega)$ for different values of the pressure in the chamber, with a tuning mass, $m_t = 240\,g$. The measures were repeated for different values of the resonance frequency obtained through the calibration procedure, both increasing and decreasing the frequency. The results are reported in Figure 4 and 5. As expected, the Quality Factor increases at the increase of the resonance frequency and shows the expected parabolic dependence, $Q(f_o) = a \cdot f_o^2$.

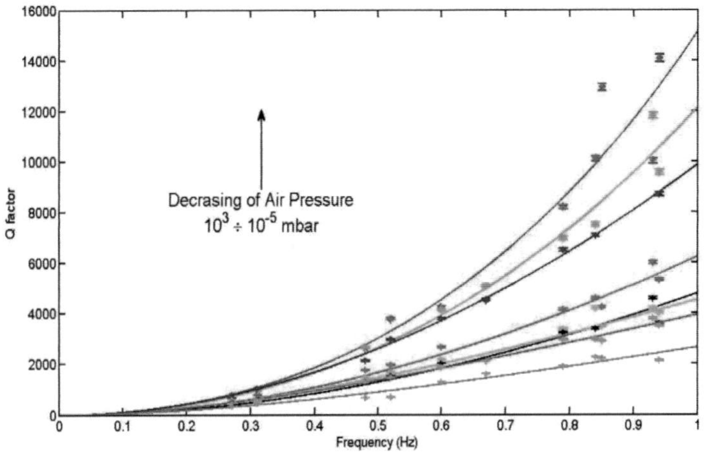

Figure 4. Quality Factor vs. Resonance Frequency

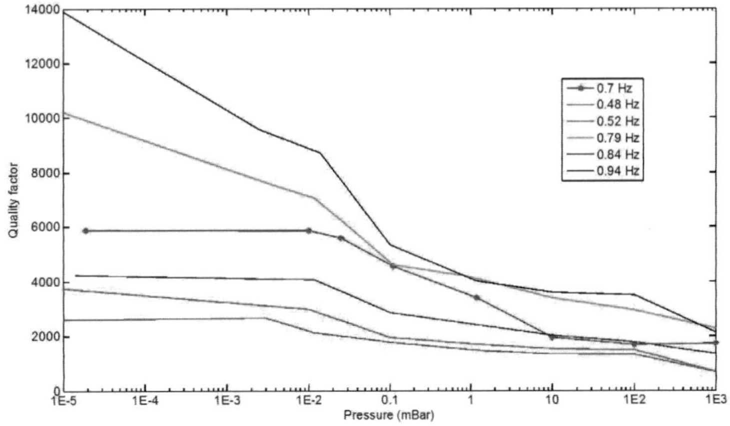

Figure 5. Quality Factor vs. Air Pressure

Figure 4 shows that already at the atmospheric pressure and at the prototype design natural resonance frequency ($f_o \approx 0.721\,Hz$) the value of the quality factor is $Q > 1800$, thus demonstrating that the UNISA sensor perfectly fits for applications in air. Moreover, a quality Factor, $Q \approx 6000$ was measured for the prototype at the design natural resonance frequency with a moderate vacuum ($p = 10^{-5}\,mbar$), reaching values up to ≈ 14000 for a natural resonance frequency of $f_o = 0.94\,Hz$. These results demonstrate, instead, that this sensor works very well both in vacuum, as expected, and in air (that is the main goal of the FP monolithic new design). It is again relevant to underline that for this monolithic Aluminium (7075-T6) prototype values of $Q > 100$ are measured also for resonance frequencies below $100\,mHz$.

Finally, we report the readout noise that is the best UNISA Horizontal Seismometer sensitivity, assuming that the thermal noise is much lower. The measurements were made with the FP central mass clamped to the frame, in air and with no thermal stabilization both for the optical lever (with PSD photodiodes) and for the interferometric readouts. In Figure 6 the best theoretical and experimental sensitivities curves are shown at ($T = 300\,K$), assuming the FP tuned at a resonance frequency ($f_o = 100\,mHz$).

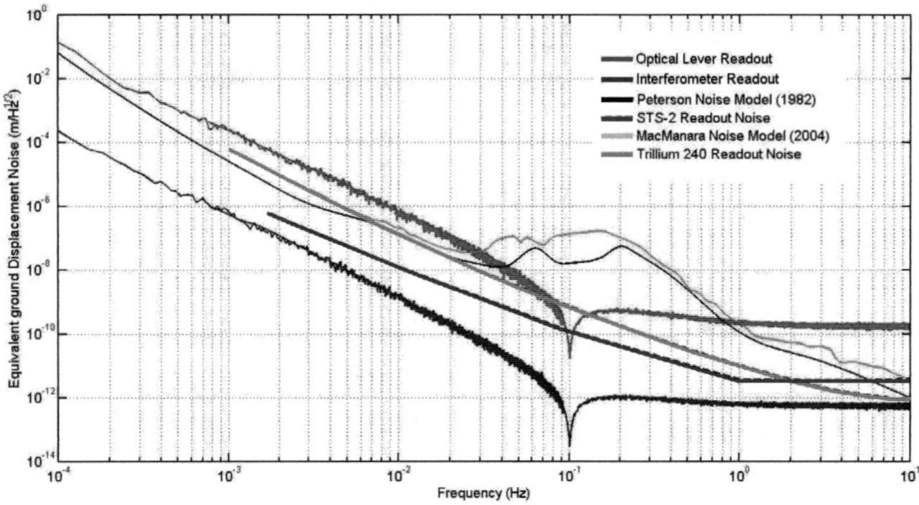

Figure 6. UNISA Horizontal Seismometer Readout Noise

5. Applications of the UNISA Horizontal Seismometer

Experimental applications of the FP UNISA monolithic seismometer have started since 2009. Preliminary tests of FP monolithic sensors have been performed in 2009 at DUSEL, in the Homestake Mine, where all the necessary upgrades were defined to improve sensitivity and band [33].

Then, for long term tests on the field we had to choose a very quiet site (from environmental and seismic noise point of view), but equipped as well as a facility (in terms of power supply, network, etc) and very easily reachable for all the necessary adjustments and tuning procedures. The obvious choice was the Gran Sasso INFN National Laboratory. A test facility in a quiet location at the INFN Gran Sasso National Laboratory was built for this task. It consists in a thermally insulated box, whose dimensions are 2 m (length) $\times$ 1 m (width) $\times$ 1.80 m (height), equipped with power supply and network for remote monitoring. It hosts thermally insulated boxes at the bottom, containing the UNISA horizontal seismometers, and placed on granite slabs fixed to the ground (Figure 7). The whole station guarantees the thermal stability of the sensors. In fact, although the environmental temperature outside the whole station changes at most of 2^oC along the year, no correlation effects between temperature changes and FP output have been highlighted. In Figure 8 the first noise spectrum measured by the UNISA

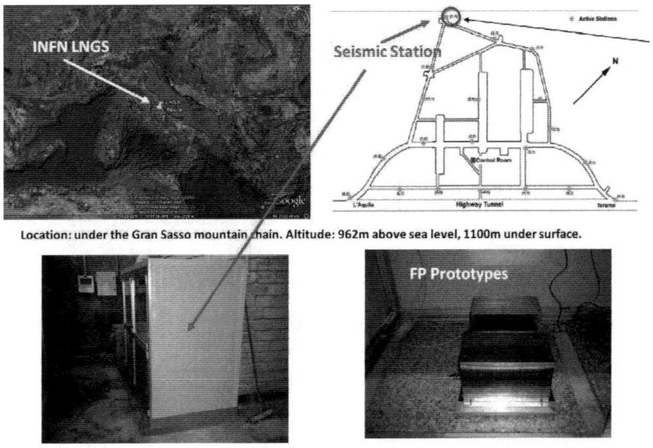

Figure 7. Gran Sasso Seismic Station

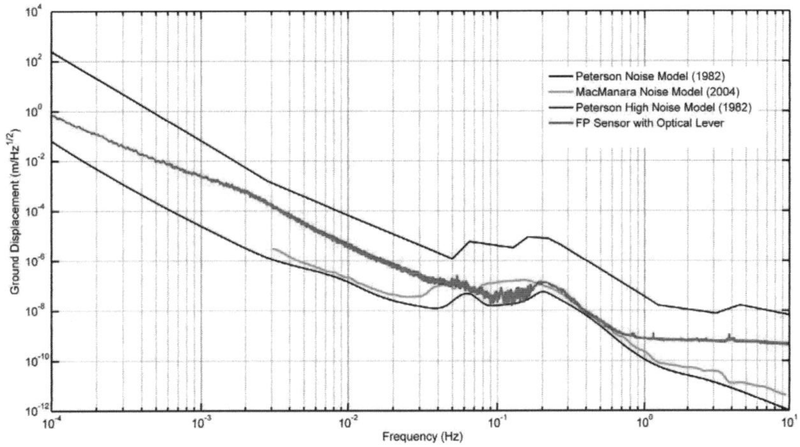

Figure 8. Seismic Noise measured at the INFN Gran Sasso National Laboratory

Horizontal Seismometer. This spectrum, based on a temporal acquisition of about 1130 hours data, shows that the FP UNISA horizontal monolithic sensors, although not yet optimized in terms of resonance frequency, $f_o = 150\,mHz$, acquisition dynamics and optical lever as readout have useful measurement bands for the site that already spans from $50\,mHz \div 500\,mHz$.

The quality of the data as sensor and its sensitivity let us think that is can useful also as sensor in the control system of mechanical suspensions. For this task preliminary tests are in progress on a TAMA like suspension operational in the INFN Napoli Virgo Laboratory.

6. Conclusions

in this paper a new implementation of monolithic horizontal sensor, developed at the University of Salerno, based on the Folded Pendulum architecture, configurable both as seismometer and as accelerometer has been described. The UNISA Horizontal Seismometer shows a large low-frequency band ($10^{-6} \div 10\,Hz$), high sensitivity ($10^{-12} m\sqrt{Hz}$ in the band $0.1 \div 10\,Hz$) and high quality factor in air ($Q > 1500$), largely better than all the previous Folded Pendulum implementations. Moreover its monolithic implementation, coupled with a full tunability of its resonance frequency ($70\,mHz \div 1.2\,Hz$) together with an integrated laser optical readout, guarantees both compactness, robustness and immunity to environmental noises, making this sensor suitable for a large number of scientific applications, also in high vacuum and cryogeny. Applications of this sensor are already started in the field of geophysics, including the study of seismic and newtonian noise for characterization of suitable sites for future underground interferometric detectors of gravitational waves.

Acknowledgments
We acknowledge the Director of the INFN Laboratori Nazionali del Gran Sasso, Prof. Lucia Votano. We acknowledge Ing. Benedetto Gallese and Ing. Piergiogio Aprili for their very professional technical support and useful suggestions on the site.

References

[1] Grote H, for the LIGO Scientific Collaboration 2008 *Class. and Quant. Grav.* **25** 114043.
[2] Abbott B, et al. the LIGO Scientific Collaboration 2009 *Reports on Progress in Physics* **72** 076901.
[3] Arai K, et al., the TAMA Collaboration 2008 *Journal of Physics: Conference Series* **120** 032010.
[4] Accadia T, et al., the VIRGO Collaboration 2011 *Class. and Quant. Grav.* **28** pp.114002-1-114002-10.
[5] The Advanced LIGO Team 2011 *Advanced LIGO Reference Design*, LIGO-M060056.
[6] Harry G M, at al., the LIGO Scientific Collaboration 2010 *Class. and Quant. Grav.* **27** 084006.
[7] The VIRGO Collaboration 2009 *Virgo Technical Report VIR-0027A-09.*
[8] Punturo M, et al., 2010 *Class. and Quant. Grav.* **27** pp.194002-1 - 194002-12.
[9] Abernathy M, et al., ET Scienze Team, 2011 *The Einstein Gravitational Wave Telescope Conceptual Design Study*, ET-0106C-10, issue 4, June 28.
[10] Cella G, Cuoco E, et al 1998 *Class. and Quant. Grav.* **15** p.3339.
[11] Nakayama Y, et al. 2004 *Performances test of STS-2 seismometers with various data loggers*, in proceedings of IWAA2004, CERN, Geneva, 4-7 October 2004.
[12] http://www.nanometrics.ca/products/trillium-240.
[13] Fergusson E S 1962 *US Nat. Museum Bull.* **228** p.185.
[14] Liu J, Ju L, Blair D G 1997 *Phys. Lett. A* **228** pp.243-249.
[15] Acernese F, De Rosa R, Giordano G, Romano R, Barone F 2008 *Rev. Sci. Instrum.* **79** 074501.
[16] Acernese F, Giordano G, Romano R, De Rosa R, Barone F 2010 *Nucl Instrum. and Meth. A* **617** pp.457-458.
[17] Acernese F, De Rosa R, Garufi F, Giordano G, Romano R, Barone F 2010 *Journ of Phys. Conf. Series* **228** p.012035.
[18] Bertolini A, et al. 2006 *Nucl. Instr. and Meth.* **556** pp.616-623.
[19] Acernese F, De Rosa R, Giordano G, Romano R, Barone F 2007 *Proc. of SPIE Int. Conf. on Remote Sensing for Environmental Monitoring, GIS Applications, and Geology VII*, Vol. 6749, p.67491I.
[20] Smith T S, Badami V G, Dale J S, Xu Y 1997 *Rev. Sci. Instrum.* **68** pp.1474-1484.
[21] Tseytlin M Y 2002 *Rev. Sci. Instrument* **73** p.3363.
[22] Barone F, Calloni E, De Rosa R, Di Fiore L, Fusco F, Milano L, Russo G 1994 *Appl. Optics* **33** pp.1194-1197.
[23] Barone F, Calloni E, Di Fiore L, Grado A, Milano L, Russo G 1995 *Appl. Optics* **34** pp.8100-8105.
[24] Zumberge M A, Berger J, Dzieciuch M A, Parker R L 2004 *Appl. Optics* **43** pp.771-775.
[25] Barone F, De Rosa R, Milano L, Qipiani K 2002, *Proc. of* 9th *SPIE Int. Conf. on Remote Sensing for Environmental Monitoring, GIS Applications, and Geology*, Vol. 4886, pp. 614-621.
[26] Acernese F, Barone F, De Rosa R, Milano L, Qipiani K 2003 *Proc. of* 20th *SPIE Int. Symp. on Smart Structures and Materials*, Vol. 5050, pp.342-349.
[27] Zumberge M A, Berger J, Otero J, Wielandt, E 2010 *Bull. Seism. Soc. of America* **100** pp.598-605.
[28] Barone F, Giordano G, 2011 *Low frequency folded pendulum with high mechanical quality factor, and seismic sensor utilizing such a folded pendulum International application published under the patent cooperation treaty (PCT)* WO 2011/004413 A3.
[29] Fan S, Cai Y, Wu S, Luo J, Hsu H 1999 *Physics Letters A* **256** 132-140.
[30] Wu S, Fan S, Luo J 2002 *Rev. Sci. Instrum.* **73** pp.2150-2156.
[31] Bertolini A, DeSalvo R, Fidecaro F, Takamori A, *IEEE Trans. on Geosci. And Rem. Sens.* **44** pp.273-276.
[32] Takamori A, Bertolini A, DeSalvo R, Araya A, Kanazawa T, Shinohara M 2011 *Meas. Sci. Technol.* **22** p.115901.
[33] Acernese F, De Rosa R, Desalvo R, Garufi F, Giordano G, Harms J, Mandic V, Sajeva A, Trancynger T, Barone F 2010 *Journ. of Phys. Conf. Series* **228** p.012036.

High accuracy measurement of the quantum efficiency using radiation pressure

Kazuhiro Agatsuma[1], Takumi Mori[2], Stefan Ballmer[3], Giulia DeSalvo[4], Shihori Sakata[5], Erina Nishida[6], Seiji Kawamura[7]

[1] National Astronomical Observatory of Japan, Mitaka, Tokyo 181-8588, Japan
[2] Department of Advanced Materials Science, University of Tokyo, Kashiwa, Chiba 277-8582, Japan
[3] Department of Physics, Syracuse University, NY 13244, USA
[4] University of California, Berkeley, CA 94720, USA
[5] Center for Photonic Innovation, University of Electro-Communications, 1-5-1 Chofugaoka, Chofu, Tokyo 182-8585, Japan
[6] Ochanomizu University Graduate School of Humanities and Sciences, Bunkyo, Tokyo 112-8610, Japan
[7] Institute for Cosmic Ray Research, University of Tokyo, Kashiwa, Chiba 277-8582, Japan

E-mail: `kazuhiro.agatsuma@nao.ac.jp`

Abstract. Preliminary investigations of a novel method to measure the laser power accurately using the radiation pressure are reported here. We aim to measure the laser power within one percent error to then obtain an accurate quantum efficiency (QE) of a photodiode. Since the typical error of QE is still a few percent due to the uncertainty of measured laser power, an accurate measurement of the laser power contributes a precise estimation of the QE. Our experimental setup is a suspended Michelson interferometer, where one of the pendulums is small, consisting of a 20-mg mirror and 10-um fiber. The motion of this small mirror is very sensitive to changes in radiation pressure. Due to this, the number of photons in the incident (intensity modulated) laser beam can be counted accurately by measuring displacement of the mirror. We set up the apparatus, and have found a suitable frequency band for the accurate measurement. Displacement caused by the radiation pressure was observed using the feedback signal.

1. Introduction

According to a recent study about squeezing of quantum noise have made great progress and will contribute to the development of essential technology for future gravitational wave detectors [1, 2]. It is significant for squeezing experiments to make the quantum efficiency (QE) of photo diodes (PDs) high because a squeezing state is easily degraded by a small amount of optical losses including QE [3]. According to a resent study, $95(\pm 2)\%$ of QE is the main limitation of a squeezing level [4]. A PD with an expected QE of close to 99% is about to be constructed [5] and it will be the main limitation of a squeezing level. Even if high QE close to 99% is achieved, it cannot yet be measured with sufficient accuracy. Uncertainty of QE affects the measurement of squeezing level. A typical error of QE is still a few percent owing to the uncertainty of measured laser power. Our objective is to measure laser power within an error of one percent, and then to measure QE accurately. In order to achieve this we are constructing an opto-mechanical power meter, which can count the number of photons by measuring displacement of a mirror pushed by laser photons.

2. Theory

The QE of a PD is defined as the efficiency with which light power is converted to current; $QE = N_{\rm e}/N_{\rm P}$. Here, $N_{\rm e}$ is the number of electrons in the output current of the PD and $N_{\rm P}$ is the number of photons in the input light. A precise measurement of $N_{\rm P}$ leads to a high accuracy of QE.

When there is an opto-mechanical system with an input laser and a mirror, $N_{\rm P}$ can be related to the displacement of the mirror shaken by input photons. We can find the displacement in the frequency domain by modulating the input power. Let us consider a photon that is reflected by a mirror. The mirror is pushed by the momentum of the photon. The equation of motion is written as $m\ddot{x} = 2h\nu/c$, where m is the mass of the mirror, x the displacement of the mirror, h the Planck constant, ν the frequency of the photon, and c the speed of light. In the frequency domain, the equation of motion is represented as

$$d\tilde{X} = \frac{2P_{\rm m}\cos\phi}{mc\omega^2}. \tag{1}$$

Here, $d\tilde{X}$ is the Fourier transformation of x, ω the angular frequency of the intensity modulation of $P_{\rm m}$, and ϕ the incident angle of the input laser to the mirror. $P_{\rm m}$ is translated to the power at DC using the modulation index, $\alpha_{\rm m}$, as $P_{\rm m} = \alpha_{\rm m}P$. Also, $N_{\rm P}$ is related to a continuous-wave laser with the power of P as $N_{\rm P} = P/(h\nu)$.

When the Michelson interferometer is used as a displacement sensor, we can obtain $d\tilde{X} = dV_{\rm PD}\lambda/(2\pi V_{\rm pp})$. $dV_{\rm PD}$ is the output voltage from the PD, and $V_{\rm pp}$ is the peak-to-peak voltage when the arm length is changed over a half wave length. A linear response is guaranteed as long as the so-called 'mid-fringe' state is kept. This equation in combination with equation(1) yields

$$P = \frac{mc\lambda\omega^2}{4\pi V_{\rm pp}\alpha_{\rm m}\cos\phi}dV_{\rm PD}. \tag{2}$$

Note that $dV_{\rm PD}$ includes the effect from control gain. One implication of equation(2) is that measuring the opto-mechanical response gives an accurate measurement of light power.

3. Experimental Setup

As shown in Fig.(1) and Fig.(2), our main apparatus is a simple Michelson interferometer (MI) with a tiny mirror suspended by a small pendulum. The displacement of this tiny mirror is sensitive to changes in radiation pressure, so that the number of photons from the incident laser beam can be counted accurately by measuring the displacement of the mirror.

The laser has a nominal power of 500 mW. The laser path is divided using a polarized beam splitter and a half-wave plate into two paths that are used for the MI and for the input power modulation. That is why we can see displacement of the tiny mirror using the MI as the tiny mirror is fluctuated by the modulated radiation pressure. The power modulation is provided using an acousto-optic modulator (AOM). The MI is housed in a vacuum tank, which is put on vibration isolation stacks, but, for the moment, this experiment is performed in air.

A characteristic of our apparatus is to use a tiny mirror whose weight is only 20 mg and diameter is 3 mm as an end mirror for one arm of the MI. The tiny mirror is suspended by a silica fiber with the diameter of 10 um. The fiber is glued on the mirror using UV cured resin. The suspension is the double pendulum with eddy current damping at the middle mass [6, 7]. The other arm of the MI comprises a large mirror whose weight is 50 g and diameter is 1 inch. The large suspension is also a double pendulum with eddy current damping. This mirror is controlled using four sets of coil-magnet actuators.

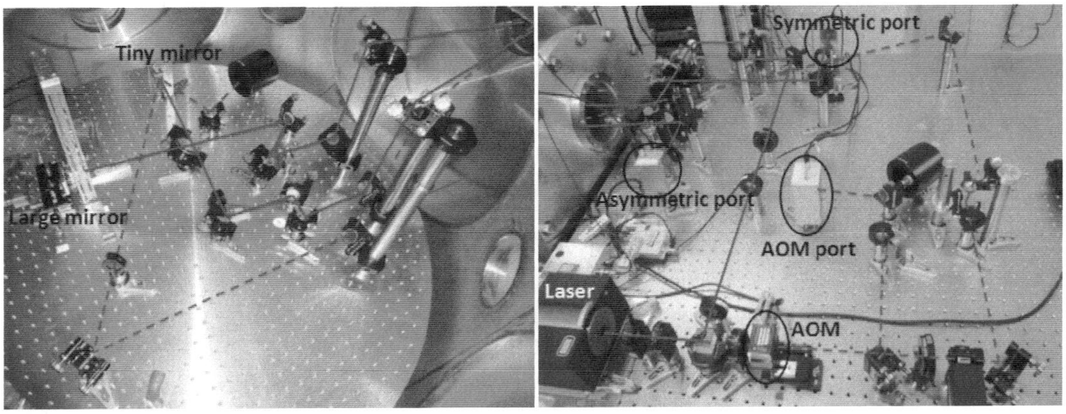

Figure 1. Optical layout.
The solid red line indicates a beam trace for the Michelson interferometer and the dotted red line is that for an intensity modulation by AOM. There are three photo detectors; Symmetric port, Asymmetric port and AOM port on an optical table.

Figure 2. Suspension systems.
Left: Tiny mirror pendulum. The tiny mirror is suspended by a 10-um silica fiber. The middle mass of 20 mg is suspended by a 10-um tungsten wire. Right: Large mirror pendulum. The large mirror and its optical bracket is suspended by 50-um tungsten wires. The intermediate mass of 60 g is suspended by 70-um tungsten wires.

4. Sensitivity and Control

In order to measure the radiation pressure response within an error of one percent, it is necessary for the signal to be above the noise floor by nearly two orders of magnitude. Figure(3) shows a superposition of the sensitivity of the MI and calculated displacement of the tiny mirror by radiation pressure modulation when the displacement noise was measured in the air. To have a sensitivity with wide range and fine resolution, data was taken in three frequency ranges (10-100 Hz, 100-800 Hz and 800-6400 Hz) and then combined. Displacement by radiation pressure is estimated using equation(1) with an intensity modulation of 50 mW (peak-to-peak). The modulation index of the AOM is about 17%, so 17% of the incident laser power can be used as an intensity modulation. The nominal power of our laser is 500 mW. If we use 400 mW for the modulation path, then we will get a modulation of 68 mW. Taking some dispersion loss into account, 50 mW is chosen as a typical value of this experiment. We found that this was sufficient to measure the radiation pressure response as the dominant signal, especially in the range between 200 Hz and 500 Hz.

The Michelson Interferometer was locked at the mid-fringe using the symmetric port signal. Contrasts are about 95% in the symmetric port and asymmetric port as well. As shown in Fig.(4), the unity gain frequencies are 100 Hz, 130 Hz and 190 Hz, corresponding to 30 degrees,

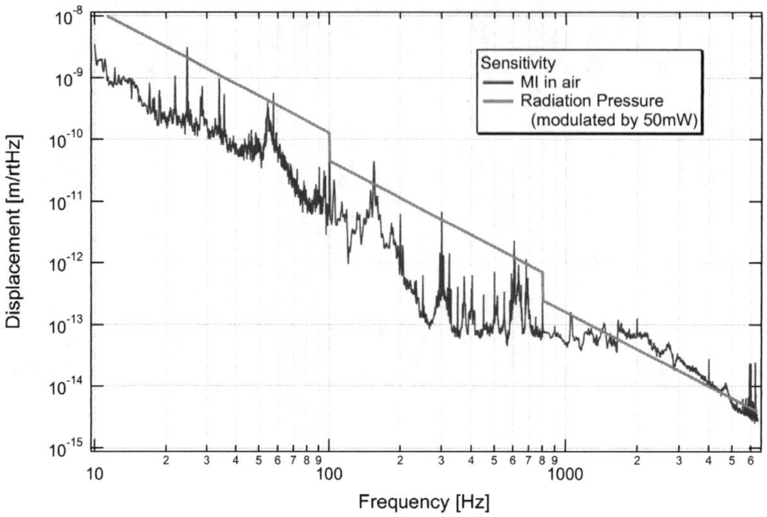

Figure 3. The sensitivity of MI and radiation pressure estimation.
The green line indicates a radiation pressure estimation. Note that jumps of the line at 100 Hz and 800 Hz are caused by differences of the integration time. Integration time of each frequency bin is 16 s below 100 Hz, 2 s between 100 Hz and 800 Hz, and 0.25 s above 800 Hz.

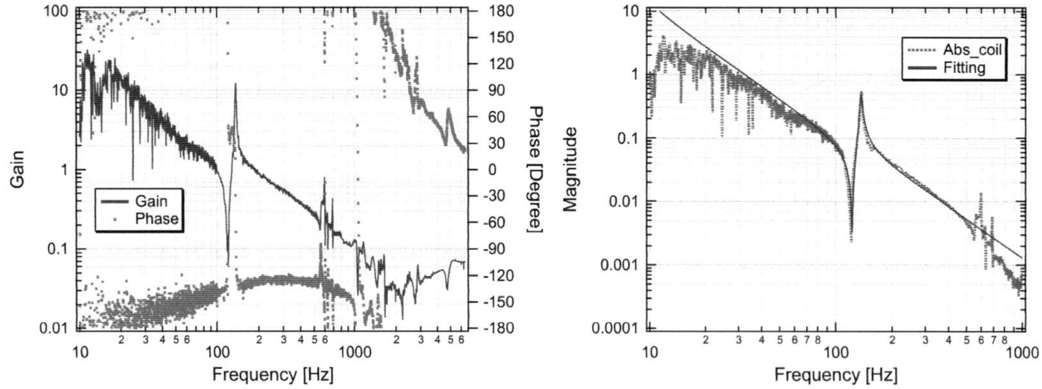

Figure 4. Measured transfer functions.
Left: Open loop transfer function of the MI servo. Right: Transfer function from the coil-magnet actuator to the photo detector at the symmetric port, which is fitted using data between 80 Hz and 500 Hz.

220 degrees and 50 degrees of phase margins, respectively. A transfer function from the coil-magnet actuator to photo detector is translated to the actuator response using a fitting function of f^{-2}. There is a structure like the parasitic resonance around 130 Hz. This might be a coupling motion with the pitch mode of the large mirror. There are no structures between 200 Hz and 500 Hz, so that fine measurements are possible in this region. The stability of the control is still not sufficient. Disorder below 100 Hz of Fig.(4) comes from shortage of the measurement

average. When unsteady large noise (mainly seismic noise) occurs, the locking of MI easily fails.

From the sensitivity and control, we can now expect an accurate measurement between 200 Hz and 500 Hz owing to fine measurement of the transfer function, lower noise, and higher stability.

5. Radiation Pressure

We compared feedback signals when the modulated incident laser power was ON using AOM at 430 Hz, and when it was OFF. There was no explicit difference between them except for the peak at 430 Hz as shown in Fig.(5). We confirmed that this peak can be moved by changing the modulation frequency. Thus the radiation pressure is observed in the feedback signal. However, the scattering effect has not yet been investigated.

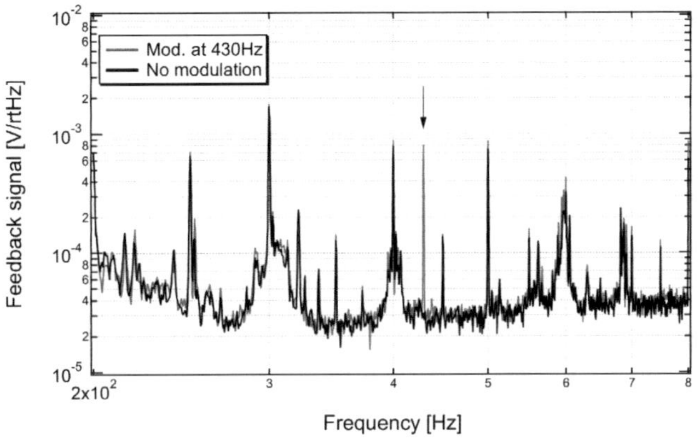

Figure 5. The feedback signal with and without modulated input laser.
The black arrow indicates the peak signal by modulated radiation pressure at 430Hz.

6. Summary and Future Plan

We are constructing a new kind of power meter to measure the fine QE of a PD, and have obtained promising results. We found that the frequency region between 200 Hz and 500 Hz is suitable for an accurate measurement. In that region, the transfer function can be measured accurately and noise of the MI is less than the modulation signal by nearly two orders. The modulation signal of the radiation pressure was observed by feedback signal.

The next step is a detailed examination of the accuracy of each measurement. We will investigate the errors due to measurement of mass (tiny mirror), deviation from free mass approximation, scattering and absorption from the mirrors, intensity noise of the laser, and deviation from mid-fringe lock. Noise hunting and suppression should broaden the frequency band where accurate measurement is possible. Beam centering at a mirror spot will suppress a coupling with a pitch mode, and then increase stability of the lock.

7. Acknowledgments

This work is supported by a Grant-in-Aid for Scientific Research from the Japan Society for the Promotion of Science (09J03624), and Young Scientists B No. 21740201.

References

[1] The LIGO Scientific Collaboration 2011 *Nature Physics, Advance Online Publication* 11 September
[2] Schnabel R, Mavalvala N, McClelland D E, and Lam P K 2010 *Nature Communications* **1** 121
[3] Kimble H J, Levin Y, Matsko B A, Thone S K and Vyatchanin P S 2002 *Phys. Rev. D* **65** 022002
[4] Vahlbruch H 2008 *Phys. Rev. Lett.* **100** 033602
[5] Vahlbruch H 2008 Phd Thesis p.63
[6] Sakata S, Miyakawa O, Nishizawa A, Ishizaki H and Kawamura S 2010 *Phys. Rev. D* **81** 064023
[7] Sakata S et al 2008 *Jour. of Phys: Conf. Ser.* **122** 012020

Journal of Physics: Conference Series **363** (2012) 012003 doi:10.1088/1742-6596/363/1/012003

Status Report of the Schenberg Gravitational Wave Antenna

O D Aguiar[1], J J Barroso[1], N C Carvalho[1], P J Castro[1], C E Cedeño M[1], C F da Silva Costa[1], J C N de Araujo[1], E F D Evangelista[1], S R Furtado[1], O D Miranda[1], P H R S Moraes[1], E S Pereira[1], P R Silveira[1], C Stellati[1], N F Oliveira Jr[2], Xavier Gratens[2], L A N de Paula[2], S T de Souza[2], R M Marinho Jr[3], F G Oliveira[3], C Frajuca[4], F S Bortoli[4], R Pires[4], D F A Bessada[5], N S Magalhães[5], M E S Alves[6], A C Fauth[7], R P Macedo[7], A Saa[7], D B Tavares[7], C S S Brandão[8], L A Andrade[9], G F Marranghello[10], C B M H Chirenti[11], G Frossati[12], A de Waard[12], M E Tobar[13], C A Costa[14], W W Johnson[14], J A de Freitas Pacheco[15], G L Pimentel[16]

[1] INPE – Divisão de Astrofísica, São José dos Campos, SP, Brazil,
[2] Universidade de São Paulo, Instituto de Física, São Paulo, SP, Brazil,
[3] Instituto Tecnológico de Aeronáutica, São José dos Campos, SP, Brazil,
[4] Instituto Federal de São Paulo, São Paulo, SP, Brazil,
[5] Universidade Federal de São Paulo, Diadema, SP, Brazil,
[6] Universidade Federal de Itajubá, Itajubá, MG, Brazil,
[7] Universidade Estadual de Campinas, Instituto de Física, Campinas, SP, Brazil,
[8] Universidade Estadual de Santa Cruz, Ilhéus, BA, Brazil,
[9] Instituto de Aeronáutica e Espaço, São José dos Campos, SP, Brazil,
[10] Universidade Federal de Bagé, Bagé, RS, Brazil,
[11] Universidade Federal do ABC, Santo André, SP, Brazil,
[12] Leiden University, Kammerlingh Onnes Laboratory, Leiden, The Netherlands,
[13] University of Western Australia, Perth, Australia,
[14] Louisiana State University, Baton Rouge, USA,
[15] Observatoire de la Côte dAzur, Nice, France,
[16] Princeton University, Princeton, USA.

E-mail: odylio.aguiar@inpe.br, barroso@plasma.inpe.br, nataliac@das.inpe.br, castro@plasma.inpe.br, cecedenom@das.inpe.br, filipe.dasilva@das.inpe.br, jcarlos.dearaujo@inpe.br, edgard@das.inpe.br, sergio@das.inpe.br, oswaldo@das.inpe.br, pedromoraes@das.inpe.br, duducosmo@das.inpe.br, patrick@das.inpe.br, cstellati@uol.com.br, nei@if.usp.br, xavier@if.usp.br, leandroifusp@yahoo.com.br, turano@if.usp.br, marinho@ita.br, nandago@ita.br, frajuca@gmail.com, bortoli.fabio@gmail.com, ricardopires29@gmail.com, dbessada@gmail.com, nadjasm@gmail.com, alvesmes@unifei.edu.br, fauth@ifi.unicamp.br, romacedo@aei.mpg.de, asaa@ime.unicamp.br, btavares@ifi.unicamp.br, claudiosoriano.uesc@gmail.com, andrade@iae.cta.br, gfrederico.unipampa@ufpel.edu.br, cecilia.chirenti@ufabc.edu.br, giorgio@leidencryogenics.com, info@milli-K.com, michael.tobar@uwa.edu.au, ccosta@phys.lsu.edu, johnson@ligo.phys.lsu.edu, pacheco@oca.eu, gpimente@Princeton.EDU

Abstract. Here we present a status report of the Schenberg antenna. In the past three years it has gone to a radical upgrading operation, in which we have been installing a 1K pot dilution refrigerator, cabling and amplifiers for nine transducer circuits, designing a new suspension and vibration isolation system for the microstrip antennas, and developing a full set of new transducers, microstrip antennas, and oscillators. We are also studying an innovative approach, which could transform Schenberg into a broadband gravitational wave detector.

1. Introduction

The Mario SCHENBERG Gravitational Wave Detector [1][2][3][4], which has a spherical resonant antenna, started commissioning operation on the 8th of September 2006 [5][6]. It involves collaboration between many Brazilian institutions such as INPE, USP, ITA, IFSP, UNIFESP, UNIFEI, UNICAMP, UESC, IAE, UFABC, UNIPAMPA, and CBPF, and also some foreign universities as the Leiden University, UWA, LSU, and OCA. FAPESP (the São Paulo State Foundation for Research), CAPES, CNPq, and MCT have been supporting this project.

The Schenberg antenna is the only spherical antenna equipped with a set of parametric transducers for gravitational wave (gw) detection.

In figure 1 is shown a schematic view of it. The Schenberg CuAl6% antenna has a diameter of 65 cm and weighs 1.15 ton. It is kept in a vacuum, isolated from mechanical noises. It has nine small holes on its surface for up to nine transducers, six of which follow the truncated icosahedron configuration proposed by Johnson and Merkowitz [7]. These transducers will monitor their fundamental modes of vibration [8]. When coupled to the antenna, the transducer–sphere system will work as a mass–spring system with three modes, where the first will be constituted by the antenna effective mass (287.5 kg), the second will be constituted by the mechanical structure of the transducer (~ 25 g), and the third one will be constituted by a membrane (~ 2 mg) that will close the transducer microwave cavity and modulate it around 3.2 kHz. Because there are five quadrupole modes in the sphere, each with an effective mass of 287.5 kg, which is one forth of the sphere mass, the total mass involved in the detection is actually larger than the total spherical antenna mass. This is impressive compared to the effective mass for the fundamental longitudinal mode of bars, which is half of the bar antenna mass. This is one of the reasons why a sphere has a cross section higher for gravitational waves than the one for a bar (the other reason is because its mass is larger then the equivalent bar, tuned to the same wave frequency).

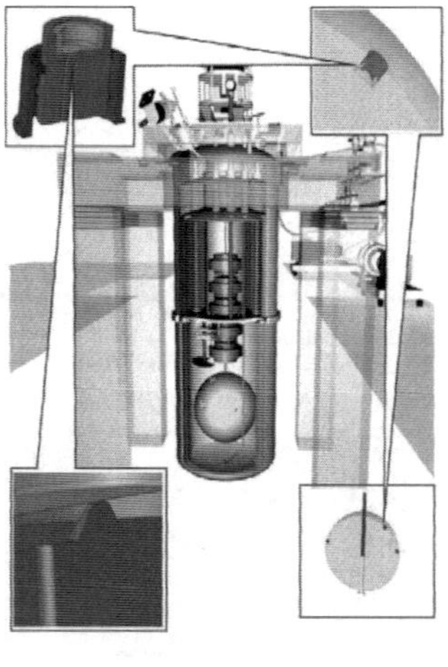

Figure 1. Mario Schenberg detector. The resonant mass (sphere) is kept in a vacuum, isolated from mechanical noises. Nine parametric transducers will monitor their fundamental modes of vibration [8].

Journal of Physics: Conference Series **363** (2012) 012003 doi:10.1088/1742-6596/363/1/012003

Its standard quantum limit sensitivity [9] corresponds to a strain noise power spectral density of $\sim 10^{-22}$ $Hz^{-1/2}$ [10]. If it reaches its design sensitivity, it could operate in coincidence with the Dutch Mini-GRAIL antenna [11] and some long baseline laser interferometer detectors [12], searching for high frequency events in the 3.1–3.3 kHz frequency bandwidth, such as core collapse in supernova events, neutron stars going to hydrodynamical instability, quakes and oscillations of neutron stars (f modes) induced by the falling of matter in binary systems, excitation of the first quadrupole normal mode of 4–9 solar-mass black holes [13], and coalescence of neutron stars and/or black hole systems of 4–9 solar-masses, among the 'classical' sources we have been studying. We also can speculate on the possibility of searching for some 'exotic' sources (if they exist), such as rotation of bosonic or strange matter stars at 1.6 kHz and the inspiralling of mini-black hole binaries. The Schenberg detector can also be useful to test the scalar component in alternative gravity theories by monitoring the antenna monopole excitation mode [14][15][16].

2. The status of the detector

We performed a few runs during the period of 2006 and 2008, testing the system with just three initial designed transducers (figure 2).

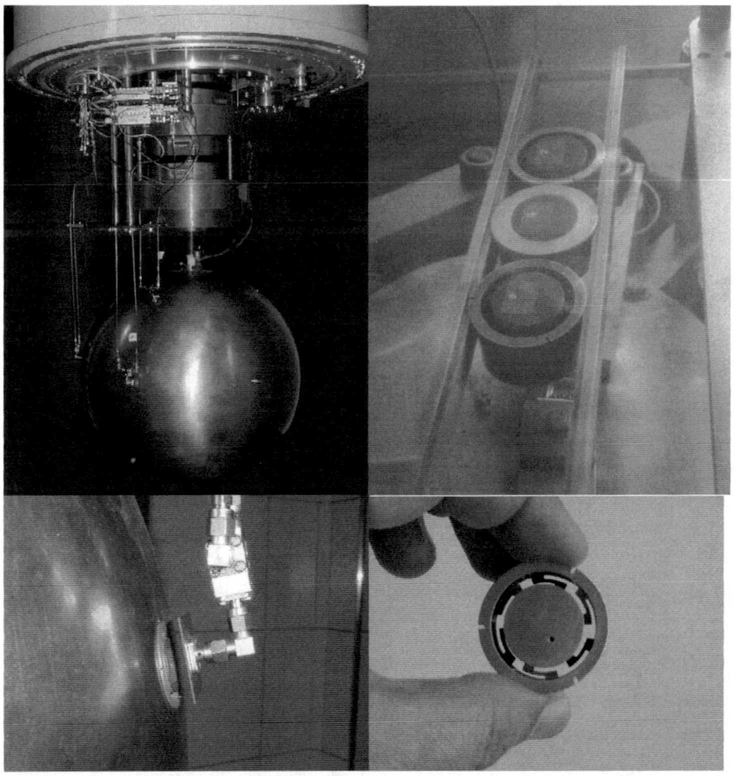

Figure 2. (Top left) The position on the sphere surface of the three initial designed transducers for the 2006-2007-2008 runs can be seen. It is also possible to see the cabling lines going and coming from these transducers and the three cryogenic microwave amplifiers installed close to the bottom wall of the liquid helium reservoir; (Top right) The three initial designed transducers inside a sputtering machine; (Bottom left) Both the carrier and the modulated signal were sent back and forth within the transducer by pairs of microstrip antennas. No cables touched the sphere in order to avoid the introduction of seismic noise; (Bottom right) One of the two first designed transducers that had a resonant mode tuned to the sphere's quadrupole frequency. We decided to abandon this design because we measured mechanical Qs for the sphere modes of only $3x10^4$, instead of the 2 million we were expecting.

Since 2008, we have been upgrading the detector, designing a new suspension and vibration isolation system for the cabling and microstrip antennas, designing a complete new set of transducers, and installing a dilution refrigerator's 1K pot.

We also measured the electrical Q of many different superconducting klystron cavities at 4.2 K using a liquid helium dewar and an Agilent 8722/ES vector network analyzer. Q as high as ~ 3×10^5 were recorded.

The various transducer designs we have tried in the period 2008-2011 are shown in figures 3, 4, and 5. The major problems to be overcome in these designs were: the optimization of the mechanical Q of the transducer modes (intermediate and membrane); the optimization of the electrical Q of the microwave superconductor cavity; and the electromagnetic coupling with this cavity.

Figure 3. (Top and middle) Various views of the parts of the second designed transducer; (Bottom) The second designed transducer assembled. We succeeded to achieve electrical Qs as high as 3×10^5 with this design, but the coupling to the cavities depended on a transmission line of 2 cm, which was presenting construction problems. We decided to remove this transmission line, mounting the microstrip antennas on the same side as the microwave cavity.

Figure 4. Drawings of the third and forth designed transducers. (Left): A view of the third designed transducer is shown. Only one prototype niobium piece was made of it, which presented very bad mechanical Qs (below 1000); (Right) The body of the forth designed transducer. We constructed 11 pieces of it, all made of niobium. However, the procedure to introduce these transducers to the sphere cavities altered the elastic behavior of the transducer arms, changing both resonant frequency as mechanical Q (to values as low as the ones from the previous design).

Each transducer in the fifth design is composed of: - a molybdenum piece, which includes the first mechanical mode springs and the part that goes attached to the CuAl (6%) spherical antenna holes by thermal differential contraction (by cooling down), - a niobium cover, which forms most of the superconductor klystron cavity, - a silicon membrane of a few milligrams, which oscillates at the sphere quadrupole frequency and which is the sensor element of the microwave superconductor cavity, - the niobium film deposited by sputtering on this silicon membrane, - twelve M2 (metric-two-millimeter) stainless steel screws, which keep the niobium cover united to the molybdenum piece.

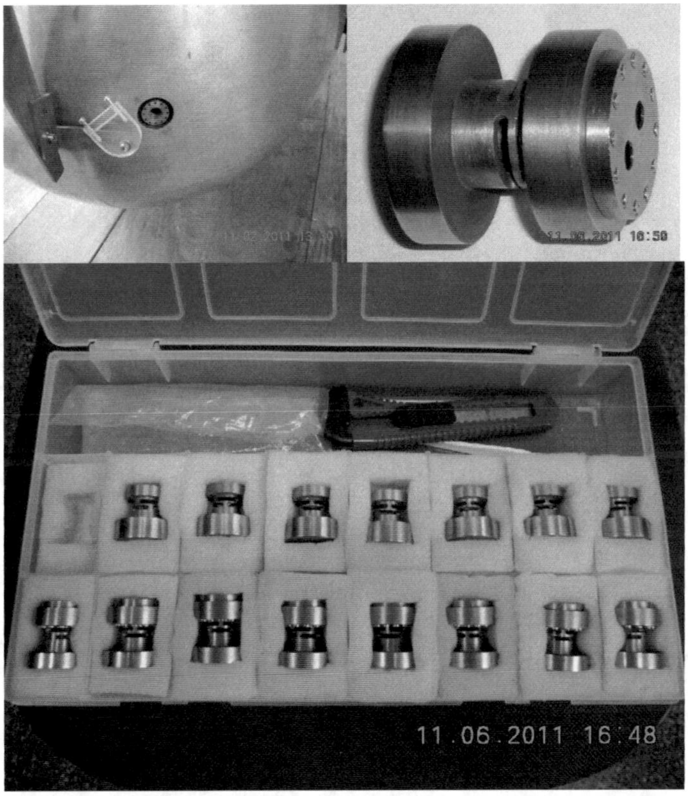

Figure 5. Photos of the transducers built using the fifth design.

We developed a sapphire oscillator, which operates at 77 K and will replace the current one thereby providing better performance in the next run. The phase noise was improved from -100 dBc/Hz [17] to -135 dBc/Hz @ 3.2 kHz [18] from the carrier. The schematic view and pictures of the setup for testing microwave sapphire cavities are shown in figure 6.

All cabling for a full set of transducers, following a new electronic set up, was installed and heat sunk.

In order to put or remove the bottom cryogenic shells we always had to use the hydraulic system to raise or lower the sphere cryostat. However, a new set of transducers with better sensitivity would not afford a movable "testbed", so we needed to stop moving the sphere cryostat up and down. In order to do that and still be able to put and remove the bottom shells, we had to immobilize the antenna to a high position, bringing it to the floor level (which also facilitate the assembling of transducers), so large tubes were constructed as "columns" for the detector concrete base (figure 7).

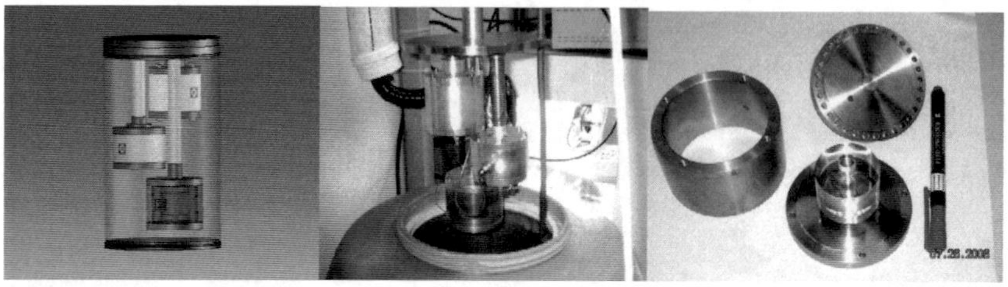

Figure 6. Schematic view and pictures of the setup for testing microwave sapphire cavities. These cavities will be used in the next run.

Figure 7. In order to immobilize the sphere cryostat at a high position (about 1.5 meters up), large tubes were constructed as "columns" for the detector concrete base, bringing the antenna to the floor level. A wood floor was constructed under it, making the transducer assembly work easier. Two "swimming pool" ladders were installed for access to the top of the detector.

A Schematic view of the electronics for the next run configuration is shown in figure 8. We removed all circulators, isolators and hybrids from the cryogenic circuit. The price to pay was to double the number of probes accessing the cavity (one for sending the carrier signal and one for bringing the modulated signal to the criogenic amplifier). We also did not cancel the carrier to the cryogenic amplifier. Instead, we decreased the carrier power to a value below -80 dBm in order to guarantee the best noise performance of this amplifier [19].

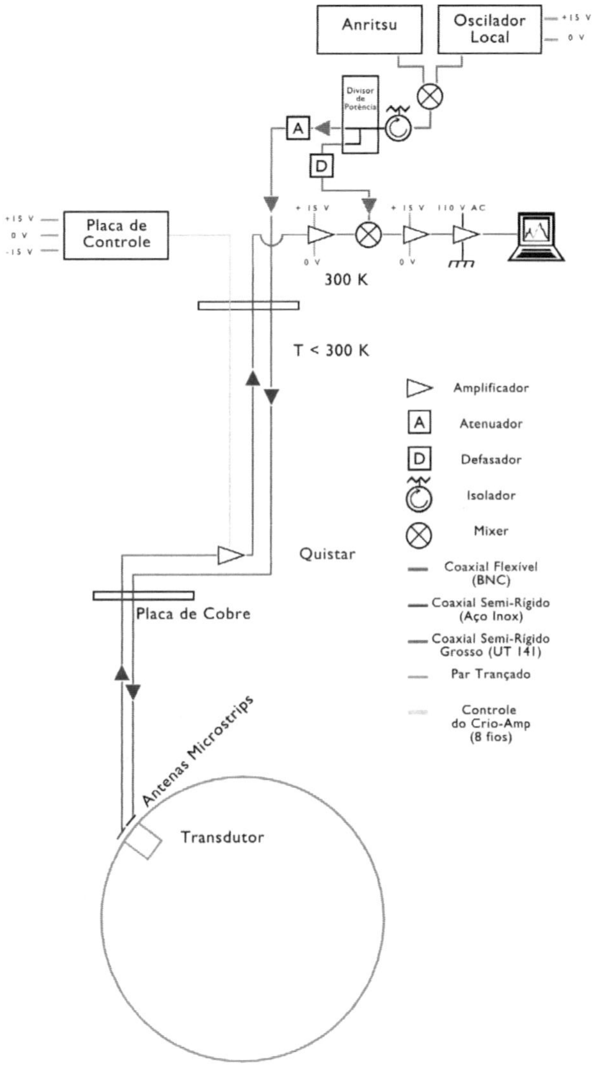

Figure 8. Schematic view of the electronics for the next run configuration.

We hope to start a run with the new set of resonant transducers soon.

We are also developing the project for an ultra-high sensitivity non-resonant *nanogap* (with a gap of one nanometer) transducer. In doing so, we want to verify if the Schenberg antenna can become a wideband gravitational wave detector through the use of an ultra-high sensitivity non-resonant transducer constructed by the application of nanotechnology. Details of this work can be found elsewhere [20][21].

3. Final Remarks

We have already succeeded to obtain electrical Qs of $3\text{x}10^5$ for some tested cavities and we have also constructed an oscillator with a phase noise of -130 dBc/Hz @ 3.2kHz [22]. If we now succeed to assemble the 3-micron gap for these cavities and to optimally couple the probes to the cavity, we should reach the design sensitivity with a thermodynamical temperature of 1K, and so entering in science mode, hopefully already during the planned 2012 runs.

Reaching the designed sensitivity, we will try to contribute to the gravitational wave astronomy with information about the wave direction and polarization of some of the astrophysical sources mentioned in the introduction.

If we improve the electrical Q to one million and decrease the thermodynamical temperature to 50mK, we could reach the standard quantum limit sensitivity. In doing so, we will try to study the behavior of the spherical antenna as a macroscopic quantum oscillator, trying to perform quantum non-demolition measurements (QND) [23][24] and exploring the independence of its five quadrupole modes.

Acknowledgments

This work has been supported by FAPESP (under grants No. 1998/13468-9 and 2006/56041-3), CNPq, CAPES and MCT/INPE. I thank the referee for all comments and suggestion made.

References

[1] Aguiar O D *et al* 2002 *Class. Quantum Grav.* **19** 1949
[2] Aguiar O D *et al* 2004 *Class. Quantum Grav.* **21** S457
[3] Aguiar O D *et al* 2005 *Class. Quantum Grav.* **22** S209
[4] Aguiar O D *et al* 2006 *Class. Quantum Grav.* **23** S239
[5] Aguiar O D *et al* 2008 *Class. Quantum Grav.* **25** 114042
[6] Costa, C A *et al* 2008 *Class. Quantum Grav.* **25** 184002
[7] Johnson, W W and Merkowitz, S M 1993 *Phys. Rev. Lett.* **70** 2367
[8] Coccia, E, Lobo, J A and Ortega, J A 1995 *Phys. Rev.* D **52** 3735
[9] Thorne, K S et al 1978 *Phys. Rev. Lett.* **40** 667
[10] Tobar, M E, Ivanov, E N and Blair, D G 2000 *Gen. Rel. Grav.* **32** 1799
[11] De Waard *et al* 2005 *Class. Quantum Grav.* **22** S215
[12] Vinet, J-Y 2010 *Research in Astron Astrophys.* **10** 956
[13] Costa C A, Aguiar O D and Magalhães N S 2004 *Class. Quantum Grav.* **21** S827
[14] Forward R L 1971 *Gen. Rel. Grav.* **2** 149
[15] Eardley D M, Lee D L, Lightman A P, Wagoner R V and Will C M 1973 *Phys. Rev. Lett.* **30** 884
[16] Bianchi M, Coccia E, Colacino C N, Fafone V and Fucito F 1996 *Class. Quantum Grav.* **13** 2865
[17] Andrade, L A *et al* 2009 *Microwave and Optical Tech. Lett.* **51** 1120
[18] Furtado, S R *et al* (in preparation) 2012
[19] Ivanov E N, Hartnett, J G, and Tobar, M E 2000 *IEEE Trans. Ultrason., Ferroelect., Freq. Contr.* **47** 1526
[20] Pimentel, G L *et al* 2008 *J. Phys. Conf. Series* **122** 012028
[21] Aguiar *et al* 2009 *Int. J. Modern Phys. D* **18** 2317
[22] Furtado, S R 2009 Ph.D. Thesis at INPE, not published
[23] Braginsky, V B, Vorontsov, Y I and Thorne, K S 1980 *Science* **209** 547
[24] Thorne, K S *et al* 1979 "The Quantum Limit for Gravitational-Wave Detectors and Methods of Circumventing It" In: "Sources of Gravitational Waves", ed. L L Smarr, Cambridge University Press, Cambridge, USA, 49

Newtonian noise and ambient ground motion for gravitational wave detectors

M.G. Beker[1], J.F.J. van den Brand[1,2], E. Hennes[1], D.S. Rabeling[1,2]

[1]National Institute for Subatomic Physics Nikhef, Science Park 105, 1098 XG Amsterdam, The Netherlands

[2]VU University Amsterdam, de Boelelaan 1081, 1081 HV Amsterdam, The Netherlands

E-mail: `M.Beker@Nikhef.nl`

Abstract. Fluctuations of the local gravitational field as a result of seismic and atmospheric displacements will limit the sensitivity of ground based gravitational wave detectors at frequencies below 10 Hz. We discuss the implications of Newtonian noise for future third generation gravitational wave detectors. The relevant seismic wave fields are predominately of human origin and are dependent on local infrastructure and population density. Seismic studies presented here show that considerable seismic noise reduction is possible compared to current detector locations. A realistic seismic amplitude spectral density of a suitably quiet site should not exceed 0.5 $\text{nm}/\sqrt{\text{Hz}}\,(\text{Hz}/f)^2$ above 1 Hz. Newtonian noise models have been developed both analytically and by finite element analysis. These show that the contribution to Newtonian noise from surface waves due to distance sources significantly reduces with depth. Seismic displacements from local sources and body waves then become the dominant contributors to the Newtonian fluctuations.

1. Introduction

The present interferometric gravitational wave (GW) detectors in the USA (LIGO) and Europe (Virgo) will undergo a major upgrade within the next few years. At the same time KAGRA is being realized in Japan [1]. By 2015 these 'second generation' detectors are expected to come online, with a sensitivity about 10 times better than that of initial designs [2–4]. Increasing the sensitivity of these instruments is a huge experimental endeavor. It will require the suppression of a number of noise sources, including some noise contributors that have previously been considered insignificant. Newtonian noise (NN) is one such noise term that may limit the sensitivity of these advanced detectors in the low frequency band (up to 10 Hz), and is certainly a limiting noise source for 'third generation' detectors that are currently in conceptual design phase.

GW detectors are designed to measure the spatial fluctuations induced by gravitational radiation from astro-physical phenomena. Seismic isolation systems, such as Virgo's super-attenuators [5], have been developed to suppress the mechanical coupling of seismic motion to unwanted vibrations of the interferometer components. However, the resulting transients in the local gravitational field produce Newtonian forces on the interferometer test masses and subsequently inject noise at the interferometer output.

A European project to establish the conceptual design for a third generation GW detector has come to a conclusion this year [6]. The resulting gravitational observatory is coined Einstein

Telescope (ET). ET will improve on the sensitivity of second generation detectors by about another factor of 10 with NN expected to limit performance below 10 Hz. To better understand the problems associated with NN a comprehensive understanding of the expected seismic wave fields is imperative. As part of the ET design study we have conducted a world-wide seismic noise investigation that involved seismic data taking at various sites in Europe and across the globe, as well as the development of analytical and finite element models. Seismic motion in the frequency range of interest predominately originate from human or wind activity. This motivates the choice of third generation detectors, such as ET, to be built underground. Other sources of NN include atmospheric density fluctuations and movement of near-by objects such as people and cars [7]. The latter is a controllable source and atmospheric NN is expected to be less significant than the seismic contribution for underground locations. Here we focus only on seismic NN.

2. Newtonian noise

For a given mass distribution described by a density function $\rho(\mathbf{r},t)$, the acceleration $\mathbf{a}$, experienced by a test mass located at $\mathbf{y}$ is given by

$$\mathbf{a}(\mathbf{y},t) = G\int_V \rho(\mathbf{r},t)\frac{\mathbf{r}'}{|\mathbf{r}'|^3}d^3\mathbf{r}, \tag{1}$$

where $\mathbf{r}' = \mathbf{r} - \mathbf{y}$ and G is the universal gravitational constant. Newtonian noise is associated with density fluctuations of the surrounding geology. When the ground is displaced by a seismic wave field, $\vec{\xi}(\mathbf{r},t)$, density perturbations are produced according to $\delta\rho = -\nabla\rho_0(\mathbf{r})\vec{\xi}(\mathbf{r},t)$. The resulting Newtonian noise term can be written as

$$\delta\mathbf{a}_{NN}(\mathbf{y},t) = G\int_V (\nabla\otimes\frac{\mathbf{r}'}{|\mathbf{r}'|^3})\vec{\xi}(\mathbf{r},t)\rho_0(\mathbf{r})d^3\mathbf{r}, \tag{2}$$

where $\rho_0(\mathbf{r})$ represents the equilibrium density of the medium. The effects of NN on GW detectors have been studied previously [8–10] where the authors present coinciding analytical estimates of NN for surface detectors. Based on these models, it is expected that the sensitivity of Advanced LIGO and Advanced Virgo will be limited around 10 Hz, the lower bound of the detector measurement bandwidth. ET proposes to push the lower bound down to 1.7 Hz, in this way placing more stringent constraints on the level of acceptable seismic noise. To achieve such a broad-band instrument, ET will consist of two separate interferometers, each optimized for respective high and low frequency detections. Fig. 1a shows the projected noise curve for Einstein Telescope's low frequency detector [11]. Newtonian noise is expected to be the major contributor to noise between 2 - 6 Hz. Below 2 Hz it will be limited by ground vibrations coupling mechanically through the conceived seismic isolation systems.

It is important to note that NN is a direct Newtonian coupling between the interferometer test masses and the surrounding geology. It can therefore not be shielded or otherwise mechanically suppressed. In addition it cannot be measured by any instrument other than the interferometer itself. Therefore, finding a suitably quiet seismic environment is the first step in reducing the effects of NN. The level of seismic noise assumed in Fig. 1a is derived from seismic studies and set to 0.5 $\mathrm{nm}/\sqrt{\mathrm{Hz}}\ (\mathrm{Hz}/f)^2$ above 1 Hz. A site with lower seismicity would simply reduce the NN contribution. A reduction by a factor of 3 would, on average be sufficient to let other noise sources, such as suspension thermal and quantum noise dominate. Subtraction techniques are under development [12, 13] that would closely monitor the surrounding geology and predict the resulting NN for each test mass. These predictions can then be subtracted from the interferometer output.

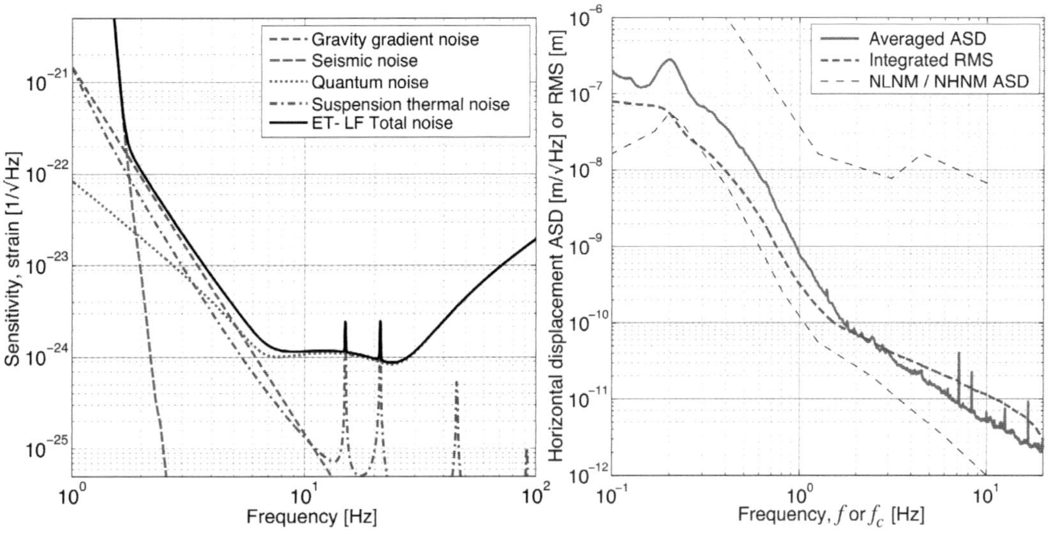

(a) (b)

Figure 1. (a) Expected noise budget for ET's low frequency interferometer [11]. Seismic noise (dashed brown curve) will limit sensitivity below 2 Hz and Newtonian noise (dashed green curve) below 6 Hz. (b) A low seismic noise environment as measured at the Canfranc underground laboratory in Spain. The horizontal displacement ASD averaged over several days (red curve) with the corresponding integrated RMS (dashed magenta curve). The dashed black curves represent the Peterson new high and low noise model spectral levels.

3. Ambient ground motion and seismic noise

As part of the ET design project studies were performed to determine seismic noise characteristics at various sites. The studies aim to explore potential locations that yield significant reduction in seismic power spectral density (PSD) in comparison with the current LIGO and Virgo sites, and thus provide a valid estimate of achievable low seismic noise environments in Europe. During the studies the geophysical aspects of underground infrastructure were also considered. Sites were chosen from varying geologies such as clay, salt and hard rock. The study includes 15 locations in 9 European countries, USA and Japan. In the USA two seismometers were installed at the Dusel 3D seismic network [14] in Lead, South Dakota. Data were also taken at the site of KAGRA, the GW detector under construction in the Kamioka mine, Japan. Close collaboration with the Royal Dutch Meteorological Institute [15] provided cross checks of data and analysis techniques at their permanent seismic observatory in The Netherlands. The sites are given in Table 1, with their depth underground, elevation above sea level and indication of general geology, while in Fig. 2 their location is indicated on a map. Two Trillium 240 seismometers were used to obtain seismic data at each site over a period of 1 - 2 weeks, being sure to always include a weekend. All the sites, with the exception of Mol, Belgium, were visited between April 1 and July 30, 2010.

We characterized the sites in terms of RMS displacement and their acceleration power spectral variation. The RMS displacement is the square root of the PSD integrated from the Nyquist frequency to some cut-off frequency, f_c [16]. The horizontal displacement amplitude spectral density (ASD) of the Spanish site and resulting integrated RMS curve are shown in Fig. 1b. The seismic noise levels for all the locations with f_c = 2 Hz are given in Table 1. A displacement ASD of 0.5 nm/$\sqrt{\text{Hz}}$ (Hz/f)^2 above 1 Hz, corresponds to an integrated RMS of 0.1 nm (at 2

Location	Depth [m]	Elevation [m]	Seismic noise [nm] (RMS, $f_c = 2$ Hz)	Geology
Spain - LSC, Canfranc	900	1600	0.070	Hard rock
Italy - Sardinia	185	205	$0.077^{\dagger}$	Hard rock
Hungary - Gyöngyösoroszi	400	400	0.082	Hard rock
	70	400	0.12	
France - LSM, Modane	1750	1000	0.10	Hard rock
Japan - Kamioka	1000	358	0.11	Hard rock
Finland - Sumiainen	0	185	0.11	Surface rock
Italy - Gran Sasso	1400	970	0.13	Hard rock
Germany - Black forest	95	850	0.20	Granite
Romania - Slănic-Prahova	190	195	$0.25^{\dagger}$	Salt
USA - Dusel, Lead, SD	1250	350	$0.64^{\dagger}$	Hard rock
	640	960	$1.12^{\dagger}$	
Germany - Moxa	35	455	0.70	Hard rock
Netherlands - Heimansgroeve	10	135	1.07	Surface rock
Belgium - Mol	230	-205	2.10	Clay
Italy - Sicily	60	-30	$9.24^{\dagger}$	Salt
Italy - Virgo, Pisa	0	5	54.52	Sedimentary soil

Table 1. Summary of the locations investigated as part of the ET seismic studies. The seismic noise RMS level is obtained from displacement spectra averaged over several days and integrated for frequencies above 2 Hz. A displacement spectrum of 0.5 nm/$\sqrt{\text{Hz}}$ $(\text{Hz}/f)^2$ above 1 Hz, provides a seismic noise RMS of 0.1 nm. For numbers denoted by a dagger (†) mining or other underground activities were underway during data taking.

Hz). The spectral variation is calculated from half hour averaged power spectra. The 90 and 10 percentiles, defined as the level under which the PSD remains for 90 and 10 percent of the time respectively, provide an indication of the variation in the seismic power. A plot of the spectral variation from a selection of sites is given in Fig. 3. The transparent color regions are bounded by the 90 and 10 percentiles while the solid curves are the mode or most common PSD value in each frequency bin. For comparison the spectral variation at the site of the Virgo GW detector is also given. It is clear that selecting a suitable location may improve seismic conditions by several orders of magnitude above 1 Hz.

Results for the Peterson new high (NHNM) and low (NLNM) noise models [17] are also shown in Fig. 3 and Fig. 1b. These models give a theoretical indication of extremely high and low ambient seismic motion respectively. The large peak between 0.1 and 0.3 Hz is known to be generated by oceanic waves. It has been shown that seismic motion from the Dusel array at these frequencies can be correlated with ocean waves from buoy data on the northern Pacific and northern Atlantic [18]. Similarly in Europe, this microseismic peak can be attributed to northern Atlantic ocean activity [19]. Italian sites suffer from an additional analogous peak around 0.5 Hz as a result of Mediterranean sea activity [20].

In the frequency range of interest, seismic activity is dominated by wind and human activity. The seismic wave fields from these sources are excited on, and propagate predominately along the surface, resulting in a reduction in seismic activity with depth [21]. The influence of near-by traffic on GW detectors has been reported before at the LIGO [22] and Virgo [23, 24] sites. This so-called anthropogenic noise is discernible in diurnal patterns in PSD values above 1 Hz, including a clear difference between weekend and weekday spectra. Similar patterns have been attributed to atmospheric effects such as wind. The presence of wind causes movement of surface

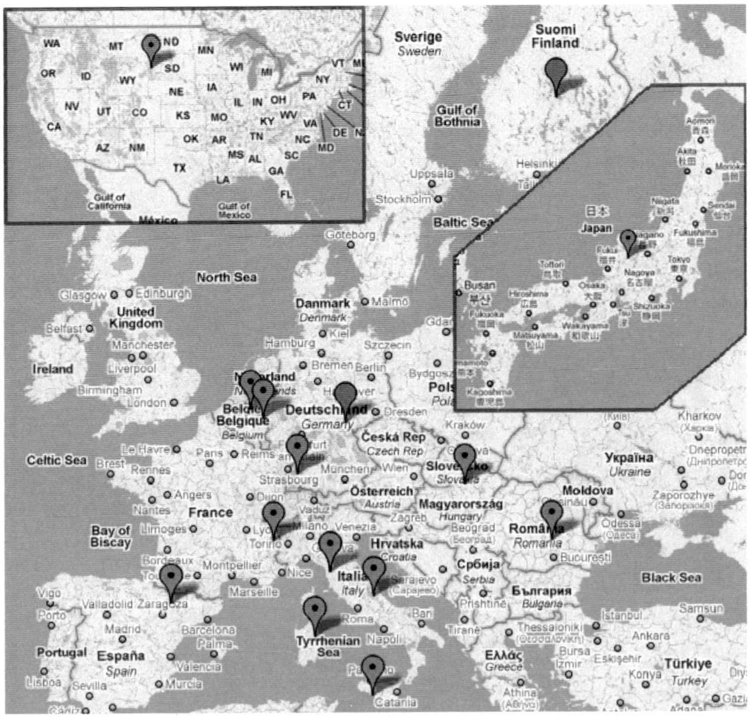

Figure 2. Map showing the locations of sites investigated during the ET site studies. Data were collected from the sites indicated by red markers with a dot, whilst third party data were used for the locations represented by blue markers.

objects, such as trees or buildings, or directly through turbulent pressures on topographical irregularities [25, 26]. The day and night ratios for various sites are shown in Fig. 4. There is an obvious dependence on day-time activity except at the Spanish site. This can be explained by the depth and remoteness of the site. It is situated in the northern Pyrenees, where the population density is just 1.38 people per km^2 in comparison with 10, 75 and 180 people per km^2 for the Sardinian, Hungarian and Romania sites respectively. The extreme ratio of the Romanian site is also due to daily mining activities in the same salt structure.

Geophysical aspects of underground construction will play a crucial role in site selection and covers a vast number of topics and considerations. Here we will restrict the discussion to a few general observations. From Table 1 we notice that the most seismically quiet sites are found in hard rock geologies, while the soft geologies such as salt and clay, feature amongst the highest seismic noise environments. In terms of construction of underground facilities, rock stability is a crucial factor, which tends to be more favorable in hard rock geologies. However, subterranean characteristics will also play a role in the transfer of seismic to Newtonian noise as we will demonstrate in the following section.

4. Newtonian noise modeling

Recent studies of NN involve mathematical and finite element (FE) analysis techniques. These attempt to describe NN as a function of depth. A major challenge of underground NN modeling is making correct assumptions about contributions from the various seismic wave types. The wave field in a homogeneous elastic medium can be expressed as a combination of surface and

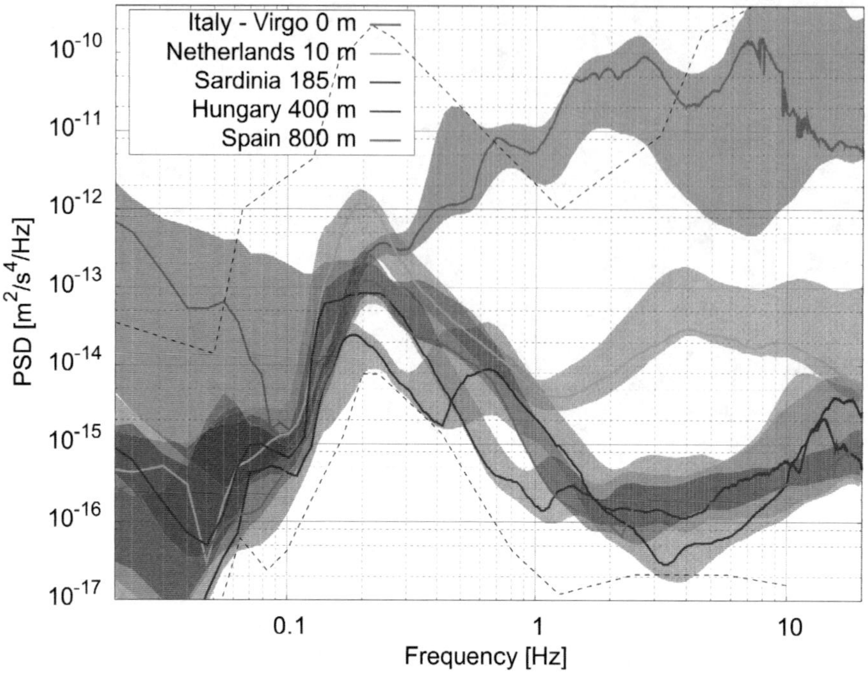

Figure 3. Spectral variation results for various sites. The transparent color regions are bounded by the 90 and 10 percentiles while the solid curves represent the mode or most common PSD value in each frequency bin. The dashed black curves indicate the Peterson new high and low noise models.

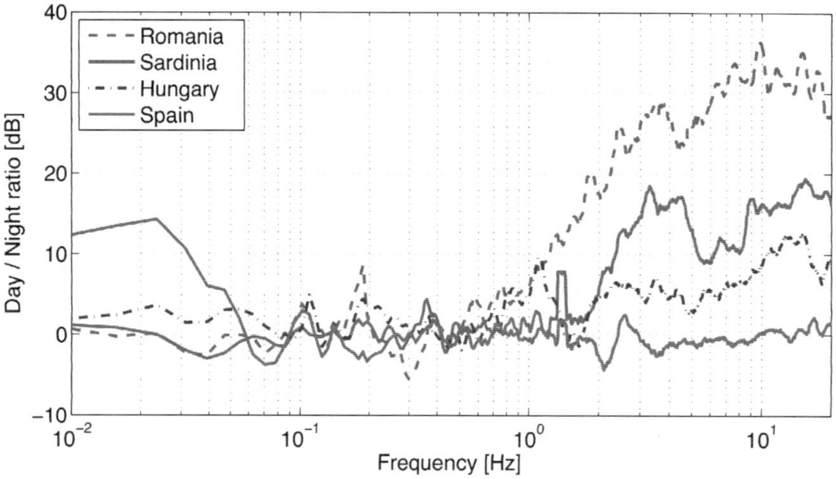

Figure 4. Midday versus midnight noise PSD ratios as a function of frequency. The variations show a strong correlation between seismic noise and human activity at frequencies above 1 Hz. The Romanian site is an extreme case due to the mining activities in the same mine during data taking.

body waves. Two types of body waves exist, pressure (P) and shear (S) waves with respective wave speeds c_P and c_S [27]. In the case of P-waves the movement of a ground particle is parallel to the direction of wave propagation. For S-waves the particle motion is perpendicular to the direction of the wave. In a medium that interfaces with air, two types of surface waves exists: Rayleigh and Love waves. Rayleigh waves produce the largest contribution to NN [9]. They are polarized perpendicular to the surface and vanish with depth.

FE modeling involves subdividing a 3D medium into small volume elements, constituting several corner points called nodes. Within each element the relevant physical parameters, like displacement and stress, are approximated by spline functions that react to a given excitation. The NN calculation is done with the numerical solution to Eq.(2), given by

$$\Delta \mathbf{a}_{NN}(\mathbf{y},t) = G\sum_i (\nabla \otimes \frac{\mathbf{r}'_i}{|\mathbf{r}'_i|^3})\vec{\xi}_i(\mathbf{r},t)m_i, \tag{3}$$

where $m_i = \rho_0 \Delta V_i$ is the mass per node and is assumed constant for all nodes within an element.

Seismic waves can roughly be characterized according to the distance to their source. For near-by or 'local' sources the wave fields are seen to propagate radially away from a single point source. Distant wave fields are excited by sources at a distance, $d \gg \lambda_P = c_P/f$, from the observer. It has been shown that for distant sources an analytical model of the seismic wave field can be constructed from a combination of plane waves [28]. The authors concluded that in such a case the contribution to NN from surface Rayleigh waves is small at depths comparable to the P-wavelength, λ_P. NN will then be dominated by body waves. Note that for hard rock at 1 Hz this involves depths of several kilometers. Similarly FE models have provided predictions for the reduction of NN as a function of depth and wave frequency as shown by the markers in Fig. 5. For comparison results from an analytical model developed by Cella [12, 29] for surface Rayleigh waves are also plotted and show qualitative agreement with the FE results.

For local sources the analytical solutions of the wave fields become more complex and involve all wave field types. In this case we rely on FE analysis, that accurately predicts seismic motion for a given excitation [30]. An example of a FE model displaying a wave field after a local vertical pulse excitation at the origin is shown in Fig. 6a. The model comprises a finite half-space (of which a quarter is shown) of a clay-like medium resulting in a P-wave speed, $c_P = 800$ m/s. The P and S waves can be seen radiating away from the source, with the fastest being the P-wave and the S-wave traveling at roughly half that speed. A Rayleigh wave propagates along the surface at a speed similar to the S-wave. Fig. 6b shows the calculated NN reduction as a function of the depth. The simulations were done along the dashed line shown in Fig. 6a. The reduction in NN as a function of depth is significantly less than expected from the previously discussed distant source Rayleigh wave models. This confirms that NN at depth is dominated by contributions from body waves. These results also clearly reveal the notion that NN suppression with depth increases with lower seismic velocities and higher frequencies.

5. Conclusions

Seismic studies have revealed that future third generation gravitational wave detectors will be able to benefit from significantly lower seismic environments, by up to several orders of magnitude in power spectral density, in comparison with current detector sites. This amounts to a conservative seismic requirement for ET of 0.5 nm/$\sqrt{\text{Hz}}$ (Hz/f)^2 above 1 Hz. At this level NN is expected to limit the interferometer sensitivity from 2 to 6 Hz, with seismic noise being the major contributor below 2 Hz. It is important to note that by simply moving to a more quiet seismic environment both NN and seismic noise will decrease. Subtraction techniques may be used to a posteriori remove part of the NN. Reduction by a factor of 3 will, on average allow suspension thermal and quantum noise to dominate the noise budget at these frequencies.

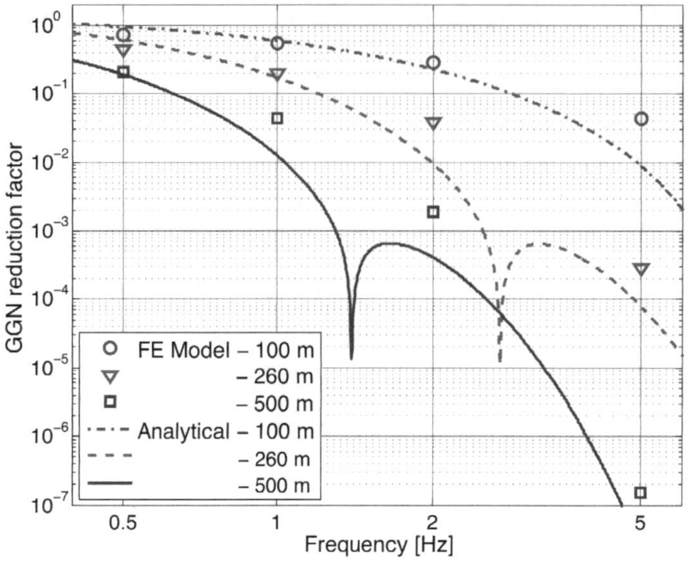

Figure 5. FE simulations of density fluctuations from plane surface waves in a clay-like medium where $c_P = 1000$ m/s. Results show NN reduction compared to surface at 100 (circles), 260 (triangles) and 500 m (squares) depth as a function of frequency. For comparison analytical results from Cella [29] of plane surface Rayleigh waves are indicated by the respective curves.

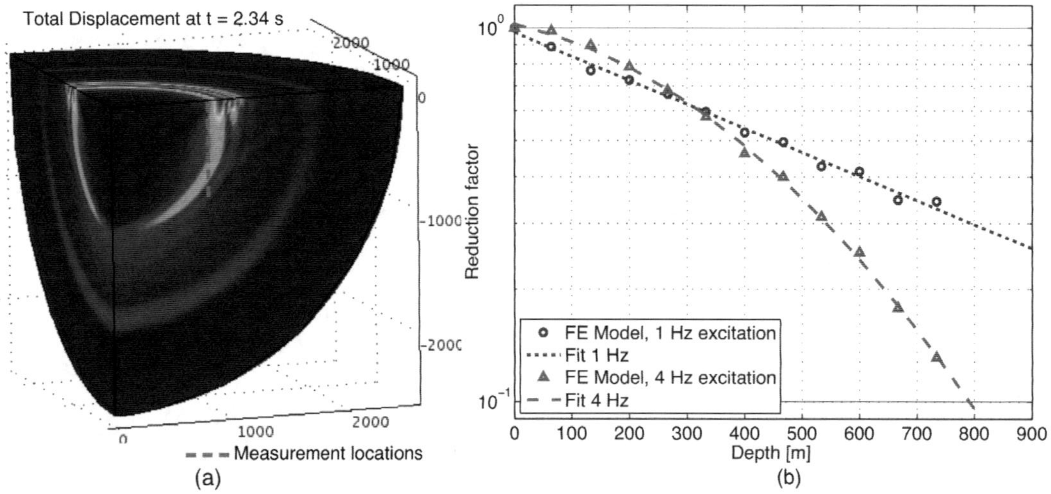

Figure 6. (a) Total displacement for a time domain simulation at 2.34 seconds after a 1 μm vertical pulse excitation at the center of the half-sphere. The P, S and Rayleigh waves can be distinguished. Ground modeled by clay-like medium with a P-wave speed, $c_P = 800$ m/s. (b) NN reduction as a function of depth for 1 and 4 Hz excitations. Simulations done at 800 m (corresponding to λ for 1 Hz waves) horizontally from the excitation point, along the dashed line in (a).

The seismic studies also show that the most quiet sites are in hard rock geologies. While this is unfavorable for NN reduction which is inversely proportional to seismic wave speed, it does provides a stable environment for the construction of large underground facilities. When considering a potential GW detector site, it is important to avoid nearby industry, busy road and train routes and highly populated areas. These anthropogenic sources are the main contributor to seismicity above 1 Hz. Moving underground provides low seismic noise and suppression of atmospheric NN. Additionally, as models have shown, NN contributions from surface waves generated by distant sources significantly attenuate with depth. The remaining NN contributions will then be due to body waves and local seismic sources.

Acknowledgments

We are grateful to many people for their valuable help during seismic data taking, in particular we thank, R. Sleeman, I. Racz, K. Kuroda, F. Ricci, R. Mircea Margineanu, R. Widmer-Schnidrig, J. Harms, R. De Salvo, A. Bettini, M. Zampaolo, E. Coccia and staff at EURIDICE. For discussions on NN we also thank G. Cella. This work has been performed with the support of the European Commission under the Framework Programme 7 (FP7) Capacities, project Einstein Telescope design study (Grant Agreement 211743). This work is part of the research programme of the Foundation for Fundamental Research on Matter (FOM), which is financially supported by the Netherlands Organisation for Scientific Research (NWO).

References

[1] Kuroda K *et al.* 2010 *Class. Quantum Grav.* **27** 084004
[2] Acernese F *et al.* 2008 *Class. Quantum Grav.* **25** 184001
[3] Smith J R for the LIGO Scientific Collaboration 2009 *Class. Quantum Grav.* **26** 114013
[4] Grote H for the LIGO Scientific Collaboration 2010 *Class. Quantum Grav.* **27** 084003
[5] Braccini S *et al.* 2005 *Astroparticle Physics* **23** 557–565
[6] Abernathy M *et al.* 2011 Einstein gravitational wave telescope conceptual design study ET-0106C-10 URL `http://tds.ego-gw.it/`
[7] Thorne K and Winstein C 1999 *Phys. Rev. D* **60** 082001
[8] Saulson P 1984 *Phys. Rev. D* **30** 732–736
[9] Hughes S and Thorne K 1998 *Phys. Rev. D* **58** 122002
[10] Beccaria M *et al.* 1998 *Class. Quantum Grav.* **15** 3339–3362
[11] Hild S *et al.* 2011 *Class. Quantum Grav.* **28** 094013
[12] Beker M *et al.* 2011 *General Relativity and Gravitation* **43** 623–656
[13] Harms J, DeSalvo R, Dorsher S and Mandic V 2009 *arXiv:0910.2774v1 [gr-qc]*
[14] Dusel, Deep Underground Science and Engineering Laboratory URL `http://www.dusel.org`
[15] KNMI, Royal Dutch Meteorological Institute URL `http://www.knmi.nl`
[16] Bialowons W *et al.* 2007 *EUROTeV-Report* **2007-011**
[17] Peterson J 1993 *U.S. Department of Interior Geological Survey* Open-File Report 93-322
[18] Harms J *et al.* 2010 *arXiv:1006.0678v1 [gr-qc]*
[19] A Friedrich F K and Klinge K 1998 *Journal of Seismology* **2** 47–64
[20] Marchetti E and Mazzoni M 2004 *Internal Virgo note* VIR-NOT-FIR-1390-261
[21] Carter J A *et al.* 1991 *Bull. Seism. Soc. of Am.* **81** 1101–1114
[22] Schofield R *et al.* 2000 *LIGO Scientific Collaboration Meeting, Hanford*
[23] Fiori I, Holloway L and Paoletti F 2003 *Internal Virgo Note* VIR-NOT-FIR-1390-251
[24] Acernese F *et al.* 2004 *Class. Quantum Grav.* **21** S433–S440
[25] Withers M *et al.* 1996 *Bull. of the Seism. Soc. of America* **86** 1507–1515
[26] McNamara D E 2004 *Bull. of the Seism. Soc. of America* **94** 1517–1527
[27] Achenbach J 1973 *Wave Propagation in Elastic Solids* (Amsterdam: North-Holland) pp 187–194
[28] Harms J, DeSalvo R, Dorsher S and Mandic V 2009 *Phys. Rev. D* **80**(12) 122001
[29] Cella G 2009 *2nd ET annual workshop, Erice* URL `http://www.et-gw.eu/2ndannualworkshop`
[30] Beker M G, van den Brand J F J, Hennes E and Rabeling D S 2010 *J. Phys.: Conf. Ser.* **228** 012034

The effect of mirror surface distortions on higher order Laguerre-Gauss modes

C Bond, P Fulda, L Carbone, K Kokeyama∗ and A Freise
School of Physics and Astronomy, University of Birmingham, Edgbaston, Birmingham, B15 2TT, UK
E-mail: czb@star.sr.bham.ac.uk

Abstract. Higher order Laguerre-Gauss (LG) beams have been proposed for use in future generation gravitational wave detectors for their potential to reduce the effects of the thermal noise of the test masses. However, it has been reported that due to the degeneracy of higher order modes using these beams will be extremely challenging. Our aim was to quantify these degeneracy effects. We present a new analytical approximation to compute the coupling between different LG modes, verified with simulation results of realistic arm cavities. This method is applied to Advanced LIGO mirror maps and used to derive requirements for mirrors for the use of the LG_{33} beam.

1. Introduction

The sensitivities of second generation gravitational wave detectors are expected to be limited by thermal noise of the test masses, in particular coating brownian thermal noise, at a range of frequencies around 100 Hz [1]. The Brownian motion of atoms in the mirror surfaces leads to excitations of the elastic modes of the mirror, which is interpreted by the interferometer as a mirror displacement, mimicking the displacement caused by a gravitational wave. It has been proposed that the use of beams with a different intensity distribution than the currently used fundamental Gaussian beam could reduce the effects of mirror thermal noise [2, 3]. Beams whose intensity is more evenly distributed over a mirror surface can more effectively average over mirror surface distortions, such as those caused by thermal noise [4]. Several different beams have been suggested, including mesa beams [5], conical beams [6] and higher order Laguerre-Gauss (LG) beams. Higher order Laguerre-Gauss beams, unlike mesa and conical beams, are compatible with spherical mirrors, which are currently used in gravitational wave detectors. Higher order LG beams therefore seem the most appropriate candidate for thermal noise reduction.

The potential of using the particular higher order beam LG_{33} in gravitational wave detectors has been investigated using simulations and table-top experiments [7, 8]. However, there is another aspect of higher order modes which must be considered. Higher-order gaussian modes are degenerate, or, in other words, there are multiple modes with the same order. The consequence of this is that an optical cavity resonant for a higher order LG mode will also be resonant for other modes of the same order. This is not the case for the currently used fundamental mode, which is unique in its mode order. This degeneracy can lead to additional optical losses. It has been shown that the LG_{33} mode, compared to the fundamental mode, could

∗Present address: Physics and Astronomy, Louisiana State University, Baton Rouge, LA 70803, USA result in an unacceptably high contrast defect at the interferometer output due to coupling into other modes via mirror surface distortions [9, 10, 11, 12]. In this paper we summarise our investigations into this coupling, focusing on the effects of mirror surface distortions on beam purity when an LG_{33} beam is injected into a high finesse cavity. A more detailed account of these investigations can be found in [13].

2. Laguerre-Gauss beams

The shape of any paraxial beam can be described by a sum of Gaussian modes. Two complete and orthogonal sets are the Hermite-Gauss (HG) modes, with rectangular symmetry, and Laguerre-Gauss (LG) modes, with circular symmetry. The Laguerre-Gauss modes are a defined by radial index $p \geq 0$ and azimuthal index l, with the helical modes given by [14]:

$$\begin{aligned} U_{p,l}(r,\phi,z) = {} & \frac{1}{w(z)}\sqrt{\frac{2p!}{\pi(|l|+p)!}}\exp\left(i\left(2p+|l|+1\right)\Psi(z)\right) \\ & \times \left(\frac{\sqrt{2}r}{w(z)}\right)^{|l|} L_p^{|l|}\left(\frac{2r^2}{w^2(z)}\right)\exp\left(-\frac{ikr^2}{2R_c(z)}-\frac{r^2}{w^2(z)}+il\phi\right) \end{aligned} \tag{1}$$

where k is the wavenumber, $w(z)$ is the beam radius, $\Psi(z)$ is the Gouy phase and $R_c(z)$ is the radius of curvature of the beam. $L_p^{|l|}$ refer to the Laguerre polynomials. The quantity $2p + |l|$ is defined as the order of the mode. The displacement seen by an interferometer due to thermal noise depends on the intensity distribution of the beams incident on the mirrors. In general higher order beams more effectively average over the mirror surface and results in a smaller effective displacement. However, using higher order beams requires smaller beam spot sizes to achieve the same low power loss due to clipping by the mirrors. The particular LG beam suggested to reduce thermal noise is LG_{33}, which results in a good reduction of thermal noise for the low clipping losses required. A plot of the intensity distribution of this higher order beam, compared to the currently used fundamental beam, is shown in figure 1. This plot shows that the intensity of the fundamental beam is concentrated around the centre of the optical axis, whereas the intensity of the LG_{33} beam is more evenly distributed around the axis. In theory

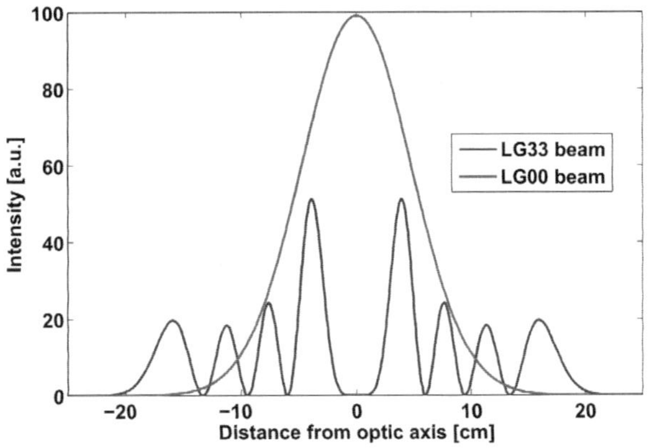

Figure 1. Plots of the intensity of the fundamental gaussian mode (LG_{00}) and the LG_{33} mode as a function of distance from the optical axis.

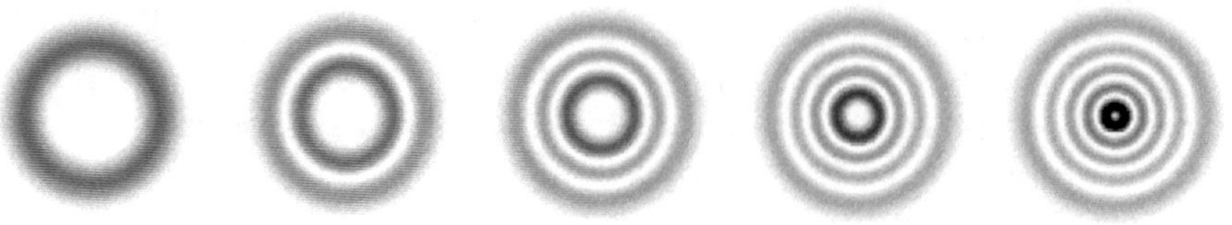

Figure 2. Intensity plots for the order 9 Laguerre-Gauss modes. From left to right; $LG_{0,\pm9}$, $LG_{1,\pm7}$, $LG_{2,\pm5}$, $LG_{3,\pm3}$ and $LG_{4,\pm1}$.

this higher order beam could reduce the thermal noise effects by a factor of 1.61 in an Advanced LIGO style interferometer [15].

2.1. Higher-order mode degeneracy

Laguerre-Gauss modes experience a phase shift of $(2p + |l| + 1)\Psi(z)$ compared to that of a plane wave. Modes of the same order $(2p + |l|)$ will accumulate the same round trip phase when circulating in the arm cavities of an interferometer and hence will be resonant for the same cavity tuning. Therefore, cavities designed for higher order modes, such as LG_{33}, will be degenerate for modes of the same order. This can become a problem if there are processes which introduce new modes to the interferometer. The imperfect nature of the mirror surfaces results in distortions of the beams interacting with them. This can be described as coupling into other modes. If this coupling results in modes which are resonant for the cavity they will have large amplitudes in the cavity.

The fundamental Gaussian mode is currently used in gravitational wave detectors. LG_{00} is the only mode of order 0 so any coupling due to mirror surface distortions results in modes of a different order. These additional modes are suppressed in the cavities. LG_{33}, on the other hand, is one of 10 order 9 modes; $LG_{0,\pm9}$, $LG_{1,\pm7}$, $LG_{2,\pm5}$, $LG_{3,\pm3}$ and $LG_{4,\pm1}$ (figure 2). Any coupling into these modes via mirror surface distortions could result in modes other than LG_{33} with large amplitudes in the arm cavities. As the surface distortions will be different in the mirrors in the different cavity arms the circulating fields will no longer have the same mode content and will not interfere as expected at the dark port. This could result in a large contrast defect.

3. Describing mirror surfaces

In an ideal world the mirrors used in gravitational wave detectors would be perfectly smooth, with a radius of curvature matching that of the incident beam. The effect of a distorted mirror surface $Z(x, y)$, with reflectivity r, on a beam $U(x, y, z)$, reflected from the surface can be described by a phase shift:

$$U_{ref}(x, y, z) = U(x, y, z)\ r \exp\left(i2kZ(x, y)\right) \tag{2}$$

where k is the wavenumber and the beam is propagating in the z direction. In terms of higher-order Laguerre-Gauss modes we can describe the beam reflected from the mirror surface as having a different mode composition to the incident beam. The coupling from a mode $U_{p,l}$ in the incident beam into a mode $U_{p',l'}$ in the reflected beam, considering a completely reflecting surface Z can be described by a coupling coefficient [16, 17]:

$$k^{Z}_{p,l,p',l'} = \int_S U_{p,l} \exp\left(2ikZ(r, \phi)\right) U_{p',l'} \tag{3}$$

Z describes the surface height of the mirror and S describes an infinite plane perpendicular to the optical axis.

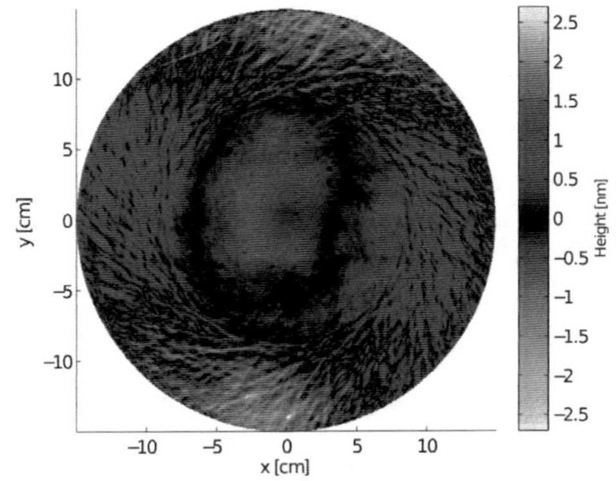

Figure 3. Surface plot of the mirror map corresponding to the Advanced LIGO end test mass ETM08. The mirror map data refers to the difference in surface height from a perfect sphere.

3.1. Mirror maps

In order to investigate the effects of surface distortions in gravitational wave detectors we require methods of representing surface distortions in mathematical models and simulations. One such method is to use mirror surface maps measured from existing mirrors designed for gravitational wave detectors. *Mirror maps* refer to an array of data detailing particular properties of a mirror over its surface. In this case mirror maps refer to the surface height of the mirror. We can use these maps to represent realistic mirror surface distortions in simulations of gravitational wave detectors. The best mirror maps of this kind currently available are those produced from mirror substrates for Advanced LIGO. In our investigations we used a typical Advanced LIGO mirror map, ETM08 [18], produced from the curved surface of an end test mass. The surface map for this mirror is shown in figure 3, where the surface refers to the deviation from a perfect sphere with radius of curvature 2249.28 m. This surface has a RMS figure error of 0.523 nm.

3.2. Zernike polynomials

Zernike polynomials provide another method of representing mirror surface distortions. Zernike polynomials form a complete set of functions which are orthogonal over the unit disc [19, 20]. Each polynomial is defined by a radial index n and azimuthal index m, with $0 \leq m \leq n$. For any index m there is an even and odd polynomial:

$$\begin{aligned} Z_n^{+m}(\rho,\phi) &= A_n^{+m}\ \cos(m\phi)R_n^m(\rho) \quad \text{even polynomial} \\ Z_n^{-m}(\rho,\phi) &= A_n^{-m}\ \sin(m\phi)R_n^m(\rho) \quad \text{odd polynomial} \end{aligned} \tag{4}$$

where $A_n^{\pm m}$ is the amplitude of the polynomial, ρ is the normalised radius, ϕ is the azimuthal angle and $R_n^m(\rho)$ is the radial function. The radial function is given by:

$$R_n^m(\rho) = \sum_{h=0}^{\frac{1}{2}(n-m)} \frac{(-1)^h(n-h)!}{h!\left(\frac{1}{2}(n+m)-h\right)!\left(\frac{1}{2}(n-m)-h\right)!}\rho^{n-2h} \tag{5}$$

for even $n-m$ and 0 otherwise. Any surface defined over a disc, such as the mirrors used in gravitational wave detectors, can be described as a sum of Zernike polynomials, with higher order (n) Zernike polynomials representing higher spatial frequencies.

4. Analytical description of mode coupling via surface distortions

Using Zernike polynomials to describe mirror surface distortions we can analyse the coupling between LG modes. Coupling via a mirror described by an individual Zernike polynomial is given by:

$$k^{n,m}_{p,l,p',l'} = \int_S U_{p,l} \exp\left(2ikZ^m_n\right) U^*_{p',l'} \tag{6}$$

As the surface distortions are small compared to the wavelength we can simplify the integral using the approximation:

$$\exp\left(2ikZ\right) \approx 1 + 2ikZ \tag{7}$$

Solving the integral [13] it can be shown that the only non-zero solutions to the approximation occur when:

$$m = |l - l'| \tag{8}$$

This is a significant result as we can already make predictions about the types of surface distortions which could lead to significant coupling between modes of the same order. The final approximation of the coupling is given by:

$$\begin{aligned} k^{n,m}_{p,l,p',l'} = A^m_n k \sqrt{p!p'!(p+|l|)!(p'+|l'|)!} \Bigg| \sum_{i=0}^{p} \sum_{j=0}^{p'} \sum_{h=0}^{\frac{1}{2}(n-m)} \frac{(-1)^{i+j+h}}{(p-i)!(|l|+i)!i!} \\ \times \frac{1}{(p'-j)!(|l'|+j)!j!} \frac{(n-h)!}{\left(\frac{1}{2}(n+m)-h\right)!} \frac{1}{\left(\frac{1}{2}(n-m)-h\right)!h!} \\ \times \frac{1}{X^{\frac{1}{2}(n-2h)}} \gamma(i+j-h+\frac{1}{2}(|l|+|l'|+n)+1, X) \Bigg| \end{aligned} \tag{9}$$

where A^m_n is the amplitude of the Zernike polynomial representing the mirror surface, $X = \frac{2R^2}{w^2}$, R is the radius of the mirror, w is the beam radius and γ refers to the lower incomplete gamma function.

4.1. Coupling for order 9 modes

We are concerned with coupling from LG_{33} into the other order 9 modes, and want to identify the particular surface distortions which will cause such couplings. Table 1 shows the required azimuthal index required to couple significantly from LG_{33} into the other order 9 modes. We can predict which particular distortions, if present in the mirror surfaces, will cause coupling into particular order 9 modes. As we have $m \leq n$ these m requirements also tell us the lowest order (n) Zernike polynomial required to couple from LG_{33} into each of the other order 9 modes.

Table 1. The azimuthal index, m, of the Zernike polynomial describing a mirror surface which results in significant coupling from an incident LG_{33} beam into the other order 9 modes.

m	2	2	4	4	6	6	8	10	12
$LG_{p,l}$ mode	2, 5	4, 1	1, 7	4, -1	0, 9	3, -3	2, -5	1, -7	0, -9

5. Simulating cavities with mirror surface distortions

The problem of higher-order mode degeneracy originates in the arm cavities of the detectors. The analytical approach taken in section 4 applies to a single reflection, whereas inside a cavity multiple reflections and couplings of the beam occur. However, for small coupling coefficients it is expected that the direct coupling from the input beam will be the dominant process in determining the mode content of the circulating beam. In this section we compare our analytical approach with simulations of realistic arm cavites.

Simulations of arm cavities were carried out to assess the purity of the beams circulating in such cavities when LG_{33} is injected into a cavity simulated with realistic mirror surface distortions. Here we refer to the purity in terms of the LG_{33} mode content of the beam. Cavity simulations were carried out in FINESSE [21, 22] based on Advanced LIGO arm cavities. The arm cavities in Advanced LIGO consist of two curved mirrors; the end test mass (ETM) and input test mass (ITM), separated by approximately 4 km. These cavities are designed for the fundamental Gaussian beam. As the LG_{33} beam has a wider intensity distribution than LG_{00} the Advanced LIGO cavity parameters result in a clipping loss of around 35%. In order to achieve smaller beam spots on the mirrors it was necessary to decrease the cavity length. A cavity length of 2802.9 m gives similar clipping to that experienced by the LG_{00} beam in the Advanced LIGO cavities, for the 30 cm aperture which the mirror maps represent and using the original mirror curvatures of 1934 m for the ITM and 2245 m for the ETM. The simulations used the Advanced LIGO mirror parameters [23, 24].

5.1. Circulating purity with Advanced LIGO mirror maps

In our simulations the Advanced LIGO mirror map ETM08 (figure 3) was used to represent surface distortions on the end test mass. In order to relate these simulations with our previous analytical approach and predict the coupling at this mirror we look at the Zernike polynomials present in the map. The Zernike content of the ETM08 mirror map is summarised in table 2. The content is ordered according to the power coupled from LG_{33} into the other order 9 modes by the individual polynmials, according to our approximation (equation 9). This shows that the astigmatism (Z_2^2) causes a large amount of coupling from LG_{33} into the other order 9 modes. This suggests that the order 9 modes LG_{25} and LG_{41} will have significant amplitudes in the cavity, as polynomials with $m = 2$ should couple largely into LG_{25} and LG_{41} (equation 8). The polynomials Z_4^2 and Z_4^4 also couple relatively large powers into the other order 9 modes, which should add to LG_{25} and LG_{41} in the case of Z_4^2, and LG_{17} and LG_{4-1} in the case of Z_4^4.

The cavity defined in section 5 was simulated with a LG_{33} input beam and the ETM08 mirror map. The field circulating in the cavity on resonance for LG_{33} was detected and is shown in figure 4. Comparison with the LG_{33} beam shown in figure 2 shows that the circulating beam is no longer pure LG_{33}. The mode content of the circulating beam is summarised in table 3. The purity of the LG_{33} beam circulating in the simulated cavity was found to be 88.6% with most

Table 2. A summary of the Zernike polynomials representing the Advanced LIGO mirror map ETM08. The polynomials are ranked according to the power they individually subtract from LG_{33} into the other order 9 modes, according to the approximation (equation 9).

Z_n^m	2, 2	4, 2	4, 4	6, 2	10, 8	other
A_n^m [nm]	0.908	0.202	0.213	0.124	0.116	-
Power [ppm]	4.66	0.331	0.0431	0.0099	0.0059	< 0.005

Table 3. A summary of the mode content of a beam circulating in a cavity simulated with a LG_{33} input beam and the Advanced LIGO mirror map ETM08 applied to the end mirror.

$LG_{p,l}$ mode	3, 3	4, 1	2, 5	4, -1	1, 7	other
Power [%]	88.6	5.70	5.02	0.333	0.313	< 0.05

of the coupled power in the other order 9 modes. We find that the two largest coupled modes in the cavity are LG_{25} and LG_{41}. This is in agreement with the predictions made from looking at the Zernike content of ETM08 mirror map (table 2). LG_{17} and LG_{4-1} also have relatively large amplitudes in the cavity. This could be due to direct coupling via $m = 4$ polynomials or coupling of LG_{25} and LG_{41} via $m = 2$ polynomials. The coupling process is complicated, but these simulation results appear to suggest that the direct coupling, as described in our mathematical model, is the dominant effect on the mode content in a cavity.

5.2. Circulating purity with improved mirrors
To provide mirror requirements for the purposes of using higher order LG beams we consider which distortions cause the most coupling. Firstly, looking at the results using the ETM08 mirror map, we suggest that the astigmatism is the dominant distortion, in terms of coupling between order 9 modes. In order to verify this the cavity was simulated again with the ETM08 mirror map but with the astigmatism removed from the map data. The circulating beam for this simulation is shown in figure 5. The LG_{33} purity of the circulating beam was found to be 99.5%, a large improvement on the original purity of 88.6%. This supports the idea that astigmatism is a significant cause of coupling between order 9 modes. The mode content of the circulating beam is summarised in table 4. Although the power in modes LG_{41} and LG_{25} has been significantly reduced they are still the dominant modes after LG_{33}. This suggests that astigmatism is not the only distortion present in this mirror which couples into these modes.

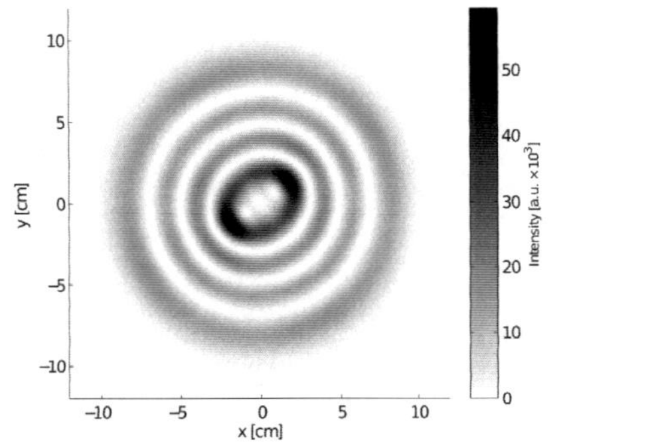

Figure 4. The field circulating on resonance in a simulated cavity with an LG_{33} input beam and the ETM08 mirror map applied to the ETM.

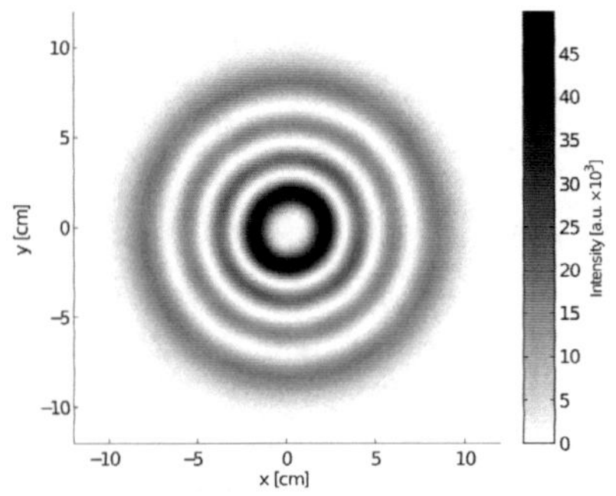

Figure 5. The field circulating in a cavity simulated with an LG_{33} input beam with the ETM08 mirror map. In this case the astigmatism has been removed from the mirror map.

Table 4. A summary of the modes circulating in a cavity simulated with a LG_{33} input beam and the Advanced LIGO mirror map ETM08, with the astigmatism removed.

$LG_{p,l}$ mode	3, 3	4, 1	2, 5	1, 7	4, -1	0, -9	other
Power [%]	99.5	0.231	0.208	0.0524	0.0165	0.0137	< 0.05

6. Mirror requirements for LG_{33}

Our investigation has shown that mirror surface distortions lead to a large degradation of the purity of LG_{33} circulating in the arm cavities. In order to use LG_{33} in gravitational wave detectors we require specific Zernike polynomials to be smaller than in the current available mirrors. The direct coupling between order 9 modes was investigated for our example mirror map, ETM08. Figure 6 illustrates the coupling from LG_{33} into the other order 9 modes from the ETM08 mirror map. In this plot the overall coupling from the Zernike polynomials representing the mirror map is approximated using equation 9, where n refers to the maximum order of the Zernike polynomials included in the approximation. This plot shows that the two modes LG_{25} and LG_{41} experience around 10 times the coupling than the other order 9 modes. The plot suggests that the coupling into these modes is not significantly increased by including higher order polynomials, i.e. the coupling into these modes is caused predominantly by low order polynomials. Therefore, reducing low order polynomials with $m = 2$, the required index for these couplings, should lead to an improvement in LG_{33} beam purity and should be the first requirement for the mirror surfaces.

Overall the ETM08 mirror map directly couples 31 ppm of the incident power from LG_{33} into other modes. Of this 6.8 ppm is coupled into the other order 9 modes. The Zernike content of the ETM08 mirror map is summarised in table 2. In this case we consider individual polynomials coupling more than 0.01 ppm into order 9 modes as causing a large amount of coupling. For this mirror these are the Z_2^2, Z_4^2 and Z_4^4 polynomials. Table 5 summarises the requirements for these

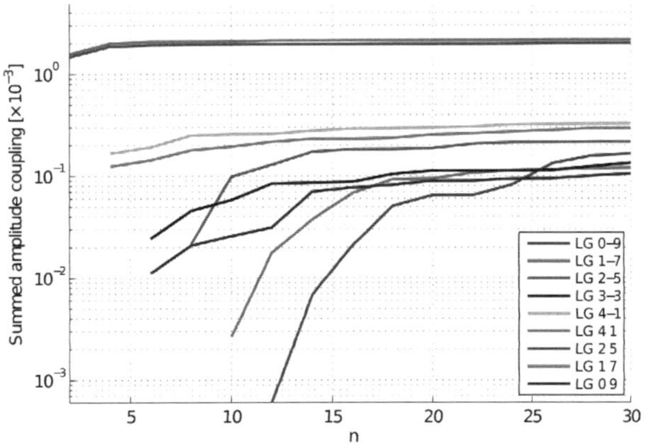

Figure 6. A plot of the approximate coupling into each order 9 mode when an LG_{33} beam is incident on a surface defined by the Advanced LIGO mirror map ETM08. The coupling is approximated from equation 9 and includes the coupling from each Zernike polynomial with order $\leq n$, weighted by the amplitude of the polynomials in the map.

Table 5. Amplitude requirements for surfaces defined by Zernike polynomials to couple 0.01 ppm from LG_{33} into the other order 9 modes.

Z_n^m polynomial	2, 2	4, 2	4, 4
Amplitude [nm]	0.042	0.035	0.100

3 polynomials to couple 0.01 ppm directly into order 9 modes. Applying these requirements to the mirror map ETM08 results in direct coupling of 19 ppm into other modes with just 0.043 ppm coupling into the other order 9 modes. The cavity defined in section 5 was simulated with this modified map, achieving a circulating beam impurity of 815 ppm. This is a very significant improvement on the original impurity of 114,000 ppm (88.6% LG_{33} purity) and a further improvement on the 5220 ppm impurity (99.5% LG_{33} purity) when the astigmatism was completely removed from the map.

We have demonstrated that a high beam purity is achievable by limiting the amplitudes of specific low order Zernike polynomials in the mirror surfaces. In order to achieve even higher beam purity the amplitudes of these distortions will have to be reduced further, as well as reducing other polynomials in the mirror surface.

7. Conclusion

We have presented an analytical approach to coupling between Laguerre-Gauss modes and used this approximation to investigate the problem of higher-order mode degeneracy in gravitational wave detectors, for the particular example of the LG_{33} mode. The condition for significant coupling between two LG modes, $m = |l - l'|$, allows us to identify which modes will have large amplitudes in the arm cavities of detectors and which polynomials should be limited to achieve high beam purity.

The performance of the LG_{33} beam was investigated using simulations of arm cavities, with Advanced LIGO mirror maps describing mirror surface distortions. It was found that mirror surface distortions significantly reduce the purity of LG_{33} circulating in a cavity, with relatively large amplitudes of LG_{25} and LG_{41} in the cavities. Our analytical approach suggested that the astigmatism of the mirror was largely responsible for the coupling into these modes, which was confirmed by the resulting increase in purity when the cavity was simulated with the astigmatism removed.

By analysing the coupling from the specific mirror map, ETM08, we derived mirror requirements (table 5) for the use of LG_{33}. Applying these to the mirror map resulted in a reduction of circulating LG_{33} impurity from 114,000 ppm to 815 ppm. This demonstrates that a high beam purity can be achieved by reducing specific low order polynomials in Advanced LIGO mirrors. However, implementing the LG_{33} beam in gravitational wave detectors will be challenging as the amplitudes of the lower order Zernike polynomials must be very small. At the time of our investigation the mirrors had not been coated, which will further add to the distortions present. However, using this analytical approach we can suggest the requirements needed to design suitable mirrors for these higher order beams.

Acknowledgments
We would like to thank GariLynn Billingsley for providing the Advanced LIGO mirror surface maps and for advice and support on using them. We would also like to thank David Shoemaker and Stefan Hild for useful discussions. This document has been assigned the LIGO Laboratory document number LIGO-P1100148.

References
[1] S. Rowan, J. Hough and D. Crooks, *Phys. Lett. A*, **347**, 25 (2005).
[2] J.-Y. Vinet, *Classical Quantum Gravity*, **24**, 3897 (2007).
[3] B. Mours, E. Tournefier and J.-Y. Vinet, *Classical Quantum Gravity*, **23**, 5777 (2006).
[4] J.-Y. Vinet, *Living Rev. Relativity*, **12**, 5 (2009), `http://www.livingreviews.org/lrr-2009-5`.
[5] E. D'Ambrosio, *Phys. Rev. D*, **67**, 102004 (2003).
[6] M. Bondarescu, O. Kogan and Y. Chen, *Phys. Rev. D*, **78**, 082002 (2008).
[7] S. Chelkowski, S. Hild and A. Freise, *Phys. Rev. D*, **79**, 122002 (2009).
[8] P. Fulda, K. Kokeyama, S. Chelkowski and A. Freise, *Phys. Rev. D*, **82**, 012002 (2010).
[9] John Miller, New Beam Shape with Low Thermal Noise (Elba 2011), `http://agenda.infn.it/contributionDisplay.py?contribId=44&sessionId=19&confId=3351`.
[10] M. Galimberti and R. Flaminio, Mirror requirements for 3rd generation GW detectors (GWADW 2010), `http://gw.icrr.u-tokyo.ac.jp/gwadw2010/program/2010_GWADW_Galimberti.pdf`.
[11] H. Yamamoto, Reduction of degenerate modes excitation in an imperfect FP cavity for LG33 beam, LIGO-T1100220-v1, LSC (2011).
[12] T. Hong, J. Miller, H. Yamamoto, Y. Chen and R. Adhikari, *Phys. Rev. D*, **84**, 102001 (2011).
[13] C. Bond, P. Fulda, L. Carbone, K. Kokeyama and A. Freise, *Phys. Rev. D*, **84**, 102002 (2011).
[14] A. Siegman, *Lasers* (University Science Books, Sausalito, California, 1986).
[15] LSC, Instrument Science White paper 2010, LIGO-T1000416-v3 (2010).
[16] A. Freise and K. Strain, *Living Rev. Relativity*, **13**, 1 (2010), `http://www.livingreviews.org/lrr-2010-1`.
[17] F. Bayer-Helms, *Appl. Opt.*, **23**, (1984) 1369-1380.
[18] G. Billingsley, Private Communication, June 2011.
[19] M. Born and E. Wolf, *Principles of Optics*, (Cambridge University Press, Cambridge, UK, 1999), 7th ed. (expanded).
[20] G-m. Dai, *Wavefront Optics for Vision Correction*, (Society of Photo-Optical Instrumentation Engineers, Washington, USA, 2008).
[21] A. Freise: 'Finesse 0.99.8: Frequency domain interferometer simulation software' (2008) Finesse manual available at `http://www.gwoptics.org/finesse/`.
[22] A. Freise, G. Heinzel, H. Lück, R. Schilling, B. Willke and K. Danzmann, *Classical Quantum Gravity*, **21**, (2004).
[23] Advanced LIGO Team, Technical Report No. LIGO-E0900041-v5, LSC (2009).
[24] Advanced LIGO Team, Technical Report No. LIGO-E0900068-v3, LSC (2009).

Development of a torsion balance for measuring charging noise

P Campsie[1], G D Hammond, J Hough, and S Rowan
SUPA[2], Institute for Gravitational Research, School of Physics and Astronomy, University of Glasgow, Glasgow, G12 8QQ, UK
E-mail: p.campsie.physics@gmail.com

Abstract.
Noise due to surface charge on gravitational wave detector test masses could potentially become a limiting low frequency noise source in future detectors. It is therefore very important that the behavior of charging noise is experimentally verified so that accurate predictions of charging noise can be made. A torsion balance that is sensitive to small forces has been constructed at the University of Glasgow in order to measure charging noise. In this article the torsion balance apparatus being developed will be described in detail. There will also be a description of the calibration of the instrument and preliminary measurements that have been taken. These measurements show that it is possible to distinguish between the surface charge and polarisation charge on a silica sample. From this measurement it was possible to estimate the surface charge on the silica disc. The remainder of the article will discuss the improvements in sensitivity that have been made which will allow initial measurements of charging noise to begin.

1. Introduction

The detection of gravitational waves still remains one of the most exciting challenges in experimental physics. A worldwide network of six interferometric detectors has been set up to detect gravitational waves generated by astrophysical events. There are three detectors in the US that form the LIGO network [1], one detector located in Italy, Virgo [2], one detector in Germany, GEO600 [3] and one detector in Japan, TAMA 300 [4].

With the beginning of the era of second generation detectors [5, 6, 7] closely approaching, a significant amount of current research in the gravitational wave community is focussed on suppressing the limiting noise sources of these complex instruments in order to maximise the chances of detection. Excess static charges on the detector test masses and suspensions could compromise the sensitivity and control

[1] corresponding author
[2] Scottish Universities Physics Alliance

of the detector and may potentially become a limiting noise source in future detectors [8]. Charging of the test masses can occur through various processes [9]: abrasion with dust as the vacuum chamber is pumped out, contact with nearby structures such as earthquake stops and during cleaning of the test masses with First Contact^{TM}[3]. Another potential source of charging is due to cosmic rays [10, 11, 12] hitting the vacuum chamber walls and showering the test masses in electrons. The excess noise caused by charging is due to charge "hopping". As electric charge moves around on the dielectric surface of the test mass it creates a fluctuating electric field which gives rise to a fluctuating force which acts on the test mass. It is assumed [13] that these fluctuating forces can be treated as a Markov process with a single correlation time, τ, and produce a power spectrum, $F^2(f)$, given by:

$$F^2(f) \approx \frac{2\langle F^2 \rangle}{\pi\tau\left(\frac{1}{\tau^2} + (2\pi f)^2\right)} \tag{1}$$

where f is frequency and $\langle F \rangle$ is the average Coulomb force. The displacement noise that arises from surface charging can be expressed as:

$$x(f) = \frac{1}{m(2\pi f)^2}\sqrt{\frac{2\langle F^2 \rangle}{\pi\tau\left(\frac{1}{\tau^2} + (2\pi f)^2\right)}} \tag{2}$$

where m is the mass of the optic. This theory has not yet been experimentally confirmed but it is very important that charging noise is measured so that accurate estimates of charging noise in future detectors can be made.

A torsion balance apparatus has been constructed at the University of Glasgow for the purpose of measuring charging noise. A description of the experimental set up, preliminary measurements and investigations into improving the instruments sensitivity will be given in this article. There will also be a brief discussion on the development of a Mark II version of the instrument which should perform even better than the instrument used for the work in this article.

2. Experimental setup

2.1. The torsion balance apparatus

The torsion balance, shown in Figure 1, consists of a light test mass, hereafter referred to as the bob of the torsion balance, made of aluminium which is suspended by a long thin tungsten fibre which is 50 μm in diameter and 0.6 m in length. The torsion bob consists of a clamping block for the fibre and four arms. There are two copper plates attached to the ends of two opposite arms which are used for position sensing and applying servo control. There are two silica discs on the other arms that are used for charge measurements. A drive signal is applied down the fibre of the torsion balance, using a Stanford SR830 lock in amplifier, in order to provide a signal that can be used for capacitive sensing. This voltage signal has an amplitude of 1 V and a frequency of 100 kHz, which is much greater than the resonant torsional frequency of the bob. The torsion balance has an observed period of 143 seconds ($\sim$ 7 mHz) which compares well

[3] http://www.photoniccleaning.com/

to the period calculated from the stiffness of the fibre and the moment of inertia of the bob.

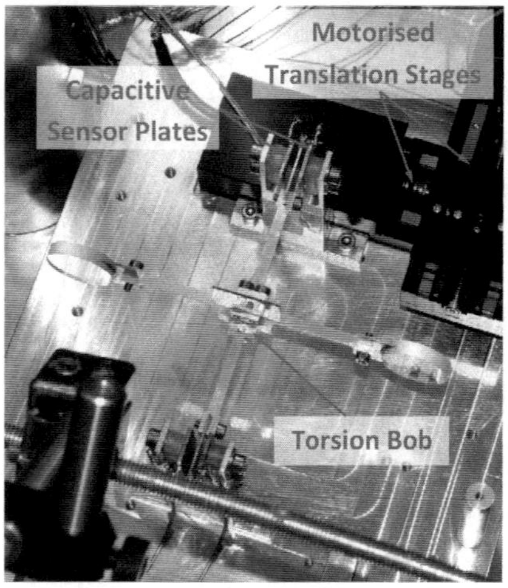

Figure 1. A picture of the torsion balance bob.

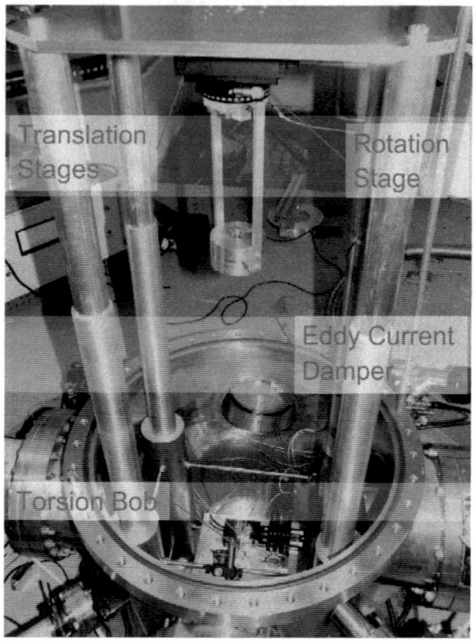

Figure 2. A picture of the vacuum tank and inner structure of the tank.

An eddy current damper [14], positioned near the top of the pendulum, is used to damp the pendulum mode of the torsion balance. The eddy current damper is composed of a copper damper disc which has fibre clamps on the top and bottom. The copper disc is enclosed in a housing made of iron which has two neodynium iron rare earth magnets positioned directly above and below the damper disc. As the pendulum mode of the torsion balance is excited, eddy currents are generated in the copper disc which damps the pendulum motion of the bob while the torsional motion of the bob remains unaffected due to the symmetry of the damping disc.

The torsion balance is housed in a vacuum chamber, shown in Figure 2, that can achieve pressures of $< 1 \times 10^{-5}$ Torr using a BOC Edwards XDS-10 dry scroll pump and a Varian Turbo-V 301 navigator pump. A Varian ConvecTorr gauge was used to monitor the pressure inside the vacuum tank down to 1×10^{-3} Torr and a Varian IMG-100 inverted magnetron gauge was used for pressures below 1×10^{-3} Torr. Inside the vacuum chamber there is an inner structure from which the bob is suspended. The position of the bob can be adjusted using two Thorlabs PT1 translation stages, in an x-y configuration, and a Thorlabs PR01 high precision rotation stage. The rotation stage can be controlled by a stepper motor whilst under vacuum. There is a fixed platform onto which the capacitive sensor plates and two Thorlabs MTS50-Z8E motorised translation stages, in an x-y configuration, are attached (shown in Figure 1). The capacitive sensor is used to monitor the position of the bob and will be discussed in detail later in this section. The motorised translation stages are used to move charged silica samples near the bob of the torsion balance in order to take a measurement. For the measurements taken in this article a small copper rod was fixed to the motorised translation stages instead of a silica sample. The copper rod was connected to a PC so that sinusoidal voltage signals could be applied to the rod using LabVIEW.

2.2. The capacitive sensor

The angular position of the bob is measured using a capacitive sensor. The copper plates that are used for position sensing and applying actuation (located on the bob) sit between two other fixed copper plates, creating what is commonly known as a capacitive bridge. A 1:1:-1 transformer, that consists of two secondary coils and one primary coil, is connected to the pair of capacitive plates that monitor the position of the torsion balance. The positive terminal of the first secondary coil is attached to one sensor plate and the negative terminal of the other secondary coil is attached to the other sensor plate. As mentioned before, the bob of the torsion balance has an AC signal applied to it in order to provide a signal for the sensor. This allows current to flow through the two secondary coils of the transformer and the position of the bob can be deduced from the difference of these currents, the result of which is an induced EMF in the primary coil. The voltage induced in the primary coil is amplified by two low noise inverting op amps (an OPA37 and an AD797) and then the amplified voltage signal is read into a Stanford SR830 lock-in amplifier where it is demodulated into a DC signal. A schematic of the capacitive sensor circuit is shown in Figure 3.

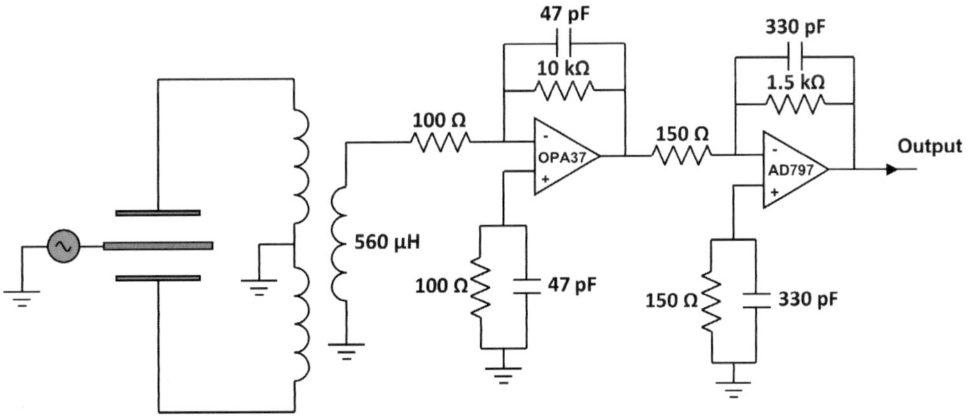

Figure 3. A schematic of the capacitive sensor circuit.

The benefit of this capacitive sensor design is that the impedance to ground, through the primary transformer windings, is much less than that due to stray capacitance. Thus all current flows through the windings. The noise of the capacitive sensor, connected to the torsion balance but with no drive signal, at 1 mHz is 1.62×10^{-4} V/$\sqrt{\text{Hz}}$ which translates into a torque noise of 6.02×10^{-14} Nm/$\sqrt{\text{Hz}}$.

2.3. The PID control

A proportional-integral-derivative (PID) servo control is used to control the position of the torsion balance. In physical terms, proportional control shortens the natural period of the torsion balance, integral control ensures that the torsion balance maintains the set-point and derivative control damps out oscillations of the torsion balance. The voltage signal demodulated by the phase sensitive detector is read onto a PC running a custom PID control program, written in LabVIEW, using a peripheral component interconnect (PCI) 6229 ADC. The program samples a predetermined number of data points from the voltage signal at a selected frequency and performs a linear least squares fit on those data points [15, 16]. The constant parameter of the least squares fit gives the most probable angular position of the torsion balance, which is used in the proportional and integral part of the control program. The gradient of the least squares fit gives the angular velocity of the torsion balance, which is used in the derivative part of the control program. After the program has performed this task it then calculates the output voltages to apply to the servo plates in order to hold the torsion balance in position. The servo control voltage, V_{servo}, is given by,

$$V_{\text{servo}} = \alpha \left(V_{\text{sensor}} - V_{\text{set}}\right) + \beta \Sigma \left(V_{\text{sensor}} - V_{\text{set}}\right) \Delta t + \gamma \frac{dV_{\text{sensor}}}{dt}, \tag{3}$$

where α, β and γ are the gains of the P, I and D controls respectively, V_{sensor} is the output voltage of the capacitive sensor and V_{set} is a programmable set-point.

3. Calibration of the torsion balance

In order to calibrate the torsion balance the bob was moved through a known angle and the resulting change in output voltage from the capacitive sensor was measured. It took 49 steps of the stepper motor to move through an angle of $1\,^{\circ}$ and thus 1 step corresponds to an angle of 356.2 μrad. The torsion balance was moved through 2.5 steps and the change in the output voltage of the capacitive sensor was measured. This was repeated for both clockwise and anti-clockwise directions until the bob had been rotated through 5 steps in both directions. A plot of voltage against rotation angle (see Figure 4) was constructed and the gradient of the best fit line through the data gave a value of 96.1 μV/μrad. Once the voltage data had been converted into an angle it was possible to convert it into a torque by dividing it by the response function of the torsion balance,

$$r(f) = \frac{1}{k\left(\left(1 - \frac{f^2}{f_0^2}\right) + \frac{i}{Q}\right)}, \tag{4}$$

where k is the stiffness of the fibre, f is frequency, f_0 is the resonant frequency of the torsion balance and Q is the quality factor of the torsion balance.

4. Initial Results

The torsion balance was set up such that a small copper rod was positioned near one of the silica discs on one arm of the torsion balance. The rod was orientated so that one of its flat circular ends was facing one of the flat circular surfaces of the silica disc. The copper rod had an AC voltage signal of amplitude 1 V and a frequency of 1 mHz applied

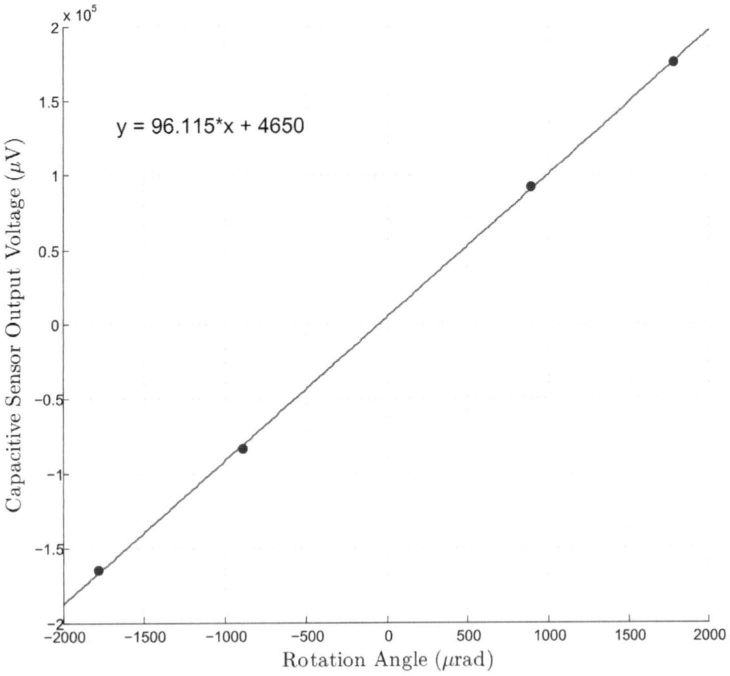

Figure 4. A plot of capacitive sensor output voltage against rotation angle. The gradient of the best fit line through the data points was used to calibrate the torsion balance in terms of an angle.

to it. Most of the surface charge was mitigated from the silica disc using a Static Clean International BR2200 rectangular static eliminator bar, however, some residual charge still remained on the disc which interacted with the copper rod. Figure 5 shows an FFT of the time series data from the capacitive sensor. The peaks at 1 mHz and 2 mHz are the 1ω and 2ω components of the drive signal respectively, while the signal at 7 mHz is the resonant frequency of the torsion bob. Two peaks arise because the force acting on the torsion balance consists of two contributions. The first is due to the force between the residual surface charge on the silica disc, Q_{surface}, and the charge on the copper rod, Q_{copper}, and the second is due to the force interaction between Q_{copper} and polarisation charges, Q_{pol}. Using Gauss's law it is possible to obtain an approximation for Q_{copper} that is related to the voltage applied to the rod (V),

$$Q_{\mathrm{copper}} = \frac{V\pi r^2 \varepsilon_0}{d} \cos \omega t, \tag{5}$$

where r is the radius of the copper rod, d is the distance between the surface of the silica disc and the rod, ω is the angular frequency of the 1 mHz voltage signal and t is time. This is then substituted into the equation that describes the force between two charged

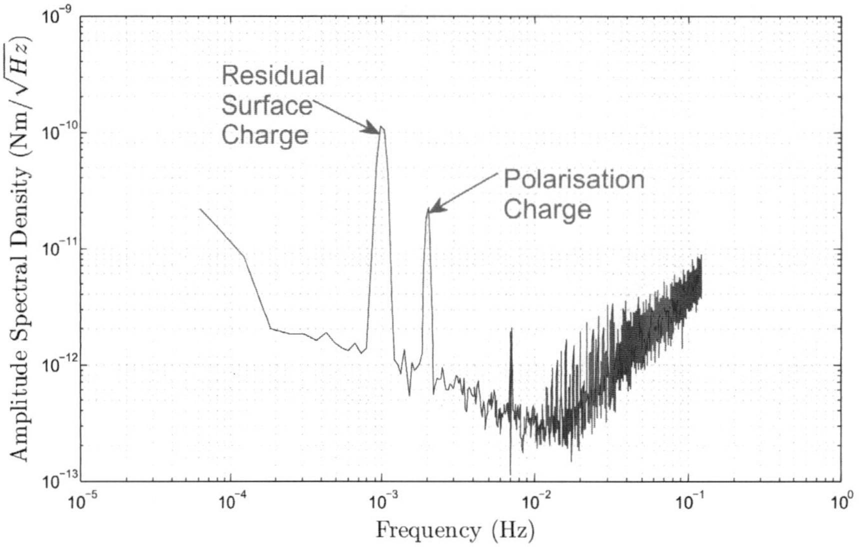

Figure 5. Amplitude spectral density of the capacitive sensor output showing a signal due to surface charge interaction at 1 mHz and a signal due to polarisation charge interaction at 2 mHz. The third peak at 7 mHz is the resonant frequency of the torsion bob.

plates to obtain an expression for the force exerted on the torsion balance,

$$F = \frac{(Q_{\text{surface}} + Q_{\text{pol}} \cos \omega t)\, Q_{\text{copper}}}{2\varepsilon_0 \pi r^2}, \tag{6}$$

which becomes,

$$F = \frac{Q_{\text{surface}} V \cos \omega t}{2d} + \frac{Q_{\text{pol}} V \cos^2 \omega t}{2d}. \tag{7}$$

If it is assumed that Q_{pol} is small compared to Q_{surface} an estimate of the surface charge on the silica disc can be calculated. Using the value for F calculated from the output voltage of the capacitive sensor a value of $\sim 2 \times 10^{-14}$ C, which gives an estimated surface charge density of 1×10^{-9} C/m^2, was obtained. This value agrees well with Kelvin probe measurements, made in a previous publication [17], of residual charge on silica discs after they had been discharged with a corona bar. With the incorporation of a Kelvin probe into this set up it will be possible to calibrate the torsion balance in terms of charge and make surface charge measurements.

5. Improving sensitivity

In order to improve the sensitivity of the instrument much effort has gone into identifying and eliminating potential noise sources. One of the first changes that was made was the arrangement of the capacitive sensor plates. Instead of having the sensor plates arranged

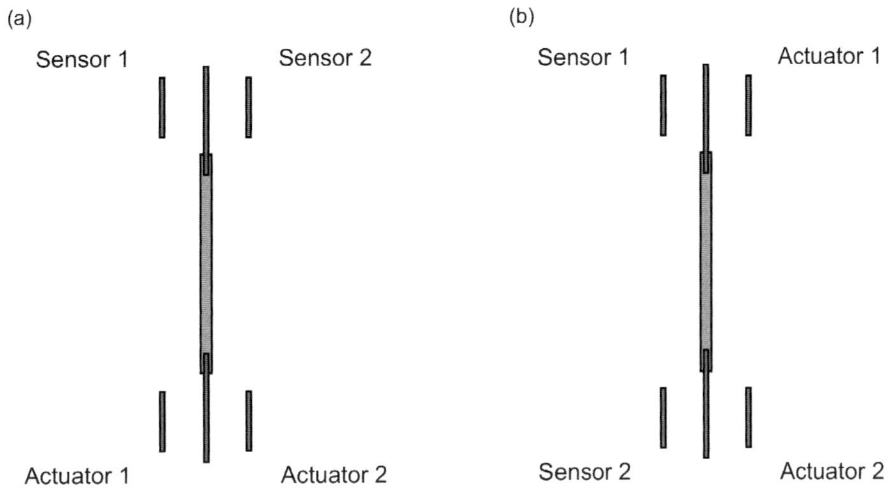

Figure 6. A diagram showing (a) the previous arrangement of the position sensor plates and (b) the new arrangement of the position sensor plates.

in the way shown in Figure 6(a) the sensor plates were changed to the set up shown in Figure 6(b). In this configuration the capacitive sensor rejects simple pendulum motion of the bob but remains sensitive to differential rotation of the bob.

A strong correlation was found between the low frequency drift of the torsion balance and temperature drift in the vacuum tank [18]. The temperature is now monitored with three PT-100 thermometers: one outside the vacuum tank, one inside the tank at the top of the tank and another inside the tank at the bottom of the tank. Temperature drift is now removed from the data computationally using a non-linear least squares fit program [19].

A simple copper shield was fabricated that screens the bob from dielectric surfaces. This proved to be very effective in reducing a vast majority of the low frequency noise.

These changes now place the torsion balance at a level of sensitivity capable of measuring charging noise. Figure 7 shows an estimate of the charging noise that would be expected from measurements of touching a silica sample with a copper rod held at a potential of 300 V. The surface charge deposited on the silica sample was $\sim 1 \times 10^{-11}\, C$ and the correlation time of the charge is assumed to be approximately the same as that measured by Ugolini [9]. Figure 7 also shows the improved torque sensitivity of the torsion balance, the capacitive sensor noise and the thermal noise limit of the instrument. Excess noise is likely due to either seismic noise or dielectric fluctuations due to the fact that there are significant contact potentials present in the apparatus.

6. Future work

In order to further improve the sensitivity of the instrument an Applied Geomechanics Inc. 755 miniature tilt sensor was bought to monitor ground tilt. If ground tilt is found

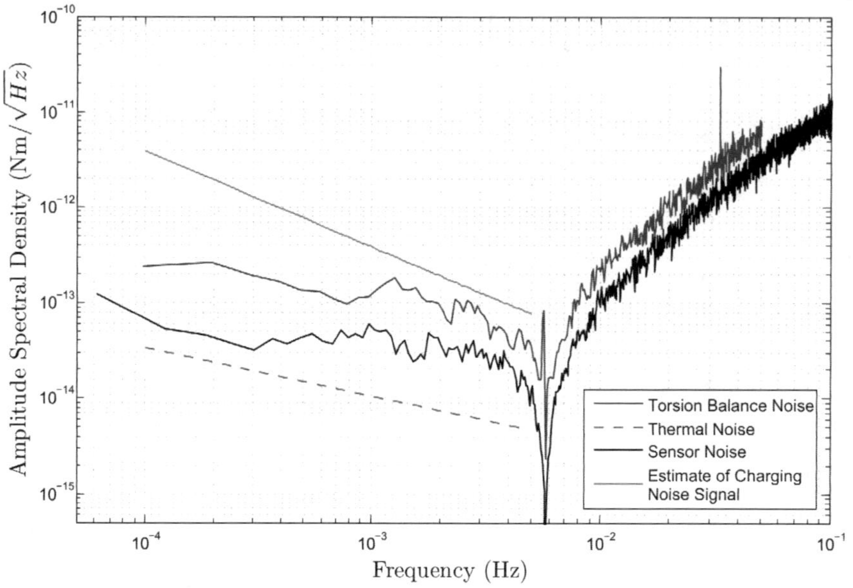

Figure 7. The torque noise of the capacitive sensor output (blue). The thermal noise limit (red) sensor noise (black) and an estimate of signal due to charging noise (green) are also featured on the plot.

to be a limiting noise source it will initially be removed computationally in the same manner as the temperature drift but it is planned that eventually a servo controlled tilt table will actively compensate for ground tilt.

It was found that effects of contact potentials [20] arising from the different metals used in the torsion bob are present in the instrument. It is therefore likely that the bob and sensor plates will be coated in gold to eliminate this [21, 22].

The torsion balance bob has been redesigned. The Mark II design will allow the majority of the bob to be grounded while the position is sensed capacitively. There are also plans to incorporate an optical method of monitoring the position of the torsion bob in the Mark II version of the instrument. An optical system should allow the bob to remain completely grounded while still maintaining high sensitivity position sensing.

7. Conclusion

A torsion balance has been constructed at the University of Glasgow that will be used to study charging noise. A direct measurement of charging noise has not been made yet and thus a measurement will allow us to confirm whether the Weiss theory of charging noise is the most appropriate description. Initial measurements have shown that it is possible to measure both the polarisation charge and surface charge present on a silica disc. With the inclusion of a Kelvin probe in the experimental set up it should be possible to make accurate measurements of the surface charge on samples of silica. This will allow accurate estimates of the signal expected from charging noise.

The limiting noise sources of the instrument have been investigated and the sensitivity of the instrument has been greatly improved to the stage where it should be possible to make a successful measurement of charging noise. This measurement will provide greater understanding of charging noise and allow accurate predictions of charging noise in advanced gravitational wave detectors to be made.

Acknowledgments

We would like to thank the physics department workshop at the University of Glasgow for manufacturing the parts for the work presented in this article and Russell Jones and Michael Perreur-Lloyd (University of Glasgow) for useful discussions on engineering the parts for this work. The authors also gratefully acknowledge the work of Iain McCrindle (University of Glasgow) who built the capacitive sensor circuitry. We are grateful for the financial support provided by Science and Technology Facilities Council (STFC), the Royal Society of Edinburgh (RSE), the Scottish Funding Council (SFC) and the University of Glasgow in the UK. Finally we would like to thank our colleagues in the LSC, the LIGO charging working group and within SUPA for their interest in this work.

References

[1] Abbott B P *et al.* 2009 *Rep. Prog. Phys.* **72** 076901
[2] Accadia T *et al.* 2011 *Class. Quantum Grav.* **28** 114002
[3] Grote H *et al.* 2010 *Class. Quantum Grav.* **27** 084003
[4] Takahashi R *et al.* 2004 *Class. Quantum Grav.* **21** S403S408
[5] Harry G M *et al.* 2010 *Class. Quantum Grav.* **27** 084006
[6] Acernese F *et al.* 2009 *Advanced Virgo Baseline Design* Internal Virgo Technical Note VIR027A09
[7] Willke B *et al.* 2006 *Class. Quantum Grav.* **23** S207
[8] Harry G M 2008 *Estimate of noise from charging in initial, enhanced and advanced LIGO* Internal LIGO Technical Note T080019-01
[9] Ugolini D, Amin R, Harry G, Hough J, Martin I, Mitrofanov V, Reid S, Rowan S and Sun K X 2008 *International Cosmic Ray Conference* vol 3 pp 1283–1286
[10] Mitrofanov V P, Prokhorov L G and Tokmakov K V 2002 *Phys. Lett. A* **300** 370–374
[11] Mitrofanov V P, Prokhorov L G, Tokmakov K V and Willems P 2004 *Class. Quantum Grav.* **21** S1083–S1089
[12] Braginsky V B, Ryazhskaya O G and Vyatchanin S P 2006 *Phys. Lett. A* **359**(2) 86–89
[13] Weiss R 1995 *Note on electrostatics in the LIGO suspensions* Internal LIGO Technical Note T960137-00
[14] Su Y, Heckel B R, Adelberger E G, Gundlach J H, Harris M, Smith G L and Swanson H E 1994 *Phys. Rev. D* **50**(6) 3614–3636
[15] Savitzky A and Golay M J E 1964 *Anal. Chem.* **36** 1627–1639
[16] Hammond G D, Speake C C, Matthews A J, Rocco E and Peña-Arellano F 2008 *Rev. Sci. Instrum.* **79**(2) 025103
[17] Campsie P, Cunningham L, Hendry M, Hough J, Reid S, Rowan S and Hammond G D 2011 *Class. Quantum Grav.* **28** 215010
[18] Adelberger E G, Stubbs C W, Heckel B R, Su Y, Swanson H E, Smith G, Gundlach J H and Rogers W F 1990 *Phys. Rev. D* **42**(10) 3267–3292
[19] Linfield G and Penny J 1999 *Numerical Methods using Matlab* 2nd ed (New Jersey: Prentice Hall)
[20] Pollack S E, Schlamminger S and Gundlach J H 2008 *Phys. Rev. Lett.* **101**(7) 071101
[21] Schlamminger S, Hagedorn C A and Gundlach J H 2010 *Phys. Rev. D* **81**(12) 123008
[22] Carbone L, Cavalleri A, Ciani G, Dolesi R, Hueller M, Tombalato D, Vitale S and Weber W J 2006 *AIP Conf. Proc* vol 873 pp 561–565

Controlling calibration errors in gravitational-wave detectors by precise location of calibration forces.

H P Daveloza[1], M Afrin Badhan[2], M Diaz[1], K Kawabe[3], P N Konverski[1], M Landry[3] and R L Savage[3]

[1] Center for Gravitational Wave Astronomy, University of Texas at Brownsville, Brownsville, Texas, USA
[2] Mount Holyoke College, South Hadley, Massachusetts, USA
[3] LIGO Hanford Observatory, Richland, Washington, USA

E-mail: hernan.daveloza@ligo.org

Abstract. We present results of finite element analysis simulations which could lead to more accurate calibration of interferometric gravitational wave detectors. Calibration and actuation forces applied to the interferometer test masses cause elastic deformation, inducing errors in the calibration. These errors increase with actuation frequency, and can be greater than 50% at frequencies above a few kilohertz. We show that they can be reduced significantly by optimizing the position at which the forces are applied. The Advanced LIGO [1] photon calibrators use a two-beam configuration to reduce the impact of *local* deformations of the test mass surface. The position of the beams over the test mass can be chosen such both the local and the *bulk* induced elastic deformation are minimized. Our finite element modeling indicates that with two beams positioned within ± 1 mm of their optimal locations, calibration errors due to test mass elastic deformation can be kept below 1% for frequencies up to 3.5 kHz. We thus show that precise control of the location of calibration forces could considerably improve calibration accuracy, especially at high frequencies.

1. Introduction

Optimizing searches for gravitational wave sources at frequencies above 1 kHz [2] [3], will require improvements in the calibration of ground-based interferometric gravitational wave detectors. Lindblom has shown that optimizing scientific benefit requires calibration accuracies on the order of 5% for first detection and 0.5% for later measurements [4].

The Photon Calibrator (Pcal) is a calibration method that uses radiation pressure from auxiliary lasers to induce reference displacements of the interferometer's test masses. The Pcal has become one of the favored methods for determination of the interferometer's response function, due to its simplicity and ability to calibrate with forces that induce displacements that are close to what is expected from candidate Gravitational Wave (GW) sources. However, Hild et al. at GEO600 [5], and Goetz et al. in LIGO [6] demonstrated that calibration forces exerted by the action of a centered Pcal beam produce local deformations that significantly change the sensed displacement at frequencies above about 500 Hz due to local elastic deformation of the test mass surface. Even though this type of local elastic deformation cannot be avoided, its effects can be mitigated by changing the configuration of the Pcal from one laser beam in the middle of the face of the test mass, to two laser beams diametrically opposed and sufficiently displaced from the center of the face such that local elastic deformations are not sensed by the interferometer. This scheme was tested and implemented in LIGO as part of the Enhanced LIGO upgrade [7] [8].

Willems [9] and Afrin Badhan et al. [10] studied another possible problem, the *bulk* elastic deformation of the End Test Masses (ETM). Their studies showed that these deformations also significantly impact the calibration accuracy, especially at high frequencies.

Our investigations show that actuation and calibration forces induce bulk deformations with spatial distributions given by the test mass normal mode shapes. The magnitude of the induced deformation increases dramatically as the frequency of excitation approaches the frequency of a given normal mode, affecting the interferometer length calibration. We incorporated the actual Initial LIGO ETM geometry and the actual physical parameters of the ETM (e.g. density, Young's modulus, Poisson's ratio, damping parameter or the inverse of the quality factor) into to our Finite Element Analysis. We also analyzed the elastic deformations of Advanced LIGO ETMs, predicting the deformations that would be produced by Pcal forces and identifying the optimal positions for the beams.

2. Elastic Theory

ETMs are made of fused silica. Although this is a stiff material, with a Young's modulus of 72.7 GPa and a Poisson's ratio of 0.16, they can experience small deformations when acted upon by calibration or actuation forces. For example, deformations on the order of 10^{-16} m induced by oscillating calibration forces of 10^{-8} N at a frequency of 3.5 kHz. Because the deformations are small, the response to the excitations induced by calibration forces can be represented with the appropriate linear combination of normal modes. We perform investigations to correctly identify these modes and their effects on the detector performance. As we will show, the amplitude in each mode, and then the deformation sensed by the interferometer, depends on the position of the calibration and actuation forces on the High-Reflective (HR) and Anti-Reflective (AR) surfaces of the ETMs.

The test masses of GW interferometers are right circular cylinders with one wedged flat surface. As introduced by Love [11], the first study of an ideal cylinder's elastic deformations was made by Pochhammer in 1876. Hutchinson, summarized these studies and presented a numerical method to calculate the normal modes of a solid cylinder [12].

GW interferometers measure the axial projection of the displacement of the surface of the test mass, $w = \vec{u} \cdot \vec{z}$, where $\vec{u}$ is the displacement vector and $\vec{z}$ is the direction parallel to the axis of the cylinder. Following Hutchinson, the solutions for this displacement are:

$$w = 2\delta J_n(\alpha r) \left\{ \begin{array}{c} -sin(\delta z) \\ cos(\delta z) \end{array} \right\} \cos(n\theta)$$

and

$$w = \alpha^2 J_n(\alpha r) \left\{ \begin{array}{c} -sin(\beta z) \\ cos(\beta z) \end{array} \right\} \cos(n\theta)$$

where J_n denotes the nth-order Bessel function and r, θ and z are the usual cylindrical coordinates. The numbers α, β and δ are related by $\alpha^2 + \delta^2 = (2\pi f)^2(1-2\nu)/2(1-\nu)$ and $\alpha^2 + \beta^2 = (2\pi f)^2$, where f is the frequency and ν is the Poisson's ratio.

The sensing by the main interferometer laser beam can be modeled as a Gaussian-weighted integral of the displacement over the entire HR surface. Due to the periodic azimuthal dependence $(cos(n\theta))$, the integrals for a centered interferometer beam will vanish except for the case of $n = 0$, in which case the azimuthal dependence vanishes. In this case we have a solution for w with only a Bessel dependence on the radial direction (the integral is performed over a surface, i.e. at $z = constant$). This mode is referred to as the *drumhead* mode.

We can reduce or eliminate the excitation of this mode by applying the forces at the nodes of this function. Associating the first minimum of $J_0(x)$ with the radius of the cylinder, a rough approximation for the ratio between the optimal radial position for the applied forces and the radius of the cylinder, is given by the ratio between the first root of $J_0(x)$ ($x \sim 2.40$), the first node, and the first root of its derivative ($x \sim 3.83$). The ratio is thus $2.40/3.83 = 0.63$.

Afrin Badhan et al. found a similar solution for the Initial LIGO ETMs and control system forces using Finite Element Analysis (FEA) [10]. To control the position of the ETM, Initial LIGO used four voice coil actuators with magnets glued to the AR surface of the ETM. Afrin Badhan modeled the ETMs as right circular cylinders of diameter $D = 250$ mm and thickness $\tau = 100$ mm with four forces applied uniformly over circles of 2 mm radius that are equally spaced displaced radially from the center of the AR surface. She found the optimum radius for the forces to be 82 mm, giving a ratio of 0.65.

The Initial LIGO ETMs are not perfect cylinders; the AR surface is wedge at 2 degree. In our analysis, see section 3 for more detail, we included this wedge and found the optimal radius to be 78.7 mm. The voice coil actuator were located at a radius of 114 mm, well outside the optimal position. Thus the contribution to the sensed motion arising from elastic deformation is in phase with the rigid body motion. The relative error, or discrepancy, in the overall sensed motion compared with the rigid body motion is plotted in Figure **1** for the actual position of the voice coil actuators (green squares) and for voice coil actuators located at the optimal radius (blue x's). For actuator at the Initial LIGO position, the induced error increases with frequency to a factor of 1.9 at 6 kHz. If the forces were located at the optimal radius, our analysis shows that the discrepancy would be less than one percent up to 6 kHz.

The measured discrepancy between the motion sensed by the interferometer and the expected free-mass motion, when the test mass is driven by the Initial LIGO voice coil actuators, has been observed [13] and is in reasonable agreement with the FEA results [10]. In order to maintain the accuracy of the interferometer calibration, the expected elastic deformation induced by the voice coil actuators must be, and is, included in the model of the interferometer response used for calibration [13].

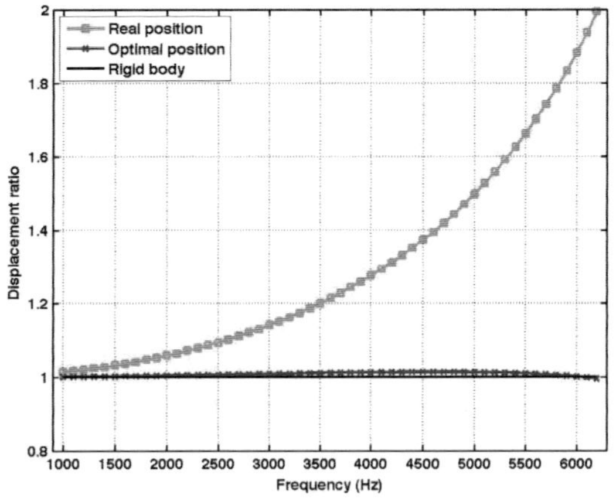

Figure 1

Ratio between total sensed motion and rigid-body motion vs. excitation frequency, for actual (114 mm) and optimal (78.7 mm) calibration force positions.

3. Finite Element Analysis

For this study we use COMSOL Multiphysics [14], a well known FEA software package. LIGO ETMs have thin dielectric film coatings to produce the HR and AR surfaces, which are not modeled in our FEA. The Initial LIGO ETM (iETM) have a diameter $D = 250.75$ mm, a maximum thickness $\tau = 99.63$ mm, and a $2°$ wedge on the AR surface. The voice coil actuators were modeled as four forces uniformly applied over circles of radius 2 mm equally spaced and displaced radially from the center of the AR surface. The Advanced LIGO End Test Masses (aETM) have $D = 340.13$ mm, $\tau = 200.2$ mm, a wedge of $0.077°$ on the AR surface, and two flats on the sides separated by 326.5 mm. The aETMs will use electrostatic drives [15] instead voice coil actuators.

LIGO ETMs are suspended as pendulums to isolate them from seismic noise. In our FEA models, we treat the ETMs as free bodies, a reasonable assumption for frequencies much larger than the ETMs pendulum resonance frequency, approximately 1 Hz. The total displacement of the HR surface is the combination of the free-mass motion and the displacement resulting from elastic deformation of the ETMs. The effective displacement measured by the interferometer is the overlap of this HR surface displacement with the main interferometer gaussian beam. The gaussian weighting for a centered interferometer beam is given by:

$$\frac{exp(-2(x^2 + y^2)/r_{ifo}^2)}{2\pi \times r_{ifo}^2}$$

where the interferometer beam spot size, r_{ifo}, is 45 mm for the iETM, and 62 mm for the aETM. The denominator is a normalization factor.

Using the elastic theory described in Section 2, we analyzed the coupling from the actuation of the photon calibrator forces to the displacement of each mode. Our analysis shows that two normal modes are most relevant for calibration frequencies below 6 kHz, the *butterfly* mode and the drumhead mode. The butterfly and drumhead mode resonant frequencies are 6612 Hz and 9221 Hz for the iETM, and 5953 Hz and 8151 Hz for the aETM, respectively. Figure **2** shows the deformation shape of the butterfly and drumhead mode for an aETM.

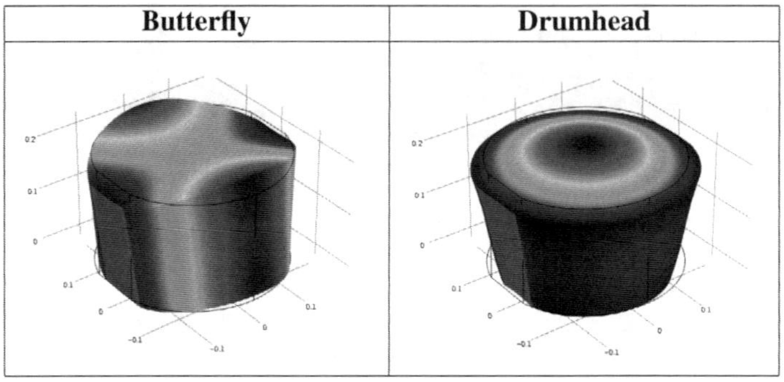

Figure 2
Advanced LIGO End Test Mass normal modes.

The drumhead mode has no azimuthal variation. The amplitude of the mode can be minimized by precise location of the actuation or calibration forces at the optimal radius, which is close to the nodal circle for the mode. The butterfly mode has a periodic variation with azimuthal angle. We already showed that for a perfect cylinder this periodic variation results in an effective zero displacement. Nevertheless, details of the actual test mass geometry, e.g. flats and wedges, affect the shapes and frequencies of the normal modes [16]. This is easy to visualize with *isosurfaces*; they are three dimensional surfaces with a constant value of the displacement of the elements. Figure **3** shows zero axial displacement *isosurfaces*, that is surfaces where $w = 0$, for an ideal cylinder and for an aETM.

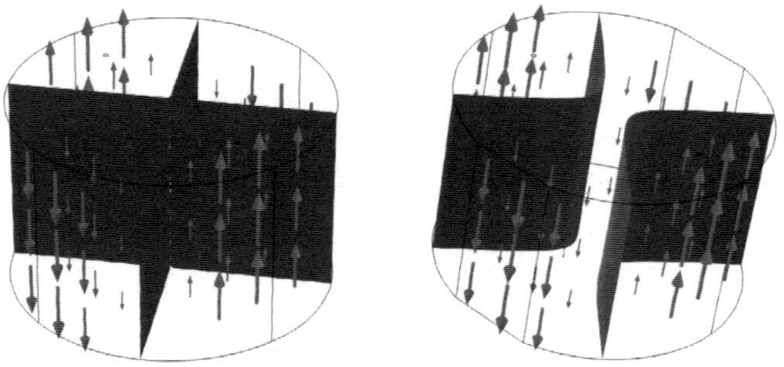

Figure 3
Butterfly mode $w = 0$ isosurfaces for an ideal cylinder (left) and an advanced LIGO End Test Mass (right). The arrows show the displacement vectors for the elements.

For ideal cylinders, the $w = 0$ isosurfaces cross the cylinder through its center, leaving an almost zero effective displacement. However, the $w = 0$ isosurfaces for the ETMs do not cross through the center of the cylinder, leaving an additional displacement that sums with the free-mass motion. Our study shows that the amplitude of the displacement produced by the elastic deformation for the butterfly mode can be minimized using three or four forces.

We also calculated the transfer function from photon calibrator forces to the displacement sensed by the interferometer to find an optimal placement of the photon calibrator beams. The Advanced LIGO Pcal forces were applied uniformly over circles of 2.5 mm radius, the approximate *spot size* for the laser beam. The discrepancy between the ideal rigid-body motion and the effective displacement measured by the interferometer is weakly dependent on the position of the interferometer beam, but strongly dependent on the positions of the driving forces. For frequencies below 4 kHz, the minimum discrepancy is observed when driving forces are positioned at 111.6 mm, giving a ratio between the optimal position and the aETM radius of 0.66. The resulting potential calibration errors are plotted versus excitation frequencies for a two-beam Pcal calibrator configuration in Figure **4**, and for a three-beam Pcal calibrator configuration in Figure **5**. For two optimally located beams the discrepancy is less than 0.1% up to 3.2 kHz and less than 1% for frequencies up to 4.3 kHz, but increases to almost 4% at 5 kHz.

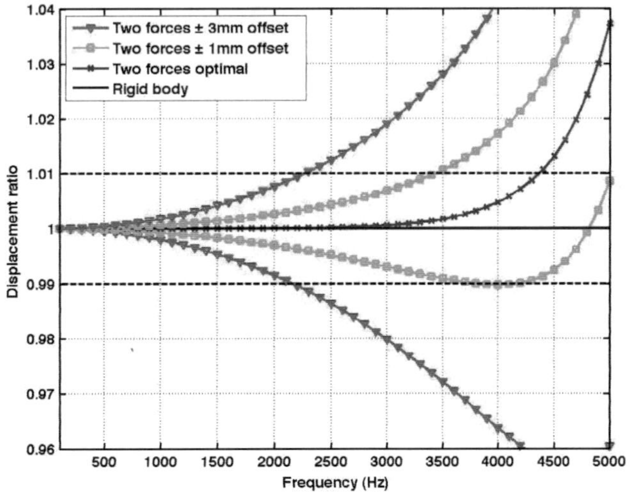

Figure 4

Advanced LIGO two-beams Photon calibrator configuration, ratio between total sensed motion and rigid-body motion vs. excitation frequency, for optimally positioned beams, 111.6 *mm, and* ± 1 *mm and* ± 3 *mm offsets from the optimal locations.*

The discrepancy increases dramatically for beams that are offset from their optimal positions. This increase in discrepancy for frequencies near and above 5 kHz can be understood by considering the increased compliance in the butterfly and drumhead deformation patterns as the excitation frequency approaches their resonance frequencies. The amplitude of the deformation related to the butterfly mode is important only for frequencies close to it's resonant frequency, 6612 Hz for the iETM and 5953 Hz for the aETM, because far away from this frequencies the shape of this mode is governed by the azimuthal periodic variation that averages to zero for a centered interferometer beam. Thus, in general, the behavior of the discrepancy is governed by the amplitude of the elastic deformation related to the drumhead mode, because it directly involves the center of the ETMs with no periodic azimuthal dependence. For beams positioned in radius smaller than the optimal radius, the displacement produced by the drumhead mode is 180° out of phase with the free-mass motion, so these motion are subtracted. On the other hand, if the beams are positioned at a radius larger than the optimal radius, the displacement due to the drumhead mode is in phase with the free-mass motion and they add, so the discrepancy between the ideal rigid-body and the effective displacement always increases with frequency. Figure **4** shows this behavior. Note that above about 4 kHz, the ratio for Pcal beams positioned at radii smaller than the optimal radius, which initially decreased, begins to increase with frequency. This behavior, which is not observed for 3 forces, is due to excitation of the butterfly deformation pattern as the drive frequency approaches its resonance frequency.

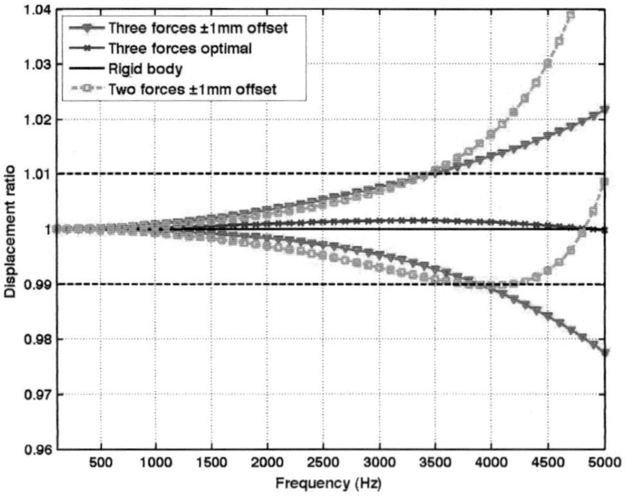

Figure 5

Advanced LIGO three-beam Photon calibrator configuration, ratio between total sensed motion and rigid-body motion vs. excitation frequency, for optimally positioned beams (112.7 mm) and ±1 mm offsets from the optimal locations.

If we now focus our attention to a discrepancy of 1%, we observe in Figure **4** that for an optimally positioned two-beam configuration this discrepancy is reached at 4.3 kHz. Meanwhile for an offset of 1 mm this discrepancy is reached at 3.5 kHz. In the case of three or four forces at the optimal radius, the discrepancy could be reduced to 0.2% for frequencies up to 5 kHz. Nevertheless, if the three or four forces are offset from the optimal locations by 1 mm, either intentionally or unintentionally due to poor positioning accuracy, the discrepancy is again 1% at 3.5 kHz, as can be seen in Figure **5**. Thus, to realize the calibration accuracies called for in Lindblom's assessment, calibration forces must be localized within ± 1 mm of their optimal locations and multiple beam configurations will be required.

We also analyzed ETMs made of sapphire, a stiffer material than fused silica and a candidate material for future GW detectors test masses due to its increased thermal conductivity. For sapphire, normal modes moves to higher frequencies, butterfly $\sim$ 9 kHz and drumhead $\sim$ 14 kHz, resulting in, compared with fused silica, smaller deformation amplitudes for frequencies below 6 kHz which is desirable for GW detection and for high-frequency calibration accuracy.

We have also carried out a preliminary investigation of the electrostatic drives that will be used to control the aETMs [17]. Advanced LIGO's strict noise requirements demand that no magnets be glued to the ETMs, as was the case for Initial LIGO. The advanced LIGO suspension system, is a quadruple pendulum, in which the bottom masses are the ETMs and reaction masses are used to exert electrostatic forces on the ETMs via electrodes patterned on the reaction masses. Preliminary studies show that the electrostatic drives could produce calibration discrepancies of about 10% at 2.5 kHz and as much as 60% at 5 kHz due to the location of the forces.

4. Conclusions

Our analysis shows that actuation and calibration forces induce bulk deformations of test masses producing a discrepancy between the ideal free-mass, rigid-body motion and the effective displacement sensed by gravitational wave interferometers. The spatial distribution of these deformations are given by a superposition of the normal modes of the test mass with amplitudes that increase dramatically as the calibration or actuation excitation frequency approaches that of the normal modes. Even when the amplitude of the butterfly spatial deformation could be minimized using three or four forces, our study shows that, unless an accuracy of ± 1 mm in locating the calibration or actuation forces is achieved, there will be no advantage in using three or four forces for frequencies below 3.5 kHz. For an aETM with two Pcal beams positioned at a radius 111.6 ± 1.0 mm it is possible to keep the discrepancy between the ideal motion and the motion sensed by the interferometer below 1% up to 3.5 kHz. For frequencies above 3.5 kHz, calibration accuracy will be improved by adopting a Pcal configuration with three or more beams in order to minimize the amplitude of the butterfly bulk deformation pattern.

Acknowledgments

We thank P. Willems and the staff of the LIGO Hanford Observatory for enlightening discussions. We also thank the National Science Foundation for support under award HRD0734800. LIGO was constructed by the California Institute of Technology and Massachusetts Institute of Technology with funding from the National Science Foundation and operates under cooperative agreement PHY-0757058. This paper has LIGO Document Number LIGO-P1100166.

References

[1] Fritschel P, et al. 2009 Advanced LIGO Systems Design *LIGO document T010075* (https://dcc.ligo.org)
[2] LIGO Scientific Collaboration 2009 *Phy. Rev. D* **80** 102002.

[3] Maggiore M 2000 Stochastic backgrounds of gravitational waves (http://arxiv.org/abs/gr-qc/0008027v1)
[4] Lindblom L 2009 *Phy. Rev. D* **80** 042005.
[5] Hild S, et al. 2007 *Class. Quantum Grav.* **24** 5681-5688.
[6] Goetz E, et al. 2009 *Class. Quantum Grav.* **26** 245011.
[7] Abbott B, et al. 2009 *Rep. Prog. Phys.* **72** 076901 (http://iopscience.iop.org/0034-4885)
[8] Adhikari R, Fritschel P, Waldman S 2006 Enhanced LIGO *LIGO document T060156* (https://dcc.ligo.org)
[9] Willems P 2008 Finite Element Modeling of Test Mass Flexure Due to OSEM Forces *LIGO document T080190* (https://dcc.ligo.org)
[10] Afrin Badhan M, et al. 2009 Analyzing Elastic Deformation of Test Masses in LIGO *LIGO document T0900401* (https://dcc.ligo.org)
[11] Love A.E.H. 1944 A treatise on the mathematical theory of elasticity *Dover*
[12] Hutchinson J R 1980 *J. of Appl. Mech.* **47** 901-908.
[13] Bartos I, et al. 2011 Frequency Domain Calibration Error Budget for LIGO in S6 *LIGO document T1100071* (https://dcc.ligo.org)
[14] www.comsol.com
[15] M. Hewitson, H. Grote, G. Heinzel, K. A. Strain, H. Ward, U. Weiland 2003 *Class. Quantum Grav.* **20** S885-S893.
[16] Miller J, et al. 2011 *Phys. Lett. A* **375** 788-794.
[17] Daveloza H, et al. 2011 aLIGO Photon Calibrator Errors due to Elastic Deformation *LIGO document G1100332* (https://dcc.ligo.org)

Journal of Physics: Conference Series **363** (2012) 012008 doi:10.1088/1742-6596/363/1/012008

The new cryogenic facility at LMA

Jérôme Degallaix, Raffaele Flaminio, Danièle Forest, Janyce Franc, Kevin Gautier, Massimo Granata, Bernard Lagrange, Christophe Michel, Nazario Morgado, Laurent Pinard, Emeline Saracco and Quentin Benoit

Laboratoire des Matériaux Avancés, 7 avenue Pierre de Coubertin, 69622 Villeurbanne, France

E-mail: j.degallaix@lma.in2p3.fr

Abstract. To support the research effort for the third generation of gravitational wave interferometers, the Laboratoire des Matériaux Avancés (LMA) at Lyon, France has developed a new cryogenic facility to characterize optics at low temperature. The new cryostat is installed in a clean room and allows samples to be cooled down to 10 Kelvin in around 12 hours.

Currently, two independent experiments have been installed in the cryostat: the measure of the optical absorption of silicon and the measurement of the coating mechanical loss. After a short presentation of the cryogenic and optical setup, preliminary results from the optical absorption experiment will be presented.

1. Motivation

While the second generation of Gravitational Wave (GW) laser interferometer is being installed, a major research effort is underway to shape the third generation of detectors. Whereas the two first generations share similar technologies, the instrument of the next generation will be likely based on different light wavelength, mirror substrate made of new material and different working temperature. For example, the interferometer of the Einstein Telescope [1] dedicated to low frequency will use a laser with a 1550 nm wavelength in conjunction with silicon test masses cooled down to a temperature of 20 K.

To design the Einstein Telescope (ET) and to help predicting its performances, knowledge of the physical properties of the material used for the mirror coating and substrate are required. Unfortunately, properties for thin layers or bulk material at low temperature are sometimes only interpolation or simply unexistent. For example, the optical absorption of silicon at 1550 nm which directly constrains the size of the suspension fiber used to evacuate the heat generated in the mirror, is still unknown (albeit supposed to be very low). An exhaustive list of missing physical data relevant to the thermal noise calculation for ET can be found in Franc et al. [2].

An important research effort focuses now on finding the most suitable materials regarding the substrate and coating for cryogenic gravitational wave detectors [3]. Mechanical loss of uncoated and coated silicon cantilevers were measured at the university of Glasgow [4, 5] as a function of the temperature whereas Q factor of bulk material were measured at the University of Jena [6, 7]. As part of the research toward LCGT [8], the optical absorption of sapphire has also been characterised at low temperature [9].

Our confidence to design successful third generation GW interferometer will be greatly improved by the characterisation of substrate and coating at low temperature. In this perspective the LMA has commissioned a new cryogenic facility in order to extend our knowledge about material at low temperature. This article aims to introduce this facility to the GW community. In a first part, the design and performances of the cryostat will be described. Then in a second part, the two first experiments at low temperature will be presented: the measurement of the coating loss angle and the optical absorption measurement.

2. The cryostat

In this section, the design and the functioning of the cryostat is described. The cryostat was custom built by Air Liquide and was installed in a clean room at LMA in September 2010.

2.1. Requirements and design

One of the main purposes of the cryostat is the measurement of the coating quality factor at low temperature. A rapid turnover of samples in the cryostat is required to test different coating formulas (with various thicknesses or dopant concentrations for example). In parallel of the coating quality factor measurement, a second independent experiment can take place: the measurement of the optical properties of silicon at low temperature. Running two experiments in parallel allows important saving on the cooling cost.

In a nutshell, while looking for a cryostat, three main requirements were derived:

- it must be able to reach a temperature of 10 K
- the low temperature state must be rapidly reached (in the order of a night time)
- the cryostat must be big enough to accommodate two experiments.

The above specifications were met with a continuous flow cryostat. The idea of using a pulse tube cryostat was disregarded since it could generate additional vibration noise. The cryostat is connected to a 100 l liquid Helium dewar and by varying the circulating flow of the liquid helium, the cooling rate can be adjusted. A copper breadboard of diameter 200 mm is mounted on top of the cryostat to install the experiments. The cryostat itself is integrated in the center of a large optical table (2 m $\times$ 1.25 m) with the copper breadboard sticking out 50 mm above the table. All the cryostat plumbing and the vacuum equipment are installed below the optical table.

Two temperature probes inside the cryostat are available: the first one is mobile and can be positioned on the sample to test, the second one is incorporated into the copper breadboard and is used for the temperature regulation. The stabilization of the temperature is achieved through a heater (maximum power of 50 W) which is also in contact with the breadboard. The heater and the temperature probes are connected to a controller where the target temperature can be entered. With the controller, we can stabilize the temperature $\pm$ 0.2 K around the desired temperature. During the experiments, the state of the cryostat (temperature and pressure) is continually monitored using Labview.

To allow the injection and extraction of probe laser beams, the vacuum chamber and each of the two internal heat shields have three two inches diameter viewports made of borosilicate glass. All the mounts of the cryostat windows were specifically designed to allow a quick and easy removal and exchange of the windows. So the cryostat could later be used with different laser wavelengths than the ones presented in this article.

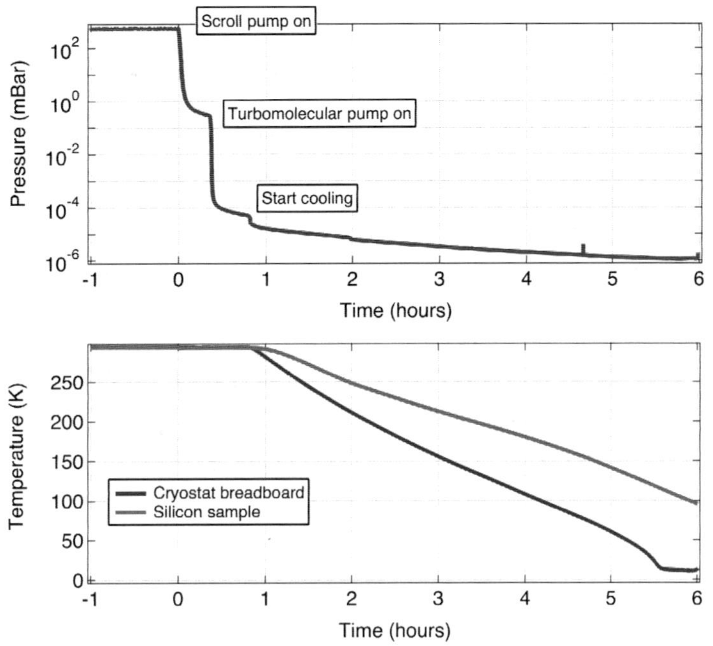

Figure 1. Evolution of pressure (upper plot) and temperature (lower plot) during the cooling of the cryostat. The pumping started at the time $t = 0$

2.2. Performances

The evolution of the pressure and temperature over time in the cryostat is presented in figure 1. The values of the two temperature probes are shown. One probe is at the level of the copper breadboard whereas the second probe is attached to a large (2 inches diameter) silicon substrate. At the beginning of the measurement the cryostat is at room pressure and temperature. Then at time $t = 0$ we started the primary scroll pump to evacuate the air in the cryostat and 20 minutes later the secondary high vacuum turbo pump was also turned on. 50 minutes after the start of the pump down, when a pressure of less than 5×10^{-4} mbar had been reached, we opened the valve to let the liquid helium circulate inside the cryostat bowels.

The temperature of the breadboard inside the cryostat started to cool immediately as shown by the blue curve in the lower plot of figure 1. The situation was different for the sample, which took longer to cool since it is not in full contact with the breadboard (as we will see in section 4). The sample took also longer to cool as it must reach the radiative equilibrium with the inner thermal shield.

By adjusting the flow rate of gaseous helium at the output of the cryostat (via an aspiration pump), the consumption of helium can be controlled. A digital flowmeter was installed to ensure an accurate monitoring and control of the helium consumption. To reach the cryogenic temperature overnight, we found the optimal flow of gaseous helium to be around 6 liter/minute.

In the cooling example presented in 1, in less than 6 hours after we close the lid of the cryostat, the breadboard is at 12 K. That is also the temperature of the cantilever used for coating loss measurement as its mount was designed to maximise the thermal link between the cantilever and the breadboard. The cooling in this example was relatively fast requiring a large

Figure 2. Photos of the copper breadboard inside of the cryostat with two experiments installed. On the left side, the 2 inches diameter silicon substrate for the measurement of the absorption can be seen, while on the right side the cantilever mounted vertically is present. Space has been reserved to add a second cantilever mount with its own excitation next to the first one.

flow of liquid helium. In normal conditions, the helium circulating rate is two times smaller and we cool our samples overnight.

3. Coating loss measurement

To measure the coating mechanical loss, a small layer of material is deposited on a silicon cantilever. This small layer has a thickness of the order of the micrometer which is enough to reduce the quality factor of the mechanical modes the cantilever. The dimensions of the cantilever are chosen for the frequency of the first mechanical mode to be around 70 Hz.

A picture of the cantilever in its mount inside the cryostat is shown in figure 2. The mount is made of copper and is bolted to the cryostat breadboard. A vacuum grease is used to maximize the thermal contact between the mount and the cryostat. The bending modes of the cantilever are excited using a contactless electrostatic drive. A 633 nm laser beam reflected by the cantilever is send to a 2 dimension position sensing device located outside the cryostat. The Q factor is directly determined by doing a fit in the time domain of the decreasing amplitude of the resonant motion of the cantilever [10].

During the measurement, the Q factor of the first three bending modes of the cantilever are measured sequentially. Since we only want to excite one mode at a time, one mode is excited only after the previous mode is no longer resonating with a large amplitude. So for each temperature, the measurements of the coating loss angle of the first three modes are repeated automatically and can take up to 2 hours per measurement (but only 5 minutes if we skip the measurement of the first mode).

The first test measurement was done on an uncoated silicon cantilever to serve as a reference. At the time of writing, we are repeating this measurement with a new cantilever and results will be presented soon.

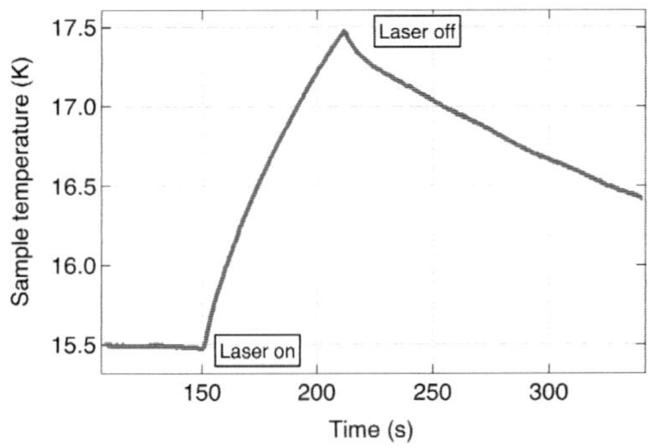

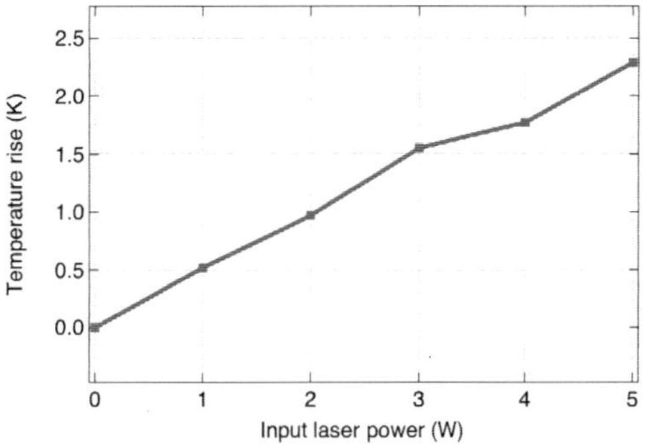

Figure 3. Evolution of the silicon sample temperature when the laser is switched on (left plot) and the measured temperature rise 50 s after the laser is switched on for different input power (right plot).

4. Optical absorption measurement

In parallel to the measurement of the silicon absorption at low temperature described thereafter, we are also doing measurement at room temperature. In that case, we use the photo-deflection technique [11]. A 30 W pump laser at 1550 nm laser is focused on the sample to create a gradient of temperature inside, while a second low power laser at 1310 nm is used to monitor the gradient of temperature created by the first laser. This technique is already intensively used at LMA on other benches with a 1064 nm pump laser to measure 2D and 3D absorption maps on large fused silica mirrors [12].

We aim also to measure the optical absorption of silicon substrate at low temperature with the LMA cryostat. At cryogenic temperature, where materials can exhibit very low specific heat, a small amount of absorbed power results in a noticeable increase of temperature. This technique of laser calorimetry has already been successfully applied to measure the optical absorption of sapphire at low temperature [9].

The silicon sample tested was manufactured by Silicon Materials [13] using the float zone growth method. This silicon has an orientation $< 100 >$, dopant type n, a purity superior to 99.9999% and a resistivity of 10 kOhm.cm. We choose one of the highest resistivity available on the market since it is linked to low optical absorption [14]. The two uncoated flat surfaces were diamond dust-dry polished giving a surface roughness of less than 0.5 nm RMS.

Inside the cryostat, the silicon sample is resting on two thin holders made of copper. The holders was designed to allow enough contact for the substrate to cool while the thermal link between the substrate and cryostat breadboard must be small enough to not dissipate immediately the heat generated from the absorbed optical power. A probe is directly attached to the sample to monitor its temperature.

The silicon sample is heated with a 30 W laser with a wavelength of 1550 nm. However, in order to avoid impacting the heat equilibrium in the cryostat, we limit the laser output power for this experiment to 5 W (power measured with an external powermeter). An example of the temperature rise when the laser is switched on is shown figure 3.

During the experiment, we keep the temperature rise of silicon sample below 3 K in order to neglect the heat lost by conduction. The rise of temperature$\Delta\, T$ during the time$\Delta\, t$ is directly proportional to the amount of heat absorbed P_{abs}:

$$P_{abs} = \frac{\Delta T}{\Delta t} C m \tag{1}$$

with C the average specific heat of the sample during the measurement and m the mass of the sample. Since the laser power incident on the cryostat is known and we can estimate the amount of light in the silicon sample (keeping in mind that each side of the uncoated silicon substrate reflects 30% of the light), we can derive the absorption per unit of length.

Using the method described above, we found the silicon absorption for a temperature between 15 K and 20 K, to be around 200 ppm/cm (+/- 15%). The power absorbed is proportional to the incident power on the sample (right plot in figure 3) indicating no nonlinear effect.

With the laser calorimetry technique, only the total absorbed power is measured, surface or substrate absorption can not be distinguished. To have an idea of the contribution of the surface absorption to the total absorbed power, samples with different lengths of the same silicon (same physical properties and made by the same manufacturer) have also been bought. A new campaign of measurement will test the effect of the sample length to the absorbed power.

The results presented in this article are preliminary and a very simplistic model has been used to derive the absorption. A more complex model using ANSYS and including the temperature dependence of the material properties and realistic thermal contact has already been developed. This model will be used to get a more accurate measurement of the absorption in the future.

5. Conclusions and perspectives

In this article, the new cryogenic facility recently installed at LMA has been presented. This cryostat is used to test coating and bulk materials at low temperature, as part of the research toward the third generation of gravitational wave detectors.

We spent the last year to learn how to master the cryostat and get familiar with low temperature technology. We are now ready to start new campaigns of measurement as using the cryostat becomes routine work. The two research directions described in this article (coating loss measurement and optical absorption measurement) will be pursued in 2012 with a more rigorous analysis than the preliminary results presented here. It is planned that new coating formulas will be tested to minimise the coating mechanical loss at low temperature. On the other front, in parallel, different samples of silicon from different manufacturer will be bought to determine which one has the lowest optical absorption.

Acknowledgments

Jérôme Degallaix is grateful for support from the European Gravitational Observatory (EGO). This work has also been performed with the support of the European Commission under the Framework Programme 7 (FP7) Capacities, project Einstein Telescope (ET) design study (grant agreement 211743).

References

[1] Hild S, Abernathy M, Acernese F, Amaro-Seoane P, Andersson N, Arun K, Barone F, Barr B, Barsuglia M, Beker M, Beveridge N, Birindelli S, Bose S, Bosi L, Braccini S, Bradaschia C, Bulik T, Calloni E, Cella G, Mottin E C, Chelkowski S, Chincarini A, Clark J, Coccia E, Colacino C, Colas J, Cumming A, Cunningham L, Cuoco E, Danilishin S, Danzmann K, Salvo R D, Dent T, Rosa R D, Fiore L D, Virgilio A D, Doets M, Fafone V, Falferi P, Flaminio R, Franc J, Frasconi F, Freise A, Friedrich D, Fulda P, Gair J, Gemme G, Genin E, Gennai A, Giazotto A, Glampedakis K, Grf C, Granata M, Grote H, Guidi G, Gurkovsky A, Hammond G, Hannam M, Harms J, Heinert D, Hendry M, Heng I, Hennes E, Hough J, Husa S, Huttner S, Jones G, Khalili F, Kokeyama K, Kokkotas K, Krishnan B, Li T G F, Lorenzini M, Lck H, Majorana E, Mandel I, Mandic V, Mantovani M, Martin I, Michel C, Minenkov Y, Morgado N, Mosca S, Mours B, MllerEbhardt H, Murray P, Nawrodt R, Nelson J, Oshaughnessy R, Ott C D, Palomba C, Paoli A, Parguez G, Pasqualetti A, Passaquieti R, Passuello D, Pinard L, Plastino W, Poggiani R, Popolizio P, Prato M, Punturo M, Puppo P, Rabeling D, Rapagnani P, Read J, Regimbau T, Rehbein H,

Reid S, Ricci F, Richard F, Rocchi A, Rowan S, Rdiger A, Santamara L, Sassolas B, Sathyaprakash B, Schnabel R, Schwarz C, Seidel P, Sintes A, Somiya K, Speirits F, Strain K, Strigin S, Sutton P, Tarabrin S, Thring A, van den Brand J, van Veggel M, van den Broeck C, Vecchio A, Veitch J, Vetrano F, Vicere A, Vyatchanin S, Willke B, Woan G and Yamamoto K 2011 *Classical and Quantum Gravity* **28** 094013 URL http://stacks.iop.org/0264-9381/28/i=9/a=094013 1

[2] Franc J, Morgado N, Flaminio R, Nawrodt R, Martin I, Cunningham L, Cumming A, Rowan S and Hough J 2009 Mirror thermal noise in laser interferometer gravitational wave detectors operating at room and cryogenic temperature arXiv:0912.0107v1 URL http://arxiv.org/abs/0912.0107v1 1

[3] Nawrodt R, Rowan S, Hough J, Punturo M, Ricci F and Vinet J Y *General Relativity and Gravitation* **43** 593–622 1

[4] Reid S, Cagnoli G, Crooks D, Hough J, Murray P, Rowan S, Fejer M, Route R and Zappe S 2006 *Physics Letters A* **351** 205 – 211 ISSN 0375-9601 1

[5] Martin I, Armandula H, Comtet C, Fejer M M, Gretarsson A, Harry G, Hough J, Mackowski J M M, MacLaren I, Michel C, Montorio J L, Morgado N, Nawrodt R, Penn S, Reid S, Remillieux A, Route R, Rowan S, Schwarz C, Seidel P, Vodel W and Zimmer A 2008 *Classical and Quantum Gravity* **25** 055005 URL http://stacks.iop.org/0264-9381/25/i=5/a=055005 1

[6] Nietzsche S, Nawrodt R, Zimmer A, Thrk M, Vodel W and Seidel P 2006 *Journal of Physics: Conference Series* **32** 445 URL http://stacks.iop.org/1742-6596/32/i=1/a=068 1

[7] Nawrodt R, Zimmer A, Koettig T, Schwarz C, Heinert D, Hudl M, Neubert R, Thrk M, Nietzsche S, Vodel W, Seidel P and Tnnermann A 2008 *Journal of Physics: Conference Series* **122** 012008 URL http://stacks.iop.org/1742-6596/122/i=1/a=012008 1

[8] Kuroda K and the LCGT Collaboration 2010 *Classical and Quantum Gravity* **27** 084004 URL http://stacks.iop.org/0264-9381/27/i=8/a=084004 1

[9] Tomaru T, Uchiyama T, Tatsumi D, Miyoki S, Ohashi M, Kuroda K, Suzuki T, Yamamoto A and Shintomi T 2001 *Physics Letters A* **283** 80 – 84 URL http://www.sciencedirect.com/science/article/pii/S0375960101001918 1, 5

[10] Martin I, Armandula H, Comtet C, Fejer M M, Gretarsson A, Harry G, Hough J, Mackowski J M M, MacLaren I, Michel C, Montorio J L, Morgado N, Nawrodt R, Penn S, Reid S, Remillieux A, Route R, Rowan S, Schwarz C, Seidel P, Vodel W and Zimmer A 2008 *Classical and Quantum Gravity* **25** 055005 URL http://stacks.iop.org/0264-9381/25/i=5/a=055005 4

[11] Jackson W B, Amer N M, Boccara A C and Fournier D 1981 *Appl. Opt.* **20** 1333–1344 URL http://ao.osa.org/abstract.cfm?URI=ao-20-8-1333 5

[12] Yan Z, Ju L, Zhao C, Gras S, Blair D G, Tokunari M, Kuroda K, Mackowski J M and Remillieux A 2006 *Appl. Opt.* **45** 2631–2637 URL http://ao.osa.org/abstract.cfm?URI=ao-45-12-2631 5

[13] Silicon Materials http://www.si-mat.com/ 5

[14] Soref R and J Larenzo J 1986 *IEEE Journal of Quantum Electronics* **22** 873–879 5

Journal of Physics: Conference Series **363** (2012) 012009 doi:10.1088/1742-6596/363/1/012009

Optimization of Mechanical Impedance Matchers for Parametric Transducers in Gravitational Wave Spherical Detectors

C. Frajuca[1], R. Pires[1], F. S. Bortoli[1], N. S. Magalhaes[2], R. M. Marinho, Jr.[3] and F. G. Oliveira[3]

[1] Sao Paulo Federal Institute of Education, Science and Technology,
Rua Pedro Vicente 625, Sao Paulo, SP 01109-010, Brazil

[2] Universidade Federal de Sao Paulo – DCET,
Rua Sao Nicolau, 221, Diadema, SP 09913-030, Brazil

[3] Technological Institute of Aeronautics,
Praça Mal. Eduardo Gomes 50, Sao Jose dos Campos, SP 12228-900, Brazil

E-mail: frajuca@gmail.com

Abstract. A resonant-mass spherical gravitational wave detector is been built at the Physics Institute of Sao Paulo University which will be part of a detection network with a similar one that is under construction in The Netherlands. The goal of such detector is to reach a sensitivity of 10^{-21} in h at frequencies close to 3200 Hz within a bandwidth of 200 Hz. Its expected sensitivity will be close to the quantum limit and for such sensitivity to be reached several parameters must be optimized. One of them is the calibration of the mechanical impedance matchers, which are mechanical oscillators assembled on the sphere with the purpose of coupling the oscillations of the sphere surface to the electromechanical transducers that are used as motion sensors. This work has the goal to optimize such mechanical coupling, consequently the detector sensitivity, using a microwave parametric transducer for which an important variable is the electrical quality factor. The study involves the simulation of the oscillation of the mechanical parts of the detector with its all 17 frequency modes while checking whether these modes behave as predicted by the theory. We found that the simulation works adequately and that most of the frequency modes investigated are in good agreement with the theory about vibration quadrupole modes in spheres. Other modes did not fit so well in the theory and we comment on this in the text.

1. Introduction

SCHENBERG is a spherical resonant-mass gravitational wave detector [1, 2, 3] being built at the Department of Materials and Mechanics of the University of Sao Paulo, Brazil. According to the theory of general relativity the sphere's quadrupole modes are the ones expected to be excited by a gravitational wave (GW) signal and theoretically they are exactly degenerate.

The sphere, with 65 cm in diameter and weighting 1.15 ton, is made of a copper-aluminum alloy [4] with 94% Cu and 6% Al. Figure 1 shows a cut view of this detector with some of its characteristics. Theoretical calculations yield a value of 3170.4 Hz for the frequency of the degenerate quadrupole modes of a bare sphere [5].

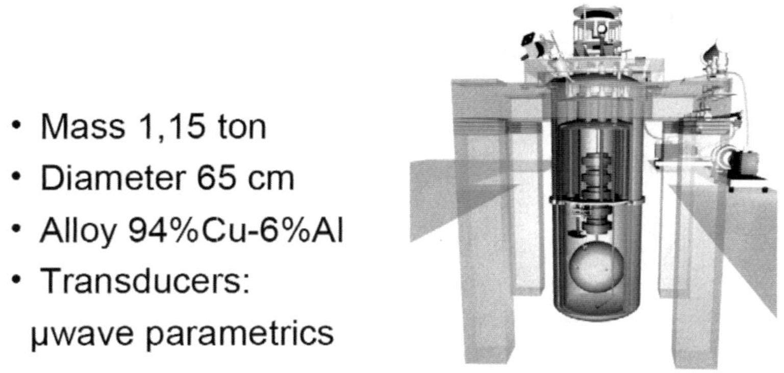

Figure 1: Schematics of the SCHENBERG detector.

The detector will have six transducers [6, 7] arranged on the sphere's surface in a half-dodecahedron distribution; these sensors will be located as if in the center of the six connected pentagons in a dodecahedron surface, following the studies by Merkowitz and Johnson [8] confirmed by Magalhaes and collaborators [9]. The transducers and the suspension hole are located in the sphere surface as indicated in figure 2. SCHENBERG's sphere modes as published in [10] were measured after holes were drilled to accommodate the suspension and future transducers. The values found were (in Hz) 3172.5, 3183.0, 3213.6, 3222.9 and 3240.0.

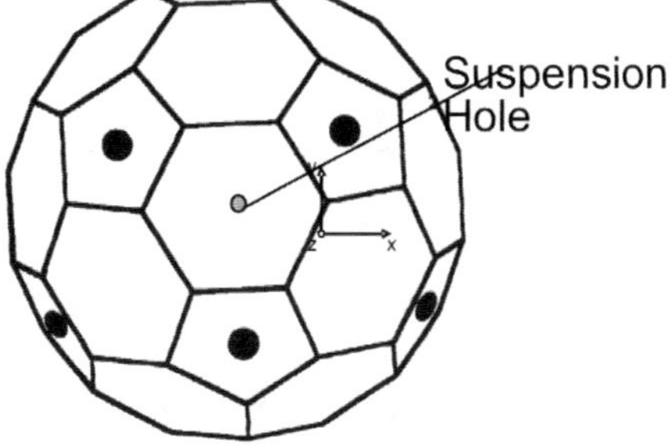

Figure 2: Position of the transducers and the suspension hole.

When the modes of the system *sphere without hole plus six transducers* are determined using a theoretic model their frequencies are grouped into three quintuplets (frequencies, in Hz, of 2842.3, 3140.8 and 3527.7) and two singlets (frequencies, in Hz, of 2936.9 and 3424.1) [5]. The quintuplets are located in the sides and at the center of the bandwidth, and the singlets are located between the quintuplets.

By analyzing the signal from such sensors the amplitudes and the direction of the incoming gravitation wave can be obtained [11, 12]. A similar detector is been built in The Netherlands, called MiniGrail [13], with a frequency close to SCHENBERG's.

The Brazilian group has decided to use as motion sensors microwave parametric transducers, like the one used in the Australian GW detector NIOBÈ [14]. In this kind of transducer a superconducting cavity is pumped with monochromatic resonant microwaves and when the size of the cavity changes due to the vibration (one of the cavity wall is connected to the sphere by the mechanical impedance matcher) two side bands are created in the microwave signal that leaves the cavity. The amplitude of the side band is proportional to the amplitude of the sphere vibration.

Between the sphere surface and the transducer an impedance matcher is placed, composed by a series of mechanical oscillators with the same frequency of the detector and with decreasing effective masses, the last of these being the microwave cavity wall itself. The function of such device is to act as a mechanical amplifier (the amplitude is amplified by the square root of the mass ratio) as well as a mechanical filter so that only signals within the relevant bandwidth are sensed by the transducer.

It is expected that a multimode mechanical impedance matcher for the transducer will have several advantages over the single mode transducer [15, 16, 17, 18, 19], either parametric, inductive or capacitive. For this reason such a matcher is planned to be used in SCHENBERG.

In order to propose improvements based on the practical tests that shall start soon, we decided to perform simulations that would allow us to estimate how changes in the parameters of the mechanical impedance matcher could influence its match to the antenna. The results of our studies are presented in the following sections.

2. The Two-Mode Mechanical Impedance Matcher

We chose a design for the impedance matcher which was simple to model with a finite element program and whose parameters could easily be changed: a mechanical matcher with a disk supported by three rings that act as springs, its second mode being a membrane located at the disk's center of mass (see figure 3). Although this is not the same shape of the actual transducer used in the experiment its matching to the sphere is physically similar, modeled by springs and masses. The design used in this simulation is expected to be used in the next detector improvement phase.

The material chosen for the impedance matcher is pure niobium with the following properties: Young's modulus: $E = 1.05 \times 10^{11}$ Pa; specific mass: $\rho = 8580.00\ kg/m^3$; Poisson's coefficient: $\nu = 0.38$.

Figure 3: Impedance matcher used in the finite elements simulation. The figure on the left shows the cut perspective and on the right is the full perspective. The membrane is located in the lower disks of the figures.

After we defined the design of the transducer we then had to determine the dimensions of the diaphragm for the first simulation. The mass of the disk was chosen to be equal to the the desired one (initially 30g) and its initial diameter was chosen as 20 mm. To find the initial thickness of the membrane we used the formula for membrane frequency found in the book by Blevins [20]. As this formula does not take into account the correct boundary conditions we used it to find only the first frequency which would then be refined in the finite element modeling. The intention behind these determinations was to tune the impedance matcher as close as possible to the frequency of the sphere quadrupole modes in a way that it would work as a resonant transducer.

3. The Simulations

3.1. The Bare Sphere

Using the finite elements program *Solid Works Simulation 2010* in all simulations, we first simulated the sphere alone based on SCHENBERG's spherical antenna as described in the first paragraph of Section 1 with no holes drilled in it. The parameters we used for the Cu-Al alloy were: Young's modulus: $E = 1.303 \times 10^{11}$ Pa; specific mass: $\rho = 8077.5$ kg/m^3; Poisson's coefficient: $\nu = 0.364$.

We found that the frequencies for the five quadrupole modes were (in Hz) 3189.7, 3190.0, 3190.8, 3191.2 and 3192.1. These numbers are within the band 3190.8 ± 1.0 Hz. The very small standard deviation (about 0.03%) indicates that the modes are degenerate as theory predicts. The small differences from the average value of 3190.8 Hz are interpreted as due to numerical limitations in the finite elements modeling. This average value differs from the theoretical prediction mentioned in Section 1 (3170.4 Hz) by 0.6%, possibly due to small differences in the values of physical parameters.

The simulations show clearly that the antenna changes its shape from a sphere into an ellipsoid during its vibration at all frequencies as is also predicted from theory.

As mentioned in Section 1, measurements of SCHENBERG's sphere modes in the presence of holes yielded an average of 3206.4 Hz with a splitting of nearly 1%. It is reasonable to speculate that the lower (less than 0.5%) average frequency obtained in the above simulation is mainly due to the fact that no holes were drawn in it. Nevertheless, the simulated frequencies and the measured ones are quite close to each other, around 3198.6 ± 0.1 Hz.

3.2. Sphere without Suspension Hole plus Six Matchers

Six impedance matchers were include in the model positioned on the sphere surface as shown in figure 4. They were located on the sphere surface as in figure 2 and were fastened to the sphere surface by the solid disk. They were aggregated into the same structure of the sphere in the finite element program.

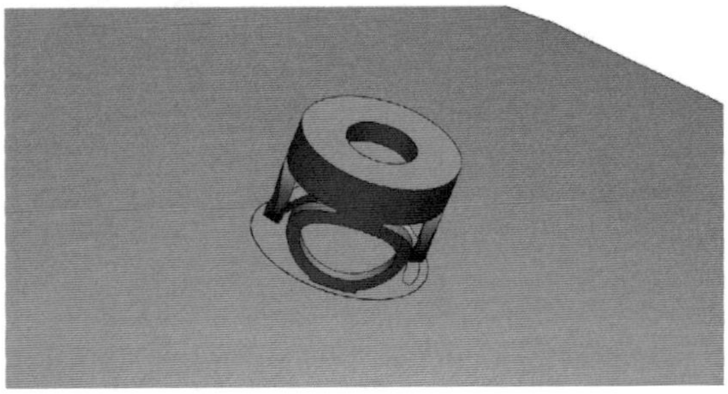

Figure 4: A two-mode impedance matcher positioned on the sphere's surface.

Then the simulation was run with six impedance matchers distributed on the full sphere, and the frequencies obtained in the simulation are shown in table 1.

Journal of Physics: Conference Series **363** (2012) 012009 doi:10.1088/1742-6596/363/1/012009

Table 1: Frequencies (in Hz) found on the simulation of the system full sphere plus six two-modes impedance matchers. The frequencies labeled *phase* correspond to the situation in which the disk and the membrane move in phase and the ones labeled *counter phase* imply that they move in phase opposition.

Mode Frequencies (Hz)	Phase Characteristics
3098.208	in phase
3098.574	in phase
3099.689	in phase
3099.953	in phase
3100.313	in phase
3101.513	in phase
3156.595	in phase
3156.725	in phase
3256.697	counter phase
3157.249	counter phase
3157.343	counter phase
3206.034	counter phase
3208.252	counter phase
3209.792	counter phase
3211.047	counter phase
3211.460	counter phase
3212.595	counter phase

The frequencies in table 1 follow the theoretical pattern presented in Section 1: three quintuplets (frequencies, in Hz, around 3099, 3156 and 3211) and two singlets (frequencies near 3101 Hz and 3206 Hz). Based on previous arguments, differences between theory and simulation are believed to be due mainly to the values of the physical parameters used and minimally due to numerical limitations.

3.3. Sphere with Suspension Hole plus Six Impedance Matchers

A suspension hole was include in the sphere drawing for a new simulation and the results are shown in table 2. The membrane thickness used was 63 μm.

The frequencies in table 2 follow a pattern similar to the case in the preceding section: three quintuplets (frequencies, in Hz, around 3082, 3138 and 3187) and two singlets (frequencies near 3085 Hz and 3184 Hz).

Table 2: Frequencies found in the simulation for the system sphere with suspension hole plus six two-modes impedance matchers. The frequencies labeled *phase* correspond to the situation in which the disk and the membrane move in phase and the ones labeled *counter phase* imply that they move in phase opposition.

Mode Frequencies (Hz)	Phase Characteristics
3080.784	in phase
3081.498	in phase
3082.119	in phase
3082.460	in phase
3083.612	in phase
3085.366	in phase
3119.850	in phase
3120.150	in phase
3038.042	counter phase
3138.157	counter phase
3153.345	counter phase
3148.637	counter phase
3186.637	counter phase
3187.147	counter phase
3187.276	counter phase
3187.722	counter phase
3189.185	counter phase

4. Conclusions

The simulation yielded patterns predicted theoretically and results close to measured values, which is a good evidence that the simulation is working properly.

Two frequency singlets and three quintuplets were identified even when we included a hole in the simulated sphere (Section 3.3). They were shifted from the values obtained in the no-hole case (Section 3.2), again in good agreement with the theory about vibration quadrupole modes in spheres since the small holes are expected to slightly change the body's mode frequencies. We found that the singlets are very close to the side quintuplets while theoretically their positions was closer to the half distance between the two nearest quintuplets. The central quintuplet is not degenerate because its average frequency is 3133,9 Hz with a standard deviation of 0,5%; this significant deviation suggests physical causes for the non-degeneracy, probably because of the presence of the suspension hole. The side quintuplets can be considered degenerate.

More work is necessary to understand why the singlets are dislocated from the position they have in the no-hole case (more symmetrical with respect to the quintuplets), and why the side quintuplets are degenerate.

The impedance matcher design presented here may be machined in a single piece and this will allow the system to have a high mechanical quality factor. Also, it is easy to change masses and frequencies of this impedance matcher by small changes in masses, membrane thickness and the rings used as springs making the tuning process easier.

In a future work we plan to include in the simulations the holes with which the transducers are attached to the sphere.

Acknowledgments
CF acknowledges FAPESP (grant # 2010/00167-4) and NSM acknowledges FAPESP (grant # 11/07731-5) for the financial support.

References

[1] O D Aguiar *et al* 2008. The Schenberg spherical gravitational wave detector: the first commissioning runs *Classical and Quantum Gravity* **25** 114042.
[2] O.D. Aguiar *et al.* The Brazilian gravitational wave detector Mario Schenberg: progress and plans. Classical and Quantum Gravity, Bristol, v. 22, p. S209-S214, 2005.
[3] C. Frajuca, K.L.Ribeiro, O.D.Aguiar, L.A. Andrade, P.J.Castro, N.S.Magalhaes and R. M. Marinho Jr, *Class. Quantum Grav.* **21**, 1107 (2004).
[4] G. Frossati et al., Proceedings *of the First International Workshop for an Omnidirectional Gravitational Radiation Observatory*, W.F.Velloso, Jr., O.D.Aguiar and N.S. Magalhaes, editors (Singapore, World Scientific, 1997).
[5] C. Stelatti. *Classification of Theories of Gravitation with the Gravitational Wave Detector Mario Schenberg*. Ph. D. thesis. (In Portuguese). Sao Jose dos Campos: ITA (2006).
[6] C. Frajuca, K.L. Ribeiro, L.A. Andrade, W.F. Velloso Jr, J.L. Melo, O. D. Aguiar and N. S. Magalhaes, *Class. Quantum Grav.* **19**, 1961 (2002).
[7] C. Frajuca, O. D. Aguiar, N. S. Magalhaes, K. L. Ribeiro and L. A Andrade, Proc. *3rd Edoardo Amaldi Conference on Gravitational Waves* (Pasadena, USA, July 1999). AIP Conf. Proc. 523 (New York, AIP) 417.
[8] W. W. Johnson and S. M. Merkowitz, *Phys. Rev. Lett.* 70, 2367 (1993).
[9] N. S. Magalhaes, O.D. Aguiar, W.W. Johnson and C.Frajuca, *Gen. Relat. Grav.* **29**, 1511 (1997).
[10] O.D. Aguiar et al., *Class. Quantum Grav.* **21**, 459 (2004).
[11] N. S. Magalhaes, W. W. Johnson, C. Frajuca and O. D. Aguiar, *MNRAS* **274**, 670 (1995).
[12] N. S. Magalhaes, W. W. Jonhson, C. Frajuca and O D. Aguiar, *Ap. J.* 475, 462 (1997).
[13] A. de Waard et al, *Class. Quantum Grav.* **21**, 465 (2004).
[14] D.G. Blair, E.N. Ivanov, M.E. Tobar, P.J. Turner, F. Kann and I.S. Heng, *Phys. Rev. Lett.* **74**, 1908 (1995).
[15] C Frajuca, N S Magalhaes, F S Bortoli and A M Horiguti 2008 Study of six mechanical impedance matchers on a spherical gravitational wave detector *Journal of Physics. Conference Series (Online)* **122** 012029 .
[16] C Frajuca and F S Bortoli 2006 Planning to improve the mechanical quality factor in the transducer impedance matchers for Mario Schenberg detector *Journal of Physics. Conference Series* **32** 315-8.
[17] C Frajuca, F S Bortoli and N S Magalhaes 2005 Resonant transducers for spherical gravitational wave detectors *Brazilian Journal of Physics* (São Paulo, SP: Brasilian Society of Physic) **35** n. 4B 1201-3.
[18] C Frajuca, F S Bortoli and N S Magalhaes 2006 Studying a new shape for mechanical impedance matchers in Mario Schenberg transducers *Journal of Physics. Conference Series* **32** 319-22.
[19] Bortli, F.S.; Frajuca, C ; Magalhaes, N.S.; Duarte, N.S. . A physical criterion for validating the method used to design mechanical impedance matchers for Mario Schenberg's transducers. Journal of Physics. Conference Series (Online), **228**, p. 012011, 2010.
[20] R.D. Blevins, Formulas for Natural Frequency and Shape (Malabar, Krieger Publ., 1979).

Review of the Laguerre-Gauss mode technology research program at Birmingham

P. Fulda, C. Bond, D. Brown, F. Brückner L. Carbone, S. Chelkowski[1], S. Hild[2], K. Kokeyama[3], M. Wang, and A. Freise

School of Physics and Astronomy, University of Birmingham
Edgbaston, Birmingham, B15 2TT, UK

E-mail: pfulda@star.sr.bham.ac.uk

Abstract. Gravitational wave detectors of the advanced generation are expected to be limited in sensitivity by thermal noise of the optics. The reduction of this noise is therefore of high importance for future detectors which aim to surpass the sensitivity of the advanced generation. A proposed method for reducing the impact of this noise is to use higher-order Laguerre-Gauss (LG) modes for the readout beam, as opposed to the currently used fundamental mode. We present here a synopsis of the research program undertaken by the University of Birmingham into the suitability of LG mode technology for future gravitational wave detectors. This will cover our previous and current work on this topic, from initial simulations and table-top LG mode experiments up to implementation in a prototype scale suspended cavity and high-power laser bench.

1. Introduction

Advanced gravitational wave detectors such as Advanced LIGO [1] and Advanced Virgo [2] are being constructed with the goal of a tenfold increase in sensitivity over the first generation of detectors [3, 4]. This is expected to be sufficient to make the first conclusive gravitational wave detection. The projected sensitivity of both Advanced LIGO and Advanced Virgo will be limited over the observation frequency band by several fundamental noise sources: seismic noise, radiation pressure noise, suspension thermal noise, shot noise, and thermal noise of the test masses themselves [5]. The dominant contribution to the total thermal noise of the test masses is expected to be from the Brownian motion of the atoms that make up the reflective coatings. This motion causes an uncertainty in the phase of the light reflected from the test mass, and will therefore appear as noise in the detector output. This thermal noise has until now not been a limiting noise source in detectors, but as progress within the field pushes other noise sources ever lower, its mitigation becomes a more urgent priority.

One way to reduce the effects of the coating Brownian thermal noise is to use a larger beam spot on the test mass surfaces, since a larger beam averages more effectively over the random motions of the surface. The coating Brownian thermal noise power spectral density has been

[1] Presently at: Jenoptik AG, Carl-Zeiss-Strasse 1, 07739, Jena, Germany
[2] Presently at: Institute for Gravitational Research, University of Glasgow, G12 8QQ, UK
[3] Presently at: Physics and Astronomy, Louisiana State University, Baton Rouge, LA 70803, USA

shown to scale with the inverse square of the Gaussian beam size parameter $w(z)$ [6]. However, as the beam spots are made larger, a larger fraction of the light power is lost over the edge of the reflective surface. This fraction is known as the *clipping loss*. If one uses a beam with a more homogeneous radial intensity distribution than the commonly used LG_{00} beam, the Brownian thermal noise level can be reduced without increasing the clipping loss. Several beams with this property have been suggested for gravitational wave detectors, including mesa beams [7], optimised conical beams [8] and higher-order LG beams [9].

Higher-order LG modes have the advantage over the other suggested beams of being compatible with spherical mirror surfaces, as currently used with the LG_{00} mode. Equation 1 shows the complex amplitude function for the helical higher-order mode set,

$$u_{p,l}^{\text{hel}}(r,\phi,z) = \frac{1}{w(z)}\sqrt{\frac{2p!}{\pi(|l|+p)!}}\; e^{(2p+|l|+1)\Psi(z)} \times \left(\frac{\sqrt{2}r}{w(z)}\right)^{|l|} L_p^{(|l|)}\left(\frac{2r^2}{w(z)^2}\right) e^{-ik\frac{r^2}{2q(z)}+il\phi}, \quad (1)$$

where p is the radial mode index, l is the azimuthal mode index, w is the beam radius, k is the wave-number, q is the complex Gaussian beam parameter, $L_p^{(|l|)}$ are the associated Laguerre polynomials, and Ψ is the Gouy phase, given by

$$\Psi(z) = \arctan\left(\frac{\lambda z}{\pi w_0^2}\right). \quad (2)$$

The order of a higher-order LG mode is given by $2p + |l|$. An alternative sinusoidal form of LG modes with a sinusoidal amplitude dependence in azimuthal angle can be used equally well. In this article we will discuss the research performed by the Birmingham group towards evaluating the feasibility of LG mode technology for implementation in detectors of the advanced generation and beyond to reduce the test mass thermal noise limit.

2. Interferometric performance simulation study of LG modes

Our initial step was to compare the interferometric performance of the LG_{33} mode to the LG_{00} mode in an optical layout similar to that proposed for the advanced detectors. By interferometric performance we mean the level of coupling of a number of variables to the measured phase ('phase noise analysis'), the ability to generate the required longitudinal and alignment control signals, and finally the maximum achievable detector sensitivity. The first consideration to ensure a fair analysis of the performance of the LG_{33} mode was the beam sizes which should be compared. The advanced detectors use larger beam sizes on the test masses than the first generation detectors, irrespective of the spatial mode used, since the limiting Brownian noise level scales with the inverse square of the beam size. The limit on beam size can be enforced either by the point at which the clipping loss becomes unacceptably high, or the point at which cavities become unacceptably close to the unstable concentric design required for maximising beam sizes. For this study we considered the limit to be enforced by the clipping loss, which we fixed at 1 ppm.

The higher-order LG modes are more spatially extended, and thus have larger clipping losses than the LG_{00} mode for a given beam spot size parameter. The ratio between LG_{00} spot size and LG_{33} spot size required to give a 1 ppm clipping loss on the same mirror is $\frac{w_{00}}{w_{33}} = 1.64$. In the rest of this section we will compare the performance of three different configurations for a symmetric 3 km cavity, to be referred to as the LG_{33}, LG_{00}^{large} and LG_{00}^{small} configurations. The LG_{33} and LG_{00}^{large} configurations both have a clipping loss of 1 ppm at the cavity mirrors, thus the beam size at the mirrors for LG_{00}^{large} configuration is factor of 1.64 larger than for the LG_{33} configuration. The LG_{00}^{small} configuration has the same spot size at the mirrors as the LG_{33} configuration, but uses a LG_{00} beam, and therefore has a lower clipping loss. The LG_{00}^{small} configuration is less realistic for comparison with the LG_{33}, for the aforementioned reason that advanced detectors will be pushing towards larger beam sizes irrespective of spatial mode used, but we found it useful to include in the study to separate the effects due beam parameters and effects due directly to mode shape.

The first control signal investigated was the longitudinal error signal for a single cavity. For a Pound-Drever-Hall (PDH) modulation/demodulation error signal generation technique [11], the resulting longitudinal control signal was identical for all three of the considered configurations. The next step was an analysis of the coupling from end mirror tilt to longitudinal phase noise in a single cavity for each configuration. Figure 1 shows the results of the investigation for a 3 km cavity based on the Advanced Virgo reference design in [10]. The LG_{00}^{small} configuration performs slightly better than the LG_{33} here, but of the two configurations with equal clipping losses, the LG_{33} configuration performs significantly better than the LG_{00}^{large} configuration. This is because the LG_{00}^{large} configuration requires the cavity to be pushed closer to the unstable concentric design in order to achieve the necessary beam sizes at the mirrors.

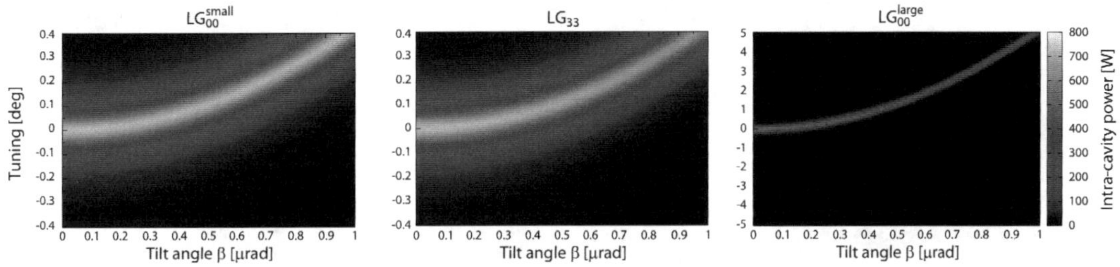

Figure 1. Intra cavity power as a function of cavity end mirror tilt angle β and longitudinal tuning ϕ for the three different configurations. The first two configurations show very similar results, but the LG_{00}^{large} configuration shows a much stronger coupling from tilt to tuning (note the larger scale on the tuning axis in the LG_{00}^{large} plot).

As well as length sensing and control, angular sensing and control is crucial to maintaining stable operation and maximum sensitivity of a gravitational wave interferometer. To investigate how the LG_{33} mode performs in this respect, an alignment scheme based on the Ward technique described in [12] was designed for the same 3 km cavities previously described. The optical layout for this scheme is shown in panel Ⓐ of figure 2. Radio frequency sidebands are added to the beam by the electro-optic modulator (EOM), and the beam is passed into the cavity via a beam splitter. The reflected beam from the cavity is split in two, and each resulting beam is passed through a telescope and detected with a quadrant photodetector. These telescopes are designed such that the Gouy phase (see equation 2) difference between the beams at each quadrant photodetector is 90°, in order to provide the maximum possible orthogonality between alignment signals from the end mirror and the input mirror. The signal from each photodetector is demodulated by mixing with the initial radio frequency, with a demodulation phase chosen such as to maximise the slope of the error signal corresponding to the mirror for which the photodetector is required to sense the alignment. In order to match the conditions under which the alignment control system will be developed in practice as closely as possible, we tuned the parameters rather than using the theoretical optimum parameters.

The alignment sensing figure of merit for each configuration can be summarized by its control matrix [13]. The elements $\sigma_{\mathrm{detector}}^{\mathrm{mirror}}$ of these control matrices are the slopes of the error signal at the working point as measured at quadrant photodetectors QPD1 and QPD2, for misalignments of the cavity input mirror IMX, and end mirror EMX, as shown in equation 3.

$$C_{\mathrm{configuration}} = \begin{pmatrix} \sigma_{\mathrm{QPD1}}^{\mathrm{IMX}} & \sigma_{\mathrm{QPD1}}^{\mathrm{EMX}} \\ \sigma_{\mathrm{QPD2}}^{\mathrm{IMX}} & \sigma_{\mathrm{QPD2}}^{\mathrm{EMX}} \end{pmatrix} \tag{3}$$

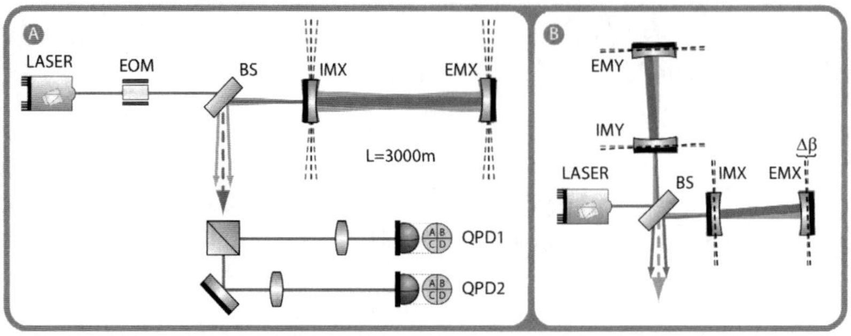

Figure 2. Two optical layouts used in the alignment analysis simulations. Ⓐ shows the single arm cavity alignment control scheme investigated, and Ⓑ shows the differential misalignment of arm cavities, for which coupling into dark port power was analysed.

The resulting control matrices for each of the three cavity configurations were as follows:

$$C_{\mathrm{LG}_{00}^{\mathrm{small}}} = \begin{pmatrix} 5.6152 & 0.0477 \\ 2.1607 & 3.5878 \end{pmatrix} = 5.6152 \begin{pmatrix} 1 & 0.009 \\ 0.385 & 0.639 \end{pmatrix} \tag{4}$$

$$C_{\mathrm{LG}_{33}} = \begin{pmatrix} 7.444 & 0.022 \\ 2.741 & 4.771 \end{pmatrix} = 7.444 \begin{pmatrix} 1 & 0.003 \\ 0.368 & 0.641 \end{pmatrix} \tag{5}$$

$$C_{\mathrm{LG}_{00}^{\mathrm{large}}} = \begin{pmatrix} 17.774 & 15.330 \\ 11.472 & 2.725 \end{pmatrix} = 17.774 \begin{pmatrix} 1 & 0.862 \\ 0.645 & 0.153 \end{pmatrix}. \tag{6}$$

An ideal control matrix would be diagonal, since the off diagonal elements correspond to the presence of information from the unwanted mirror in a given photodetector signal. In this respect, the LG_{33} and $\mathrm{LG}_{00}^{\mathrm{small}}$ configurations give control matrices which are much closer to the ideal than the $\mathrm{LG}_{00}^{\mathrm{large}}$ configuration; the information from the two mirrors is better separated in these two configurations.

The next aspect of the interferometric performance of the LG_{33} mode to be investigated was the coupling of differential arm cavity misalignment to power at the interferometer output port. If the two arm cavities are differentially misaligned, the overlap of the two beams at the central beam splitter will not be complete, leading to a change in the light power present at the output port of the interferometer, as illustrated in panel Ⓑ of figure 2. If this differential misalignment varies with time, the power measured at the dark port will also vary with time, producing a signal at the photodetector that is indistinguishable from gravitational wave signals. We obtained values for the output power enhancement due to differential misalignment for each of the three previously described cavity configurations, as shown in figure 3. Comparing the simulation results with the reference limit (around $7{\times}10^{-9}$ W) calculated from a differential arm length requirement of 10^{-15} m [14] and a dark fringe offset of 10^{-12} m [15], we see that the $\mathrm{LG}_{00}^{\mathrm{small}}$ configuration performs best, but crucially that the LG_{33} configuration also outperforms the $\mathrm{LG}_{00}^{\mathrm{large}}$ configuration. In summary, for all the interferometric performance aspects analysed, the LG_{33} configuration performs significantly better than the $\mathrm{LG}_{00}^{\mathrm{large}}$ configuration. The $\mathrm{LG}_{00}^{\mathrm{small}}$ performs as well or better than the LG_{33} configuration, but it should be recalled that this will have a significantly higher thermal noise level, and for this reason is not representative of a likely configuration for advanced detectors.

The final comparison between the performance of the LG_{00} and LG_{33} modes was in terms of the overall detector sensitivity. Several scenarios for using the LG_{33} mode in Advanced Virgo detector were evaluated in [16], but here we just discuss a scenario which compares the

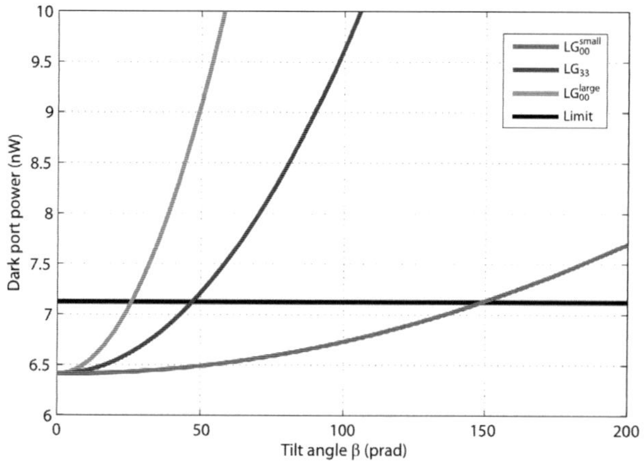

Figure 3. Dark port power as a function of differential arm cavity misalignment for the three different configurations. The black line shows the limit calculated for dark port power based on a differential arm length requirement of 10^{-15} m for a dark fringe offset of 10^{-12} m.

sensitivity of the detector with LG_{00}^{large} and LG_{33} modes, where both beams experience a 1 ppm clipping loss at the cavity mirrors. The cavity mirror curvatures, RoC, and beam sizes at the mirrors, w, for the two configurations in this scenario are shown in table 1. In each case the detector sensitivity was calculated using a version of Gravitational Wave Interferometer Noise Calculator (GWINC) [17] especially adapted for Advanced Virgo. Table 1 shows the results of the calculation for the two cases NS/NS, and BH/BH where the signal recycling detuning (SR det.)[18] was optimized for detection of signals from binary neutron star inspirals and binary black hole inspirals respectively. The figures of merit chosen were the effective detection ranges for the two signal sources, $\Gamma_{NS/NS}$ and $\Gamma_{BH/BH}$. According to these results, the LG_{33} mode provides a relative improvement of the inspiral ranges by around 20 % and 25 % for signal recycling detunings of 750 Hz and 300 Hz respectively, compared to the LG_{00} mode. This corresponds to a potential increase by up to a factor of 2 in the observable event rate of the Advanced Virgo detector by using the LG_{33} mode instead of the LG_{00} mode. The method and results of the investigation described in this section are presented in more detail in [16].

Laser mode	SR det. [Hz]	w [cm]	l_{clip} [ppm]	RoC [m]	$\Gamma_{NS/NS}$ [Mpc]	$\Gamma_{BH/BH}$ [Mpc]
LG_{00}^{large}	750	6.47	1	1523	126	900
LG_{33}	750	3.94	1	1708	148	1140
LG_{00}^{large}	300	6.47	1	1523	130	580
LG_{33}	300	3.94	1	1708	163	715

Table 1. Relevant parameters and results of the GWINC simulation analysis of an Advanced Virgo design which considers identical clipping loss l_{clip} of 1 ppm at the arm cavity mirrors for both the LG_{00}^{large} and LG_{33} modes.

3. Table-top demonstrations

Following the positive results of the simulation investigation into the interferometric performance of the LG_{33} mode, we proceeded to perform table-top experiments to verify these results. The first step was to generate LG modes with a reasonably high purity for use in the experiments. This was achieved using a Holoeye liquid-crystal on Silicon spatial light modulator (SLM), as for example in [19]. Upon reflection from the surface of the SLM, a beam can be imprinted with a transverse phase pattern determined by the user. By choosing the phase pattern to be that of a LG_{33} mode, i.e. three azimuthal phase vortices and three radial phase jumps of π for the helical case, one can achieve a partial conversion from the LG_{00} mode to the LG_{33} mode.

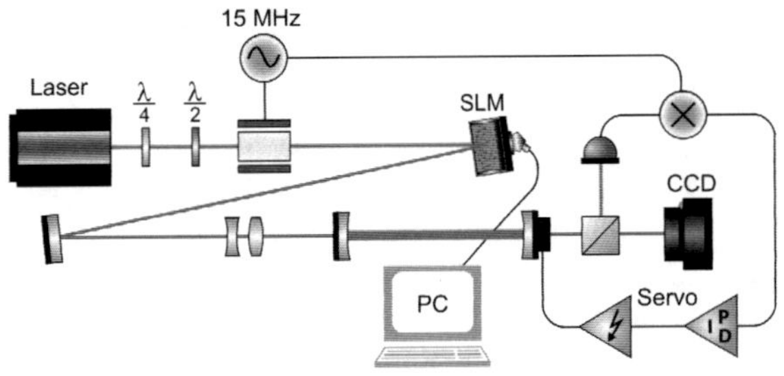

Figure 4. The experimental setup for mode cleaning a SLM generated higher-order LG beam. The input beam is converted to a higher-order LG beam by the SLM, then passed via a mode-matching telescope into the linear mode cleaner cavity. The transmitted light is used to generate an error signal which is fed back to the PZT attached to the curved end mirror to control the length of the cavity, and simultaneously imaged on the CCD camera.

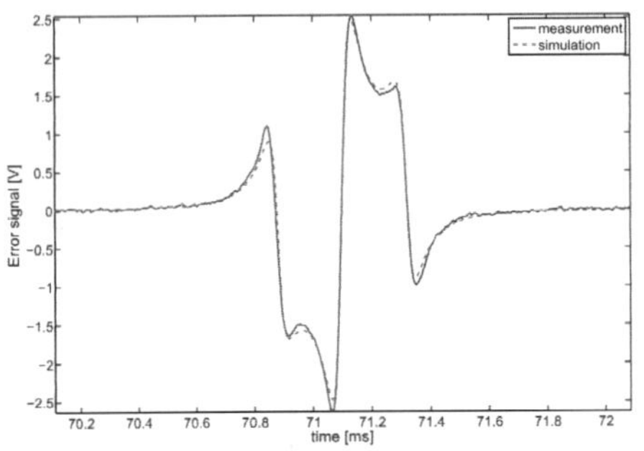

Figure 5. The blue trace and the red dashed trace show the measured sinusoidal LG_{33} beam and simulated LG_{00} PDH error signal respectively, for the linear mode cleaner cavity set up as shown in figure 4. The primary features are identical, as predicted in [16].

The experimental setup shown in figure 4 was then used to investigate the length control signals that could be generated for the resulting beam. The 1064 nm laser light was passed through an EOM to add 15 MHz sidebands to the light, for generating the PDH error signal for length control of the mode cleaner cavity. The light was then partially converted to the LG_{33} mode by interaction with the SLM, and passed into the mode cleaner cavity via a mode matching telescope. The light transmitted through the mode cleaner cavity was simultaneously measured with a photodiode and a CCD camera. The photodiode signal was mixed down with the 15 MHz signal to give the required error signal, which was then fed back to a Piezo-electric transducer attached to one of the mode cleaner mirrors, via two amplifier stages. In this way the length of the cavity was controlled to keep it in the resonant condition for any given mode order. The CCD measurement of the transmitted beam was then used for analysing the mode purity. The mode cleaner is a 2 mirror flat-concave cavity of length 21 cm and with a measured finesse of 172.

We verified the first result of the initial simulations by recording the PDH error signal generated using the LG_{33} mode, shown in figure 5. This signal was equivalent for the LG_{33} and LG_{00} modes as predicted, and therefore allowed us to control the cavity on resonance for the order 9 LG_{33} mode. Demonstrating robust and repeatable feedback control of an optical cavity was a significant step in validating the higher-order LG mode operation of interferometers.

Figure 6 shows the beams before and after transmission through the mode cleaner cavity. It is immediately apparent that the transmitted beams exhibit greater circular symmetry than the input beams, as would be expected due to the filtering out of other spatial modes. We used the CCD measurement of the transmitted beam intensity pattern to estimate its mode content. This required fitting the beam size parameter to calibrate the CCD, and subtracting

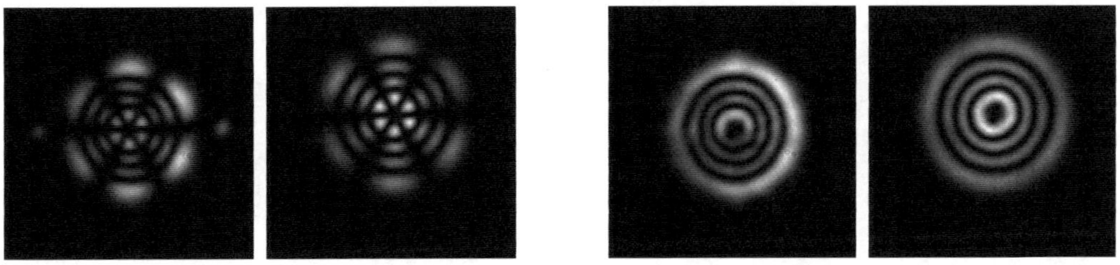

Figure 6. The measured intensity patterns of the sinusoidal (left pair) and helical (right pair) LG_{33} beams before (left of pairs) and after (right of pairs) transmission through the linear mode cleaner.

an ideal LG_{33} mode intensity pattern, with beam size given by the result of the fit, from the measured data. These residuals were then analysed by comparing the measured data with a numerical simulation of the setup which included some misalignment into the cavity, performed using the interferometer simulation Finesse [20]. For a Finesse model of the mode cleaner setup which included an input beam misalignment of -100 μrad in horizontal and 60 μrad in vertical, the residual between the transmitted mode and an ideal LG_{33} matched the measured data extremely well. By performing a modal decomposition of the transmitted field in the simulation, we estimated the mode content of the transmitted beam, as shown in table 2. The

u_{lp}^{sin} mode	3, 3	4, -1	2, -5	4, 1	2, 5	other
power	99%	0.4%	0.3%	0.1%	0.1%	$<$ 10 ppm

Table 2. Results of a modal decomposition of the transmitted sinusoidal LG_{33} beam in the numerical model of the linear mode cleaner, with an input beam misalignment of -100 μrad in the horizontal axis, and 60 μrad in the vertical axis. Most of the beam is in the sinusoidal LG_{33} mode, with the rest almost entirely found in other modes of order 9.

analysis showed that the transmitted beam was to 99 % in the LG_{33} mode for both the helical and sinusoidal cases. With an estimate of the transmitted mode purity, together with measurements of the cavity finesse and throughput efficiency, we were also able to estimate the LG_{33} mode content of the input beam to be 66 % for the helical beam, and 51 % for the sinusoidal beam. A detailed description of the method and results of this experiment can be found in [21].

4. LG mode degeneracy investigation

One important difference between higher-order LG modes and the fundamental LG_{00} mode is that only the LG_{00} mode is unique in its mode order. For each higher-order mode there exists at least one other mode of the same order. Since the mode filtering effect of optical cavities relies principally on the round trip Gouy phase difference between different mode orders, they cannot therefore filter out all other modes than the LG_{33} when on resonance for order 9. For this reason we call the other order 9 modes degenerate with the LG_{33} mode.

This degeneracy of higher-order modes has serious implications for their application in gravitational wave interferometers. In the initial simulations reported in [16] the interferometer mirrors were modelled as perfect spherical curved mirrors. In reality however, the mirrors will have some small deviations from perfect spherical surfaces which can cause coupling from the LG_{33} mode into other modes of order 9. If we consider this process occurring within the arm cavities of a gravitational wave interferometer, the degeneracy of the modes of order 9 will mean that the mode content of the circulating beam may have a significantly reduced proportion of LG_{33}, the remainder being made up principally of other order 9 modes. If the mode content of both arm cavities is different (which is likely to be the case if the coupling between modes is driven by the randomly oriented mirror surface distortions) the modal overlap at the beam splitter will be imperfect and the output port contrast will be reduced.

The method and results of a detailed numerical and analytical investigation into the effect of mirror surface distortions on the purity of LG modes within optical cavities were presented at this Amaldi meeting [22], and in [23], and another study on this topic is presented in [24]. One of the most important outcomes from the work described in [23] was the derivation of an analytical formula for predicting the amount of coupling between different LG modes upon reflection from a mirror, based on the spatial features of mirror surfaces as described by Zernike polynomial functions. In the limit that the height of the surface distortions is much smaller than the wavelength of the light, coupling between an incident mode LG_{pl} and a reflected mode $LG_{p'l'}$ is only significantly caused by Zernike polynomials Z_n^m which satisfy the condition $m = |l - l'|$. Based on this result, it was possible to propose limits on the heights of the most important lower-order Zernike polynomial present on mirror surfaces, in order to achieve a circulating mode purity of over 99.9 % [23].

5. Ongoing experimental investigations

Following the simulation studies and table-top experiments, we continue to pursue experimental tests of LG mode technology. Currently, this involves evaluating the performance of the LG_{33} mode in a suspended 10 m cavity at the Glasgow prototype facility, and the development of a high-power LG_{33} mode laser beam at the Hannover high-power laser facility.

Procession from table-top experiments to a prototype experiment with suspended mirrors is a standard feature of the development of new technologies for gravitational wave detectors, in order to ensure compatibility with realistic interferometer subsystems and requirements. Experiments at the Glasgow prototype also give us a way to investigate the LG mode degeneracy problem heretofore only looked at theoretically and numerically. Though we did not observe the detrimental effects of the mode degeneracy problem in our table-top experiment we believe this is most likely due to the relatively small beam sizes used on the mirrors, and the low cavity finesse in comparison to the advanced detector arm cavities. However, it is clear that LG modes must also work with larger beam sizes in order to provide the thermal noise benefits that make them attractive in the first place. LG modes must also be compatible with cavities of similar finesse to the advanced detector arm cavities. We have therefore been setting up an experiment at the Glasgow 10 m prototype facility which uses larger beam sizes in a higher finesse cavity.

In this experiment we will use an etched diffractive optic to facilitate the conversion from LG_{00} to LG_{33}, similar to that used in [25], rather than the SLM used in the table-top experiments. Though the SLM was very useful for prototyping purposes to investigate the effects of different phase profiles, a fixed transmissive optic allows to generate a higher mode purity with more stability than the SLM. The setup for the LG_{33} mode conversion bench is shown in figure 7. The beam is picked off from the previously used laser path (shown in purple) after a fibre output coupler, and directed with a flip mirror into the LG mode conversion area. The red path shows the LG_{33} mode conversion path, in which the beam passes through an EOM for generating the control sidebands, and the etched diffractive phaseplate. The resulting beam is then transmitted through a linear mode cleaner, which is feedback controlled using the PDH method to remain on resonance for mode order 9. This serves to increase the mode purity of the beam, as demonstrated in [21]. The transmitted beam is then passed back into the previously used laser path using another flip mirror, and on towards the suspended cavity. The green path is a phaseplate bypass path, which enables us to alternatively operate the linear mode cleaner

 doi:10.1088/1742-6596/363/1/012010

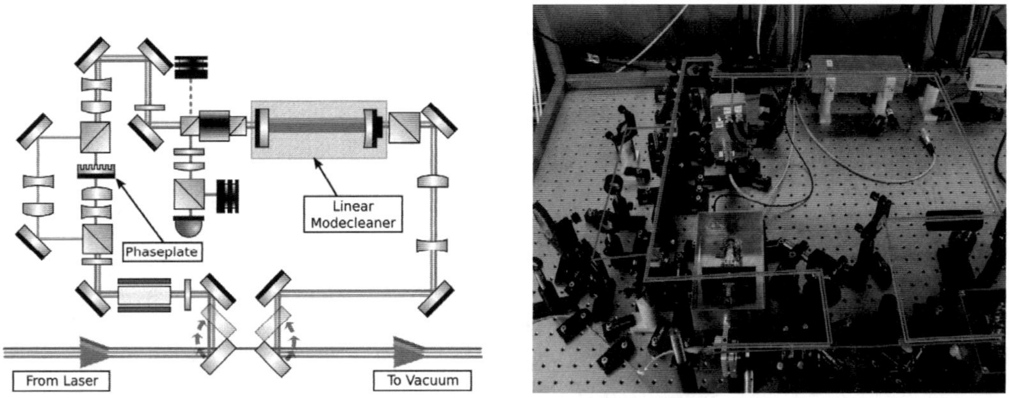

Figure 7. The laser mode conversion path on the Glasgow 10 m prototype input optics bench. The red line shows the LG_{33} mode path, and the green line shows the LG_{00} path that bypasses the phaseplate. The purple line shows the original LG_{00} laser path.

with the LG_{00} mode. Since the light transmitted by the mode cleaner must be defined by the cavity eigenmodes, the transmitted LG_{00} and LG_{33} beams should have the same alignment and mode matching relative to the suspended cavity. This enables us to make a valid assessment of the relative performance of the two modes within the suspended cavity.

The experimental work on LG modes at the Glasgow prototype is still under way, though we will soon have results about the performance of the LG_{33} mode within a cavity of finesse in the region of 600 - around the same finesse foreseen for the advanced detector arm cavities. We plan to analyse the length control signals for the cavity, as well as the transmitted beam intensity pattern, in order to ascertain the level of coupling between LG modes that occurs in the cavity. Analysis of the circulating mode purity, together with measured data of the cavity mirror surfaces, should enable us to verify the results of the theoretical investigation into the LG mode degeneracy problem described in [23]. With this information, we should be in a position to make well informed decisions on the suitability of LG mode technology for future detectors, based on the surface quality of the available optics.

Another crucial requirement for LG mode technology to be useful in future detectors is the ability to generate high-power and high-purity LG mode laser beams for the interferometer input. It is planned to use laser powers in excess of 100 W in the advanced detectors in order to bring the shot noise level down, so any prospective LG mode light source must also be able to reach comparable power levels. In collaboration with the members of the AEI in Hannover that developed the new 200 W laser system for Advanced LIGO [26], we are currently setting up an experiment to investigate the technical challenges of generating LG modes at such high powers. High-power beams from tens to hundreds of Watts will be passed through etched diffractive phaseplates. Following the example of our very first experiment on LG modes described in section 3, the LG beams that are generated will be then filtered by means of a linear mode cleaner, and their subsequent purity analysed. This work is also still at a preliminary stage, but we expect to be able to report results in the near future.

6. Summary

We have investigated the suitability of higher-order LG mode technology for use in future gravitational wave interferometers for reducing the test mass thermal noise, from many different angles. An initial simulation study showed that the longitudinal and angular control performance of the LG_{33} mode in an Advanced Virgo-like interferometer was significantly better than the LG_{00} mode with the same clipping losses at the cavity mirrors. It was also shown that increases in the observable event rates of Advanced Virgo by up to a factor of 2 could be gained by using the LG_{33} mode in place of the LG_{00} mode. Table-top experiments verified the equivalence of the LG_{33} mode and LG_{00} mode for PDH longitudinal control signal generation in an optical cavity, and demonstrated the mode cleaning of SLM generated LG_{33} beams to reach a mode purity of over 99 %. More detailed theoretical investigations into the effects of mirror surface distortions on LG mode purity within optical cavities led to the derivation of an analytical formula for the coupling between LG modes caused by specific mirror distortions. Using this formula, mirror specifications required for compatibility with the LG_{33} mode were subsequently derived. We are currently engaged in a collaborative effort with the University of Glasgow to verify the results of these theoretical investigations, by means of experiment within the 10 m prototype facility in Glasgow. We are also currently investigating the feasibility of generating high-power and high-mode purity LG_{33} beams at the Hannover high-power laser facility. Higher-order LG modes can potentially offer a significant improvement in the sensitivity of future gravitational wave interferometers: our research program will evaluate how much of that potential is likely to be realisable in practice.

Acknowledgments

We thank B. Barr, A. Bell, S. Huttner, J. Macarthur, B. Sorazu and K. Strain for providing the opportunity and helping to perform the LG mode experiments at the Glasgow 10 m prototype, and we thank C. Bogan, P. Kwee and B. Willke for the same assistance in the high-power LG modes experiment in Hannover. This document has been assigned the LIGO document number LIGO-P1100146.

References

[1] G. M. Harry and the LIGO Scientific Collaboration, 2010 *Class. Quantum Grav.* **27** 084006
[2] The Virgo Collaboration, 2009 *Tech. Rep.* VIR-027A-09
[3] B. Abbott, *et al*, 2009 *Rep. Prog. Phys.* **72** 076901
[4] F. Acernese and the VIRGO Collaboration, 2006 *Class. Quantum Grav.* **23** 635
[5] S. Rowan, J. Hough, and D. Crooks, 2005 *Phys. Lett. A* **347**, 25
[6] J-Y. Vinet, 2009 *Living Rev. Relativity* **12**, http://www.livingreviews.org/lrr-2009-5
[7] E. D'Ambrosio, 2003 *Phys. Rev. D* **67**, 102004
[8] M. Bondarescu, O. Kogan, and Y. Chen, 2008 *Phys. Rev. D* **78**, 082002
[9] B. Mours, E. Tournefier, and J.-Y. Vinet, 2006 *Class. Quantum Grav.* **23**, 5777
[10] A. Freise and M. Mantovani, 2004 *Tech. Rep.* VIR-NOT-EGO-1390-330
[11] E. D. Black,2001 *Am. J. Phys.* **69**, 79
[12] E. Morrison, D. I. Robertson, H. Ward, and B. J. Meers, 1994 *Appl. Opt.* **33**, 5041
[13] M. Mantovani and A. Freise, 2008 *J. Phys. Conf. Ser.* **122**, 012026
[14] R. Adhikari, S. Ballmer, and P. Fritschel, 2008 *LIGO Report* No. LIGO-T070236-00-D
[15] R. Abbott *et al*, 2008, *LIGO Report* No. LIGO-T070247-00-I
[16] S. Chelkowski, S. Hild and A. Freise, 2009 *Phys. Rev. D* **79**, 122002
[17] GWINC, http://lhocds.ligo-wa.caltech.edu:8000/advligo/GWINC
[18] S. Hild, H. Grote, M. Hewitson, H. Lück, J.R. Smith, K.A. Strain, B. Willke, and K. Danzmann, 2007 *Class. Quantum Grav.* **24**, 1513
[19] N. Matsumoto et al., 2008 *J. Opt. Soc. Am. A* **25**, 1642
[20] A. Freise, G. Heinzel, H. Lück, R. Schilling B. Willke and K. Danzmann, 2004 *Class. Quantum Grav.*, **21**, S1067-S1074. Finesse simulation software and manual available at http://www.gwoptics.org/finesse/
[21] P. Fulda, K. Kokeyama, S. Chelkowski and A. Freise, 2010 *Phys. Rev. D* **82**, 012002
[22] C. Bond *et al*, this conference proceedings
[23] C. Bond, P. Fulda, L. Carbone, K. Kokeyama, A. Freise, 2011 *Phys. Rev. D* **84**, 102002
[24] T. Hong, J. Miller, H. Yamamoto, Y. Chen, R. Adhikari, 2011 *Phys. Rev. D* **84**, 102001
[25] S. A. Kennedy, M. J. Szabo, H. Teslow, J. Z. Porterfield, and E. R. Abraham, 2002 *Phys. Rev. A* **66**, 043801
[26] L. Winkelmann *et al*, 2011 Appl. Phys. B **102**, 529-538

Non-equilibrium "thermal noise" of low loss oscillators

G Karapetyan[1], D Agguiaro[1], M Bonaldi[2,3], L Castellani[1], R Hajj[1], C Lazzaro[1], D Mazzaro[1], M Pegoraro[1], C Poli[1], R Thakur[1], L Conti[1]

[1] INFN, Sezione di Padova, Via Marzolo 8, I-35131 Padova, Italy
[2] Institute of Materials for Electronics and Magnetism, Nanoscience-Trento-FBK Division, I-38123 Povo, Trento, Italy
[3] INFN, Gruppo Collegato di Trento, Sezione di Padova, I-38123 Povo, Trento, Italy

gagik.karapetyan@pd.infn.it

Abstract. The RareNoise project aims at studying how ground-based gravitational wave detector are affected by the thermodynamic non-equilibrium states which are present in their experimental apparatus. We present the RareNoise experimental work which focuses on the study of the 'thermal noise' of low loss oscillators subject to steady-state thermal gradients. We also present the first results of the experimental campaign on steady-state non-equilibrium oscillators around room temperature.

1. Introduction.

Ground-based detectors of gravitational waves (GW) are sophisticated experiments to detect small distortions of the spacetime predicted by general relativity. Ground based ones consist of Michelson laser interferometers with an arm length of up to a few kilometers (VIRGO [1], LIGO and aLIGO [2], GEO600 [3]), or mechanical oscillators with masses of up to several tons [4]. To reach a detection rate of a few events per year, a sensitivity of 10^{-22} in $1msec$ is required [5]. To meet this goal several techniques are proposed and used to lower the contribution of the noise sources; among them thermal noise proves to be difficult to tackle.

Usually GW detectors are considered as equilibrium devices, hence the thermal noise is assumed to have Gaussian distribution. However this picture is not fully justified [6]: in fact non equilibrium states arise for instance in suspended mirrors due to absorbed laser power and thermal compensation technique. In the cryogenic environment planned for LCGT [7, 8] and under consideration for the 3rd generation GW detector [9] these issues are potentially more relevant. However there is not a clear understanding of thermal noise in non equilibrium states.

The RareNoise project aims at studying how the performance of ground-based gravitational wave detectors is affected by the thermodynamic non-equilibrium states which are present in their experimental apparatus [6].

2. Experimental setup.

We have developed an experimental setup to investigate the statistical properties of the thermal noise of mechanical oscillators subject to steady state thermal gradients around room temperature. The setup hosts simultaneously 3 oscillators in vacuum environment (figure 1) with low mechanical noise.

2.1 The oscillators

Each oscillator is machined from a single piece of Al5056, a material with low intrinsic mechanical losses [10]. Basically it consists of a square cross section rod kept in the vertical position, fixed at the top end and loaded by a cuboid mass at the bottom end (Fig. 2). A prototype version of such oscillators is reported in reference [11]. The length of the rod is $0.1m$, cross section is $30.25\cdot10^{-6}\ m^2$ and the cuboid mass is about $0.25kg$. These dimensions are chosen after FEM modelling, so that the first longitudinal mode of the part resonates at about $1.4kHz$. In this mode the position of the cuboid mass oscillates along the vertical axis, as the rod elongates and shrinks. The 3 oscillators are identical except for the volume of the cuboid masses: they differ by 5.5% and 11% in volume to separate the resonant frequencies of the longitudinal modes. This is done in order to avoid mechanical couplings of the oscillators. To the same bottom part of the suspension where the oscillators are fixed we also attached piezoelectric actuator to provide for mechanical excitation of the oscillators (see figure 1).

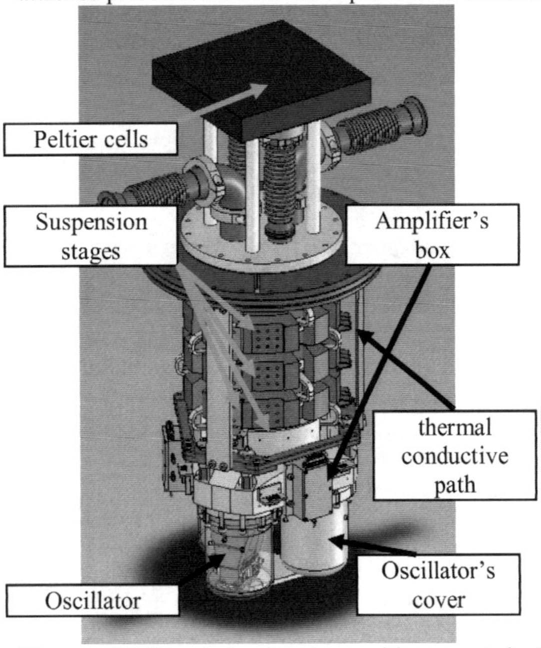

Figure 1. Suspension system with mounted oscillators. Oscillators are inserted into metallic covers for protection while in air.

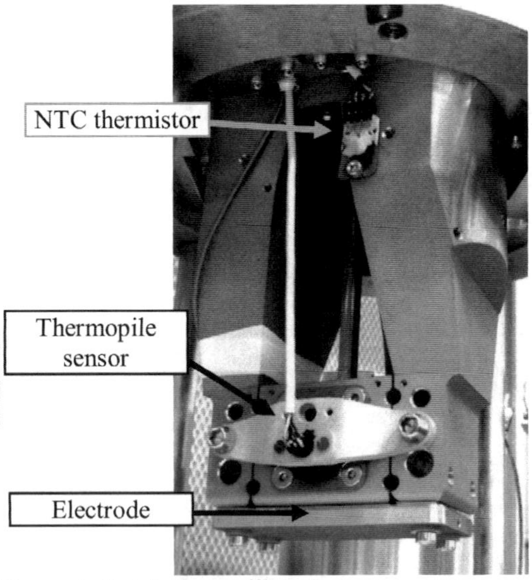

Figure 2. Inside the oscillator's cover.

2.2 The displacement readout.

To measure the oscillations of the cuboid mass each of the 3 oscillators is equipped with capacitive readout. To this end we form a capacitor by facing a Al5056 plate to the bottom surface of the cuboid mass. This plate is supported by two arms parallel to the rod and electrically insulated from them by two Teflon spacers (figure 2). The thicknesses of the spacers range between $100\mu m$ and $300\mu m$. The resulting overall capacitance ranges from $160pF$ to $290pF$. This capacitance is the sum of the capacitance C formed by the plate and the cuboid mass (dielectric medium is the vacuum) and the capacitance C_{teflon} formed by the plate and its support (dielectric medium is Teflon). From the mechanical design we estimate the ratio between these capacitances to be $\frac{C}{C_{teflon}} \approx 3.21$. The facing surfaces of the electrodes and cuboid masses are diamond turned in order to be able to setup electric fields up to 10^7 *V/m*. In fact working at constant charge the sensitivity of the readout is proportional to the electric field according to the following equation:

$$dV = E_0 \frac{dx}{\left(1 + \frac{C_p}{C}\right)}$$

- where dx is the displacement of the cuboid mass, dV is the corresponding developed voltage signal and the E_0 is the electric field across the capacitance C, and C_p is the sum of C_{teflon} and all parasitic capacitances. In order to reduce the latter we have minimized the length of the wire between the readout and the amplifier by placing it in the vacuum chamber close to the oscillator (see figure 1); we estimate the parasitic capacitances to be about 10*pF*. Each of the 3 amplifiers is custom made and adds a (single sided) noise contribution of about 6 $nV/Hz^{-1/2}$ in the frequency range of interest: from 0.3*kHz* to 2*kHz*. The amplifiers are designed as band pass filters with gain G=100 in the range of interest (see figure 3).

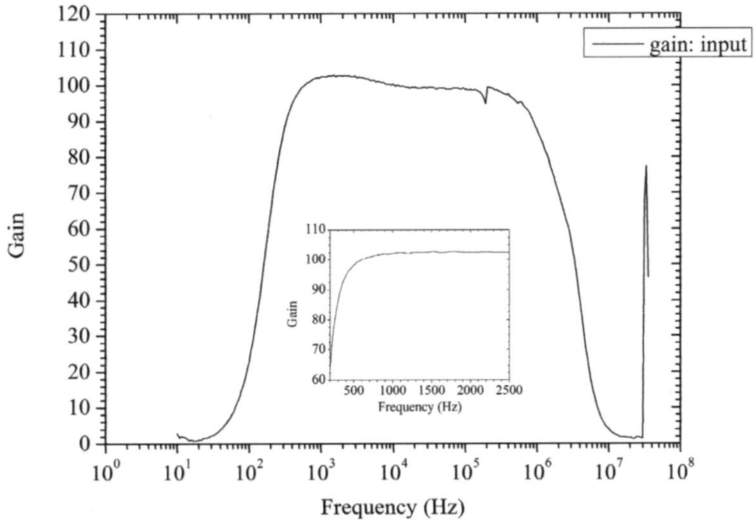

Figure 3. Voltage gain measurement one of the 3 (identical) amplifiers. The inset is a zoom in the range of 200-2500 Hz.

2.3 Mechanical suspension

To reduce mechanical noise in the setup we suspend oscillators and amplifiers by a cascade of active and passive filters; the active filter is provided by an air suspended platform which supports all the experimental apparatus. The passive suspension consists of a cascade of 4 mechanical filters effective in all directions and housed inside the vacuum chamber: a prototype version of the filters is reported in reference [11]. Overall we estimate to achieve vibration isolation of more than 200*dB* at 1.5*kHz* in all spatial directions. To avoid introducing mechanical noise we pump the vacuum chamber by vibration free ion pump. The pressure level in the chamber is below 10^{-5}*mbar*.

2.4 Thermal control

As mentioned our goal is to study the oscillator thermal noise in non equilibrium steady states due to thermal differences. Thus we need to control the temperature T_1 of the top end of the oscillator rods and the temperature T_2 of each cuboid mass. The control of T_1 is achieved by stabilizing the temperature of the aluminium mass at the end of the suspension cascade and where all oscillators are fixed to. This is achieved by feedback loop that controls the current flowing through Peltier cells placed on the top of the vacuum chamber. To reduce the thermal time constant of this loop we provide thermal conductive path in parallel to the mechanical filters. The resulting time constant is about 12 hours. To be able to set thermal differences across the oscillators we heat the cuboid masses radiatively with thermal sources. This way we are able to set thermal differences up to 15*K* by flowing about 1*W* power. To measure temperatures of the cuboid masses without adding mechanical losses we use thermopiles while for T_1 we use NTC thermistors (see figure 1,2). To improve radiative coupling we attached graphene coated plates to the two opposite faces of the cuboid masses facing respectively the thermal source and the thermopile (see figure 2).

2.5 Data acquisition.

The outputs of the amplifiers are acquired by a system that can sample simultaneously up to 4 channels at high speed and 16 channels at low speed in continuous mode. For the amplifier output we use 24-bit resolution ADC NI PXI-4462 DAQ at 8*kSamples/sec* within ±300*mV* range. All temperature sensors and environmental monitor signals are acquired at 0.1*Sample/sec* by NI 9219 24-bit universal analog inputs. The development platform of the acquisition system was chosen to be NI LabView. During the acquisition it also provides live data view from the oscillators. Data storage is arranged in proprietary binary data format: the header is the 10sec time average of each environmental signal and the body consists of *80000* samples of the amplifier signals. Data analysis is implemented with ROOT libraries [12].
The feedback control program of T_1 interacts with acquisition of the temperature sensors. It reads T1 measurement, computes the correction signal with PID algorithm for a given temperature set point and controls via Ethernet a Sorensen XTR850 DC power supply connected to the Peltier cells. The power supply to the thermal sources of the oscillators and for the thermopiles is provided by NI PXI-4110 dc power supply.

3. Experimental results.

Before starting experimental runs in non equilibrium steady states we perform test runs for diagnostics. Figure 4 is an example of the power spectral density of the amplifier output with $E_0=1.2\cdot 10^7$ *V/m*.

Up to 200*Hz* dominant noise source is mechanical noise, also amplified by suspension resonances. Above 200*Hz* the amplifier noise becomes dominant. A few peaks are still present due to mains harmonics. At about 300*Hz* transverse modes of the oscillator are present. The inset shows the zoom around the resonance of the longitudinal mode at 1348*Hz*. Red curve is lorentzian fit of the resonance:

$$f(x) = y_0 + A\frac{w}{4(x-x_c)^2+w^2}$$

where x_c is resonance frequency, w is full width at half maximum and A is the curve area times $2/\pi$. From the fit we infer the oscillator quality factor $Q=\frac{x_c}{w}=4380$. From the curve area we compute equivalent noise temperature of the oscillator as:

$$T=\frac{(2\pi)^3 A\, m\, x_c^2}{4k_B E_0^2 G^2}\left(1+\frac{C_p}{C}\right)^2$$

where m is the mass of the longitudinal mode, which we approximate as the cuboid mass 0.25*kg*. The resulting noise temperature is 230*K* which is in agreement with the temperature of 288*K* measured by the thermometers within the error bars. These are mainly due to estimation of the electric field and mode mass. This result shows that the oscillator is dominated by thermal noise. Similar results we get with the other oscillators.

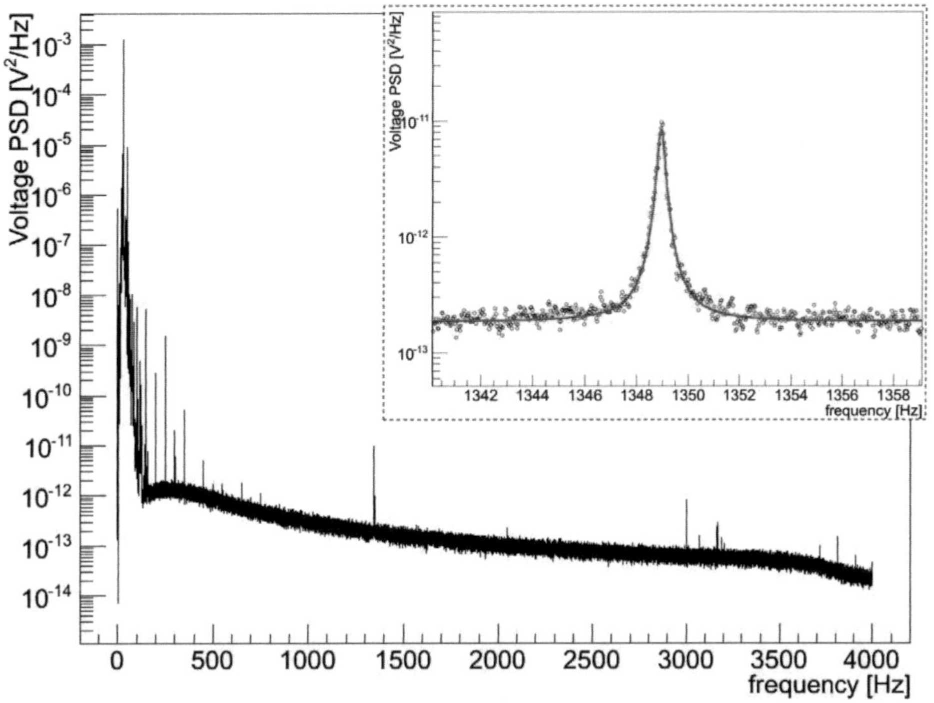

Figure 4. Power spectral density of the amplifier output for one of the oscillators in equilibrium. The inset is the zoom around the resonance of the longitudinal mode of the oscillator. The black points represent experimental data and the red solid line is the lorentzian curve that best fits the data. The decrease of the noise level above 3500*Hz* is due to the anti-alias filter in the acquisition system.

4 . Conclusions.

We have developed a system capable to make thermal noise measurements of the macroscopic mechanical oscillators around 1.4*kHz*. The thermal profiles of the oscillators can be actively controlled: this system is ready to be used for experimental campaign to thermal noise of the oscillators in non-equilibrium steady states.

References

[1] B Caron *et al* - 1997 *Class. Quantum Grav.* **14** 1461 doi:10.1088/0264-9381/14/6/011
[2] https://www.advancedligo.mit.edu/summary.html
[3] B Willke *et al* - 2002 *Class. Quantum Grav.* **19** 1377 doi:10.1088/0264-9381/19/7/321
[4] L Ju *et al* - 2000 *Rep. Prog. Phys.* **63** 1317 doi:10.1088/0034-4885/63/9/201
[5] The gravitational wave international committee roadmap – June 2010, http://gwic.ligo.org/roadmap/
[6] L. Conti *et al.* – 2010 *Class. Quantum Grav.* **27**, 084032
[7] S Miyoki *et al* - 2004 *Class. Quantum Grav.* **21** S1173 doi:10.1088/0264-9381/21/5/116
[8] T Uchiyama *et al* - 2004 *Class. Quantum Grav.* **21** S1161 doi:10.1088/0264-9381/21/5/115
[9] http://www.et-gw.eu/
[10] P. W. Adams *et al* - 1991 *Rev. Sci. Instrum.* **62**, 2461
[11] Mario Saraceni *et al* - 2010 *Rev. Sci. Instrum.* **81**, 035115
[12] http://root.cern.ch/drupal/

The AEI 10 m Prototype Interferometer frequency control using the reference cavity and its angular control

Fumiko Kawazoe[1], Gerald Bergmann[1], Alessandro Bertolini[2], Michael Born[1], Yanbei Chen[3], Alan V. Cummning[4], Liam Cunningham[4], Katrin Dahl[1], Christian Gräf[1], Giles Hammond[4], Gerhard Heinzel[1], Stefan Hild[4], Sabina H. Huttner[4], Russell Jones[4], Sina Köhlenbeck[1] Gerrit Kühn[1], Harald Lück[1], Kasem Mossavi[1], Jan Hendrik Pöld[1], Kentaro Somiya[5], A. Marielle van Veggel [4], Alexander Wanner[1], Tobias Westphal[1], Benno Willke[1], Kenneth Strain[1,4], Stefan Goßler[1], and Karsten Danzmann[1]

[1] Max-Planck-Institut für Gravitationsphysik (Albert-Einstein-Institut) und Leibniz Universität Hannover, Callinstr. 38, D–30167 Hannover, Germany
[2] National Institute for Subatomic Physics, Science Park 105, 1098 XG Amsterdam, Netherlands
[3] California Institute of Technology, Pasadena, CA 91125 USA
[4] University of Glasgow, Glasgow, G12 8QQ, United Kingdom
[5] Tokyo Institute of Technology, 2-12-1 Ookayama, Meguro-ku, Tokyo 152-8550, Japan

E-mail: fumiko.kawazoe@aei.mpg.de

Abstract. The main purpose of the AEI 10 m Prototype is to reach and eventually surpass the Standard Quantum Limit at frequencies ranging from 20 Hz to 1 kHz with a 10 m arm-length Michelson interferometer named the sub-SQL interferometer. The frequency control system uses a 20 m optical path length triangular suspended cavity named the reference cavity, with the goal of suppressing frequency noise of the input laser to a level of $\sim 10^{-4}\,\mathrm{Hz}/\sqrt{\mathrm{Hz}}$ at 20 Hz rolling off to below $6 \times 10^{-6}\,\mathrm{Hz}/\sqrt{\mathrm{Hz}}$ above 1 kHz. It is expected that tight angular control of the reference cavity's mirrors is necessary to reach this stringent requirement.

1. Introduction

In the AEI 10 m Prototype a 10 m arm-length Fabry-Perot Michelson interferometer called the sub-SQL interferometer will be installed [1]. Its objective is to first reach and finally surpass the Standard Quantum Limit (SQL) for 100 g mirror masses in a frequency band of 20 Hz to 1 kHz, with its highest sensitivity centered around 200 Hz. We aim to achieve this goal by suppressing all other noise sources well below the quantum noise, as shown in Fig. 1.

We set up several subsystems to tackle known noise sources; isolation tables with a seismic attenuation system will attenuate the ground motion by a factor of about 10^{-3} at 10 Hz in vertical and horizontal directions, a suspension platform interferometer will measure and correct the residual motions between these tables to the level of $10\,\mathrm{nrad}/\sqrt{\mathrm{Hz}}$ at 10 mHz for angular motions and $100\,\mathrm{pm}/\sqrt{\mathrm{Hz}}$ at 10 mHz for longitudinal motions [2], triple cascaded mirror suspensions with

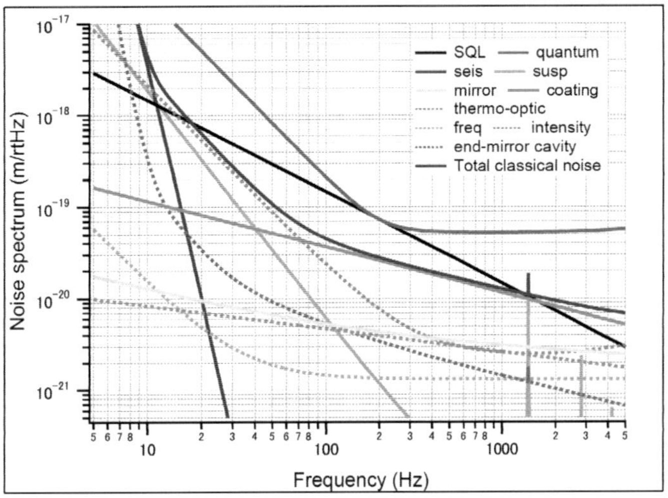

Figure 1. Design sensitivity of the sub-SQL interferometer.

an all silica monolithic last stage will minimize suspension thermal noise, and the frequency control system will suppress the laser frequency noise to a level of $\sim 10^{-4}\,\mathrm{Hz}/\sqrt{\mathrm{Hz}}$ at 20 Hz rolling off to below $6 \times 10^{-6}\,\mathrm{Hz}/\sqrt{\mathrm{Hz}}$ above 1 kHz [3].

2. Frequency Control System

Figure 2 shows a schematic layout of the reference cavity. It is a triangular cavity consisting of two flat mirrors and a curved mirror. Two flat mirrors are located close together, (the distance between them is 0.3 m) and the curved mirror is located $\sim$ 10 m away from the flat mirrors. The cavity has an optical path length of $\sim$ 20 m.

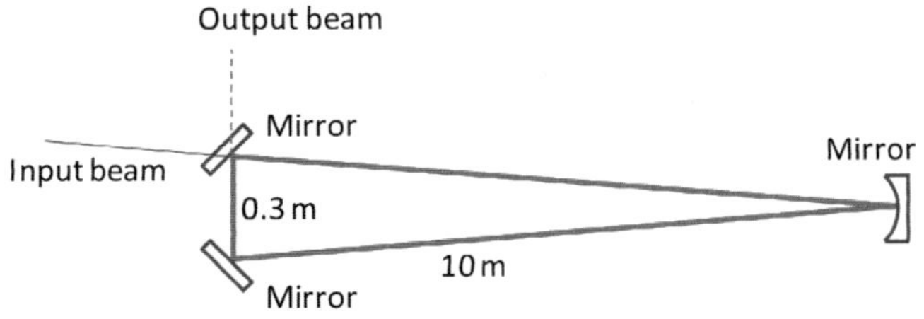

Figure 2. Schematic layout of the frequency reference cavity.

An asymmetry between the two arms of the sub-SQL interferometer can cause laser frequency noise to show up at the detection port. In our design we assume a common-mode rejection ratio of 99 %. The laser is a 2 W InnoLight Mephisto non-planar monolithic ring oscillator (NPRO), which is amplified by a four head Nd:YVO_4 laser amplifier to a power level of 35 W [4]. Assuming the well known free running noise of an NPRO ($10^4\,\mathrm{Hz}/\sqrt{\mathrm{Hz}}$ at 1 Hz rolling off as $\propto 1/f$) [5] and the above mentioned common-mode rejection ratio, and adding a safety margin of $\sim$ 10, we set the target level of the frequency noise in the input beam to be $\sim 10^{-4}\,\mathrm{Hz}/\sqrt{\mathrm{Hz}}$ at 20 Hz dropping to below $6 \times 10^{-6}\,\mathrm{Hz}/\sqrt{\mathrm{Hz}}$ at 1 kHz. Expressed in other units, this corresponds to a reference-cavity length-noise of $\sim 10^{-17}$ and $< 2 \times 10^{-19}\,\mathrm{m}/\sqrt{\mathrm{Hz}}$, at $\sim 20\,\mathrm{Hz}$ and $> 1\,\mathrm{kHz}$, respectively. In Fig. 3 the free running noise of an NPRO and the target level are compared. This stringent suppression ratio requires us to set up a frequency reference cavity, and control the laser frequency to follow the cavity length fluctuations which are designed to be much smaller than the free running noise of the laser. The widely-adopted method to use a high-finesse arm cavity of the main interferometer (in this case, the sub-SQL) as a frequency reference fails in our case, as the arms are too susceptible to the radiation pressure noise. (The masses of the suspended mirrors for the sub-SQL will be 100 g while the power inside the arms will exceed 1 kW). A straightforward way to design a frequency reference cavity is to make its optical path

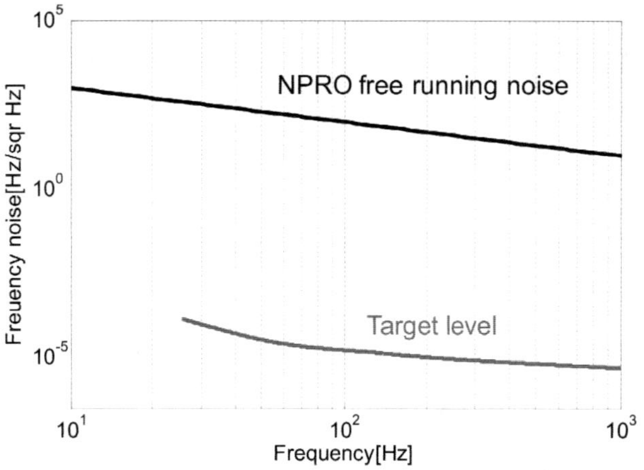

Figure 3. Target frequency noise level and a free running NPRO noise level.

length (L) long because given the minimum achievable displacement noise (ΔL), the frequency noise is proportional to $\Delta L/L$ and can be minimized by choosing the longest possible path length. The distance between two adjacent isolation tables being $\sim 10\,\mathrm{m}$ allows an optical path length of order 20 m.

The mirror mass is 850 g, the substrate is made from Suprasil 2, the diameter is 10 cm, and the thickness is 5 cm. The mass is chosen as a compromise between the suspension thermal noise and the radiation pressure noise performance, and the load/space capacity of the isolation table. Note that the suspension system for each mirror weighs approximately 15 times more than the mirror itself.

With the mirror mass fixed, the power inside the reference cavity is optimized such that the radiation pressure noise and the shot noise level are well below the target level. The cavity finesse is optimized so that the value is high enough to have a reasonable signal to shot noise ratio, yet within experimental reach.

The beam radius of $\sim 2.5\,\mathrm{mm}$ on the cavity mirrors is a compromise between the coating Brownian noise performance, and tolerance to optically induced torque and power fluctuations arising from angular fluctuations. The cavity g-factor is selected so that it is a non-degenerate cavity in which higher order spatial-modes are sufficiently suppressed. We assume that no more than 5 % of the input power is distributed over those modes. Non-degeneracy will be important for the alignment control because the control signal depends on the different resonance conditions between the fundamental mode and higher order spatial-modes.

Table 1 shows the design parameters. Note that slight modifications to the early phase design parameters that are shown in [3] were made to compensate for the unpredictable manufacturing tolerances.

Table 1. Design parameters of the reference cavity.

Parameter	Specification
Optical path length	21.2 m
Input power	0.133 W
Finesse	7305
Circulating power	232 W
Higher order spatial-mode suppression for $n + m < 6$	>1000
Cavity g-factor	0.72
Beam waist size	2.4 mm
Mirror mass	850 g

With these parameters the calculated total noise of the reference cavity is at the same level as the target in the frequency range of interest, see Fig. 4. Several individual noise sources that are either limiting or close to limiting the total noise are plotted.

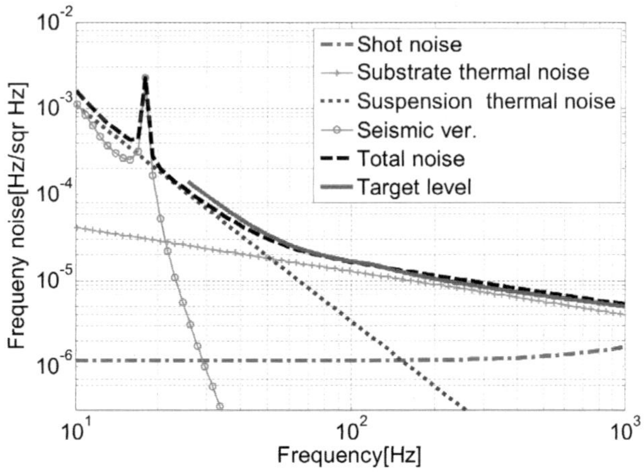

Figure 4. Noise budget of the reference cavity. The target level, the expected total noise level, and noise that are close to limiting the total noise of the reference cavity, and hence influence the cavity performance significantly, are shown.

In principle, if the NPRO frequency noise is suppressed to the target level within this frequency range our goal is achieved. In reality however, noise outside of this frequency range is expected to couple in through non-linear effects, e.g. scattering, and increase the total noise in the frequency range of interest. In order to prevent this we control the NPRO from below 1 Hz to the highest achievable frequency at which the phase lag in the control loop does not harm the control system's stability.

We take a common approach of using a Pound-Drever-Hall (PDH) technique, with several actuators that work in different frequency bands to cover the whole range: NPRO Temperature actuator, NPRO Piezoelectric Transducer(PZT), and a phase correcting Electro-optic Modulator (EOM). Figure 5 shows a block diagram for the feedback control system with the three actuators. Blue lines represent electrical signal path and the red dotted line represents the laser beam. The Laser Temperature feedback signal is picked up after the Laser PZT servo stage so that any changes to the PZT gain do not require consequential changes to the Laser Temperature gain to maintain a stable crossover condition. The bandwidth and the actuator coefficient of each

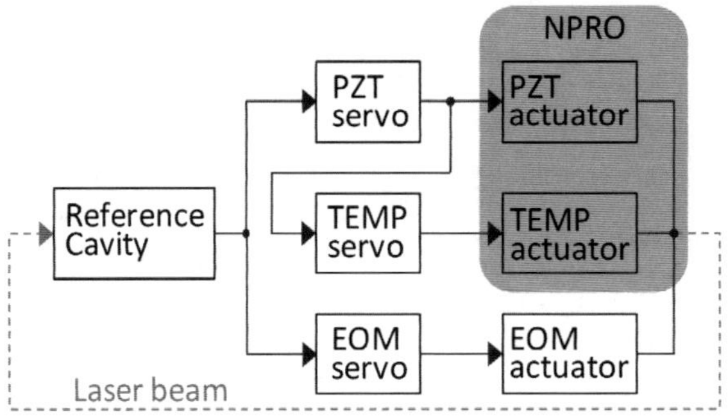

Figure 5. Block diagram for the feedback control system.

actuator are shown in Table 2. The unity gain frequency aimed at needs to include some safety margin because additional phase lag will arise from electronics, for example photo detectors, EOM driver, and post-mixer filters, that cannot be accurately predicted at this design phase. We aim at 250 kHz for the unity gain frequency at which the designed servo phase lag is less than 10 degrees.

Table 2. Design parameters of the reference cavity control.

Actuator	Bandwidth	Actuator coefficient
Laser Temperature	$< 1\,\text{Hz}$	$-3\,\text{GHz/V}$
Laser PZT	1 Hz - 10 kHz	1 MHz/V, Resonance at 200 kHz, $Q = 10$
Phase correcting EOM	10 kHz - 250 kHz	15 mrad/V

3. Angular control of the reference cavity

In order to maintain the full sensitivity, alignment degrees of freedom also need to be controlled. Otherwise they will introduce a drift in the system that will deteriorate the feedback control signal, making a robust control of the system impossible.

Three mirrors with two misalignment directions (horizontal and vertical) will give six degrees of freedom. A Differential Wavefront Sensing scheme (DWS) [6] and a Spot Position Control (SPC) will be used for controlling these degrees of freedom.

Wavefront misalignments appear as higher order spatial-modes that carry phase information that is different from that of the fundamental mode due to Gouy phase contributions. The DWS uses a PDH technique with a split photo diode on which two segments detect opposite effects of a wavefront tilt, hence by taking a difference of the demodulated signals from each segment we obtain an error signal to feed back to the servo.

Combinations of four degrees of freedom of the reference cavity (see [7]), will define the tilt between wavefronts of the promptly reflected beam and the beam leaving the cavity, i.e. DWS is sensitive to them. On the other hand the remaining two degrees of freedom do not cause detectable changes, hence the effect of the promptly reflected beam and the beam leaving the cavity is the same, and will be detected as a spot position change on a position sensitive photo detector.

Two suspended steering mirrors will be placed in front of the reference cavity to act as actuation points for alignment control loops with a control bandwidth of up to a few Hz. In this way a faster control will be possible because the cavity mirrors are suspended as triple suspensions with no actuator at the mirror level while the steering mirrors use a single suspension stage allowing a direct actuation at the mirror level. For the SPC the cavity mirrors are used as actuators, with a control bandwidth up to 0.1 Hz. At frequencies below 0.1 Hz angular alignment of the remaining mirror degrees of freedom will be used for beam centering on the mirrors.

This control speed hierarchy is subject to change because of the lack of decisive factors in this design phase, e.g. it is not yet clear if the cavity mirror motions will create larger changes in the eigenmode than the steering mirror motions. However, it will not be a problem to change the hierarchy since it will only require minor changes in the experimental setup.

Figure 6 and 7 show a schematic overview of the alignment control of the DWS part and the SPC part, respectively. A modulation frequency of $\sim 8\,\mathrm{MHz}$ is chosen so that the sidebands are

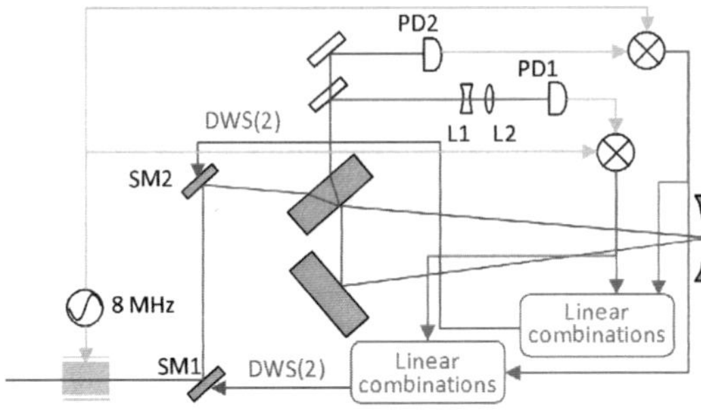

Figure 6. Schematic overview of the DWS part of the alignment control. (SM: steering mirrors, PD: quadrant photo detectors, L: lenses)

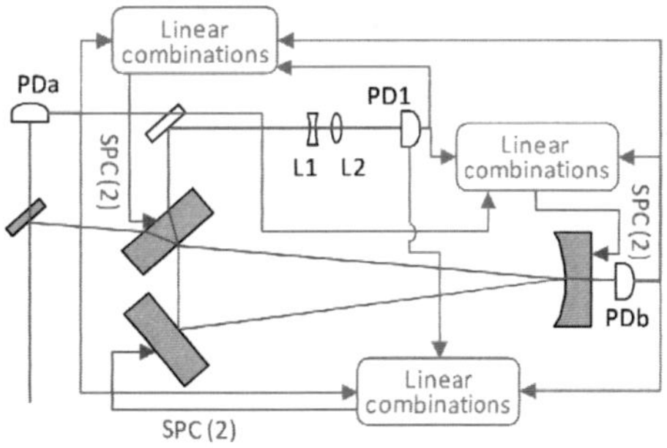

Figure 7. Schematic overview of the SPC part of the alignment control.

not resonant inside the cavity. For the SPC three photo detectors will detect six error signals and by taking proper linear combinations these signals will be fed back to the three cavity mirrors, and will be nulled via the DWS control. For the DWS, two photo detectors will create four demodulated signals and by taking proper linear combinations these signals will be fed back to the two steering mirrors, labeled SM1 and SM2. The number of degrees of freedom in the signals are denoted by the numbers in round brackets. L1 and L2 are widely called Gouy lenses, and are necessary for obtaining misalignment signals similar to those in the far field but at a much shorter distance than otherwise required. Figure 8 shows the overall layout of the reference cavity. The laser beam is sent into the vacuum tank through a fiber. The fiber and a Pre-Mode-Cleaner will give sufficient spatial mode filtering. Part of this spatially filtered beam will be sent into the reference cavity. All the detection systems will be placed on the detection bench located outside the vacuum tanks.

4. Summary

The Frequency Control System is expected to suppress the frequency noise to a level of $\sim 10^{-4}\,\mathrm{Hz}/\sqrt{\mathrm{Hz}}$ at 20 Hz rolling off to below $6 \times 10^{-6}\,\mathrm{Hz}/\sqrt{\mathrm{Hz}}$ above 1 kHz. In order to maintain the full sensitivity a tight alignment control of the cavity is required. The overall control design has been developed and will be tested shortly.

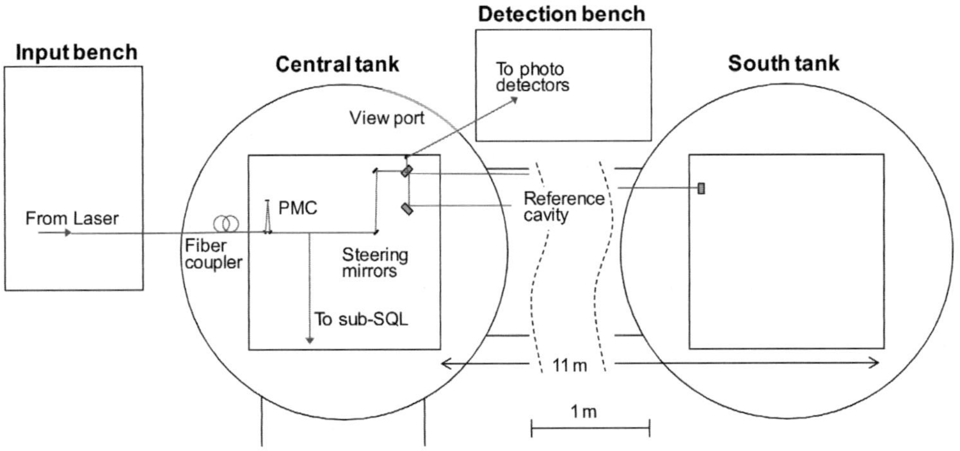

Figure 8. Optical layout of the reference cavity.

References

[1] Gossler S *et al* 2010 *Class. Quantum Grav.* **27** 084023
[2] Dahl K *et al* 2010 *J. Phys.: Conf. Series* **228** 012027
[3] Kawazoe F *et al* 2010 *J. Phys.: Conf. Series* **228** 012028
[4] Frede M, Schulz B,Wilhelm R, Kwee P, Seifert F, Willke B and Kracht D 2007 *Opt.Express* **15** 459
[5] Kwee P and Willke B 2008 *Appl.Opt.* **47** 6022
[6] Morrison E, Meers B J, Robertson D I and Ward H 1994 *Appl. Opt.* **33** 5041
[7] Kawazoe F, Schilling R and Lück H 2011 *J. Opt.* **13** 055504

Status of the GEO 600 squeezed-light laser

Alexander Khalaidovski, Henning Vahlbruch, Nico Lastzka, Christian Gräf, Harald Lück, Karsten Danzmann, Hartmut Grote and Roman Schnabel

Max-Planck-Institut für Gravitationsphysik (Albert-Einstein-Institut) and
Institut für Gravitationsphysik, Leibniz Universität Hannover
Callinstr. 38, 30167 Hannover, Germany

E-mail: Roman.Schnabel@aei.mpg.de

Abstract. In the course of the high-frequency upgrade of GEO 600, its optical configuration was extended by a squeezed-light laser [1]. Recently, a non-classically enhanced measurement sensitivity of GEO 600 was reported [2]. In this paper, a characterization of the squeezed-light laser is presented. Thereupon, the status of the integration into GEO 600 is reviewed, focussing on the sources of optical loss limiting the shot noise reduction by squeezing at the moment. Finally, the possibilities for a future loss reduction are discussed.

1. Introduction

The German-British project GEO 600 [3] is part of an international network of kilometre-scale Michelson-type laser interferometers [4, 5] searching for gravitational wave (GW) signals at audio frequencies in a band of 10 Hz – 10 kHz. While at frequencies below a few hundred Hertz the measurement sensitivity is, at the moment, still limited by seismic, thermal and technical noises, at higher frequencies photon shot noise constitutes the limiting factor (for an extended review of quantum effects in GW interferometers see [6]). Currently, GEO 600 undergoes an upgrade program to GEO - HF, aiming at a measurement sensitivity improvement by about an order of magnitude at high frequencies [7, 8].

A gravitational wave signal is 'translated' by the laser interferometer into a variation of the optical power at the signal output port. A limit for the possible measurement accuracy is set by the Poissonian distribution of the number of photons arriving within a time interval at the interferometric photo detector. Laser GW interferometers are operated close to their *dark fringe*, which means that due to destructive interference almost no light leaves the signal port and all optical power is reflected back to the laser. However, zero-point (vacuum) fluctuations of the electro-magnetic field are coupled into the interferometer's signal port and generate a signal that is not distinguishable from a GW signal. A possibility to reduce this *shot noise*, as was first proposed by Caves in [9], is the use of *squeezed light* [10, 11]. Such a light field has a non-classical noise distribution with a reduced uncertainty in one of the field quadratures, when compared to a vacuum state. Injected from the signal port, the squeezed state replaces the vacuum state, thereby directly reducing the interferometer's quantum noise.

Since the first observation of squeezed light in the 1980s [12–14], the degree of squeezing has been constantly improved [15–17], recently reaching a value of almost 13 dB [18]. In the GW detection band, the generation of squeezing remained an unsolved problem for a long time. It was not until 2004 that squeezing at audio frequencies down to 280 Hz could be generated for the first time [19], soon followed by the creation of a coherent control scheme [20, 21] that allowed the observation of frequency-independent squeezing over the entire Earth-bound gravitational wave detection band from

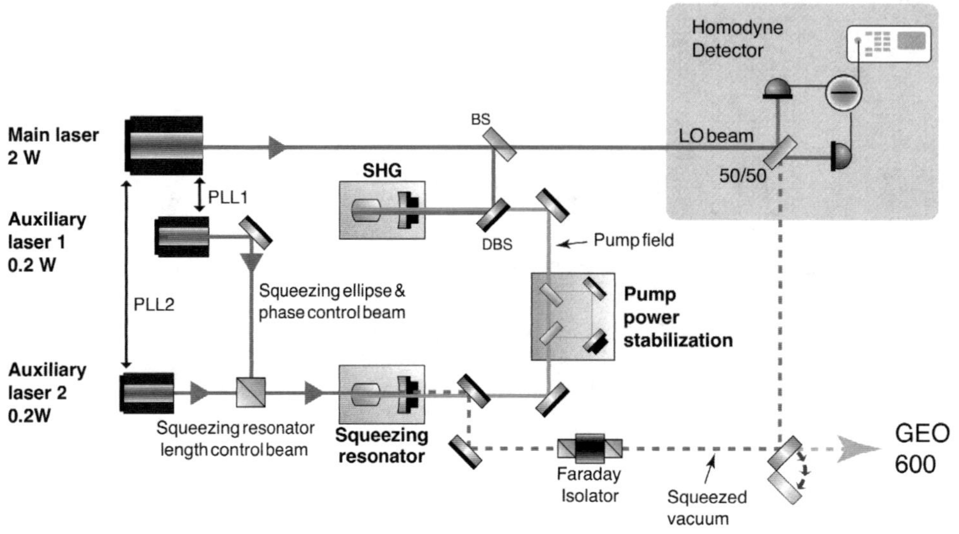

Figure 1. Simplified sketch of the optical layout of the squeezed-light laser. Auxiliary laser 1 is used to stabilize the length of the squeezing resonator and auxiliary laser 2 allows control of the phase of the squeezed field. The two auxiliary lasers are phase locked (PLL) to the main 2 W laser. For characterization, a balanced homodyne detector is available. BS: beam splitter, DBS: dichroic beam splitter, LO: local oscillator, PLL: phase-locked loop.

10 Hz to 10 kHz [22]. The implementation of squeezed light was tested in a table-top Michelson-type interferometer [22, 23] as well as in a suspended GW prototype detector [24]. Finally, a squeezed-light laser for GEO 600 was constructed [1], first results of a squeezing-improved measurement sensitivity were recently published in [2]. In this paper, a characterization of the squeezed-light laser is presented along with the current status of the implementation in GEO 600.

2. Optical layout

Figure 1 shows a schematic of the optical layout of the squeezed-light laser. To maintain clarity, only the core components are shown. Altogether, the experiment consists of approximately 130 components which are situated on a 135 cm $\times$ 113 cm optical breadboard. A detailed discussion of the single subsystems and of the employed control scheme is provided in the References [1, 25].

The experiment is driven by a monolithic non-planar Nd:YAG ring laser (NPRO) with 2 W of single-mode output power at 1064 nm. One part of this beam is frequency up-converted in a second-harmonic generator (SHG), which uses 7% MgO doped $LiNbO_3$ as nonlinear $\chi^{(2)}$ medium. This 532 nm *pump beam* is filtered by a mode-cleaning travelling-wave resonator (not shown in Fig. 1). Thereby, high-frequency phase noise that has been shown to diminish the maximal degree of squeezing achievable [26, 27] is attenuated. Subsequently, the pump beam is injected into the squeezed-light source that consists of a periodically poled potassium titanyl phosphate (PPKTP) crystal placed in a standing-wave hemilithic cavity.

For use in a gravitational wave detector the degree of squeezing must be highly stable in the long term and consequently the pump beam power must also be highly stable. The dependence of the degree of squeezing on the pump power value is discussed in detail in [28]. A Mach-Zehnder interferometer is employed to stabilize the pump beam intensity. The scheme ensures the squeezing value to be stable on long-term timescales, first characterization results are presented in [28].

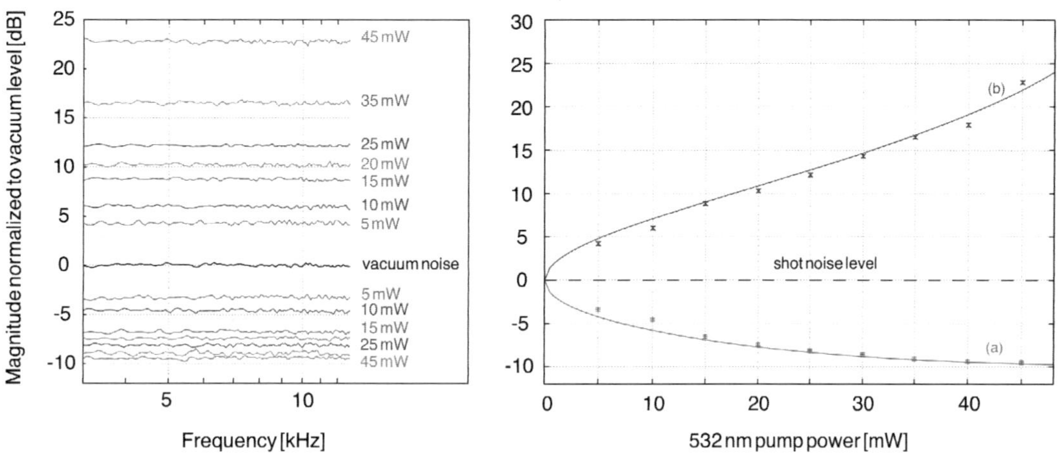

Figure 2. Left: Squeezed and anti-squeezed variances measured at different pump powers. A threshold power of 61 mW was estimated from the measurements. For clarity, only a small frequency range of 3 kHz – 12.8 kHz is shown. Right: the blue and red points show the squeezing and anti-squeezing measured at 5 kHz, as shown in the left graph. Traces (a) and (b) show the theoretical prediction of the expected anti-squeezing and squeezing levels, respectively, taking into account an overall loss of 10.5 %.

The generation of squeezed states at audio frequencies from 10 Hz to 10 kHz requires the implementation of a control scheme that avoids the introduction of technical laser noise to the squeezed vacuum state. The basic idea is to remove all noisy fields that could by means of interference introduce noise to the squeezed field at the frequencies of interest [19, 29, 30]. The *coherent control scheme* [20, 21] employed uses two auxiliary optical frequencies that are provided by two additional NPRO laser sources. The frequency of the auxiliary units is locked to the main laser that is in turn phase-locked to the GEO 600 master laser when operating the squeezed-light laser at the detector site.

For characterization of the squeezed-light laser, an on-board diagnostic homodyne detector is available. A small fraction of the main 1064 nm beam, that is also filtered by a ring mode-cleaner, is used as a local oscillator (LO) beam.

3. Characterization of the squeezed-light laser

The left part of Figure 2 shows squeezing and anti-squeezing values measured for different 532 nm pump powers. The values are normalized to the noise level of a vacuum state that can be measured when the signal port of the diagnostic homodyne detector is blocked (black trace). It is worth noting that very low 532 nm pump powers are sufficient to produce a high degree of squeezing. At a power of 5 mW (corresponding to about 15 mW intra-cavity power), the quantum noise was reduced by more than 3 dB, equivalent to a factor of two in noise power reduction. The degree of squeezing reached a maximal value of 9.5 dB at a pump power of 45 mW. This value is more than an order of magnitude below the pump power levels necessary to reach correspondent squeezing values with similar cavity designs using $MgO{:}LiNbO_3$ as nonlinear medium [15, 16].

In the right part of Figure 2, the measured values (taken at a frequency of 5 kHz) are compared with the theoretical expectations. The squeezed and anti-squeezed noise powers in decibel [26]

$$V_{\mathrm{s/a}} = 10 \cdot \log_{10}(R_{\mathrm{s/a}}) \tag{1}$$

with the variances

$$R_{\text{s/a}} = 1 \mp \eta_{\text{opt}} \frac{4\sqrt{P/P_{\text{th}}}}{(1 \pm \sqrt{P/P_{\text{th}}})^2 + 4\kappa^2} \tag{2}$$

depend on the pump power P as well as on optical loss. In this expression, a and s and the upper and lower signs denote the anti-squeezing and the squeezing, respectively, η describes the total detection efficiency and P_{th} is the OPO threshold power. The normalized frequency κ that is a function of cavity parameters is not relevant for the presented discussion, since frequencies much smaller than the cavity linewidth are considered.

The different loss sources during the generation and detection processes are:

- The escape efficiency $\rho = T/(T + L)$ of the squeezed-light resonator. Here, $T = 0.08$ is the power transmittance of the output coupling mirror and L is the intra-cavity loss. The total loss per round-trip, due to the non-perfect AR-coating of the crystal facet as well as to the residual power absorption of the nonlinear crystal, is estimated to be $L \approx 0.04$, thus resulting in $\rho \approx 0.95$.
- The total propagation loss suffered by the squeezed field on the way from the squeezing resonator to the homodyne detector. The critical components in this path are two polarizing beam splitters (PBSs) as well as a Faraday rotator, while the contribution of the super-polished lenses is negligible. In an independent measurement, the propagation efficiency was estimated to be $\xi \approx 0.98$.
- The fringe visibility $\mathcal{V}$ at the homodyne detector, contributing quadratically, was measured to range between 0.985 and 0.9875.
- The quantum efficiency of the homodyne photo diodes. To minimize this loss, custom-made high-efficiency diodes are used. The diameter of the active area is $500\,\mu$m. Equivalent diodes have already been used in high-level squeezing experiments, as e. g. reported in [15, 16], and have been estimated to have a quantum efficiency of $\eta_{\text{hmd}} > 0.99$.

A consideration of all possible sources results in a total detection efficiency of $\eta_{\text{tot}} \approx 0.895$. This value corresponds very well to the measured squeezing and anti-squeezing levels, as shown in Figure 2. From the measurements, an OPO threshold power of $P_{\text{th}} \approx 61$ mW was estimated.

It is important to note that the homodyne detector is merely a diagnostic device which is by-passed when the squeezed-light laser is used in GEO 600. Hence, the homodyne visibility loss is not relevant for a later operation and has to be corrected for. Thus, a total loss of merely 7 % can be deduced.

The initial requirement formulated for the squeezed-light laser was a squeezing value of 10 dB. When operated at a pump power of 35 mW, more than 10 dB of squeezing are available for the injection into the dark GEO 600 port. The advantage of this operation point is that with a corresponding anti-squeezing value of about 17 dB, the generated state is still relatively pure. If the pump power is increased further, the gain in squeezing is marginal while the anti-squeezing increases rapidly, leading to an increased sensitivity to phase noise. The highest squeezing value available is 11.3 dB (with about 23 dB of anti-squeezing) when using a pump power of 45 mW. The high squeezing factor was achieved over almost the entire frequency band of Earth-bound instruments extending from 10 Hz up to 10 kHz [1]. The long-term stability of the device was discussed in [28]. Altogether, this means that the requirements initially formulated for the GEO 600 squeezed-light laser have been fully achieved.

4. The squeezed-light laser in GEO 600

A sketch of the squeezing-input stage at the GEO 600 detector site is shown in Fig. 3. The squeezing injection scheme is realized through the combination of a Faraday rotator (FR) and a polarizing beam splitter: The squeezed field is reflected at the PBS between the output mode-cleaner (OMC) and the signal-recycling mirror (MSR), experiences a 45 degree polarization rotation passing the FR and another 45 degree rotation after having been reflected at the MSR. The total polarization rotation of 90 degrees allows the squeezed beam to be transmitted by the PBS together with the output beam of the

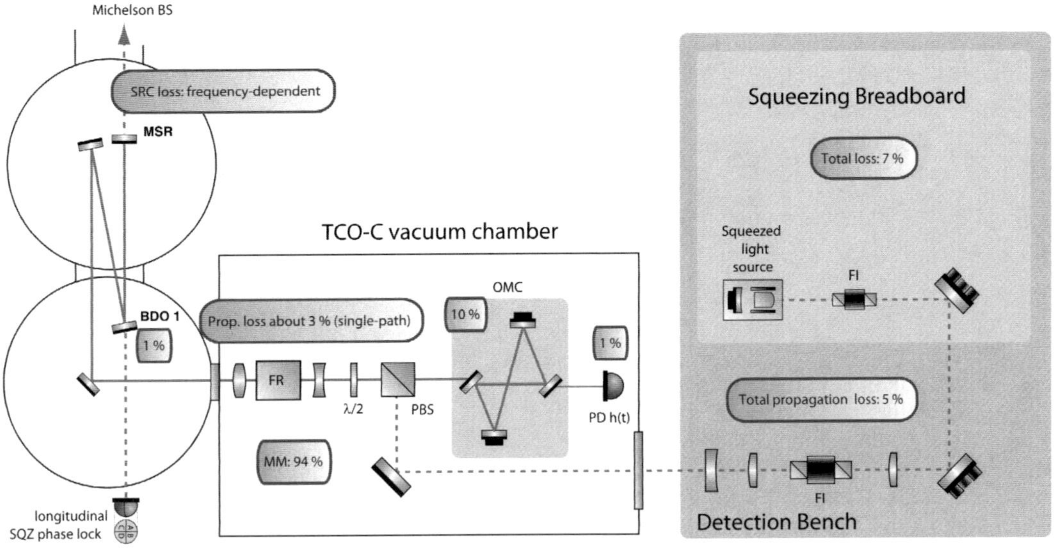

Figure 3. Sketch of the optical layout of the GEO 600 detection stage with squeezing input. Auxiliary optics are omitted for clarity reasons. The text boxes show estimations of the optical loss experienced by the squeezed field. BDO 1: suspended steering mirror with $T = 1\%$, FI: Faraday isolator, FR: Faraday rotator, MM: mode-matching, PD h(t): GW signal photo detector, SRC: signal-recycling cavity.

interferometer (carrying the GW signal) and to arrive at the GW signal photo detector after a transmission through the OMC. The squeezed-light laser is situated on the so-called *detection bench*, an optical table located close to the vacuum chamber that contains the squeezing input PBS, the OMC and the GW photo detector (this vacuum chamber is referred to as *TCO-C* in the following).

At the output of the squeezing breadboard, up to 11.3 dB of squeezing is available for injection into GEO 600. However, different loss sources appearing at various stages of the squeezed-light injection have to be considered. In the end, they limit the maximal sensitivity improvement achievable with squeezed light. In the following, the optical losses experienced by the squeezed field are summarized, the individual contributions are illustrated in Fig. 3.

- The total loss on the squeezing breadboard is, as stated above, about 7 %, taking into account an estimated escape efficiency of 95 % and a propagation loss of 2 %.
- During the implementation of squeezing, it was discovered that an additional Faraday isolator was required to protect the interferometer from back-scattering as well as the squeezed-light source from a bright beam coming from the interferometer. The additional Faraday isolator adds 44 dB of isolation but contributes 4 % to the propagation loss. The total propagation loss on the detection bench is therefore about 5 %.
- The expected single-path propagation loss inside TCO-C, due to imperfect AR coatings and absorption of the PBS, the Faraday rotator and the waveplate can be estimated to be about 3 %.
- 1 % is transmitted at one of the suspended steering mirrors (BDO_1 in Fig. 3) for error signal generation (This loss also contributes twice due to the double-path of the squeezed beam).
- A frequency-dependent loss is introduced by the signal-recycling cavity (SRC). While at frequencies outside the SRC linewidth it acts as a mirror, at low frequencies the squeezed field is coupled into the interferometer and experiences the intra-cavity loss. Figure 4 shows the expected

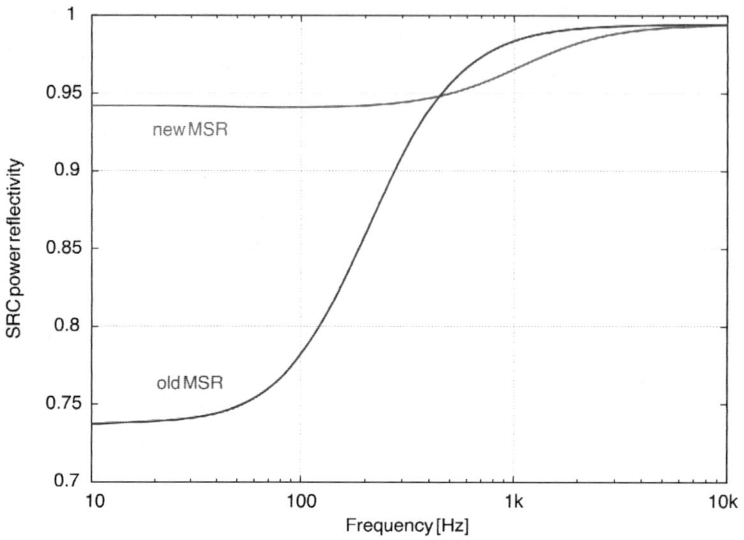

Figure 4. FINESSE simulation of the signal-recycling cavity's frequency-dependent power reflectivity with the old (blue trace) and new (red trace) MSR. The bandwidth of the SRC was increased from 220 Hz to 1100 Hz.

power reflectivity of the SRC for the squeezed field. For this, the FINESSE file of GEO 600 published in [32] was used.
In the course of the GEO - HF upgrade, the MSR reflectivity was reduced from 98.1 % to 90 % . As far as the squeezed field is concerned, the higher bandwidth leads to a slightly increased loss value at kHz-frequencies. The expected effect is shown in the blue trace of Fig. 4.

- The best mode-matching to the SRC and to the OMC measured was about 94 %.
- From the design parameters, the optical loss for a single transmission through the OMC was estimated to be below 1 %. During characterization, however, a much higher value of about 10 % was measured. This loss could not be reduced further by cleaning and probably arises from a combination of imperfect optical surfaces and dielectric coatings deviating from the design specifications.
- Before the GEO - HF upgrade, a Perkin Elmer C30642 photo diode was used. It was substituted by a custom-made high-efficiency photo diode, as already used in the diagnostic homodyne detector. The diameter of the active area is 3 mm. The quantum efficiency was measured to be 8 % higher than for the Perkin Elmer PD and is expected to be similar to the 500 μm diodes used in previous squeezing experiments, namely about 99 %.

The total loss for the squeezing, being the sum of all contributions mentioned above, is therefore estimated to a value of 32 % at kHz-frequencies and up to 50 % at frequencies much lower than the SRC linewidth. It is worth noting that since the SRC linewidth is about 220 Hz, the interferometer sensitivity at those frequencies is no longer limited by quantum noise. Figure 5 shows the expected squeezing and anti-squeezing values for the squeezed-light laser operated with a pump power of 35 mW. Traces (a – b) refer to the squeezing and anti-squeezing measured at the diagnostic homodyne detector. Traces (c – d) correct those values for the homodyne detector loss and therefore show the squeezing or anti-squeezing available for injection at the squeezing breadboard output. Traces (e – f) finally give an estimation of the amount of squeezing or anti-squeezing present at the GEO 600 GW photo detector. Traces (g – h) illustrate the effect of the MSR exchange, leading to an increased SRC linewidth.

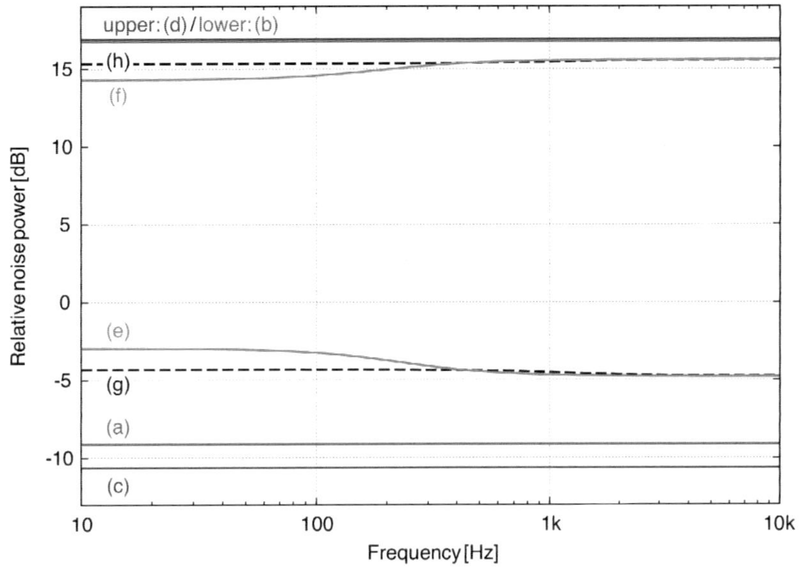

Figure 5. Simulated squeezing and anti-squeezing values prior to and after the injection into GEO 600. A value of 35 mW was assumed for the 532 nm pump power. Traces (a – b): measuring at the diagnostic homodyne detector. Traces (c – d): squeezing and anti-squeezing available at the output of the squeezing breadboard, corrected for the homodyne detector losses. Traces (e – f): frequency-dependent squeezing and anti-squeezing values at the GEO 600 GW photo detector with the old MSR. Traces (g – h): same with the new MSR.

5. Future work

The long-term goal to be achieved with squeezed-light injection is a desired reduction of the observatory noise by 6 dB at quantum-noise limited frequencies and a permanent squeezing contribution during the operation of GEO - HF. After the first successful enhancement of the observatory sensitivity by squeezed light, the achievement of both intents seems to be feasible in the near future. Regarding the squeezing factor, the following loss sources need to be addressed:

- The optical propagation loss on the detection bench is mainly due to the currently used Faraday isolator with a suboptimal performance in this respect as well as to polarization optics. The use of a high-throughput Faraday isolator – as already used on the squeezing breadboard – will allow a reduction of the propagation loss. Apart from the Faraday isolator, merely super-polished optics with a high-performance AR-coating (as used on the squeezing breadboard) will be used, thus reducing loss due to surface scattering and to residual AR-coating reflection.
- Lenses with a super-polished surface and a low residual reflection at the AR coating (as already used in the squeezed-light laser setup) will be used inside TCO-C.
- Up to now, the largest single loss contribution arises from the transmission through the OMC. In the future, the OMC optics will be exchanged. From the specifications of the single optics, the total transmission loss (mainly due to coating and surface imperfections and to a 100 ppm residual transmission of one of the mirrors for monitoring purposes) was estimated to be below 1 %.

An important issue to be addressed is the automatic alignment of the squeezed beam to the interferometer. Without this control loop, the suspended interferometer optics, though locked with respect to each other, are free swinging with respect to the squeezed beam injection path. This introduces a time-dependent, unknown effective loss for the squeezing injection. The generation of auto-alignment error signals is currently under investigation.

6. Conclusion

In this paper, we presented a characterization of the GEO 600 squeezed-light laser. The highest directly observed degree of squeezing is 9.5 dB. Thus, more than 11 dB of squeezing is available for the injection into the signal port of GEO 600. Furthermore, individual sources of optical loss experienced by the squeezed field in the course of generation, propagation in GEO 600, and the final photo-electric detection at the output port were discussed. The presented possibilities to reduce the total optical loss suggest an even stronger non-classical sensitivity improvement of GEO 600 by squeezed light to be feasible in the near future.

Acknowledgments

This work has been supported by the international Max Planck Research School (IMPRS) and the cluster of excellence QUEST (Centre for Quantum Engineering and Space-Time Research).

References

[1] Vahlbruch H, Khalaidovski A, Lastzka N, Gräf C, Danzmann K and Schnabel R 2010 *Class. Quantum Grav.* **27** 084027
[2] Abadie J *et al.* 2011 *Nature Phys.* **7** 962
[3] Grote H 2010 *Class. Quantum Grav.* **27** 084003
[4] Abbott B P *et al.* 2009 *Rep. Prog. Phys.* **72** 076901
[5] Acernese F *et al.* 2008 *Class. Quantum Grav.* **25** 114045
[6] Schnabel R, Mavalvala N, McClelland D E and Lam P K 2010 *Nat. Commun.* **1** 122
[7] Willke B *et al.* 2006 *Class. Quantum Grav.* **23** S207
[8] Lück H *et al.* 2010 *J. Phys. Conf. Ser.* **228** 012012
[9] Caves C M 1980 *Phys. Rev. D* **23** 1693
[10] Yuen H P 1976 *Phys. Rev. A* **13** 2226
[11] Gerry C C and Knight P L 2004 *Cambridge University Press*
[12] Slusher L E, Hollberg L W, Yurke B, Mertz J C and Valley J F 1985 *Phys. Rev. Lett.* **55** 2409
[13] Wu L-A, Xiao M and Kimble H J 1987 *J. Opt. Soc. Am. B* **4** 1465
[14] Pereira S F, Xiao M, Kimble H J and Hall J L 1988 *Phys. Rev. A* **38** 4931
[15] Vahlbruch H, Mehmet M, Chelkowski S, Hage B, Franzen A, Lastzka N, Goßler S, Danzmann K and Schnabel R 2008 *Phys. Rev. Lett.* **100** 033602
[16] Mehmet M, Vahlbruch H, Lastzka N, Danzmann K and Schnabel R 2010 *Phys. Rev. A* **81** 013814
[17] Mehmet M, Ast S, Eberle T, Steinlechner S, Vahlbruch H and Schnabel R 2011 *Opt. Express* **19** 25763
[18] Eberle T, Steinlechner S, Bauchrowitz J, Händchen V, Vahlbruch H, Mehmet M, Müller-Erbhardt H and Schnabel R 2010 *Phys. Rev. Lett.* **104** 251102
[19] McKenzie K, Grosse N, Bowen W P, Whitcomb S E, Gray M B, McClelland D E and Lam D E 2004 *Phys. Rev. Lett.* **93** 161105
[20] Vahlbruch H, Chelkowski S, Hage B, Franzen A, Danzmann K and Schnabel R 2006 *Phys. Rev. Lett.* **97** 011101
[21] Chelkowski S, Vahlbruch H, Danzmann K and Schnabel R 2007 *Phys. Rev. A* **75** 043814
[22] Vahlbruch H, Chelkowski S, Danzmann K and Schnabel R 2007 *New J. Phys.* **9** 371
[23] McKenzie K, Shaddock D A, McClelland D E, Buchler B C and Lam P K 2002 *Phys. Rev. Lett.* **88** 231102
[24] Goda K, Miyakawa O, Mikhailov E E, Saraf S, Adhikari R, McKenzie K, Ward R, Vass S, Weinstein A J and Mavalvala N 2008 *Nature Phys.* **4** 472
[25] Khalaidovski A 2011 *Gottfried Wilhelm Leibniz Universität Hannover* PhD thesis
[26] Takeno Y, Yukawa M, Yonezawa H and Furusawa A 2007 *Optics Express* **15** 4321
[27] Franzen A, Hage B, DiGuglielmo J, Fiurášek J and Schnabel R 2006 *Phys. Rev. Lett.* **97** 150505
[28] Khalaidovski A, Vahlbruch H, Lastzka N, Gräf C, Danzmann K, Grote H and Schnabel R 2011 *accepted for publication at Class. Quantum Grav.* arXiv:1109.3731
[29] Bowen W P, Schnabel R, Treps N, Bachor H-A and Lam P K 2002 *J. Opt. B* **4** 421
[30] Schnabel R, Vahlbruch H, Franzen A, Chelkowski S, Grosse N, Bachor H-A, Bowen W P, Lam P K and Danzmann K 2004 *Opt. Commun.* **240** 185
[31] Drever R W P, Hall J L, Kowalski F V, Hough J, Ford G M, Munley A J and Ward H 1983 *Appl. Phys. B* **31** 97
[32] Hild S 2007 *Gottfried Wilhelm Leibniz Universität Hannover* PhD thesis

Phase effects in Gaussian beams on diffraction gratings

D Lodhia, F Brückner, L Carbone, P Fulda, K Kokeyama[1] and A Freise

School of Physics and Astronomy, University of Birmingham, Edgbaston, Birmingham B15 2TT, UK

E-mail: dl@star.sr.bham.ac.uk

Abstract. Diffraction gratings have been proposed as replacements for transmissive optical elements in the next generation of gravitational wave detectors. However, they couple additional alignment noise to phase noise, and current models are based on unrealistic plane-wave expansion theories. There is a need for a description of grating-related phase noise which is compatible with standard interferometer tools. In this paper we investigate the grating-related phase shift by presenting a fully analytical Gaussian model for the phase accumulation of a displaced beam when diffracted from a grating. We consider a first-order modal decomposition as the method employed by simulation tools for off-axis beams. We show that the phase distribution of a typical displaced beam and a decomposed beam is accurate to within 3.9×10^{-8} radians. However, we find that the grating-related phase noise is not present, and this is further validated experimentally by the absence of a phase shift in beams with different modes. The phase noise must therefore be implemented manually into existing interferometer simulation tools.

1. Introduction

Currently a worldwide network of ground-based gravitational-wave detectors such as Advanced LIGO [1], Advanced VIRGO [2] and GEO HF [3] are being upgraded to form extremely sensitive second-generation laser interferometric observatories. Based on the experience gained through the first-generation operation, researchers are confident in obtaining the first direct detection of gravitational waves once the second-generation detectors are up and running. Simultaneously, efforts are being made to yield new concepts to further increase detector sensitivities, allowing for an evolution from gravitational-wave detection to gravitational-wave astronomy. One promising approach for the next-generation of interferometers is to replace conventional partly-transmissive optics such as beam splitters and cavity couplers by dielectric diffraction gratings. By employing an all-reflective interferometer setup, the severe limitation of thermal effects caused by absorption of laser light can be significantly reduced. Moreover, their use enables for a broader choice of opaque substrate materials with a potential for superior mechanical properties. Given these two benefits towards a thermal noise reduction, diffraction gratings are promising to become key elements for future generations of gravitational-wave detectors.

Based on the initial concept of all-reflective interferometry [4], various groups around the world have presented experimental proof of their feasibility and compatibility with standard

[1] Presently at: Physics and Astronomy, Louisiana State University, Baton Rouge, LA 70803, USA

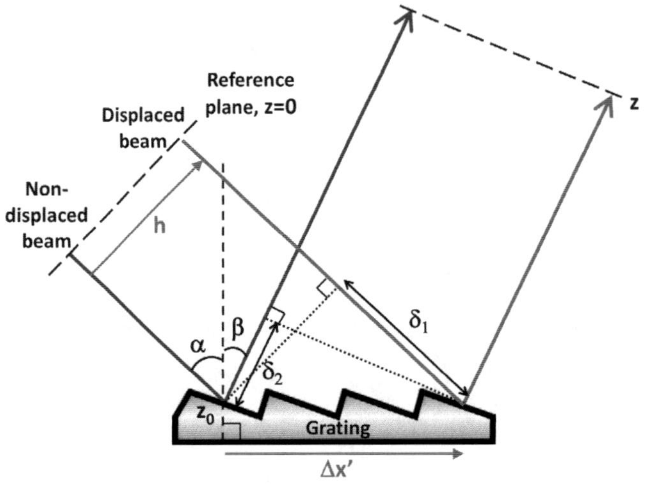

Figure 1. Diffraction of light into the m-th diffraction order when the grating is displaced by amount $\Delta x'$ relative to the beam. The angle conventions imply the incidence angle α to be positive while the diffraction angles β_m can be both negative and positive. A grating displacement $\Delta x'$ (corresponding to a parallel beam shift h) leads to an output optical path length change of ΔP according to Eq. (2). For clarity only one diffraction order m is considered.

interferometric techniques. This work includes the demonstration and characterisation of 2-port [5] and 3-port cavity couplers [6, 7], replacing conventional cavity input mirrors, as well as a full operation of a 4-port reflectively coupled Michelson interferometer [8]. Very recently, a concept for reflective coupling without the need of adding a multilayer coating was proposed which would further reduce thermal effects associated with the coating stack itself [9]. However, due to the broken symmetry of light deflection when using diffraction gratings, their use implies an additional coupling from alignment noise to output phase noise [10, 11]. This effect is illustrated in Fig. 1 where an incident beam with vacuum wavelength λ is diffracted from a grating with period d into the m-th diffraction order. Following the grating equation

$$\sin\alpha + \sin\beta_m = \frac{m\lambda}{d}, \tag{1}$$

the diffraction angle β_m of a certain functional output coupling port will be different from the incident angle α, with an exception for the zero-order specular order. From a simple geometrical consideration it is obvious that a slight displacement $\Delta x'$ of either the grating or the beam relative to each other causes an optical path length shift of $\Delta P = \delta_1 + \delta_2$ (where δ_2 is negative). This shift can be related to the displacement $\Delta x'$ via the grating equation (1) [11] to give:

$$\Delta P = \delta_1 + \delta_2 = -\Delta x' \frac{m\lambda}{d}. \tag{2}$$

Given this phase noise effect, the use of gratings results in more challenging requirements for the suspension and isolation systems for optical components compared to conventional interferometric settings with a natural symmetry of light reflection. Recently, we proposed an advanced readout of the ports of a grating coupler which gives a suppression of phase noise originating from lateral grating displacement, resulting in a factor of 20 relaxation in the lateral displacement suspension requirement at 10 Hz [12]. In order to be able to evaluate these requirements in great detail we want to use realistic interferometry simulation tools such as FINESSE [13] which are based on a Gaussian modal decomposition description of the laser beams. However, until now grating-related phase noise effects were solely investigated using simple and non-realistic plane-wave models, which are simplified models of Gaussian beam optics and therefore incompatible with such simulation tools.

In this paper we present a fully analytical Gaussian model of the phase accumulation of a beam when diffracted from a grating. The beam displacement by a small amount Δx relative to the grating (which is of course equivalent to a grating displacement by the same amount) is considered by means of a first-order modal decomposition of the beam being the standard approach for off-axis beams in simulation tools. We track and analyse the phase behaviour of both the displaced and non-displaced beams using Gaussian beam optics. For beam displacements in the order of the grating period, Eq. (2) predicts a phase change of up to 2π between a non-displaced and displaced beam, yet we find that this phase shift is absent from Gaussian modal decomposition. This means that the implementation of grating phase noise effects into existing simulation tools requires ad-hoc simulation code. In order to test whether the treatment of this subject by a decomposed beam is justified we experimentally investigated the difference of the phase change between a fundamental and first-order beam imprinted by a diffraction from a grating. The experiment was carried out using a 600 grooves/mm grating in the first order within a Mach-Zehnder setup.

2. Theoretical phase effects in Gaussian beams

The aim of this section is to examine the phase accumulation of a non-displaced and displaced Gaussian beam as they both travel from one reference plane, undergo grating diffraction, and reach a second reference plane. The displaced beam is then decomposed into zero-order and first-order modes, after which the phase accumulation for all three cases can be analysed.

We refer to Fig. 1 to illustrate the setup. The grating is orientated such that it lies in the x'-y' plane, where the grooves lie parallel to the y'-axis. The beam propagates along the z-axis, and only changes to the beam parameters in the x-z direction of the beam are considered. Note that the coordinate system of the beam (x, y, z) is rotated by the angle α relative to the coordinate system of the grating (x', y', z').

The grating displacement $\Delta x'$ is related to the beam displacement h and the angle of incidence α using the following relation:

$$h = \Delta x' \cos\alpha. \tag{3}$$

2.1. Gaussian optics

We begin by studying the description of Gaussian beams in more detail. Hermite-Gauss modes have the general form:

$$E(x, y, z) = \sum_{nm} a_{nm}(x, y, z) u_{nm}(x, y, z) e^{-ikz}. \tag{4}$$

The normalised Hermite-Gauss function $u_{nm}(x, y, z)$ describes the transverse spatial distribution of the beam as it varies slowly with z and is defined as:

$$\begin{aligned} u_{nm}(x, y, z) &= \left(2^{n+m-1} n! m! \pi\right)^{\frac{1}{2}} \frac{1}{\omega(z)} \exp\big(i(n+m+1)\Psi(z)\big) H_n\left(\frac{\sqrt{2}x}{\omega(z)}\right) \\ &\times \; H_m\left(\frac{\sqrt{2}y}{\omega(z)}\right) \exp\left(-i\frac{k(x^2+y^2)}{2R_C(z)} - \frac{x^2+y^2}{\omega^2(z)}\right), \end{aligned} \tag{5}$$

where H_n and H_m are Hermite polynomials, $\omega(z)$ is the beam size, $R_C(z)$ is the radius of curvature of the beam wavefronts, and $\Psi(z)$ is the Gouy phase. Unless otherwise specified, the beam waist, ω_0, will always be located at the grating, i.e. where $z = z_0$.

In general, an offset beam is displaced in both the x and y directions. Due to the symmetry of the system, we consider the displacement of the beam for only one degree of freedom, along the x-axis. The normalised Hermite-Gauss function, $u_n(x, z)$ in x becomes:

$$u_n(x, z) = \left(\frac{2}{\pi}\right)^{\frac{1}{4}} \left(\frac{\exp\big(i(2n+1)\Psi(z)\big)}{2^n n! \omega(z)}\right)^{\frac{1}{2}} H_n\left(\frac{\sqrt{2}x}{\omega(z)}\right) \exp\left(-i\frac{kx^2}{2R_C(z)} - \frac{x^2}{\omega^2(z)}\right). \tag{6}$$

At the waist, the Gouy phase is zero. In addition, $R_C \to \infty$ and therefore the R_C term in Eq. (6) can be ignored. Since $H_0 = 1$, a zero-order mode where $n = 0$ can be described at the waist position in the following form:

$$u_0(x, z_0) = \left(\frac{2}{\pi}\right)^{\frac{1}{4}} \frac{1}{\sqrt{\omega_0}} \exp\left(-\frac{x^2}{\omega_0^2}\right). \tag{7}$$

A first-order mode with $n = 1$ at the waist position is given as:

$$u_1(x, z_0) = \left(\frac{2}{\pi}\right)^{\frac{1}{4}} \frac{1}{\sqrt{2\omega_0}} \left(\frac{2\sqrt{2}x}{\omega_0}\right) \exp\left(-\frac{x^2}{\omega_0^2}\right). \tag{8}$$

Using Eqs. (7) and (8), we obtain a simple relationship between the zero-order and first-order modes at the beam waist:

$$u_1(x, z_0) = \frac{2x}{\omega_0} u_0(x, z_0). \tag{9}$$

2.2. Beam translation and modal decomposition

We now consider the description of a translated and modally decomposed beam in Gaussian terms. Due to the symmetry of the setup, we can assume that the beam is translated along the x-axis, rather than a translation of the grating along the x'-axis [10]. If we introduce a displacement to a fundamental beam by an amount h, we can describe the Hermite-Gauss function of the translated beam as:

$$u_0^t(x, z_0) = \left(\frac{2}{\pi}\right)^{\frac{1}{4}} \frac{1}{\sqrt{\omega_0}} \exp\left(-\frac{(x-h)^2}{\omega_0^2}\right), \tag{10}$$

where the superscript t indicates a translation. By expanding Eq. (10), we can substitute in Eq. (7). Since typical grating displacements are small compared to the beam waist, we can use the approximation $h/\omega_0 \ll 1$ and apply a first-order Taylor expansion to obtain the expression

$$u_0^d(x, z_0) = u_0(x, z_0) + \frac{h}{\omega_0} u_1(x, z_0), \tag{11}$$

with the superscript d to denote decomposition. From this it is clear that the translated beam is composed of zero-order and first-order mode terms; we can therefore deduce that a translated zero-order mode beam can be described by a decomposition into higher order modes.

2.3. Phase terms

Although we have established an expression to describe a displaced beam by means of modal decomposition, we need to verify if the same applies to the phase of the beam. In order to observe how accurately a decomposed beam describes a laterally displaced beam, we need a further understanding of the phase terms when propagated away from the waist, i.e. $z \neq z_0$. From Eqs. (4) and (6), we find three contributions to the overall phase: $\exp(-ikz)$, $\exp\left(i\frac{1}{2}\Psi\right)$

and $\exp\left(-i\frac{kx^2}{2R_C}\right)$. Using the general form, $\exp(-i\theta)$, the phase of a beam, θ, at any given point in the x-z plane is described as

$$\theta_{f,t,d} = kz - \tfrac{1}{2}\Psi + \phi_{f,t,d}, \tag{12}$$

where the subscripts f, t and d correspond to the fundamental non-translated, translated and decomposed beams respectively, and Ψ is the Gouy phase. The common factor of $kz - \frac{1}{2}\Psi$ can be omitted to leave $\phi_{f,t,d}$, defined for each beam as follows:

$$\phi_f = \frac{kx^2}{2R_C}, \tag{13}$$

$$\phi_t = \frac{k(x-h)^2}{2R_C}, \tag{14}$$

$$\phi_d = \frac{kx^2}{2R_C} - \varphi. \tag{15}$$

The φ term in Eq. (15) arises from the fact that the beam in Eq. (11) is a superposition of modes. By expanding and simplifying Eq. (11), we find that the remaining factor consists of a sum of terms, which is subsequently treated as a complex number. We use the relation $e^{i\varphi} = (\cos\varphi + i\sin\varphi)$ to reach the expression:

$$\varphi = \arctan\left(\frac{\sin\Psi}{\cos\Psi + \left(\frac{\omega\,\omega_0}{2xh}\right)}\right). \tag{16}$$

Using Eqs. (13)-(16), the phase distribution for each beam at a given position z can be plotted against the radial distance from the central optical axis, x, as depicted in Fig. 2. The phase distribution of a displaced beam (green) is the same shape as that of a non-translated beam (blue) but is simply shifted by an amount h along the x-axis, as expected. More importantly, the decomposed beam (red) also shows the same distribution and the same shift of h. However, this decomposed beam deviates away from the displaced beam along the z-axis, giving rise to negative phase. We can clarify this by examining the expression given in Eq. (16): the nature of the equation constricts the phase distribution such that the phase equals zero when $x = 0$ and $x = 2h$, hence the reason why the red and blue traces cross at $x = 0$ where the phase is zero.

We can also examine the behaviour of these phase distributions for differing values of h. Figure 3 shows the phase difference between a translated beam and a decomposed beam for increasing h, where h is given as a ratio of the waist size, $\omega_0 = 10\,\text{mm}$. We immediately see that as h increases, the phase of the decomposed beam deviates further away from that of a translated beam since we are violating the approximation ($h/\omega_0 \ll 1$) made to obtain Eq. (11) and in turn Eq. (16).

2.4. Effects from astigmatism

Returning to the original setup, we pose the question: does diffraction of the beam affect the phase distribution upon reflection? When a beam is reflected off a grating parallel to the x-z plane, the angles of incidence and reflection are different, i.e. $\alpha \neq \beta$. Consequently, only the beam parameters along the x-axis change whilst those along the y-axis remain the same, and an elliptical beam spot is produced. This astigmatism gives rise to a different waist size along the x-axis for the diffracted beam, $\omega_{0_r}^x$, and in terms of the incident beam waist, ω_{0_i}, is given by

$$\omega_{0_r}^x = \omega_{0_i}\frac{\cos\beta}{\cos\alpha}, \tag{17}$$

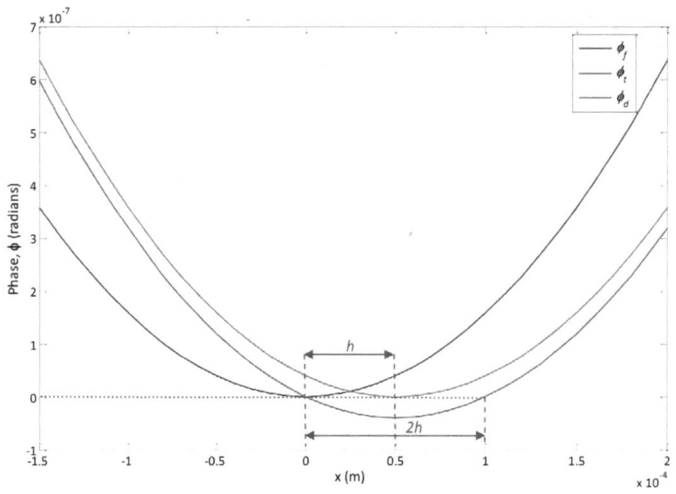

Figure 2. Phase distribution plots of a fundamental non-translated beam (blue), translated beam (green) and a decomposed beam (red). Only the x-direction is considered, and the following values are assumed: $\lambda = 10^{-6}$ m, $\omega_0 = 10$ mm, $z = 0.5$ m and $h = 0.05$ mm.

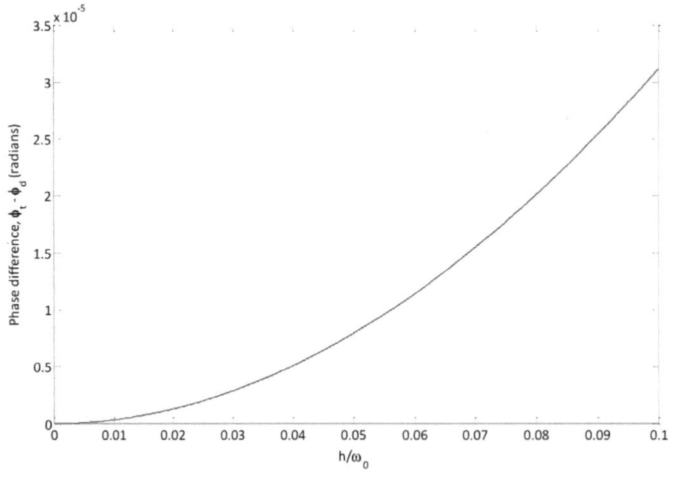

Figure 3. Difference in phase between a translated and decomposed beam as a function of h/ω_0. Increasing displacement gives rise to a further deviation in phase of the decomposed beam from the translated beam due to the approximation $h/\omega_0 \ll 1$.

where ω_{0_i} is the waist of the incident beam in both x and y-directions. Note that $\omega^y_{0_r} = \omega_{0_i}$.

From Eq. (1), we let $d = 1666$ nm and $\alpha = 10°$ to give $\beta_1 = 27.7°$. If $\omega_{0_i} = 10$ mm, then according to Eq. (17), $\omega^x_{0_r} = 8.99$ mm. Using these parameters as an example, we can plot and analyse the phase distributions after diffraction, as seen in Fig. 4. In addition to the fundamental non-translated (blue), translated (green) and decomposed (red) beams in the x-direction, we include the phase of a fundamental non-translated beam in the y-direction (blue dashed) for comparison. Notice that although the phase distribution of the diffracted beam is more narrow along x than y, the phase of the translated and decomposed beams still exhibit the same 'shifting' behaviour as seen in Fig. 2.

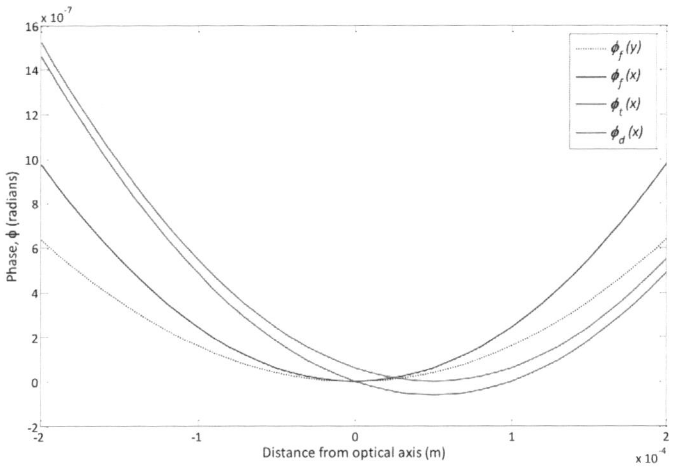

Figure 4. Phase distribution plots upon grating reflection along the x-axis for a fundamental non-translated beam (blue), translated beam (green) and a decomposed beam (red), and for a fundamental non-translated beam along the y-axis (blue dashed). The following parameters are assumed: $\lambda = 10^{-6}$ m, $\omega_{0_i} = 10$ mm, $\omega_{0_r}^x = 8.99$ mm, $z = 0.5$ m and $h = 0.05$ mm.

2.5. Analytical results

In reality, the amount of displacement a beam experiences is extremely tiny, typically less than 0.5% of the beam waist size. In the example used, a displacement of 0.05 mm for a beam with a waist size of 10 mm gives a difference in phase of only 3.9×10^{-8} radians. This suggests that the resulting phase from modal decomposition is very close to the phase of a translated beam, and that modal decomposition is a good approximation for a small beam displacement. The beam is affected by the grating through the effects of astigmatism, yet we found that astigmatism does not change the behaviour of the phase of the beam. In other words, the phase shift for a translated or modally decomposed beam relative to the non-displaced beam is unaffected by grating diffraction. However, taking Eq. (2) into account, we expected to see a substantial phase change of up to 2π between a non-displaced and a displaced beam for displacements in the order of the grating period (i.e. $\Delta x' = d$) [11], yet we find no evidence of this. We reach the conclusion that the modal decomposition model does not contain the observed grating related noise, and this noise must therefore be implemented manually into existing simulation tools.

3. An experimental demonstration of mode-switching

We continue to investigate the validity of the modal decomposition model in an attempt to determine experimentally whether or not phase changes exist between different orders of mode when subjected to grating diffraction. Any 'shift' in phase between modes would contradict with modal decomposition calculations (which suggest that the phase of a diffracted beam is independent of its mode), but it may possibly explain the absence of periodic phase shifts.

3.1. Experimental setup

We test the possibility of phase changes between zero-order and first-order modes using the Mach-Zehnder setup shown in Fig. 5. The laser beam passes through a series of modematching lenses and steering mirrors before entering the three-mirror mode-cleaner (MC). The MC can be tuned by means of a piezoelectric transducer (PZT) to allow any chosen mode to resonate and pass through. The input beamsplitter (BS) splits the beam into two arms of equal lengths. A ruled grating, with $d = 1666.7\,\mathrm{nm}$, was placed in one of the arms, with reflection in the first order. It is worth noting that for the purpose of this work, the grating is fixed in position and not translated in any direction. A second PZT is located in the other arm, and both beams recombine at the output BS, creating two superimposed beams. Photodetectors (PDs) are situated at each output beam, denoted as the 'east port' and 'south port'.

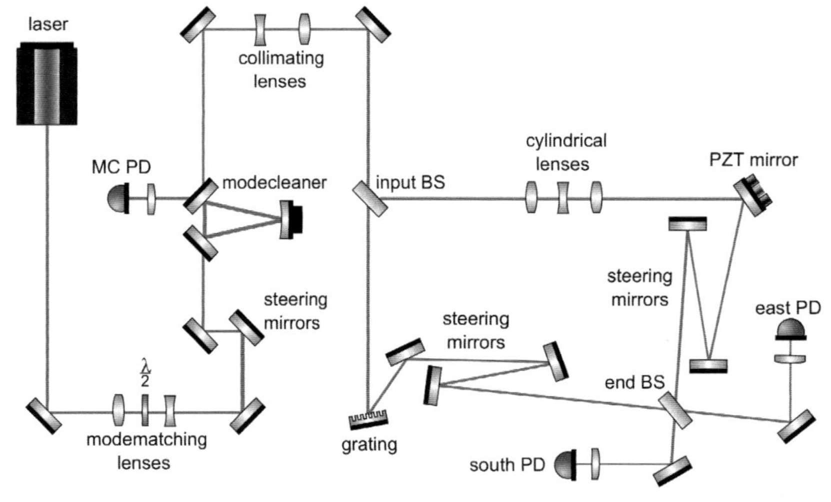

Figure 5. Layout of a grating Mach-Zehnder experiment. The mode-cleaner was supplied with a square-ramp signal, locking to zero-order and first-order mode resonances alternately. The diffraction grating is situated in one arm of the Mach-Zehnder, whilst the other arm is modulated to create fringe signals at the output.

3.2. Dual-mode locking system

The Pound-Drever Hall scheme [14] is used to provide feedback control for the MC. A combination of steering mirrors are used to isolate and increase the resonance peaks for zero- and first-order modes. A square wave signal is injected into the MC, allowing the PZT to ramp back and forth between two very specific positions at a frequency of 3 Hz. These positions were determined by the amplitude of the square-wave signal, which in turn was dependent on the distance between the resonant peaks of the two modes. Once the amplitude was determined (in this case 2.1 V), we were able to successfully lock to zero-order and first-order modes alternately.

Ramping the PZT in the Mach-Zehnder arm causes tiny changes in arm length and results in alternating phases of constructive and destructive interference, detected by the output PDs as fringes. Note that when the east port PD senses constructive interference, the south port PD

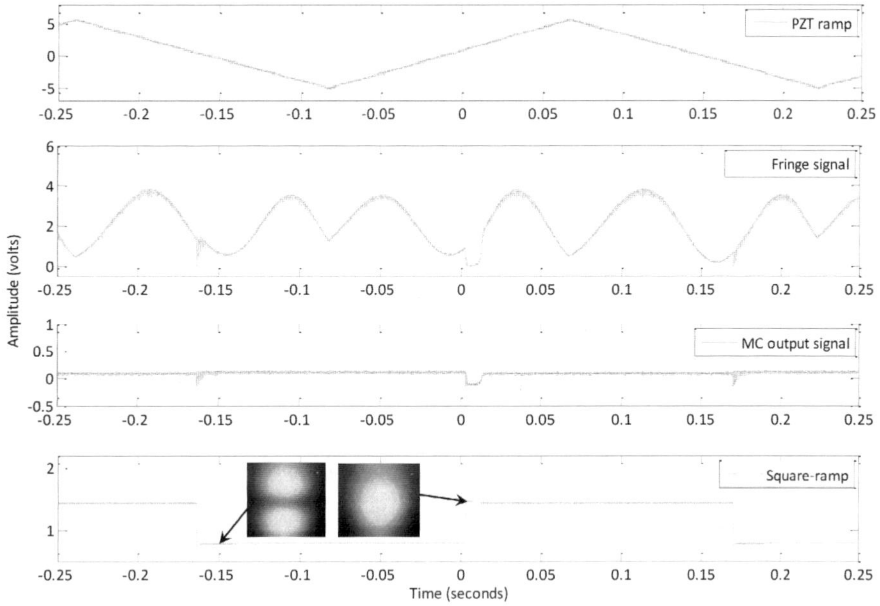

Figure 6. Fringe pattern during alternate locking on zero-order and first-order modes. From top to bottom are: main PZT ramp for the Mach-Zehnder (purple); fringe signal from the east port PD (red); output PD signal from the MC (blue); square-wave ramp signal to alternate the PZT in the MC (green). The maximum and minimum of the square ramp locks to the zero-order and first-order modes respectively. As the system switches between the two modes, no shifts are visible in the fringe signal. The slight fluctuations in the fringe signal coincide with those seen in the PD signal, caused by the electronics stabilising between each lock.

sees destructive interference, and vice versa. If the two beam modes contained different phases, we would expect to see a shift in the fringe signal at the instant when the ramp stepped and down. Instead, we found in our preliminary results that as the modes alternated, the fringe waveform remained continuous and no shifting was present, as shown in Fig. 6.

It should be noted that the breaks visible in the fringes coincide with the step-up/step-down of the ramp. They are caused by the electronics attempting to lock to a new mode each time. After a brief moment, the system stabilises into a locked system, and the fringes signal continues the waveform. For this reason, the square-wave ramp is set to a frequency of 3Hz - a higher frequency forces the electronics to destabilise more frequently and results in very noisy fringe signals, yet if the frequency is too low, it is difficult to lock the modecleaner at both resonances. The breaks in fringe symmetry are due to the main PZT changing direction during ramping (peaks and troughs of the triangular wave).

3.3. Experimental results

At first glance it appears that the fringes continue their waveform between mode-switching, implying that phase changes do not occur between zero-order and first-order modes upon grating diffraction. There is no doubt, however, that there are certain constraints with the setup described, such as the reaction of the electronics for locking, compromising the square-wave ramp frequency and sensitivity of the photodetectors. These could affect the level of detail with which the fringe signals can be examined, and we continue to investigate this area of research.

4. Conclusion and future scope

With the purpose of providing a description of grating-related phase noise compatible with standard interferometer simulation tools, we analytically investigated such effects using a Gaussian beam model. Similar to the respective approach in simulation, we took into account grating displacement, or equivalently beam displacement, by means of a modal decomposition of a displaced beam into zero and first-order modes. We showed that for the assumption of small displacements, the phase distributions of a displaced beam and decomposed beam match remarkably well with a deviation of only 3.9×10^{-8} radians for typical beam and propagation parameters. However, the diffraction from a grating does not affect the phase distribution of non-displaced and displaced beams as predicted, apart from introducing astigmatism. The Gaussian beam model does not automatically contain the previously observed phase shift, leading to the conclusion that simulation tools will require a separate input to account for phase from grating translation. We also investigated the possibility of phase differences within the modal decomposition model using a Mach-Zehnder setup. Our obtained data suggests there is no difference in phase change between both modes which corresponds to a mode-independent diffraction phase change. This experimental activity is currently in progress as we intend to improve the level of detail of the preliminary but promising measurements reported here.

Acknowledgments

This document has been assigned the LIGO document number LIGO-P1100149.

References

[1] Harry G 2010 *Classical and Quantum Gravity* **27** 084006

[2] Accadia T, Acernese F, Antonucci F, Astone P, Ballardin G, Barone F, Barsuglia M, Basti A, Bauer T, Bebronne M *et al.* 2011 *Classical and Quantum Gravity* **28** 114002

[3] Willke B, Ajith P, Allen B, Aufmuth P, Aulbert C, Babak S, Balasubramanian R, Barr B, Berukoff S, Bunkowski A *et al.* 2006 *Classical and Quantum Gravity* **23** S207

[4] Drever R 1996 *Proceedings of the Seventh Marcel Grossman Meeting on recent developments in theoretical and experimental general relativity, gravitation, and relativistic field theories* vol 1 p 1401

[5] Bunkowski A, Burmeister O, Clausnitzer T, Kley E, Tünnermann A, Danzmann K and Schnabel R 2006 *Journal of Physics: Conference Series* vol 32 (IOP Publishing) p 333

[6] Bunkowski A, Burmeister O, Beyersdorf P, Danzmann K, Schnabel R, Clausnitzer T, Kley E and Tünnermann A 2004 *Optics Letters* **29** 2342–2344

[7] Britzger M, Friedrich D, Kroker S, Brückner F, Burmeister O, Kley E, Tünnermann A, Danzmann K and Schnabel R 2011 *Applied Optics* **50** 4340–4346

[8] Friedrich D, Burmeister O, Bunkowski A, Clausnitzer T, Fahr S, Kley E, Tünnermann A, Danzmann K and Schnabel R 2008 *Optics Letters* **33** 101–103

[9] Kroker S, Käsebier T, Brückner F, Fuchs F, Kley E and Tünnermann A 2011 *Optics Express* **19** 16466–16479

[10] Wise S, Quetschke V, Deshpande A, Mueller G, Reitze D, Tanner D, Whiting B, Chen Y, Tünnermann A, Kley E *et al.* 2005 *Physical Review Letters* **95** 13901

[11] Freise A, Bunkowski A and Schnabel R 2007 *New Journal of Physics* **9** 433

[12] Hallam J, Chelkowski S, Freise A, Hild S, Barr B, Strain K, Burmeister O and Schnabel R 2009 *Journal of Optics A: Pure and Applied Optics* **11** 085502

[13] Freise A, Heinzel G, Lück H, Schilling R, Willke B and Danzmann K 2004 *Classical and Quantum Gravity* **21** S1067

[14] Drever R, Hall J, Kowalski F, Hough J, Ford G, Munley A and Ward H 1983 *Applied Physics B: Lasers and Optics* **31** 97–105

Development of a high power optical cavity for optomechanical QND experiment

T Mori[1,2], K Agatsuma[2], S Ballmer[3], S Sakata[4], O Miyakawa[5], S Kawamura[5], A Furusawa[6] and N Mio[7]

[1] Department of Advanced Materials Science, The University of Tokyo, 5-1-5 Kashiwanoha, Kashiwa, Chiba 277-8561, Japan
[2] National Astronomical Observatory of Japan, 2-21-1 Osawa, Mitaka, Tokyo 181-8588, Japan
[3] Department of Physics, Syracuse University, NY 13244, USA
[4] Center for Photonic Innovation, University of Electro-Communications, 1-5-1 Chofugaoka, Chofu, Tokyo 182-8585, Japan
[5] Institute for Cosmic Ray Research, University of Tokyo, 5-1-5 Kashiwanoha, Kashiwa, Chiba 277-8582, Japan
[6] Department of Applied Physics, School of Engineering, The University of Tokyo, 7-3-1 Hongo, Bunkyo-ku, Tokyo 113-8656, Japan
[7] Photon Science Center, School of Engineering, The University of Tokyo, 2-11-16 Yayoi, Bunkyo-ku, Tokyo 113-8656, Japan

E-mail: npfazy@gmail.com

Abstract. In order to observe quantum radiation pressure noise and reduce it by measuring the ponderomotively squeezed light on a table-top experiment, we are developing a laser interferometer with Fabry-Perot cavities with very small suspended mirrors. As a preliminary setup, we have constructed a Fabry-Perot cavity of finesse 1300 with a suspended mirror of 20 mg. The cavity was locked stably at low laser power for which the classical radiation pressure caused little effect on the dynamics of the small mirror. For the stable operation of this cavity with higher laser power, a technique to control the motion of the small mirror, especially its yaw motion, is necessary. We describe that the motion can be stabilized through the radiation pressure of light inside the cavity, by controlling the motion of the front mirror of a Fabry-Perot cavity properly.

1. Introduction

For the detection of gravitational waves (GWs) from violent astronomical events such as a supernova explosion or a coalescence of neutron-star binaries, large laser interferometers of km-scale have been constructed and are working worldwide. Since the expected GW signals are quite weak, noise reduction techniques are of particular interest. In current laser interferometer GW detectors, the sensitivity is limited by the quantum phase fluctuations of the photons of the laser at frequencies above several hundred Hz [1]. This noise is called the shot noise, and the equivalent displacement noise caused by the shot noise is inversely proportional to the square root of the laser power. Therefore, a higher laser power is needed to lower the shot noise limited sensitivity [2]. It is theoretically expected that another aspect of the quantum fluctuations of the laser arises as the laser power increases, namely the radiation pressure acting on suspended mirrors. The noise is called the quantum radiation pressure noise, and arises from the quantum photon-number fluctuations. The quantum radiation pressure noise has not been observed experimentally yet. However, in the near future, the sensitivity of the GW detectors will be totally limited by these two noise sources [3, 4]. Techniques to reduce these quantum noises are necessary to achieve a better sensitivity.

The key to overcome the quantum noises is that the shot noise and the quantum radiation pressure noise arise from different quadrature component of a light field, namely phase fluctuations and photon number fluctuations. Therefore if these quadrature phases are correlated, one fluctuation can be reduced at the sacrifice of the other. One method that utilizes the radiation pressure as the squeezing mechanism or a variational readout has been proposed [5]. This squeezing is called the ponderomotive squeezing. The quantum radiation pressure noise can be reduced by extracting this squeezing with homodyne detection.

We are developing a system that will enable the observation and reduction of quantum radiation pressure noise based on ponderomotive squeezing. The realization of such a system is of importance for laser interferometer GW detectors for some reasons. For instance, it will allow to verify the theory of quantum fluctuations in laser interferometers. Since no one has observed the quantum radiation pressure noise experimentally yet, the measurement itself has become a subject of growing interest. Further, to demonstrate a reduction of it is also important for the verification of the theory. In addition to this, the system will be useful as a test bench for advanced techniques regarding GW detectors.

2. Experimental design

This section describes the conceptual design of the system to observe and reduce the quantum radiation pressure noise. The main interferometer is a Fabry-Perot Michelson interferometer. In order to amplify the effect caused by the radiation pressure, the mirrors of the interferometer should be small and the finesse of the cavity should be high. We set the mass of the end mirror of the cavity to 20 mg. The main parameters of the cavity are as follows; the finesse of the cavity $\mathcal{F}$ is 10000, the laser power injected to the Fabry-Perot Michelson interferometer P is 120 mW. This leads to the design sensitivity to observe the quantum radiation pressure of about 10^{-16} m/$\sqrt{\text{Hz}}$ at 300 Hz and 10^{-17} m/$\sqrt{\text{Hz}}$ at 1 kHz [8]. The displacement noise spectrum due to the quantum fluctuations can be described as [5, 6, 7]

$$h_{\text{quantum}}(f) = \sqrt{\frac{h_{\text{SQL}}^2}{2}\left(\kappa(f) + \frac{1}{\kappa(f)}\right)}, \tag{1}$$

where $\mathcal{F}$ is the finesse the assumptions made here are;

$$\kappa(f) \equiv \frac{4P\omega_0}{r_{\text{F}}(1-r_{\text{F}})^2 m}\left(\frac{t_{\text{F}}\mathcal{F}}{(2\pi f)\pi c}\right)^2, \tag{2}$$

$$h_{\text{SQL}}(f) \equiv \sqrt{\frac{4\hbar}{m(2\pi f)^2}}. \tag{3}$$

Here ω_0 and c are the angular frequency and the speed of light, respectively, m is the reduced mass of the mirrors of the cavity, r_{F} and t_{F} are the amplitude reflectivity and transmissivity of the front mirror of the cavity, respectively. The reflectivity of the end mirror is set to unity. The first and second term in Eq. (1) describe the quantum radiation pressure noise and the shot noise, respectively. By extracting the ponderomotive squeezing, these two noises can be partially canceled at a certain frequency. It is expected that the radiation pressure noise can be suppressed by 6 dB with the actual losses of the system taken into consideration [8].

In the design, other noises such as the thermal noise of the mirror or the suspension, laser classical noise, or seismic noise are much smaller than the quantum noise .

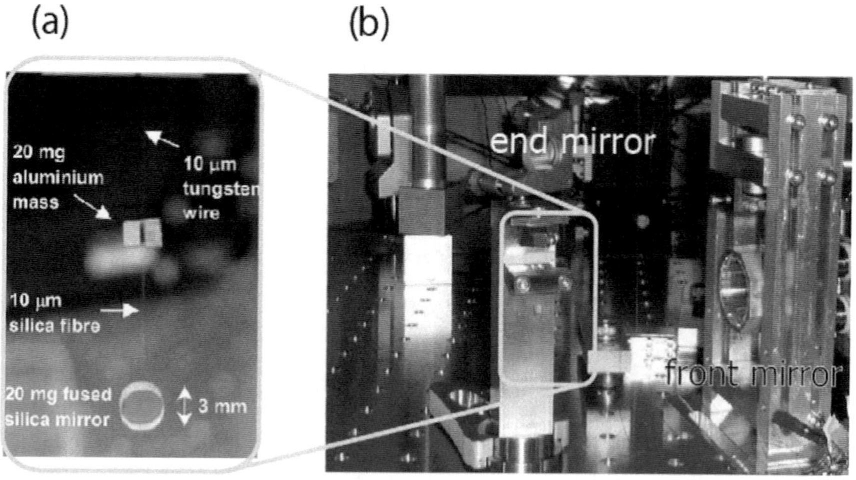

Figure 1. (a) Picture of the suspension with the mirror of 20 mg. The mirror is suspended by a silica fiber of 10 μm diameter. (b) Picture of one inch front mirror and 3 mm end mirror forming a Fabry-Perot cavity with a length of 80 mm. The laser comes from right to left.

3. Current experimental setup

As a preliminary setup, we have constructed a Fabry-Perot cavity of finesse 1300 with a suspended mirror of 20 mg in advance to the cavity of finesse 10000 in order to develop the control system. In the following, the main aspects of the current experimental setup are briefly described.

3.1. Small suspended mirror

Fig. 1(a) shows the suspension system of the small mirror, which is realized as a double pendulum. The mirror of 20 mg is suspended by a silica fiber of 10 μm diameter. A thin silica fiber is needed to lower the thermal noise of the fiber since silica has high mechanical Q factor. The middle mass is made of pure aluminum and also has a mass of 20 mg. The middle mass is surrounded by some small magnets in a way that its motion can be damped by eddy-current losses. Note that in Fig. 1(a) the magnets are removed. The middle mass is suspended by a 10 μm tungsten wire.

3.2. Suspended one inch mirror with coil-magnet actuators

As the front mirror of the cavity, a suspended pendulum with a one inch mirror has been constructed (see right side of Fig. 1(b)). The middle mass of this pendulum is made of copper, and is damped by eddy-current losses. Four magnets are attached to the back of the mirror holder, with each being surrounded by a coil. The motion of the mirror can be controlled by a force to the magnets exerted by a magnetic field that is produced by the coils.

3.3. Fabry-Perot cavity

The Fabry-Perot cavity consists of these two suspended mirrors, as shown in Fig. 1(b). The radius of curvature of the front mirror is one meter and the small mirror is flat. The length of the cavity is 80 mm.

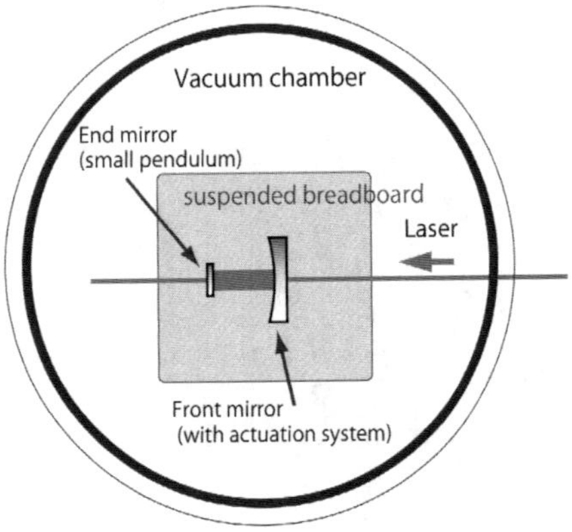

Figure 2. Schematic view of the optical configuration. Core optics are set on a suspended breadboard in a vacuum chamber.

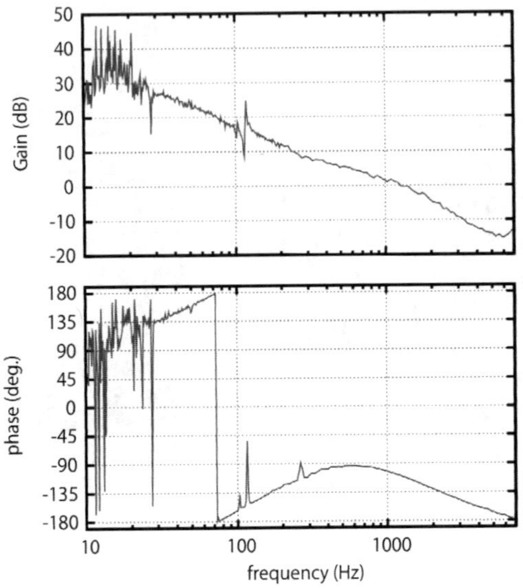

Figure 3. Open loop transfer function of the length control.

3.4. Seismic isolation system

Since the motion of the small mirror is easily excited by external vibration, an additional seismic attenuation system is necessary. Therefore, the mirrors that compose the Fabry-Perot cavity were set on the suspended breadboard of 40 cm × 40 cm. The schematic view of the system is shown in Fig. 2. This breadboard was suspended by a double pendulum, giving sufficient attenuation of the seismic motion.

4. Single Fabry-Perot operation

We have succeeded in the stable locking of the single Fabry-Perot cavity owing to the seismic attenuation system described above. The input laser power was low so that the radiation pressure caused little effect on the dynamics of the small mirror; the cavity could be operated stably when the laser power inside the cavity was less than about 200mW. The length of the cavity was controlled by the Pound-Drever-Hall locking scheme [9]. The open loop transfer function of the feedback servo is shown in Fig. 3. The unity gain frequency was at 1.3 kHz with a phase margin of about 70 degrees, which resulted in the stable locking.

The displacement noise spectrum sensitivity shown in Fig. 4 was estimated from the feedback signal applied to the coil-magnet actuators. The sensitivity was about $10^{-15}\,\mathrm{m}/\sqrt{\mathrm{Hz}}$ at 1 kHz which is about two orders of magnitude higher than the expected quantum radiation pressure noise level. The sensitivity was limited by the frequency noise of the laser which was estimated to be about $5 \times 10^3/f\ \mathrm{Hz}/\sqrt{\mathrm{Hz}}$. This noise will be suppressed in a next step by a frequency stabilization of the laser using a reference cavity.

5. Control system of the angular motion of the mirror

The cavity is stable only when the input laser power is low enough not to excite the motion of the small mirror through the radiation pressure. Since the g-factor of the cavity is positive, the radiation pressure works as a negative spring for angular motion [10]. At high laser power, the angular antispring effect caused by the radiation pressure prevents from stable operation of the

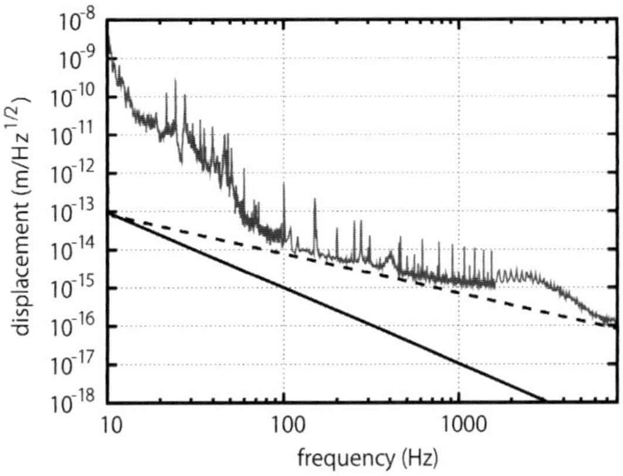

Figure 4. Measured displacement sensitivity of the Fabry-Perot interferometer. The solid blue line is the target radiation pressure noise level, and the dotted black line is the estimated frequency noise of the laser.

cavity. In Fig. 5(a) it is sketched how this instability occurs. When the rotational motion of the small mirror is excited slightly, the optical axis of the cavity moves. Since the rotational motion caused by the radiation pressure of the cavity with new optical axis is the same direction as the initial rotation, the motion is excited. This happens when the g-factor of the cavity is positive which is the case of this experiment. This effect was measured quantitatively with the small mirror used in this experiment [11].

For the stable operation of this cavity with higher laser power, an additional system to deal with this instability is required. Here we propose a technique to control the angular motion of the small mirror through the radiation pressure of photons inside the cavity. The schematic explanation of this control is shown in Fig. 5(b). When the small mirror rotates, the optical axis of the cavity moves. Therefore, the movement can be measured by detecting the transmitted light with a quadrant photo detector (QPD). The signal is then fed back to the angular motion of the front mirror in order to move the optical axis of the cavity such that the optical axis coincides with the center of the small mirror again. In order to obtain the signal which is just proportional to the rotation of the small mirror at the QPD, a lens was inserted in front of the QPD. Furthermore, in high laser power operation, the radiation pressure pushes the small mirror to the initial position and prevents the rotation when the control of the rotational motion is worked. Here the radiation pressure can work as angular spring for the end mirror instead of antispring.

The angular instability caused by the radiation pressure at full laser power (~ 400 W inside the cavity) is calculated to be about 1000 times larger than the mechanical restoring force of the torsion mode of the small mirror. To maintain the stability, the control bandwidth of about 1 kHz is needed. The experiment to verify them is now under way.

6. Summary

Toward the observation of the quantum radiation pressure noise and the reduction of it by measuring the ponderomotively squeezed light, we are developing a laser interferometer with Fabry-Perot cavities with very small suspended mirrors. The prototype cavity of the Finesse 1300 was locked stably at low laser power. The size of the mirror is the smallest, as far as the authors' knowledge, as a suspended mirror working as an optical cavity. For the stable

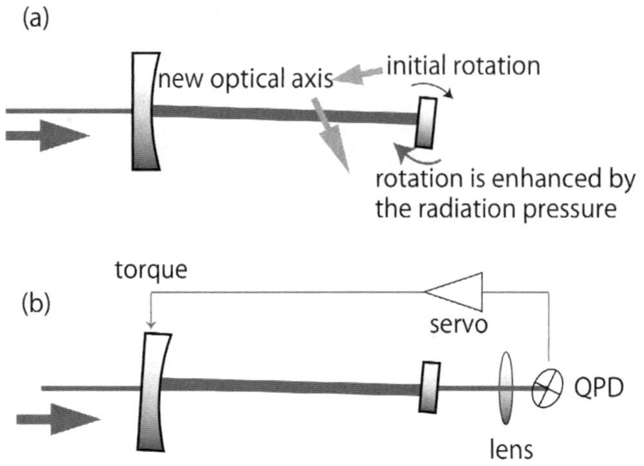

Figure 5. (a) Schematic view of how the radiation pressure inside the cavity works as an angular antispring. The rotational motion of the small mirror can be excited by the radiation pressure. (b) Schematic view of how to control the yaw motion of the small mirror through the radiation pressure.

operation of this cavity with higher laser power, a technique to control the motion of the small mirror is necessary. The method of the control through the radiation pressure of photons inside a cavity by controlling the motion of the front mirror of a Fabry-Perot cavity is proposed. The experiment to verify this is now under way.

Acknowledgments
T Mori is a Research Fellow of the Japan Society for the Promotion of Science. This work is supported by a Grant-in-Aid for Scientific Research from the Japan Society for the Promotion of Science (09J03624), and partly supported by the Global Center of Excellence for Physical Sciences Frontier, the University of Tokyo. This work is also supported by Grant-in-Aid for Scientific Research (B) (23340077).

References
[1] Abbott B P *et al* 2009 *Rep. Prog. Phys.* **72** 076901
[2] Ohmae N, Takeno K, Moriwaki S and Mio N 2008 *Appl. Phys. Exp.* **1** 092601
[3] Kawamura S 2010 *Class. and Quant. Grav.* **37** 084001
[4] Kuroda K *et al* 2010 *Class. and Quant. Grav.* **27** 084004
[5] Kimble H J, Levin Y, Matsko B A, Thone S K and Vyatchanin P S 2002 *Phys. Rev. D* **65** 022002
[6] Caves M C and Schumaker L B 1985 *Phys. Rev. A* **31** 3068
[7] Schumaker L B and Caves M C 1985 *Phys. Rev. A* **31** 3093
[8] Sakata S *et al* 2008 *Jour. of Phys: Conf. Ser.* **122** 012020
[9] Drever R W, Hall J L, Kowalski F V, Hough J, Ford G M, Munley A J and Ward H 1983 *Appl. Phys. B* **31** 97
[10] Sidles J A and Sigg D 2006 *Phys. Let. A* **354** 167
[11] Sakata S, Miyakawa O, Nishizawa A, Ishizaki H and Kawamura S 2010 *Phys. Rev. D* **81** 064023

Thermal effects and their compensation in Advanced Virgo

A. Rocchi[1], **E. Coccia**[1,2], **V. Fafone**[1,2], **V. Malvezzi**[1], **Y. Minenkov**[1], **L. Sperandio**[1,2]

[1]INFN Sezione di Roma Tor Vergata, I-00133 Rome, Italy
[2]University of Rome Tor Vergata and INFN, I-00133 Rome, Italy

E-mail: alessio.rocchi@roma2.infn.it

Abstract. Thermal effects in the test masses of the gravitational waves interferometric detectors may result in a strong limitation to their operation and sensitivity. Already in initial LIGO and Virgo, these effects have been observed and required the installation of dedicated compensation systems. Based on CO_2 laser projectors, the thermal compensators heat the peripheral of the input test masses to reduce the lensing effect. In advanced detectors, the power circulating in the interferometer will increase, thus making thermal effects more relevant. In this paper, the concept of the compensation system for Advanced Virgo is described.

1. Introduction

In an interferometric GW detector, the amount of allowable circulating probe power is limited by the nonzero optical absorption in the substrate and coatings of the test masses. The dependence from temperature of the refraction index and the thermal expansion coefficient of optical materials ensure that temperature gradients, induced by the absorption of the Gaussian-profiled probe light, will result in nonuniform optical path length distortions, which will affect both the controllability and length sensitivity of the instrument [1, 2].

Thermal effects have already been observed in Virgo [3] and LIGO [4] and required the installation of Thermal Compensation Systems [5] (TCS). Advanced detectors [6, 7] will be characterized by a higher circulating power (from 20 kW in the initial interferometers to 700 kW in the second generation detectors) and thermal effects will become even more relevant.

Absorption of optical power in the test mass of GW interferometric detectors predominantly occurs in the high reflectivity coating. The absorbed power is converted into heat, producing a gradient of temperature inside the substrate. Two different effects originate from the heating of the test mass:

- nonuniform optical path length distortions (thermo-optic effect, also termed thermal lensing) mainly due to the temperature dependency of the index of refraction.
- change of the profile of the high reflective surface, due to thermal expansion (thermo-elastic deformation) in both input and end test masses.

In the case of first generation detectors, the latter effect is negligible (as shown in figure 2): the main thermal effect is the thermal lensing, affecting the Power Recycling Cavity (PRC). In presence of this effect, which changes the cavity mode, the input laser no longer matches the

PRC and the coupling coefficient between the laser TEM_{00} and the cavity TEM_{00} becomes less than one. This leads to a decrease of the recycling cavity gain and thus in the sidebands power. Since sidebands are used to keep the interferometer locked, thermal lensing affects the possibility to operate the detector at high input powers. The ultimate consequence is a loss of signal-to-noise ratio at high frequencies due to the increase of shot noise. It is possible to compare the amount of thermal effects in Virgo and Advanced Virgo. The computations have been made for 1 ppm coating absorption, that leads to a total absorbed power of about 20 mW and 600 mW for Virgo (25 W input power) and Advanced Virgo (125 W input power) respectively. In figure 1, the comparison between the optical path length increase is shown. We can see that the optical path length increase expected in Advanced Virgo at a distance of 5 cm (one Gaussian beam radius) is about 470 nm, while it is about 30 nm in Virgo, at the same distance. Figure 2 compares the thermo-elastic deformations of the high reflectivity face of the Fabry-Perot cavity mirrors for the two detectors.

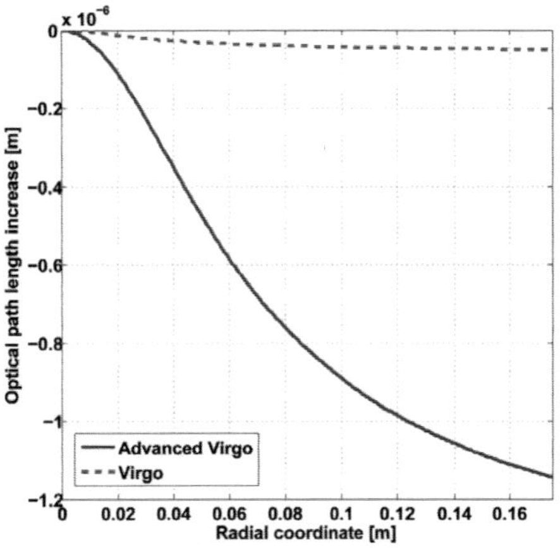

Figure 1. Optical path length increase in Advanced Virgo (blue) and Virgo (green) ITM due to coating absorption.

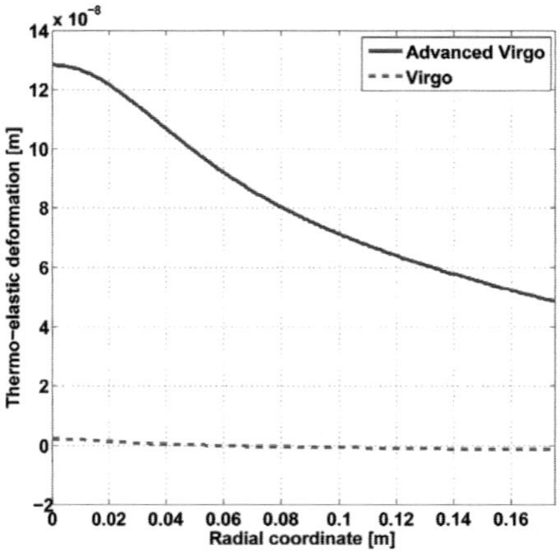

Figure 2. Thermo-elastic deformation of the HR surface in Advanced Virgo (blue) and Virgo (green).

As a consequence, in second generation detectors, besides the thermal lensing, the thermo-elastic deformation will also be relevant, due to the much higher circulating power in the Fabry-Perot cavities. Thermal expansion will change the profile of the high reflectivity surface: a bump will raise in the center of the test mass faces, making their surface profile non-spherical and the Radius of Curvature (RoC) will increase by about 44 m. The cavity will become less concentric, and the spot sizes at the mirrors will shrink. To maintain the arm cavity mode structure, it will be then necessary to control the radii of curvature of all test masses within ± 2 m from the initial RoC.

In advanced detectors, TCS will need to compensate for both effects, acting on both input and end test masses.

2. Thermal compensation in Advanced Virgo

Thermal compensation in Advanced Virgo needs to cope with the increase of circulating power in the optical cavities and with the improvement of the detector sensitivity. The main consequence of the first point is that, as already stated above, TCS will also need to correct the RoC increase in all test masses. The outcome of the sensitivity improvement is that it is no more possible to shine the input test masses directly with the CO_2 laser, as it is done in initial detectors [5]. The displacement noise introduced by the intensity fluctuations of the CO_2 laser would spoil the detector sensitivity in the frequency band 50 Hz - 100 Hz. To make TCS compliant with Advanced Virgo noise requirement, the relative intensity noise of the CO_2 laser should be reduced to the level of $10^{-8}/\sqrt{\text{Hz}}$ at 50 Hz, one order of magnitude below what it is possible to achieve with the present technology. This implies the need of an additional transmissive optic, named Compensation Plate (CP) to act on with the compensating beam. The CPs are placed in the recycling cavity, where the noise requirements are less stringent by a factor of $\pi/2F$ than in the Fabry-Perot cavity.

The CPs will be shined with the appropriate heating pattern to compensate distortions in the recycling cavity, while to control the radii of curvature of all the test masses, ring-shaped resistive heaters (RH) will be used. This scheme also allows to reduce the coupling between the two degrees of freedom (lensing and RoC), so to have a control matrix as diagonal as possible.

Following the considerations above, the conceptual actuation scheme of the compensation system foreseen for advanced detectors is shown in figure 3. A pictorial view of the input payload, with the ITM, RH and CP, is shown in figure 4.

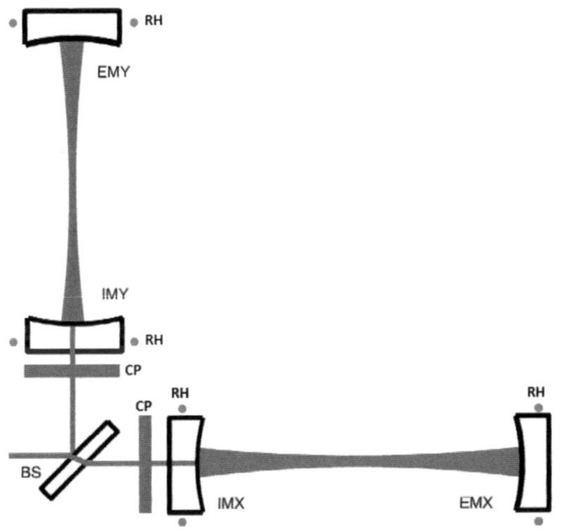

Figure 3. Actuation scheme of the Advanced Virgo TCS: blue rectangles represent the CPs while the green dots around the test masses are the ring heaters.

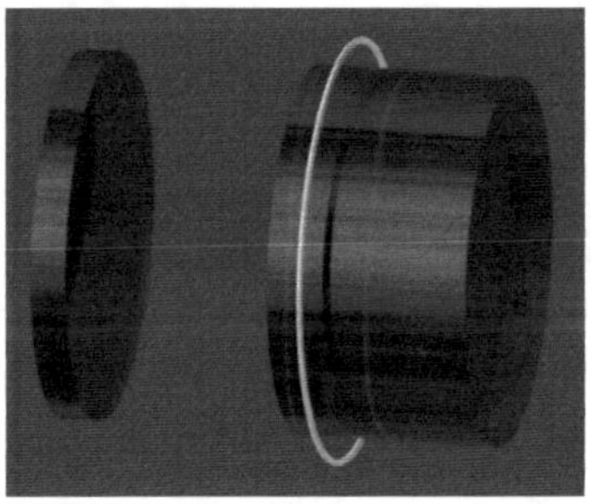

Figure 4. Picture of the input payload, comprising the ITM with the RH and the CP.

The thickness of the CP has been optimized by minimizing the heat escaping from its barrel and taking into account the need to accumulate enough optical path length. The distance between CP and ITM is 20 cm, this allows to minimize the radiative coupling between the two optics. In fact, the heated CP radiates heat towards the test mass. The heating of the TM is uniform, but since the barrel of the input mirror radiates away a part of the heat, a radial temperature gradient is established. This gives rise to an increase of optical path length that adds to the thermal lensing.

The position of the RH along the barrel of the TM is such as to maximize its efficiency (see section 3 for more details).

A useful way to picture the effect of a thermal distortions is the fractional power scattered out from the TEM_{00} mode [8], termed "coupling losses" L:

$$L = 1 - A^* A, \tag{1}$$

where

$$A = \frac{\langle E_0 | E \rangle}{\langle E_0 | E_0 \rangle} = \frac{\langle E_0 | e^{i\phi(x,y)} | E_0 \rangle}{\langle E_0 | E_0 \rangle} = 2\pi \int_0^{r_{ITM}} e^{i\phi(r)} |E_0(r)|^2 r dr. \tag{2}$$

E_0 represents the undisturbed cavity field before being subjected to the phase distortion $\phi(x, y) = 2\pi/\lambda \cdot OPL(r)$, E is the distorted field and $OPL(r)$ is the optical path length in the substrate. The last equality in equation 2 holds if cylindrical symmetry is assumed. A phase distortion in a cavity acts to scatter power out of the fundamental mode, and thus out of the cavity, and can be viewed as a simple coupling loss term.

With no thermal compensation, in Advanced Virgo, the coupling losses would amount to approximately $5 \cdot 10^5$ ppm. For a comparison, in Virgo, the losses due to thermal lensing are of the order of 10^4 ppm. The Advanced Virgo TCS needs to reduce to coupling losses at least by a factor of 10^3 (corresponding, roughly speaking, to a maximum peak-to-valley residual optical path length increase of about 8 nm) to allow the correct operation of the detector at design sensitivity.

Current interferometers use axicon-based optical projectors to convert a CO_2 laser Gaussian beam into an annular beam [5]. This heating pattern is not adequate for the advanced detectors, since it can reduce the residual coupling losses only by a factor of 10^2, while at least a factor of 10^3 is required. Studies on the optimization of the heating pattern, made with Finite Element Model (FEM), have shown that it is possible to reduce the residual coupling losses to about 6 ppm, thus leading to a reduction factor of about 10^5, with about 18 W of CO_2 power shined on the compensation plate.

Different methods for the generation of the optimal heating pattern are being investigated.

3. Test masses RoCs tuning

The need to control the radius of curvature of the test mass in GW interferometric detectors has already been faced in the past: the GEO detector [9] used a ring heater to change the RoC of one of the two test masses [10]. The GEO RH is placed on the back of the mirror, radiatively coupled with the optic.

Compensation and control of the test mass high reflectivity (HR) surfaces will be accomplished in Advanced Virgo with the same technique. The TCS baseline design considers four ring heaters, one around each test mass. The input mirror RH also provides limited compensation of thermo-optic effect in the recycling cavities. Unlike the GEO heater, the Advanced Virgo RH will be equipped with a reflecting shield to maximize the amount of power reaching the test mass.

The efficiency of the ring heater to change the RoC also depends on the position of the RH along the barrel of the test mass. In order to study the RH dynamics, an ANSYS coupled thermal-structural FEM has been developed, modeling a radiating ring placed around the barrel of the TM at different distances from the HR surface. At each position, the model calculates the RoC of the TM as a function of the RH power. It is found that, for a 20 cm thick test mass, the power required to recover the cold state RoC (1420 m) is minimized when the position of the RH is at about 18 cm from the TM HR surface.

The engineered design of the ring heater for Advanced Virgo is in progress, taking also into account two important constraints: high temperature operation and UHV compatibility. Moreover, since the last stage of the suspension system will use coil-magnet actuators for the control of the mirrors, it is necessary to avoid any stray magnetic field generated by the ring heater. In fact, the coupling of the RH magnetic field to the actuators would introduce displacement noise in the detector, limiting its sensitivity.

4. Measurement of thermal effects

The aberrations induced by absorptions will be sensed by several complementary techniques. The amplitude of the lensing will appear in some ITF channels, such as the power stored in the radio frequency sidebands. These are scalar quantities that can only give a measurement of the amount of power scattered into higher order modes. Phase cameras [11] will sense the intensity distribution and phase of the fields in the recycling cavity. However, control of thermal effects using error signals coming from such detectors has not been demonstrated yet.

Therefore, each optic with a significant thermal load will be independently monitored. The HR face of each test mass will be monitored in reflection for deformation. The input test mass/compensation plate phase profile will be monitored on reflection on-axis from the recycling cavity side. The TCS control loop will then use a blend of all the signals from the different channels, as shown in figure 5.

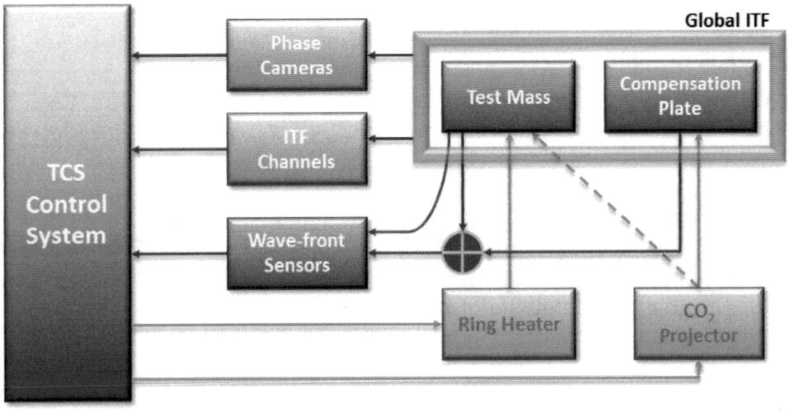

Figure 5. Conceptual scheme of the TCS control loop.

The ITM-CP phase profile dedicated sensors consist of a Hartmann Wavefront Sensor (HWS), and a probe beam whose wavefront contains the thermal aberration information to be sensed. The working principle of the device is shown in figure 6.

The Hartmann sensor selected for Advanced Virgo is that already developed and characterized on test bench experiments and in the Gingin High Optical Power Test Facility for the measurement of wavefront distortion [12].

This sensor [13] has been demonstrated to have a shot-to-shot reproducibility of $\lambda/1450$ at 820 nm, which can be improved to $\lambda/15500$ with averaging, and with an overall accuracy of $\lambda/6800$.

Due to the requirement on the maximum residual optical path length peak-to-valley of 8 nm, the precision of the wavefront sensing must be of the order of 0.8 nm, corresponding to $\lambda/1000$ @ 800 nm (to be compared with the measured noise level of $\lambda/15500$). This indicates that the selected Hartmann sensor is both sufficiently sensitive and accurate for the measurements of absorption-induced wavefront distortions in advanced GW interferometers. A prototype of the sensor is in the Tor Vergata Laboratories for further characterization and integration into the test facility for the development of the Advanced Virgo Thermal Compensation System.

5. Conclusions

Advanced Virgo aims to achieve a factor of 10 sensitivity improvement over Virgo and requires pushing the limits of technology on several frontiers. Thermal Compensation System will play a

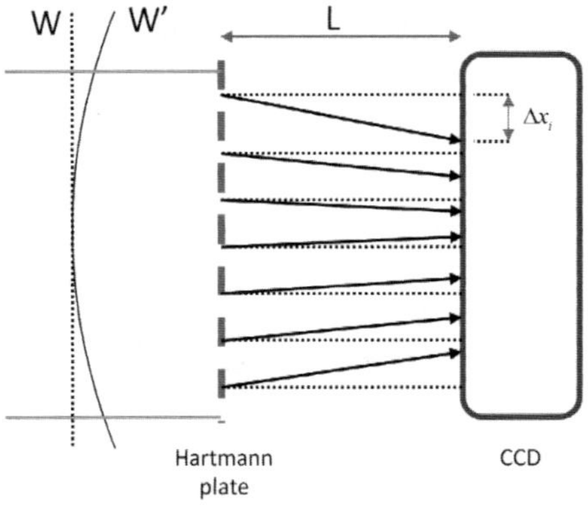

Figure 6. Working principle of the Hartmann Wavefront Sensor: an aberrated wavefront W' is incident on a Hartmann plate. The resulting rays propagate a distance L, normal to the wavefront, and are incident on a CCD. The new spot position, x'_i, is measured and compared to a reference spot positions, x_i, determined using a non-aberrated wavefront W. The wavefront gradient in the ith position is given by $\frac{\partial \Delta W}{\partial x} = \frac{\Delta x_i}{L}$.

crucial role in correcting optical path length distortions in the recycling cavities and in allowing the detector to reach its design sensitivity. High precision beam shaping of the CO_2 laser beam and high sensitivity wavefront sensors will be implemented to face this challenge, moving TCS towards adaptive optical systems.

Acknowledgments
This work has been performed with the support of the European Gravitational Observatory (collaboration conventions EGO-DIR-95-2008 and EGO-DIR-95-2008), of the European Commission (Framework Programme 7, project Einstein Telescope (ET) design study, grant agreement 211743) and of the Italian Ministero dell'Istruzione, dell'Universitá e della Ricerca (PRIN2007NXMBHP).

Virgo and Advanced Virgo are joint projects of the Centre National de la Recherche Scientifique (CNRS) and the Istituto Nazionale di Fisica Nucleare (INFN).

We thank R. Simonetti for his precious technical assistance.

References

[1] P. Hello and J.-Y. Vinet, *J. Phys. I France* **3**, 717-732 (1993).
[2] P. Hello and J.-Y. Vinet, *Phys. Lett. A* **178**, 351-356 (1993).
[3] F. Acernese et al., *J. Phys.: Conf. Ser.* **120**, 032007 (2008).
[4] B. P. Abbott et al., *Rep. Prog. Phys.* **72**, 076901 (2009).
[5] T. Accadia et al., A Thermal Compensation System for the Gravitational Wave Detector Virgo, *Proceedings of the 12th Marcell Grossmann Meeting* ed T Damour, R T Jantzen and R Ruffini (Singapore: World Scientific) (2011).
[6] T. Accadia et al., Advanced Virgo, *Proceedings of the 46th Rencontres de Moriond* (2011), in press.
[7] G.M. Harry (LIGO Scientific Collaboration), *Class. Quantum Grav.* **27**, 084006 (2010).
[8] P. Hello, *Eur. Phys. J. D* **15**, 373-383 (2001).
[9] H. Grote (LIGO Scientific Collaboration), *Class. Quantum Grav.* **27**, 084003 (2010).
[10] H. Luek et al., *Class. Quantum Grav.* **21**, S985-S989 (2004).
[11] K. Goda, D. Ottaway, B. Connelly, R. Adhikari, N. Mavalvala, and A. Gretarsson, *Opt. Lett.* **29**, 1452 (2004).
[12] Y. Fan et al., *Rev. Sci. Instrum.* **79**, 104501 (2008).
[13] A.F. Brooks et al., *Opt. Express* **15**(16), 10370-10375 (2007).

Amaldi 9 and NRDA 2011 IOP Publishing
Journal of Physics: Conference Series **363** (2012) 012017 doi:10.1088/1742-6596/363/1/012017

Search for a Stochastic Gravitational-wave Background with Torsion-bar Antennas

Ayaka Shoda[1], Masaki Ando[2], Kenshi Okada[1], Koji Ishidoshiro[3], Wataru Kokuyama[1], Yoichi Aso[1], Kimio Tsubono[1]

[1]Department of Physics, The University of Tokyo, Bunkyo-ku, Tokyo 113-0033, Japan
[2]Department of Physics, Kyoto University, Kitashirakawa Oiwake-cho, Sakyo-ku, Kyoto 606-8502, Japan
[3]Institute of Particle and Nuclear Studies, High Energy Accelerator Research Organization (KEK), 1-1 Oho, Tsukuba, Ibaraki 305-0801, Japan

E-mail: `shoda@granite.phys.s.u-tokyo.ac.jp`

Abstract. We performed a simultaneous observational run with prototypes of Torsion-bar Antenna (TOBA) and searched for a stochastic gravitational waves (GW) background. TOBA is a new type of GW detector which measures differential rotation of two test-mass bars caused by tidal force from GWs. It fundamentally has a good sensitivity at lower frequencies, such as 0.1 – 1.0 Hz. The prototype has a 20-cm test mass bar which is levitated by the pinning effect of a superconductor. The data was taken from 1:00 am to 10:00 am on March 11th 2011, at Tokyo and Kyoto in Japan. As a result, we did not detect a stochastic GW background with false alarm rate of 5 %, and set an upper limit on a stochastic GW background. Our 95 % confidence upper limit is $\Omega_{\rm gw} h_0^2 < 1.2 \times 10^{19}$ at 0.06 – 0.9 Hz, where $\Omega_{\rm gw}$ is the GW energy density per logarithmic frequency interval in units of the critical density and h_0 is the Hubble constant per 100 km/sec/Mpc. We had established the simultaneous observation and the analysis pipeline with two TOBAs, and set an upper limit at a wider frequency band.

1. Introduction

Recently, many gravitational wave (GW) detectors have been developed. Though GWs have not been detected yet, once detected, they will provide us new aspects of universe. Especially, a stochastic GW background is a cosmologically interesting target. It will provide us much information about phenomena occurred in the universe shortly after its birth, such as the inflation. Therefore, it has been eagerly searched for with a number of experiments.

The ground-based laser interferometer GW detectors, LIGO [1] and Virgo [2], have set a 95 % confidence upper limit on a stochastic GW background of in the frequency band around 100 Hz [3]. A pair of synchronous recycling interferometers set an upper limit at 100 MHz [4]. The cryogenic bar detectors, Explorer and Nautilus, had done at 907 Hz [5]. Also, the ALLEGRO resonant-bar detector and LIGO Livingston had set an upper limit at 917 Hz [6]. At much lower frequency band, the Doppler tracking with the Cassini spacecraft at 10^{-6} – 10^{-3} Hz [7], the pulsar timing by PSR B1855+09 at 10^{-9} – 10^{-7} Hz [8], and measurement of cosmic microwave background has established an upper limit at 10^{-18} – 10^{-16} Hz [8][9].

So far, no upper limit had been set at 0.01 – 1 Hz. The space GW detectors, LISA [10] and DECIGO [11], are planned to be launched and explore this frequency band. However, it is very challenging to launch more than three satellites and construct laser interferometer between the satellites.

Then, we designed a novel GW detector, "Torsion-bar Antenna", called TOBA [12]. It is fundamentally sensitive to GWs below 1 Hz even on the ground. We have already developed prototypes of TOBA with one 20-cm long test mass bar. By a single prototype TOBA, the first upper limit on a stochastic GW background had been set at 0.2 Hz [13].

However, it is difficult to judge whether a stochastic GW background is present or not by a single detector. We can mention about the presence of a stochastic GW background only if the signal exceeds the expected noise level. Therefore, it is necessary to take cross-correlation of multiple detectors in order to search a stochastic GW background. In this paper, we report on the simultaneous observational run and the results obtained by the cross-correlation analysis with two prototypes of TOBA placed at Tokyo and Kyoto in Japan. Here we demonstrated the two-detector correlation with TOBA and evaluate the its capabilities for low-frequency GW observations.

2. Torsion-bar Antenna

2.1. Principle of Torsion-bar Antenna

TOBA has two test mass bars suspended at their centers. (See figure 1.) When they are arranged orthogonal to each other, the bars rotates differentially by the tidal force caused by GWs. Their differential angular fluctuation is monitored by laser interferometric sensors.

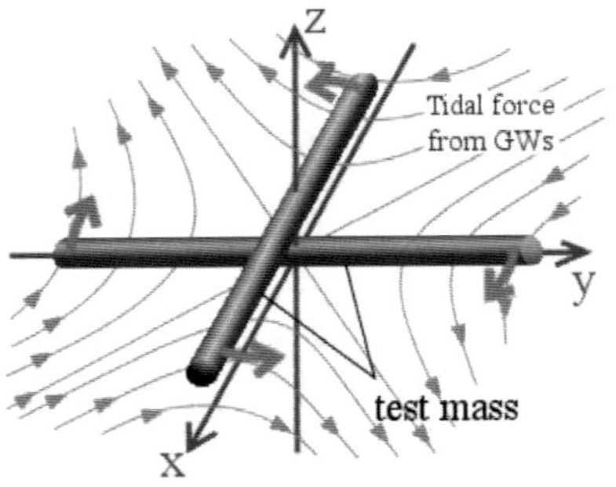

Figure 1. A conceptual drawing of TOBA.

Let us consider a test mass bar is arranged along x axis, and suspended so that they rotate around the z axis. (Figure 1.) When the GW come along z axis, the bars are affected by the tidal force from GWs. The angular fluctuation θ obeys the equation of motion:

$$I\ddot{\theta} + \gamma\dot{\theta} + \kappa\theta = F_{\mathrm{gw}}(t) \tag{1}$$

where I is the moment of inertia of the test mass, γ and κ are the damping constant and the spring constant around z axis, and F_{gw} is the torque caused by the GW. Assuming that the antenna is smaller enough than the wavelength of the GW, and that GW has $\times$-polarization, the angular fluctuation is simply represented as

$$\tilde{\theta}(f) = \frac{q_\times}{2I}\tilde{h}_\times(f) \tag{2}$$

above the resonant frequency of the suspension $f_0 = 1/(2\pi)\sqrt{\kappa/I}$, where $q_\times = q_{12} = q_{21}$ is the dynamic quadrupole moment of the test mass, and $h_\times$ is the amplitude of the $\times$-polarized GW [12].

TOBA and laser interferometric GW antennas has sensitivity to GWs above the resonant frequency. The point is that the resonant frequency in the rotational degree of freedom is on the order of a few mHz while the resonant frequency of the pendulum is around 1 Hz. This is the reason that a TOBA is fundamentally sensitive to GWs below 1 Hz even on the ground.

2.2. Prototype Torsion-bar Antenna

We have developed prototypes of TOBA. Pictures of the prototype TOBA is shown in figure 2.

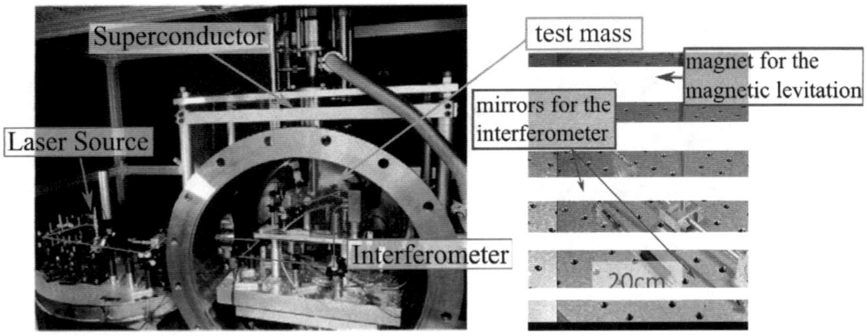

Figure 2. Pictures of the prototype TOBA detector. The left is the picture of the all experimental system. The Laser beam comes from the left hand side of the vacuum tank and composes the interferometer in order to monitor the rotation of the test mass bar. At the top of the vacuum tank, a superconductor is set for the magnetic levitation. The right picture is of a test mass bar. It has a magnet whose magnetic field is pinned by the superconductor and mirrors for the interferometer.

It has a 20-cm test mass bar, which is levitated by the flux pinning effect of a superconductor. This magnetic suspension let the test mass free to rotate while it provide large suspension force. In this case, its rotational resonance frequency is about 5 mHz. The rotation of the test mass bar is read by a laser Michelson interferometer. A laser source has a wavelength of 1064 nm and an output power of 40 mW. The beam goes to the mirrors attached at the both ends of the test mass, thus the differential change in the two beam path lengths is proportional to the rotation angle. The test mass is controlled by coil-magnet actuators. To compose the actuators, the test mass has two magnets of $\phi 1$ mm at the each end of the bar.

We have almost the same prototype TOBAs at University of Tokyo and Kyoto University. We performed simultaneous observational run with them.

3. Observation

The observation is performed from 1:00 am – 10:00 am at March 11, 2011 at Tokyo and Kyoto in Japan. The latitude and longitude of Tokyo is 35.71° N and 139.76° E, and of Kyoto is 35.03° N and 135.78° E. The Tokyo site is about 370 km far from the Kyoto site. (See figure 3) The both test mass bars are oriented to the north-to-south direction. An overlap reduction function which represents the difference of the response to GWs between two detectors is shown in figure 4 [14]. Note that the overlap reduction function of two TOBAs is calculated in the same way as two interferometers since the antenna pattern of TOBA is the same. At lower frequency than 1 Hz, the overlap reduction function is almost one, which means that the two detectors respond to GWs almost equally.

We recorded pulse per second (PPS) signals from GPS as well as feedback and error signals during the observation. The time shift between the two cites are adjusted after the observation

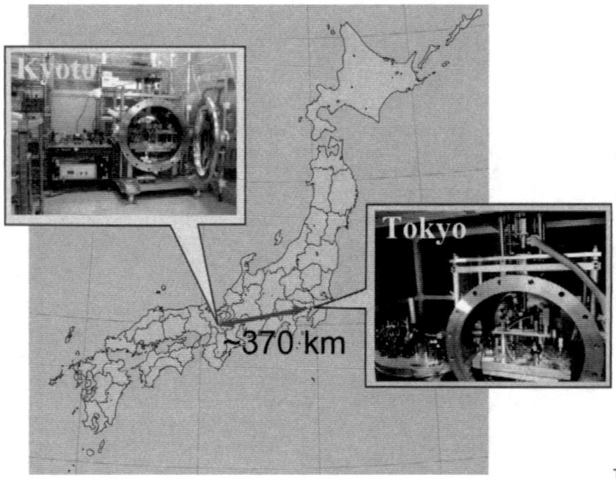

Figure 3. The location of Tokyo and Kyoto in Japan. Kyoto cite is about 370 km far west from Tokyo cite.

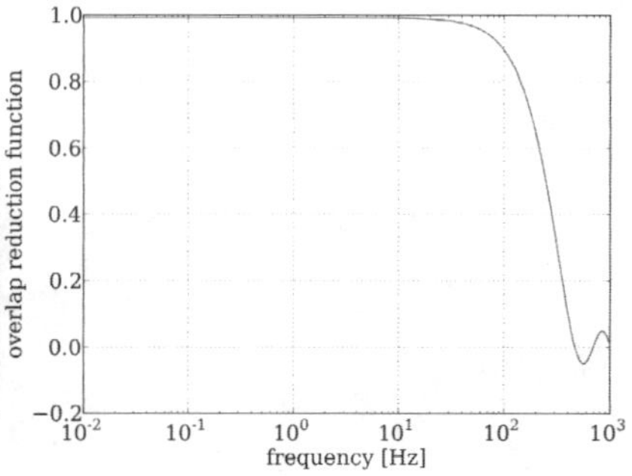

Figure 4. The overlap reduction function of two prototypes of TOBA at Tokyo and Kyoto. It is almost 1 below 10 Hz.

so that the PPS signals rise at the same position in the time series data. We also used the on-line calibration signal which is 8.7 Hz sine wave in order to monitor the fluctuation of the control open-loop gain. The on-line calibration signal is injected into the feedback signal, and the real-time gain is calculated by comparing signal before the injection and signal after the injection.

The equivalent strain noise spectra of two detectors are shown in fig 5. Their sensitivity is limited by the seismic noise at higher frequency than 0.1 Hz, and by the magnetic coupling noise at lower frequency than 0.1 Hz [13] [15].

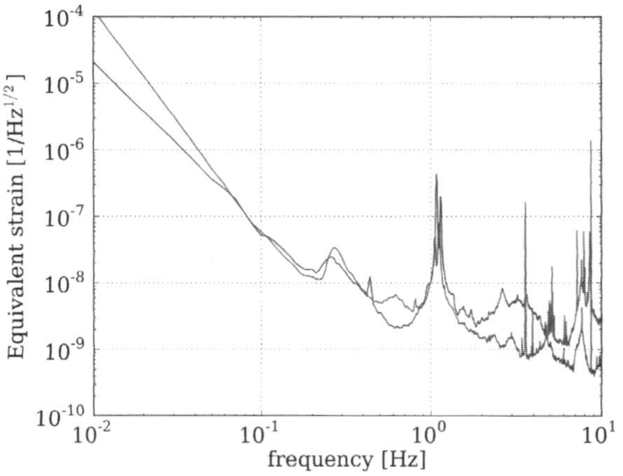

Figure 5. The equivalent strain noise spectra of two detectors. Blue line is the strain of the detector in Tokyo, and green line is the strain of the detector in Kyoto.

4. Analysis

4.1. A spectrum of a stochastic gravitational wave background

Our target, a stochastic GW background, is considered to be a superposition of GWs from astrophysical sources and from cosmological sources. Here, we assume that a stochastic GW background is stationary, unpolarized, and isotropic. Also, the amplitude of a stochastic GW background h_{ij} is assumed to have random variables, and we characterize the amplitude by its ensemble average of power spectrum density (PSD) $S_{\rm gw}(f)$ written as

$$S_{\rm gw}(f) = \frac{3H_0^2}{10\pi^2} f^{-3}\Omega_{\rm gw}(f). \tag{3}$$

Here, H_0 is the Hubble constant [16], and $\Omega_{\rm gw}(f)$ is defined as

$$\Omega_{\rm gw}(f) = \frac{f}{\rho_c}\frac{d\rho_{\rm gw}}{df}, \tag{4}$$

where $\rho_c = 3c^2H_0^2/8\pi G$ is the critical energy density of the universe and $\rho_{\rm gw} = (c^2/32\pi G)\langle \dot{h}_{ij}\dot{h}^{ij}\rangle$ is the gravitational-wave energy density [17].

4.2. The principle of the cross correlation analysis

The equivalent strain output $s_i(t)$ is written as

$$s_i(t) = n_i(t) + h_i(t), \tag{5}$$

where i donates the index of the two detectors. $n_i(t)$ and $h_i(t)$ is the detector's noise in equivalent strain and the signal due to a stochastic GW background. The cross correlation between the two outputs is

$$\begin{aligned} Y &= \int_{-T/2}^{T/2} dt \int_{-T/2}^{T/2} dt' s_1(t)s_2(t')Q(t-t') \\ &\sim \int_{-\infty}^{+\infty} df \tilde{s}_1^*(f)\tilde{Q}(f)\tilde{s}_2(f), \end{aligned} \tag{6}$$

where T is the observation time and $Q(t)$ is a real filter function called an optimal filter, which is decided so that signal-to-noise ratio is maximized. Here we assume that T is long enough to be approximated as $T \to +\infty$ and that $Q(t)$ converges on 0 when t is large. Note that the integrated interval has to be restricted for the analysis with the actual discrete output data.

The signal-to-noise ratio of Y, which is equal to $\langle Y\rangle/\sigma_Y$, is maximized when

$$\tilde{Q}(f) = N\frac{\gamma(f)}{P_1(f)P_2(f)f^3}, \tag{7}$$

where $\gamma(f)$ is the normalized overlap reduction function and P_i $(i=1,2)$ is the power spectrum density of the i-th detector's output. N is a normalization factor set so that $\langle Y\rangle = \Omega_{\rm gw}h_0^2T$:

$$N = \frac{20\pi^2}{3H_{100}^2}\left(\int_{-\infty}^{+\infty} df \frac{\gamma^2(|f|)}{f^6P_1(|f|)P_2(|f|)}\right)^{-1}. \tag{8}$$

Here, $H_{100} = 100$ km/sec/Mpc and $h_0 = H_0/H_{100}$ is the normalized Hubble constant.

If you want to know more details, see [17].

4.3. Main analysis

The process of the main analysis is divided into four parts: data selection, calculation of the cross correlation, signal detection, and calculation of the upper limit. Before calculating the cross correlation, we have to remove the data which badly affects the result, such as glitches. Then, we decide whether a stochastic GW background is present or not according to the cross correlation value calculated with survived data. When it is judged that a stochastic GW background is absent, the upper limit on a stochastic GW background is calculated.

4.3.1. Data selection Ideally, the noise levels of the two detectors should be stationary through the whole observation. However, there are some glitches or non-stationary noise in the time series data, which will affect the result. Therefore we remove the data where the noise level is bigger.

First, the time series data is divided into several segments. (See figure 6.) Then, we remove the segments in which the data is noisy, and calculate the cross correlation only with the survived segments. Here, it is important to distinguish the noise from a stochastic GW background, otherwise we may remove the segments where a stochastic GW background is happened to be bigger. In this case, the frequency band where the sensitivity is best is used for the calculation of the cross correlation, and at other frequencies, the sensitivity is assumed to be limited by the noises. Therefore, we judge whether the data is noisy or not according to the whitened RMS at the frequencies excluding the analyzed frequency band.

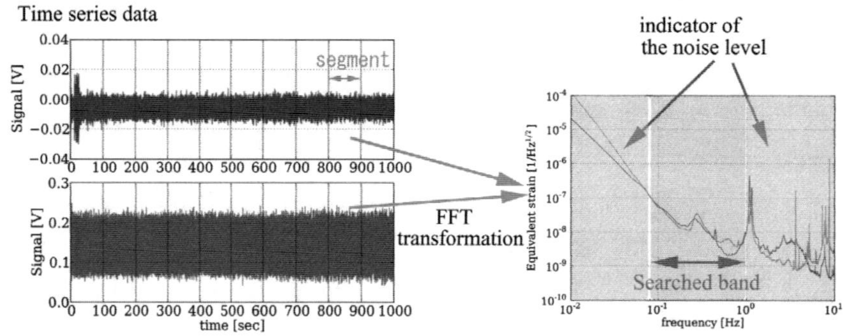

Figure 6. Data selection method. The blue and green line on the left graph is the time series data from the detector at Tokyo and Kyoto respectively. These data streem is devided into the segments whose length is 100 seconds (this length is decided by the parameter tuning described later) and transformed into the frequency domain. Then, the whitened RMS is calculated at the frequency band colered green at right graph. The analyzed frequency band, which is the region colored red, is not included into the noise level indicators.

4.3.2. Calculating cross correlation The cross correlation value Ys are calculated with each survived segments according to the equation 6.

Here we restrict the analyzed frequency band, since the frequency band where the sensitivity is worse does not produces a good result. In order to decide the analyzed frequency band, we used the optimal filter $\tilde{Q}(f)$. The optimal filter is larger when the sensitivity to a stochastic GW background is better. Therefore, the frequency band where the optimal filter is biggest is chosen for the analyzed frequency band. (See figure 7)

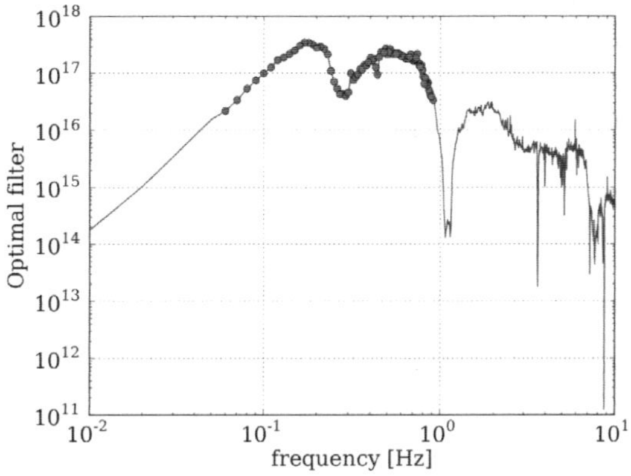

Figure 7. The blue line is the optimal filter. The red dots are the frequency bins which are chosen to be analyzed.

4.3.3. Signal detection According to Ys calculated at each segments, we decide whether a stochastic GW background is present or not by the Neyman-Pearson criterion:

- if $\langle Y \rangle \geq z_\alpha$, a stochastic GW background is present.
- if $\langle Y \rangle < z_\alpha$, a stochastic GW background is absent.

Here, $\langle Y \rangle$ is the average of Ys calculated with the survived segments and α is the false alarm rate. This detection threshold z_α is set so that

$$\alpha = \int_{z_\alpha}^{+\infty} dyp(y|0), \tag{9}$$

where $p(y|0)$ is the probability distribution of $\langle Y \rangle$ when a stochastic GW background is absent.

In this case, we calculate the distribution of $\langle Y \rangle$ with time shifted data, which is considered to be relative to the probability distribution of $\langle Y \rangle$ when a stochastic GW background is absent. The distribution is shown in figure 8.

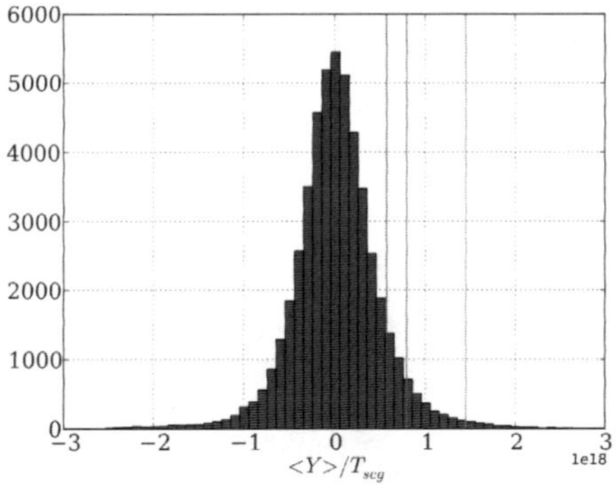

Figure 8. The distribution of $\langle Y \rangle$ with time shifted data. The sky blue, red, and green lines are the detection threshold z_α where $\alpha = 0.01, 0.05,$ and0.1 respectively.

4.3.4. Setting an upper limit When we cannot detect a stochastic GW background, we set an upper limit on a stochastic GW background. An upper limit means the amplitude of a stochastic GW background which we can detect if it would come to these detectors with the same noise levels.

In order to calculate it, we use the actual data set and mock signals. We make a mock signal of a stochastic GW background with a certain amplitude according to equation 3 (see figure 9), and inject it into the observational data. Then, we perform the same analysis described above with the injected data. This process is repeated many times and we compute the rate at which we detect the mock signal, which is called detection efficiency. Figure 10 shows the detection efficiency at variable amplitude of a injected signals. The detection efficiency is equal to the confidence level to the upper limit. Therefore, for example, 95 % confidence upper limit is the amplitude of a mock signal at which the detection efficiency is 95 %.

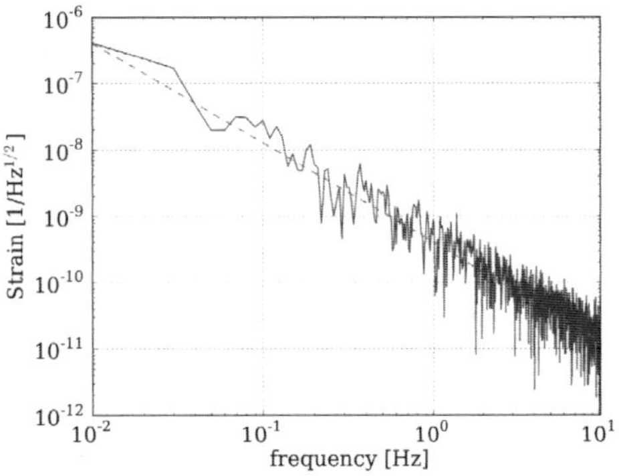

Figure 9. The strain of the mock signal of a stochastic GW background where $\Omega_{\rm gw}h_0^2 = 10^{18}$. It is basically a Gaussian noise, but it is multiplied by $f^{-3/2}$.

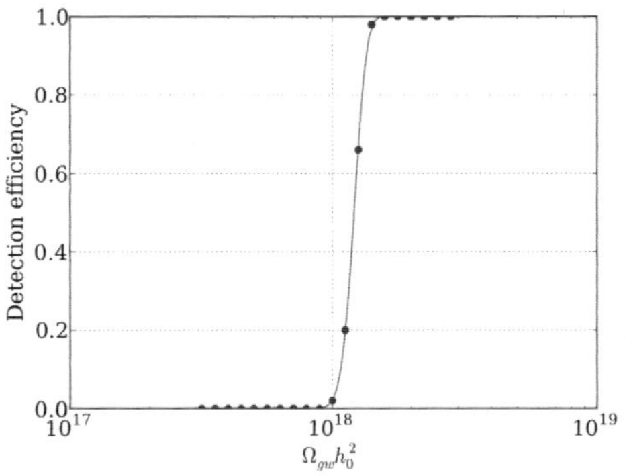

Figure 10. The detection efficiency. The blue dots are calculated probability and the blue line is an error function fitted to the blue dots. This detection efficiency is equal to $1 - \beta$, where β is the false dismissal rate.

4.4. Parameter tuning

In the process of the main analysis, there are some parameters whose optimal values depend on the data quality. Therefore, we performed pre-analysis and tuned these parameters. Here the time shifted data is used in order to prevent the result from being intentionally good.

The parameters to be tuned are the length of segments, the frequency band used as the indicator of the noise level at data selection, the amount of the segments which is removed, and the bandwidth of the analyzed frequencies.

Then, A flow chart of the whole analysis process is shown in figure 11.

5. Result

As a result of the parameter tuning, the length of the segment is set as 100 sec, i.e., the frequency resolution df is 0.01 Hz. We calculate RMS as the indicator of noise level at 0.01 – 0.05 and 1.0 – 10.0 Hz. Each 50 % of the segments are removed by the data selection, which results in about 14,000-second effective observation time. The analyzed frequency band is adjusted to 0.06 – 0.9 Hz. The width of searched frequency band is 0.85 Hz.

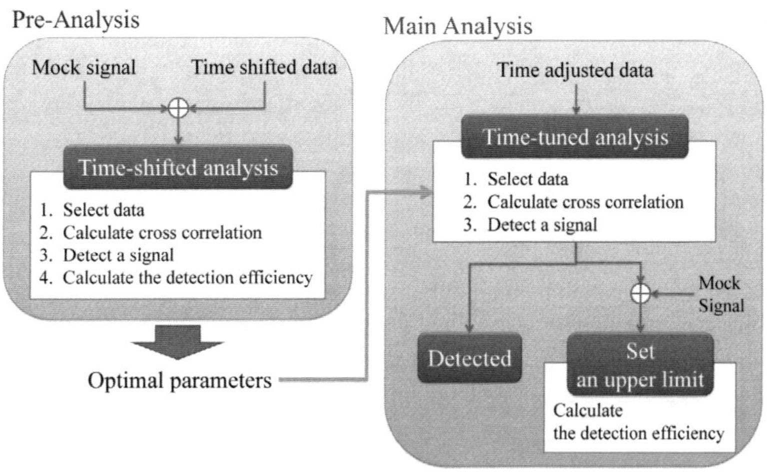

Figure 11. The flow chart of the analysis process. The process is roughly devided into two parts: the pre-analysis for the parameter tuning and the main analysis. We use the time-shifted data at pre-analysis and decide optimal parameters. Then, we perform the main analysis with time-tuned data and detect a GW signal or set an upper limit on a stochastic GW background.

When the data is analyzed with these parameters, the detection threshold with false alarm rate 1 % for $\langle Y \rangle / T$ is $z_{0.01} = 1.2 \times 10^{19}$. And calculated $\langle Y \rangle / T$ with time adjusted data is -1.4×10^{17}, which is smaller than $z_{0.01}$. Therefore, we concluded that a stochastic GW background is absent. Moreover, we calculated that 95 % confidence upper limit with false alarm rate 1 % is $\Omega_{\rm gw} h_0^2 \leq 1.2 \times 10^{19}$.

This result is worse than the upper limit calculated with a single TOBA by K. Ishidoshiro [13]. It is considered to be because of a large seismic noise due to the earthquake occurred at May 11th, 2011. This observational run was started about 13 hours before the earthquake, and there were some foreshocks during the observation. Now we are planning to perform the simultaneous observational run again before long.

6. Summary and future plan

We searched a stochastic GW background with the prototypes of TOBA by the cross correlation method. TOBA is the novel type of GW detector with good sensitivity below 1 Hz unlike other ground-based detectors such as large interferometers. The prototype has 20-cm test mass bar, and levitated by the magnetic force of the superconductor and the magnet. Its sensitivity is $h \sim 10^{-8} [1/\sqrt{\rm Hz}]$ at 0.3 Hz.

We demonstrated the simultaneous observational run and cross correlation analysis with two prototype TOBAs at Tokyo and Kyoto, and established the cross correlation analysis scheme with TOBAs. As a redult of the analysis for detect a stochastic GW background, we cannot detect a stochastic GW background. Our 95% confidence upper limit is $\Omega_{\rm gw} h_0^2 \leq 1.2 \times 10^{19}$.

While we set a new upper limit at wider frequency band than previous result, this result is not good because the seismic noise is bigger than before. This is considered to be due to the fore-shocks due to the big earthquake occurred in Japan at May 11th, 2011. Therefore, we will perform the simultaneous observational run again. It is expected that we can derive a better result.

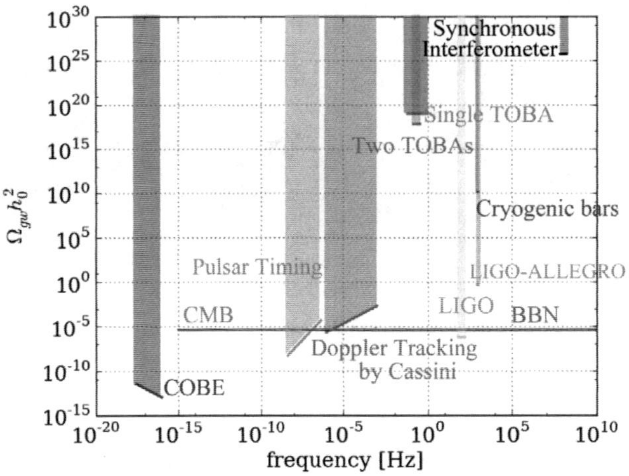

Figure 12. Upper limits established by TOBAs and other previous observations. This result does not exceed the previous result because of the seismic noise. However, we extended the analyzed frequency band.

References

[1] Abbott B P *et al.* 2009 *Reports on Progress in Physics* **72** 076901
[2] Acernese F *et al.* 2008 *Journal of Optics A: Pure and Applied Optics* **10** 064009
[3] Abbott B *et al.* 2009 *Nature* **460** 990–994
[4] Akutsu T *et al.* 2008 *Phys. Rev. Lett.* **101** 101101
[5] Astone P *et al.* 1999 *Astron. Astrophys.* **351** 811–814
[6] Abbott B *et al.* (LIGO Scientific Collaboration and ALLEGRO Collaboration) 2007 *Phys. Rev. D* **76**(2) 022001 URL `http://link.aps.org/doi/10.1103/PhysRevD.76.022001`
[7] Armstrong J, Iess L, Tortora P and Bertotti B 2003 *Astrophys. J.* **599** 806
[8] Maggiore M 2000 *Phys. Rep.* **331** 283–367
[9] Smith T, Pierpaoli E and Kamionkowski M 2006 *Phys. Rev. Lett.* **97** 21301
[10] Danzmann K 1996 *Class. Quantum Grav.* **13** A247–A250
[11] Kawamura S *et al.* 2011 *Class. Quantum Grav.* **28** ISSN 0264-9381
[12] Ando M *et al.* 2010 *Phys. Rev. Lett.* **105** ISSN 0031-9007
[13] Ishidoshiro K *et al.* 2011 *Phys. Rev. Lett.* **106** ISSN 0031-9007
[14] Allen B and Romano J 1999 *Phys. Rev. D* **59** 102001
[15] Ishidoshiro K 2009 *Search for low-frequency gravitational waves using a superconducting magnetically-levitated torsion antenna* Ph.D. thesis University of Tokyo
[16] Spergel D *et al.* 2003 *Astrophys. J.* **148** 175–194 ISSN 0067-0049
[17] Maggiore M 2008 *Gravitational Waves Volume I Theory and Experiments* (Oxford University Press)

Thermal tuning the optical cavity for 3 mode interaction studies using a CO_2 laser

Sunil Susmithan, Chunnong Zhao, Fang Qi, Li Ju and David Blair

School of Physics, The University of Western Australia, 35 Stirling Highway, Crawley, WA 6009, Australia

E-mail: susmis01@student.uwa.edu.au

Abstract. Three mode interactions could induce parametric instability in advanced gravitational wave detectors with high optical power circulating in the cavities. One of the conditions for parametric instability to occur is when the cavity frequency difference between fundamental mode and the high order mode matches the test mass acoustic mode frequency. The optical mode spacing is a function of cavity g-factor (radius of curvature). At the Gingin High Optical Power Facility, we have an 80 meter optical cavity particularly designed for studying high optical power effects in advanced gravitational wave detectors such as parametric instabilities. Here we present the recent results of thermal tuning the cavity g-factor by heating the test mass surface with a CO_2 laser to investigate the 3-mode interactions. Observation of test mass thermal noise peaks above 160 kHz enhanced by 3 mode interaction is presented.

1. Introduction

Advanced gravitational wave detectors are aiming to detect frequent gravitational wave signals at least once a year, with the sensitivity able to reach a distance of 200MPc [1]. To achieve this sensitivity, the detectors will use extremely high optical power inside the arm cavities to reduce quantum shot noise limit. However, high optical power, together with low optical and mechanical loss test masses could induce parametric instability in a gravitational wave detector. At Gingin high optical power facility, an 80m optical cavity is used to investigate parametric instabilities and method for parametric instability control. Currently, spontaneous parametric instability has not been observed due to the low power level in the instrument. However, 3-mode interaction was observed [2] and possible control strategies was proposed [3,4,5]. Here we present further studies of 3-mode interaction using CO_2 laser thermal tuning method.

1.1. Three Mode Parametric Interaction and Parametric Instability

Gravitational wave detectors consist of a pair of Fabry-Perot (FP) cavities [6]. Thermal motion of the test mass will scatter the incoming fundamental optical mode of frequency (ω_0) to higher order optical modes. The higher order optical modes provide a back action force on the test mass mirrors through radiation pressure creating a resonance condition which leads to further amplification of the higher order optical mode. If the beating frequency between a resonant higher order optical mode (ω_h) and the fundamental optical mode matches the acoustic mode frequency (ω_m) of the test mass mirror, as well as when the optical mode shape has a good spatial overlap with the acoustic mode then, 3-mode interaction will occur. This either amplifies or dampen the mechanical motion of the test mass. Amplification of the mechanical mode could lead to parametric instability, whereas damping leads to the cooling of the mechanical mode.

A quantitative discussion on parametric instability can be described using parametric gain R [7]. Instability occurs when the value of R is greater than 1. The gain can be defined as

$$R = \frac{4P_c Q_1 Q_m}{m_{eff} L c \omega_m^2} \cdot \frac{\Lambda}{1 + (\frac{\Delta\omega}{\delta})^2} \tag{1}$$

where P_c is the power of the laser circulating within the cavity, Q_1 and Q_m are the quality factors of higher order optical mode and the acoustic mode, m_{eff} is the effective mass of the mirror acoustic mode, L is the length of the cavity, c is the velocity of light, ω_m is the frequency of the acoustic mode, $\Delta\omega$=$|\omega_0$-$\omega_h|$-ω_m , δ is the half-line width of the higher order mode and Λ is the spacial overlap factor between the optical and the acoustic mode.

With a reasonably good spatial overlap, parametric gain R will be maximum at $\Delta\omega$=$|\omega_0$-$\omega_h|$-$\omega_m = 0$, i.e the frequency difference between the higher order optical mode and the fundamental mode becomes equal to the acoustic mode of the test mass. The frequency difference (mode gap) of the fundamental mode and the high order mode is given by

$$|\omega_0 - \omega_h| = \omega_m \tag{2}$$

$$\omega_0 - \omega_h = \frac{c}{L}[cos^{-1}\sqrt{(1 - \frac{L}{R_1}) \cdot (1 - \frac{L}{R_2})}] \tag{3}$$

where R_1 and R_2 are the radius of curvature of the input test mass (ITM) and the end test mass (ETM) respectively and L is the length of the cavity. It can be seen that mode gap can be tuned by changing the radius of curvature of the cavity mirrors as shown in eqn.(3). By heating the test mass mirror, we change the radius of curvature of the test mass as shown in fig.1. We could either tune the $|\omega_0$-$\omega_h|$ away from ω_m to reduce parametric instability in the case of R > 1, or tune $|\omega_0$-$\omega_h|$ close to ω_m to cool the test mass mode.

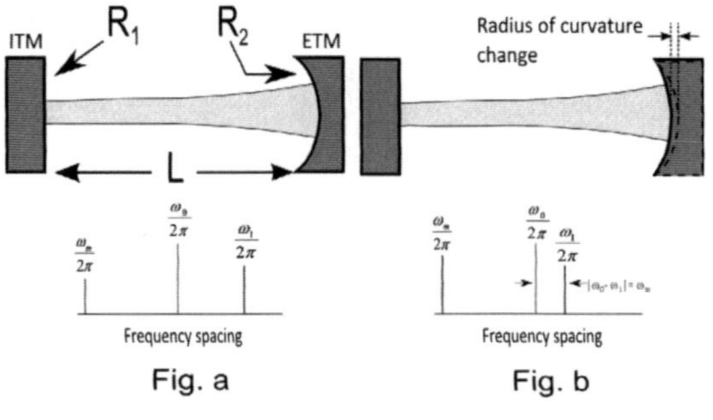

Figure 1. Fig.a Frequency spacing between the modes when the cavity is cold. Fig.b Thermal tuning of the ETM changes the radius of curvature of the mirror. This change in radius of curvature tune the spacing between the optical frequencies to match the acoustic mode frequency.

We use a CO_2 laser to perform a localized heating to create a deformation on the surface of the test mass which changes the radius of curvature as shown in the figure 1.

2. Experimental setup for thermal tuning

Experimental setup is shown in figure 2. A stabilized CO_2 laser heats one of the test masses and creates a localized deformation. We use two diagnostic tools to measure the radius of curvature change of the heated test mass. They are (i) Hartmann sensor and (ii) a beam profile CCD camera. Table I shows the optical cavity parameters. Calculation using equation (2) shows that

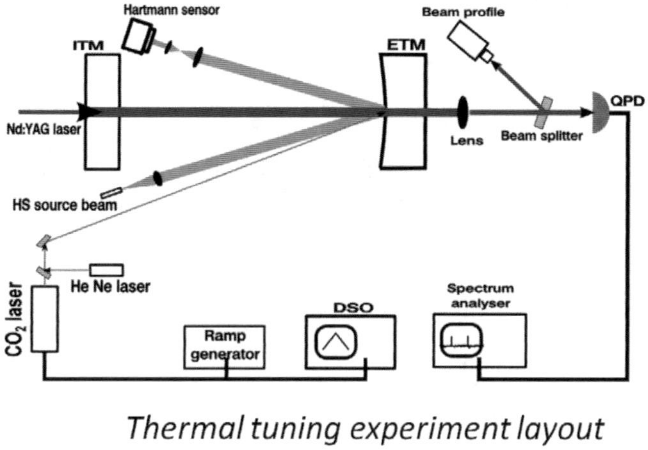

Figure 2. Experimental setup of thermal tuning of the cavity

the frequency difference ($\Delta\omega$) between the fundamental mode TEM_{00} and the first higher order optical mode TEM_{01} is 1.298 MHz ($f = \frac{\Delta\omega}{2\pi} = 206kHz$).

PARAMETERS	VALUES
Test mass material	Sapphire
ROC R_1 ITM	Flat
ROC R_2 ETM	720m
Length of the cavity L	77m
Q_1 quality factor of optical mode	1.88x10^{11}
Q_m quality factor of acoustic mode	2x10^6
m_{eff} effective mass of the acoustic mode	0.25 kg
ω_m acoustic mode frequency	181.6 kHz
Λ overlap factor	1.67
R Parametric gain	0.1

Using FEA simulation, an acoustic mode at 181.6 kHz is identified. This acoustic mode has a good spatial overlap with the first order optical mode. By thermal tuning the radius of curvature of the mirror we can adjust the frequency spacing between the optical modes to match the acoustic mode frequency as shown in figure 1(a) and figure 1(b). To tune the frequency spacing from 206 kHz to 181.6 kHz, a radius of curvature change of 1.5 km is needed. The heating power and spot size of the heating beam needed for thermal tuning is estimated using the following equation:

$$R_c = \frac{4 \cdot \pi \cdot k \cdot a^2}{2 \cdot \alpha \cdot P} \tag{4}$$

where R_c is the radius of curvature change of the heated test mass, k is the thermal conductivity of the test mass, a is the heating spot size, α is coefficient of thermal expansion of the material and P is the laser heating power.

2.1. Hartmann sensor

Our Hartmann sensor [8] consists of a metal plate 16x16mm with uniformly distributed holes of $151\mu m$ in diameter and hole to hole spacing $429\mu m$. The plate is kept at a distance of 10mm from the CCD pixel photodiode array. A super luminescent laser emitting diode (SLED) source beam is collimated before illuminating the test mass, and the reflected beam is demagnified and

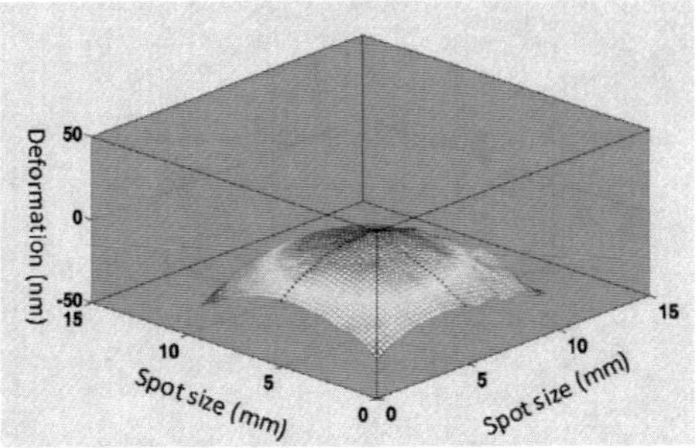

Figure 3. Hartmann sensor: Deformation of ETM at the place heated using a CO_2 laser. The spot size falling on the Hartman sensor is a de-magnified image of the mirror surface illuminated using a Hartmann source (HS) beam.

collimated to fall on the Hartmann sensor using a telescopic arrangement as shown in figure 2. The CCD read the position of the illuminated holes and compares these positions with any new position that has formed during the deformation. Data obtained from the pixels are processed and the localized deformation of the test mass due to CO_2 heating is measured. A typical front surface deformation during CO_2 heating measured by the Hartmann sensor is shown in figure 3.

2.2. Beam profile camera

A beam profile camera measures the leaking transmitted light from the locked Fabry-Perot cavity. The image on the camera is proportional to the spot size of the beam on the ETM.

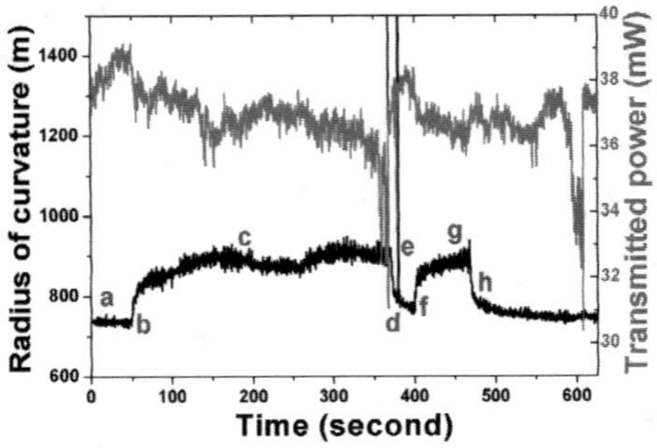

Figure 4. Transient radius of curvature change measured using CCD camera (lower curve). The upper curve is the transmitted power. Region a to b shows the mirror before heating, b denotes the instant when CO_2 laser is switched ON, c and g denotes the equilibrium state attained by the mirror during heating, d and h denotes the region when the CO_2 laser is switched OFF and e denotes when the cavity is unlocked.

Using the Spiricon software we measure the horizontal and vertical crossection of the spot on the camera. Figure 4 shows the variation in the spot size as ETM is heated with the CO_2 laser.

3. Results

Figure 5(a) and figure 5(b) shows the spectrum of the acoustic mode of 181.6 kHz without and with CO_2 thermal tuning. A frequency gap of 181.6 kHz of the optical cavity, corresponds to a radius of curvature of 923m of the ETM. CO_2 laser heating of the ETM showed that at 650mW heating power, the acoustic thermal noise peak value is maximum as shown in figure 5(c). The value of heating power is consistence with the radius of curvature obtained from equation 4.

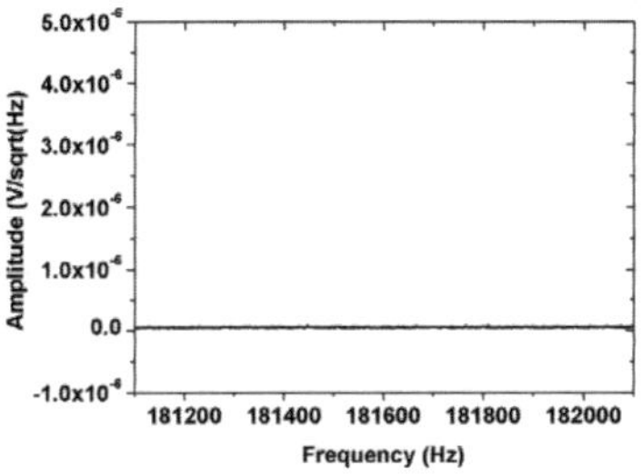

(a)

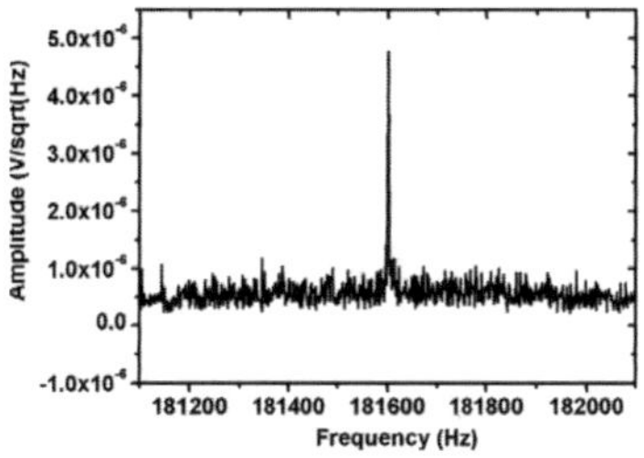

(b)

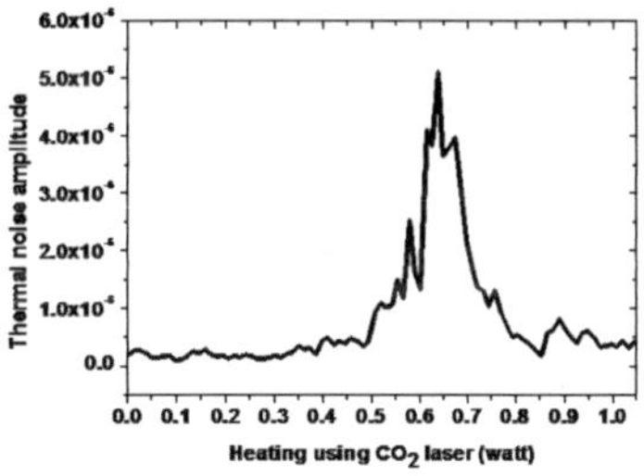

(c)

Figure 5. (a) Frequency spectrum observed without heating (b) Thermal noise peak of 181.6 kHz is observed due to the 3 mode interaction, (c) Variation of thermal noise peak of 181.6 kHz with CO_2heating power.

4. Conclusion

We have observed 3-mode interactions with controlled thermal tuning using a CO_2 laser. The experimental result is consistent with the theoretical calculation. The CO_2 laser thermal tuning provides an effective method for 3-mode interaction study and possibly for parametric instability control. This research is supported by the Australian Research Council and is part of the Australian Consortium for Gravitational Astronomy.

5. References

[1] S.A.Hughes *et.al* Tech. Rep. P010029-00-D, LIGO Project, 2001. http://www.ligo.caltec.edu/docs/P/P010029-00.pdf

[2] C. Zhao *et.al* 2008 *Phys. Rev.* A **78** 023807
[3] Y.H. Fan *et.al* 2010 *Class. Quantum Grav.* **27** 084028
[4] S. Gras, D.G. Blair, L. Ju 2008 *Phys. Lett.* A **372** 1348
[5] L. Ju *et.al* 2009 *Class. Quantum Grav.* **26** 015002
[6] A.E. Siegman, Lasers (University Science, 1986)
[7] V.B. Braginsky, S.E. Strigin, S.P. Vyatchanin 2001 *Phy. Lett.* A **287** 331
[8] A. Brooks Ph. D thesis, The University of Adelaide (2007)

Testing Chern-Simons modified gravity with observations of extreme-mass-ratio binaries

P. Canizares[1], J.R. Gair[2] and C.F. Sopuerta[3]

[1,3]Institut de Ciències de l'Espai (CSIC-IEEC), Facultat de Ciències, Campus UAB, Torre C5 parells, Bellaterra, 08193 Barcelona, Spain.
[1,2]Institute of Astronomy, Madingley Road, Cambridge, CB30HA, United Kingdom.

E-mail: [1]pcm@ast.cam.ac.uk,[2]jrg23@cam.ac.uk, [3]sopuerta@ieec.uab.es

Abstract. Extreme-Mass-Ratio Inspirals (EMRIs) are one of the most promising sources of gravitational waves (GWs) for space-based detectors like the Laser Interferometer Space Antenna (LISA). EMRIs consist of a compact stellar object orbiting around a massive black hole (MBH). Since EMRI signals are expected to be long lasting (containing of the order of hundred thousand cycles), they will encode the structure of the MBH gravitational potential in a precise way such that features depending on the theory of gravity governing the system may be distinguished. That is, EMRI signals may be used to test gravity and the geometry of black holes. However, the development of a practical methodology for computing the generation and propagation of GWs from EMRIs in theories of gravity different than General Relativity (GR) has only recently begun. In this paper, we present a parameter estimation study of EMRIs in a particular modification of GR, which is described by a four-dimensional Chern-Simons (CS) gravitational term. We focus on determining to what extent a space-based GW observatory like LISA could distinguish between GR and CS gravity through the detection of GWs from EMRIs.

1. Introduction

It is known that surrounding the central MBH of many quiescent galaxies there are around $10^7 - 10^8$ stars forming a cusp/core (see e.g. [1]). As a consequence of mass segregation and large scattering encounters between stars in the cluster, the largest stellar-mass compact objects (SCOs), like neutron stars, $m \approx 1.4M_\odot$, white dwarfs, $m \approx 0.6M_\odot$, and stellar-mass black holes, $m \approx 1 - 50M_\odot$, may be perturbed onto an orbit that passes close enough to the central MBH and become gravitationally bound to it. It is expected that the capture of a SCO by a MBH may be a frequent phenomenon in the Universe. The MBHs of interest for a GW observatory like LISA [2] have masses in the range $M_\bullet \sim 10^5 - 10^7 M_\odot$. In addition, there are observations suggesting the existence of Intermediate-Mass Black Holes (IMBHs), whose masses are in the range $M_\bullet \sim 10^2 - 10^4 M_\odot$ (see e.g. [3]). These may form EMRI-like systems, either an IMBH-MBH inspiral or an SCO-IMBH system. Space-based detectors like LISA will be sensitive to the inspiral of an IMBH into a MBH, whereas ground-based detectors like the Einstein Telescope [4] will be sensitive to the inspiral of SCOs into IMBHs.

Once the SCO is captured, the binary loses energy and angular momentum through the emission of GWs. This drives the inspiral of the SCO until it plunges into the MBH. Due to the fact that the mass of the SCO is much smaller than the mass of the MBH, with typical mass ratios lying in the range $\mu = m/M_\bullet \sim 10^{-7} - 10^{-3}$, the inspiral can be modelled as a slow adiabatic process in which the orbital time scale is much shorter than the timescale on which the orbit shrinks.

During the last year of evolution before plunge, the orbit of the SCO spends of the order of 10^5 cycles in the strong field region of the MBH, mapping its geometric structure (which can be characterized in terms of mass and current multipolar moments). This map is encoded in the shape and timing of the GWs emitted [5]. If we can decode this map, EMRI detections will allow us to study different aspects of the MBHs located at the centers of galaxies, the stellar dynamics around galactic centers, etc. In particular, EMRI GW signals may be very useful to test the spacetime geometry of MBHs [6, 7, 8, 9, 10, 11, 12, 13] and even alternative theories of gravity (see, e.g. [14, 15, 13, 16, 17, 18]), which is the subject of this article.

Due to the myriad of alternative theories of gravity available, the question that arises here is which kind of theory do we choose to test? Alternatively, what kind of phenomena can we expect to test using EMRI observations? In this paper, we will focus on Dynamical Chern-Simons Modified Gravity (DCSMG) which is a 4D effective theory of gravity that arises in certain low-energy limits of *string theory* and can also arise in *Loop Quantum Gravity* (see [19] for a review of DCSMG). DCSMG has been proposed as an explanation for the cosmic baryon asymmetry [20] and makes predictions for the polarisation of the cosmic microwave background [21]. In addition, it has potentially observable signatures in EMRI GW observations [22].

In this paper we summarize the necessary techniques that we have developed to estimate the capability of a LISA-like GW detector to distinguish between GR and DCSMG using GW signals from EMRIs, extending and improving the results of [22]. A more detailed account of these developments will appear elsewhere [23]. Due to the large difference between the masses of the components in the binary, an EMRI signal can be modelled accurately using black hole perturbation theory (see e.g. [24]), in which the SCO is represented as a particle orbiting in a given MBH spacetime background. In this framework, at the lowest order of approximation the orbit of the particle is a geodesic of the MBH spacetime, but radiation reaction (RR) effects, i.e. the effects that arise from the interaction of the SCO with its own gravitational field, accelerate the trajectory and then drive the inspiral of the SCO. Depending on the theory of gravity that describes the system, the RR effects drive the inspiral in different ways .

This paper is organised as follows. In Sec. 2 we summarize the formulation of our EMRI model in DCSMG. In Sec. 3 we review the basics of signal analysis employed in our study. Finally, in Sec. 4, we present results concerning whether a GW detector like LISA will be able to discriminate between an EMRI in GR and one in DCSMG.

Throughout this paper, we follow the conventions of Misner, Thorne and Wheeler [25]: Greek letters will represent spacetime indexes and Latin letters in the middle of the alphabet $i, j, \ldots$, will represent spatial indexes only. Partial derivatives are denoted by $\partial_\alpha h = \partial h/\partial x^\alpha = h_{,\alpha}$, while, for any quantity h, covariant derivatives are denoted by $\nabla_\alpha h = h_{;\alpha}$. In addition, we use the Einstein summation convention and geometrised units, i.e. $G = c = 1$ and the gravitational constant is given by $\kappa = 1/(16\pi G)$. Finally, we use a bar, e.g. $\bar{h}$, to denote background quantities.

2. Formulation of our EMRI model

We approximate our system following the procedure described in [22], in which an EMRI is modelled as a binary composed of a point particle moving in a modified Kerr MBH background geometry. That is, the MBH geometry is slightly modified as a consequence of the CS gravitational modifications to GR. The motion of the particle, at the lowest order of approximation, is given by a geodesic of this CS-modified MBH geometry. The action describing the system is given by:

$$S = \kappa S_{EH} + \frac{\alpha}{4} S_{CS} + \frac{\beta}{2} S_\vartheta + S_{\rm mat}\,, \qquad (1)$$

where the different terms are:

$$S_{EH} = \int d^4x\sqrt{-\mathrm{g}}R\,, \tag{2}$$

$$S_{CS} = \int d^4x\sqrt{-\mathrm{g}}\,\vartheta\,{}^*RR\,, \tag{3}$$

$$S_{\vartheta} = -\int d^4x\sqrt{-\mathrm{g}}\,[\,\mathrm{g}^{\mu\nu}(\nabla_\mu\vartheta)(\nabla_\nu\vartheta)]\,, \tag{4}$$

$$S_{\rm mat} = \int d^4\sqrt{-\mathrm{g}}\,\mathcal{L}_{\rm mat}\,, \tag{5}$$

which represent the Einstein-Hilbert action [Eq. (2)], the CS gravitational correction [Eq. (3)], the CS scalar field action [Eq. (4)], and the action of the matter degrees of freedom [Eq. (5)]. The coupling constant, α, is associated with the Pontryagin density, ${}^*RR := \frac{1}{2}\epsilon^{\gamma\delta\mu\nu}R^{\alpha}_{betau\mu\nu}R^{\beta}{}_{\alpha\gamma\delta}$, which is coupled to the CS scalar field, ϑ. The coupling constant, β, determines the strength of the coupling of the CS scalar field with the stress-energy distribution.

Following [26], we can construct the modified MBH geometry and the CS scalar field using the slow-rotation ($a/M_\bullet \ll 1$, where $M_\bullet$ and $a = |\mathbf{S}/\mathbf{M}_\bullet|$ are the MBH mass and spin respectively) and weak-coupling approximation. Then, the spacetime metric, $\mathrm{g}_{\mu\nu}$, and the CS scalar field, ϑ, can be written as the sum of a background and a perturbative quantity characterized by the perturbative parameter υ:

$$\mathrm{g}_{\mu\nu} = \bar{\mathrm{g}}_{\mu\nu} + \upsilon h_{\mu\nu} + \mathcal{O}(\upsilon^2)\,, \tag{6}$$

$$\vartheta = \bar{\vartheta} + \upsilon\vartheta + \mathcal{O}(\upsilon^2)\,. \tag{7}$$

The MBH metric has the following non-zero components in Boyer-Lindquist-like coordinates (t, r, θ, ϕ) [26]:

$$\bar{\mathrm{g}}_{tt} = -\left(1 - 2M_\bullet\frac{r}{\rho^2}\right)\,, \tag{8}$$

$$\bar{\mathrm{g}}_{rr} = \frac{\rho^2}{\Delta}\,, \tag{9}$$

$$\bar{\mathrm{g}}_{\theta\theta} = \rho^2\,, \tag{10}$$

$$\bar{\mathrm{g}}_{\phi\phi} = \frac{\Sigma}{\rho^2}\sin^2\theta\,, \tag{11}$$

$$\bar{\mathrm{g}}_{t\phi} = \left[\frac{5}{8}\frac{\xi}{M_\bullet^4}\frac{M_\bullet^5}{r^4}\left(1 + \frac{12M_\bullet}{7r} + \frac{27M_\bullet^2}{10r^2}\right) - 2M_\bullet^2\frac{r}{\rho^2}\right]\frac{a}{M_\bullet}\sin^2\theta\,, \tag{12}$$

where $\Sigma = (r^2 + a^2)^2 - a^2\Delta\sin^2\theta$, $\rho^2 = r^2 + a^2\cos^2\theta$ and $\Delta = r^2 f + a^2$, with $f = 1 - 2M_\bullet/r$. The CS parameter ξ is given by

$$\xi := \frac{\alpha^2}{\beta\kappa}\,. \tag{13}$$

Notice that the only metric component that contains the CS parameter ξ is $\bar{\mathrm{g}}_{t\phi}$ and that it behaves as r^{-4} for $M_\bullet/r \ll 1$, decaying faster than the rest of metric contributions and becoming negligible at large distances. Under the preceding approximations, the CS scalar field is given by [26]:

$$\bar{\vartheta} = \frac{5}{8}\frac{\alpha}{\beta}\frac{a}{M_\bullet}\frac{\cos\theta}{r^2}\left(1 + \frac{2M_\bullet}{r} + \frac{18M_\bullet^2}{5r^2}\right)\,. \tag{14}$$

The numerical evolution of the system comprises two stages, namely the computation of the SCO trajectory around the MBH and then the computation of the corresponding waveform based on this trajectory. To generate the trajectory, it has been shown in [22] that in the test-mass limit the SCO must follow geodesics of the CS deformed MBH background. It turns out that the metric [Eqs. (8)-(12)] has almost identical symmetry properties as the Kerr metric (at least at the level of approximation used for its derivation). As a consequence, the geodesic motion of the particle is obtained by solving the following set of Ordinary Differential Equations (ODEs):

$$\dot{t} = \dot{t}_K + L_z \delta g_\phi^{CS}\,, \tag{15}$$

$$\dot{\phi} = \dot{\phi}_K - E\delta g_\phi^{CS}\,, \tag{16}$$

$$\dot{r}^2 = \dot{r}_K^2 + 2EL_z f \delta g_\phi^{CS}\,, \tag{17}$$

$$\dot{\theta}^2 = \dot{\theta}_K^2\,, \tag{18}$$

where E is the energy and L_z is the z-component of the orbital angular momentum of the particle per unit mass. Here, a dot denotes differentiation with respect to the proper time and $(\dot{t}_K,\, \dot{r}_K, \dot{\theta}_K,\, \dot{\phi}_K)$ are the corresponding expressions for geodesics in a Kerr MBH background:

$$\rho^2 \dot{t}_K = -\left(a^2 E\sin^2\theta - aL_z\right) + \left(r^2+a^2\right)\left(\left(r^2+a^2\right)\frac{E}{\Delta} - \frac{aL_z}{\Delta}\right), \tag{19}$$

$$\rho^2 \dot{\phi}_K = -\left(aE - \frac{L_z}{\sin^2\theta}\right) + (r^2+a^2)\frac{aE}{\Delta} - \frac{a^2 L_z}{\Delta}\,, \tag{20}$$

$$\rho^4 \dot{r}_K^2 = \left[(r^2+a^2)E - aL_z\right]^2 - \Delta\left[\mathcal{Q} + (aE - L_z)^2 + r^2\right]\,, \tag{21}$$

$$\rho^4 \dot{\theta}_K^2 = \mathcal{Q} - \cot^2\theta L_z^2 - a^2\cos^2\theta(1-E^2)\,, \tag{22}$$

where $\mathcal{Q}$ denotes the Carter constant per unit mass squared. The correction δg_ϕ^{CS} appearing in Eqs. (15)-(18) is the CS correction, which is given by:

$$\delta g_\phi^{CS} = \frac{\xi a}{112\, r^8 f}\left(70r^2 + 120 r M_\bullet + 189 M_\bullet^2\right)\,. \tag{23}$$

In order to avoid the presence of turning points in the integration of these equations of motion, we introduce two new angle coordinates, ψ and χ, associated with the radial and polar motion respectively, which are defined by the equations:

$$r = \frac{pM_\bullet}{1+e\cos\psi}\,, \qquad \cos^2\theta = \cos^2\theta_{\rm min}\cos^2\chi\,, \tag{24}$$

where p and e are the semilatus rectum and the eccentricity of the orbit respectively, and $\theta_{\rm min}$ is the minimum value of the polar angle θ in the orbit. This value is related to the angle of inclination of the orbit, $\theta_{\rm inc}$ (see Table 1), through

$$\theta_{\rm min} = \frac{\pi}{2} - \theta_{\rm inc}\,. \tag{25}$$

A given orbit can be characterized in terms of the constants of motion $(E, L_z, \mathcal{Q})$ or, alternatively, in terms of the orbital parameters $(e, p, \theta_{\rm inc})$. To transform between these two sets of constants we use the equations at the turning points of the radial and polar coordinates. At $\theta = \theta_{\rm min}$ we have that $\dot{\theta} = 0$, leading to an expression for $\mathcal{Q}$ [22]:

$$\mathcal{Q} = \cos^2\theta_{\rm min}\left[\frac{L_z^2}{\sin^2\theta_{\rm min}} + a^2(1-E^2)\right]\,. \tag{26}$$

The additional two equations needed are those that define the radial turning points, that is, the pericenter, $r_{\rm peri}$, and apocenter, $r_{\rm apo}$, which we relate to (e, p) in the usual way:

$$r_{\rm peri} = \frac{pM_\bullet}{1+e}\,, \qquad r_{\rm apo} = \frac{pM_\bullet}{1-e}\,. \tag{27}$$

Eq. (26) together with the equations $\dot{r}(r_{peri}) = \dot{r}(r_{apo}) = 0$ allow us to relate the values of the constants $(E, L_z, \mathcal{Q})$ to those of the orbital parameters $(e, p, \theta_{\rm inc})$.

Until now we have only considered geodesic motion. Constructing waveforms from geodesic trajectories can lead to a a confusion problem in the sense that it is possible that an EMRI waveform in GR matches an EMRI waveform in DCSMG but with different parameters, given that the motion in both cases is going to be characterized by just three fundamental frequencies of motion. This degeneracy is overcome when RR effects are taken into account, since these tend to break such confusion, as the RR effects drive the inspiral in different ways depending on the theory of gravity that describes the EMRI system. Nevertheless, it was shown in [22] that, at leading order, the GW emission in DCSMG takes the same form as in GR. So, as a first approximation we describe RR effects in DCSMG using expressions from GR. In particular we use the fluxes for the constants of motion $(E, L_z, \mathcal{Q})$ provided in [27]. In principle one might think that this would lead to exactly the same inspiral evolution in DCSMG and GR, but it is not the case due to the different relations between the constants of motion $(E, L_z, \mathcal{Q})$ and the orbital parameters $(e, p, \theta_{\rm inc})$.

To compute the EMRI waveforms in DCSMG, we first solve the ODEs given in Eqs. (15)-(18) for the angle variables $(\psi(t), \chi(t), \varphi(t))$ (see e.g. [28]). The scheme for evolving a particular EMRI in DCSMG can be summarized as follows: For a given set of initial orbital parameters, $(e^{(0)}, p^{(0)}, \theta^{(0)}_{\rm inc})$, we find the associated initial constants of the motion, $(E^{(0)}, L^{(0)}_z, \mathcal{Q}^{(0)})$, which are different from the ones that one would obtain in GR for the same orbital parameters. Then, the time evolution of the "constants" of motion, $(\dot{E}^{(0)}, \dot{L}^{(0)}_z, \dot{\mathcal{Q}}^{(0)})$, the RR effect, is computed using the fluxes of [27]. From the values of the constants of motion, $(E^{(0)}, L^{(0)}_z, \mathcal{Q}^{(0)})$, their evolution due to RR, $(\dot{E}^{(0)}, \dot{L}^{(0)}_z, \dot{\mathcal{Q}}^{(0)})$, and the value of the geodesic radial period, T_r (the coordinate time to go from the apocenter to the pericenter and back again to apocenter), we obtain the values of the new constants of motion, $(E^{(1)}, L^{(1)}_z, \mathcal{Q}^{(1)})$. Finally, from the new values of the constants of motion we find the new values of the orbital parameters $(p^{(1)}, e^{(1)}, \theta^{(1)}_{\rm inc})$. This algorithm is iterated along the whole EMRI evolution to obtain the SCO orbit around the MBH. The gravitational waveform is then generated using the multipolar formalism given by Thorne [29] up to quadrupolar order, by identifying the Boyer-Lindquist coordinates with spherical polar coordinates in a pseudo flat-space.

3. Signal analysis formalism

In this section we briefly review the formalism of signal analysis that we use in this paper. For our calculations we use the LISA noise sensitivity curve (see e.g. [30]) to represent a possible future space-based GW detector. At low-frequencies the response of the LISA detector can be decomposed as the response of a pair of independent, right-angle Michelson detectors, which we label by $\alpha = I, II$ in the standard way. Each of these data streams is a time series, $s_\alpha(t)$, which is a superposition of a possible gravitational wave signal $h_\alpha(t)$ and noise $n_\alpha(t)$, $s_\alpha(t) = h_\alpha(t) + n_\alpha(t)$. Under the assumptions that the noise in the detector is stationary, Gaussian and uncorrelated between the two data streams, and that the spectral density of the noise is the same in the two channels, the Fourier components of the noise, $\tilde{n}_\alpha(f)$, would have a Gaussian probability distribution and

$$\langle\, \tilde{n}_\alpha(f)\, \tilde{n}_\beta(f')^*\,\rangle = \frac{1}{2}\,\delta_{\alpha\beta}\,\delta(f-f')\,S_n(f)\,, \tag{28}$$

where a tilde denotes the Fourier transform, an asterisk denotes complex conjugation, angle brackets denote the ensemble expectation value, and $S_n(f)$ is the one-sided noise spectral density of the data stream, which is the same for both channels.

The vector space of signals has a natural inner product associated with $S_n(f)$:

$$(\mathbf{a}|\mathbf{b}) = 2\sum_{I,II}\int_0^\infty \frac{\tilde{a}^*_\alpha(f)\tilde{b}_\alpha(f)+\tilde{a}_\alpha(f)\tilde{b}^*_\alpha(f)}{S_n(f)}\mathrm{d}f\,. \tag{29}$$

The assumption of Gaussianity then means that the probability that the noise, $\mathbf{n}$, has a particular realisation, $\mathbf{n}_0$, has the form

$$p(\mathbf{n}=\mathbf{n}_0) \propto e^{-(\mathbf{n}_0|\mathbf{n}_0)/2}\,, \tag{30}$$

and hence the likelihood that a signal $\mathbf{h}$ is present in the data stream $\mathbf{s}$ is proportional to $\exp(-(\mathbf{s}-\mathbf{h}|\mathbf{s}-\mathbf{h})/2)$. The 'best-fit' waveform will be the one that maximises $(\mathbf{s}|\mathbf{h})$ and, thus, provides the maximum likelihood parameter estimate. The expected signal-to-noise ratio (SNR) when filtering with the correct waveform is

$$\mathrm{SNR} = \frac{(\mathbf{h}|\mathbf{h})}{\mathrm{rms}(\mathbf{h}|\mathbf{n})} = \sqrt{(\mathbf{h}|\mathbf{h})}\,, \tag{31}$$

where the second equality follows from the fact that the expectation value of $(\mathbf{a}|\mathbf{n})(\mathbf{b}|\mathbf{n})$ is $(\mathbf{a}|\mathbf{b})$ [31].

A given template family $\mathbf{h}(t;\vec{\lambda})$ will depend on a set of parameters $\vec{\lambda}$. If we expand $\exp(-(\mathbf{s}-\mathbf{h}|\mathbf{s}-\mathbf{h})/2)$ around the best-fit parameters, $\vec{\lambda}_0$, by writing $\vec{\lambda}=\vec{\lambda}_0+\delta\vec{\lambda}$, we can derive a probability distribution function for the parameter uncertainties

$$p(\delta\vec{\lambda}) = \mathcal{N}\exp\left(-\frac{1}{2}\Gamma_{jk}\delta\lambda^j\delta\lambda^k\right)\,, \tag{32}$$

where $\mathcal{N}=\sqrt{\det(\Gamma/2\pi)}$ is the normalization factor and Γ_{ij} is the *Fisher information matrix* [32]

$$\Gamma_{jk} = \left(\frac{\partial\mathbf{h}}{\partial\lambda^j}\bigg|\frac{\partial\mathbf{h}}{\partial\lambda^k}\right)\,. \tag{33}$$

The variance-covariance matrix for the waveform parameters is therefore given by the inverse of the Fisher matrix

$$\langle\delta\lambda^j\delta\lambda^k\rangle = \left(\Gamma^{-1}\right)_{jk} + O(1/\mathrm{SNR}) \tag{34}$$

and so we can estimate the precision with which we will be able to measure a particular parameter, λ_i, using LISA observations by computing the component Γ^{-1}_{ii} of this inverse matrix.

In our simulations, the space of EMRI waveforms in DCSMG, $\{h(t;\lambda^i)\,|\,i=1,\ldots,N\}$ is characterised by the following 15 parameters (see Table 1 for a description of these parameters):

$$\vec{\lambda} = \{M_\bullet, a, \mu, e_0, p_0, \theta_0^{\mathrm{inc}}, \hat{\xi}, \theta_S, \phi_S, \theta_K, \phi_K, D_L, \psi_0, \chi_0, \phi_0\}\,. \tag{35}$$

The Fisher matrices for EMRI waveforms have very large condition numbers (the ratio of the largest to the smallest eigenvalues) and so we use an LU decomposition to invert them, where the Fisher matrix [Eq. (34)] is written as the product of a lower triangular matrix and an upper triangular matrix [33]. In addition, to verify that the error estimates that we obtain are reliable, we employ the maximum-mismatch criterion proposed in [34].

4. Results

The main goal of our study is to determine how well a GW detector like LISA could estimate the parameters of an EMRI system (see Table 1) in DCSMG and in particular how well it could constrain the CS parameter ξ and hence distinguish between an inspiral in DCSMG and one in GR. To that end, we have followed the approach of [22] to model the motion of the SCO in the modified DCSMG MBH geometry, but this time we have included RR effects. To evolve the inspiral we have followed the algorithmic scheme described above. Once the trajectory of the SCO has been computed, we obtain the waveforms by using the multipolar formalism described by Thorne in [29], but only keeping the quadrupolar term. To compute the Fisher matrix, the waveform derivatives, $\partial_i \mathbf{h} = \partial \mathbf{h}/\partial \lambda^i$, are obtained numerically through a five-point finite difference rule. Finally, the response of the detector and the Doppler modulation for a detector like LISA are computed using the expressions in [30] and in [35].

Table 1. Summary of the EMRI parameter space employed in our analysis. The angles (θ_S, ϕ_S) and (θ_K, ϕ_K) are spherical coordinates and t_0 stands for the initial time in the computation. For the parameters with physical dimensions, we indicate the units used in round brackets.

$M_\bullet$	MBH mass.
$a = \lvert\mathbf{S}\rvert/M_\bullet$ $(M_\bullet^2)$	MBH spin.
$\mu = m/M_\bullet$	Mass-ratio.
e_0	Eccentricity of the particle orbit at t_0.
p_0	Semilatus rectum at t_0.
$\theta_0^{\rm inc}$	Inclination of the orbit at t_0.
$\hat{\xi} = \xi \times S$ $(M_\bullet^6)$	Product of the Chern-Simons parameter and the spin.
θ_S	Polar angle with respect to the ecliptic.
ϕ_S	Azimuthal angle with respect to the ecliptic.
θ_K	Spin polar angle with respect to the ecliptic.
ϕ_K	Spin azimuthal angle with respect to the ecliptic.
D_L (Gpc)	Distance to the EMRI from the ecliptic baricentre.
ψ_0	Phase angle associated with the radial motion at t_0.
χ_0	Phase angle associated with the polar motion at t_0.
ϕ_0	Boyer-Lindquist azimuthal coordinate, ϕ, at t_0.

Table 2. Parameters of the different systems used in our work, where we have fixed $p = 10$, $\theta_{\rm inc} = 0.85$, $\theta_S = 1.1$, $\phi_S = 0.3$, $\theta_K = 1.4$, $\phi_K = 0.25$, $D_L = 1\,{\rm Gpc}$, $\psi_0 = 0.25$, $\chi_0 = 1$, and $\phi_0 = 0.1$.

System	$M_\bullet$	$a/M_\bullet$	μ	e_0	$\hat{\xi}/M_\bullet^6$
A	5×10^5	0.5	2×10^{-5}	0.5	10^{-2}
B	5×10^5	0.5	2×10^{-5}	0.5	0
C	5×10^5	0.25	2×10^{-5}	0.5	10^{-2}
D	5×10^5	0.25	2×10^{-5}	0.5	0
E	10^6	0.25	10^{-5}	0.32	10^{-2}
F	10^6	0.25	10^{-5}	0.32	0

We have computed parameter measurement error estimates for different systems in DCSMG and the analogous systems in GR (that is, setting $\xi = 0 = \delta\xi$). In Table 2, we show the parameters of the different (systems) used in this analysis, where the distance to the source is given in Gpc, the angles are given in radians and the rest of the parameters are given in MBH mass units, while the MBH mass itself is given in solar masses. The direction of the MBH spin is taken to be parallel to the z-direction of the EMRI reference frame.

As a test of our numerical set-up, in Table 3 we compare our numerical estimates in GR with the ones obtained in [30]. Since the parameter space employed in our analysis differs with the one chosen in [30], we just compare the results obtained for the parameters common to both works and, as we can see from Table 3, the two methods produce very similar results. However, as it is shown in the table, our estimate for the error in the MBH spin is two orders of magnitude smaller than the one obtained in [30]. In our analysis, we include the effect of the MBH spin directly in the geodesic equations, whereas in [30] the effect of the MBH spin is introduced as a post-Newtonian correction in their *post-Keplerian* equations of motion. Our analysis therefore includes higher order spin corrections which may explain some of the improvement that our results show. A similar improvement in spin measurement precision was observed in [33] who used a similar method based on exact Kerr geodesic trajectories.

Table 3. Parameter accuracy estimates for the inspiral of a $10M_\odot$ SCO into a $10^6M_\odot$ MBH. It is assumed that we have observed the last year of inspiral and that the system is an EMRI described by GR, i.e., $\xi = 0$. We compare the results we have obtained with the ones obtained for the same parameters by Barack and Cutler [30].

Parameter	GR (this work)	GR ([30])
$\Delta \ln M_\bullet$	8.6×10^{-4}	9.2×10^{-4}
$\Delta a/M_\bullet$ $(M_\bullet)$	9.5×10^{-6}	6.3×10^{-4}
$\Delta \ln \mu$	1.4×10^{-4}	9.2×10^{-5}
Δe	2.7×10^{-5}	2.8×10^{-4}
$\Delta \ln D_L{}^{(a)}$, $\Delta \ln (\mu/D_L)^{(b)}$	$6.5 \times 10^{-2\,(a)}$	$3.7 \times 10^{-2\,(b)}$

We have considered two different kind of EMRIs, namely one composed of a MBH with $M_\bullet = 5 \times 10^5 M_\odot$ and a SCO with mass $m = 10M_\odot$, and the other one composed of a MBH with mass $M_\bullet = 10^6 M_\odot$ and a SCO with mass $m = 10M_\odot$. For the former, we consider an observation lasting for the six months before plunge (SNR= 35) and for the latter we consider an observation lasting one year before plunge (SNR= 40). In Table 4, we show the error estimates obtained in our simulations. We typically find that an EMRI observation can determine the extrinsic parameters, namely $\vec{\lambda}_{\text{extrinsic}} = \{\theta_S, \phi_S, \theta_K, \phi_K, D_L, \psi_0, \chi_0, \phi_0\}$, to a fractional error of $\sim 10^{-2}$, while the intrinsic parameters: $\vec{\lambda}_{\text{intrinsic}} = \{M_\bullet, a, \mu, e_0, \theta_0^{\text{inc}}\}$, are typically measured with a fractional error of $\sim 10^{-4}, 10^{-5}, 10^{-5}, 10^{-5}, 10^{-5}$ respectively. For the CS parameter $\hat{\xi}$, we find that in the best case it can be measured with a fractional error $\sim 10^{-1}$. These results suggest that LISA might be able to distinguish between GR and DCSMG using certain EMRI observations. Notice that in this paper we are using the concept of the LISA mission as a reference and, as it is shown in [23], the results obtained are also valid for the updated eLISA or NGO (New Gravitational-Waves Observatory) [36] by normalising to a fixed SNR. In addition, we find that for a given value of $\hat{\xi}$, it will be better estimated for systems with smaller MBH mass, $M_\bullet$, and spin, a. This is not surprising, since the leading-order term that contains the CS parameter in the DCSMG metric [Eq. (12)] is $\sim \hat{\xi}/M_\bullet^6 (M_\bullet/r)^6$. By decreasing the MBH mass with fixed $\hat{\xi}$, this term becomes larger at fixed $r/M_\bullet$ and by decreasing the MBH spin this term is larger relative to the Kerr $g_{t\phi}$ metric coefficient.

Due to the relatively high computational cost of these calculations, a full (i.e., including all the

Table 4. Error estimates for half a year of evolution (SNR= 35) for Systems A, B, C, and D; and for one year of evolution (SNR= 40) for Systems E and F.

Parameter	A	B	C	D	E	F
$\Delta \ln M_\bullet$	1.0×10^{-4}	9.8×10^{-3}	7.8×10^{-4}	7.6×10^{-4}	8.6×10^{-4}	8.4×10^{-4}
$\Delta a \ \ (M_\bullet)$	1.5×10^{-5}	9.5×10^{-6}	7.5×10^{-6}	6.3×10^{-6}	1.8×10^{-5}	9.1×10^{-6}
$\Delta \ln \mu$	1.8×10^{-4}	3.9×10^{-5}	5.7×10^{-5}	2.4×10^{-5}	3.2×10^{-4}	1.2×10^{-4}
Δe_0	9.2×10^{-5}	9.2×10^{-5}	8.4×10^{-5}	8.4×10^{-5}	1.9×10^{-5}	1.7×10^{-5}
$\Delta \theta_0^{\mathrm{inc}}$	2.8×10^{-5}	1.7×10^{-5}	2.6×10^{-5}	2.2×10^{-5}	8.1×10^{-5}	3.2×10^{-5}
$\Delta \ln \hat{\xi}$	2.1		4.4×10^{-1}		5.1	
$\Delta \theta_S$	1.8×10^{-2}	1.7×10^{-2}	1.5×10^{-2}	1.4×10^{-2}	1.1×10^{-2}	1.1×10^{-2}
$\Delta \phi_S$	1.7×10^{-2}	1.6×10^{-2}	1.5×10^{-2}	1.5×10^{-2}	2.1×10^{-2}	2.0×10^{-2}
$\Delta \theta_K$	2.8×10^{-2}	2.8×10^{-2}	2.5×10^{-2}	2.5×10^{-2}	2.3×10^{-2}	2.3×10^{-2}
$\Delta \phi_K$	7.2×10^{-2}	7.2×10^{-2}	6.5×10^{-2}	6.5×10^{-2}	8.6×10^{-2}	8.6×10^{-2}
$\Delta D_L \ \ (Gyr)$	2.7×10^{-2}	2.6×10^{-2}	2.2×10^{-2}	2.3×10^{-2}	1.8×10^{-2}	1.9×10^{-2}
$\Delta \psi_0$	9.8×10^{-2}	7.1×10^{-2}	7.2×10^{-2}	6.0×10^{-2}	1.0×10^{-2}	8.6×10^{-2}
$\Delta \chi_0$	8.6×10^{-2}	8.5×10^{-2}	7.7×10^{-2}	7.7×10^{-2}	9.2×10^{-2}	9.0×10^{-2}
$\Delta \varphi_0$	8.6×10^{-2}	8.3×10^{-2}	7.6×10^{-2}	7.5×10^{-2}	1.1×10^{-1}	1.1×10^{-2}

parameters) error estimation takes approximately three days using a single processor machine. For this reason, we have only studied a small number of points in the parameter space of all possible EMRI systems that we might detect. In this sense, these results could be improved by performing a more exhaustive study of the parameter space and also using a full Monte Carlo sampling. In addition, we could have considered longer gravitational waveforms, that cover the entire length of the LISA mission ($\sim$ 2–5 years), which would be expected to improve our ability to measure the system parameters.

5. Conclusions

In this work we have presented a first study aimed at assessing the ability of a GW detector like LISA to discriminate between GR and DCSMG by performing parameter estimation studies of EMRIs in DCSMG. We have computed the waveforms generated by a SCO orbiting in a MBH geometry modified by CS corrections. The parameter error estimates have been computed using a Fisher matrix analysis. To validate our computational framework, we have studied a typical EMRI system in GR and we have found agreement between our results and previous results found in the literature. We have applied these parameter estimation studies to estimate the ability of LISA to distinguish between GR and DCSMG by estimating the CS parameter ξ. To that end, we have performed simulations of the GWs emitted during the last six months before plunge of an EMRI system for which the gravitational wave emission is at frequencies in the most sensitive part of the LISA band. Our results indicate that, for certain EMRI systems, a detector like LISA may be able to discriminate between GR and DCSMG. We have also seen that the error in estimating ξ decreases with the MBH mass and spin. We are now performing a more exhaustive study of the parameter space of possible EMRIs to better understand if these results hold in general [23].

Another subject that deserves further study is the question of the systematic errors that would arise by using GR waveform templates to detect EMRIs that are actually described by DCSMG. This can be done by using the formalism developed recently by Cutler and Vallisneri [37], which allows for the estimation of the magnitude of the model errors. Finally, the present work can be extended to other GW detectors, for instance, to study Intermediate-Mass-Ratio Inspiral detection with the future Einstein Telescope.

Acknowledgments
We would like to thank Michele Vallisneri for helpful discussions. PCM is supported by a predoctoral FPU fellowship of the Spanish Ministry of Science and Innovation (MICINN). JG's work is supported by the Royal Society. CFS acknowledges support from the Ramón y Cajal Programme of the Ministry of Education and Science of Spain, from a Marie Curie International Reintegration Grant (MIRG-CT-2007-205005/PHY) of the 7th European Community Framework Programme, from contract AYA-2010-15709 of the MICINN, and from contract 2009-SGR-935 of the Catalan Agency for Research Funding (AGAUR). We acknowledge the computational resources provided by the Barcelona Supercomputing Centre (AECT-2011-3-0007) and CESGA (contract CESGA-ICTS-200).

References

[1] Amaro-Seoane P and Preto M 2011 *Class. Quant. Grav.* **28** 094017 (*Preprint* 1010.5781)
[2] LISA (Laser Interferometer Space Antenna), http://www.esa.int, http://lisa.jpl.nasa.gov
[3] Amaro-Seoane P, Gair J, Freitag M, Miller C M, Mandel I, Cutler C and Babak S 2007 *Class. Quant. Grav.* **24** R113 (*Preprint* astro-ph/0703495)
[4] ET (Einstein Telescope), http://www.et-gw.eu/
[5] Ryan F D 1995 *Phys. Rev. D* **52** 5707
[6] Collins N A and Hughes S A 2004 *Phys. Rev.* **D69** 124022 (*Preprint* gr-qc/0402063)
[7] Glampedakis K and Babak S 2006 *Class. Quant. Grav.* **23** 4167–4188 (*Preprint* gr-qc/0510057)
[8] Barack L and Cutler C 2007 *Phys. Rev.* **D75** 042003 (*Preprint* gr-qc/0612029)
[9] Yunes N, Sopuerta C F, Rubbo L J and Holley-Bockelmann K 2008 *Astrophys. J.* **675** 604–613 (*Preprint* 0704.2612)
[10] Brink J 2008 *Phys. Rev.* **D78** 102001 (*Preprint* 0807.1178)
[11] Vigeland S J and Hughes S A 2010 *Phys. Rev.* **D81** 024030 (*Preprint* 0911.1756)
[12] Hughes S A 2010 (*Preprint* 1002.2591)
[13] Sopuerta C F 2010 *GW Notes* **4**(4) 3–47 (*Preprint* 1009.1402)
[14] Hughes S A 2006 *AIP Conf. Proc.* **873** 233–240 (*Preprint* gr-qc/0608140)
[15] Schutz B F 2009 *Class. Quant. Grav.* **26** 094020
[16] Babak S, Gair J R, Petiteau A and Sesana A 2011 *Class. Quant. Grav.* **28** 114001 (*Preprint* 1011.2062)
[17] Vigeland S, Yunes N and Stein L 2011 *Phys. Rev. D* **83** 104027 (*Preprint* 1102.3706)
[18] Gair J R and Yunes N 2011 (*Preprint* 1106.6313)
[19] Alexander S and Yunes N 2009 *Phys. Rep.* **480** 1
[20] Alexander S H, Peskin M E and Sheikh-Jabbari M M 2006 *Phys. Rev. Lett.* **96** 081301 (*Preprint* hep-th/0403069)
[21] Lue A, Wang L M and Kamionkowski M 1999 *Phys. Rev. Lett.* **83** 1506 (*Preprint* astro-ph/9812088)
[22] Sopuerta C F and Yunes N 2009 *Phys. Rev. D* **80** 064006 (*Preprint* 0904.4501)
[23] Canizares P, Gair J R and Sopuerta C F 2012 *to be submitted*
[24] Drasco S and Hughes S A 2006 *Phys. Rev. D* **73** 024027 (*Preprint* gr-qc/0509101)
[25] Misner C W, Thorne K and Wheeler J A 1973 *Gravitation* (San Francisco: W. H. Freeman & Co.)
[26] Yunes N and Pretorius F 2009 *Phys. Rev. D* **79** 084043 (*Preprint* 0902.4669)
[27] Gair J R and Glampedakis K 2006 *Phys. Rev. D* **73** 064037
[28] Drasco S and Hughes S A 2004 *Phys. Rev. D* **69** 044015 (*Preprint* astro-ph/0308479)
[29] Thorne K S 1980 *Rev. Mod. Phys.* **52** 299
[30] Barack L and Cutler C 2004 *Phys. Rev. D* **69** 082005 (*Preprint* gr-qc/0310125)
[31] Cutler C and Flanagan E E 1994 *Phys. Rev. D* **49** 2658 (*Preprint* gr-qc/9402014)
[32] Fisher R A 1935 *Roy. Stat. Soc. J.* **98** 35
[33] Huerta E A and Gair J R 2009 *Phys. Rev. D* **79** 084021 (*Preprint* 0812.4208)
[34] Vallisneri M 2008 *Phys. Rev. D* **77** 042001
[35] Cornish N J and Larson S L 2003 *Class. Quant. Grav.* **20** 163 (*Preprint* arXiv:gr-qc/0206017)
[36] Amaro-Seoane P, Aoudia S, Babak S, Binetruy P, Berti E *et al.* 2012 (*Preprint* 1201.3621)
[37] Cutler C and Vallisneri M 2007 *Phys. Rev. D* **76** 104018

Journal of Physics: Conference Series **363** (2012) 012020 doi:10.1088/1742-6596/363/1/012020

Gravitational waves from strongly magnetised neutron stars

Paul D Lasky[1], **Burkhard Zink and Kostas D Kokkotas**

Theoretical Astrophysics, Institute for Astronomy and Astrophysics, University of Tübingen, Auf der Morgenstelle 10, Tübingen 72076, Germany

E-mail: `paul.lasky@unimelb.edu.au`

Abstract. Strongly magnetised neutron stars are prime candidates for multi-messenger astronomy given their proximity and regular, energetic flaring events. We present non-linear, ideal MHD simulations of strongly magnetised neutron stars in general relativity that are models for post-flare internal dynamics of these stars. In particular, magnetic field instabilities are used to trigger global reconfigurations of the magnetic field, which in turn excites both fluid and Alfvén modes throughout the star. These simulations are discussed in the context of gravitational wave emissions and detectability for ground-based gravitational wave detectors.

1. Introduction

Soft-gamma repeaters (SGRs) and anomalous X-ray pulsars (AXPs) are classes of neutron stars, collectively known as *magnetars*, that contain the most extreme magnetic fields in the Universe. These exotic objects exhibit sporadic bursting and flaring events, which are commonly associated to dynamics of the magnetic field. The strongest of these flares has been considered in the literature as possible multi-messenger gravitational wave sources, which has prompted targeted gravitational wave searches [1–5]. Presently, these searches have placed upper limits on gravitational wave energies of 1.4×10^{49} erg for the f-mode (i.e. in the kHz frequency range), and 3.5×10^{44} erg for white noise around $100 - 200$ Hz.

Comparatively little theoretical work has been done on the topic of magnetar flares as a viable source of gravitational radiation. In part this is due to a reasonably poor understanding of the internal dynamics of the objects during and immediately following a flare. Ioka [6] first investigated this question by looking at the maximum gravitational wave energy released by a change in moment of inertia induced by a dynamical rearrangement of the core magnetic field inside the star. They were able to place an upper limit of about 10^{49} erg under ideal conditions, including optimistic values of the internal magnetic field. More recently, Corsi & Owen [7] found similar values to be possible under more generic conditions, still tapping into the full energy reservoir associated with an instantaneous change in the magnetic potential energy of the star. In contrast, Levin & van Hoven [8] did not find f-mode detection to be very likely in the near future. Their model was based on some mechanism external to the star triggering the internal f-modes.

In this work we present details of recent theoretical work aimed at understanding these gravitational wave emissions caused by a large-scale rearrangement of the internal magnetic

[1] Current Address: School of Physics, University of Melbourne, Parkville, VIC 3010, Australia

field [9]. In particular, we utilise magnetic field instabilities to trigger global reconfigurations of the magnetic field, and determine the gravitational wave output for such models. Recently, Ciolfi *et al.* [10] performed numerical simulations of the same hydromagnetic simulations as our earlier paper [11], concluding that giant flares *could* give rise to observable gravitational radiation. However, they utilised a stellar model with a surface magnetic field strength some sixty times the strongest observed to reach these conclusions. We have shown [9] that the gravitational wave energy emitted in these events is a highly nonlinear function of the magnetic field strength, implying f-mode detection is unlikely.

This article provides a conglomeration of our two recent papers [9, 11]. In section 2 we briefly describe our model and numerical method. In section 3 we discuss the hydromagnetic instability that we elucidated in [11], and in section 4 we discuss the gravitational wave output and detectability of the signal.

2. Model

We study the time evolution of the ideal MHD equations in general relativity under the Cowling approximation utilising the three-dimensional GRMHD code THOR [12, 13] and her sister GPU code HORIZON [14]. The MHD portion of the code utilises the conservative formalism outlined in [15], with hyperbolic divergence cleaning employed following the prescription of ref. [16]. In the following simulations we use a Cartesian grid with 120^3 grid points. Resolution studies have been completed using 90^3, 150^3 and 200^3 grid points, finding consistent phenomenology. The outer boundary of our star is located approximately 1.4 times the stellar radius (at the closest point), and we adopt Dirichlet boundary conditions for the evolution of the magnetic field at this outer boundary. We have performed various tests with different boundary conditions and again found consistent phenomenology. For further specifics on the code and the numerical method we refer the reader to a forthcoming article [17].

The spectral code LORENE is utilised to create initial conditions, which are self-consistent solutions of the Einstein-Maxwell field equations in ideal MHD with purely poloidal magnetic fields[18]. Our fiducial model is a non-rotating star with polytropic equation of state relating the pressure, P, and rest-mass density, ρ, through $P = K\rho^\Gamma$, with $\Gamma = 2$ and $K = 100$. This gives a stellar model with gravitational mass $M = 1.3\,\mathrm{M}_\odot$ and equatorial radius $R = 12.6\,\mathrm{km}$.

3. Hydromagnetic Instabilities

Purely poloidal fields are known to be dynamically unstable to the kink instability from local, linear studies in Newtonian physics [19]. Such studies have been confirmed with global, linearised numerical evolutions [20] and in full non-linear studies [21]. Such configurations were also shown to be unstable in fully nonlinear relativistic calculations in Lasky *et al.* [11] – some details of which we repeat here.

The modal structure of the instability is extracted by performing a Fourier decomposition of the B_ϕ component of the magnetic field on a ring in the equatorial plane. We compute complex weighted averages

$$C_m\left(B_\phi\right) = \frac{1}{2\pi}\int_0^{2\pi} B_\phi\left(\varpi,\,\phi,\,z=0\right)\mathrm{e}^{im\phi}d\phi, \qquad (1)$$

where $\varpi = \sqrt{x^2+y^2} = \mathrm{const.}$ lies in the initial equatorial plane of the magnetic field. In figure 1 we plot the evolution of the $m = 1,\ldots,4$ modes for $C_m\left(B_\phi\right)$ for an evolution with surface magnetic field $B_{\rm surf} = 8.8\times10^{15}\,\mathrm{G}$. These curves are extracted at a radius of $\varpi = 0.6\varpi_\star$, where $\varpi_\star$ is the stellar radius.

An instability can be seen in all modes after the first couple of Alfvén crossing times. Each mode grows exponentially by approximately six orders of magnitude over many subsequent Alfvén crossing times. Saturation of the modes is seen to occur after about 75 ms, at which

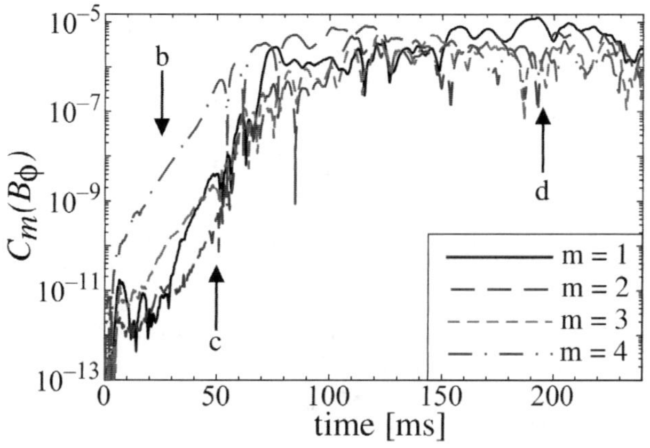

Figure 1. Evolution of $C_m(B_\phi)$ as a function of time for a model with initial surface magnetic field of $B_{\rm surf} = 8.8\times10^{15}$ G, yielding an Alfvén crossing time of 5.0 ms. The arrows represent the times of the three-dimensional snapshots plotted in figure 2. This figure was originally published in [11].

point the simulation evolves to a pseudo-equilibrium state. We have evolved such simulations to 400 ms $\equiv 80\,t_A$, where t_A is the Alfvén crossing time, with little variation in the equilibrium configuration following the first hundred or so milliseconds. The equilibrium configurations are discussed in more detail in refs. [11, 17].

In figures 2 we present three-dimensional plots of the magnetic field at various instances throughout the evolution. The instability acts near the neutral line, which is the line where $B = 0$ around the equatorial plane. In our simulations this is located at approximately two-thirds of the stellar radius. For clarity we have plotted red magnetic field lines seeded near the neutral line in the equatorial plane. Also plotted are black field lines seeded in the equatorial plane interior to the neutral line. The blue volume rendering is an isopycnic surface of $\rho = 0.37\rho_c$, where ρ_c is the central rest-mass density. This surface lies at a radius of approximately 50% of the stellar radius, which is well inside the neutral line.

Figure 2a shows the initial data imported from the LORENE spectral code. The domain of our grid is larger than that plotted here, and field lines are truncated at the surface of the star for clarity in the figure.

Figure 2b shows the evolution after 25 ms $\equiv 5\,t_A$. The onset of the "sausage" or "varicose" mode [19], involving a change in the cross-sectional area of a flux tube around the neutral line, is clearly visible. This is strongest in the $m = 4$ mode, which is a result of the transient excitation at the beginning of our simulation (for more details we refer the reader to ref. [11]). It is worth noting that, while this transient reduces with increasing grid resolution, the presence of the varicose mode is an inherent characteristic of the system.

Figure 2c shows the point at which the "kink" instability begins to visually dominate the system, which is after 50 ms $= 10\,t_A$. The kink mode acts perpendicularly to the gravitational field, in accordance with the prediction of ref. [19]. While one can still see the presence of the varicose mode in this snapshot, this point represents the non-linear development of the instability where the change in field structure is of similar order to the background field.

Figure 2d shows the simulation after 195 ms $= 39\,t_A$. This is a typical snapshot many Alfvén timescales after the non-linear saturation of the unstable modes. In some ways, this figure is consistent with the "twisted-torus" configurations seen in the non-linear evolutions of refs. [22, 23], and in the semi-analytic equilibrium calculations of refs. [24, 25]. Figure 2d shows about half of the star is well approximated by a twisted-torus, however the remainder of the star also exhibits large non-axisymmetric structures.

A way of classifying variation from the initial state is through calculating the relative energy in the toroidal and poloidal components of the magnetic field. In figure 3 we plot the evolution

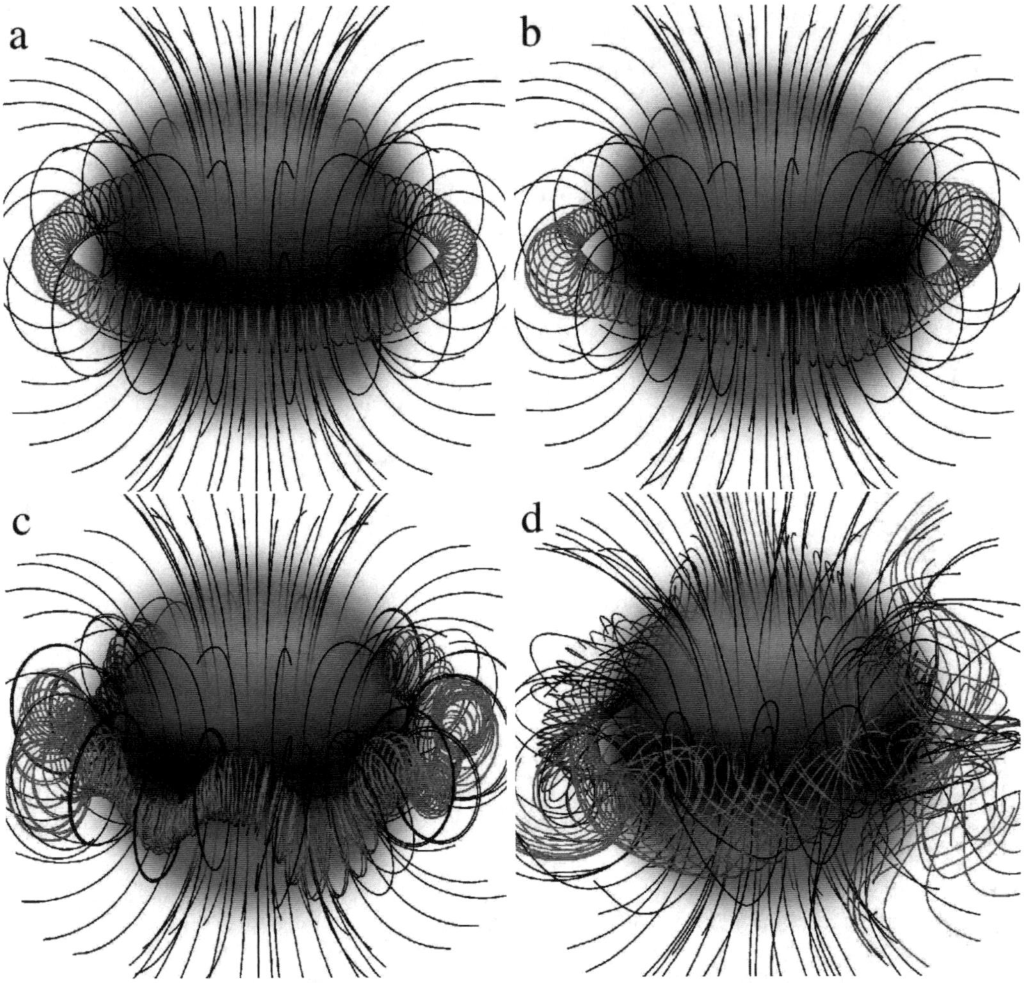

Figure 2. Time evolution of model in figure 1 with an Alfvén crossing time of 5.0 ms. The figures are (a) $t = 0$ ms, (b) $t = 25$ ms, (c) $t = 50$ ms and (d) $t = 195$ ms. The red magnetic field lines are seeded on the equatorial plane close to the neutral line to more clearly visualise the instabilities, while the black magnetic field lines are seeded on the equatorial plane interior to the neutral line. The volume rendering is an isopycnic surface at 37% of the central rest-mass density which is shown only to provide contrast with the field lines. A full movie of this simulation lasting 400 ms can be seen at www.tat.physik.uni-tuebingen.de/~tat/grmhd. These figures were originally published in [11].

of the ratio of poloidal magnetic energy, E_p, to total magnetic energy, E, calculated inside the stellar radius. Numerous models are plotted here with polar magnetic fields between $B_{15} = 6$ to 55, where $B_{15} = B_{\rm surf}/10^{15}$ G. The temporal evolution is here normalised to the Alfvén crossing time of the system, showing that this is the dominant timescale for the evolution of the instability. It is worth stressing here that it takes between twenty to thirty Alfvén crossing times for the instability in the system to saturate. It can be seen from this plot that the equilibrium configurations are generically characterised by magnetic fields with $E_p/E \sim 0.65$. The relevance of these results for neutron star physics are discussed in more detail in refs. [11, 17]

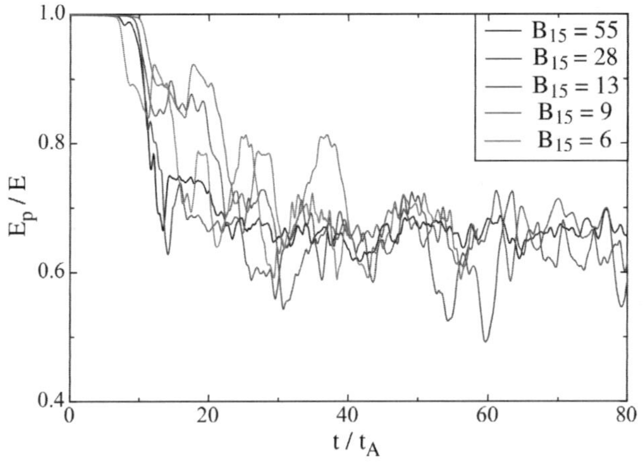

Figure 3. Evolution of the ratio of poloidal magnetic energy, E_p, to total magnetic energy, E, for models with varying magnetic field strengths, from $27\,B_{15}$ to $3.2\,B_{15}$, where $B_{15} = B_{\rm surf}/10^{15}$G. The evolutions here are normalised to the Alfveń crossing time of the system.

4. Gravitational Wave Emissions

The preceding simulations provide an invaluable tool for predicting gravitational wave emissions from magnetised neutron stars. In particular, we utilise the hydromagnetic instabilities to mimic a global magnetic field reconfiguring that may occur during, or immediately following, a magnetar flare. While the entire magnetic field may not rearrange in a realistic magnetar, rearranging the entire field gives us an *upper limit* of gravitational wave excitation in one of these events.

Our simulations exhibit negligible gravitational wave strain (i.e. $h_\times \lesssim 10^{-27}$ measured at 10 kpc) for the initial period of the simulation. During the non-linear phase of the instability, typically around $20\,t_A$ into the evolution, the gravitational wave strain grows until the instability saturates, at which point the strain signal remains roughly constant. More details of this can be seen in Zink *et al.* [9]. It is worth noting that the dominant mechanism for damping the f-mode is through gravitational wave emissions, which occurs on a timescale of order $100-200$ ms [26, 27]. As we are working in the Cowling approximation, we do not see such damping, and hence the gravitational wave amplitude remains nearly constant throughout the evolution of the system.

In figure 4 we show the gravitational wave amplitude as a function of surface magnetic field strength. The approximate values and error bars for the strains are derived from the time variation in the $h_\times$ signals (which are similar in amplitude to h_+) after the instability has saturated. We find an approximate power-law relation between h and $B_{\rm surf}$ given by

$$h \approx 1.1 \times 10^{-27} \left(\frac{10\,kpc}{d}\right) \left(\frac{B_{\rm surf}}{10^{15}\,{\rm G}}\right)^{3.3} . \tag{2}$$

It is worth noting that the 3.3 exponent is an approximate value we find from this particular set of simulations. In general, this number will vary with the equation of state and magnetic field configuration, however an exponent of approximately 3 does agree with theoretical predictions. This will be explored in detail in a future article [17].

Most of the energy in the aforementioned signal is in the f-mode. Assuming a gravitational wave damping time of approximately 100 ms [26, 27], we find a corresponding power-law relation

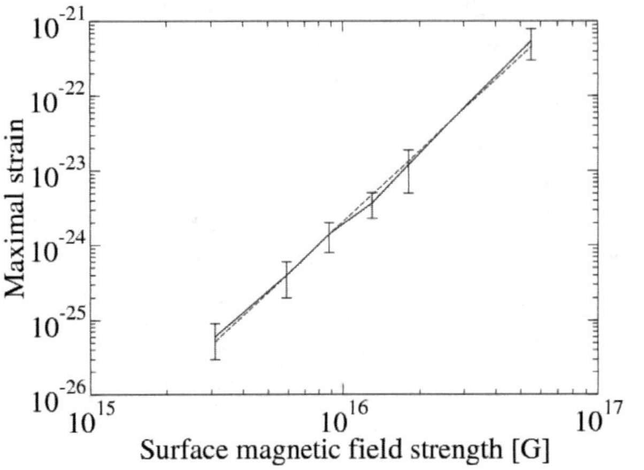

Figure 4. Gravitational wave strain as a function of surface magnetic field strength at the stellar pole. The signal amplitude is found to depend strongly on the magnetic field strength, ranging from almost 10^{-21} for our most extreme model down to less than 10^{-25} for models closer to realistic magnetar strengths. The dashed line is a power-law fit to the data – equation 2. This figure was originally published in [9].

for the energy emitted in gravitational wave radiation

$$E_{gw} \approx 1.5 \times 10^{36} \left(\frac{B_{\text{surf}}}{10^{15}\,\text{G}} \right)^{6.5} \text{erg}. \tag{3}$$

We can see from the above two relations that the gravitational wave amplitude and energies are highly non-linear functions of the surface magnetic field strength. Moreover, for typical magnetar field strengths of order 10^{15} G, one finds strains *below* 10^{-25} and energies lower than 10^{40} erg for a source at 10 kpc, even if we assume a catastrophic global restructuring of the field to be associated with a giant flare.

We provide the signal-to-noise ratios for different detector sensitivity curves in figure 5. Here we have plotted the signal amplitude, $\sqrt{T}|\tilde{h}(f)|$, where T is the damping time of the oscillation and $\tilde{h}(f)$ is the Fourier transform of $h_\times$, as a function of the frequency for various magnetic field strength models. At approximately 1.8 kHz we show the signal amplitude for the f-mode as excited by the hydromagnetic instability assuming a damping time of $50\,\text{ms} \leq T \leq 200\,\text{ms}$. In the lower part of the spectrum we plot the other maximal mode seen in the Fourier transform, assuming a damping time between $10\,\text{ms} \leq T \leq 1\,\text{s}$. Over this we plot the entire spectrum (assuming $T = 100\,\text{ms}$) for two models with $B_{\text{surf}} = 8.8 \times 10^{15}$ G and $B_{\text{surf}} = 1.8 \times 10^{16}$ G. Finally, we also plot the root of the noise power spectral density, $\sqrt{|S_h(f)|}$, as a function of frequency for the LIGO, AdvLIGO and ET detectors [28].

The amplitude signal-to-noise ratio, defined by $\sqrt{T}|\tilde{h}(f)|/\sqrt{|S_h(f)|}$, can be read off the graph as the ratio between the signal and the noise curve for the respective detector. The main conclusion from this figure is this: Assuming a giant flare is associated with a catastrophic large-scale rearrangement of the core magnetic field, *the gravitational wave signal associated with f-modes are not observable with present or near-future gravitational wave observatories* [9].

We do note the presence of lower frequency modes in the 10–100 Hz region of the spectrum. These are Alfvén modes, whereby the restoring force is provided by the magnetic field itself. The dominant damping mechanism for these modes is largely unknown, although it is generally expected that they will last significantly longer than the f-mode signal. For this reason we have plotted them in figure 5 with damping times between 10 ms and 1 s, although we note they may be excited for significantly longer than this. These modes are an exciting new prospect for gravitational wave detection [9], which we discuss in significantly more detail in a forthcoming article [17].

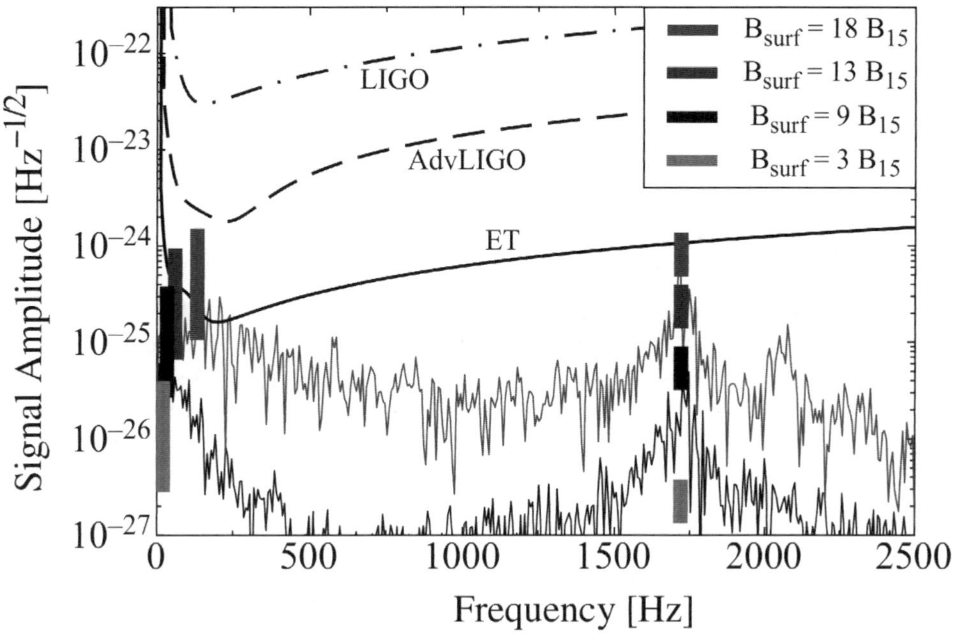

Figure 5. Signal amplitude, $\sqrt{T}|\tilde{h}(f)|$ against oscillation frequency. The coloured boxes on the right represent the maximum for the f-mode assuming a constant periodic source lasting between 50 and 200 ms. The coloured boxes on the left represent the maximum mode seen in the Fourier transform for any given frequency below the f-mode frequency assuming a constant periodic source lasting between $10\,ms$ and $1\,s$. These scale with $\sqrt{T}$, implying one can easily extrapolate to alternative values of the damping time. We have further plotted the entire spectrum (assuming a damping time of 100 ms) for both the $B_{\rm surf} = 1.8 \times 10^{16}$ G model (blue line) and $B_{\rm surf} = 8.8 \times 10^{15}$ G model (black line). This figure was originally published in [9].

5. Conclusion

We have performed three-dimensional, general relativistic magnetohydrodynamic simulations of neutron stars endowed with initially purely poloidal magnetic field geometries. This was achieved by evolving initially self-consistent solutions of the Einstein-Maxwell equations under the assumption of ideal MHD and the Cowling approximation. Purely poloidal magnetic field configurations are intrinsically unstable to the kink instability, which acted to reconfigure our internal magnetic field structure. This reconfiguring enabled us to investigate three separate phenomena: (i) the nature of the instability itself, (ii) the quasi-equilibrium configurations achieved as steady-state solutions and (iii) the gravitational wave emission from such magnetic field rearrangement. These have each been addressed in two previous papers [9, 11], and will be broached in more detail in a forthcoming article [17].

The most pertinent conclusions to be garnered from this work are two-fold: Firstly, that gravitational wave emissions associated with f-modes caused by the magnetic field rearrangement in a magnetar are not observable using current technologies. This is consistent with the conclusion of Levin & van Hoven [8] who argue that an *external* rearrangement of the magnetic field will not cause sufficient gravitational radiation for detection. The second, and significantly more speculative conclusion, is that lower frequency radiation – i.e. in the 10's to

100's of Hz range – may be detectable. The existence of modes at these frequencies is particularly intriguing given the detectors are most sensitive in this regime, and that these modes may have significantly longer damping times which would act to further increase their detectability.

Acknowledgments
We are indebted to Kostas Glampedakis for obtusely irreverent banter and occasional scientific insight. This work was supported by the SFB/Transregio 7 on "Gravitational Wave Astronomy" by the DFG. PL was supported by the Alexander von Humboldt Foundation. Computations were performed on the Multi-modal Australian ScienceS Imaging and Visualisation Environement MASSIVE (www.massive.org.au).

References

[1] Baggio L, Bignotto M, Bonaldi M, Cerdonio M, Conti L, de Rosa M, Falferi P, Fortini P, Inguscio M, Liguori N, Marin F, Mezzena R, Mion A, Ortolan A, Prodi G A, Poggi S, Salemi F, Soranzo G, Taffarello L, Vedovato G, Vinante A, Vitale S and Zendri J P 2005 *Phys. Rev. Lett.* **95** 081103
[2] Abbott B and et al 2007 *Phys. Rev. D* **76** 062003
[3] Abbott B and et al 2008 *Phys. Rev. Lett.* **101** 211102
[4] Modestino G and Pizzella G 2011 *Phys. Rev. D* **83** 062004
[5] Abadie J and *et al* 2011 *Astrophys. J.* **734** L35
[6] Ioka K 2001 *Mon. Not. R. Astron. Soc.* **327** 639
[7] Corsi A and Owen B J 2011 *Phys. Rev. D* **83** 104014
[8] Levin Y and van Hoven M 2011 *Mon. Not. R. Astron. Soc.* **418** 659
[9] Zink B, Lasky P D and Kokkotas K D 2012 *Phys. Rev. D* **85** 024030 arXiv:1107.1689
[10] Ciolfi R, Lander S K, Manca G M and Rezzolla L 2011 *Astrophys. J.* **736** L6 arXiv:1105.3971
[11] Lasky P D, Zink B, Kokkotas K D and Glampedakis K 2011 *Astrophys. J.* **735** L20
[12] Zink B, Schnetter E and Tiglio M 2008 *Phys. Rev. D* **77** 103015
[13] Korobkin O, Abdikamalov E B, Schnetter E, Stergioulas N and Zink B 2011 *Phys. Rev. D* **83** 043007 arXiv:1011.3010
[14] Zink B 2011 HORIZON: Accelerated General Relativistic Magnetohydrodynamics arXiv:1102.5202
[15] Antón L, Zanotti O, Miralles J A, Marti J M, Ibánez J M, Font J A and Pons J A 2006 *Astrophys. J.* **637** 296
[16] Anderson M, Hirschmann E W, Liebling S L and Neilsen D 2006 *Class. Quantum Grav.* **23** 6503
[17] Lasky P D, Zink B and Kokkotas K D 2012 in Preparation
[18] Bocquet M, Bonazzola S, Gourgoulhon E and Novak J 1995 *A&A* **301** 757
[19] Markey P and Tayler R J 1973 *Mon. Not. R. Astron. Soc.* **163** 77
[20] Lander S K and Jones D I 2011 *Mon. Not. R. Astron. Soc.* **412** 1730
[21] Braithwaite J 2007 *A&A* **469** 275
[22] Braithwaite J and Nordlund A 2006 *A&A* **450** 1077
[23] Braithwaite J 2009 *Mon. Not. R. Astron. Soc.* **397** 763
[24] Yoshida S, Yoshida S and Eriguchi Y 2006 *Astrophys. J.* **651** 462
[25] Ciolfi R, Ferrari V, Gualtieri L and Pons J A 2009 *Mon. Not. R. Astron. Soc.* **397** 913
[26] Lindblom L and Detweiler S L 1983 *Astrophys. J. S.* **53** 73
[27] Andersson N and Kokkotas K D 1998 *Mon. Not. R. Astron. Soc.* **299** 1059
[28] Sathyaprakash B S and Schutz B F 2009 *Living Rev. Relativity* **12** 2

Approximate Waveforms for Extreme-Mass-Ratio Inspirals: The Chimera Scheme

Carlos F. Sopuerta[1] **and Nicolás Yunes**[2,3,4]

[1]Institut de Ciències de l'Espai (CSIC-IEEC), Facultat de Ciències, Campus UAB, Torre C5 parells, Bellaterra, 08193 Barcelona, Spain.
[2]Department of Physics, Montana State University, Bozeman, MT 59717, USA.
[3]Department of Physics and MIT Kavli Institute, 77 Massachusetts Avenue, Cambridge, MA 02139, USA.
[4]Princeton University, Physics Department, Princeton, NJ 08544, USA.

E-mail: sopuerta@ieec.uab.es,nyunes@physics.montana.edu

Abstract. We describe a new kludge scheme to model the dynamics of generic extreme-mass-ratio inspirals (EMRIs; stellar compact objects spiraling into a spinning supermassive black hole) and their gravitational-wave emission. The *Chimera* scheme is a hybrid method that combines tools from different approximation techniques in General Relativity: (i) A multipolar, post-Minkowskian expansion for the far-zone metric perturbation (the gravitational waveforms) and for the local prescription of the self-force; (ii) a post-Newtonian expansion for the computation of the multipole moments in terms of the trajectories; and (iii) a BH perturbation theory expansion when treating the trajectories as a sequence of self-adjusting Kerr geodesics. The EMRI trajectory is made out of Kerr geodesic fragments joined via the method of osculating elements as dictated by the multipolar post-Minkowskian radiation-reaction prescription. We implemented the proper coordinate mapping between Boyer-Lindquist coordinates, associated with the Kerr geodesics, and harmonic coordinates, associated with the multipolar post-Minkowskian decomposition. The Chimera scheme is thus a combination of approximations that can be used to model generic inspirals of systems with extreme to intermediate mass ratios, and hence, it can provide valuable information for future space-based gravitational-wave observatories, like LISA, and even for advanced ground detectors. The local character in time of our multipolar post-Minkowskian self-force makes this scheme amenable to study the possible appearance of transient resonances in generic inspirals.

1. Introduction

The modeling of gravitational wave (GW) sources is key for the success of GW astronomy. Detectors require precise theoretical waveforms to extract and characterize signals buried in the noise. One of the main GW sources for space-based observatories, like the *Laser Interferometer Space Antenna* (LISA) [1, 2], that requires accurate templates for detection and analysis are extreme-mass-ratio inspirals (EMRIs) [3]. These events consist of a small compact object (SCO), such as a stellar-mass black hole (BH) or a neutron star, spiraling in a generic orbit into a spinning, (super)massive black hole (MBH), in the regime where GW emission is relevant. In such inspirals, the SCO spends up to millions of cycles in close orbits around the MBH, possibly with large pericenter velocities and eccentricities, sampling the strong gravitational field of the MBH.

The relevant EMRI phase for data analysis is the inspiral, during which the number of cycles accumulated scales with the inverse of the mass ratio $q = m_\star/M_\bullet$ [$\mathcal{O}(10^{-4})$-$\mathcal{O}(10^{-6})$]. During EMRI plunge and merger, the accumulated number of cycles scales with the MBH mass ($M_\bullet$) only. Moreover, the signal-to-noise ratio (SNR) is in the range $\mathcal{O}(10)$-$\mathcal{O}(10^2)$ for EMRIs at realistic distances, and it scales linearly with the total number of cycles. Therefore, the contribution of the merger/plunge phase to the SNR is a fraction $\mathcal{O}(m_\star/M_\bullet)$ relative to the total inspiral contribution. In addition, the EMRI ringdown phase is not detectable, as the SCO barely perturbs the MBH geometry as it plunges and crosses the MBH's event horizon.

As a consequence of the large number of GW cycles contained in EMRIs, they can provide a detailed map of the MBH geometry, allowing us to extract the EMRI physical parameters with high precision [4]. This information is crucial to test the Kerr hypothesis [5, 6, 7, 8, 9, 10] and constrain modified gravity theories (see, e.g. [11, 12, 10, 13]). EMRI observations will also have an important impact on astrophysics (for a review see [3]).

The extreme mass ratios involved in the problem also lead to the appearance of two different spatial and temporal scales. The two different spatial scales are due to the different MBH and SCO sizes, $m_\star/M_\bullet \ll 1$. The two different temporal scales are the orbital one(s) and the one associated with *radiation reaction*, $T_{\rm Orbital}/T_{\rm RR} \sim m_\star/M_\bullet \ll 1$. Due to this, a fully relativistic numerical evolution of EMRIs is currently unfeasible.

A framework to model EMRIs that exploits the extreme mass ratios involved is BH perturbation theory, where one treats the SCO as a small perturbation of the MBH background geometry. In this context, the inspiral is modeled as due to a local *self-force*. This force is composed of the metric perturbations generated by the SCO, regularized by eliminating divergences due to the SCO's point particle description. The SCO's motion is then governed by the MiSaTaQuWa equation of motion, derived in [14, 15]. This equation is the foundation of a self-consistent scheme to describe EMRIs, coupling this equation of motion to the partial differential equations that describe the perturbations produced by the SCO.

At present, the gravitational self-force has been computed for the case of a non-rotating MBH using time-domain techniques [16, 17]. Progress is currently being made toward the calculation of more astrophysically relevant (spinning) EMRIs [18] (see also [19, 20] for progress in the computations of the scalar self-force in Kerr). Given the number of cycles contained produced by an EMRI in one year and the present complexity of self-force calculations, we cannot expect to generate complete waveform template banks by means of full self-force calculations. Instead, it has been proposed that the goal of these studies should be to understand all the details of the structure of the self-force, so that one can formulate efficient and precise algorithms (phenomenological models) to create the waveforms needed for LISA data analysis, perhaps complementing existent approximation schemes.

In this paper, we summarize the results of [21] where a new *approximate* scheme was proposed to model EMRIs: the Chimera framework. This scheme can be thought of as a new *kludge* model that combines different approximation methods, including BH perturbation theory, the multipolar post-Minkowskian formalism of Blanchet and Damour [22] and post-Newtonian theory. The Chimera scheme could be used for qualitative descoping studies, i.e. to determine the accuracy to which parameters could be extracted given an eLISA EMRI detection as a function of detector properties. The Chimera scheme could also be used to study how certain resonances, found by Flanagan and Hinderer [23], affect such parameter estimation and detection. This would need the inclusion of conservative and second-order dissipative pieces in the self-force. Chimera waveforms, however, would require improvements before they can be used as realistic templates in data analysis. The Chimera scheme is indeed amenable to such improvements, which will be the focus of future work. Other works in which EMRI orbits are evolved through a self-force can be found in [24, 25].

2. Ingredients of the Chimera Scheme

2.1. MBH Geometry and Geodesic Motion

The geometry of the MBH is described by the Kerr metric. In Boyer-Lindquist (BL) coordinates, $(x^{\mu}_{\rm BL}) = (t, r, \theta, \phi)$, the line element is given by

$$ds^2 = -dt^2 + \frac{\rho^2}{\Delta}dr^2 + \rho^2 d\theta^2 + (r^2+a^2)\sin^2\theta\, d\phi^2 + \frac{2M_\bullet r}{\rho^2}(dt - a\sin^2\theta\, d\phi)^2\,, \tag{1}$$

where $\rho^2 = r^2 + a^2\cos^2\theta$ and $\Delta = r^2 - 2M_\bullet r + a^2 = r^2 f + a^2$, with $f = 1 - 2M_\bullet/r$. The function Δ has two roots: $r_\pm \equiv M_\bullet \pm \sqrt{M_\bullet^2 - a^2}$; the positive one r_+ ($\geq r_-$) coincides with the location of the event horizon. The Kerr geometry is a stationary, axisymmetric and vacuum solution to the Einstein equations, which has has an additional symmetry described by a Killing tensor.

In the limit where the SCO has zero mass ($m_* \to 0$), it follows timelike geodesics of the MBH geometry which, in terms of proper time, can be parameterized as $z^\mu(\tau) = (t(\tau), r(\tau), \theta(\tau), \phi(\tau))$. Given the symmetries of the Kerr metric, each of these geodesics can be fully characterized by three constants of motion (per SCO unit mass): The energy (stationarity), E; the angular momentum in the MBH spin direction (axisymmetry), L_z, and the *Carter constant* (Killing tensor symmetry), Q (or $C \equiv Q - (L_z - aE)^2$). Using these symmetries we can completely separate the geodesic equations so that they can be written in the form:

$$\rho^2\frac{dt}{d\tau} = \frac{1}{\Delta}\left(\Sigma^2 E - 2M_\bullet a r L_z\right)\,, \tag{2}$$

$$\rho^4\left(\frac{dr}{d\tau}\right)^2 = \left[\left(r^2+a^2\right)E - aL_z\right]^2 - \left(Q + r^2\right)\Delta\,, \tag{3}$$

$$\rho^4\left(\frac{d\theta}{d\tau}\right)^2 = C - \cot^2\theta L_z^2 - a^2\cos^2\theta\left(1 - E^2\right)\,, \tag{4}$$

$$\rho^2\frac{d\phi}{d\tau} = \frac{1}{\Delta}\left[2M_\bullet\, a\, rE + \frac{L_z}{\sin^2\theta}\left(\Delta - a^2\sin^2\theta\right)\right]\,, \tag{5}$$

where $\Sigma^2 \equiv (r^2+a^2)^2 - a^2\Delta\,\sin^2\theta$. For the purposes of EMRI modeling we restrict ourselves to bound geodesics (orbits), which can also be characterized in terms of orbital elements: the eccentricity e, the semilatus rectum p, and the inclination $\theta_{\rm inc}$. The first two are defined in terms of the turning points (extrema) of the radial motion, the pericenter and apocenter ($r_{\rm peri}$ and $r_{\rm apo}$), by the Newtonian relations: $r_{\rm peri} = pM_\bullet/(1+e)$ and $r_{\rm apo} = pM_\bullet/(1-e)$. Likewise, the inclination angle $\theta_{\rm inc}$ is associated with the turning point of the polar motion, i.e. the minimum of θ ($\theta_{\rm min} \in [0, \pi/2]$), via the relation: $\theta_{\rm inc} = \mathrm{sign}(L_z)\,(\pi/2 - \theta_{\rm min})$. Another useful definition of the orbital inclination angle is: $\cos\iota = L_z/\sqrt{L_z^2 + C}$.

Alternatively, the orbits can also be characterized in terms of three *fundamental* frequencies (see e.g. [26, 27, 28]) with respect to the BL time t (however, given a set of fundamental frequencies, there is no necessarily a unique orbit associated with it [29]): Ω_r, associated with the radial motion (from periapsis to apoapsis and back); Ω_θ, associated with polar motion; and Ω_ϕ, associated with azimuthal motion. Precessional orbital effects are due to mismatches between these frequencies, and they can be used to decompose, among other things, the GWs in a Fourier expansion [27].

2.2. Chimera Osculating Trajectories

The foundations of the gravitational backreaction of a moving massive body on its own trajectory were laid down in the papers by Mino, Tanaka, and Sasaki [14] and Quinn and Wald [15]. These works established that the trajectory of a point mass can be understood as geodesic motion in the background geometry modified by a local force, the *self-force*, given in terms of the gradients of the *regularized*[1] first-order metric perturbations. Then, the equation of motion for the SCO, the MiSaTaQuWa equation, is

$$\frac{d^2 z^\alpha}{d\tau^2} + \Gamma^\alpha_{\mu\nu} u^\mu u^\nu = m_\star^{-1} F^\alpha_{\rm SF}\,, \tag{6}$$

where the self-force, $F^\alpha_{\rm SF}$, is given by

$$F^\alpha_{\rm SF} = -\frac{1}{2} m_\star \left(\mathrm{g}^{\alpha\lambda} + u^\alpha u^\lambda \right) u^\mu u^\nu \left(2\nabla_\mu h^{\rm R}_{\nu\lambda} - \nabla_\lambda h^{\rm R}_{\mu\nu} \right)\,, \tag{7}$$

and $h^{\rm R}_{\mu\nu}$ denotes the regularized metric perturbations. This equation can be also reinterpreted as a geodesic equation for a point particle in the perturbed geometry $\mathrm{g}_{\alpha\beta} + h^{\rm R}_{\alpha\beta}$ [31]. In order to evolve the SCO's trajectory subject to the self-force in a self-consistent way, we use the method proposed in [32], which is a relativistic extension of the *osculating orbit method.* The main idea is to model the SCO's worldline as a sequence of geodesics tangent to the worldline at any given point. The transition is smooth since EMRI trajectories remain close to a geodesic for a significant amount of time. In principle, to carry out this transition we need to take into account all the parameters that characterize the *instantaneous* geodesic (the *principal* orbital elements) and the position of the SCO within that geodesic (the *positional* orbital elements). In the initial version of the Chimera scheme [21] we only consider the dissipative effects of the self-force, i.e. those that only affect to principal orbital elements. Then, the evolution of $(E, L_z, C/Q)$ is given by

$$\frac{dE}{d\tau} = -\zeta^{(t)}_\alpha a^\alpha_{\rm SF}\,, \qquad \frac{dL_z}{d\tau} = \zeta^{(\phi)}_\alpha a^\alpha_{\rm SF}\,, \tag{8}$$

$$\frac{dQ}{d\tau} = 2\,\xi_{\alpha\beta} u^\alpha a^\beta_{\rm SF}\,, \qquad \frac{dC}{d\tau} = \frac{dQ}{d\tau} + 2\,(aE - L_z)\left(\frac{dL_z}{d\tau} - a\frac{dE}{d\tau}\right)\,, \tag{9}$$

where the SCO self-acceleration $a^\alpha_{\rm SF}$ is related to the self-force by $a^\alpha_{\rm SF} = m_\star^{-1} F^\alpha_{\rm SF}$, and where $\zeta^{(t)}_\alpha$ and $\zeta^{(\phi)}_\alpha$ are the timelike and axial Killing vectors respectively, and $\xi_{\alpha\beta}$ is the Killing tensor.

2.3. Multipolar Post-Minkowskian Self-Acceleration

Since the self-force is not known for generic Kerr orbits, we approximate it here via a multipolar, post-Minkowskian expansion (see e.g. [22, 33, 34]). We begin by recasting the post-Minkowskian metric perturbation in the far field as the Kerr metric of the MBH $\mathrm{g}^{\rm K,H}_{\mu\nu}$ in harmonic coordinates (we shall discuss the issue of coordinates in next section), plus perturbations induced by the SCO. The latter can be decomposed into conservative (*time-symmetric*) and dissipative (*time-asymmetric*) perturbations $h^{\rm RR}_{\mu\nu}$. Neglecting the former, we then have

$$\mathrm{g}^{\rm PM}_{\mu\nu} = \mathrm{g}^{\rm K,H}_{\mu\nu} + h^{\rm RR}_{\mu\nu}\,, \tag{10}$$

The metric perturbations $h^{\rm RR}_{\mu\nu}$ are the quantities we use to describe the *radiation-reaction* effects, and hence they are our approximation to the regularized metric perturbations $h^{\rm R}_{\alpha\beta}$ of Eqs. (6) and (7). The rate of change of the principal orbital elements in then described by Eq. (8), with the force of Eq. (7) and the regularized metric perturbation of Eq. (10).

The regular metric perturbation is a function of certain time-asymmetric scalar, $V_{\rm RR}$, and vector, $V^i_{\rm RR}$, radiation-reaction potentials, which to lowest order can be written as

$$h^{\rm RR}_{tt} = 2\,V_{\rm RR}\,, \qquad h^{\rm RR}_{ti} = -4\,V^i_{\rm RR}\,, \qquad h^{\rm RR}_{ij} = 2\,V_{\rm RR}\,\delta_{ij}\,, \tag{11}$$

[1] For a point mass, the first-order metric perturbation is singular at the particle location, and a regularization prescription [30] must be employed to separate out the piece responsible of the gravitational backreaction.

and the expressions for these potentials themselves are [22, 35]

$$V_{\rm RR}(t_{\rm H},\boldsymbol{x}_{\rm H}) = -\frac{1}{5}x^{ij}_{\rm H}M^{(5)}_{ij}(t_{\rm H}) + \frac{1}{189}x^{ijk}_{\rm H}M^{(7)}_{ijk}(t_{\rm H}) - \frac{1}{70}\boldsymbol{x}^2_{\rm H}x^{ij}_{\rm H}M^{(7)}_{ij}(t_{\rm H}) + \mathcal{O}(x^{\rm H}_{ij}M^{(9)}_{ij}), \quad (12)$$

$$V^i_{\rm RR}(t_{\rm H},\boldsymbol{x}_{\rm H}) = \frac{1}{21}\hat{x}^{<ijk>}_{\rm H}M^{(6)}_{jk}(t_{\rm H}) - \frac{4}{45}\epsilon_{ijk}x^{jl}_{\rm H}S^{(5)}_{kl}(t_{\rm H}) + \mathcal{O}(x^{ijk}_{\rm H}M^{(8)}_{ij}), \quad (13)$$

where $(x^\alpha_{\rm H}) = (t_{\rm H}, x^i_{\rm H})$ are spacetime harmonic coordinates, ϵ_{ijk} is the Levi-Civita symbol, and

$$\hat{x}^{<ijk>}_{\rm H} \equiv x^{ijk}_{\rm H} - \frac{3}{5}\boldsymbol{x}^2_{\rm H}\delta^{(ij}x^{k)}_{\rm H}\,. \quad (14)$$

is the symmetric trace-free (STF) projection of the multi-index quantity $x^{ijk} = x^i x^j x^k$. The quantities $M^{(n)}_{ij}$, $M^{(n)}_{ijk}$, and $S^{(n)}_{ij}$ are the nth-time-derivative of the STF mass quadrupole, mass octopole and current quadrupole multipole moments of the source. The first term in $V_{\rm RR}$ [Eq. (12)] corresponds to the well-known Burke-Thorne radiation reaction potential [36].

The Chimera local self-force is completely specified once we prescribe how the multipole moments depend on the orbital trajectories. This can be achieved by asymptotically matching a PN and a post-Minkowskian solution in some buffer zone where both approximations are valid. The current Chimera implementation employs the leading-order expressions for these multipoles. As such, we are not consistently keeping a given PN order, but instead using leading-order expressions for the multipole moments, while keeping several such moments in the expansions. To lowest order, the mass moments are given by

$$M_{ij} = \eta\, m\, z_{<ij>}\,, \qquad M_{ijk} = \eta\, \delta m\, z_{<ijk>} \quad (15)$$

and the current quadrupole moment is

$$S_{ij} = \eta\, \delta m\, \epsilon_{kl<i}z_{j>}{}^k \dot{z}^l\,, \quad (16)$$

where $\eta = m_\star M_\bullet/(m_\star + M_\bullet)^2$ is the symmetric mass ratio and $\delta m = M_\bullet - m_\star$ is the mass difference. To higher order, these moments become more complicated as there are now non-linear contributions from the non-reactive potentials (i.e. non-linear contributions from the background) as well as tail and memory contributions.

2.4. From Boyer-Lindquist to Harmonic Coordinates

From the previous discussion it is clear that the MBH metric must be written in harmonic coordinates in order for all terms to be in the same coordinate system and to be consistent with other ingredients of the Chimera scheme, in particular the waveform generation procedure that we describe in next section. Nevertheless, there are parts of the scheme that are easier to implement in BL coordinates, such as the integration of the geodesic equations. Therefore, it is crucial to find a map from BL to harmonic coordinates.

Harmonic coordinates refer to coordinate systems, $\{x^\alpha_{\rm H}\}$, that satisfy $\Box x^\alpha_{\rm H} = 0$, where $\Box$ is the D'Alembertian operator: $\Box \equiv {\rm g}^{\alpha\beta}\nabla_\alpha\nabla_\beta$. In [37], based on work by Ding (see Ref. [37] for references on this), a transformation to harmonic coordinates was found. Following [37], we can map the Kerr metric from BL coordinates (t, r, θ, ϕ) to harmonic ones $(t_{\rm H}, x_{\rm H}, y_{\rm H}, z_{\rm H})$, and the converse, via the coordinate transformations

$$
\begin{aligned}
t_{\rm H} &= t\,, & t &= t_{\rm H}\,,\\
x_{\rm H} &= \sqrt{(r-M_\bullet)^2 + a^2}\,\sin\theta\cos[\phi - \Phi(r)]\,, & r &= M_\bullet + \left[\frac{r^2_{\rm H}-a^2}{2} + \sqrt{\left(\frac{r^2_{\rm H}-a^2}{2}\right)^2 + a^2 z^2_{\rm H}}\right]^{1/2},\\
y_{\rm H} &= \sqrt{(r-M_\bullet)^2 + a^2}\,\sin\theta\sin[\phi - \Phi(r)]\,, & \theta &= \arccos\left(\frac{z_{\rm H}}{r-M_\bullet}\right),\\
z_{\rm H} &= (r - M_\bullet)\,\cos\theta\,, & \phi &= \Phi(r) + \arctan\left(\frac{y_{\rm H}}{x_{\rm H}}\right),
\end{aligned}
\quad (17)
$$

where $r_H \equiv (x_H^2 + y_H^2 + z_H^2)^{1/2}$ and the angle function $\Phi(r)$ is given by

$$\Phi(r) = \frac{\pi}{2} - \arctan\left\{\frac{\frac{r-M_\bullet}{a} + \Omega(r)}{1 - \frac{r-M_\bullet}{a}\,\Omega(r)}\right\}, \quad \text{with } \Omega(r) = \tan\left[\frac{a}{2\sqrt{M_\bullet^2 - a^2}}\ln\left(\frac{r-r_-}{r-r_+}\right)\right]. \quad (18)$$

2.5. Waveform Generation

Given the orbital evolution in harmonic coordinates one can construct the gravitational waveforms through multipolar, post-Minkowskian expressions in terms of a sum over multipole moments (see e.g. [38] for a review). This prescription is fully specified once expressions for the radiative moments are given in terms of derivatives of the orbital trajectory. Such identification, however, is difficult, as these moments are defined in the *far-zone* and have no knowledge of the *source* multipole moments, defined in the *near zone*. In this initial version of the Chimera scheme, we only consider the leading-order contributions to the multipoles moments, i.e. we identify the source and radiative moments. Then, in a transverse-traceless gauge and keeping multipole moments that include both the mass hexadecapole and the current octopole (thus keeping contributions one order higher than traditional kludge waveforms), we have:

$$h_{ij}^{\mathrm{TT}} = \frac{2}{r}\ddot{M}_{ij}^{\mathrm{STF}} + \frac{2}{3r}\left[\dddot{M}_{ijk}n^k + 4\epsilon^{kl}{}_{(i}\ddot{S}_{j)k}n_l\right]^{\mathrm{STF}} + \frac{1}{6r}\left[\ddddot{M}_{ijkl}n^k n^l + 6\epsilon^{kl}{}_{(i}\dddot{S}_{j)km}n^l n^m\right]^{\mathrm{STF}}. \quad (19)$$

The expressions for these multipole moments has been given in Eqs. (15) and (16), except for the mass hexadecapole and current octopole multipoles, which are given respectively by

$$M_{ijkl} = \eta\, m\, z_{<ijkl>}, \qquad S_{ijk} = \eta\, m\, \epsilon_{lm<i}z_{jk>}{}^{l}\,\dot{z}^m. \quad (20)$$

The plus and cross-polarized projections can then be constructed via

$$h_{+,\times} = e^{ij}_{+,\times}h_{ij}^{\mathrm{TT}}, \quad (21)$$

where $e^{ij}_{+,\times}$ are the plus- and cross-polarization tensors. Finally, the observables, i.e. the GW response function, are given by a projection of the plus- and cross-polarized waveform with the beam pattern functions of the detector. For a detector like LISA, there are two such functions, $F_{+,I/II}$ and $F_{\times,I/II}$ (see, e.g. [39, 4]) and the response is thus

$$h \equiv \frac{\sqrt{3}}{2\,D_L}\sum_{A=I,II}\left(F_{+,A}\,h_+ + F_{\times,A}h_\times\right), \quad (22)$$

where D_L is the luminosity distance from the source to the observer and the prefactor of $\sqrt{3}/2$ is due to the triangular arrangement of the LISA spacecraft constellation.

3. Numerical Implementation and some Numerical Results

One of the main ingredients of the scheme is the integration of the geodesic equations (2)-(5). The resulting equations for the radial and polar BL coordinates, Eqs. (3) and (4) respectively, have turning points (at pericenter and apocenter in the case of the radial coordinate, and at the location of the orbital inclination angle in the case of the polar coordinate). At these points we have either $\dot{r} = 0$ or $\dot{\theta} = 0$ and the ODE solvers may encounter convergence problems. To avoid these problems, we use angle variables associated with radial and polar BL coordinates via

$$r = \frac{pM_\bullet}{1 + e\cos\psi}, \qquad \cos^2\theta = \cos^2\theta_{\min}\cos^2\chi. \quad (23)$$

For convenience, we parametrize the trajectory in terms of the BL coordinate time t, which is also a time harmonic coordinate, instead of the proper time τ. Then, the direct result of the ODE integration is the sequence $(t, \psi(t), \chi(t), \phi(t))$ for $t \in [t_{\rm ini}, t_{\rm end}]$.

The main challenge in the numerical implementation of the Chimera scheme is the evaluation of the time derivatives of the different quantities involved. To understand this problem let us focus on the computation of our multipolar, post-Minkowskian self-force. The radiation reaction potentials $V_{\rm RR}$ and $V^i_{\rm RR}$ in (12) and (13) depend on up to the seventh time derivative of the mass quadrupole and octopole moments and up to the fifth time derivative of the current quadrupole moment. But since one also needs certain time derivatives of these potentials [see Eq. (7)] one must compute up to the eighth time derivative of the mass quadrupole and octopole moments and the sixth time derivative of the current quadrupole moment.

In principle, the computation of these derivatives can be done analytically by using repeatedly the equations of motion, Eqs. (2)-(5). The problem is that we need the time derivatives of the trajectory in harmonic coordinates, and to pass from the ODE angles $(\psi(t), \chi(t), \phi(t))$ to harmonic coordinates we first need to use Eq. (23) to go from these angles to BL coordinates, and then Eqs. (17) to go from BL to harmonic coordinates. Therefore, we would need to obtain analytically higher time-derivatives of the ODE angles (up to eighth order), which involves using Christoffel symbols of the Kerr metric and several of their derivatives, and also to differentiate several times Eqs. (23) and (17). In practice, this makes the analytical computations unfeasible, even using modern computer algebra systems.

Therefore, one needs to resort to numerical evaluations of these derivatives. First, notice that the trajectory, the velocity, and the acceleration can be computed with high accuracy directly from the integration of the ODEs in Eqs. (2)-(5). Then, using analytical expressions, we can directly obtain the second time derivatives of the mass quadrupole and octopole moments and the first time derivative of the current quadrupole moment. Thus, we just need to compute numerically up to six additional time derivative of the mass moments (i.e. starting from their second time derivative) and up to five additional time derivative of the current quadrupole moment (i.e. starting from its first time derivative).

Computing numerical derivatives is a subtle task, which we tried to implement in different ways. For instance, we experimented with many finite difference rules (involving from a few to more than 20 evaluation points), as well as other generic numerical differentiation techniques (such as numerical interpolation or Chebyshev differentiation). We found that, in order to obtain decent performance, a substantial amount of fine-tuning was necessary. To overcome this difficulty, we looked for a method better adapted to our problem. Notice that multipole moments are functionals of the trajectories, which are piecewise Kerr bounded geodesic, and bounded geodesics can be characterized by three fundamental frequencies (in the generic case). Then, an arbitrary functional of Kerr orbits, say $f[\psi, \chi, \phi](t)$, can be expanded in a multiple Fourier series of these frequencies as [27]

$$f[\psi, \chi, \phi](t) = \sum_{k,m,n} f_{k,m,n} e^{-i\,(k\Omega_r + m\Omega_\theta + n\Omega_\phi)\,t}\,, \qquad (24)$$

where (k, m, n) are integers running from $-\infty$ to $+\infty$ and $f_{k,m,n}$ are complex coefficients such that $f_{-k,-m,-n} = \bar{f}_{k,m,n}$. There are three special cases in which this expansion is simplified: (i) Circular equatorial orbits; (ii) Equatorial non-circular orbits; (iii) Circular non-equatorial orbits. In case (i), the Fourier series contains only a single frequency, the azimuthal one Ω_ϕ. In case (ii) and (iii), there are two independent frequencies, (Ω_r, Ω_ϕ) and $(\Omega_\theta, \Omega_\phi)$ respectively.

By modeling the required multipole moments as an expansion of the form of Eq. (24), using a standard least-square fitting algorithm, time derivatives are then computed via

$$f^{(N)}[\psi,\chi,\phi](t) = \sum_{k,m,n} f^N_{k,m,n} e^{-i\,(k\Omega_r + m\Omega_\theta + n\Omega_\phi)\,t}\,, \tag{25}$$

where $f^N_{k,m,n} = (-i)^N (k\Omega_r + m\Omega_\theta + n\Omega_\phi)^N f_{k,m,n}$. We tested this technique with different types of functions and we have found that it is very robust and provides very high accuracy even for the highest derivatives. For instance, the sixth time derivative is typically accurate to one part in 10^5, which is more than enough for our purposes.

In order to illustrate the strengths of the Chimera scheme, we present results for a generic, eccentric and inclined orbit. The evolution of all the constants of motion/orbital elements for many orbital periods is shown in Figure. 1. While the large scale evolution looks very smooth, the shorter scale evolution in the subplots shows patterns with periods that match the orbital ones. These patterns are a consequence of using a local in time self-force and do not appear in evolutionary schemes based on flux averaging over a certain number of orbital periods. We can also appreciate from the global evolution that all quantities decay in time except for the inclination, which grows. However, if we look at the evolution over a few orbital periods, e.g. for the eccentricity, we can see that it can grow locally in time although the global tendency is to decay. The Chimera scheme leads to rich evolutionary patterns due to its local-in-time character, which also makes it a very valuable tool to investigate questions like transient resonances [23].

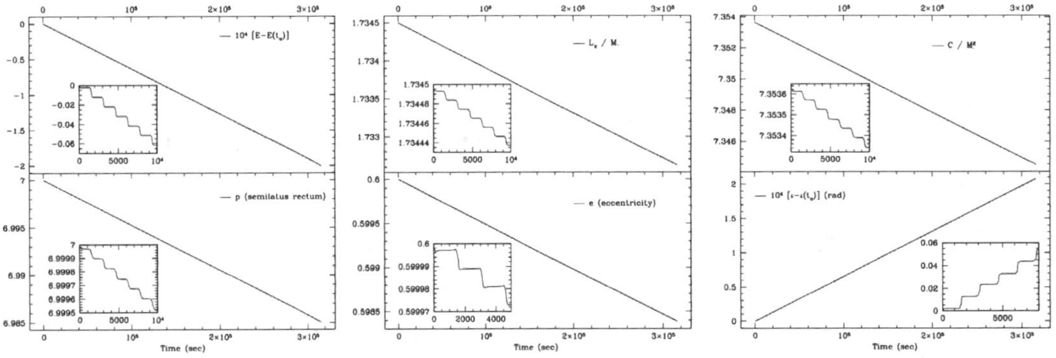

Figure 1. Evolution (for a total time of 10^{-2} yrs) of an eccentric and inclined inspiral of a system characterized by: $M_\bullet = 10^6 M_\odot$, $a/M_\bullet = 0.98$, and $q = 10^{-5}$. The plots show the evolution of the following quantities: Energy E (top left), actually $10^4[E - E(t_o)]$, where $E(t_o) = 0.9575513$; angular momentum along the spin axis, L_z (top centre); Carter constant C (top right); semi-latus rectum, p (bottom left), with $p(t_o) = 7$; eccentricity, e (bottom centre), with $e(t_o) = 0.6$; and inclination angle ι (bottom right), actually $10^4[\iota - \iota(t_o)]$, where $\iota(t_o) = 57.39$ deg. All plots contain subplots where the detailed evolution during a few orbital periods is shown.

4. Conclusions and Discussion

We have presented the main ingredients of the Chimera scheme, a framework initially designed for the modeling of EMRI dynamics and GW emission, but in principle also adaptable to intermediate-mass ratio systems. This scheme combines ingredients from the multipolar, post-Minkowskian formalism and black hole perturbation theory to evolve a non-geodesic wordline (with respect to the MBH geometry) and to construct waveforms. The orbits are built as a sequence of local geodesics, whose orbital elements evolve according to a local self-force that we approximate via a multipolar, post-Minkowskian expansion. The leading-order term of the self-force corresponds to the well-known Burke-Thorne radiation-reaction potential. A crucial ingredient in this construction is the mapping from BL coordinates, which we use to locally integrate the orbits, to harmonic coordinates, required both by the multipolar, post-Minkowskian self-force and by the GW multipolar expansion. Once we have trajectories in harmonic coordinates, it is straightforward to build waveforms.

The Chimera scheme can be improved in different ways. One possibility is to use more accurate, beyond leading-order, expressions for the different multipole moments. Although these PN corrections are in general not known for generic spinning binaries, they should certainly improve the accuracy of the Chimera evolutions for certain orbits. Another way to improve the scheme would be to introduce conservative corrections to the background, for instance as it is done in the EOB formalism [40, 41, 42].

A more detailed and exhaustive validation of the Chimera scheme would include, as a first step, a comparison of the evolution of the constants of motion with those obtained by solving the Teukolsky equation [43, 44]. One has to be careful in doing this comparison given that the latter employs averages over several cycles, while the Chimera fluxes are computed locally at the SCO's location. Once the fluxes have been validated, one should compare the waveforms themselves. An overlap study would determine the level of agreement between them.

The Chimera scheme might also be useful to model IMRIs and even systems with moderate mass ratios. Recent comparisons between self-force, PN, and numerical relativity computations have concluded that by replacing the mass ratio q with the symmetric mass ratio η, the self-force predictions compare quite well with numerical relativity and PN predictions in the comparable-mass range [45, 46]. These results may allow for a simpler description of IMRIs, which otherwise would require either long, full numerical computations or higher-order perturbative computations, or a combination of both. In the Chimera scheme, $q \to \eta$ is actually mostly built in already, since the mass ratio information enters through the definition of the multipole moments. For these, we already use general binary expressions, instead of effective one-body ones with mass $qM_\bullet$.

The ability to compute approximate local quantities at the location of the SCO, in particular the multipolar, post-Minkowskian self-force, makes the Chimera scheme an interesting tool to study certain local behavior believed to exist in EMRIs. For example, Flanagan and Hinderer [23] have recently reported that certain rapid changes in orbital elements can arise for generic EMRIs when the orbital frequencies become commensurate. During these rapid changes, it has been suggested that EMRI waveforms may present *glitches*. Reliable quantitative predictions for these glitches are needed to assess the importance of this effect for gravitational wave astronomy. In this sense, we could compare the effect of these glitches in waveforms as computed from an averaged-scheme (such as the Teukolsky one) and a local one (such as the Chimera scheme).

Acknowledgments

CFS acknowledges support from the Ramón y Cajal Programme of the Spanish Ministry of Education and Science, by a Marie Curie International Reintegration Grant (MIRG-CT-2007-205005/PHY) within the 7th European Community Framework Programme, from contract AYA-2010-15709 of the Spanish Ministry of Science and Innovation (MICINN), and contract 2009-SGR-935 from the Catalan Agency for Research Funding (AGAUR). NY acknowledges support from NSF grant PHY-1114374, as well as support provided by the National Aeronautics and Space Administration through Einstein Postdoctoral Fellowship Award Number PF0-110080, issued by the Chandra X-ray Observatory Center, which is operated by the Smithsonian Astrophysical Observatory for and on behalf of the National Aeronautics Space Administration under contract NAS8-03060. NY also acknowledges support from NASA grant NNX11AI49G, under sub-award 00001944. We acknowledge the computational resources provided by the

Barcelona Supercomputing Centre (AECT-2011-3-0007) and CESGA (CESGA-ICTS-200).

References

[1] Danzmann K and Rudiger A 2003 *Class. Quant. Grav.* **20** S1–S9
[2] Prince T 2003 *American Astronomical Society Meeting* **202** 3701
[3] Amaro-Seoane P *et al.* 2007 *Class. Quant. Grav.* **24** R113–R169 (*Preprint* astro-ph/0703495)
[4] Barack L and Cutler C 2004 *Phys. Rev.* **69** 082005 (*Preprint* gr-qc/0310125)
[5] Collins N A and Hughes S A 2004 *Phys. Rev.* **D69** 124022 (*Preprint* gr-qc/0402063)
[6] Glampedakis K and Babak S 2006 *Class. Quant. Grav.* **23** 4167–4188 (*Preprint* gr-qc/0510057)
[7] Barack L and Cutler C 2007 *Phys. Rev.* **D75** 042003 (*Preprint* gr-qc/0612029)
[8] Vigeland S J and Hughes S A 2010 *Phys. Rev.* **D81** 024030 (*Preprint* 0911.1756)
[9] Hughes S A 2010 (*Preprint* 1002.2591)
[10] Sopuerta C F 2010 *Gravitational Wave Notes* **4**(4) 3–47 (*Preprint* 1009.1402)
[11] Hughes S A 2006 *AIP Conf. Proc.* **873** 233–240 (*Preprint* gr-qc/0608140)
[12] Schutz B F 2009 *Class. Quant. Grav.* **26** 094020
[13] Babak S, Gair J R, Petiteau A and Sesana A 2011 *Class. Quant. Grav.* **28** 114001 (*Preprint* 1011.2062)
[14] Mino Y, Sasaki M and Tanaka T 1997 *Phys. Rev.* **D55** 3457–3476 (*Preprint* gr-qc/9606018)
[15] Quinn T C and Wald R M 1997 *Phys. Rev.* **D56** 3381–3394 (*Preprint* gr-qc/9610053)
[16] Barack L and Sago N 2009 *Phys. Rev. Lett.* **102** 191101 (*Preprint* 0902.0573)
[17] Barack L and Sago N 2010 *Phys. Rev.* **D81** 084021 (*Preprint* 1002.2386)
[18] Shah A G, Keidl T S, Friedman J L, Kim D H and Price L R 2011 *Phys. Rev.* **D83** 064018 (*Preprint* 1009.4876)
[19] N. Warburton N and Barack L 2011 *Phys. Rev.* **D83** 124038 (*Preprint* 1103.0287)
[20] Dolan S R, Wardell B and Barack L 2011 *Phys. Rev.* **D84** 084001 (*Preprint* 1107.0012)
[21] Sopuerta C F and Yunes N 2011 *Phys.Rev.* **D84** 124060 (*Preprint* 1109.0572)
[22] Blanchet L and Damour T 1984 *Phys. Lett.* **A104** 82–86
[23] Flanagan E E and Hinderer T 2010 (*Preprint* 1009.4923)
[24] Warburton N, Akcay S, Barack L, Gair J R and Sago N 2011 (*Preprint* 1111.6908)
[25] Diener P, Vega I, Wardell B and Detweiler S 2011 (*Preprint* 1112.4821)
[26] Schmidt W 2002 *Class. Quant. Grav.* **19** 2743 (*Preprint* gr-qc/0202090)
[27] Drasco S and Hughes S A 2004 *Phys. Rev.* **D69** 044015 (*Preprint* astro-ph/0308479)
[28] Fujita R and Hikida W 2009 *Class. Quant. Grav.* **26** 135002 (*Preprint* 0906.1420)
[29] Barack L and Sago N 2011 *Phys. Rev.* **D83** 084023 (*Preprint* 1101.3331)
[30] Detweiler S and Whiting B F 2003 *Phys. Rev.* **D67** 024025 (*Preprint* gr-qc/0202086)
[31] Detweiler S 2001 *Phys. Rev. Lett.* **86** 1931–1934 (*Preprint* gr-qc/0011039)
[32] Pound A and Poisson E 2008 *Phys. Rev.* **D77** 044013 (*Preprint* 0708.3033)
[33] Iyer B R and Will C M 1993 *Phys. Rev. Lett.* **70** 113–116
[34] Iyer B R and Will C M 1995 *Phys. Rev.* **D52** 6882–6893
[35] Blanchet L 1997 *Phys. Rev.* **D55** 714–732 (*Preprint* gr-qc/9609049)
[36] Burke W L 1971 *J. Math. Phys.* **12** 401–418
[37] Abe M, Ichinose S and Nakanishi N 1987 *Progress of Theoretical Physics* **78** 1186–1201
[38] Blanchet L 2006 *Living Rev. Relativity* **9** 4 (*Preprint* gr-qc/0202016)
[39] Cutler C 1998 *Phys. Rev.* **D57** 7089–7102
[40] Buonanno A and Damour T 1999 *Phys. Rev.* **D59** 084006 (*Preprint* gr-qc/9811091)
[41] Buonanno A and Damour T 2000 *Phys. Rev.* **D62** 064015 (*Preprint* gr-qc/0001013)
[42] Damour T 2010 *Phys. Rev.* **D81** 024017 (*Preprint* 0910.5533)
[43] Hughes S A 2000 *Phys. Rev.* **D61** 084004
[44] Hughes S A 2001 *Phys. Rev.* **D64** 064004
[45] Barack L, Damour T and Sago N 2010 *Phys. Rev.* **D82** 084036 (*Preprint* 1008.0935)
[46] Le Tiec A, Mroue A H, Barack L, Buonanno A, Pfeiffer H P *et al.* 2011 *Phys. Rev. Lett.* **107** 141101 (*Preprint* 1106.3278)

Amaldi 9 and NRDA 2011 IOP Publishing
Journal of Physics: Conference Series **363** (2012) 012022 doi:10.1088/1742-6596/363/1/012022

Multimessenger Sources of Gravitational Waves and High-energy Neutrinos: Science Reach and Analysis Method

B Baret[1], I Bartos[2,†], B Bouhou[1], E Chassande-Mottin[1], A Corsi[3], I Di Palma[4], C Donzaud[1,5], M Drago[6], C Finley[7], G Jones[8], S Klimenko[9], A Kouchner[1], S Márka[2], Z Márka[2], L Moscoso[1,*], M Alessandra Papa[4], T Pradier[10], G Prodi[6], P Raffai[2,11], V Re[12], J Rollins[3], F Salemi[4,13], P Sutton[8], M Tse[2], V Van Elewyck[1] and G Vedovato[14]

[1]AstroParticule et Cosmologie (APC), CNRS: UMR7164-IN2P3-Observatoire de Paris-Université Denis Diderot-Paris VII-CEA: DSM/IRFU, France
[2]Department of Physics, Columbia University, New York, NY 10027, USA
[3]LIGO Laboratory, California Institute of Technology, Pasadena, CA 91125, USA
[4]Albert-Einstein-Institut, Max-Planck-Institut für Gravitationsphysik, D-30167 Hannover, Germany
[5]Université Paris-sud, Orsay, F-91405, France
[6]INFN, Gruppo Collegato di Trento and Università di Trento, I-38050 Povo, Trento, Italy
[7]Oskar Klein Centre & Dept. of Physics, Stockholm University, SE-10691 Stockholm, Sweden
[8]Cardiff University, Cardiff CF24 3AA, UK
[9]University of Florida, Gainesville, FL 32611, USA
[10]University of Strasbourg & Institut Pluridisciplinaire Hubert Curien, Strasbourg, France
[11]Eötvös University, Institute of Physics, 1117 Budapest, Hungary
[12]INFN, sezione di Roma Tor Vergata, I-00133 Roma Italy
[13]Leibniz Universität Hannover, D-30167 Hannover, Germany
[14]INFN, Sezione di Padova, I-35131 Padova, Italy
[*] Deceased
[†] Corresponding author. Email: ibartos@phys.columbia.edu

Abstract. Sources of gravitational waves are often expected to be observable through several messengers, such as gamma-rays, X-rays, optical, radio, and/or neutrino emission. The simultaneous observation of electromagnetic or neutrino emission with a gravitational-wave signal could be a crucial aspect for the first direct detection of gravitational waves. Furthermore, combining gravitational waves with electromagnetic and neutrino observations will enable the extraction of scientific insight that was hidden from us before. We discuss the method that enables the joint search with the LIGO-Virgo-IceCube-ANTARES global network, as well as its methodology, science reach, and outlook for the next generation of gravitational-wave detectors.

The observation of multimessenger sources of gravitational waves (GWs) and high-energy neutrinos (HENs) is on our doorstep due to recent and near-future development of large-scale observatories. Several GW observatories have been in operation in the past years, including LIGO [1], Virgo [2] and GEO [3]. These detectors will be upgraded to second generation detectors within the next few years. Another advanced GW detector, LCGT [4] in Japan is

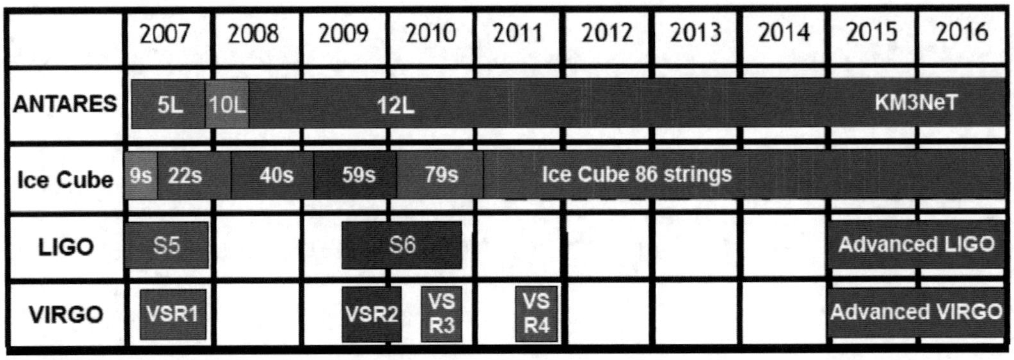

Figure 1. Time chart. Planned data-taking periods for detectors involved in the GW+HEN search.

under construction. LIGO may build a third observatory in India [5] soon after the two US-based detectors. Several HEN observatories are currently in operation, such as IceCube [6], a cubic-kilometer detector at the geographic South Pole, and ANTARES [7] in the Mediterranean sea. ANTARES is being upgraded to a cubic-kilometer detector called KM3NeT in the following years [8]. A third HEN detector operating at the lake Baikal is also planned to be upgraded to a km^3 volume [9].

The gravitational-wave – high-energy neutrino (GW+HEN) working group of the LIGO, Virgo, IceCube and ANTARES detectors[1] has been formed to facilitate cooperation and data-sharing between participating institutions. Collaboration is further aided by past and planned coincident data-taking periods, as we show in Figure 1.

The first published result of the GW+HEN working group involved the analysis of the coincident time window within which one can expect the observation of coincident GW and HEN signals from a common source [10]. The analysis focused on gamma-ray bursts (GRBs) as arguably most promising multimessenger sources. The obtained conservative, ~ 500 s long time window took into account processes motivated by current GRB models. The duration of these process was determined based on electromagnetic observational data. The considered processes and their duration are shown in Figure 2.

An important milestone in the work on common GW+HEN sources was the first observational constraint obtained by Bartos *et al.* [11]. They explored the reach of current and planned experiments in constraining the population of multi-messenger sources of GWs and HENs. They derived constraints on the rate of GW and HEN transients based on independent observations by the initial LIGO and Virgo GW detectors and the partially completed IceCube (40-string) HEN detector. Further, they compared the estimated reach of joint GW+HEN searches using advanced GW detectors and the completed km^3 IceCube detector to the reach of independent searches using the same detectors. They conclude that, while the main advantage of joint searches is increased sensitivity for the actual detection of sources, joint searches will provide better constraints than independent observations if, upon non-detection they result in an increased exclusion distance by at least a factor $\sim f_b^{1/3}$ compared to independent searches, where f_b is the neutrino beaming angle. The projected population constraint from joint searches with advanced GW and HEN detectors is shown in Figure 5 (right).

As a next important milestone, the baseline analysis has been designed [12] to search for

[1] In order of greatest linear dimension

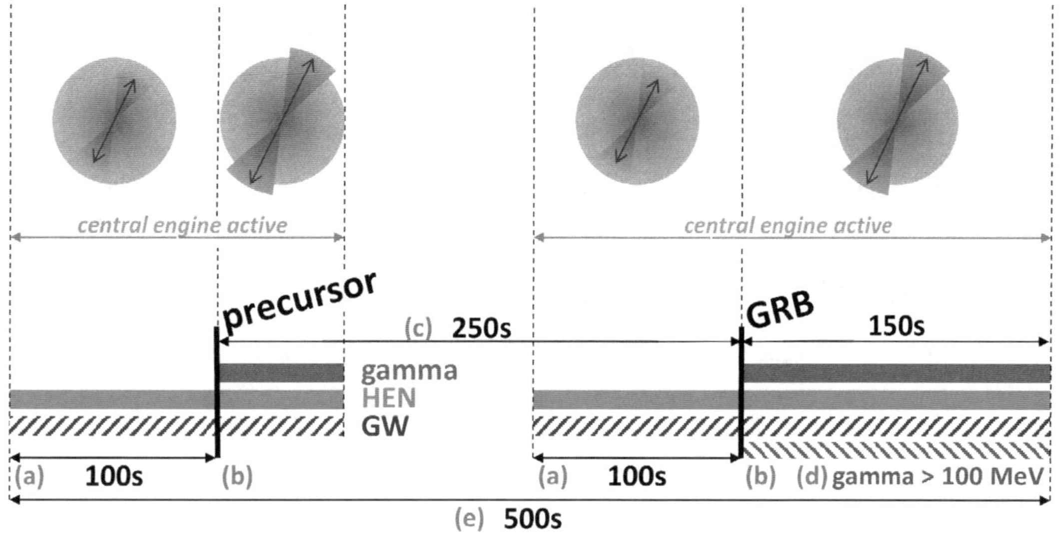

Figure 2. Summary of upper bound of GRB emission process durations taken into account in the total GW+HEN coincidence time window (courtesy of [10]). (a) active central engine before the relativistic jet has broken out of the stellar envelope; (b) active central engine with the relativistic jet broken out of the envelope; (c) delay between the onset of the precursor and the main burst; (d) duration corresponding to 90% of GeV photon emission; (e) time span of central engine activity.

common sources. The baseline joint analysis takes into account the significance of the GW and HEN signals, calculated based on the excess energy in the GW datastream (see see e.g. [13]) and the reconstructed neutrino energy and neutrino flux (i.e. number of coincident neutrinos). The analysis also takes into account the directional probability distributions of the GW and HEN signals, as well as the *a priori* source distribution using the observed distribution of blue luminosity in the universe[2]. An example set of spatial probability distributions used by the analysis, as well as their combination, are shown in Figure 3. A parallel search is performed where the blue-luminosity distribution is ignored in order to search for sources not connected to blue luminosity, e.g. galactic sources. The joint search will use the coincidence time window derived by Baret *et al.* [10]. It will consider individual signal candidates, as well as an ensemble of weak, sub-threshold signals that could not be detected individually. In the case of no detection, results will be used to obtain upper limit estimates of multimessenger GW+HEN sources. A schematic flow diagram of the baseline analysis is shown in Figure 4. Projected results for the search using initial LIGO-Virgo detectors (S5/VSR1 science run) with the partial IceCube detector with 22 strings are shown in Figure 5 (left). These upper limit projections were obtained by Baret *et al.* [10] following the calculation of Bartos *et al.* [11], using the different detector parameters.

We are close to the milestone of finishing thê first coincident search for GWs and HENs for the initial LIGO-Virgo (S5/VSR1 science runs) and the partial ANTARES detector in its 5-string configuration. The analysis uses the directional distribution and the time of arrival of HENs to trigger a GW follow-up analysis, similar to the analysis used for GW follow-up searches of

[2] I.e. the analysis assumes that the source distribution follows the blue-luminosity distribution of galaxies.

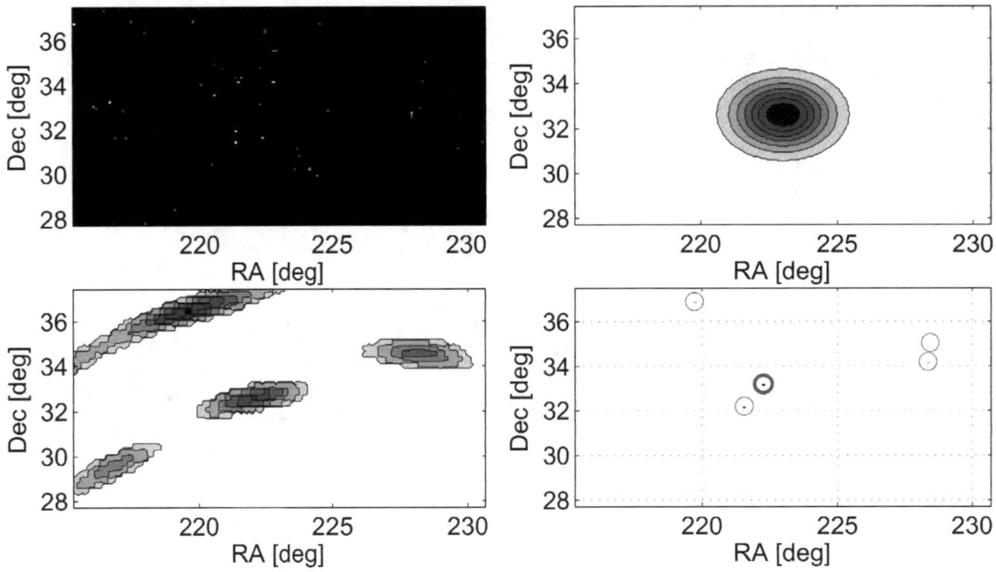

Figure 3. Example likelihood distributions as parts of the joint likelihood ratio: weighted galaxy distribution (upper left), HEN directional probability distribution function (PDF) (upper right), GW PDF (lower left) and joint PDF (lower right), all shown in arbitrary units. On the joint PDF plot, every galaxy for which the joint PDF is non-zero is circled for visibility. Courtesy of [12].

GRBs (e.g. [14]). The search uses the conservative coincidence time window of 500 s of Baret *et al.* [10]. There are ~200 neutrino triggers from ANTARES, most of which are detected by digital optical modules (DOMs) on two strings, while 13 neutrinos are detected by DOMs on three strings. The detailed method and the first scientific results will be published soon.

One can consider specific source emission models to interpret the reach of the projected population upper limits [12]. Taking two typical HEN emission models, Figure 6 shows the corresponding population upper limits as functions of GW emission.

The first observational results and projected population estimates indicate that multimessenger searches for gravitational waves and high-energy neutrinos will be an exciting and fruitful field in the near future.

Acknowledgement
We are grateful for the generous support of Columbia University in the City of New York and the National Science Foundation under cooperative agreements PHY-0847182. The authors are grateful for support from the Swedish Research Council (VR) through the Oskar Klein Centre. The authors thank the European Union FP7 (Marie Curie European Reintegration Grant NEUTEL-APC 224898). We acknowledge the financial support of the National Science Foundation under grants PHY-0855044 and PHY-0855313 to the University of Florida, Gainesville, Florida. The authors acknowledge financial support from the French Agence Nationale de la Recherche (contract ANR-08-JCJC-0061-01) and from the EU FP7 (Marie Curie Reintegration Grant). LIGO was constructed by the California Institute of Technology and Massachusetts Institute of Technology with funding from the National Science Foundation under cooperative agreement PHY-0757058. This paper has document number LIGO-P1100193.

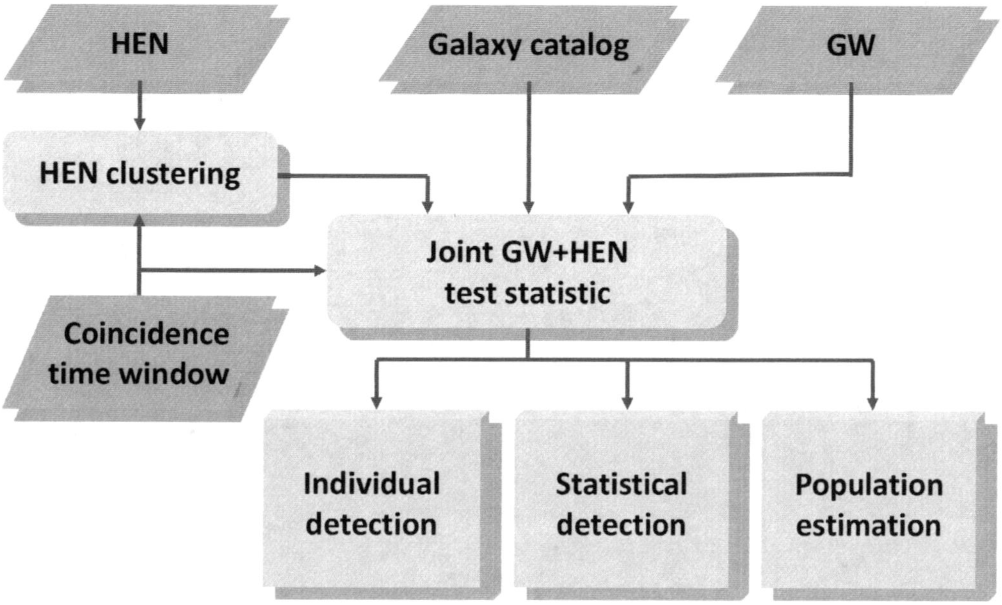

Figure 4. Schematic flow diagram of the baseline joint GW+HEN search pipeline [12]. This pipeline is used for the search with initial LIGO-Virgo (S5/VSR1) + IceCube (22 strings).

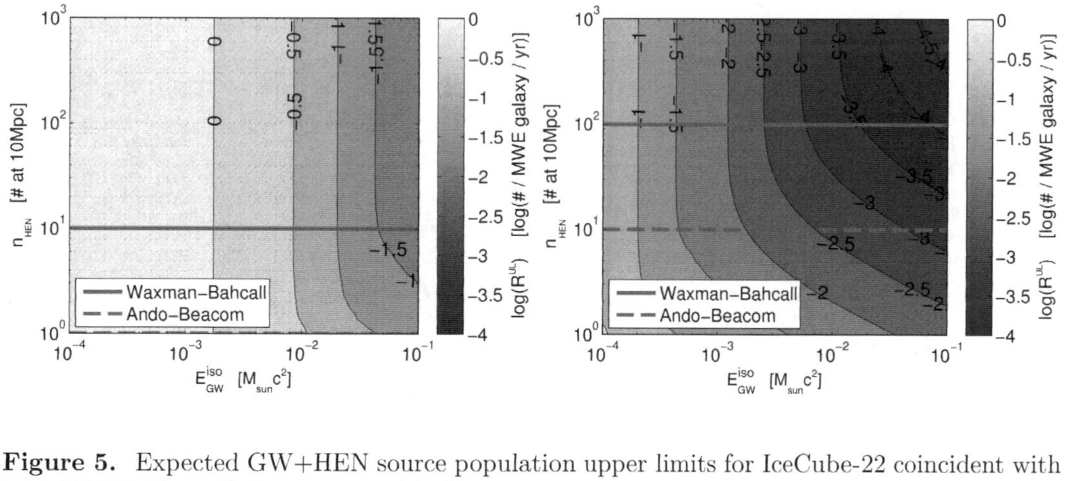

Figure 5. Expected GW+HEN source population upper limits for IceCube-22 coincident with initial LIGO-Virgo (left; courtesy of [12]) and IceCube-86 coincident with advanced the LIGO-Virgo detectors (right; courtesy of [15]), with one year of coincident measurement time. The results take into account the blue luminosity-weighted galaxy distribution. The x-axis represents the GW energy output of a standard source. The y-axis represents the number of detected neutrinos from a standard source at 10 Mpc. The color scale shows the obtained source rate upper limit R^{UL} in logarithmic units of number of sources per (Milky Way equivalent) galaxy per year. On both plots, the two horizontal lines (scaled for detector sensitivity) show the Waxman-Bahcall emission model [16] (higher) and the HEN emission model of Ando and Beacom [17] for reverse shocks in mildly relativistic supernova jets / choked GRBs (lower).

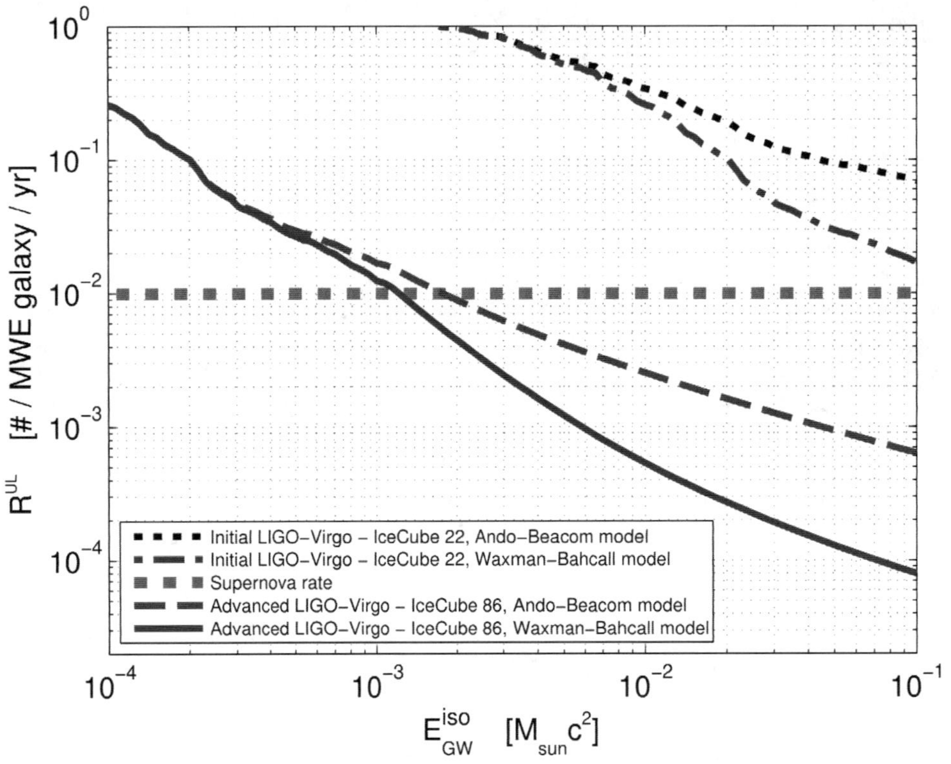

Figure 6. Expected GW+HEN source population upper limits for anticipated HEN emission from two emission models, as functions of isotropic-equivalent GW emission energy E_{GW}^{iso} (courtesy of [12]). Results are shown both for measurements with the initial LIGO-Virgo detectors and the IceCube-22 detector (dashed line), as well as for the advanced LIGO-Virgo detectors and the IceCube-86 detector (solid line), both with one year of coincident measurement time. For comparison, the galactic supernova rate is shown (dotted line).

References

[1] B. P. Abbott *et al.*, Rep. Prog. Phys. **72**, 076901 (2009).
[2] T. Accadia *et al.*, Classical Quantum Gravity **28**, 114002 (2011).
[3] H. Grote and the LIGO Scientific Collaboration, Classical Quantum Gravity **27**, 084003 (2010).
[4] K. Kuroda and the LCGT Collaboration, Classical Quantum Gravity **27**, 084004 (2010).
[5] LIGO-India, http://www.gw-indigo.org/.
[6] J. Ahrens *et al.*, Astroparticle Physics **20**, 507 (2004).
[7] M. Ageron *et al.*, Nuclear Instruments and Methods in Physics Research A **656**, 11 (2011), 1104.1607.
[8] M. de Jong, Nucl. Instrum. Methods Phys. Res., Sect. A **623**, 445 (2010).
[9] A. Avrorin *et al.*, Nucl. Instrum. Methods Phys. Res., Sect. A **626-627**, S13 (2011).
[10] B. Baret *et al.*, Astropart. Phys. **35**, 1 (2011).
[11] I. Bartos, C. Finley, A. Corsi, and S. Márka, Phys. Rev. Lett. **107**, 251101 (2011).
[12] B. Baret *et al.*, ArXiv (2011).
[13] J. Abadie *et al.*, Phys. Rev. D **81**, 102001 (2010).
[14] B. P. Abbott *et al.*, Astrophys. J. **715**, 1438 (2010).
[15] I. Bartos, C. Finley, A. Corsi, and S. Márka, ArXiv e-prints (2011), 1108.3001.
[16] D. Guetta, D. Hooper, J. Alvarez-Mun˜Iz, F. Halzen, and E. Reuveni, Astropart. Phys. **20**, 429 (2004).
[17] S. Ando and J. F. Beacom, Phys. Rev. Lett. **95**, 061103 (2005).

Scientific Benefit of Enlarging Gravitational Wave Detector Networks

Qi Chu[1,2], **Linqing Wen**[1,2],**David Blair**[1]

[1]Australian International Gravitational Research Centre, School of Physics, University of Western Australian, 35 Stirling Hwy, Crawley, WA 6009, Australia
[2]ICRAR-Fairway M468, School of Physics, University of Western Australian, Crawley, WA 6009, Australia

E-mail: chicsheep@gmail.com, linqing.wen@uwa.edu.au

Abstract. Localising the sources of gravitational waves (GWs) in the sky is crucial to observing the electromagnetic counterparts of GW sources. The localisation capability is poor by a single GW detector yet can be improved by adding more detectors to the detector network. In this paper we review recent studies on scientific benefits of global detector networks and focus on their localisation capability. We employ Wen-Chen's formula to compare this merit of current and future detector networks for localising gravitational wave bursts. We find that the addition of a new detector located in Japan, or India, or Australia will increase angular resolution 3~5 fold with respect to current LIGO-Virgo network, and that the angular resolution improvement by adding a single detector in Australia is comparable to that achieved by adding detectors in both India and Japan. A six-site network achieves a 11-fold improvement in angular resolution compared with the existing three-site network.

1. Introduction

The first detection of gravitational waves (GW) is likely to take place within the coming decade. It will start a new era for GW astronomy. The Laser Interferometer Gravitational Wave Observatory (LIGO)[1] in the USA has achieved its initial sensitivity goal and the upgraded configuration, advanced LIGO (advLIGO) is expected to be operational in the next 4~5 years with a 10-fold improvement in sensitivity. During the same timescale, advanced Virgo[2], an upgrade of the current Virgo detector in Italy, is scheduled to achieve a similar sensitivity. Another advanced detector is under construction in Japan–KAmioka GRAvitational wave detector (KAGRA)[3]. Although LIGO-Australia (AIGO)[4] was not approved in 2011, LIGO-India (IndIGO)[5] has now been proposed, and as an alternative international collaboration for AIGO is now under discussion.

There are many well-known sources that can generate GWs, such as compact binaries system, stellar core collapse and supernovae, or pulsars. The advanced detectors are expected to detect at least several compact binary coalescence sources per year. The coalescence of neutron star/neutron star (NS/NS) is widely considered to be the progenitor of short hard gamma ray bursts (SHBs) and observation by GW detectors could tell us whether the events are indeed the cause of SHBs. It is critical to localise the event source accurately to allow the observation of electromagnetic counterparts. In this paper we will first review recent studies on the scientific benefits of global detector networks, then narrow down our discussion on localisation capability of detector networks.

2. Recent studies on scientific benefits of detector networks

A global detector network excels a single detector mainly in the aspects that it can detect the signal more robustly and measure the source location more accurately. A LIGO report[6] gave an extensive review by comparing two possible detector networks: a 4 detector network using two detectors at Hanford (USA), one at Livingston (USA), and one at Cascina (Italy) and a new network obtained by moving one Hanford detector to the proposed AIGO site. The networks are denoted HHLV and AHLV (see Table 1 for detector labels). Comparisons were based on considerations of source localisation, source parameter estimation capacity, network sensitivity and false alarm rate, and how local environmental perturbations would affect the network. Schutz[7], assuming that coherent network analysis is the default method for detection, summarised the properties of detector networks into three criteria: a) triple detection rate, b) sky coverage, and c) angular precision. Schutz considered the possibility that individual detectors might be off-duty and characterised the detection rate of a given network to be the detection volumes of all sub-networks weighted by their on-duty probability. The sky coverage is often computed as sky reach above a certain threshold of a network property. The LIGO report used the network sensitivity for this property while Schutz used antenna power patterns.

The work of measuring how accurately an array of detectors can estimate source location has long been appreciated. So far two strategies have been developed. One is to perform Monte-Carlo simulations on posterior parameter sampling and calculate the variations with true values[8, 9, 10]. The other is to derive an analytical expression of localisation accuracy (or angular resolution, expressed as an angular area) directly. The difficulty in such a direct derivation will be illustrated in detail in section 3. Based on triangulation for source localisation, Fairhurst[11] presumed that source localisation is primarily related to timing information when the signal arrives at different detectors. Therefore he used the Fisher matrix to obtain time accuracies by which he derived a form of localisation accuracy. Wen and Chen[12] give a more generalized form of localisation accuracy appropriate for any unbiased localisation method. This deviation considered a full parameter space and marginalised the unnecessary parameters except source location parameters. Vitale and Zanolin[13] employed the asymptotic expansions to give a frequentist assessment of parameter estimation errors. The results of the above three strategies are comparable, and agree well with Monte-Carlo simulations given that signal-to-noise ratio (SNR) is relatively high[14]. A thorough study of the Fisher matrix for assessment of GW parameter estimation has been done by Vallisneri. This gives an SNR threshold above which the Fisher matrix is reliable[15]. Schutz[7] gave a figure of merit for angular resolution from a simplification of Wen-Chen's work. In this paper, we compare the performance of different detector networks with the aid of the Wen-Chen approach.

3. Parameter measurement

3.1. Detection and parameter estimation

In practice the detection and parameter estimation of GW signals are treated separately, as detecting a signal and solving all parameters of the signal simultaneously is not sensible due to immense computational cost. The purpose of detection is to recognise GW signals from noisy data efficiently and reliably. Much effort has been paid to reduce the complete form of the signal to make templates for detection. The most popular detection algorithm - matched filtering, uses templates defined on a reduced parameter space (the intrinsic parameter space) of the signal and time-frequency analysis techniques map the multi-dimensional parameter space to the 2-dimensional space.

For the matched filtering detection method, the estimation of intrinsic parameters are performed at the time of detection to maximise the SNR criterion. The remaining extrinsic parameters can be estimated using the maximum likelihood (ML) estimator. Advanced estimators like the maximum a posteriori estimator(MAP) or the Bayesian estimator are more reliable but are more difficult to compute.

We first take a quick look at the background on estimators. Let X be a random variable with probability function $f(x;\theta)$ which depends on an unknown parameter θ. The joint probability of a random sample from X is given by $\prod_1^n f(x;\theta)$, which is also known as the likelihood function for θ. An estimator of θ is any statistics denoted by $\Theta = s(x_1, ..., x_n)$ and is itself a random variable. When $E(\Theta) = \tilde{\theta}$ (E is referred to expected value, $\tilde{\theta}$ is the true value), this estimator is said to be unbiased. A maximum likelihood estimator which maximise the likelihood function is an unbiased estimator.

One can understand that although detection and estimation can be treated separately, parameter estimation depends on the presence of the signal. In other words, the error of estimation is related to the confidence level of the presence. The SNR of an observation can be a criterion to reflect the presence of a signal. Higher SNR indicates the presence of signal is more likely and therefore the estimated parameter values are very likely to fall within the parameter space where signal is present. The statistical error is then well approximated by the inverse of the Fisher information which will be shown below.

3.2. Fisher information matrix and Cramér-Rao bound

The Fisher information matrix(or Fisher matrix) of θ, where θ has more than one parameter, is defined as:

$$\Gamma_{ij} = E\left[\left(\frac{\partial}{\partial\theta_i}\log f(x;\theta)\right)\left(\frac{\partial}{\partial\theta_j}\log f(x;\theta)\right)\right] \tag{1}$$

According to the Cramér-Rao bound[16], the lower bound of the variance of any unbiased estimator is determined by the inverse of the Fisher matrix: $Var(\Theta) \geq \Gamma^{-1}$. Using this fact, we can obtain the best precision of an unbiased estimator of any parameter. Here the focus of our paper is on angular position θ, ϕ. Wen and Chen(2010) have derived geometrical expressions for angular resolution in several practical situations. In the case of detecting short signals (in which time the motion of the detector network is negligible), the general expression of angular resolution is

$$\Delta\Omega = \frac{4\sqrt{2}\pi c^2}{\sqrt{\sum_{J,K,L,M}\Delta_{JK}\Delta_{LM}|(\mathbf{r}_{KJ}\times\mathbf{r}_{ML})\cdot\mathbf{n}|^2}} \tag{2}$$

In this formula, J, K, L, M are detector indices; $\Delta\Omega$ is the angular resolution; $\Delta_{IJ} = \langle\Omega d_I|\Omega d_J\rangle\delta_{IJ} - \langle\Omega d_I|\partial_{\lambda_k}d_J\rangle\mathbf{B}_{\mathbf{kl}}^{-1}\langle\partial_{\lambda_l}d_J|\Omega d_J\rangle$, where d_I is the Fourier transform of noise weighted waveform at detector I, $\langle|\rangle$ is complex vector inner product, δ_{IJ} is the Kronecker delta, and $B_{kl} = \langle\partial_{\lambda_k}\mathbf{d}|\partial_{\lambda_l}\mathbf{d}\rangle$; $\mathbf{r}$ is positional vector of connecting detectors; $\mathbf{n}$ is the source direction vector.

Two special cases of short signals have been considered in [12] and also in this paper. For one case is the waveform and the arrival time of the signal are both unknown. This we refer to as the worst case. The best case is the one when the waveform is known and the arrival time is unknown. In this scenario, Δ_{IJ} can be simplified as $\Delta_{IJ} = -\frac{\xi_I\xi_J}{\sum_I\xi_I}$ for $I \neq J$, where $\xi_I \equiv \langle\Omega d_I|\Omega d_I\rangle$. We will use angular resolution for the best case and the worst case to compare the localisation capability of a range of detectors.

4. Results and discussion

The GW detector networks that we consider comprises combinations of three established interferometers: LIGO-Virgo (LHV), one partially funded: KAGRA (J), and two possible ones:

Table 1: Properties of detectors. Longitude is positive toward east and latitude positive toward north. The locations for L,H,V,J can be found on corresponding detector websites. The location of AIGO is set in Gingin,Western Australia. The location of IndIGO is found from[7]. There are two interferometers in Hanford, USA and the symbol "HH" is used when both detectors are considered.

Detector	Label	Longitude	Latitude	Sensitivity
LIGO Livingston, USA	L	−90.77°	30.56°	advL[1]
LIGO Hanford, USA	H	−119.41°	46.45°	advL
Virgo, Italy	V	10.5°	43.63°	advV[2]
KAGRA, Japan	J	137.18°	36.25°	advL
IndIGO, India	I	74.05°	19.09°	advL
AIGO, Australia	A	115.87°	−31.95°	advL

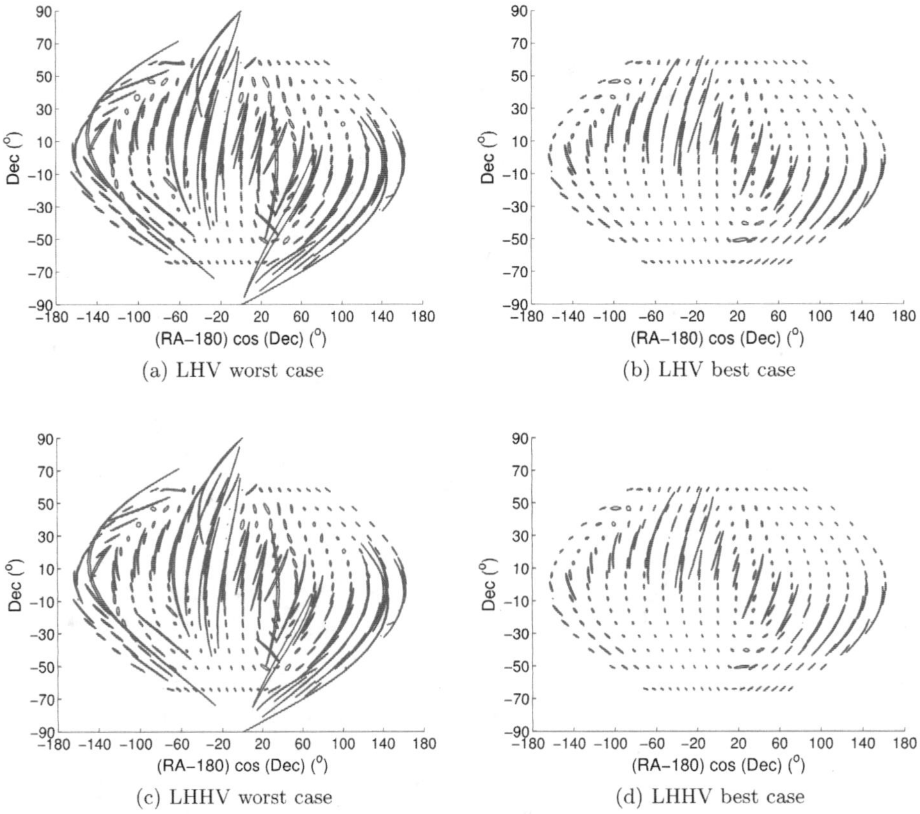

(a) LHV worst case

(b) LHV best case

(c) LHHV worst case

(d) LHHV best case

Figure 1

IndIGO (I) or AIGO (A). Other first generation GW interferometers detectors such as TAMA in Japan and GEO600 in Germany will not be discussed in this paper.

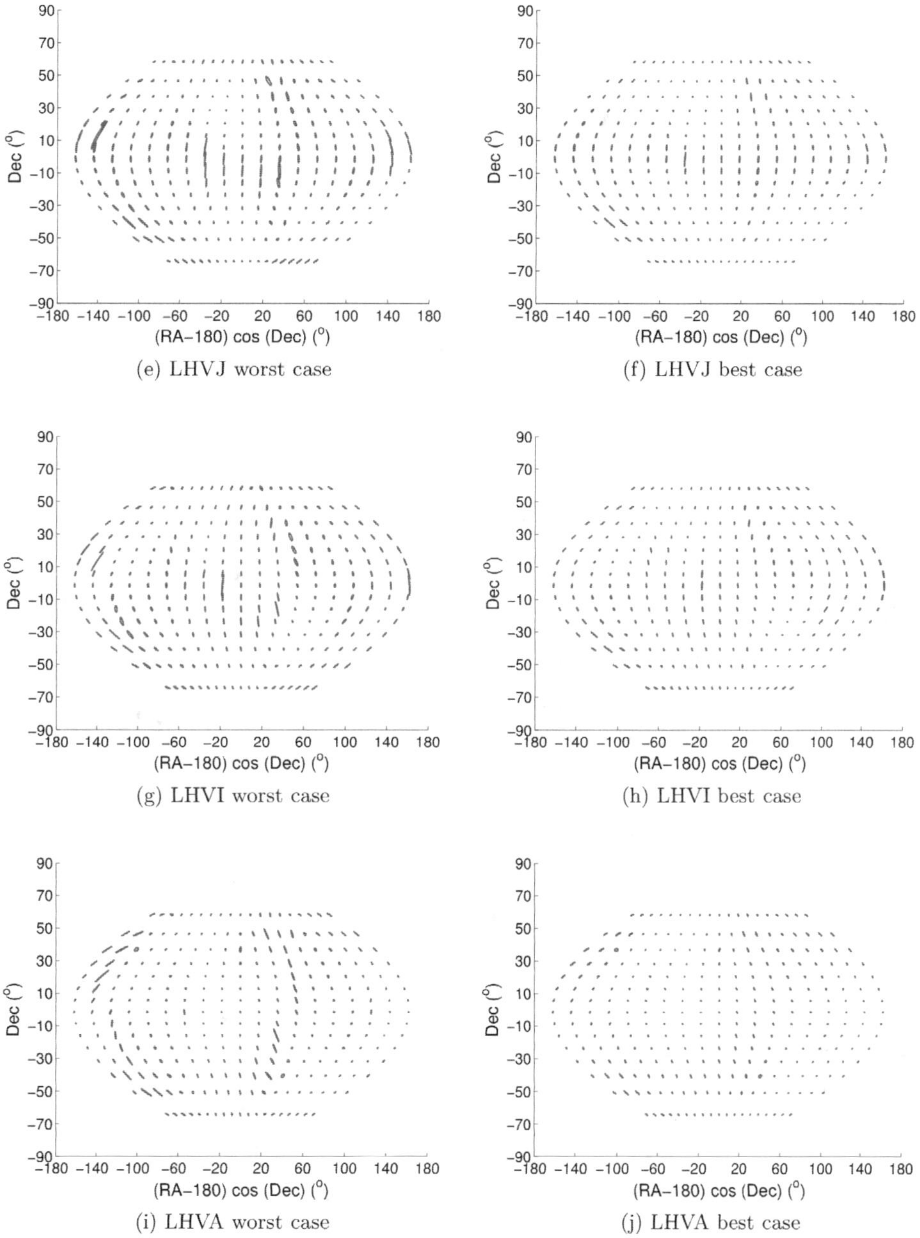

(e) LHVJ worst case (f) LHVJ best case

(g) LHVI worst case (h) LHVI best case

(i) LHVA worst case (j) LHVA best case

Figure 1

The GW signal for experiments here is the GW burst from supernova(U11 model)[17] with central frequency 600Hz. We adjust the distance of the signal so that the optimal network SNR of LHV is 10. Note that Eq.(2) obtained from the Fisher matrix is the lower bound of angular resolution. Therefore what we discuss in this paper is the best localisation capability of the detector networks.

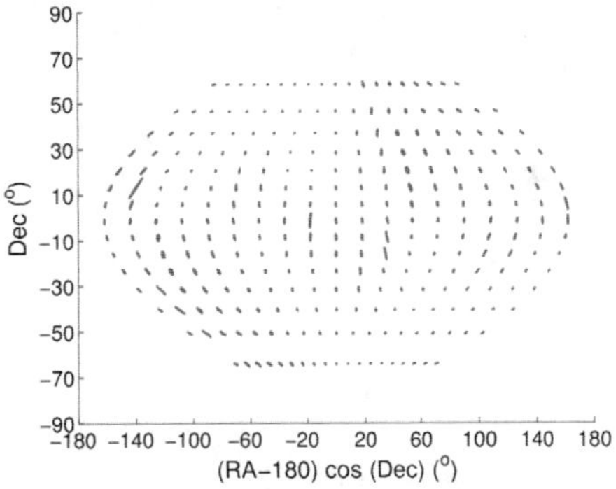

(k) LHVJI worst case

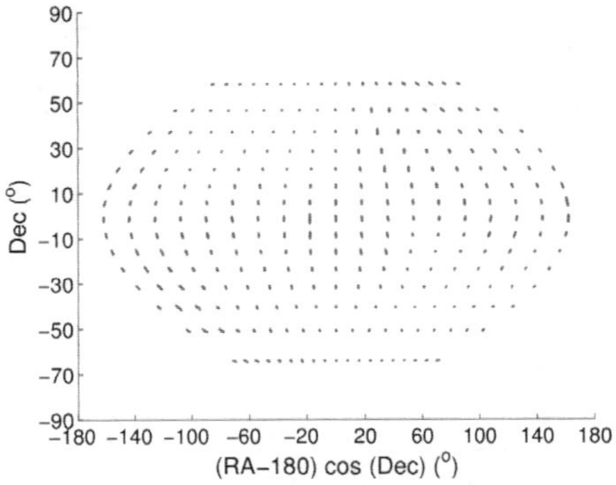

(l) LHVJI best case

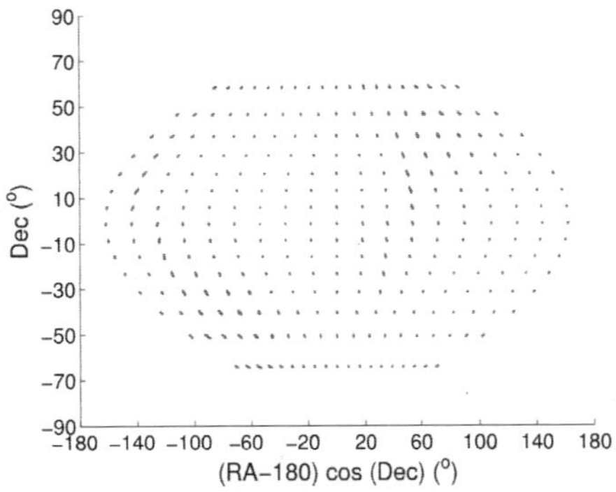

(m) LHVJIA worst case

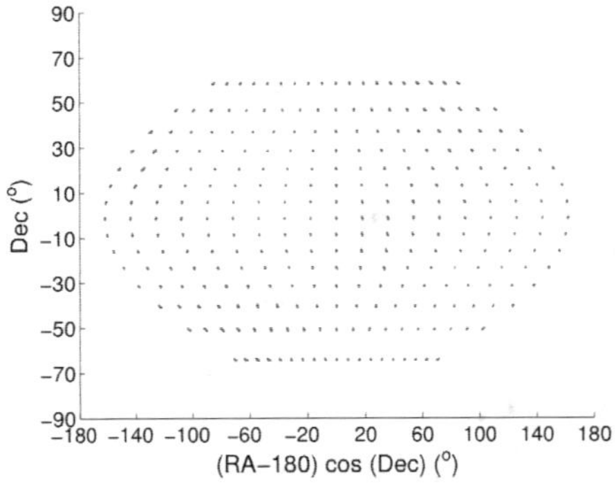

(n) LHVJIA best case

Figure 1: All sky-map error ellipses of 7 detector networks. They are LHV, LHHV, LHVJ, LHVI, LHVA, LHVJI, and LHVJIA. All angular areas are in 95% confidence interval.

We present our results in the term of all sky maps which show the distribution of error ellipses for point source located uniformly over the sky. For each detector network, we apply both two cases (explanations of best case and worst case can be found in Section 3).

Observation of the various all-sky maps gives a qualitative estimate of network performance in the best and worst cases as shown in Fig. 1. To obtain quatitative characteristics we perform accumulative distributions for the angular area as shown in Fig. 2. From Fig.1 and Fig.2, we can see the following:

(i) If a signal is formed along the plane formed by detector network LHV, then it is hard to locate the signal accurately by this network.

(ii) The improvement of angular resolution if an extra detector is added at the same place is negligible. The angular resolution of LHVJ and LHVI are comparable and so is LHVA and LHVJI. The reason why AIGO can improve source localisation significantly lies in that it can form the largest baselines with existing detectors and is roughly in antipodal position with LIGO.

Our results are consistent with Schutz[7] who used a simplified version of Wen-Chen formula, and with Vitale and Zanolin[13] who used a frequentist strategy, and with Fairhust[11] who took advantage of timing information, and with Nissanke *et.el*[9] who employed Monte-Carlo sampling. The results in Fig. 2 show the great improvement obtained by extending the network but also emphasize on outstanding benefit of the AIGO detector, as discussed further below.

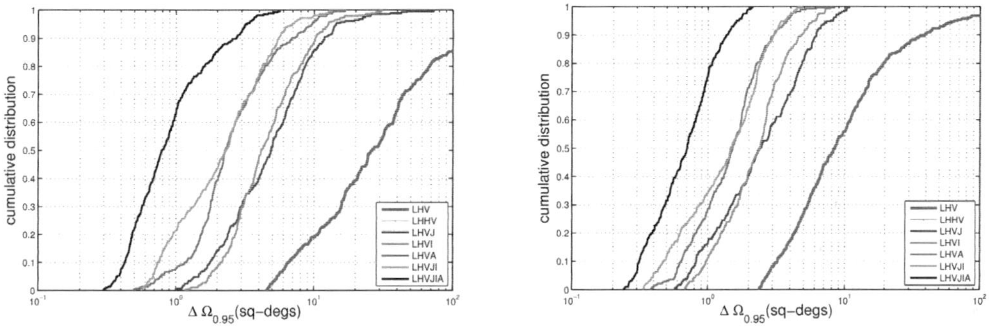

(a) (b)

Figure 2: Cumulative distribution of error ellipses at 95% confidence in worst case (a) and in best case (b) of 7 detector networks. The best case is most likely to occur in reality which the signal waveform is unknown but the arrival time is unknown. In the best case scenario, 50% of the area of error ellipses will be below 8 deg^2 for LHV/LHHV networks, and goes down to around 2.5 deg^2 for LHVJ/LHVI networks, further down to less than 1.5 deg^2 for LHVA/LHVJI networks, achieving its best at around 0.7 deg^2 for the LHVJIA network. The worst-case figure depicts more significant improvement of angular resolution when extending detector networks.

5. Conclusion

We have reviewed recent work on the scientific benefits of extending global detector networks. In particular we discussed the angular resolution of future networks. Our results show that the addition of new sites in Japan (KAGRA) or in India (IndIGO) will improve the angular resolution about 3-fold compared to the current LHV network. However, adding a new detector in Australia (AIGO) alone has similar angular resolution performance to KAGRA and IndIGO combined. AIGO gives a 5-fold better angular resolution compared with the LHV network. The best network is LHVJIA which includes LIGO Livingston, LIGO Hanford, Virgo, KAGRA, IndIGO, and AIGO. It has a 11-fold improvement compared with LHV, achieving angular resolution of 0.7 deg^2 for 50% of sources, and 0.3 deg^2 for 10% of sources. In future work we will repeat this analysis for compact binary coalescence sources and consider other performance measures for the networks.

Acknowledgement

We are grateful to Jean-Charles Dumas for correcting English expressions in this paper. This work is supported by the Australian Research Council Discovery Grants and Future Fellowship program.

References

[1] Shoemaker D Advanced ligo anticipated sensitivity curves Tech. Rep. LIGO-T0900288-v3
[2] Collaboration T V 2009 Advanced virgo baseline design note vir–027a–09 may 16 The Virgo Collaboration
[3] Kuroda K 2010 *Class. Quantum Grav.* **27** 084004
[4] Blair D G *et al.* 2008 *Journal of Physics Conference Series* vol 122 p 012001
[5] URL `www.gw-indigo.org`

[6] Weiss 2010 Report of the committee to compare the scientific cases for two gravitational-wave detector networks: (ahlv) australia, hanford, livingston, virgo; and (hhlv) two detectors at hanford, one at livingston, and virgo Tech. rep.
[7] Schutz B F 2011 *Class. Quantum Grav.* arXiv:1102.5421v2
[8] Abadie J *et al.* 2010 *Class. Quantum Grav.* **27** 173001
[9] Nissanke S, Sievers J, Dalal N and Holz D 2011 *Astrophys. J.* **739** 99
[10] Klimenko S *et al.* 2011 *Phys. Rev. D* **83** 102001
[11] Fairhust S 2011 *Class. Quantum Grav.* **28** 105021
[12] Wen L and Chen Y 2010 *Phys. Rev. D* **81** 082001
[13] Vitale S and Zanolin M 2011 *Phys. Rev. D* **84**
[14] Balasubramanian R *et al.* 1996 *Phys. Rev. D* **53** 3033–3055
[15] Vallisneri M 2008 *Phys. Rev. D* **77** 042001
[16] Cramer H 1946 *Mathematical Methods of Statistics* (Princeton University Press)
[17] Baiotti L, Pietri R D, Manca G M and Rezzolla L 2007 *Phys. Rev. D* **75** 044023

Journal of Physics: Conference Series **363** (2012) 012024 doi:10.1088/1742-6596/363/1/012024

Noise monitor tools and their application to Virgo data

T. Accadia[1], F Acernese[2ac], M Agathos[3a], P Astone[4a], G Ballardin[5], F Barone[2ac], M Barsuglia[6], A Basti[7ab], Th S Bauer[3a], M Bebronne[1], M Bejger[8c], MG Beker[3a], M Bitossi[7a], M A Bizouard[9a], M Blom[3a], F Bondu[10b], L Bonelli[7ab], R Bonnand[11], V Boschi[7a], L Bosi[12a], B Bouhou[6], S Braccini[7a], C Bradaschia[7a], M Branchesi[13ab], gabriel chardin T Briant[14], A Brillet[10a], V Brisson[9a], T Bulik[8b], H J Bulten[3ab], D Buskulic[1], C Buy[6], E Calloni[2ab], B Canuel[5], F Carbognani[5], F Cavalier[9a], R Cavalieri[5], G Cella[7a], E Cesarini[13b], O Chaibi[10a], E Chassande-Mottin[6], A Chincarini[15], A Chiummo[5], F Cleva[10a], E Coccia[16ab], P-F Cohadon[14], C N Colacino[7ab], J Colas[5], A Colla[4ab], M Colombini[4b], A Conte[4ab], J-P Coulon[10a], E Cuoco[5], S D'Antonio[16a], V Dattilo[5], M Davier[9a], R Day[5], R De Rosa[2ab], G Debreczeni[17], W Del Pozzo[3a], L Di Fiore[2a], A Di Lieto[7ab], M Di Paolo Emilio[16ac], A Di Virgilio[7a], A Dietz[1], M Drago[18ab], G Endrőczi[17], V Fafone[16ab], I Ferrante[7ab], F Fidecaro[7ab], I Fiori[5], R Flaminio[11], L A Forte[2a], J-D Fournier[10a], J Franc[11], S Franco[9a], S Frasca[4ab], F Frasconi[7a], M Galimberti[11], L Gammaitoni[12ab], F Garufi[2ab], M E Gáspár[17], G Gemme[15], E Genin[5], A Gennai[7a], A Giazotto[7a], R Gouaty[1], M Granata[6], C Greverie[10a], G M Guidi[13ab], J-F Hayau[10b], A Heidmann[14], H Heitmann[10], P Hello[9a], G Hemming[5], P Jaranowski[8d], RJG Jonker[3a], M Kasprzack[5,9a], I Kowalska[8b], A Królak[8ae], N Leroy[9a], N Letendre[1], T G F Li[3a], N Liguori[18ab], M Lorenzini[13a], V Loriette[9b], G Losurdo[13a], E Majorana[4a], I Maksimovic[9b], V Malvezzi[16a], N Man[10a], M Mantovani[7ac], F Marchesoni[12a], F Marion[1], J Marque[5], F Martelli[13ab], A Masserot[1], C Michel[11], L Milano[2ab], Y Minenkov[16a], M Mohan[5], N Morgado[11], A Morgia[16ab], S Mosca[2ab], B Mours[1], L Naticchioni[4ab], F Nocera[5], L Palladino[16ac], C Palomba[4a], F Paoletti[7a,5], R Paoletti[7a], M Parisi[2ab], A Pasqualetti[5], R Passaquieti[7ab], D Passuello[7a], G Persichetti[2ab], F Piergiovanni[13ab], M Pietka[8d], L Pinard[11], R Poggiani[7ab], M Prato[15], G A Prodi[18ab], M Punturo[12a], P Puppo[4a], D S Rabeling[3ab], I Rácz[17], P Rapagnani[4ab], V Re[16ab], T Regimbau[10a], F Ricci[4ab], F Robinet[9a], A Rocchi[16a], L Rolland[1], R Romano[2ac], D Rosińska[8cf], P Ruggi[5], B Sassolas[11], D Sentenac[5], L Sperandio[16ab], R Sturani[13ab], B Swinkels[5], M Tacca[5], L Taffarello[18c], A P M ter Braack[3a], A Toncelli[7ab], M Tonelli[7ab], O Torre[7ac], E Tournefier[1], F Travasso[12ab], G Vajente[7ab], J F J van den Brand[3ab], C Van Den Broeck[3a], S van der Putten[3a], M Vasuth[17], M Vavoulidis[9a], G Vedovato[18c], D Verkindt[1], F Vetrano[13ab], A Viceré[13ab], J-Y Vinet[10a], S Vitale[3a],

H Vocca[12a], R L Ward[6], M Was[9a], K Yamamoto[18bd], M Yvert[1], A Zadrożny[8e], J-P Zendri[18c]

[1]Laboratoire d'Annecy-le-Vieux de Physique des Particules (LAPP), Université de Savoie, CNRS/IN2P3, F-74941 Annecy-Le-Vieux, France
[2]INFN, Sezione di Napoli [a]; Università di Napoli 'Federico II'[b] Complesso Universitario di Monte S.Angelo, I-80126 Napoli; Università di Salerno, Fisciano, I-84084 Salerno[c], Italy
[3]Nikhef, Science Park, Amsterdam, the Netherlands[a]; VU University Amsterdam, De Boelelaan 1081, 1081 HV Amsterdam, the Netherlands[b]
[4]INFN, Sezione di Roma[a]; Università 'La Sapienza'[b], I-00185 Roma, Italy
[5]European Gravitational Observatory (EGO), I-56021 Cascina (PI), Italy
[6]Laboratoire AstroParticule et Cosmologie (APC) Université Paris Diderot, CNRS: IN2P3, CEA: DSM/IRFU, Observatoire de Paris, 10 rue A.Domon et L.Duquet, 75013 Paris - France
[7]INFN, Sezione di Pisa[a]; Università di Pisa[b]; I-56127 Pisa; Università di Siena, I-53100 Siena[c], Italy
[8]IM-PAN 00-956 Warsaw[a]; Astronomical Observatory Warsaw University 00-478 Warsaw[b]; CAMK-PAN 00-716 Warsaw[c]; Białystok University 15-424 Białystok[d]; IPJ 05-400 Świerk-Otwock[e]; Institute of Astronomy 65-265 Zielona Góra[f], Poland
[9]LAL, Université Paris-Sud, IN2P3/CNRS, F-91898 Orsay[a]; ESPCI, CNRS, F-75005 Paris[b], France
[10]Université Nice-Sophia-Antipolis, CNRS, Observatoire de la Côte d'Azur, F-06304 Nice[a]; Institut de Physique de Rennes, CNRS, Université de Rennes 1, 35042 Rennes[b], France
[11]Laboratoire des Matériaux Avancés (LMA), IN2P3/CNRS, F-69622 Villeurbanne, Lyon, France
[12]INFN, Sezione di Perugia[a]; Università di Perugia[b], I-06123 Perugia,Italy
[13]INFN, Sezione di Firenze, I-50019 Sesto Fiorentino[a]; Università degli Studi di Urbino 'Carlo Bo', I-61029 Urbino[b], Italy
[14]Laboratoire Kastler Brossel, ENS, CNRS, UPMC, Université Pierre et Marie Curie, 4 Place Jussieu, F-75005 Paris, France
[15]INFN, Sezione di Genova; I-16146 Genova, Italy
[16]INFN, Sezione di Roma Tor Vergata[a]; Università di Roma Tor Vergata, I-00133 Roma[b]; Università dell'Aquila, I-67100 L'Aquila[c], Italy
[17]RMKI, H-1121 Budapest, Konkoly Thege Miklós út 29-33, Hungary
[18]INFN, Gruppo Collegato di Trento[a] and Università di Trento[b], I-38050 Povo, Trento, Italy; INFN, Sezione di Padova[c] and Università di Padova[d], I-35131 Padova, Italy

E-mail: `elena.cuoco@ego-gw.it`

Abstract.
The understanding of noise in interferometric gravitational wave detectors is fundamental in terms of both enabling prompt reactions in the mitigation of noise disturbances and in the establishment of appropriate data-cleaning strategies. Monitoring tools to perform online and offline noise analysis in areas such as transient signal detection, line identification algorithms and coherence are used to characterise the Virgo detector noise. In this paper, we describe the framework into which these tools are integrated - the Noise Monitor Application Programming Interface (NMAPI) - and provide examples of its application.

1. Introduction

The detection of gravitational wave (GW) signals is one of the big challenges of astrophysical research. The interferometric detectors [1] can detect gravitational waves by measuring the relative displacement of mirrors suspended at the end points of optical cavities. Virgo [2], the 3km-long French-Italian interferometer, located in Italy, is one of these detectors. The LIGO [3] laboratory, in the USA, operates two 4km-long detectors, while a 600m-long detector is operated by GEO [4] in Germany.

For all of them, the detector noise, which limits their sensitivity to gravitational wave signals, is a major challenge to face. As a result, understanding detector noise and the identification of physical noise sources and the ways in which they affect the Dark Fringe (DF) channel [1], is a major goal. The DF is sensitive to gravitational waves and is calibrated [5] to produce an equivalent strain signal. It is important to be able to follow the noise behaviour, not only to enable prompt reactions in the mitigation of noise disturbances, but also to help in the development of appropriate data-cleaning strategies. Before the last Virgo Scientific run (VSR4, June 3th - September 5th 2011) a new framework for detector noise monitoring was implemented. The idea was to have tools (Noise Monitors, sometime referred to as NM) able to process data that were either aquired in real-time by the online data acquisition system [6] (these are referred to as online data) or data archived on disk (offline data). The results of this processing are then archived in MySQL databases or similar storage solutions.

From these sources, users would be able to produce plots based on available data, including those that had been recently acquired. The aim was to give, to both scientists and the commissioning crew, instruments that are able to easily provide a snap-shot of noise behaviour at any given moment. Tools to monitor transient noise signals [10] and spectral lines [13], as well as to check coherence with signals from auxiliary sensors monitoring external noise sources, were included in a common framework, the Noise Monitor Application Programming Interface (NMAPI) [8]. One of the main requirements for NMAPI was to make the integration of new noise monitors as straightforward as possible.

This paper is organised as follows: in section 2 the NMAPI framework is described; in section 3 the monitoring tools used during VSR4 are briefly detailed; section 4 looks at the noise spectral lines catalogue application (LinesDB); while in section 5 a typical NMAPI use case is explained. In section 7 a possible evolution strategy for the NMAPI project is discussed.

2. Noise Monitor Application Programming Interface (NMAPI)

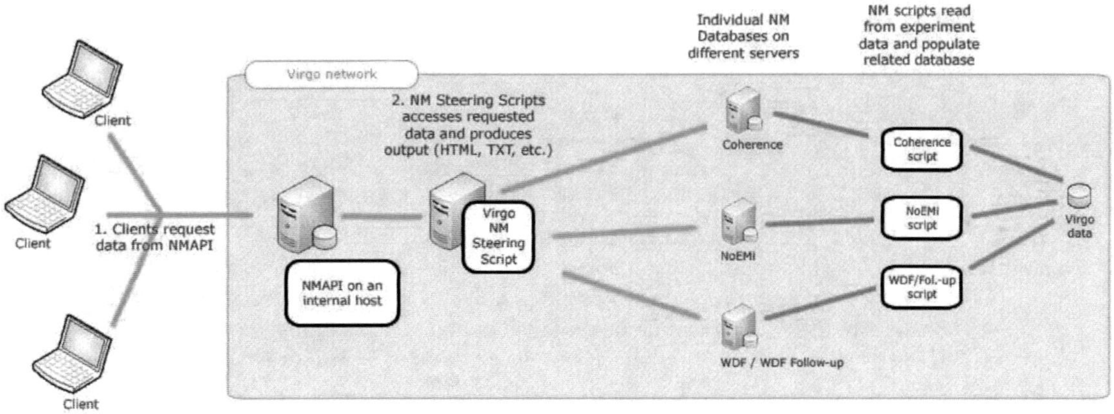

Figure 1. A simple representation of the NMAPI architecture. The noise analysis algorithms produce results that are archived in MySQL databases on different servers. A client requests HTML pages from the NMAPI host, which queries the relevant database, retrieving the requested data and producing the corresponding HTML pages.

[1] Channel refers to a time series obtained following conversion of an analague data stream to a digital data stream.

NMAPI provides a framework into which different noise monitors can be integrated and then configured, enabling users with little knowledge of web development or web programming languages and standards, to interface their scripts to an Internet audience. The application enables authenticated users to not only access information, but also to produce results, e.g. HTML pages or text files, based upon bespoke search criteria specified by them via the user interface. This mode of producing results dynamically means that, not only are users able to produce highly-configurable results, but storage requirements are negligible, given that files are created on-the-fly and only the summary plots are stored on disk. For example, in a first version of the Coherence tool the plots showing the coherence between channels, available to users, were saved to disk. With the NMAPI solution it was possible to reduce disk space usage by 98%.

Noise Monitor Administrators use the tool to build HTML forms, specific to a single Noise Monitor, which may then in turn be used by authenticated users to launch the Steering Script related to that Noise Monitor. The Steering Script receives the arguments passed to it by the HTML form POST array and produces the relevant results as required, e.g. text-format datasets or plots. In addition, each Noise Monitor provides a summary page, configurable by the related Administrator, which provides general information, which may be amended by standard users in terms of both date and related channel. NMAPI also provides an in-built debugging system, which can be used to monitor and adjust Steering Scripts and HTML forms as required. In general, NMAPI is designed to be as flexible as possible. It enables Administrators to concentrate their time and effort on providing a first-class script, rather than having to worry about how to launch it via a web interface. The degree of potential customisation available means that it is possible to provide script launch forms that are appropriate to each individual Noise Monitor. In Figure 1 a simple representation of the NMAPI architecture is displayed.

3. Noise monitors

Before VSR4 began, we integrated the following monitors into the NMAPI framework:

- *Wavelet Detection Filter (WDF)*: WDF finds triggers [2] associated to transient signal events, which could potentially be due to noise, in the detector data. In particular, it analyses data in the time domain, where a whitening [9] procedure is applied and, following a wavelet transform (using different banks of wavelets), it looks for an excess of power with respect to background noise due to transient signals. It identifies the triggers and assigns a start time for the event, calculates its signal-to-noise ratio, frequency content and duration [10]. This is done online on the DF signal and on a set of selected auxiliary channels. All of the parameters that characterise the events are stored in a MySQL database.
- *Noise Events Miner (NoEMi)*: NoEMi is the Virgo-dedicated software for noise line monitoring and identification. Based on the algorithms implemented for the continuous wave (CW) search, it analyses, on a daily basis, the GW channel and a subset of the auxiliary channels, identifying the noise lines in the various channels and looking for coincident lines between the Dark Fringe and the auxiliary channels. A line tracker algorithm tracks the lines over time, making the follow-up of the non-stationary lines easier. A detailed description of the NoEMi software is given in [13]. Three Noise Monitors have been developed using the NoEMi output data, and have been integrated into the NMAPI:
  - *NoEMi Peakmaps*: the first step of the NoEMi process is the identification of the peaks from the power spectral density of each channel. The peak maps are stored in a dedicated database, from which they can easily be queried and displayed thanks to the NMAPI infrastructure.

[2] We define a trigger as an event caused by a generic transient signal and which can be identified by features such as the event start time, its duration, its spectral content and its signal-to-noise ratio.

  - *LinesDB*: identified spectral lines are archived into the NoEMi lines database, which stores all of the information relating to the lines reconstructed by the NoEMi line tracker algorithm. An interface that allows users to easily query information from the lines database (LinesDB) is part of the NMAPI tool suite and is described in more detail in Section 4.
  - *NoEMi Spectra*: NoEMi also stores the daily-averaged frequency spectra of the calibrated data. The spectra have a resolution of 10 mHz. The NMAPI allows users to access the spectra in a given time interval and calculate the resulting average, comparing it with other time intervals.

- *Coherence*: estimates the averaged (over 15 minutes) coherence between the DF signal and all of the (approximately 1000) auxiliary channels. It runs periodically (two or three times per hour) and archives results in a MySQL database. The NMAPI framework makes it possible to retrieve data from the database via a web interface or to request coherence plots between the DF and a selected acquired channel, producing them on-the-fly. This script can also be called by NoEMi to further investigate the coincidence between lines (as explained in the following section).

4. LinesDB

LinesDB is the web interface to the lines database, the archive of noise lines identified in the Virgo data by the NoEMi framework. The database is updated in real-time during the NoEMi data-taking periods, and is linked to a metadata table, which is filled with user-defined information, such as the type of instrument used in measurement and general information relating to the line, once the source of the disturbance is understood.

The LinesDB interface allows users to easily search lists of archived lines that meet specific criteria, for instance lines found in a given time or frequency interval. It also makes it possible to set thresholds on dynamic parameters [13] and even on the lines coincident with a subset of the auxiliary channels. It is also possible to use the metadata as additional filtering criteria, for example making it possible to distinguish between 'known' (i.e. lines for which a metadata description is available) and still 'unknown' lines, or to select lines containing specific descriptions.

Figure 2 shows the web page, including the interface used to query the lines database, with an example query (in this example the search looks for the list of identified 'violin modes', the mirror suspension resonances). Figure 3 presents the result of the query. The line list is formatted in a table which shows, for each line, the mean values of the dynamic parameters, the list of the most coincident auxiliary channels (constantly updated by NoEMi) and the line description (part of the line metadata). The list of coincident auxiliary channels is itself a link to the Coherence Noise Monitor. By clicking on an auxiliary channel name, the Coherence script is launched, which queries the Coherence database and displays the coherence between the Dark Fringe and the selected auxiliary channel in the time and frequency ranges of the line.

5. How NMAPI works

An example of how NMAPI can help in noise hunting and how the links between the different noise monitors operate is presented here.

In this example, a user examines the LinesDB summary page, which contains a list of all of the lines identified and archived by NoEMi (see Figure 4). An inspection of the results shows a line of particular interest (for example the 18.6 Hz). A look at the line plot in Figure 5 confirms that the line is worth examining in further detail, it being a non-stationary line, which shows a step down in the frequency after 30 days.

The user studies the history of the disturbances in the Dark Fringe channel and the coincident auxiliary channels versus time, using the calendar tool available in the NoEMi PeakMaps

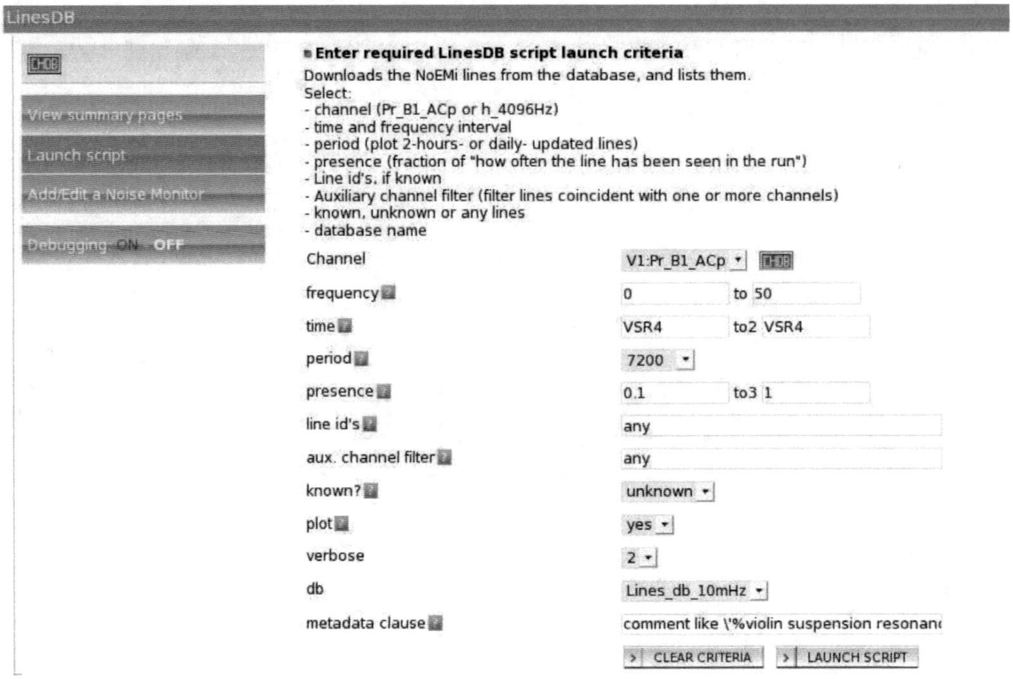

Figure 2. The web interface used to query the lines database. Among the query parameters it is possible to choose between 'known' and 'unknown' lines, to define the frequency and time ranges, as well as to use filters based on the available metadata information.

Mean Freq. (Hz)	Freq. Range (Hz)	First/Last Seen - Presence	Mean Persist.	Mean CR	log(Mean Ampli)	Aux. channels	Comment on source
2749.995	2749.795-2750.024	2011-06-02_22/ 2011-09-06 - 1.00	0.4	10.4	-13.3	Em_MABDMC02(28.3%) Em_MABDWE01(26.3%) Em_MABDNE01(21.9%) Em_MABDCE01(21.5%)	violin suspension resonance (6th harmonic)
2760.015	2759.990-2760.057	2011-06-02_22/ 2011-09-06 - 0.98	0.2	8.6	-13.5		violin suspension resonance (6th harmonic)
2761.345	2761.315-2761.373	2011-06-02_22/ 2011-09-06 - 0.97	0.2	8.8	-13.5		violin suspension resonance (6th harmonic)

Figure 3. A LinesDB query result. For a description of the displayed parameters (Persistency, CR, amplitude, etc.) see [13].

summary pages, in order to choose a day of interest. Via the NM script interface, the user launches the steering script, obtaining a detailed peak-map or line plot for a specific channel (Figure 6). The user confirms the results by looking at the Coherence tool (Figure 7), which displays the coherence between the Dark Fringe and the selected auxiliary channel in the given time and frequency ranges. Once the source of the disturbance is understood (in this example, the 18.6 Hz line was caused by the motor of an air conditioning machine in the Virgo Central

Mean Freq. (Hz)	Freq. Range (Hz)	First/Last Seen - Presence	Mean Persist.	Mean CR	log(Mean Ampli)	Aux. channels	Comment on source
17.004	16.994-17.052	2011-04-25/ 2011-09-05_22 - 1.00	1.0	133.8	-8.9		Global Control (Gc_MICH) - Calibration Line
18.612	18.501-18.721	2011-04-25/ 2011-09-05_22 - 0.98	0.5	26.5	-10.6	Em_SETODE01(98.1%) Em_AC_EIB(97.2%) Em_SE_CryoT(96.7%) Em_SETCSWI01(96.1%) Em_SEBDCE01(96.1%) Em_SETCSNI02(94.2%) Em_SE_EIB_04(92.2%) Em_SE_LB_04(92.1%) Em_SE_BrewIN)(92.0%) Em_ACBDCE01(91.4%) Em_SEDBDL03(84.0%) Em_SEBDCE08(82.0%) Em_SEBDCE06(72.2%) Em_ACTCSNI(57.9%)	

Figure 4. List of spectral lines identified by NoEMi

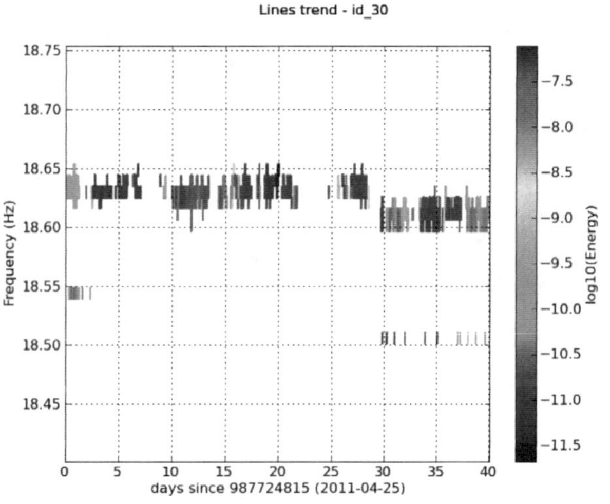

Figure 5. Wandering line at 18.6Hz. It has a step down in the frequency after 30 days

building, which was producing seismic vibrations of the floor that subsequently entered the data), using the LinesDB interface, the user can insert related metadata, adding a description to the line information and providing qualitative information. This information might regard, for instance, which sub-system of the interferometer caused the disturbance, how the line was measured, which mitigation actions were undertaken, and so on.

6. A real-life example: the VSR4 Vela bump

As already mentioned, the NM infrastructure was first used in the last Virgo science run, VSR4. Despite the short run duration (3 months), the framework proved to be a useful instrument in the detection of noise disturbances and helped to provide fast feedback to scientists, in order to help them to identify and mitigate noise sources.

The best example of noise monitoring and hunting is of a broad disturbance detected by the NoEMi tool within the first few days of the run. The disturbance affected the ITF sensitivity at the expected GW emission frequency of the Vela pulsar (22.4 Hz), hence the name 'Vela bump'.

Thanks to the frequency resolution of the NoEMi databases, we observed that the disturbance

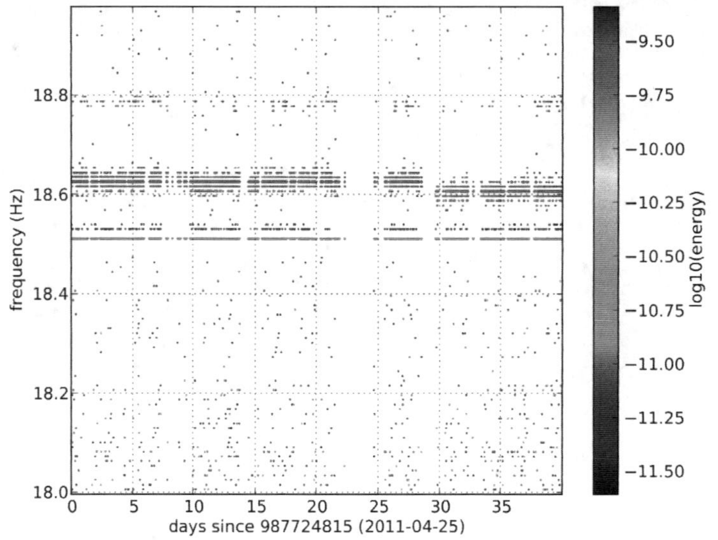

Figure 6. Correlation with auxiliary channels

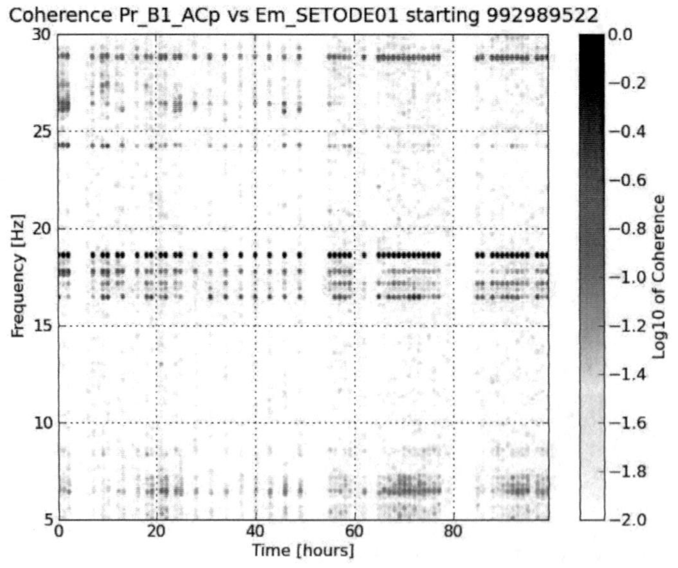

Figure 7. Coherence plot: coherence values between DF signal and one seismic monitoring signal in a frequency-time map

consisted of a dense comb of lines (see Figure 8), and, after deeper investigation, the source of the lines was understood: they were generated by the non-linear coupling of a set of calibration and control lines (sinusoidal signals injected into the interferometer for calibration and control purposes). The noise was moved away from the Vela pulsar frequency by shifting the frequencies of the calibration lines. The effect of the mitigation is visible in Figure 9.

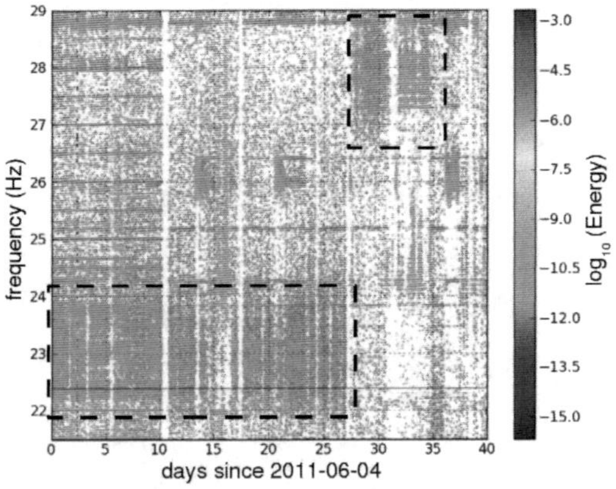

Figure 8. Time-frequency plot of the peak maps at the Vela frequency region. The noise bump was highlighted in the dashed rectangles before and after the change of the frequencies of the calibration lines.

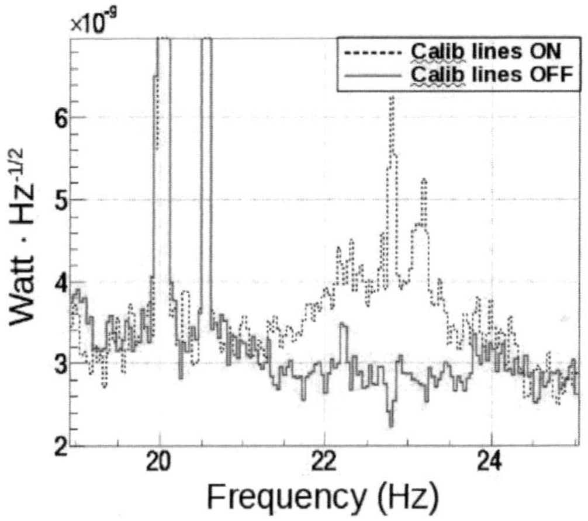

Figure 9. Effect of the calibration signal switch off on the noise power spectral density.

More details on the Vela bump are given in [13].

7. NMAPI evolution

Over the next short- and long-term periods, several developments and upgrades of NMAPI are planned. These include minor modifications, to render NMAPI ever-simpler to use from a user perspective, as well as more complex and detailed developments, designed with the improvement of NMAPI back-end operations in mind. Here we look at each of the upgrades individually.

- *Interface developments.* While NMAPI enables Noise Monitor Administrators to create bespoke script-launch forms, we have seen during testing and usage phases, that there are some occasions upon which users really need to be able to be even more flexible than usual when attempting to produce datasets and results. In order to meet this requirement, the possibility to enter conditional script syntax directly to an HTML textarea, will be provided.
- *Database developments.* In terms of the NMAPI database itself, the possibility to tag data (as VSR3, VSR4, etc.) in an extensible and modular format, will be added.
- *Improvement of cross-application linkages and communication.* NMAPI already incorporates linkages with another Virgo PHP environment application in the form of the Channels Database [14], a source of regularly updated information on channels available within the Virgo Data Acquisition system. The aim is to increase the inter-operability and communication between applications within this environment and NMAPI via the use of a dedicated connection-handling database (CDB) [15].
- *Use of additional data sources and result cross-correlation.* Discussions have taken place regarding the use of NMAPI to interrogate data produced by other interferometers (GEO, LIGO). In the event of additional data sources being used by a central NMAPI installation, a tagging system, enabling data to be recognised by given criteria, would be fundamentally important. In addition, questions relating to data location, input-output and connectivity issues, as well as usage of reduced datasets would also have to be addressed. Integration of

data and results produced by other similar detectors will also facilitate cross-correlation of advanced detector results.

- *Distributed NMAPI.* With Advanced Virgo [16] and LIGO [17] in mind and significant increases in channel examination foreseen for several of the Noise Monitors already interfaced to NMAPI, it will be necessary to move to a distributed version of the application. At present, Steering Scripts are managed by a PHP semaphore and launched by NMAPI directly on the NMAPI local host, as and when conditions allow. However, given increased usage and larger amounts of data to process, the restriction to the use of a single host could become a serious limit. Therefore, a move to a distributed system, D-NMAPI [18], is foreseen.

8. Conclusion

The NMAPI framework, into which the noise monitoring tools were integrated during the last Virgo scientific run, has been described. Online monitoring is fundamental for noise detector characterisation, in order to be able to act promptly on the interferometer and cure possible problems, as well as to assist in the development of data-cleaning strategies for the astrophysical detection groups. An example of how NMAPI was successfully used during VSR4 to identify an excess of noise in the region of the Virgo sensitivity in which the gravitational signal emitted by the Vela pulsar could be detected, has also been presented. Thanks to this it was possible to identify the noise source and to remove the problem, leading to an improved sensitivity in that region in the VSR4 data-taking period.

We plan to further enrich the NMAPI framework, in order to allow for a heavier computational load, by distributing the requested scripts and using a database cluster to collect the processed data of the Advanced generation of interferometric detectors.

9. References

[1] Cella G & Giazotto A 2011 Interferometric gravity wave detectors *Rev. Sci. Instrum.* **82** 101101
[2] Acernese *et al* 2008 *Class. Quantum Grav.* **25** 114045
[3] Abbott B *et al* 2009 *Rep. Prog. Phys.* **72** 076901
[4] Grote H *et al* 2010 *Class. Quantum Grav.* **27** 084003
[5] Accadia T *et al* 2011 Calibration and sensitivity of the Virgo detector during its second science run *Class. Quantum Grav.* **28** 025005
[6] Acernese F et al. Data Acquisition System of the Virgo Gravitational Waves Interferometric Detector Nuclear Science, IEEE Transact, 55, 252-232, (2008)
[7] `http://www.mysql.com`
[8] Colla A, Cuoco E & Hemming G, 2011 Noise Monitor Application Programming Interface (NMAPI) Software Requirements (VIR-0226B-11)
[9] Cuoco E *et al* On-line power spectra identification and whitening for the noise in interferometric gravitational wave detector 2011 *Class. Quantum Grav.* **18** 17271751
[10] Cuoco E, A new wavelet-based method for transients detection. Efficiency with respect the whitening algorithms (VIR-NOT-EGO-1390-308),
[11] Carbognani F, Colla A, Cuoco E, Hemming G, 2011 Lines Database Web Interface Software Requirements, Virgo internal note (VIR-0227A-11)
[12] Accadia T *et al* 2011 The seismic Superattenuators of the Virgo gravitational waves interferometer Journal of Low Frequency Noise, Vibration and Active Control **30**
[13] Accadia T *et al* 2011 The NoEMi framework *These proceedings*
[14] Hemming G, Verkindt D Virgo Channels List Data Base (VIR-0079A-08)
[15] Hemming G Connections Database Project Proposal,Thecnical report (VIR-0225A-11)
[16] Accadia T et al. (Virgo collaboration) Plans for the upgrade of the gravitational wave detector Virgo: Advanced Virgo, to be published in Proceedings of Twelfth Marcel Grossman Meeting on General Relativity, edited by Thibault Damour, Robert T Jantzen and Remo Ruffini, World Scientific, Singapore, 2012
[17] Harry G M (for the LIGO Scientific Collaboration) Advanced LIGO: the next generation of gravitational wave detectors Class. Quantum Grav. 27 084006 (2010)
[18] Colla A, Cuoco A, Hemming G Distributed NMAPI requirements (in preparation)

A new glitch-rejection algorithm forged in the spherical harmonic basis

Mark Edwards, Patrick J. Sutton

School of Physics and Astronomy, Cardiff University, Queens Buildings, CF24 3AA, Cardiff, United Kingdom

E-mail: `mark.edwards@astro.cf.ac.uk`

Abstract. The noise in the output of a gravitational-wave interferometer is known not to be due to stationary Gaussian processes. In fact, even after whitening, it will contain spikes of very large power (called *glitches*), due to environmental disturbances, instabilities in detector systems, and other factors. These glitches can sometimes mimic, and, due to their loudness, obfuscate real signals close in time. In this article we outline a new method of discriminating these glitches from gravitational-wave signals which works on the principle of the spherical harmonic decomposition of the correlation of the data streams in the detector network. It is intrinsically a coherent all-sky method (working, as it does, in the spherical harmonic domain) which can produce a time-series showing the glitchiness of the underlying data. We demonstrate this spherical-harmonic technique using real interferometer data and compare it to another glitch-rejection method in current use.

1. Introduction

Detecting or reducing the effect of glitches in the output of gravitational-wave interferometers is an important part of any search for transient gravitational-wave signals [1, 2]. A loud glitch is nothing but a nuisance; it can mimic a signal when coincident with glitches in the other data streams, it can wash out the effect of a real signal present, and it can affect background estimation too. Thus, there is a need to identify the presence of glitches accurately and quickly without giving false readings - mischaracterising a signal as a glitch or vice versa. This article will outline a flexible, and simple, discriminator that can be used to categorise transients as signals or glitches. Specifically, we perform a spherical harmonic decomposition of the sky map constructed from the pairwise correlation of the whitened data streams from the detector network. We find that the ratio of the power in the $\ell > 0$ modes to that in the $\ell = 0$ mode provides a powerful discriminator for rejecting background glitches. We demonstrate this technique using real interferometer data from the fourth LIGO Science Run [3]. We also compare it to another glitch-rejection technique used in coherent searches for gravitational-wave bursts.

2. Gravitational-wave burst searches

We focus on searches for gravitational-wave *bursts*; transient signals whose waveforms are not known *a priori*. (Our method does not require knowledge of the waveform, but it should be equally useful in a matched-filter search where the waveform is known.) Burst searches are usually carried out using a coherent analysis, in which the data from a network of detectors is combined in amplitude and phase to form one or more statistical measures of the likelihood of a signal being present. These statistics are commonly computed as time-frequency maps, where candidate signals are identified as clusters of pixels of large value; see Figure 1 for an example. One commonly used search pipeline is COHERENT WAVEBURST (cWB) [4], which computes a constrained likelihood measure of the gravitational wave signal-to-noise ratio, and combines this with a measure of the correlation across the network to reject or down-weight glitches. X-PIPELINE [5] computes time-frequency maps of the energy in the reconstructed plus, cross and null [7] polarisations of a gravitational wave, and rejects glitches based on the relative contribution of the detector cross-correlations to these energies [6]. Similarly, the STAMP algorithm [8, 9] uses the auto-power in each detector to discriminate between Gaussian noise (possibly including a gravitational wave) and glitches.

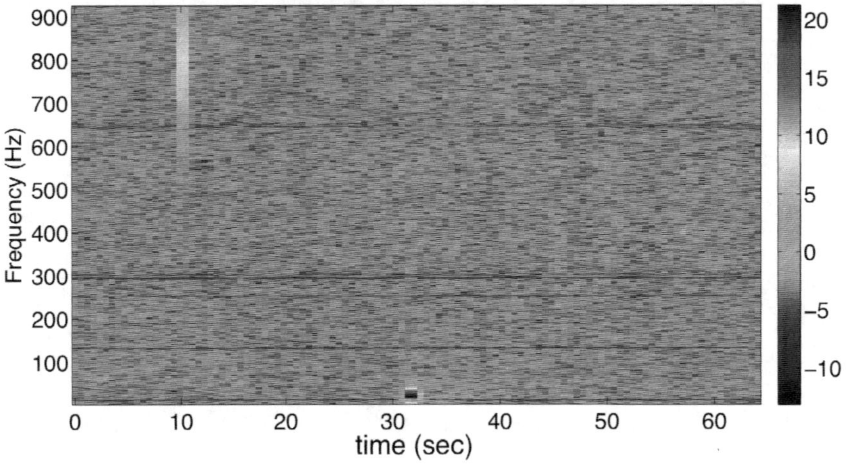

Figure 1. A time-frequency map of whitened data; to the upper left we see a short-duration white noise burst (WNB) signal, to the lower middle a glitch. The color scale is logarithmic in coherent network power, as defined in equation 4.

3. The Spherical Radiometer pipeline

A common problem with the coherent analysis is its speed. It takes time to form a skymap of coherent energy - each direction is usually calculated independently, so the higher the resolution required, the longer the generating time. To improve the speed of all-sky, all-time burst searches, we have been exploring a fundamentally new approach: spherical radiometry. This approach takes advantage of the fact that sky maps in gravitational-wave searches show strong correlations over large angular scales in a pattern closely tied to the network geometry. Computing sky maps indirectly through their spherical harmonics avoids much of the redundant calculations.

The core of our spherical radiometer pipeline, SPHRAD, is a set of fast cross-correlator codes written by Cannon [10]. This engine transforms the problem of computing correlations between data streams into the spherical harmonic domain, allowing correlation between detectors in a network to be performed extremely quickly. These codes are wrapped by the Ω pipeline [11];Ω finds, loads, and whitens the data, optionally generating simulated glitch and injection signals (for testing). We also use the suite of post-processing codes in Ω to process the output event triggers.

3.1. Output

To generate the cross-correlation for each detector pair in our network, we need to compute the integral

$$\xi_{1,2}(\hat{s}) = \int_{t-T/2}^{t+T/2} g_1(t' - \vec{r}_1 \cdot \hat{s}) g_2(t' - \vec{r}_2 \cdot \hat{s}) dt' \,, \tag{1}$$

where we follow the notation of [10], so $\vec{r}_1$ and $\vec{r}_2$ are the position of each detector, and $\hat{s}$ is the sky direction of interest. We can compute this integral efficiently by using the fact that the dependence on sky position, $\hat{s}$, can be separated out. We assume that the data $g_{1,2}(t)$ has been regularly sampled and can be represented as a vector, $\vec{g}$. This means we can write

$$\vec{g}(\hat{s}) = \mathbb{T}(\vec{r} \cdot \hat{s}) \cdot \vec{g} \,, \tag{2}$$

where $\mathbb{T}(\vec{r} \cdot \hat{s})$ is the matrix that retards the time series according to the sky location. The integral (1) can now be written as an inner product

$$\begin{aligned} \xi_{1,2}(\hat{s}) &= \frac{1}{N} \vec{g}_1^T(\hat{s}) \cdot \vec{g}_2(\hat{s}) \\ &= \frac{1}{N} \vec{g}_1^T \cdot \mathbb{T}^T(\vec{r}_1 \cdot \hat{s}) \cdot \mathbb{T}(\vec{r}_2 \cdot \hat{s}) \cdot \vec{g}_2 \,. \end{aligned} \tag{3}$$

Note that the dependence on sky position, $\hat{s}$, has been decoupled from the data, $\vec{g}_i$. By using an Earth-fixed coordinate system for $\hat{s}$, Fourier transforming $\vec{g}_i$ and expanding $\mathbb{T}$ to finite order in spherical harmonics we can efficiently calculate this integral over the sky (for more details see [10]). This calculation must be performed for each pair of detectors that we have in our network, so the total coherent network power is therefore given by

$$N_{\text{total}} = \underbrace{\xi_{1,2} + \xi_{1,3} + \xi_{2,3}}_{\text{cross-correlation}} + \underbrace{\xi_{1,1} + \xi_{2,2} + \xi_{3,3}}_{\text{auto-correlation}} \tag{4}$$

which we use as our statistic when we are forming time-frequency and time-harmonic maps.

For each 'time bin' (our FFT length), the detector data is whitened, Fourier transformed and fed into the correlator. The output is a 2-dimensional 'grid' (see Figure 2) - the spherical harmonic ex pansion for each frequency of interest. Along one axis we have the frequency span of our data, along the other we have the spherical harmonic coefficients, c_{lm}, of that expansion (the coefficients are ordered according to SPHARMKIT [12]).

Similarly, if we sum along the coefficients, we get the power at *each frequency* for that time bin, averaged over the whole sky (as we are dealing with spherical harmonics). If we then generate this for each of the time bins in our segment (usually 64 or 128 seconds, divided into 1 second bins), we get a coherent time-frequency map. Here though, the value in the time-frequency pixel is the power in the correlations for our detector network, N_{total}, defined by equation (4) above. See Figure 1 for an example time-frequency map.

Similarly, if sum along the frequencies for each spherical harmonic coefficient, we get a single set of coefficients for *all frequencies*, relating to the power over the sky. Generating this for each time bin in our segment we get a time-harmonic map. Each set of coefficients tell us something about the distribution of power over the sky for that time (see Figure 3). This 'dual view' of the data is very important, and is one of the major advantages of decomposing the integral into spherical harmonics.

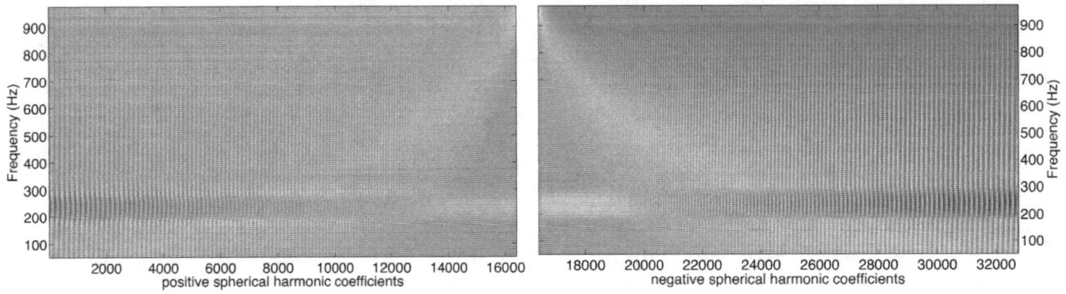

Figure 2. Frequency-spherical harmonic map in log-scale colouring for the time bin containing an injected sine-Gaussian. The injection is clearly visible as a horizontal line at $f = 235$ Hz, the central frequency of the injection.

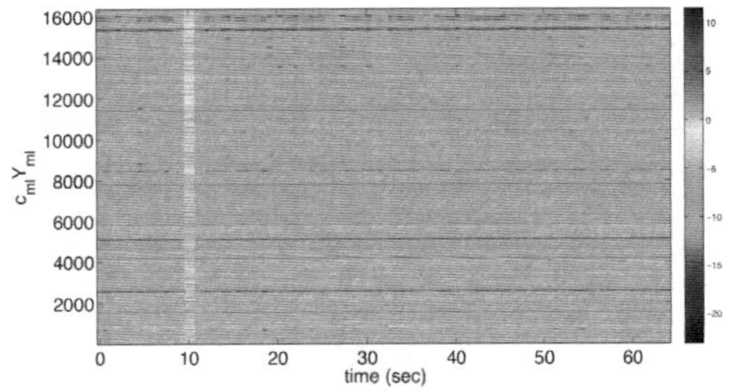

Figure 3. Time-harmonic map for a segment containing a glitch and an injection. Only the injection is visible (the vertical line at 10s) - compare to Figure 1)

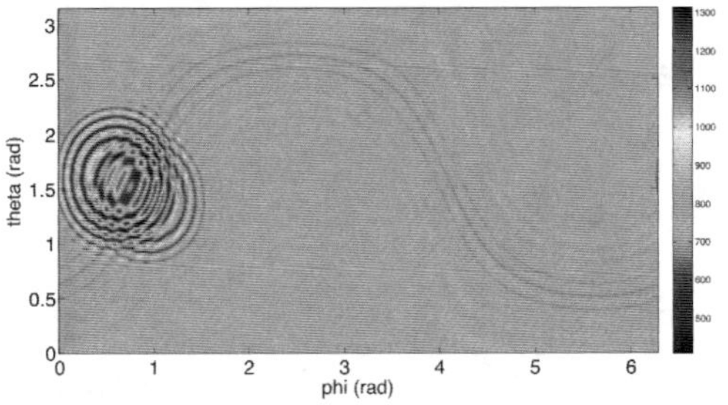

Figure 4. Skymap of the coefficients corresponding to the time bin containing the injection, see Figure 3 at 10s

3.2. Skymaps

For easier interpretation, the spherical harmonic coefficients can be transformed into the image domain using a Fast Spherical Harmonic Transform (FSHT), which is the spherical harmonic equivalent of the Fast Fourier Transform. The FSHT is defined in [13] and found in [12]. Transforming a set of coefficients into the image domain using the FSHT generates a skymap showing the localisation of power in the sky; see Figure 4 for an example. This allows for the formation of any statistic that can be constructed from the correlation between the detectors in the network. For the Livingston, Hanford, Virgo (*HLV*) network, the cross-correlations *HL*, *HV*, *LV* are available, as well as the auto-correlations of *H*, *L* and *V*. Thus, it is possible to easily construct (for instance) the coherent energy in the plus polarisation (equation 5; see [5]), by multiplying the correlation of each detector pair, $\tilde{d}_{ij}$, by the appropriate antenna factors, e^+_{ij}, in the image domain and summing:

$$E_+ = \sum |e^+ \cdot d|^2 \tag{5}$$

$$\begin{aligned}
&= e^+_{11} \cdot \tilde{d}_{11} \cdot e^+_{11} \cdot \tilde{d}^*_{11} + e^+_{11} \cdot \tilde{d}_{11} \cdot e^+_{12} \cdot \tilde{d}^*_{12} + e^+_{11} \cdot \tilde{d}_{11} \cdot e^+_{13} \cdot \tilde{d}^*_{13} + \\
&\quad e^+_{21} \cdot \tilde{d}_{21} \cdot e^+_{21} \cdot \tilde{d}^*_{21} + e^+_{21} \cdot \tilde{d}_{21} \cdot e^+_{22} \cdot \tilde{d}^*_{22} + e^+_{21} \cdot \tilde{d}_{21} \cdot e^+_{23} \cdot \tilde{d}^*_{23} + \\
&\quad e^+_{31} \cdot \tilde{d}_{31} \cdot e^+_{31} \cdot \tilde{d}^*_{31} + e^+_{31} \cdot \tilde{d}_{31} \cdot e^+_{32} \cdot \tilde{d}^*_{32} + e^+_{31} \cdot \tilde{d}_{31} \cdot e^+_{33} \cdot \tilde{d}^*_{33}
\end{aligned} \tag{6}$$

Similar calculations can be carried out for the cross polarisation and null stream [5, 7, 14], or any other statistic that can be written as the sum of pairwise cross- and auto-correlations between detectors.

3.3. Projections

The time-frequency and time-harmonic maps derive from a higher order map, so they contain the same total power. The distribution is different however, which implies something about the underlying structure. For simulated signals injected with a prescribed central frequency and peak time, the time-frequency map lights up for the duration and bandwidth expected. Similarly, for glitches, the time-frequency map also lights up the pixels around the expected time and frequency (see Figure 1). In contrast, the time-harmonic map shows structure along the coefficients at the time of an injection, but nothing at the time of a glitch (see Figure 3).

This leads to the conjecture that the spherical harmonic coefficients could be used as some form of glitch discriminator, and a quick aside to think about how details on the sky map relate to the spherical harmonic coefficients should convince why this is so.

3.4. Sky localisation

Using a network of gravitational wave detectors to localise an event is a process of triangulation - the time that the event is seen in each detector gives you a measure of its location [15, 16].

- **Behaviour with signal:** In this case, there is a preferred time delay for each detector that will match the signal up, corresponding to a preferred sky direction. In the time-frequency representation, we have essentially summed over the whole sky, so direction/time delays are not important. However, in the time-harmonic representation a signal is resolved to a patch of the sky meaning that power will be present in most of the coefficients - generally none will stand out more than any other.
- **Behaviour with glitch:** There is no preferred time-delay for a glitch, as there is nothing *coherent* for it to couple to in the other detector streams. And, as the time-frequency representation is independent of the time delay, we see similar behaviour to that with a signal. In spherical harmonic terms, the zeroth component, $c_{00}Y_{00}$, corresponds to the amount of energy present over the whole sphere, while the higher order terms relate to fine details. Thus, the zeroth component is a measure of amount of incoherent energy there is in the skymap, which directs relates to the glitch energy, as this is mostly present in the $c_{00}Y_{00}$ coefficient.

4. Spherical harmonic statistic

The new coherent statistic works on the principle that we can decouple the 'incoherent' and 'coherent' energy using the decomposition into spherical harmonics. Using the nomenclature c_{lm} to represent the spherical harmonic coefficients, the $|c_{00}|$ coefficient corresponds to the incoherent energy, while the sum of the rest of the coefficients corresponds to the coherent energy. For a signal (which has a preferred sky location), all of the coefficients have some measure of power, however, for a glitch, most of the energy is located in the c_{00} coefficient.

The algorithm is as follows:

(i) Decompose a coherent skymap into the spherical harmonic domain using the FSHT (if required).
(ii) Sum all of the $\ell > 0$ coefficients, $\sum |c_{lm}|$, as a measure of coherent energy.
(iii) Divide the sum by $|c_{00}|$, the measure of incoherent energy.
(iv) Repeat for all time bins in a data segment. Any bin that rises above the background is a signal, any falling below a glitch. See Figure 5 for an example.

Essentially, for each time bin we calculate:

$$\mathbf{Z_{ratio}} = \frac{\sum_{l=1}^{l_{max}} \sum_{m=-l}^{l} |c_{lm}|}{|c_{00}|} \tag{7}$$

where l_{max} is defined using a combination of baseline length and the data sample frequency (see [10] for more details). Figure 5 shows the statistic for data from the 4th LIGO science run (S4).

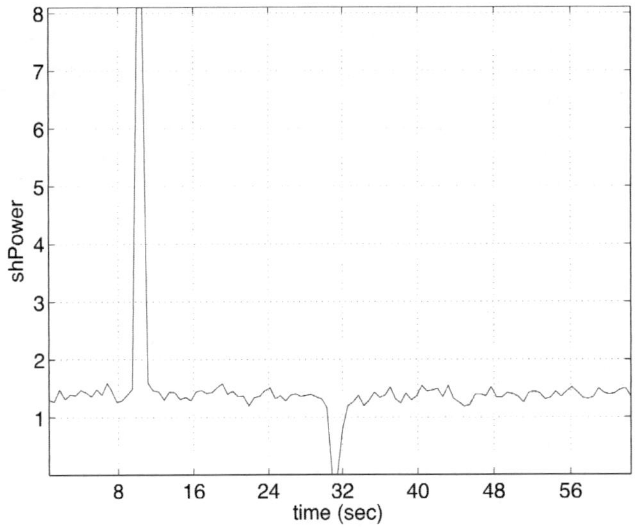

Figure 5. The $\mathbf{Z_{ratio}}$ statistic of equation (7) for the 64 seconds of data displayed in Figures 1 and 3). The injected gravitational-wave signal rises well above the background noise at t = 10 s, while the glitch drops below at t = 31s.

5. Results

To test the robustness of the new statistic, it was applied to both simulated and real data, sampled at 2048 Hz, and using the Hanford-Livingston-Virgo (HLV) network. White Noise Bursts (WNB) containing 2 independent polarisations of approximately equal power were injected into the data with a central frequency of 300 Hz, a bandwidth of 50 Hz and a duration of 0.1 seconds. A range of root-sum-square injection amplitudes were used, from 3.7×10^{-23} $s^{-1/2}$ to 1.5×10^{-19} $s^{-1/2}$, defined as:

$$h_{RSS} = \sqrt{\int_{-\infty}^{\infty} (|h_{+}(t)|^2 + |h_{\times}(t)|^2)dt} \tag{8}$$

where $h_{+}(t)$ and $h_{\times}(t)$ are the plus and cross-polarisation strain functions, and h_{RSS} has units of $Hz^{-1/2}$.

5.1. Simulated noise data

We analysed approximately one month of simulated noise data using the SPHRAD pipeline. Obviously, this data is actually Gaussian and does not contain any glitches, so, we used a very simple glitch model to generate a population of roughly one glitch per hour for the analysis. This consisted of a family of sine-Gaussians with random parameters (central frequency, duration, peak time, amplitude) inserted into the simulated data. In the course of the analysis, all of the standard coherent energies are generated (plus, cross, null and standard likelihood), as well as a timeline of the amount of power in the spherical harmonic coefficients. This allows us to generate scatter plots of the coherent/incoherent energy for each trigger.

In the following scatter plots, I_{sh} corresponds to the power in the zeroth spherical coefficient (i.e. $|c_{00}|$, the incoherent energy), while E_{sh} corresponds to the sum of the $\ell > 0$ coefficients. The energy in the null stream (i.e. the projection of the data in which a gravitational wave signal should be cancelled, E_{null} and I_{null}), is computed using the cross-product of the orthonormal antenna factors for the cross and plus-polarisation, see [5].

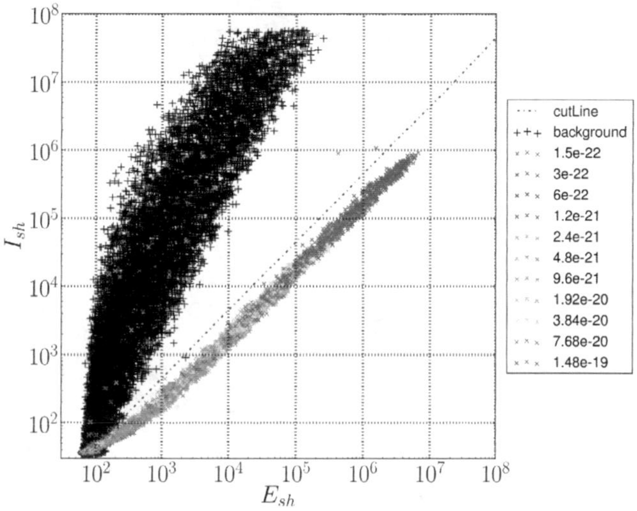

Figure 6. Scatter plot of the coherent spherical harmonic energy, E_{sh}, against the incoherent, I_{sh}, for simulated data.

Figure 6 shows such a scatter plot. The bifurcation pattern is the defining feature of these type of plots, and shows good separation between the black crosses of the background (generated using the standard time slides method), and the simulated signals.

5.2. S4 data

In order to test the statistic on real data containing a real glitch population, the same analysis was run over a month of S4 data [3] when three detectors were taking science data. In reality, the network used to take this data was the Hanford-Livingston network (H1L1H2), whereas the network we are interested in is Hanford-Livingston-Virgo (H1L1V1). To simulate this, we 'relabeled' the H2 data stream as V1 (i.e. once the H2 data was loaded, treat it as if it had been generated by the Virgo detector). As we were only interested in using the data as a source of noise and glitches, this was sufficient. We note, however, that in reality the H1 and H2 detectors were co-located; meaning noise between the data streams could be correlated, while we are treating them as independent.

Once again, there is a clear separation between the background and the injected signals (Figure 7), but notice that the angle between them has reduced. This is a common feature of

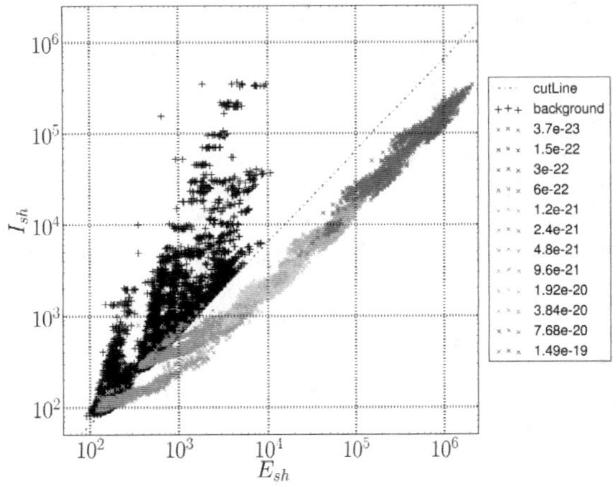

Figure 7. Scatter plot of the coherent part of the statistic versus the incoherent part for S4 data.

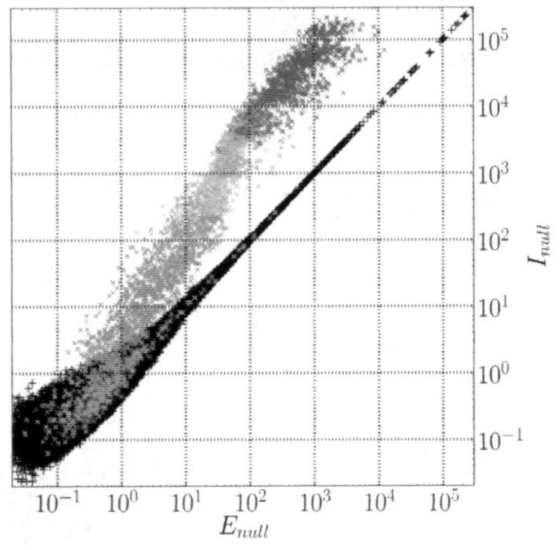

Figure 8. Scatter plot of the coherent and incoherent null energy for S4 data.

these plots when using real data and appears to be related to the amount of correlation in the noise between different detectors. This is more of an issue here, because there will be coupling between H1 and 'V1' (actually the relabelled H2).

In Figure 8 we see a scatter plot of the null energy calculated by SPHRAD. It is clear that the spherical harmonic power compares favourably to this - the clear separations, the well defined background etc. But one thing to note here is the null energy is calculated for a single sky direction, whereas the spherical harmonic power is all sky, and we get it as a by-product of the way the pipeline works. It can be generated exceptionally quickly and be used as a first pass discriminator of signal/glitch.

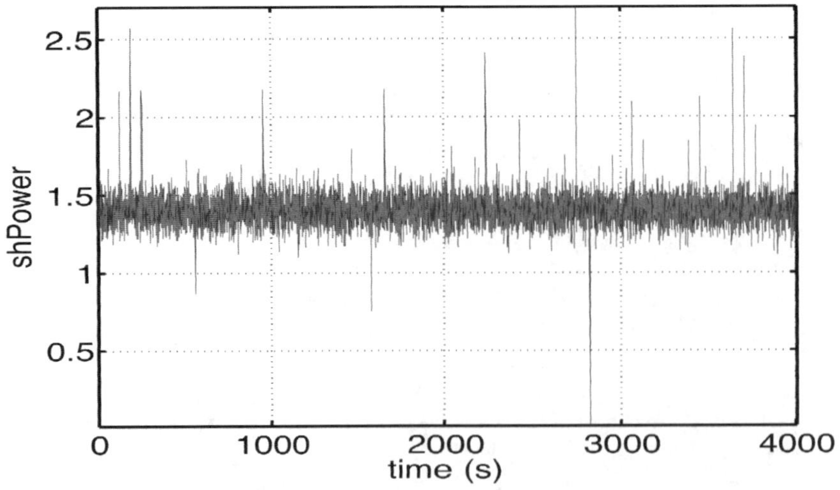

Figure 9. The SH statistic for 4000 seconds of data starting at GPS time 793502657.

To demonstrate, Figure 9 shows the plot of spherical harmonic power, $\mathbf{Z_{ratio}}$, for a large stretch of data. There is a horizontal band at $\mathbf{Z_{ratio}} \approx 1.4$ which is used to discriminate glitches and signals. Here, the injected signals have a hrss of 2.4×10^{-21} $s^{-1/2}$ and can seen rising above the band, while the three large, clear, glitches can be seen dropping below the band (at approximately 550s, 1600s and 2900s).

6. Conclusions

The spherical harmonic statistic has a 'clean' background, pushing up against the line at approximately $I_{sh} = 0.8E_{sh}$. This means that the parameters of a cut can easily be calculated and applied, it is merely a straight line. As can be seen from comparing Figures 6 and 7 to 8, there is good separation between the injections and the background. What these plots do not demonstrate is the computational cost associated with calculating the cuts - the method of [5] has to loop over different parameters (generating the curve in the log-log space) and polarisations, while the SPHRAD statistic simply has to generate a straight line. Another major advantage of using the spherical harmonic information as a discrimintor is that if the sky pointing statistic fails to select the correct sky location, it has no effect, whereas the standard coherent energy can be calculated incorrectly as it is highly dependent on getting the timing information correct.

In summary, we believe this this statistic has all the advantages of the standard coherent statistics, but very few of their disadvantages. Timing errors do not have a significant effect on its value, and it strongly discriminates between signals and glitches as shown by Figures 5 and 9.

7. Acknowledgments

We would like to thank Kipp Cannon for many conversations and much advice on his fast correlator.

References

[1] McIver J for the LIGO Scientific Collaboration and the Virgo Collaboration 2011, Data quality studies of enhanced interferometric gravitational wave detectors
[2] Blackburn L *et al.* 2008 Class. Quant. Grav. **25**, 184004 [arXiv:0804.0800 [gr-qc]].
[3] Abbott B et al. [LIGO Scientific Collaboration] 2007 *Class. Quant. Grav.* **24** 5343-5369
[4] S. Klimenko, I. Yakushin, A. Mercer and G. Mitselmakher, Class. Quant. Grav. **25**, 114029 (2008) [arXiv:0802.3232 [gr-qc]].
[5] Sutton P, Jones G, Chatterji S, Kalmus P, Leonor I, Poprocki S, Rollins J, Searle A, Stein L, Tinto M and Was, M 2010 *New J. Phys.* **12** 053034
[6] S. Chatterji et al 2006 Coherent network analysis technique for discriminating gravitational-wave bursts from instrument noise *Phys. Rev. D* **74** 082005
[7] Wen L and Schutz B F 2005 Coherent network detection of gravitational waves: the redundancy veto *Class. Quant. Grav.* **22** S1321-S1335
[8] Thrane E, Kandhasamy S, Ott C D, Anderson W G, Christensen N L, Coughlin M W, Dorsher S, Giampanis S, Mandic V, Mytidis A, Prestegard T, Raffai P, and Whiting B 2011 Long gravitational-wave transients and associated detection strategies for a network of terrestrial interferometers *Phys. Rev. D* **83** 083004
[9] Prestegard, T., Thrane, E., Christensen, N. L., Coughlin, M. W., Hubbert, B., Kandhasamy, S., MacAyeal, E., and Mandic, V. 2011 arXiv:1111.1631
[10] Kipp C. Cannon 2007 An efficient algorithm for computing the time resolved full sky cross power in an interferometer with omnidirectional elements *Phys. Rev. D* **75** 123003
[11] J Abadie *et al.* 2010 All-sky search for gravitational wave bursts in the first joint LIGO-GEO-Virgo run *Physical Review D* **81** 10200
[12] Fast Fourier Analysis Group at Dartmouth *Fast Fourier Analysis on Groups* `http://www.cs.dartmouth.edu/~geelong/sphere/`
[13] Driscoll J R and Healy D 1994 Computing Fourier transforms and convolutions on the 2-sphere. *Adv. in Appl. Math.* **15** 202 - 250
[14] Gürsel Y and Tinto M 1989 Near optimal solution to the inverse problem for gravitational-wave bursts *Phys. Rev. D* **40** 38843938

[15] Stephen Fairhurst 2011 Source localization with an advanced gravitational wave detector network *Class. Quantum Grav.* **28** 105021
[16] S. Klimenko *et al.*, Phys. Rev. D **83** (2011) 102001 [arXiv:1101.5408 [astro-ph.IM]].

Amaldi 9 and NRDA 2011 IOP Publishing
Journal of Physics: Conference Series **363** (2012) 012026 doi:10.1088/1742-6596/363/1/012026

Recent searches for gravitational-wave bursts associated with magnetar flares with LIGO, GEO, and Virgo

Raymond Frey for the LIGO Scientific Collaboration and the Virgo Collaboration

Dept of Physics, University of Oregon, Eugene, OR USA 97403

E-mail: rayfrey@uoregon.edu

Abstract. Energetic electromagnetic flares from magnetars – highly magnetized neutron stars – are associated with sudden rearrangements of the mechanical and/or magnetic configurations of the star, which can give rise to mechanical oscillations, some of which may be strong radiators of gravitational waves. General arguments have indicated that gravitational-wave bursts associated temporally with (giant) flares from galactic magnetars may be observable with ground-based gravitational wave detectors. After discussing the expectations based on the astrophysical models, we present results from several campaigns to search for such bursts using the first generation of LIGO, GEO, and Virgo detectors over the period 2005-2009, emphasizing the most recent results. No detections have been made, and we present astrophysically informed limits. Finally, we discuss prospects for progress.

1. Introduction

Understanding neutron star structure is an important astrophysical objective. Perturbations of isolated neutron stars followed by their mechanical relaxation have provided some insights. The electromagnetic emissions associated with such events have provided valuable insights into neutron star astrophysics. However, the observations in the electromagnetic spectrum have so far been unable to provide decisive information to address questions of neutron star structure. Gravitational-wave observations, on the other hand, can potentially probe the large-scale structure of neutron stars and hence may play an important role in the advancement of neutron-star astrophysics.

The astrophysical sequence considered here involves magnetars — neutron stars with magnetic fields $\sim 10^{15}$ G — which undergo a rapid reconfiguration heralded by the emission of (giant) x-ray flares. The perturbation of the star is likely to excite mechanical modes, some of which may couple strongly to gravitational-wave (GW) emission. In the searches presented here, we assume the modes which are relevant to GW emission are damped in times of ~ 1 s or less [1]. Because the potential mechanisms for GW emission are uncertain, we do not assume any specific GW waveform in the search. An additional element of the GW search we employ relies on the expectation that the x-ray flare is nearly coincident with the GW emission. Hence, we use a triggered search strategy, as described below. The knowledge of the event time and position on the sky allows an improvement in sensitivity relative to an untriggered, all-sky search. Finally, the vast majority of detected flares associated with magnetars are galactic. The relative proximity of the magnetars, combined with the associated energetics of their rearrangements, make them viable targets in a search for a first GW detection.

The searches discussed here were performed using data from the LIGO [2], GEO [3], and Virgo [4] gravitational-wave detectors from 2004 to 2009. During this period, the detectors were operating at or near the sensitivity associated with their "initial" (or first-generation) configurations. The searches are reported in Refs. [5], [6], and [7].

1.1. Astrophysical context

Models for x-ray flares from galactic magnetars are based on a handful of well-measured episodes. The objects associated with these magnetar emissions may be termed soft gamma-ray repeaters (SGRs) or anomalous x-ray pulsars (AXPs). (The typical energy of the burst photons of ~ 100 KeV can be alternatively described as (hard) x-rays or (soft) gamma rays.) The prevailing scenario [8] posits that a neutron star crust cracking event is associated with the rearrangement of the magnetic field, which gives rise to the flare. Two classes of field rearrangement have been considered: internal mechanisms based on the large-scale rearrangement of the interior magnetic field, and external mechanisms which likely involve magnetic reconnections in the crust and magnetosphere. The total luminosity of giant flares and the time structure of their subsequent quasi-periodic oscillations are consistent with this picture. The energetics of these episodes can be used to estimate the possible energy content of associated gravitational-wave bursts. Such calculations were recently performed [9] for several models, resulting in available energies in the range 10^{45} to 10^{48} erg for neutron stars of conventional composition and about 10^{48} to 10^{50} erg for neutron stars which include quark matter or baryon-meson matter.

While the energetics are promising, efficient gravitational-wave production requires the excitation of quadrupolar mechanical modes. Much attention has thus been given to the quadrupolar f-modes. In fact, such modes are thought to damp primarily via GW radiation with a damping time ~ 200 ms [10]. Low-order f-modes in neutron stars have frequencies in the range 1-3 kHz. The excitation of f-modes from the standard giant flare mechanisms involving internal [11] and external [12] magnetic field rearrangement has recently been considered. Both calculations give poor f-mode coupling, indicating that even advanced GW detectors will be unlikely to detect gravitational waves via this mechanism. On the other hand, lower frequency modes are also likely to be excited, perhaps more strongly. For example, torsional modes (~ 100 Hz) are found [12] to couple well to flares. The viability of low-frequency modes for potential gravitational-wave observations is very uncertain and a topic of ongoing investigations.

Given the astrophysical uncertainties, with few observational constraints, we have elected to perform searches which do not assume a GW waveform. If no significant signal is observed, we evaluate upper limits using the available astrophysical guidance. In particular, we consider generic white noise bursts with frequencies ~ 100 Hz to evaluate the sensitivity to low-frequency modes and damped sinusoids with frequencies > 1 kHz to evaluate the sensitivity to f-modes.

1.2. Flare triggers

Figure 1 indicates the magnetars and their x-ray bursts. Most bursts trigger a GW search, as described in the next section. The six galactic magnetars shown represent those which produced significant flare activity during the LIGO/GEO/Virgo running periods discussed here. All detections were made by satellites of the IPN network [13]. Especially notable events are indicated. The SGR 1806−20 giant flare [14] was the most luminous of these emissions. In fact, with an energy estimated [15] to be $(3.7\pm0.9)\times10^{46}$ erg, it is the most luminous transient galactic event recorded. The corresponding GW search is described in [5]. (For complete citations for the flare detections and other astrophysical data relevant to the present analyses, see Refs. [5], [6], and [7].) SGR 1900+14 released a "storm" of flares over a 30 s period. The GW search is described in Refs. [5] and [6]. The emission from AXP 1547−5409 was discovered as

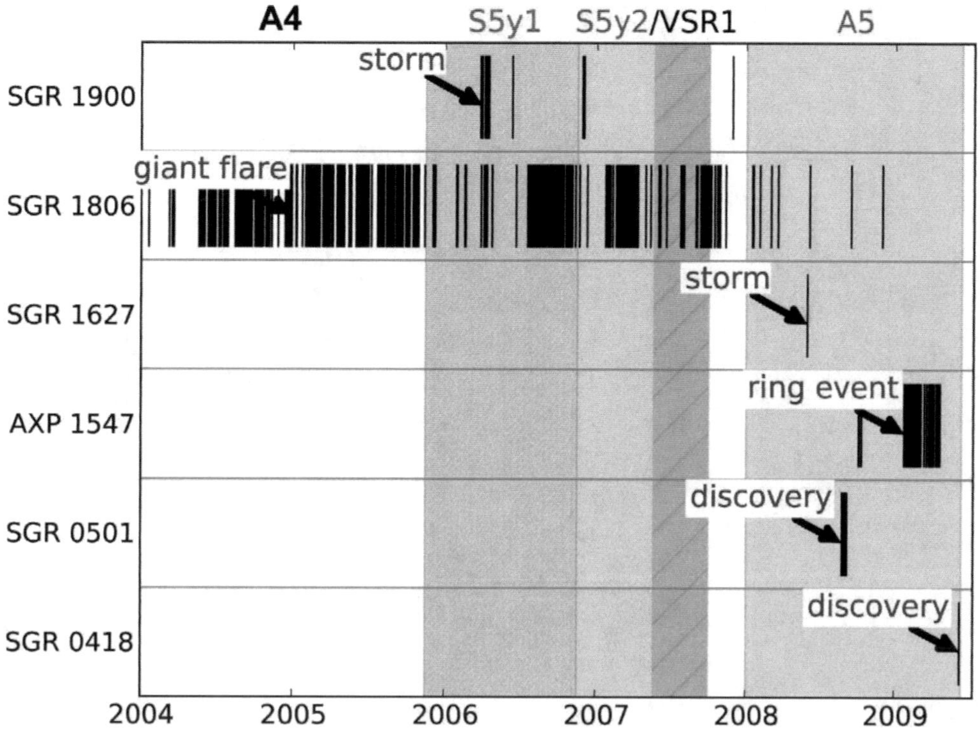

Figure 1. History of the x-ray bursts used as triggers in the GW burst searches. Each vertical black line represents an x-ray burst. The LIGO and GEO running periods A4, S5, and A5, and Virgo running period VSR1, are indicated.

expanding rings of x-rays scattered from dust sheets surrounding the magnetar. Its estimated energy [16] of 10^{44} to 10^{45} erg is comparable to a giant flare. The magnetars SGR 0501+4516 and SGR 0418+5729 were first revealed by flare activity in 2008 and 2009, respectively. SGR 0501+4516, at a distance of about 1 kpc, is also notable for being the nearest of these objects. Besides the flares associated with the SGR 1900+14 storm and giant flare episodes, the GW searches corresponding to the remaining flares are described in Ref. [7]. In all, roughly 1.5×10^3 electromagnetic bursts have been analyzed for associated GW bursts.

2. Methodology

2.1. Detectors and data

All three initial LIGO detectors were employed at various periods for the searches described here: the 4 km Hanford (H1) and Livingston (L1) interferometers, and the 2 km Hanford (H2) interferometer. The 3 km Virgo interferometer was employed for much of 2007. The GEO 600 interferometer was also crucial for many of the analysis epochs. Power spectral density curves are shown for all of these detectors in Figure 2. These represent a snapshot of the typical detector noise, in units of gravitational wave strain per $\sqrt{Hz}$, during the magnetar searches. It is worth noting that, although the LIGO detectors have better sensitivity at lower frequency ($\sim$ 100 Hz), all detectors are comparable in the f-mode regime ($>$ 1 kHz).

Data taking epochs are categorized by detector sensitivity and other factors. The LIGO

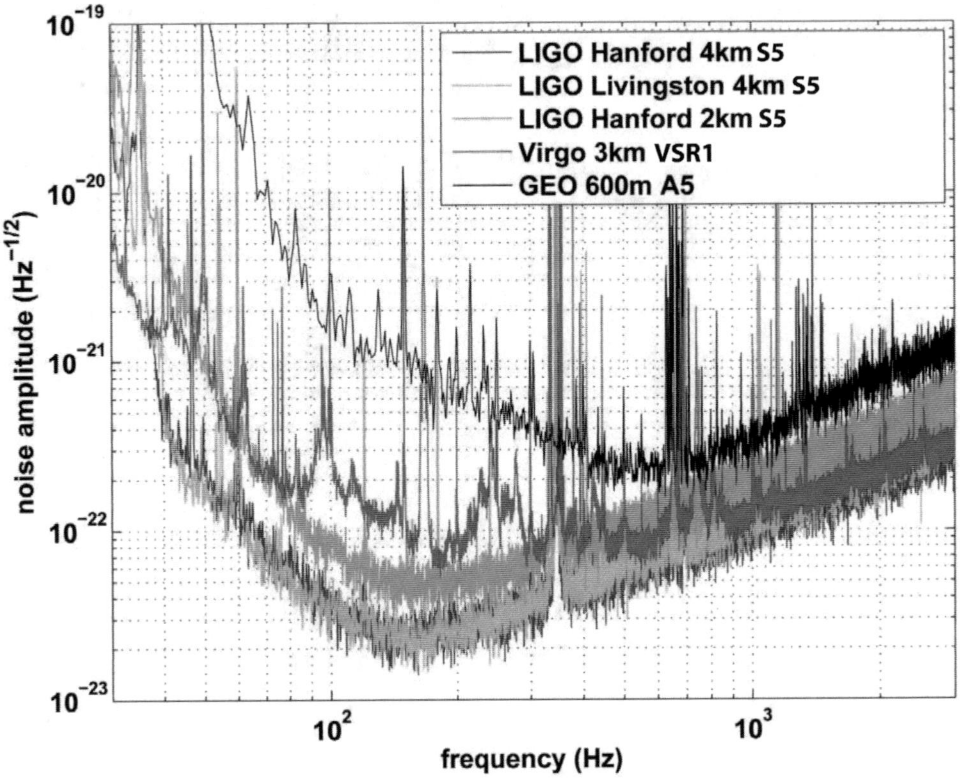

Figure 2. Power spectral density of detector noise in units of strain/$\sqrt{\text{Hz}}$. The curves represent typical performance of the detectors used for the magnetar searches. Curves from top to bottom at 100 Hz are GEO, Virgo, LIGO H2, LIGO H1, and LIGO L1.

two-year S5 science run took place after the detectors reached their design sensitivity in the initial-LIGO configuration. GEO also took data throughout S5. The first Virgo science run VSR1 began in 2007, and this data was used in the search for the S5/VSR1 period. Virgo approached design sensitivity in its initial configuration subsequent to the periods used for this search. A unique feature of the magnetar searches is that data was also analyzed outside of the designated science running periods. During these periods, termed *astrowatch*, data taking, typically at night, is interleaved with detector development and commissioning work. Because of the short stretches of data required for an externally triggered search like that employed here, the astrowatch searches are viable. These are labeled A4 and A5 in Figure 1. The typical LIGO amplitude sensitivity for A4 was roughly three times worse than for S5. GEO was operating with good uptime for the A5 period, and was joined only by the LIGO H2 interferometer. For most of this search, at least two interferometers were operating.

2.2. Analysis overview

The data segments used to search for gravitational-wave bursts are defined by the times of the detected magnetar x-ray bursts. The electromagnetic and GW bursts are expected to arrive nearly simultaneously. We also assume, as discussed above, that the GW signals should damp in less than 1 s. Hence, an "on-source" window of 4 s centered on the x-ray burst trigger is chosen. This should be large enough to include most of the GW signal energy and also not be sensitive to systematic errors in the burst timing measurement. The background is determined from 1000 s of off-source data on both sides of the trigger. When a different x-ray burst falls in the off-source, that 4 s segment is masked out.

The data are analyzed [17] using an excess power statistic applied to time-frequency pixels. We have checked that the excess power algorithm is insensitive to waveform, provided that the GW energy mostly falls within the on-source window. After pixel clustering, a list of significant "events" is formed. The algorithm has been constructed to properly weight excess power for general multi-detector networks. The off-source data is analyzed identically to the on-source, and a false-alarm rate (FAR) is determined. Signal candidates are identified when a loudest on-source event has FAR smaller than a predetermined threshold. Candidates are then scrutinized to determine if they were subject to any known instrumental problems or significant environmental disturbances.

For every electromagnetic burst considered, if no viable candidate results, then an upper limit is formed in the following way. GW waveforms are injected into 4 s off-source data segments and analyzed for excess power, as above. For a given injected GW strain amplitude, the loudest event in each 4 s segment is found and the fraction of injections louder than the loudest on-source event is determined. When this fraction is 90%, the strain upper limit is determined. The confidence level in strain is related to the isotropic GW energy release by

$$E_{\rm GW} = 4\pi d^2 \frac{c^3}{16\pi G} \int_{-\infty}^{\infty} \left(\dot{h}_+^2 + \dot{h}_\times^2\right) dt. \tag{1}$$

where h_+ and $h_\times$ are the strain amplitudes and d is the magnetar distance. A 90% confidence level upper limit calculated in this way is written $E_{\rm GW}^{90\%}$.

Unlike the search algorithm, which does not presume a GW waveform morphology, the determination of upper limits, outlined above, must assume something for the signal injections. We have chosen two classes of injected waveforms which are informed by the astrophysics:

(i) f-modes. We use damped sinusoids (ring-downs) with frequency in the range 1-3 kHz and damping time 200 ms. We consider both linearly and circularly polarized cases.

(ii) low-frequency modes. We use white-noise bursts (WNB) of duration 10-100 ms and bands ranging from 100-200 Hz to 100-1000 Hz.

As can be seen from Figure 2, because of the better sensitivity of the LIGO detectors at lower frequency, we expect the WNB injections to produce better limits for a given x-ray burst.

3. Search results

One event survived as a GW signal candidate. It was the loudest event in the S5+S5/VSR1+A5 search [7] and the only event with FAR below the predetermined threshold of 6.9×10^{-5} Hz. It corresponds to an a x-ray burst which occurred when the H2 detector was operating in the A5 astrowatch period. However, the follow-up of this candidate revealed that it coincided with a large glitch recorded in a magnetometer. Its origin was an AC electrical power disturbance. Having discarded this candidate, upper limits were calculated for this and all other bursts.

It is interesting to note that a refinement of the search procedure was applied for the SGR 1900+14 "storm" episode (see Figure 1). Because of their proximity in time, the x-ray burst on-source times were "stacked" [18,6] together. This resulted in an upper limit approximately an order of magnitude better than the previous result [5].

The full set of upper limits are given in Refs. [5], [6], and [7]. Table 1 provides a few representative examples of the best limits. As discussed above, two classes of waveforms were used. The column labeled "1090 Hz" results from circularly polarized, damped sinusoidal

Table 1. Best upper limits for selected magnetar events. The "1090 Hz" column represents f-mode inspired limits. The "100-200 Hz" column is for white-noise bursts, representing unspecified low-frequency modes.

	distance (kpc)	event	$E_{\rm GW}^{90\%}$ (erg), 1090 Hz	$E_{\rm GW}^{90\%}$ (erg), 100-200 Hz
SGR 0501+4516 [7]	1	2008 discovery	1×10^{47}	3×10^{44}
AXP 1547−5408 [7]	4	2009 bright flare	4×10^{48}	6×10^{45}
SGR 1806−20 [5]	10	2004 giant flare	3×10^{50}	5×10^{47}
SGR 1900+14 [6]	10	2006 storm	1×10^{48}	2×10^{45}

waveforms, inspired by f-modes. The damping time is 200 ms. The central frequency of 1090 Hz is the lowest chosen frequency for this class of waveforms. It is expected that this would produce the best f-mode limit since the detector sensitivities worsen for higher frequencies. As discussed above, there remains much uncertainty for low-frequency modes. However, it is interesting to gauge the sensitivity to such modes using white-noise bursts. The column labeled "100-200 Hz" is for such bursts in the 100-200 Hz band and with duration 100 ms. Again, this band corresponds to the best LIGO sensitivity.

The nominal magnetar distances given in Table 1 are those used to form the energy upper limits using Equation (1). The measured distances are typically not precise and in some cases different measurements disagree. Rivising the limits to other distances is easily accomplished using d^2 scaling.

4. Discussion

While the search for GW bursts in association with energetic electromagnetic bursts from magnetars using data from initial GW detectors has yielded no detections, it is interesting to consider the astrophysical relevance of the calculated upper limits, as represented by the examples given in Table 1. As discussed above, current magnetar models imply that f-modes are unlikely to be excited by (giant) flares to the extent that they would result in detectable GWs. Nevertheless, one sees that the best limits of $\sim 10^{47}$ erg are comparable to the estimates [9] based on energy available to GW bursts.

Excitations resulting in GW bursts with frequency content $\sim$ 100 Hz are possible, but astrophysical expectations are currently uncertain. Nevertheless, we note that our best limits $\sim 10^{45}$ erg are comparable to the electromagnetic energy of the giant flares. Improved astrophysical understanding of potential lower-frequency modes would be welcome, as this represents a promising avenue for GW detection.

GW detections would likely provide a huge advance in the astrophysical understanding of magnetars and neutron stars. The biggest factors available for improving the likelihood of detections using the types of searches discussed here are (1) discovery of a very nearby (< 1 kpc) magnetar; and (2) the realization of advanced GW detectors: The factor ~ 10 improvement in sensitivity for Advanced LIGO and Virgo gives a factor 10^2 in energy sensitivity.

Acknowledgments

The authors gratefully acknowledge the support of the United States National Science Foundation for the construction and operation of the LIGO Laboratory, the Science and Technology Facilities Council of the United Kingdom, the Max-Planck-Society, and the State of Niedersachsen/Germany for support of the construction and operation of the GEO600 detector, and the Italian Istituto Nazionale di Fisica Nucleare and the French Centre National de la Recherche Scientifique for the construction and operation of the Virgo detector. The authors also gratefully acknowledge the support of the research by these agencies and by the Australian Research Council, the International Science Linkages program of the Commonwealth of Australia, the Council of Scientific and Industrial Research of India, the Istituto Nazionale di Fisica Nucleare of Italy, the Spanish Ministerio de Educación y Ciencia, the Conselleria d'Economia Hisenda i Innovació of the Govern de les Illes Balears, the Foundation for Fundamental Research on Matter supported by the Netherlands Organisation for Scientific Research, the Polish Ministry of Science and Higher Education, the FOCUS Programme of Foundation for Polish Science, the Royal Society, the Scottish Funding Council, the Scottish Universities Physics Alliance, The National Aeronautics and Space Administration, the Carnegie Trust, the Leverhulme Trust, the David and Lucile Packard Foundation, the Research Corporation, and the Alfred P. Sloan Foundation.

References

[1] Benhar O, Ferrari V, and Gualtieri L 2004 *Phys Rev* **D70** 124015
[2] Abbott B, et al. 2009 *Rep. Prog. Phys.* **72** 076901
[3] Grote H and the LIGO Scientific Collaboration, 2010 *Class. Quantum Grav.* **27** 084003
[4] Accadia T, et al. 2011 *Class.Quantum Grav.* **28** 114002
[5] Abbott B, et al. 2008 *Phys. Rev. Lett.* **101** 211102
[6] Abbott B, et al. 2009 *Astrophys. J.* **701** L68
[7] Abadie J, 2011 *Astrophys. J.* **734** L35
[8] Thompson C and Duncan R C 1995 *MNRAS* **275** 255
[9] Corsi A and Owen B J 2011 *Phys Rev* **D83** 104014
[10] Lindblom L and Detweiler S L, 1983 *Astrophys. J. Suppl.* **53** 73
[11] Zink B, Lasky P D, and Kokkotas K D 2011 arXiv:1107.1689
[12] Levin Y and van Hoven M 2011 arXiv:1103.0880
[13] IPN, `http://ssl.berkeley.edu/ipn3`
[14] Palmer D M, et al. 2005 *Nature* **434** 1107
[15] Hurley K, et al. 2005 *Nature* **434** 1098
[16] Tiengo A, et al. 2010 *Astrophys. J.* **710** 227
[17] Kalmus P, et al. 2007 *Class. Quant. Grav.* **24** S659
[18] Kalmus P, et al. 2009 *Phys Rev* **D80** 042001

Progress on the Low-Latency Inspiral Gravitational Wave Detection algorithm known as SPIIR

Shaun Hooper[1,2], Linqing Wen[1,2], Chad Hanna[3], Kipp Cannon[4], Drew Keppel[5], David Blair[1], Shin-Kee Chung[1,2], Leo Singer[6], Yanbei Chen[6]

[1]Australian International Gravitational Research Centre, School of Physics, University of Western Australia, 35 Stirling Hwy, Crawley, WA 6009, Australia
[2]ICRAR-Fairway M468, School of Physics, The University of Western Australia, Crawley, WA 6009, Australia
[3]Perimeter Institute of Theoretical Physics
[4]Canadian Institute of Theoretical Astrophysics
[5]Albert-Einstein-Institut
[6]California Institute of Technology

E-mail: shaun.hooper@uwa.edu.au

Abstract. Low-latency event triggers to signify the presence of gravitational waves from coalescing binaries will be required to make prompt electromagnetic follow-up observations of electromagnetic counterparts. We present the recent progress made on implementing the time-domain low-latency detection algorithm known as summed parallel infinite impulse response (SPIIR) filtering into a real gravitational wave search pipeline.

1. Introduction

Neutron star binary mergers are currently the leading model of the progenitors of short hard gamma-ray bursts (short GRBs) [1, 2]. A very short delay (from 0.1 seconds to hundreds of seconds [3, 4]) is thought to happen between the final GW emission and the onset of the GRB. Presently the electromagnetic emission of the GRB event is not well understood. A prompt emission in X-ray and optical wavelengths followed by a delayed afterglow of cascading wavelengths is expected to be related to the initial GRB. For long GRBs (those where the burst duration is >2s) it is known that there is a prompt optical component that may occur tens to hundreds of seconds after the initial burst [5]. It is expected that for short GRBs there is also a prompt optical component (and possibly afterglow) happening on short time scales. Low-latency detection of the associated GW will be required in order to make follow-up observations of the electromagnetic counter-parts.

The current strategy to search for the existence of inspiral waveforms in the detector data is based on matched filtering [6]. This method, based on Wiener optimal filtering, is a correlation of an expected inspiral waveform template and the detector data, weighted by the inverse noise spectral density of the detector [7]. In order to save computational costs, this correlation is performed in the frequency domain, via a Fourier transform of a finite segment of detector data. Although computationally efficient, this naturally incurs a latency due to the fact that data must be collected before the Fourier transform can be applied. A method known as *Multi-Band Template Analysis* (MBTA) is currently under development to reduce the latency by splitting the matched filtering over two frequency bands [8]. However this is still a frequency domain method. Another low-latency method known as *Low-Latency On-line Inspiral Data* analysis (LLOID) can also give low-latency triggers in the time domain by first down-sampling the incoming data into multiple streams and then applying time domain finite impulse response (FIR) filters [9]. The computational cost of this pipeline is reduced by decreasing the number of templates via singular value decomposition [10]. In both cases, the computational cost scales with the length of the gravitational waveform. Advanced LIGO detectors will have improved bandwidth, especially at the low-frequency end. This means that the gravitational waveform length will increase, and consequently the computational cost will also increase.

The authors have previously [11, 12] introduced a new method to detect inspiral signals in the time domain using infinite impulse response (IIR) filters. This method works by approximating an inspiral waveform by a summation of time shifted exponentially increasing sinusoids. Each sinusoid can be searched for by applying a single pole IIR filter. In this manner the IIR filter acts as a narrow bandpass filter responding to a peak characteristic frequency. In order to search for the template a bank of single-pole IIR filters is applied in parallel. When each appropriately delayed IIR filter output is added coherently, the total output approximates the matched filter output of the exact waveform. We call this the *summed parallel infinite impulse response* (SPIIR) method.

In this proceeding we will discuss the current status and recent progress on the implementation of the low-latency method known as SPIIR.

2. The SPIIR Method

The *summed parallel infinite impulse response* (SPIIR) method has been previously been outlined in [11] and its mathematical concepts detailed in [12]. In this section, we will briefly go over the core parts of the SPIIR method.

The SPIIR method is designed as a low-latency time domain replacement to the optimal search strategy, matched filtering, commonly used in inspiral searches [6]. The authors have previously shown that is it possible to approximate the inspiral gravitational waveform components $h_{c,s}(t)$ by a summation of damped sinusoids u_l [11, 12],

$$h_{c,s}(t) = A(t)e^{i\phi(t)} \simeq U(t) = \sum_l u_l(t) = \sum_l b_{0,l} e^{(\gamma_l + i\omega_l)(t-t_l)} \Theta(t_l - t). \tag{1}$$

It is then possible to search for each sinusoid using the discrete single pole IIR filter

$$y_{k,l} = a_{1,l} y_{k-1} + b_{0,l} x_{k-d_l}. \tag{2}$$

Where the index k denotes discretely sampled times of t, namely $t_k = k\Delta t$. Figure 1 shows a schematic of the data processing pipeline for a one IIR filter. This is exactly equivalent to a cross-correlation of the damped sinusoids u_l where

$$a_{1,l} = e^{-(\gamma_l + i\omega_l)} \tag{3}$$

and d is a delay term equivalent to t_l. Here γ_l represents a damping factor and ω_l the peak frequency of the l-th IIR filter. The IIR filter (2) is a narrow band pass filter, centered on the frequency ω_l. As is shown in [11, 12], the summation of a *bank* of M IIR filters run in parallel, each with a different b_0, d, γ, ω, can approximate the optimal matched filter output,

$$z_k \simeq 2\Delta t \sum_{l=1}^{M} y_{k,l}. \tag{4}$$

for each template. The matched filter output z divided by a normalization constant gives the signal to noise ratio (SNR).

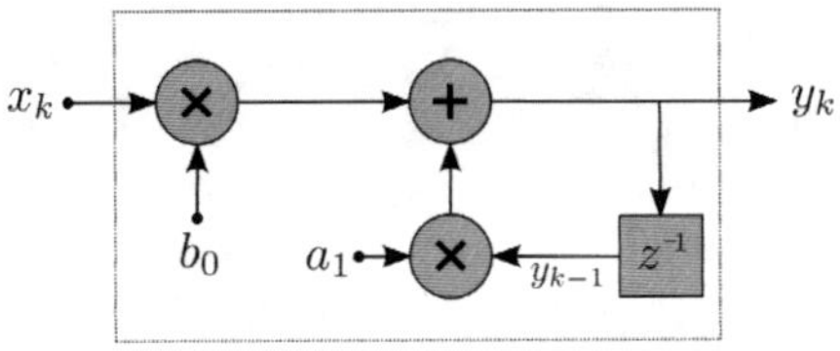

Figure 1. A signal processing schematic showing the flow of data through a single-pole IIR filter. The input, x_k is multiplied by a complex constant b_0, then added to the previous output that has been multiplied by another complex constant a_1, resulting in the current output y_k. It should be noted that this filter, in principle, should be have been run forever.

3. Progress

3.1. Advanced LIGO template bank

In order to implement the SPIIR method for the advanced detector era [13, 14], the first necessary step was to produce IIR banks for each template in the entire template bank. We choose to focus on the $1 - 3M_{\odot}$ mass range. Using the predicted advanced LIGO noise curve [15] with a lower cut off of 15 Hz we obtain 21890 mass-pairs in the given mass range with a minimal match of 97%. For each mass-pair, we produce a restricted second post Newtonian (2PN) waveform and find a bank of M IIR filters using the method outlined in [11]. We calculated the inner product between the exact 2PN waveform h and the IIR bank impulse response U to see if the impulse response approximates the waveform. We call the value of this inner product the overlap. Please see section IIIA of [11] for greater detail as to how the overlap is calculated. The overlap in all 21890 cases was above 99.2% (see figure 2). This step was necessary to assure that the SPIIR method would be reasonably capable of recovering detections all templates in the mass range $1 - 3M_{\odot}$.

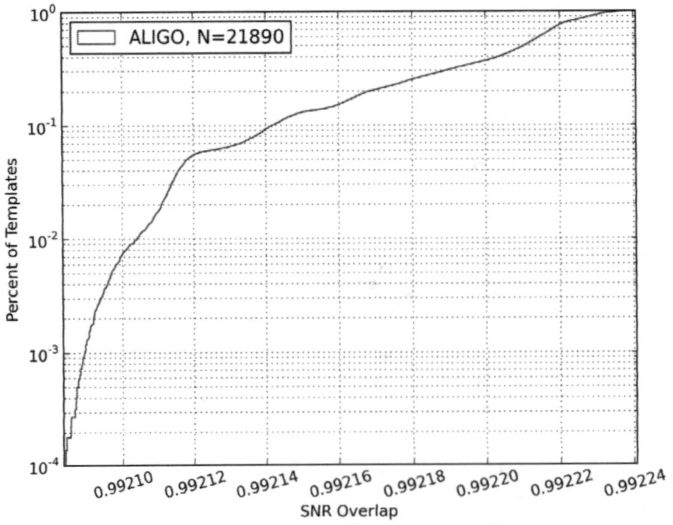

Figure 2. A cumulative distribution plot of the template SNR overlaps. The SNR overlaps is defined as the normalized inner product of the exact 2PN waveform, and the impulse response of the IIR bank.

3.2. Implementation with `gstlal`
Recent work by the authors has focused on the implementation of the SPIIR method via integration with an existing gravitational wave detection framework, `gstlal` [9]. `gstlal` is the combination of the LIGO Algorithm Library (LAL) and the open-source real-time multimedia handling software GStreamer [16]. Since processing gravitational wave data is very similar to processing audio data, GStreamer's stock signal processing elements, such as re-samplers and filters can be used as a powerful framework with which to create new gravitational wave search pipelines.

This software library is an ideal platform to integrate the SPIIR method, as there currently exists a pipeline that focuses on low-latency detection of inspiral signals; the LLOID pipeline [17]. The LLOID pipeline consists of many stock GStreamer "elements" along with gravitational wave specific elements connected together to take raw detector data, whiten the data, and then matched filtering the data via the SVD method [10]. Once the SNR time series is produced, the pipeline then passes time segment "buffers" to an element that generates standard trigger files.

Because of the relative maturity of `gstlal`, the authors have focused on replicating the LLOID pipeline, but with the core processing element replaced with an "IIR bank" element. In order to implement the SPIIR method into the larger structure of `gstlal`, a core "element" code was written, which is now a part of the `gstlal` project. The element we wrote is capable of taking the single incoming channel of data and applying a set of N IIR banks simultaneously (one for each template) thereby producing N channel outputs. Each template IIR bank is a vector of M a_1's, b_0's and d's.

3.3. Multi-rate filtering
In order to reduce computational costs, the authors have worked to implement a very similar multi-rate scheme that exists in the LLOID pipeline. The motivation for this is as follows; IIR filters with characteristic frequencies ω much lower than the native sample rate need only be run at a sample rate of twice (or slightly more) ω. In order to achieve this, the coefficients a_1 and b_0, as well as d must be redefined for the new sample rate. This comes at the cost of first down-sampling the whitened detector data, as well as up-sampling the sub IIR bank output to the original sample rate. However preliminary theoretical calculations show that the additional computational costs of re-sampling are negligible compared the savings made from running the IIR filters at a reduced rate.

For each template's IIR bank, the M IIR filters are divided into different sample rate bins. For each sample rate bin, there will be a different *sub* IIR bank that must be filtered. Figure 3 shows the basic outline of the multi-rate pipeline.

The main goal of the SPIIR method is to have a time domain method that uses less computational resources than the tradiational time-domain matched filter (based on FIR filtering) and the frequency domain matched filter for low latencies. In [12], the authors outline the theoretical computational cost of the standard SPIIR scheme and multi-rate SPIIR scheme, and compare the cost with both the time domain and frequency domain matched filter. Figure 4 shows that if an extremely low latency is required, the SPIIR method can be faster than both time domain and frequency domain matched filters.

4. Future Work
As the advanced detectors won't be online until at least 2014, the authors are currently planning to test the `gstlal` SPIIR method in the next engineering run, scheduled for January 2012. The engineering run is proposed to test important software infrastructure that will be used in the advanced detector era. As a preliminary preparation for this experiment, the pipeline will be tested on using detector data from previous science runs. The triggers generated from this preliminary experiment will be compared with other methods, such as the LLOID pipeline.

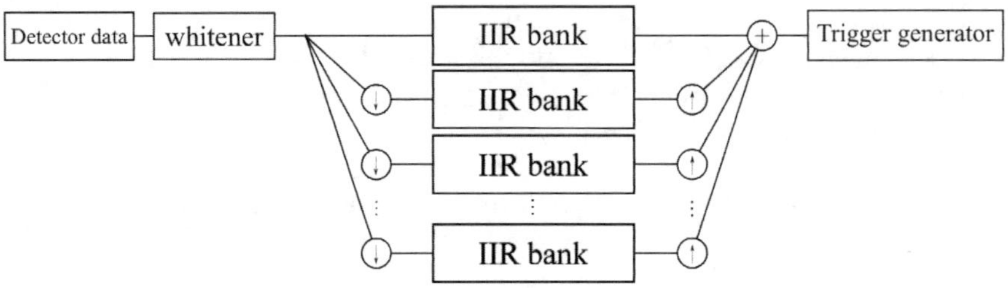

Figure 3. Basic data flow chart of the `gstlal` SPIIR method. After detector data is acquired, it is then whitened. Next the whitened data is sent to several down-samplers. Each down-sampler reduces the sample rate to an appropriate rate, as well as applying a low-pass filter. The data is then passed to IIR bank processing elements, where a bank of IIR filters runs on the data. The output of the IIR banks is the up-sampled to the highest sample rate, and added to the output of the other up-samplers. The SNR stream is then passed to a trigger generator, which compiles a database of events.

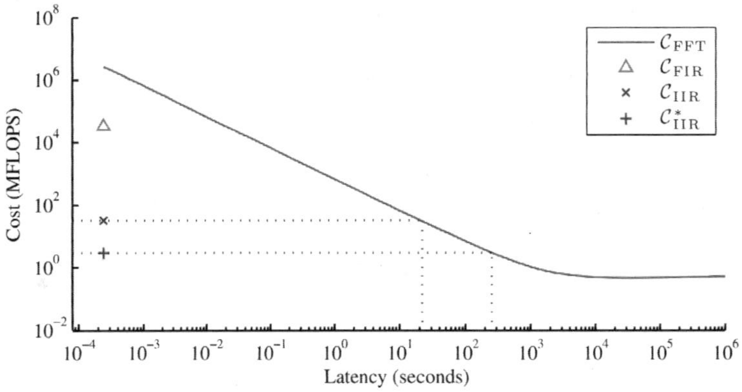

Figure 4. Theoretical computational cost of the frequency-domain (FFT) matched filter (solid line), time-domain (FIR) matched filter (triangle), IIR method (cross) and multi-rate IIR method (plus) to filter a representative $1.4 + 1.4\,\mathrm{M}_\odot$ template in advanced LIGO. The frequency domain filter requires segmenting, and is therefore a function of latency. Time domain methods have inherently sample time latencies.

As a side project to the `gstlal` implementation, the authors are also working on a GPU based SPIIR development (see [18] for details). The parallelization of the SPIIR method executed on GPU's is expected to result in significant computational speed improvements.

A further reduction in computational costs are expected to be made by the introduction of the template interpolation scheme introduced in [12]. This method works by reducing the complete, *fine*, template bank into a reduced *coarse* bank. It can be shown that different frequency bands of the coarse bank templates can be manipulated to produce all templates in the fine bank. This frequency dividing, much like the MBTA method, is ideally executed using the sub IIR banks. Although [12] demonstrates this procedure for Newtonian waveforms, we plan to extend this idea to post-Newtonian waveforms.

References

[1] Fox D B, Frail D A, Price P A, Kulkarni S R, Berger E, Piran T, Soderberg A M, Cenko S B, Cameron P B,

Gal-Yam A, Kasliwal M M, Moon D, Harrison F A, Nakar E, Schmidt B P, Penprase B, Chevalier R A, Kumar P, Roth K, Watson D, Lee B L, Shectman S, Phillips M M, Roth M, McCarthy P J, Rauch M, Cowie L, Peterson B A, Rich J, Kawai N, Aoki K, Kosugi G, Totani T, Park H, MacFadyen A and Hurley K C 2005 *Nature* **437** 845–850 (*Preprint* arXiv:astro-ph/0510110)
[2] Nakar E 2007 *Phys. Rep.* **442** 166–236 (*Preprint* arXiv:astro-ph/0701748)
[3] van Putten M H P M 2009 (*Preprint* 0905.3367)
[4] Zhang B and Mészáros P 2004 *International Journal of Modern Physics A* **19** 2385–2472 (*Preprint* arXiv:astro-ph/0311321)
[5] Klotz A, Boër M, Atteia J L and Gendre B 2009 AJ **137** 4100–4108 (*Preprint* 0902.0898)
[6] Abbott B P, Abbott R, Adhikari R, Ajith P, Allen B, Allen G, Amin R S, Anderson S B, Anderson W G, Arain M A and et al 2009 Phys. Rev. D **79** 122001 (*Preprint* 0901.0302)
[7] Wainstein L A and Zubakov V D 1962 *Extraction of Signals from Noise* (Prentice-Hall)
[8] Buskulic D, for the LIGO Scientific Collaboration and the Virgo Collaboration 2010 *Classical and Quantum Gravity* **27** 194013–+
[9] https://www.lsc-group.phys.uwm.edu/daswg/projects/gstlal.html
[10] Cannon K, Chapman A, Hanna C, Keppel D, Searle A C and Weinstein A J 2010 *Phys. Rev. D* **82** 044025
[11] Hooper S *et al.* 2011 (*Preprint* 1108.3186)
[12] Luan J, Hooper S, Wen L and Chen Y 2011 (*Preprint* 1108.3174)
[13] Harry G M, for the LIGO Scientific Collaboration 2010 *Classical and Quantum Gravity* **27** 084006
[14] Acernese F *et al.* 2009 https://tds.ego-gw.it/itf/tds/file.php?callFile=VIR-0027A-09.pdf
[15] https://www.lsc-group.phys.uwm.edu/daswg/projects/lal/nightly/docs/html/LALAdvLIGOPsd_8c_source.html
[16] http://gstreamer.freedesktop.org/
[17] Cannon K *et al.* 2011 (*Preprint* 1107.2665)
[18] Liu Y 2011 *J. Comput. Phys.* (in preparation)

Towards a generic test of the strong field dynamics of general relativity using compact binary coalescence: Further investigations

T.G.F. Li[1], W. Del Pozzo[1], S. Vitale[1], C. Van Den Broeck[1], M. Agathos[1], J. Veitch[1,2], K. Grover[3], T. Sidery[3], R. Sturani[4,5], A. Vecchio[3]

[1]Nikhef – National Institute for Subatomic Physics, Science Park 105, 1098 XG Amsterdam, The Netherlands
[2]School of Physics and Astronomy, Cardiff University, Queen's Buildings, The Parade, Cardiff CF24 3AA, United Kingdom
[3]School of Physics and Astronomy, University of Birmingham, Edgbaston, Birmingham B15 2TT, United Kingdom
[4]Dipartmento di Scienze di Base e Fondamenti, Università di Urbino, I-61029 Urbino, Italy
[5]INFN, Sezione di Firenze, I-50019 Sesto Fiorentino, Italy

Abstract.
In this paper we elaborate on earlier work by the same authors in which a novel Bayesian inference framework for testing the strong-field dynamics of General Relativity using coalescing compact binaries was proposed. Unlike methods that were used previously, our technique addresses the question whether *one or more* 'testing coefficients' (*e.g.* in the phase) parameterizing deviations from GR are non-zero, rather than all of them differing from zero at the same time. The framework is well-adapted to a scenario where most sources have low signal-to-noise ratio, and information from multiple sources as seen in multiple detectors can readily be combined. In our previous work, we conjectured that this framework can detect *generic* deviations from GR that can in principle not be accomodated by our model waveforms, on condition that the change in phase near frequencies where the detectors are the most sensitive is comparable to that induced by simple shifts in the lower-order phase coefficients of more than a few percent (~ 5 radians at 150 Hz). To further support this claim, we perform additional numerical experiments in Gaussian and stationary noise according to the expected Advanced LIGO/Virgo noise curves, and coherently injecting signals into the network whose phasing differs structurally from the predictions of GR, but with the magnitude of the deviation still being small. We find that even then, a violation of GR can be established with good confidence.

1. Introduction

The theory of General Relativity (GR) is currently the most commonly accepted theory describing space-time and gravitation. The theory has been accurately tested within the weak-field, stationary regime (*e.g.* through Solar System tests [1]), and no deviations from GR have been conclusively found. However, by the very nature of GR, gravitation is a dynamical phenomenon, a key aspect being the prediction of gravitational waves (GW) [2, 3]. The first clue towards this dynamical aspect came with the discovery of the Hulse-Taylor binary pulsar [4]. Its orbital motion is in agreement with the assumption that GWs carry away energy and angular momentum as predicted by GR. This test, as well as similar test performed on other binary pulsars that were subsequently discovered [5, 6], however still involve indirect observations of GWs. Furthermore, even for the most relativistic binary pulsar that is currently known (PSR J0737-3039), one has $GM/(c^2R) \simeq 4.4 \times 10^{-6}$ (where M is the total mass, and R is the orbital separation) and a typical orbital speed of $v/c \simeq 2 \times 10^{-3}$. The observed binary pulsars are still very far from merger. However, given access to a sufficiently large volume of space, one should find compact binaries in the final stages of inspiral. At the nominal last stable orbit, these will have a separation of $R = 6GM/c^2$, $GM/(c^2R) = 1/6$ and $v/c = 1/\sqrt{6}$. The process of inspiral has been modelled in detail in the context of the so-called post-Newtonian (PN) approximation (*cf.* [7] and references therein). The direct detection of the gravitational wave signals from such binaries would enable us to probe the genuinely strong-field, dissipative dynamics of GR.

Compact binary systems that are close to merger are in fact among the primary targets for kilometer-sized interferometric detectors. These include the Virgo detector in Italy [8, 9], the two LIGO detectors in the United States [10] and GEO600 in Germany [11]. Both Virgo and LIGO are in the process of being upgraded to Advanced Virgo and Advanced LIGO, which are expected to be completed around 2015 [12, 13, 14, 15]. Furthermore, an interferometer named KAGRA (formerly know as LCGT) in Japan is in a planning stage [16], and a detector in India [17] is also being considered. With the current estimates for the advanced detectors, the rate of detection of inspiralling compact binaries is expected be around a few tens per year [18].

Quite a number of alternative theories to GR have been discussed in the literature, and the accuracy with which some of these could be probed with gravitational waves has been studied within the Fisher matrix formalism; for scalar-tensor theories, see [19, 20, 21, 22, 23, 24], for a varying Newton constant [25], for modified dispersion relation theories (commonly referred to as 'massive gravity') [23, 24, 26, 27, 28, 29, 30], for violations of the No Hair Theorem [31, 32, 33, 34, 35], for violations of Cosmic Censorship [35, 36], and for parity violating theories [37, 38, 39, 40]. Practical Bayesian methods for performing tests of GR on actual gravitational wave data when they become available include the work by Del Pozzo *et al.* [41] in the context of massive gravitons, that of Cornish *et al.* [42], which employed the so-called parameterized post-Einsteinian (ppE) waveform family [43, 44], and that of Gossan *et al.* [45] which focussed on the ringdown signal.

A proposal by Arun *et al.* [46, 47, 48] is to measure various quantities within the phase and to check their consistency with the predictions of GR. This could lead to a very generic test, in that one would not be looking for particular (classes of) alternative theories. However, so far its viability was only explored through Fisher matrix studies.

Inspired by the method of Arun *et al.*, the authors of the present paper developed a new Bayesian framework, with the following features [49]:

- Contrary to previous Bayesian treatments such as [42, 45], it addresses the question "Do *one or more* testing parameters characterizing deviations from GR differ from zero?" as opposed to "Do *all* of them differ from zero?" In practice this comes down to testing a number of auxiliary hypotheses, in each of which only a subset of the set of testing parameters is allowed to be non-zero;
- Precisely because in most of the auxiliary hypotheses a smaller set of testing parameters is used, this method will be more suited to a scenario where most sources have low signal-to-noise ratio, as we expect to be the case with Advanced LIGO/Virgo;
- As with most Bayesian methods, information from multiple sources can easily be combined;
- The framework is not tied to any particular waveform family or even any particular part of the coalescence process. However, in [49] we focussed on the inspiral part and chose the testing parameters to be shifts in the lower-order inspiral phase coefficients, as we will also do here.

Besides establishing a theoretical framework, [49] also showed results for a few simple example deviations from GR. In particular, it was illustrated how the method can be sensitive to deviations which in principle cannot be accomodated by the model waveforms. In fact, it is reasonable to assume that the technique will be able to pick up *generic* deviations from GR, on condition that their effect on the phasing is of the same magnitude as that of a simple shift in one or more of the low-order phasing coefficients of the standard post-Newtonian waveform. More precisely, as long as the change in the phase at frequencies where the detectors are the most sensitive is comparable to the effect of a shift of a few percent at $(v/c)^3$ beyond leading order (corresponding to ~ 5 radians at $f = 150$ Hz), we expect the deviation from GR to be detectable by our method. In this paper, we will show some striking examples to provide further support for this claim.

Subsequent sections of this paper are structured as follows. In section 2 we recall the theory and the implementation of the method introduced in [49]. Section 3 shows results from simulations done with some specific examples of modifications to the waveform phase. A discussion and conclusions are presented in section 4.

2. Method

2.1. Bayesian Inference

At the heart of the method we proposed in [49] lies the question "to what degree do we believe GR is the correct theory describing the detected signals?" This question is best answered within the framework of Bayesian model selection [50]. The cornerstone of Bayesian analysis is the comparison between the probabilities of two hypotheses given the available data. This is quantified by the *odds ratio*

$$O^i_j = \frac{P(\mathcal{H}_i|d, \mathrm{I})}{P(\mathcal{H}_j|d, \mathrm{I})}, \tag{1}$$

where $\mathcal{H}_i, \mathcal{H}_j$ are the hypotheses of the models to be compared, d represents the data and I is the relevant background information. Using Bayes' theorem, we can then write this odds ratio as

$$O^i_j = \frac{P(d|\mathcal{H}_i, \mathrm{I})}{P(d|\mathcal{H}_j, \mathrm{I})}\frac{P(\mathcal{H}_i|\mathrm{I})}{P(\mathcal{H}_j|\mathrm{I})} = B^i_j \frac{P(\mathcal{H}_i|\mathrm{I})}{P(\mathcal{H}_j|\mathrm{I})}. \tag{2}$$

The odds ratio is thus the product of two ingredients. The first factor is the ratio of the so-called *evidences*, $B^i_j = \frac{P(d|\mathcal{H}_i,\mathrm{I})}{P(d|\mathcal{H}_j,\mathrm{I})}$, which is also known as the *Bayes factor.* The evidence (also called the marginal likelihood) for *e.g.* the hypothesis $\mathcal{H}_i$ is given by

$$P(d|\mathcal{H}_i, \mathrm{I}) = \int d\vec{\theta}\, p(\vec{\theta}|\mathcal{H}_i, \mathrm{I})\, p(d|\vec{\theta}, \mathcal{H}_i, \mathrm{I}) \tag{3}$$

where $\vec{\theta}$ are the parameters associated with the hypothesis $\mathcal{H}_i$, and $p(\vec{\theta}|\mathcal{H}_i, \mathrm{I})$ is the prior probability distribution of the parameters.

The second factor in Eq. (2) is the ratio of the prior probabilities, $\frac{P(\mathcal{H}_i|\mathrm{I})}{P(\mathcal{H}_j|\mathrm{I})}$, and is often referred to as *prior odds.* It should be noted that the prior probability distribution is uniquely determined by the prior information.

The assignment of the prior probability will be further explained in subsection 2.3. Details on the calculation of the Bayes factor can be found in subsection 2.6.

2.2. Waveform model

Before moving on to define the odds ratio for the problem at hand, let us explain the model waveforms used in this paper. In the inspiral regime of compact binary coalescing systems, the waveforms are accurately described by the post-Newtonian approximation. This approximation describes important quantities such as the energy and the flux as expansions in powers of v/c, where v is the characteristic velocity of the binary system.

To illustrate our method we will use the analytic, frequency domain TaylorF2 waveform model [54, 55], which is implemented in the LIGO Algorithms Library in the following way [56]:

$$h(f) = \frac{1}{D}\frac{\mathcal{A}(\theta,\phi,\iota,\psi,\mathcal{M},\eta)}{\sqrt{\dot{F}(\mathcal{M},\eta;f)}} f^{2/3}\, e^{i\Psi(\mathcal{M},\eta;f)}, \tag{4}$$

where D is the distance, (θ, ϕ) the sky position in the detector frame, and (ι, ψ) the orientation of the orbital plane with respect to the direction to the line of sight. $\mathcal{M}$ is the so-called chirp mass, and η is the symmetric mass ratio; in terms of the component masses (m_1, m_2) one has $\eta = m_1 m_2/(m_1 + m_2)^2$ and $\mathcal{M} = (m_1 + m_2)\,\eta^{3/5}$. The phase $\Psi(\mathcal{M}, \eta; f)$ takes the form

$$\Psi(\mathcal{M},\eta;f) = 2\pi f t_c - \phi_c - \pi/4 + \sum_{i=0}^{7}\left[\psi_i + \psi_i^{(l)}\ln f\right] f^{(i-5)/3}, \tag{5}$$

with t_c and ϕ_c the time and phase at coalescence, respectively. Central to the method are the phase coefficients ψ_i and $\psi_i^{(l)}$. These either have a functional dependence on $(\mathcal{M}, \eta)$ as predicted by GR (*cf.* [48] for the explicit expressions), or are allowed to deviate from the value predicted by GR. In Eq. (4), the 'frequency sweep' $\dot{F}(\mathcal{M}, \eta; f)$ is itself an expansion in powers of the frequency f with mass-dependent coefficients. Note that $\dot{F}$ is related to the phase Ψ and we could in principle allow it to deviate from the GR prediction. However, for stellar mass binaries and with advanced detectors, we do not expect to be particularly sensitive to sub-dominant contributions to the amplitude [36], so we will keep the function $\dot{F}$ fixed to its GR expression.

We note that in the case of binary neutron stars, which are the sources we will in fact focus on, TaylorF2 waveforms are likely to already suffice for a first test of GR. Indeed, in the relevant mass range, TaylorF2 has a match and fitting factor close to unity with Effective One-Body waveforms modified for optimal agreement with numerical simulations [57]. Spins are unlikely to be very important in this case. One might worry about finite size effects, but as shown in [58], even for the most extreme neutron star equations of state and for sources as close as 100 Mpc, advanced detectors will not be sensitive to these at frequencies below $\sim$ 400 Hz; hence one could cut off the recovery waveforms at 400 Hz, in which case the loss in SNR would be less than a percent. However, if one also wanted to test GR using systems composed of a neutron star and a black hole, or two black holes, then dynamical spins [51], sub-dominant signal harmonics [52, 36] and merger/ringdown [53] would become important, and in that case more sophisticated waveform models will be called for. The latter is currently the subject of investigation.

2.3. Defining the odds ratio

Focussing our attention on deviations from GR in the phase of the measured waveform, *i.e.* keeping the amplitude fixed to its GR-predicted value, we consider within the Bayesian model selection framework the following two hypotheses:

- $\mathcal{H}_{\rm GR}$: The waveform has a phase with the functional dependence on $(\mathcal{M}, \eta)$ as predicted by GR;
- $\mathcal{H}_{\rm modGR}$: One or more of the phase coefficients do not have the functional dependence on $(\mathcal{M}, \eta)$ as predicted by GR.

The GR hypothesis, $\mathcal{H}_{\rm GR}$, is the hypothesis that our GR waveform model (TaylorF2 in this case) correctly describes the signal originating from the inspiral of two compact objects. Ideally, $\mathcal{H}_{\rm modGR}$ would simply have been the negation of $\mathcal{H}_{\rm GR}$. However, *a priori*, deviations from GR can occur an infinite number of ways. What we will argue is that for the core question that we want to address, *i.e.* whether or not the observed phase deviates from GR, it will be sufficient to allow for a *limited* set of possible deviations in the recovery waveforms. For TaylorF2, we take the set of deviations only to be within the known phase coefficients $\{\psi_0, \psi_1, \psi_2, \ldots, \psi_N\}$.

To date, the TaylorF2 phase has ten known phase coefficients ($\psi_0, \ldots, \psi_7$, and two additional coefficients $\psi_5^{(l)}$ and $\psi_6^{(l)}$ associated with logarithmic contributions). Here we will not use ψ_0 as a variable coefficient; even so, if one were to consider all the subsets of the set of remaining coefficients, one would have to take into account $2^9 - 1 = 511$ ways in which a deviation can occur. Apart from this being computationally demanding, we do not expect to be sensitive to the highest-order coefficients; hence it makes sense to limit oneself to all the subsets of

$$\{\psi_1, \psi_2, \ldots, \psi_{N_T}\}, \tag{6}$$

where N_T is the number of phase coefficients one chooses to consider. We thus allow one or more of the coefficients $\{\psi_1, \psi_2 \ldots, \psi_{N_T}\}$ to vary freely, instead of following the functional dependence on $(\mathcal{M}, \eta)$ as predicted by GR. The choice of N_T will be in part be influenced by the required generality of the test, measurability of phase coefficients, and computational limitations.

Finally, we quantify our belief in whether one or more phase coefficients deviate from GR by means of auxiliary hypotheses $H_{i_1 i_2 \ldots i_k}$, which are defined as follows:

> $H_{i_1 i_2 \ldots i_k}$ is the hypothesis that the phasing coefficients $\psi_{i_1}, \ldots, \psi_{i_k}$ do not have the functional dependence on $(\mathcal{M}, \eta)$ as predicted by General Relativity, but all other coefficients ψ_j, $j \notin \{i_1, i_2, \ldots, i_k\}$ *do* have the dependence as in GR.

It is important to note that by definition of the hypotheses $H_{i_1 i_2 \ldots i_k}$, they are mutually, logically disjoint, *i.e.*, $H_{i_1 i_2 \ldots i_k} \wedge H_{j_1 j_2 \ldots j_l}$ is always false for $\{i_1, i_2, \ldots, i_k\} \neq \{j_1, j_2, \ldots, j_l\}$.

For a signal to be inconsistent with GR, we require that one or more phase coefficients deviate from GR. In terms of hypotheses, we are thus interested in the logical 'or' of the sub-hypotheses, $H_{i_1 i_2 \ldots i_k}$, defined above. With this in hand, we can now define $\mathcal{H}_{\rm modGR}$ to be:

$$\mathcal{H}_{\rm modGR} = \bigvee_{i_1 < i_2 < \ldots < i_k; k \leq N_T} H_{i_1 i_2 \ldots i_k}. \tag{7}$$

The odds ratio for N_T coefficients is given by:

$$^{(N_T)}O_{\rm GR}^{\rm modGR} = \frac{P(\mathcal{H}_{\rm modGR}|d, \mathrm{I})}{P(\mathcal{H}_{\rm GR}|d, \mathrm{I})} = \frac{P(\bigvee_{i_1 < i_2 < \ldots < i_k; k \leq N_T} H_{i_1 i_2 \ldots i_k}|d, \mathrm{I})}{P(\mathcal{H}_{\rm GR}|d, \mathrm{I})}, \tag{8}$$

Using the fact that the auxiliary hypotheses are mutually, logically disjoint, one can write

$$P(\bigvee_{i_1 < i_2 < \ldots < i_k; k \leq N_T} H_{i_1 i_2 \ldots i_k}|d, \mathrm{I}) = \sum_{k=1}^{N_T} \sum_{i_1 < i_2 < \ldots < i_k} P(H_{i_1 i_2 \ldots i_k}|d, \mathrm{I}). \tag{9}$$

Applying Bayes' theorem, one finds

$$^{(N_T)}O_{\rm GR}^{\rm modGR} = \sum_{k=1}^{N_T} \sum_{i_1 < i_2 < \ldots < i_k} \frac{P(H_{i_1 i_2 \ldots i_k}|\mathrm{I})}{P(\mathcal{H}_{\rm GR}|\mathrm{I})} B_{\rm GR}^{i_1 i_2 \ldots i_k}, \tag{10}$$

where

$$B_{\rm GR}^{i_1 i_2 \ldots i_k} = \frac{P(d|H_{i_1 i_2 \ldots i_k}, \mathrm{I})}{P(d|\mathcal{H}_{\rm GR}, \mathrm{I})}. \tag{11}$$

At this point, one has to set the values for the relative prior probabilities, $\frac{P(H_{i_1 i_2 \ldots i_k}|\mathrm{I})}{P(\mathcal{H}_{\mathrm{GR}}|\mathrm{I})}$. When one is devoid of prior information as to which of the test coefficients are inconsistent with GR, one can choose to invoke total ignorance and assign to each an equal weight, *i.e.*

$$\frac{P(H_{i_1 i_2 \ldots i_k}|\mathrm{I})}{P(\mathcal{H}_{\mathrm{GR}}|\mathrm{I})} = \frac{P(H_{j_1 j_2 \ldots j_l}|\mathrm{I})}{P(\mathcal{H}_{\mathrm{GR}}|\mathrm{I})} \quad \text{for any } k, l \leq N_T. \tag{12}$$

Despite the choice of total ignorance, however, one more quantity needs to be set. The overall relative prior, $\frac{P(\mathcal{H}_{\mathrm{modGR}}|\mathrm{I})}{P(\mathcal{H}_{\mathrm{GR}}|\mathrm{I})}$, describes the prior belief in whether GR is the correct theory or not. The choice of this quantity is left to the reader. For convenience, however, we write

$$\frac{P(\mathcal{H}_{\mathrm{modGR}}|\mathrm{I})}{P(\mathcal{H}_{\mathrm{GR}}|\mathrm{I})} = \alpha. \tag{13}$$

As will become apparent below, α will end up being just an overall scaling of the odds ratio. Later on, for the purposes of showing results, we will set $\alpha = 1$.

The equality (13), together with (12) and the logical disjointness of the $2^{N_T} - 1$ hypotheses $H_{i_1 \ldots i_k}$ implies

$$\frac{P(H_{i_1 i_2 \ldots i_k}|\mathrm{I})}{P(\mathcal{H}_{\mathrm{GR}}|\mathrm{I})} = \frac{\alpha}{2^{N_T} - 1}. \tag{14}$$

In terms of the $H_{i_1 i_2 \ldots i_k}$, the odds ratio can then be written as

$${}^{(N_T)}O^{\mathrm{modGR}}_{\mathrm{GR}} = \frac{\alpha}{2^{N_T} - 1} \sum_{k=1}^{N_T} \sum_{i_1 < i_2 < \ldots < i_k} B^{i_1 i_2 \ldots i_k}_{\mathrm{GR}}. \tag{15}$$

Up to an overall prefactor, the odds ratio is thus a straightforward average of the Bayes factors from the individual sub-hypotheses, $H_{i_1 i_2 \ldots i_k}$.

2.4. Multiple sources

Although the detection rate for compact binary coalescences is still rather uncertain, we expect advanced instruments to detect several events per year [18]. It is therefore important to take advantage of multiple detections to provide tighter constraints on the validity of GR.

The extension of the odds ratio to include observations from several independent sources can be found in [41, 49]. Here we simply state the result, referring the interested reader to these papers for details. If one assumes $\mathcal{N}$ independent measurements and the events are labelled by A, one can write the odds ratio as

$${}^{(N_T)}\mathcal{O}^{\mathrm{modGR}}_{\mathrm{GR}} = \frac{\alpha}{2^{N_T} - 1} \sum_{k=1}^{N_T} \sum_{i_1 < i_2 < \ldots < i_k} \prod_{A=1}^{\mathcal{N}} {}^{(A)}B^{i_1 i_2 \ldots i_k}_{\mathrm{GR}}, \tag{16}$$

where

$${}^{(A)}B^{i_1 i_2 \ldots i_k}_{\mathrm{GR}} = \frac{P(d_A|H_{i_1 i_2 \ldots i_k}, \mathrm{I})}{P(d_A|\mathcal{H}_{\mathrm{GR}}, \mathrm{I})}, \tag{17}$$

with d_A being the data associated to the Ath detection.

2.5. Noise

From a theoretical point of view, the data favours the hypothesis $\mathcal{H}_{\rm modGR}$ compared to the hypothesis $\mathcal{H}_{\rm GR}$ when $\mathcal{O}_{\rm GR}^{\rm modGR} > 1$. The relative degree of belief in the two hypotheses is encapsulated in the magnitude of the odds ratio. However, in the case of advanced ground-based detectors, the signals will be buried deep inside the noise. This introduces the problem that the noise itself can mimic the effect of a deviation from GR that is non-negligible.

Hence we need to study the effect of noise on the odds ratio. For this purpose, we constructed a so-called *background*, *i.e.* a distribution of log odds ratios from a large number of catalogues, collectively denoted by κ, of simulated signals consistent with $\mathcal{H}_{\rm GR}$ and embedded within noise. The background distribution $P(\ln \mathcal{O}_{\rm GR}^{\rm modGR}|\kappa, \mathcal{H}_{\rm GR}, {\rm I})$ of log odds ratios for catalogues of GR sources can be seen as the blue dotted histogram in the right hand panel of Fig. 2 below.

In the advanced detector era, one will only have access to a single catalogue of detected *foreground* events. The associated measured log odds ratio should subsequently be compared with the background distribution in order to quantify our belief in a deviation from GR. To do this, in Sec. 3.3 we will introduce a maximum tolerable false alarm probability β, which together with the background distribution sets a threshold $\ln \mathcal{O}_\beta$ for the measured log odds ratio to overcome.

For specific examples of GR violations, we will want to know how likely it is that the catalogue of foreground sources will have an odds ratio that is above threshold. For this reason we will also simulate large numbers of foreground catalogues, collectively denoted by κ'. For a given false alarm rate β, one can then calculate what fraction of the simulated foreground catalogues has a log odds ratio above the associated threshold $\ln \mathcal{O}_\beta$; this fraction we will call the *efficiency*.

2.6. Implementation

A few remarks have to be made regarding the implementation of the aforementioned method. First, we use $N_T = 3$ testing parameters $\{\psi_1, \psi_2, \psi_3\}$. The varying of these phase coefficients was parameterised in the following fashion:

$$\psi_i = \psi_i^{\rm GR}(\mathcal{M}, \eta)\, \left[1 + \delta\chi_i\right], \tag{18}$$

with $\psi_i^{\rm GR}(\mathcal{M}, \eta)$ the functional form of the dependence of ψ_i on $(\mathcal{M}, \eta)$ according to GR, and the dimensionless $\delta\chi_i$ is a fractional shift in ψ_i. The 0.5PN case, ψ_1, however cannot be implemented in a similar way, as GR predicts $\psi_1^{\rm GR} = 0$. Instead, deviations from $\psi_1^{\rm GR}$ are modelled as

$$\psi_1^{\rm GR}(\mathcal{M}, \eta) = 0 \quad \rightarrow \quad \frac{3}{128\eta}(\pi \mathcal{M}\, \eta^{-3/5})^{-4/3} \delta\chi_1, \tag{19}$$

and the interpretation of a fractional shift is not adequate; rather, $\delta\chi_1$ is related to the magnitude of the deviation itself.

For the computation of the odds ratio defined in Eq. (15) and Eq. (16), one needs to compute the relevant Bayes factors via the evidences. In high-dimensional problems, brute force methods to calculate the integral in Eq. (3) are computationally too expensive. One can, however, make use of more efficient methods to make this calculation computationally feasible. In this paper, we resort to an algorithm called Nested Sampling [59]. More specifically, an implementation tailored to ground-based observations of coalescing binaries by Veitch and Vecchio [60, 61, 62] was used.

Both the model waveforms and the Nested Sampling algorithm were appropriately adapted from existing code in the LIGO Algorithms Library [56].

3. Results

In this section, we want to lend further support to the claim in [49] that the method is in principle sensitive to deviations that are not considered within the model waveforms, as long as the phase shift at $f \sim 150$ Hz, where the detectors are the most sensitive, is comparable to, say, a shift of $\delta\chi_3 \sim$ (a few) $\times 10^{-2}$ at 1.5PN order (corresponding to a shift in the overall phase of ~ 5 radians at the given frequency). To this end we use two heuristic examples where the change in phase of the signals cannot be accomodated by the model waveforms, yet the deviation from GR turns out to be detectable.

The first example, in subsection 3.1, considers an additional term in the phase associated with a power of frequency which *itself* depends on the total mass of the system. This power is chosen in such a way that within the range of total masses we consider, the frequency dependence of the anomalous contribution varies from effectively being 0.5PN at the lower end to 1.5PN at the higher end. Clearly, our model waveforms are in no way designed to capture such a deviation from GR. The second case, in subsection 3.2, considers a deviation at a PN order (2PN) that is higher than the orders at which we allow phase coefficients to vary in our model waveforms (0.5PN, 1PN, and 1.5PN).

After presenting the main results, we study the effects of the number of detected sources on our confidence in a deviation from GR. For this investigation we use the example in subsection 3.2.

From Fisher matrix analyses, it has been shown that the phase coefficients in Eq. (5) are best measured as the total mass of the system goes down [46, 47, 48]. Therefore, the signals were chosen to originate from neutron stars with masses between $1\,M_\odot$ and $2\,M_\odot$. For such systems, it has been shown that contributions from the spin interactions and the sub-dominant signal harmonics are negligible, and merger/ringdown do not have a significant impact [36, 63, 64].

The aim is to simulate the situation at Advanced Virgo and LIGO as realistically as possible. We have assumed an advanced detector network with detectors at Hanford and Livingston, both with the Advanced LIGO noise curve [65], and a detector at Cascina with the Advanced Virgo noise curve [14]. Three data streams were produced, containing stationary, Gaussian noise coloured by these respective noise curves, to which simulated signals were added. Events were placed uniformly in volume (*i.e.* probability density proportional to D^2, where D is the luminosity distance), between 100 Mpc and 400 Mpc, to reflect the estimates of the number of detectable sources and the appropriate horizon distance. A lower cut-off of 8 was imposed on the *network* SNR, defined as the quadrature sum of the individual detector SNRs, so as to be consistent with the LIGO/Virgo minimum for an event to be claimed as detected.

The waveforms were chosen to go up to 2PN in phase both for the injected signals and the model waveforms. The test coefficients were taken to be ψ_1, ψ_2 and ψ_3, so that the hypothesis $\mathcal{H}_{\rm modGR}$ contains $2^{N_T} - 1 = 7$ logically disjoint sub-hypotheses.

The priors given to the deviations $\delta\chi_i$ were chosen to be flat and centered around zero, with a total width of 0.5. The priors on the remaining parameters were taken to be the same as in [60], with the exception that the distance is allowed to go up to 1000 Mpc.

It should be stressed that the choice of waveform approximant, test coefficients, and priors on the deviations were, to a large extent, arbitrary. In the advanced detector era, one would seek to perform the most general test that computational resources will allow. This will include the most accurate waveforms available at that time, the highest number of test coefficients one can handle, and the least restrictive priors that are in accordance with our prior information at that moment.

3.1. A deviation with a mass dependent power of frequency

In our first example, the signals are given a deviation in the phase that has a mass dependent frequency power. Specifically, the deviation is of the form:

$$\Psi^{\rm GR}(\mathcal{M},\eta;f) \to \Psi^{\rm GR}(\mathcal{M},\eta;f) + \frac{3}{128\eta}(\pi M f)^{-2+M/(3M_\odot)}, \tag{20}$$

where M denotes the total mass of the binary system. We note that for a system with component masses in the middle of our range, $(1.5, 1.5)\,M_\odot$, the change in phase at $f = 150$ Hz is about the same as for a 10% shift in ψ_3. More precisely, for these masses the change in $\Psi(\mathcal{M}, \eta; 150\,\text{Hz})$ is 13.3 radians, to be compared with the 12.8 radians change induced by a constant 10% shift in ψ_3.

In order to assess the statistics of the odds ratio, a large number or signals were simulated, with the parameter distribution explained above. For each of the signals, we calculated the odds ratio as defined in Eq. (15). The distribution of the odds ratio as a function of SNR can be seen in Fig. 1. The separation between 'foreground' and 'background' is more or less complete already below SNR ~ 15.

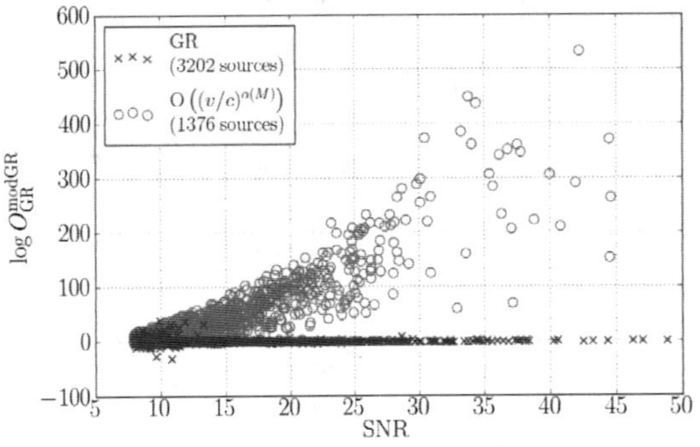

Figure 1. Distribution of log odds ratios as a function of the optimal network SNR. The crosses are for sources whose GW emission is in accordance with GR, while the circles are for sources with a deviation as in Eq. (20). For the GR sources, the log odds ratios are concentrated tightly around zero, while for sources with the deviation in phase, they increase as a function of SNR. This is in agreement with the expectation that parameter estimation improves as the SNR increases. As our parameter estimation becomes better, deviations become more pronounced, which increases our confidence in a deviation from GR.

In the top panel of Fig. 2, the odds ratios for the sources with a deviation from GR are compared with a 'background' distribution (in the sense defined above). Next we collected sources into 'catalogues' of 15 sources each and computed the *combined* odds ratio of Eq. (16) for all of these catalogues; the distribution of these odds ratios for 'background' and 'foreground' are shown in the bottom panel of Fig. 2. Clearly, the ability to combine information from multiple sources is a powerful tool in increasing one's confidence in a violation of GR. For the given deviation, a violation of GR can be established with near-certainty.

3.2. A deviation at a higher PN order than the testing coefficients

Our testing coefficients are ψ_1, ψ_2, ψ_3, so that the model waveforms can only have shifts in PN phase contributions up to 1.5PN order. To show that we can nevertheless be sensitive to anomalies at higher PN order, we now consider signals with a constant shift at 2PN. We note in passing that theories with quadratic curvature terms in the action tend to introduce extra contributions at 2PN [66, 67, 68].

Thus, we consider injections with

$$\psi_4 = \psi_4^{\mathrm{GR}}(\mathcal{M}, \eta)\,[1 + \delta\chi_4]\,, \tag{21}$$

where the magnitude is set to be $\delta\chi_4 = 0.2$. For comparison, at $f = 150$ Hz and for a system with component masses $(1.5, 1.5)\,M_\odot$, the change in the phase caused by such a deviation is

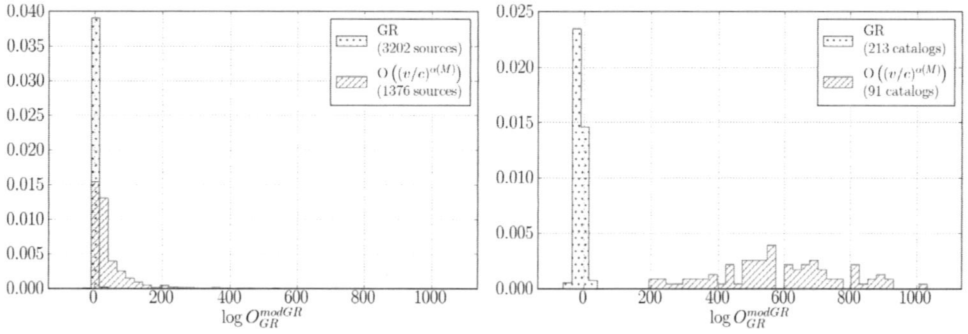

Figure 2. Left: Normalised distributions of log odds ratios for individual sources, where the signals are in accordance with $\mathcal{H}_{\mathrm{GR}}$ (blue dotted) or have a shift with a mass dependent frequency behaviour given in Eq. (20) (red striped). Right: Normalised distributions of logs of the combined odds ratios for the same injections as at the top, but collected into independent catalogues of 15 sources each. The effect of combining sources is to separate the distribution of GR injections and anomalous injections, increasing one's confidence in a deviation from GR.

comparable to the one caused by a negative relative shift in ψ_3 of 3.5% (namely, the shift in $\Psi(\mathcal{M}, \eta, f)$ at $f = 150$ Hz is ~ 4.5 radians).

Fig. 3 shows the odds ratio as a function of the optimal SNR, both for GR injections and anomalous ones. This time, as opposed to the example considered in subsection 3.1, separation between the odds ratios of signals with the deformations of Eq. (21) and the noise induced distribution of odds ratios for GR injections becomes apparent at SNR $\sim$ 20. This can be attributed the fact that the deviations are in more subdominant contributions to the phase compared to the case considered earlier.

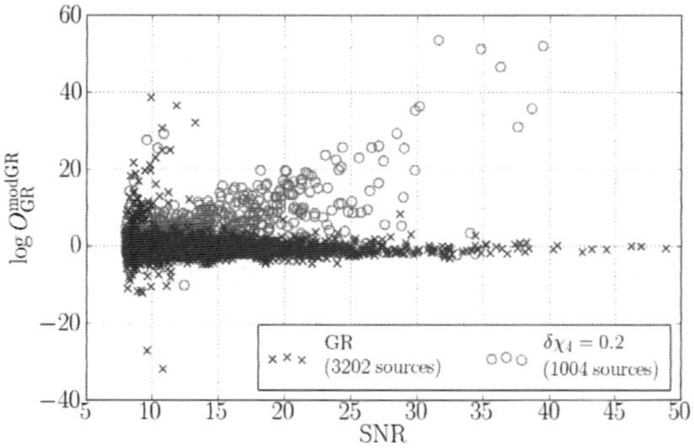

Figure 3. The distribution of odds ratio as a function of the optimal network SNR for GR injections (crosses) and injections with deviations from GR as in Eq. (21) (circles). For the anomalous injections, the odds ratio increases as a function of SNR.

In Fig. 4 we show the odds ratio for individual sources and for random catalogues with 15 sources each. For individual sources, the separation between the background and the foreground is present but weak. However, when one assumes random catalogues of 15 sources each, the separation becomes very significant. This further illustrates the importance of combining information from multiple sources.

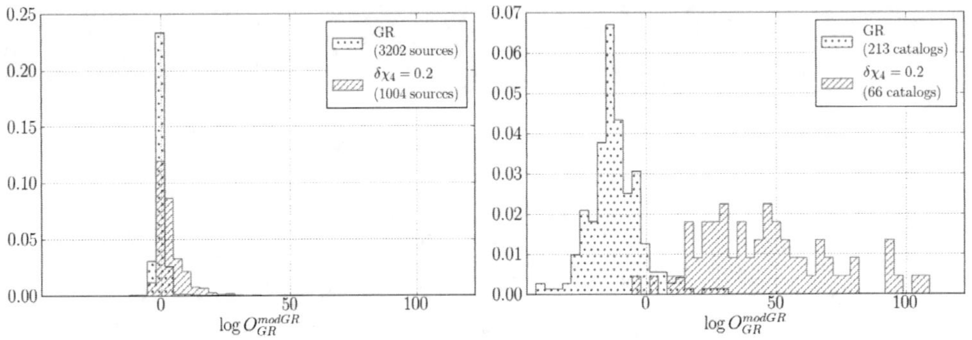

Figure 4. Left: normalized distribution of log odds ratios for individual sources, where the signals are in accordance with $\mathcal{H}_{\rm GR}$ (blue dotted) or have a deviation of the form given in Eq. (21) (red striped). Right: normalised distribution of logs of the combined odds ratios for the same signals as at the top, but randomly arranged in catalogues of 15 sources each. The effect of combining sources is in this case is profound. Only a small difference between background and foreground is visible when considering individual sources. For catalogues of 15 sources, the differentiation becomes significant.

3.3. Effect of catalogue size

We have seen in the previous section that constructing the odds ratio from a catalogue of sources greatly increases the confidence in a deviation from GR. However, the rates of binary inspiral observed in the advanced detector era are highly uncertain. It is therefore instructive to study the effect of the catalogue size on our confidence in detecting a deviation.

To characterise such a confidence, we introduce the concept of *efficiency*. Assume one has two distributions of log odds ratios: The *background distribution* of log odds ratio obtained when the simulated catalogues, collectively denoted by κ, are in agreement with $\mathcal{H}_{\rm GR}$, $P(\ln \mathcal{O}_{\rm GR}^{\rm modGR}|\kappa, \mathcal{H}_{\rm GR}, I)$, and the *foreground distribution* obtained when the simulated catalogues κ' adhere to some alternative theory, $P(\ln \mathcal{O}_{\rm GR}^{\rm modGR}|\kappa', \mathcal{H}_{\rm alt}, I)$. Now choose a maximum tolerable *false alarm probability* β. This sets a threshold $\ln \mathcal{O}_\beta$ for the measured log odds ratio to overcome, as follows:

$$\beta = \int_{\ln \mathcal{O}_\beta}^{\infty} P(\ln \mathcal{O}|\kappa, \mathcal{H}_{\rm GR}, {\rm I})\, d\ln \mathcal{O}\,. \tag{22}$$

We now define the *efficiency*, ζ, as the fraction of foreground with a false alarm probability of β or less, *i.e.* the portion that lies above the threshold $\ln \mathcal{O}_\beta$:

$$\zeta = \int_{\ln \mathcal{O}_\beta}^{\infty} P(\ln \mathcal{O}|\kappa', \mathcal{H}_{\rm alt}, {\rm I})\, d\ln \mathcal{O}. \tag{23}$$

The efficiency can be viewed as the chance that if there is a deviation from GR corresponding to $\mathcal{H}_{\rm alt}$, the catalogue of sources that is actually detected will have a log odds ratio above threshold, *i.e.* that it will have a false alarm probability of β or less. Note that with these definitions, the efficiency is independent of the overall prior odds ratio α in Eqs. (13) and (16), as it corresponds to the same shift of $\ln \alpha$ in the background distribution, the threshold $\ln \mathcal{O}_\beta$, and the foreground distribution.

In Fig. 5, we show the efficiency ζ for the example shown in subsection 3.2 as a function of the catalogue size, for $\beta \in \{0.32, 0.05, 0.01\}$. Which sources are placed together in a catalogue is determined randomly. To understand the statistical fluctuations in the efficiency when collecting sources into catalogues in different ways, for the same set of signals we considered 5000 random orderings in which the signals are combined into catalogues. The resulting median and the 68% confidence levels are shown as the central curve and the error bars, respectively.

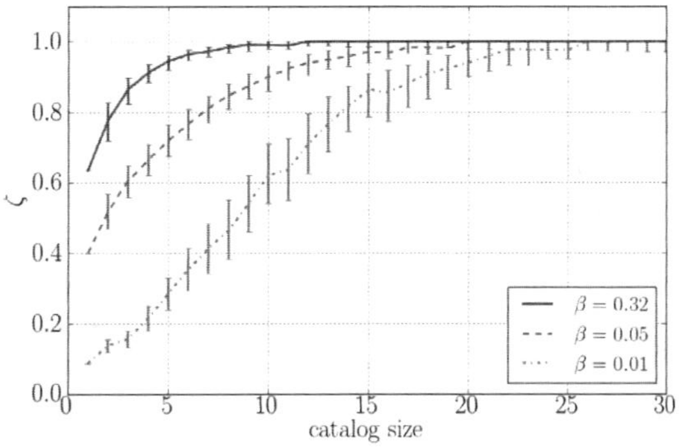

Figure 5. The efficiency at a fixed false alarm rate as function of the catalogue size for the example described in subsection 3.2 ($\delta\chi_4 = 0.2$). 5000 random orderings of the same set of sources were split into catalogues. The mean (central curve) and the 68% confidence intervals (error bars) are plotted. The efficiency rises sharply as a function of the number of sources, underscoring the importance of coherently considering all detected signals events.

As can be seen in Fig. 5, the acceptance probability rises sharply as a function of the catalogue size. This underscores the importance of considering all the detected source in a coherent fashion, as was explained in subsection 2.4. Even though a single detection might not yield confidence in a deviation from GR, coherently adding information from multiple sources can rapidly increase this confidence.

To put the numbers in Fig. 5 into perspective, the predicted rate for binary inspiral in the so-called 'realistic' case is 40 per year [18].

4. Conclusions and discussion

We have given two striking examples to support the claim that our method proposed in [49] can distinguish deviations that are not captured by the limited model waveforms, as long as the phase shift in the frequency range where the detectors are the most sensitive ($f \sim 150$ Hz) is comparable to one caused by a shift of at least $\delta\chi_3 \sim$ (a few) $\times 10^{-2}$, *i.e.*, ~ 5 radians.

In the first example, signals were studied that have a deviation in the phase with a mass dependent power of frequency, effectively ranging from 0.5PN to 1.5PN as the total mass is varied from the lowest to the highest value we consider. The magnitude of the effect was such that at $f = 150$ Hz, the change in phase (~ 13 radians) was about the same as that induced by a constant relative shift $\delta\chi_3 = -0.1$. The odds ratio for individual sources already showed confidence that such deviations can be measured. When sources were combined into catalogues of 15 each, the confidence in having detected a deviation improved drastically, and a deviation of this kind will be measurable with a false alarm probability of essentially zero.

We further showed results for signals with a shift in the 2PN phase coefficient, ψ_4. Setting $\delta\chi_4 = 0.2$, the induced change in phase at $f = 150$ Hz is comparable to a constant shift $\delta\chi_3 = -0.035$ at 1.5PN, namely 4.5 radians. The choice of a modification at 2PN was inspired by corrections to the phase if one considers a modified Einstein-Hilbert action containing terms that are quadratic in the Riemann tensor, as calculated in [66, 67, 68]. As can be seen from Fig. 5, the efficiency for a maximum false alarm probability of 1% is essentially unity for catalogues comprising more than 25 sources.

Lastly, we investigated the effect of the catalogue size on our confidence in detecting a deviation. In general, this confidence rises sharply with the number of sources in the catalogue, underscoring the necessity to combine information from multiple sources in the advanced detector era.

Finally, we want to mention some necessary future developments. First and foremost, the most accurate waveforms will need to be incorporated in order to distinguish between genuine effects predicted by GR, and possible deviations from GR. Especially for systems consisting of two black holes, or a neutron star and a black hole, these waveforms will need to include dynamical spins, sub-dominant signal harmonics, residual eccentricity, a description of the merger and ringdown, *etc.* The development of waveforms including these effects is ongoing [69, 70, 71, 72, 73, 74, 75, 76, 77, 78, 79]. Furthermore, the effects of realistic detector noise, instead of idealised stationary, Gaussian noise, need to be studied in detail.

Once the advanced detectors have reached their design sensitivities and a number of detections have been made (*e.g.* using template-based searches [80]), a test of General Relativity using compact binary coalescences could go as follows. Starting from the best available GR waveforms, one introduces parameterised deformations in phase as well as in amplitude, leading to disjoint hypotheses $H_{i_1 i_2 \ldots i_k}$, the logical 'or' of all of which is $\mathcal{H}_{\rm modGR}$. Next, many injections are performed of GR waveforms into real or realistic data and collected into 'catalogues' to establish a background distribution for the log odds ratio $\ln \mathcal{O}_{\rm GR}^{\rm modGR}$, and a suitable threshold $\ln \mathcal{O}_\beta$ is set below which a deviation from GR will not be accepted. Then $\ln \mathcal{O}_{\rm GR}^{\rm modGR}$ is computed for the catalogue of sources that were actually detected. If this number is above threshold, a violation of GR is likely.

The number of testing parameters one can consider will be limited, mainly by the computational restrictions one will have in the advanced detector era. Our method is meant, first and foremost, to establish whether or not a violation of GR is plausible, *of whichever kind* and *not* mainly to pinpoint the eventual alternative theory of gravity responsible for the GW signal, nor to estimate the parameters of the alternative model, as it is unlikely that low-SNR signals as those expected for the advanced stage of LIGO/Virgo will enable a detection of a GR deviation *and* an identification of its nature. However, once a deviation is found, a follow-up investigation can be performed with our inference method in an attempt to find out its precise nature by trying different alternatives to GR, *i.e.* using waveforms inspired by specific (families of) alternative theories of gravity. A version of the so-called parameterised post-Einsteinian waveform family [43, 42] could be useful in this respect. In this regard we recall that our framework is not tied to any particular waveform family.

The results of [49], and the further investigations presented here, motivate the construction of a full data analysis pipeline based on the method we have presented. Although much work remains to be done on the data analysis side, the advanced detectors will enable us to go well beyond the tests performed using the observed binary pulsars, and give us our very first empirical access to the genuinely strong-field dynamics of space-time.

Acknowledgements

TGFL, WDP, SV, CVDB and MA are supported by the research programme of the Foundation for Fundamental Research on Matter (FOM), which is partially supported by the Netherlands Organisation for Scientific Research (NWO). JV's research was funded by the Science and Technology Facilities Council (STFC), UK, grant ST/J000345/1. KG, TS and AV are supported by Science and Technology Facilities Council (STFC), UK, grant ST/H002006/1. The work of RS is supported by the EGO Consortium through the VESF fellowship EGO-DIR-41-2010.

It is a pleasure to thank N. Cornish, B.R. Iyer, B.S. Sathyaprakash and N. Yunes for their valuable comments and suggestions. The authors would also like to acknowledge the LIGO Data Grid clusters, without which the simulations could not have been performed. Specifically, these include the computing resources supported by National Science Foundation awards PHY-0923409 and PHY-0600953 to UW-Milwaukee. Also, we thank the Albert Einstein Institute in Hannover, supported by the Max-Planck-Gesellschaft, for use of the Atlas high-performance computing cluster.

References

[1] C. Misner, K. Thorne, and J. Wheeler, *Gravitation* (WH Freeman & co, 1973).
[2] C. M. Will, Living Reviews in Relativity **9** (2006).
[3] B. Sathyaprakash and B. F. Schutz, Living Reviews in Relativity **12** (2009).
[4] R. A. Hulse and J. H. Taylor, The Astrophysical Journal Letters **195**, L51 (1975).
[5] M. Burgay *et al.*, Nature **426**, 531 (2003).
[6] M. Kramer *et al.*, Science **314**, 97 (2006).
[7] L. Blanchet, Living Reviews in Relativity **9** (2006).
[8] F. Acernese *et al.*, Classical and Quantum Gravity **25**, 184001 (2008).
[9] T. Accadia *et al.*, Classical and Quantum Gravity **28**, 114002 (2011).
[10] B. P. Abbott *et al.*, Reports on Progress in Physics **72** (2009).
[11] H. Grote and the LIGO Scientific Collaboration, Classical and Quantum Gravity **27**, 084003 (2010).
[12] G. M. Harry and the LIGO Scientific Collaboration, Classical and Quantum Gravity **27**, 084006 (2010).
[13] `http://www.ligo.caltech.edu/advLIGO/`,
[14] Virgo Collaboration, Advanced Virgo Baseline Design, Report No. VIR027A09, 2009 (unpublished).
[15] `http://wwwcascina.virgo.infn.it/advirgo/`,
[16] K. Kuroda and the LCGT Collaboration, Classical and Quantum Gravity **27**, 084004 (2010).
[17] B. Sathyaprakash, Scientific Benefits of LIGO-India, LIGO Scientific Collaboration Report No. G1100991, 2011 (unpublished).
[18] J. Abadie *et al.*, Classical and Quantum Gravity **27**, 173001 (2010).
[19] C. M. Will, Phys. Rev. D **50**, 6058 (1994).
[20] T. Damour and G. Esposito-Farèse, Phys. Rev. D **58**, 042001 (1998).
[21] P. D. Scharre and C. M. Will, Phys. Rev. D **65**, 042002 (2002).
[22] C. M. Will and N. Yunes, Classical and Quantum Gravity **21**, 4367 (2004).
[23] E. Berti, A. Buonanno, and C. M. Will, Phys. Rev. D **71**, 084025 (2005).
[24] E. Berti, A. Buonanno, and C. M. Will, Classical and Quantum Gravity **22**, S943 (2005).
[25] N. Yunes, F. Pretorius, and D. Spergel, Phys. Rev. D **81**, 064018 (2010).
[26] C. M. Will, Phys. Rev. D **57**, 2061 (1998).
[27] K. G. Arun and C. M. Will, Classical and Quantum Gravity **26**, 155002 (2009).
[28] A. Stavridis and C. M. Will, Phys. Rev. D **80**, 044002 (2009).
[29] D. Keppel and P. Ajith, Phys. Rev. D **82**, 122001 (2010).
[30] E. Berti, J. Gair, and A. Sesana, (2011), arXiv:1107.3528v2.
[31] E. Berti, V. Cardoso, and C. M. Will, Phys. Rev. D **73**, 064030 (2006).
[32] S. A. Hughes, AIP Conference Proceedings **873**, 233 (2006).
[33] L. Barack and C. Cutler, Phys. Rev. D **75**, 042003 (2007).
[34] J. R. Gair, C. Li, and I. Mandel, Phys. Rev. D **77**, 024035 (2008).
[35] I. Kamaretsos, M. Hannam, S. Husa, and B. Sathyaprakash, (2011), arXiv:1107.0854v2.
[36] C. Van Den Broeck and A. S. Sengupta, Classical and Quantum Gravity **24**, 1089 (2007).
[37] S. Alexander, L. S. Finn, and N. Yunes, Phys. Rev. D **78**, 066005 (2008).
[38] N. Yunes and L. S. Finn, Journal of Physics: Conference Series **154** (2009).
[39] C. F. Sopuerta and N. Yunes, Phys. Rev. D **80**, 064006 (2009).
[40] N. Yunes, R. O'Shaughnessy, B. J. Owen, and S. Alexander, Phys. Rev. D **82**, 064017 (2010).
[41] W. Del Pozzo, J. Veitch, and A. Vecchio, Phys. Rev. D **83**, 082002 (2011).
[42] N. Cornish, L. Sampson, N. Yunes, and F. Pretorius, (2011), arXiv:1105.2088.
[43] N. Yunes and F. Pretorius, Phys. Rev. D **80**, 122003 (2009).
[44] N. Yunes and S. A. Hughes, Phys. Rev. D **82**, 082002 (2010).
[45] S. Gossan, J. Veitch, and B.S. Sathyaprakash, (2011), arXiv:1111.5819
[46] K. G. Arun, B. R. Iyer, M. S. S. Qusailah, and B. S. Sathyaprakash, Classical and Quantum Gravity **23**, L37 (2006).
[47] K. G. Arun, B. R. Iyer, M. S. S. Qusailah, and B. S. Sathyaprakash, Phys. Rev. D **74**, 024006 (2006).
[48] C. K. Mishra, K. G. Arun, B. R. Iyer, and B. S. Sathyaprakash, Phys. Rev. D **82**, 064010 (2010).
[49] T. G. F. Li *et al.*, (2011), arXiv:1110.0530v1.
[50] E. Jaynes and G. Bretthorst, *Probability theory: the logic of science* (Cambridge Univ Pr, 2003).

[51] L. Blanchet, T. Damour, B. R. Iyer, C. M. Will, and A. G. Wiseman, Phys. Rev. Lett. **74**, 3515 (1995).
[52] L. Blanchet, G. Faye, B. R. Iyer, and S. Sinha, Classical and Quantum Gravity **25**, 165003 (2008).
[53] Y. Pan *et al.*, (2011), arXiv:1106.1021v1.
[54] K. Thorne, *300 years of gravitation* (Cambridge University Press, Cambridge, England, 1987).
[55] B. S. Sathyaprakash and S. V. Dhurandhar, Phys. Rev. D **44**, 3819 (1991).
[56] `https://www.lsc-group.phys.uwm.edu/daswg/projects/lalsuite.html`.
[57] A. Buonanno, B. R. Iyer, E. Ochsner, Y. Pan, and B. S. Sathyaprakash, Phys. Rev. D **80**, 084043 (2009).
[58] T. Hinderer, B.D. Lackey, R.N. Lang, and J.S. Read, Phys. Rev. D **81**, 123016 (2010).
[59] J. Skilling, AIP Conference Proceedings **735**, 395 (2004).
[60] J. Veitch and A. Vecchio, Phys. Rev. D **78**, 022001 (2008).
[61] J. Veitch and A. Vecchio, Classical and Quantum Gravity **25**, 184010 (2008).
[62] J. Veitch and A. Vecchio, Phys. Rev. D **81**, 062003 (2010).
[63] C. Van Den Broeck and A. S. Sengupta, Classical and Quantum Gravity **24**, 155 (2007).
[64] S. Vitale and M. Zanolin, Phys. Rev. D **82**, 124065 (2010).
[65] D. Shoemaker, Advanced LIGO anticipated sensitivity curves, LIGO Scientific Collaboration Report No. T0900288, 2009 (unpublished).
[66] N. Yunes and L. C. Stein, Phys. Rev. D **83**, 104002 (2011), arXiv:1101.2921v2.
[67] L. C. Stein and N. Yunes, Phys. Rev. D **83**, 064038 (2010), arXiv:1012.3144v3.
[68] K. Yagi, L. C. Stein, N. Yunes, and T. Tanaka, (2011), arXiv:1110.5950v1.
[69] E. Barausse and A. Buonanno, (2011), arXiv:1107.2904v2.
[70] P. Ajith *et al.*, Classical and Quantum Gravity **24**, S689 (2007).
[71] P. Ajith *et al.*, Phys. Rev. Lett. **106**, 241101 (2011).
[72] L. Santamaria *et al.*, Phys. Rev. D **82**, 064016 (2010).
[73] A. Buonanno and T. Damour, Phys. Rev. D **59**, 084006 (1999).
[74] A. Buonanno and T. Damour, Phys. Rev. D **62**, 064015 (2000).
[75] T. Damour, P. Jaranowski, and G. Schäfer, Phys. Rev. D **62**, 084011 (2000).
[76] T. Damour, Phys. Rev. D **64**, 124013 (2001).
[77] A. Buonanno, Y. Chen, and T. Damour, Phys. Rev. D **74**, 104005 (2006).
[78] R. Sturani *et al.*, Journal of Physics: Conference Series **243** (2010).
[79] R. Sturani *et al.*, (2010), arXiv:1012.5172v2.
[80] S. Babak, R. Balasubramanian, D. Churches, T. Cokelaer, and B. Sathyaprakash, Classical and Quantum Gravity **23**, 5477 (2006).

Pulsar Timing Arrays: No longer a Blunt Instrument for Gravitational Wave Detection

Andrea N. Lommen
Franklin and Marshall College, 415 Harrisburg Pike, Lancaster, PA 17604
E-mail: andrea.lommen@fandm.edu

Abstract. Pulsar timing now has a rich history in placing limits on the stochastic background of gravitational waves, and we plan soon to reach the sensitivity where we can detect, not just place limits on, the stochastic background. However, the capability of pulsar timing goes beyond the detection of a background. Herein I review efforts that include single source detection, localization, waveform recovery, a clever use of a "time-machine" effect, alternate theories of gravity, and finally studies of the noise in our "detector" that will allow us to tune and optimize the experiment. Pulsar timing arrays are no longer "blunt" instruments for gravitational-wave detection limited to only detecting an amplitude of the background. Rather they are shrewd and tunable detectors, capable of a rich and dynamic variety of astrophysical measurements.

1. Introduction

Pulsars are basically celestial clocks, and as such, can be used to construct a Galactic-scale gravitational wave detector using the same concept as ground-based interferometric detectors, i.e. one looks for phase changes in the arrival of the signal at the vertex station, in this case, earth. The length scales of our detector 'arms' (1000 light years) as compared to the length of ground-based arms (4km) allow us to probe a different gravitational-wave frequency regime (nHz), a complement to the ground-based kHz regime (Yardley et al. 2010). For almost 30 years pulsar timers have been putting limits on the energy density of the stochastic background using pulsar timing (Romani & Taylor 1983; Stinebring et al. 1990; Kaspi, Taylor, & Ryba 1994; Lommen 2001; Jenet et al. 2006; van Haasteren et al. 2011; Yardley et al. 2011). They point out that at some moment in the future, we will detect rather than limit the stochastic background. This moment is predicted to be sometime within this decade (Demorest et al. 2009; Verbiest et al. 2009).

In the last 10 years the field of gravitational-wave detection using pulsars has matured, and we are now considering much more than just the background of gravitational waves. We are demonstrating that very precise work on specific sources can be done, and that we need to 'tune' this detector in order to maximize our sensitivity to these sources. This manuscript briefly reviews these efforts, and is organized as follows. In §2 I give some more details about the concept and current thought behind using pulsar timing arrays (PTAs) to detect gravitational waves. In the subsequent sections I review the work that shows that PTAs can be (§3) directional detectors, (§4) used to recover the gravitational waveform, (§5) used to recover information about the source at some time in past, (§6) used to measure luminosity distance to gravitational-wave sources, (§7) used to test alternate theories of gravity, and (§8) characterized as a formal 'detector' using measurements of their noise. Finally, in §9 I summarize the ways in which a PTA is no longer a 'blunt' instrument for gravitational-wave detection, but rather a tunable, pointable, and adjustable detector that can be used to gain very specific astrophysical information about the gravitational-wave source being detected.

2. An overview of the concept of gravitational-wave detection using pulsars.

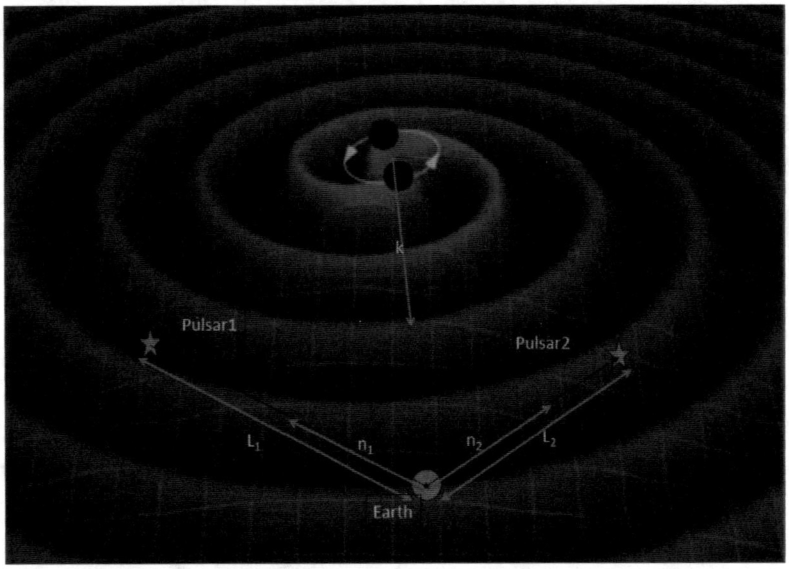

Figure 1. (Adapted from NASA) Schematic of a gravitational wave from a black hole binary impinging on a 2-pulsar pulsar timing array. When proper length scales are used the gravitational waves are nearly planar on the scale of the earth-pulsar systems, but §6 discusses the possibility of measuring their curvature.

Assume a gravitational wave is propagating through space in direction $\hat{k}$ apparently due to some distant source such as a supermassive binary black hole (see figure 1). The gravitational wave changes the curvature of the space-time along which the electromagnetic wave is traveling, and as such induces a change in the time that the pulse arrives at earth. The size of the change at time t for a pulse from pulsar j, $\tau_{\rm GW}(\hat{k},t)_j$, is given by

$$\tau_{GW}(\hat{k},t)_j \;= F^{+}(\hat{k},\hat{n}_j)g_{+}(t,L_j,\hat{k}\cdot\hat{n}_j) + F^{\times}(\hat{k},\hat{n}_j)g_{\times}(t,L_j,\hat{k}\cdot\hat{n}_j), \tag{1}$$

for a TT-gauge gravitational-wave metric perturbation with form $h_+\mathbf{e}_+ + h_\times\mathbf{e}_\times$. $\hat{n}_j$ is a unit vector pointing to pulsar j, L_j is the distance to that pulsar. $F^{+/\times}$ are geometric functions of $\hat{k}$ and $\hat{n}$ which we omit here for brevity but can be found in Burt, Lommen, & Finn (2011). Functions g_+ and $g_\times$ are integrals of h_+ and $h_\times$ as follows (Finn & Lommen 2010):

$$g_{(+/\times)}(t,L_j,\hat{k}_j\cdot\hat{n}^j) = \int_0^{L_j} h_{+/\times}\left(t-(1+\hat{k}\cdot\hat{n}_j)(L_j-\lambda)\right)d\lambda. \tag{2}$$

Note that we are using geometrized units where $c = G = 1$. Following Finn & Lommen (2010) we assume that a function f exists for which

$$df_{+/\times}(u)/du = h_{+/\times}(u). \tag{3}$$

For a plane wave we can then do the integral as follows:

$$g_{(+/\times)}(t, L_j, \hat{k}_j \cdot \hat{n}^j) = \frac{f_{+/\times}(t)}{1 + \hat{k} \cdot \hat{n}_j} - \frac{f_{+/\times}(t - (1 + \hat{k} \cdot \hat{n}_j)L_j)}{1 + \hat{k} \cdot \hat{n}_j}. \tag{4}$$

The first term is the so-called 'earth term', and the second the 'pulsar term'(Jenet et al. 2004). The pulsar term is delayed from the earth term by $(1 + \hat{k} \cdot \hat{n}_j)L_j$ which amounts to thousands of years in most cases.

Pulsar timers attempt to measure τ_{GW} by measuring the difference between the expected arrival time of the pulse and the measured arrival time of the pulse. This difference is called a 'residual.' The residuals that we measure are certainly not entirely due to passing gravitational waves, but to a variety of effects including the interstellar medium (van Straten 2006; You et al. 2007; Hemberger & Stinebring 2008), measurement noise, and calibration errors (Verbiest et al. 2009). The gravitational-wave signature, however, has a distinct feature that none of the other sources of noise can produce: part of the gravitational-wave signal, the earth term, is correlated among all the pulsars. In other words, the earth term describes a response that all pulsar timing residuals will exhibit at the same moment, independent of their distances. This characteristic can be used to distinguish gravitational waves from other sources of residual (Jenet et al. 2005).

The pulsar term, however, is not correlated among the pulsars. For each pulsar, the amplitudes of both terms will be modulated by the direction of the pulsar with respect to the gravitational-wave source. In addition, the temporal signature in the pulsar term is delayed by an amount which depends upon the distance to that pulsar. The distance is generally not known to better than 10% which corresponds to hundreds of gravitational wavelengths, and as such, it is basically a randomizer of the phase. One must confront this fact when attempting to recover the original waveform and direction of the gravitational waves, as I discuss in the following section.

However, before discussing PTAs as single source directional detectors, let us first consider the combined effect of many sources. This is actually the situation we expect in the universe at large - hundreds of thousands of galaxy pairs merging, with their central black holes eventually coming together to form black hole binaries. Those black hole binaries create a stochastic background of gravitational waves that we expect to be able to detect in pulsar timing.

Though no stochastic background has been detected thus far, many authors have used pulsar timing to limit the energy density of gravitational waves in the universe (Stinebring et al. 1990; Kaspi, Taylor, & Ryba 1994; Lommen & Backer 2001; Jenet, Creighton, & Lommen 2005; van Haasteren et al. 2011). The most strict limit, that placed by van Haasteren et al. (2011) places an upper limit on A equal to 6×10^{-15} where $h_c = A\left(\frac{f}{\mathrm{yr}^{-1}}\right)^{\alpha}$ and $\alpha = -2/3$. which is at the boundary of existing models of galaxy merger rates (see Jaffe & Backer 2003; Sesana, Vecchio, & Colacino 2008, and references therein). Wen et al. (2011) use this to constrain the coalescence rate of supermassive black-hole binaries. Van Haasteren et al. (2011) demonstrate that their analysis will eventually yield not just the amplitude, but also the slope and perhaps even the shape of the spectrum.

Thus, limiting the stochastic background has been a necessary first step for the field, but there is much to do beyond that. PTAs can do much more than detect a single number such as the amplitude of the stochastic background. They can be fashioned into a multi-faceted precision instrument for gravitational-wave detection.

Armed with an understanding of the earth and pulsar terms, we can now proceed to discuss the challenges of the analysis using PTAs for gravitational-wave detection. Basically, most methods of detection seek to capitalize on the coherence of the earth term without being unduly hindered by the incoherence of the pulsar term. In the next section I review ways in which analyses of single sources go about seeking the same advantage.

3. PTAs as directional detectors

As shown above, there are two terms in the response of a PTA: the earth term and pulsar term. Given the location and waveform of a gravitational-wave source, the earth term in the equation above is completely known. We can turn that statement around and say that if we could somehow distinguish the earth term separately from the pulsar term, and could then measure the amplitudes of the earth term in all our pulsars, we could determine the location and waveform of the gravitational-wave source. In most cases, we cannot separate those two terms. However, it turns out to be possible to determine the direction of the source using various means described below. The methods below all differ in the ways they have treated the pulsar term.

Finn & Lommen (2010) looked at bursts of gravitational waves, sources whose duration is shorter than the data span. Pulsar timing data spans are tens of years, so a burst could be a month-long source. They put forth that by looking at bursts, one can ignore the pulsar term because it is unlikely to enter the dataset in a human lifetime. Note that because the distance between the earth and the pulsar is hundreds or thousands of light years, the delay between the earth term and the pulsar term is hundreds or thousands of years. Thus, any source that produces a burst of coherent response in all the pulsars (i.e. the earth term) will inflict its pulsar term on the same pulsars many hundreds of years later. Finn & Lommen (2010) were able to localize a strong source to less than 1 square degree. For a moderate source it was hundreds of square degrees. For a weak source it was thousands of square degrees.

Sesana and Vecchio (2010) looked at continuous gravitational-wave sources such as binary black holes, where one must include the pulsar term, and they are able to localize the gravitational-wave source to within 40 square degrees for a 100-pulsar array and a signal to noise ratio (S/N) of 10. That number is big because they assumed they could not know the distances to the pulsars, so the 100% variation of the pulsar term makes it very hard to pin down the direction as described at the beginning of this section. Ellis, Jenet, & McLaughlin (2012) has achieved similar results working on a continuous wave pipeline.

Corbin & Cornish (2010) claim they can actually search over and recover the pulsar distances from the chirp signal. They have assumed Gaussian noise for the timing residuals which we may in fact have (see section on detector characterization below), but their technique has not been studied in the presence of red noise. They are able to localize the source to less than 3 square degrees for strong sources. Lee et al. (2011) point out that if the timing parallax (a distance measurement independent of the chirped signal) is estimated at the same time as the GW parameters are estimated then the pulsar term can be used as a great asset in increasing signal strength in single source cases. They carefully predict the statistical uncertainty that PTAs can expect to achieve in determining characteristics of gravitational-wave sources such as orbital inclination angle, source position, frequency, and amplitude. They predict that PTA source localization ability will range from a radian down to several microradians depending on the strength of the source.

Before I summarize these localization results, let me point out that all the above work was done on simulated data in preparation for using real data. However, work has been done on single gravitational-wave sources using real data. Yardley et al. (2010) has used pulsar timing data to limit the number of coalescing binary systems of a given chirp mass as a function of redshift.

My summary for single source localization is as follows. We can localize well for strong burst sources, i.e. less than 1 square degree. For continuous wave sources it is more difficult because one cannot ignore the pulsar term. Without knowing the pulsar distances, it is not clear that one can do better than 40 square degrees, but there is hope for getting the pulsar distances either by mitigating the red noise in pulsars (see section on detector characterization below) so that the distance can be found from the chirp signal (Corbin & Cornish 2010), or by getting the distances from other means such as timing or VLBI parallax (Lee et al. 2011). One interesting thing to note is that if one had an eccentric black-hole-binary instead of a circular one, it would be a repetitive burst source, and the localization would be much easier.

4. PTAs as waveform recoverers

PTAs are potentially able to do much more than detect and localize a gravitational-wave source. They may be able to actually recover the gravitational waveform. This means that we would not just be able to pinpoint a source, but also actually decode the structure of the object that created it. Things like masses, spins, and orientations are encoded in these waveforms. Direct observations would enable us to learn about the underlying structure of the black holes.

Finn & Lommen (2010) demonstrate that one can recover the waveform of the gravitational-wave source provided the signal to noise ratio of the source in the data is above about 1/10 in each of about 30 pulsars. They employ a likelihood method that does not rely on any input template 'bank.' This could be used to recover astrophysical parameters of the source such as inclination and polarization angles (Sesana & Vecchio 2010).

5. PTAs as time machines

Mingarelli et al. (2011) show that when detecting a binary gravitational-wave source using pulsar timing, one can detect the source at two different moments in its orbit, because the earth and pulsar terms represent two different moments in its evolution. If the black holes are spinning, for example, and there is spin-orbit precession in the system, the PTA will measure two moments in its spin-orbit evolution at the same time. They can therefore probe higher order relativistic corrections in (v/c), including the effect of spin-orbit coupling beyond the Newtonian approximation to the dynamics.

6. PTAs as measurers of luminosity distance

Deng & Finn (2011) demonstrate that the curvature of the waveform (seen in Figure 1) can be used to measure a gravitational-wave parallax, and thereby obtain the luminosity distance to sources approaching or exceeding 100 Mpc. This would serve as an important independent measurement of distance to black hole binaries and other gravitational-wave sources.

7. PTAs as detectors of alternate theories of Gravity

In General Relativity, there are two transverse gravitational-wave polarization modes. However, alternate theories of gravity can include up to 6 gravitational-wave polarization modes. Lee, Jenet, & Price (2008) find that if these extra polarizations exist, they could be detected in 5 years given 40-60 pulsars. Chamberlin & Siemens (2012) show that sensitivity to the vector and scalar-longitudinal modes can increase dramatically for pulsar pairs with small angular separations. For example, the J1853+1303/J1857+0943 pulsar pair, with an angular separation of about 3°, is about 10,000 times more sensitive to a longitudinal component of the stochastic background than to the transverse components. Detecting an extra gravitational-wave polarization would provide the first evidence for the violation of General Relativity.

Also in General Relativity, gravitational waves travel at the speed of light, and the graviton is therefore massless. However, in some alternate theories of gravity the graviton has mass, and the gravitational wave is therefore dispersive. Lee et al. (2010) estimate that it should be possible to detect this dispersion using PTAs. In particular they conclude that massless gravitons can be distinguished from gravitons heavier than 3×10^{-22} eV with 5 years of bi-weekly observations of 60 pulsars with a pulsar rms timing accuracy of 100 ns. This is not as good as limits placed currently by Goldhaber & Nieto (1974) using galaxy cluster observations but the two methods are independent and pulsar timing would represent an important substantiation of the cluster results.

8. Characterizing the Detector Noise

As we make this transition to thinking about PTAs as shrewd and tunable detectors, one of the things we have had to confront is that we must characterize the noise in the detector just as any other experiment would (See for example Cuocco et al. 2001). In our case the "detector" is the collection of pulsars. Shannon & Cordes (2010), Osłowski et al. (2011), Jenet, Armstrong, & Tinto (2011), Perrodin et al. (2012), and Finn (2012) have all sought to characterize the noise in the pulsars. Shannon & Cordes (2010), and Osłowski et al. (2011) concentrate on determining what the ultimate sensitivity limit will be due to intrinsic variability in the pulsars. Perrodin et al. (2012) using the Cholesky transform as put forth by Coles et al. (2011) concentrates on characterizing the current noise spectrum, while Finn (2012) uses a Bayesian analysis to identify the parameters of a noise model that best describe the timing noise statistics. Jenet, Armstrong, & Tinto (2011) give a full error budget for pulsar timing considering various instrumental, propagation, and other fundamental sources of noise. The effort in all cases is complicated by the fitting of the pulsar model to the time-of-arrival (TOA) data and by the fact that the pulsars are sampled at irregular intervals. The former renders stationary noise un-stationary, and the latter precludes the use of simple FFTs for finding the spectrum.

Once the noise in the detector is characterized, we can optimize the detector. Lee et al. (2012) demonstrated the advantage of optimization which in our case primarily consists of adjusting the amount of time spent on each pulsar, and showed that it only helps appreciably in the case of white noise. Red noise has been identified in many pulsar data sets (Verbiest et al. 2009; van Haasteren et al. 2011), but recent techniques such as those by Demorest et al. (2012) which remove interstellar medium effects seem to significantly reduce the "redness" of the data. In any case the possibility of "tuning" the detector seems very promising.

9. Summary

PTAs have been long regarded as our hope for detecting gravitational waves in the nanoHertz band, but up until recently people have thought that the best one could hope for would be a detection plus one or two numbers parameterizing the stochastic background in gravitational waves in that regime (e.g., the spectral index and overall amplitude). We have started to be able to think about PTAs as detectors, with noise budgets, and various modes that we can tune to suit the source in which we are most interested. I have pointed to work here that is being done not only to detect gravitational waves, but to use them for very particular goals including gravitational-waveform recovery, parameter estimation for individual gravitational-wave sources, locating a source on the sky, distinguishing between theories of gravity, measuring luminosity distance to gravitational-wave sources, and also using the delayed pulsar term as something of a time machine, where we get a glimpse of the source as it existed in the past, and the present, all at once.

Acknowledgments

I would like to thank Sydney Chamberlin, Justin Ellis, Kejia Lee, Dick Manchester, Chiara Mingarelli, Willem van Straten, Daniel Yardley, Joris Verbiest, and the entire detection group in NANOGrav, in particular Sydney Chamberlin, Sam Finn, Xavi Siemens, and Delphine Perrodin for very useful discussions.

References

Burt, B. J., Lommen, A. N., & Finn, L. S. 2011, *ApJ* , 730, 17

Chamberlin, S., & Siemens, X. 2012, submitted to ApJ

Coles, W., Hobbs, G., Champion, D. J., Manchester, R. N., & Verbiest, J. P. W. 2011, *MNRAS* , 418, 561

Corbin, V., & Cornish, N. J. 2010, ArXiv e-prints

Cuoco, E., Losurdo, G., Calamai, G., Fabbroni, L., Mazzoni, M., Stanga, R., Guidi, G., & Vetrano, F. 2001, *Phys. Rev. D* , 64, 122002

Demorest, P., Lazio, J., Lommen, A., Archibald, A., Arzoumanian, Z., Backer, D., Cordes, J., Demorest, P., Ferdman, R., Freire, P., Gonzalez, M., Jenet, R., Kaspi, V., Kondratiev, V., Lazio, J., Lommen, A., Lorimer, D., Lynch, R., McLaughlin, M., Nice, D., Ransom, S., Shannon, R., Siemens, X., Stairs, I., Stinebring, D., Reitze, D., Shoemaker, D., Whitcomb, S., & Weiss, R. 2009, In astro2010: The Astronomy and Astrophysics Decadal Survey, ArXiv Astrophysics e-prints, p. 64

Demorest, P., Ferdman, R., Gonzalez, M., Nice, D., Ransom, S., Stairs, I., Arzoumanian, Z., Cordes, J., Finn, L., Freire, P., Jenet, F., Kaspi, V., Lazio, J., Lommen, A., Lorimer, D., McLaughlin, M., Perrodin, D., Shannon, R., Siemens, X., & Stinebring, D. 2012, ApJ, in preparation

Deng, X., & Finn, L. S. 2011, *MNRAS* , 414, 50

Ellis, J., Jenet, F., & McLaughlin, M. 2012, in preparation

Finn, L. S. 2012, in preparation

Finn, L. S., & Lommen, A. N. 2010, *ApJ* , 718, 1400

Goldhaber, A. S., & Nieto, M. M. 1974, *Phys. Rev. D* , 9, 1119

Hemberger, D. A., & Stinebring, D. R. 2008, *ApJ Lett.* , 674, L37

Jaffe, A. H., & Backer, D. C. 2003, *ApJ* , 583, 616

Jenet, F. A., Lommen, A., Larson, S. L., & Wen, L. 2004, *ApJ* , 606, 799

Jenet, F. A., Hobbs, G. B., Lee, K. J., & Manchester, R. N. 2005, *ApJ Lett.* , 625, L123

Jenet, F. A., Creighton, T., & Lommen, A. 2005, *ApJ Lett.* , 627, L125

Jenet, F. A., Hobbs, G. B., van Straten, W., Manchester, R. N., Bailes, M., Verbiest, J. P. W., Edwards, R. T., Hotan, A. W., Sarkissian, J. M., & Ord, S. M. 2006, *ApJ* , 653, 1571

Jenet, F. A., Armstrong, J. W., & Tinto, M. 2011, *Phys. Rev. D* , 83, 081301

Kaspi, V. M., Taylor, J. H., & Ryba, M. F. 1994, *ApJ* , 428, 713

Lee, K., Jenet, F. A., Price, R. H., Wex, N., & Kramer, M. 2010, *ApJ* , 722, 1589

Lee, K. J., Jenet, F. A., & Price, R. H. 2008, *ApJ* , 685, 1304

Lee, K. J., Wex, N., Kramer, M., Stappers, B. W., Bassa, C. G., Janssen, G. H., Karuppusamy, R., & Smits, R. 2011, *MNRAS* , 414, 3251

Lee, K. J., Bassa, C. G., Janssen, G. H., Karuppusamy, R., Kramer, M., Smits, R., , & Stappers, B. W. 2012, *MNRAS* , accepted

Lommen, A. 2001, PhD Thesis, UC Berkeley

Lommen, A. N., & Backer, D. C. 2001, *ApJ* , 562, 297

Mingarelli, C. M. F., Grover, K., Sidery, T., Smith, R. J. E., & Vecchio, A. 2011, in preparation

Osłowski, S., van Straten, W., Hobbs, G. B., Bailes, M., & Demorest, P. 2011, *MNRAS* , 418, 1258

Perrodin, D., Jenet, F., Lommen, A. N., Finn, L. S., Demorest, P. B., Ferdman, R. D., Gonzalez, M. E., Nice, D., Ransom, S., & Stairs, I. H. 2012, in preparation

Romani, R. W., & Taylor, J. H. 1983, *ApJ Lett.* , 265, L35

Sesana, A., & Vecchio, A. 2010, *Phys. Rev. D* , 81, 104008

Sesana, A., Vecchio, A., & Colacino, C. N. 2008, *MNRAS* , 390, 192–209

Shannon, R. M., & Cordes, J. M. 2010, *ApJ* , 725, 1607

Stinebring, D. R., Ryba, M. F., Taylor, J. H., & Romani, R. W. 1990, Physical Review Letters, 65, 285

van Haasteren, R., Levin, Y., Janssen, G. H., Lazaridis, K., Kramer, M., Stappers, B. W., Desvignes, G., Purver, M. B., Lyne, A. G., Ferdman, R. D., Jessner, A., Cognard, I., Theureau, G., D'Amico, N., Possenti, A., Burgay, M., Corongiu, A., Hessels, J. W. T., Smits, R., & Verbiest, J. P. W. 2011, *MNRAS* , 414, 3117

van Straten, W. 2006, *ApJ* , 642, 1004

Verbiest, J. P. W., Bailes, M., Coles, W. A., Hobbs, G. B., van Straten, W., Champion, D. J., Jenet, F. A., Manchester, R. N., Bhat, N. D. R., Sarkissian, J. M., Yardley, D., Burke-Spolaor, S., Hotan, A. W., & You, X. P. 2009, *MNRAS* , 400, 951

Wen, Z. L., Jenet, F. A., Yardley, D., Hobbs, G. B., & Manchester, R. N. 2011, *ApJ* , 730, 29

Yardley, D. R. B., Hobbs, G. B., Jenet, F. A., Verbiest, J. P. W., Wen, Z. L., Manchester, R. N., Coles, W. A., van Straten, W., Bailes, M., Bhat, N. D. R., Burke-Spolaor, S., Champion, D. J., Hotan, A. W., & Sarkissian, J. M. 2010, *MNRAS* , 407, 669

Yardley, D. R. B., Coles, W. A., Hobbs, G. B., Verbiest, J. P. W., Manchester, R. N., van Straten, W., Jenet, F. A., Bailes, M., Bhat, N. D. R., Burke-Spolaor, S., Champion, D. J., Hotan, A. W., Oslowski, S., Reynolds, J. E., & Sarkissian, J. M. 2011, *MNRAS* , 414, 1777

You, X. P., Hobbs, G., Coles, W. A., Manchester, R. N., Edwards, R., Bailes, M., Sarkissian, J., Verbiest, J. P. W., van Straten, W., Hotan, A., Ord, S., Jenet, F., Bhat, N. D. R., & Teoh, A. 2007, MNRAS, 378, 493

Towards optimal integration of external information in a multi-trigger population study analysis

Marc E. Normandin[1,2,3], **Soumya D. Mohanty**[1,4]

[1]The University of Texas at Brownsville, 80 Fort Brown St, Brownsville, TX, 78520
[2]The University of Texas at San Antonio, One UTSA Circle, San Antonio, TX, 78249

E-mail: [3]`marc.normandin@ligo.org`, [4]`mohanty@phys.utb.edu`

Abstract. We present an algorithm for population study which optimally uses information obtained from the electromagnetic (or particle) counterpart of gravitational wave (GW) events. Existing methods do not associate suitable weight factors to the triggers besides trivial ones like a hard cutoff on flux. However, the assignment of weights needs to take into account the background astrophysical distribution to be estimated. This is done using a likelihood-based approach where electromagnetic (or particle) and gravitational wave data are incorporated into a common likelihood before the parameters of a given population distribution are estimated. We present preliminary results from this method using simulated data corresponding to simplified models of GW and electromagnetic detectors.

1. Introduction

Consider a set of triggers, such as gamma-ray bursts (GRBs), with which we would like to test the association of gravitational waves (GWs). This can be done by suitably combining data from multiple triggers, such that the signal-to-noise ratio (SNR) of the association is increased. This idea, called population study, was developed and analyzed quantitatively in the context of GWs and GRBs by Finn, Mohanty and Romano in [1] (FMR).

An improved version of the FMR method was derived in [2] from first principles using a likelihood-based approach. In [3] it was shown that a population study analysis can significantly increase our belief in the association of GWs with a set of triggers when we do not have a single-trigger detection from that set, or our belief in such a detection is weak at best. Population study based observational results have been reported in [4, 5, 6, 7].

There is significant room for improvement in the sensitivity of population study methods. One expects the biggest improvement from applying suitable weights to the triggers such that the ones likely to contribute mostly to noise are suppressed in the analysis. Given an astronomical model, this may be done by using additional information available for the triggers.

In this paper we introduce the first steps towards such a method, which is to be extended for use in realistic burst searches. To demonstrate that this method suppresses noisy triggers and incorporates non-GW information, we apply it to a simple toy model of a triggered search. The results from this simple exercise support our expectations about how this method may work in realistic cases.

The rest of this paper is organized as follows. In section 2 we introduce our method and then derive the combined likelihood function. Section 3 gives an implementation of the method for the toy model, the results of which are given in section 4. In section 5 we give our conclusions and thoughts on future work.

2. The method

Consider a set of measurements, S, consisting of the information we know about a collection of trigger events. The trigger events are assumed to be samples from a population parameterized by Γ_1 and Γ_2, where Γ_1 is defined to be the set of parameters related to the GWs, and Γ_2 is defined to be the set of parameters related to the redshift distribution. In practice, there will be more information available about some trigger events than others.

In this paper we assume that each and every trigger event has an associated GW detection statistic, $\hat{CC}$, and estimated sky location, $\hat{\theta}, \hat{\phi}$, where a $\hat{A}$ denotes a measured value of the quantity A. In addition, trigger events may or may not have an associated measured redshift, $\hat{Z}$. The set of information that we will use to estimate the population parameters is of the form $S = \{\{\hat{CC}_1, \hat{Z}_1, \hat{\theta}_1, \hat{\phi}_1\}, \{\hat{CC}_2, \hat{\theta}_2, \hat{\phi}_2\}, ...\}$, where, in this example, the first trigger event has a measured redshift and the second does not. Thus, our method incorporates trigger events with varying information content.

We begin the derivation by assuming that the trigger events are statistically independent of one another. This is the case with GRBs, which are widely separated in space, but may not be applicable to triggers from the same object, such as soft-gamma-repeaters (SGRs) [8]. The probability of the combined event, $P(S;\ \Gamma_1, \Gamma_2)$, is then simplified as the multiplication of the probability of the individual trigger events. There are then two different probability types that need to be derived: When the redshift is known, and when it is not. We begin with the former in 2.1, and the latter in 2.2. The combined likelihood function is then derived in 2.3.

2.1. Case I: Trigger event with known redshift, $\hat{Z}$

The likelihood function for the data associated with a single trigger can be expressed as follows. Here, Z, θ, ϕ denote the true "redshift", and location on the sky, and $\hat{CC}$ denotes the value of the statistic used for GW detection. (Here Z is simply a scaling factor for the observed GW signal amplitude).

$$P(\hat{CC}, \hat{Z}, \hat{\theta}, \hat{\phi};\ \Gamma_1, \Gamma_2) = \int d\theta\ d\phi\ dZ\ P(\hat{CC}, \hat{Z}, \hat{\theta}, \hat{\phi}, Z, \theta, \phi;\ \Gamma_1, \Gamma_2) \quad (1)$$

$$= \int d\theta\ d\phi\ dZ\ P(\hat{CC} \mid \hat{\theta}, \hat{\phi}, Z, \theta, \phi;\ \Gamma_1) \, P(\hat{Z} \mid Z) P(Z;\ \Gamma_2) P(\hat{\theta}, \hat{\phi}, \theta, \phi). \quad (2)$$

Note that $P(\hat{Z} \mid Z)$ is not part of the population model. It is determined by the measurement error of the method for measuring redshift from the electromagnetic information.

2.2. Case II: Trigger event with unknown redshift

The probability for a trigger that does not have associated redshift information, $\hat{Z}$, is derived similarly to the previous case.

$$P(\hat{CC}, \hat{\theta}, \hat{\phi};\ \Gamma_1, \Gamma_2) = \int d\theta\ d\phi\ dZ\ P(\hat{CC}, \hat{\theta}, \hat{\phi}, Z, \theta, \phi;\ \Gamma_1, \Gamma_2) \quad (3)$$

$$= \int d\theta\ d\phi\ dZ\ P(\hat{CC} \mid \hat{\theta}, \hat{\phi}, Z, \theta, \phi;\ \Gamma_1) P(Z;\ \Gamma_2) P(\hat{\theta}, \hat{\phi}, \theta, \phi). \quad (4)$$

We see that the expression is similar to the case for known redshift, except that the $P(\hat{Z} \mid Z)$ term is understandably absent.

2.3. The likelihood function

We can now write down the loglikelihood function for the set of measurements, S:

$$\log(P(S;\ \Gamma_1,\Gamma_2)) = \log(P(\hat{CC}_1,\hat{\phi}_1,\hat{\theta}_1,\hat{Z}_1,\hat{CC}_2,\hat{\phi}_2,\hat{\theta}_2,\hat{CC}_3,\hat{\phi}_3,\hat{\theta}_3,\hat{Z}_3,\ldots\ ;\ \Gamma_1,\Gamma_2)) \quad (5)$$

$$= \sum_{m=1}^{M} \log(P(\hat{CC}_m,\hat{\phi}_m,\hat{\theta}_m,\hat{Z}_m\ ;\ \Gamma_1,\Gamma_2)) + \sum_{n=M+1}^{N} \log(P(\hat{CC}_n,\hat{\phi}_n,\hat{\theta}_n\ ;\ \Gamma_1,\Gamma_2)). \quad (6)$$

The first summation is over those trigger events with a measured redshift, which are labelled 1 to M. The second is over those trigger events lacking a measured redshift, which are labelled $M+1$ to N. Thus there are a total of N trigger events whose information is used for the combined loglikelihood.

Once a specific model is chosen for the population, as well as a model for the measured redshift given the true redshift, then the population parameters can be estimated by maximizing the likelihood function just obtained. This is illustrated in the next section.

Note especially that the expressions derived in the present case take into account the presence and absence of redshift information, but the procedure can similarly be performed for other types of information such as electromagnetic flux and neutrino flux.

3. Implementation

To demonstrate and explore the method we choose a simplified toy model: The GW waveforms are sine-gaussians with parameterization $\Gamma_1 = \{A\}$, where A is the intrinsic GW amplitude. The distribution of the triggers on the sky is a delta function centered at a fixed sky location. This reduces the computational demands as well as allows us to understand the impact of missing information better.

The redshift distribution is taken to be a scaled (across redshift) beta distribution [9] parameterized by $\Gamma_2 = \{\alpha,\beta\}$, where the α and β values are positive real values that define a specific member of the beta distribution family. Examples of this distribution are shown in Figures 1 and 3, where in both cases it has been scaled to be from $Z = 1$ to $Z = 4$. This distribution was chosen because it is finitely bounded, and so we can be sure that no random trigger will ever have a redshift lower than some chosen minimum. This allows a clean test to be made of the population without contamination from extra loud triggers. Finally, for some choices of Γ_2, the beta distribution resembles predicted astrophysical redshift distributions in the sense that it has a single mode round which it is asymmetric.

The measured sky location, $(\hat{\theta},\hat{\phi})$, is assumed to be normally distributed about the true sky location. The measured redshift, $\hat{Z}$, if known, is equal to the true redshift (which means that the probability $P(\hat{Z} \mid Z)$ is a delta function). Equation (2) is simplified by integrating over the delta function, which gives

$$P(\hat{CC},\hat{Z},\hat{\theta},\hat{\phi};\ \Gamma_1,\Gamma_2) = \int P(\hat{CC} \mid \hat{Z},\hat{\theta},\hat{\phi},\theta,\phi;\ \Gamma_1)P(\hat{Z};\ \Gamma_2)\ P(\hat{\theta},\hat{\phi},\theta,\phi)\ d\theta\ d\phi. \quad (7)$$

The set of trigger information, S, is simulated by picking random samples from the toy model described above for some fixed choice of Γ_1,Γ_2.

The $\hat{CC}$ detector statistic was chosen to be the cross-correlation between two GW detectors. Both detectors have white, Gaussian noise with variance given by $\sigma_n^2 = 1$. We appealed to the law of large numbers, which resulted in the conditional cross-correlation probability distribution, $P(\hat{CC} \mid \hat{\theta},\hat{\phi},Z,\theta,\phi;\ \Gamma_1)$, being approximately a normal distribution with variance, $\sigma_{\rm CC}^2$, and mean, μ_{CC}, given by

$$\mu_{CC} = \langle h_1, h_2 \rangle \quad (8)$$

$$\sigma^2_{CC} = (\langle h_1, h_1 \rangle + \langle h_2, h_2 \rangle)\, \sigma_n^2 + K\sigma_n^4, \quad (9)$$

where K is the number of time samples per time series, h_1 is the GW time series from detector 1, h_2 is the GW time series from detector 2, $\langle ..., ... \rangle$ is the dot product.

The sine-gaussians have a frequency of 250 Hz, and the standard deviation of the Gaussian envelope is 0.003 seconds. Each GW detector sampled at 4096 Hz. Each strain time series consisted of 650 simulated time samples. Based on the sky location, the two time series were shifted relative to one another and then the cross-correlation was taken in a window consisting of 256 window samples ($K = 256$).

The cross-correlation SNR (CSNR) is defined to be the mean of the cross-correlation in the presence of a signal, divided by the standard deviation of the cross-correlation in the absence of a signal (only noise). Applying the definition to the implementation given above results in

$$\text{CSNR} = \frac{A^2}{d^2} \frac{\langle q_1,\ q_2 \rangle}{\sqrt{K\sigma_n^2}}, \quad (10)$$

where $q_1 = h_1/(A/d)$, and $q_2 = h_2/(A/d)$. Thus the maximum CSNR will be at the closest distance, and where $\langle q_1,\ q_2 \rangle$ is greatest. Since we used a scaled beta function for the population distance distribution, $P(z;\ \Gamma_2)$, no trigger in the study has a CSNR greater than some CSNR_{max}, which we treat as study variable in the results that follow.

Finally, the numerical integration itself was performed over a fine grid in redshift, Z. For the probabilities themselves, we binned their pdf values so that they could be used on the integration grid. This explains, for example, the non-zero value at $Z = 1$ in Figure 1, and for the non-zero value at $Z = 4$ in Figure 3.

The implementation was programmed in C99, and used the Message Passing Interface (MPI) in order to reduce the computation times when run on a computer cluster. The processing times often take hours due to the loglikelihood involving integrals, and the need to recompute the integrals during the maximization stage when estimating parameters.

4. Results

We sought to evaluate how well our method performs in reconstructing the GW signal amplitude parameter, $\Gamma_1 = A$, which is related to the energy of the GWs. This should depend on a number of factors, including the number of triggers composing the set of information, S, the loudest possible trigger as specified by CSNR_{max}, and the number of known versus unknown redshift triggers. To explore our method in terms of these factors, we ran a number of separate analyses where each analysis differed by one of the factors mentioned.

In the discussion so far we have not specified Γ_2, which determines the population's redshift distribution, $P(z;\ \Gamma_2)$. Since this population parameter plays such a large role, we used two different values for it. The first, where $\Gamma_2 = \{2, 4\}$, gives a population redshift distribution that has most triggers with low redshift, as shown in Figure 1. The second, where $\Gamma_2 = \{4, 2\}$, gives a population redshift distribution where most of the triggers are at high redshifts, as shown in Figure 3. The key qualitative difference between the two is that the latter will have more triggers buried in the noise than the former. Using these two extremes allows us to better qualitatively explore the method.

To obtain statistics for the error in estimation we performed each unique analysis 4096 times, and so obtained 4096 estimates of Γ_1 for each specific analysis. For example, in one analysis we had $\Gamma_2 = \{2, 4\}$, $\text{CSNR}_{max} = 1.5$, and the number of triggers equal to 64. We then performed that analysis 4096 times, and so obtained 4096 estimations of Γ_1, which we denote by $\overline{\Gamma_1}_n$, where n ranges from 1 to 4096. From those 4096 estimations we then computed one relative mean squared error (RMSE), which we then plotted as a single data point in Figure 4.

The RMSE was calculated in terms of Γ_1,

$$\text{RMSE} = \frac{1}{\Gamma_1}\sqrt{\frac{1}{N-1}\sum_{n=1}^{N}(\overline{\Gamma_{1n}} - \Gamma_1)^2} \tag{11}$$

where we had $N = 4096$, and where each Γ_1 is associated with a correpsonding CSNR_{max}.

The full analysis results are presented in two separate figures. Figure 2, is for analyses with $\Gamma_2 = \{2, 4\}$, while Figure 4 is for analyses with $\Gamma_2 = \{4, 2\}$. We expect that the RMSE will be lower in general for the former, than the latter, due to the former having more triggers at lower redshifts and so having a correspondingly stronger population signature. We also expect that knowing the redshift will result in a lower RMSE over not knowing it.

As qualitatively expected, the results show that the RMSE decreases with increasing number of triggers, and with increasing CSNR_{max}. Importantly, we see that incorporating redshift information into the parameter estimation results in a lower RMSE in all of the cases studied. Thus, the method behaves as expected in that it produces a better parameter estimation when the extra redshift information is included. Since we used a toy model for the simulations, and because we are aiming to show that our method can successfully incorporate non-GW information, in this case redshift, into the parameter estimation, we do not focus on the specific values obtained.

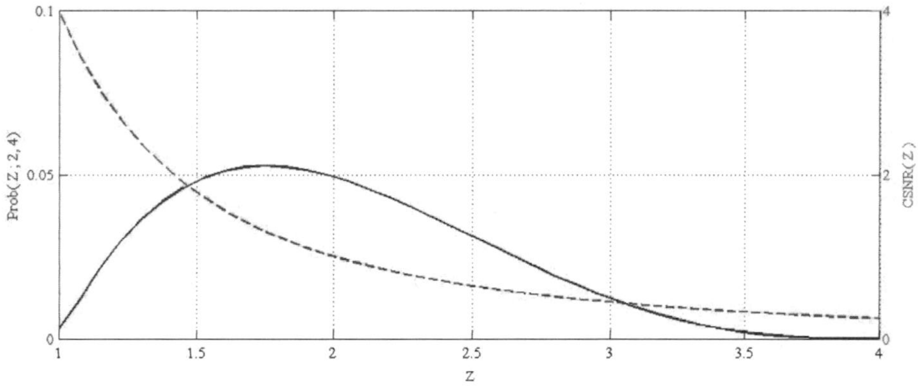

Figure 1. Shown in the solid line is an example redshift distribution, $P(z;\ \Gamma_2)$, where we have used a scaled beta distribution with $\Gamma_2 = \{2, 4\}$. The GW amplitude parameter, Γ_1, has been chosen such that at the minimum redshift possible, which is $Z = 1$, there will be a $\text{CSNR}_{max} = 4$. In this way, we guarantee that in the simulations there will not be any trigger with a larger CSNR than 4. The CSNR for the different redshifts is shown in the dashed curve, and uses the axes on the right. We see that the maximum CSNR occurs at the smallest Z.

5. Conclusions

The main question we ask at this stage is whether our proposed method can correctly take non-GW information into account. To determine this we constructed a toy model on which we tested our method. The results obtained show that the error in estimating the strength of the gravitational waves was lower when redshift information was incorporated, which is what is qualitatively expected, and demonstrates that this approach is on the right track.

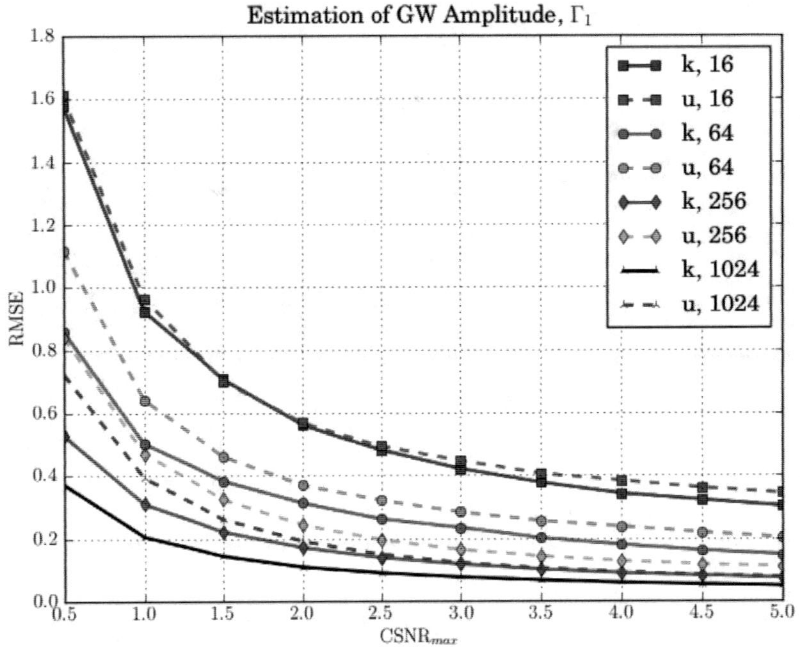

Figure 2. Shown are the errors in estimating the GW amplitude parameter Γ_1 versus CSNR_{max}, where we have fixed $\Gamma_2 = \{2, 4\}$ and whose corresponding redshift distribution is shown in Figure 1. Each series shown differs in the number of triggers in the set (denoted by the number in the series label), and whether or not the redshift information was known (indicated by the 'k'), versus unknown (indicated by the 'u').

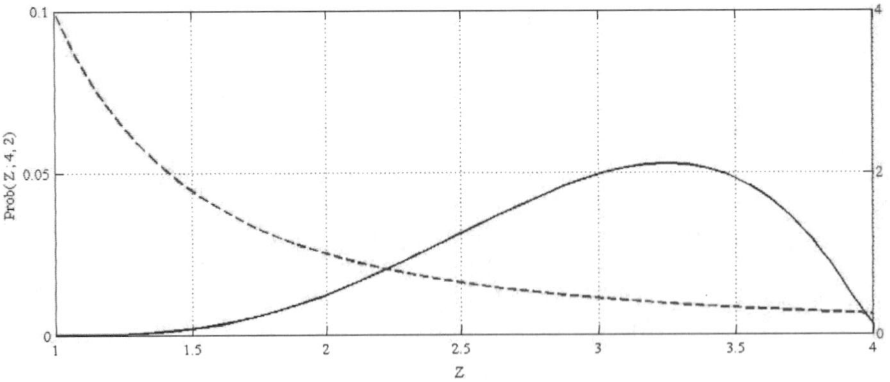

Figure 3. Shown is an example plot of redshift distribution, which has been parameterized the same as that for Figure 1, with the only exception that this time we have $\Gamma_2 = \{4, 2\}$. The shape of the distribution is now peaked more strongly towards higher redshifts. Thus triggers whose redshift is generated from this distribution will tend to have a lower CSNR, than those generated from the distribution shown in Figure 1.

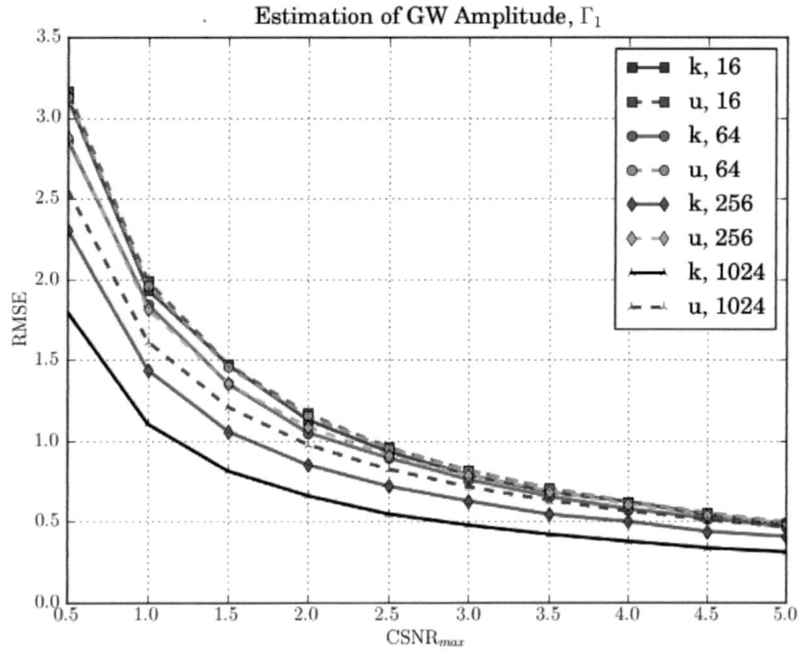

Figure 4. Shown are the errors in estimating the GW amplitude parameter Γ_1 versus CSNR$_{max}$, where we have fixed $\Gamma_2 = \{4, 2\}$ and whose corresponding redshift distribution is shown in Figure 3. Each series shown differs in the number of triggers in the set (denoted by the number in the series label), and whether or not the redshift information was known (indicated by the 'k'), versus unknown (indicated by the 'u').

In this present study, we used redshift as the non-GW information, but the method itself is not limited to redshift. It can also incorporate other non-GW information such as neutrino fluxes, for instance. Also, we estimated Γ_1, which is related to the energy of the GWs, but the method can also be used to estimate Γ_2, which parameterizes the population's redshift distribution.

The important point is the existence of a gain in performance when non-GW information is included, rather than the actual magnitude obtained, as the approximations used to construct the toy model will not be the same as those used for a full population study.

The computations involved in our method can be time consuming. Due to the fact that the likelihoods derived involve integrals, Equations (2) and (4), the numerical computations often require hours of computation. The parallel nature of the method should lend itself well to an implementation using graphics programming units, which can greatly reduce the computation times for parameter estimation.

In the future we intend to use a GW network statistic, instead of GW cross-correlation, which we expect to perform better. Also, we plan to explore *simulated maximum likelihood* methods [10], as they can better handle the larger parameter space of full astrophysical models.

Acknowledgments

This work is supported by NSF grant PHY-0855371.

References

[1] L. S. Finn, S.D. Mohanty, J. Romano, "Detecting an association between gamma ray and gravitational wave bursts," *Phys. Rev. D* **60** 121101 (1999)

[2] S.D. Mohanty, "Improving the sensitivity of searches for an association between gamma-ray bursts and gravitational waves," *CQG* **22** S1349 (2005)

[3] S.D. Mohanty, "Population study of gamma ray bursts: detection sensitivity and upper limits," *CQG* **23** S723 (2006)

[4] P. Astone, D. Babusci, M. Bassan, P. Carelli, E. Coccia, C. Cosmelli, S. D'Antonio, V. Fafone, F. Frontera, G. Giordano, C. Guidorzi, A. Marini, Y. Minenkov, I. Modena, G. Modestino, A. Moleti, E. Montanari, G.V. Pallottino, G. Pizzella, L. Quintieri, A. Rocchi, F. Ronga, L. Sperandio, R. Terenzi, G. Torrioli, M. Visco, "Cumulative analysis of the association between the Gravitational Wave detectors NAUTILUS and EXPLORER and the Gamma Ray Bursts detected by BATSE and BEPPOSAX", arXiv:astro-ph/0408544 (2004).

[5] P. Astone *et al*, "Search for correlation between GRB's detected by BeppoSAX and gravitational wave detectors EXPLORER and NAUTILUS", *Phys. Rev. D* **66** 102002 (2002).

[6] B. Abbott *et al*, "Search for gravitational-wave bursts associated with gamma-ray bursts using data from LIGO Science Run 5 and Virgo Science Run 1," *Astrophys. J.* **715** 1438 (2010).

[7] J. Abadie *et al*, "Search for gravitational-wave inspiral signals associated with short Gamma-Ray Bursts during LIGO's fifth and Virgo's first science run," *Astrophys. J.* **715** 1453 (2010).

[8] B. Abbott *et al*, "Stacked Search for Gravitational Waves from the 2006 SGR 1900+14 Storm," *Astrophys. J.* **701** L68-L74 (2009).

[9] Johnson N L, Kotz S and Balakrishnan N *Continuous Univariate Distributions vol 1* (New York: Wiley-Interscience)

[10] W. Jank, "Efficient simulated maximum likelihood with an application to online retailing," *Statistics and Computing* **16** 2 (2006)

Performance of a Chirplet-based analysis for gravitational-waves from binary black-hole mergers

Satya Mohapatra[†], **Zachary Nemtzow**[†], **Éric Chassande-Mottin**[‡] **and Laura Cadonati**[†]

[†] Physics Department, University of Massachusetts, Amherst MA 01003 (USA)
[‡] APC, Univ Paris Diderot, CNRS/IN2P3, CEA/Irfu, Obs de Paris, Sorbonne Paris Cité (France)

E-mail: satya@physics.umass.edu

Abstract. The gravitational wave (GW) signature of a binary black hole (BBH) coalescence is characterized by rapid frequency evolution in the late inspiral and merger phases. For a system with total mass larger than 100 $M_{\odot}$, ground based GW detectors are sensitive to the merger phase, and the in-band whitened waveform is a short-duration transient lasting about 10-30 ms. For a symmetric mass system with total mass between 10 and 100 $M_{\odot}$, the detector is sensitive instead to the inspiral phase and the in-band signal has a longer duration, between 30 ms - 3 s. *Omega* is a search algorithm for GW bursts that, with the assumption of locally stationary frequency evolution, uses sine-Gaussian wavelets as a template bank to decompose interferometer strain data. The local stationarity of sine-Gaussian waveforms induces a performance loss for the detection of lower mass BBH signatures, due to the mismatch between template and signal. We present the performance of a modified version of the Omega algorithm, *Chirplet Omega*, which allows a linear variation of frequency, to target BBH coalescences. The use of Chirplet-like templates enhances the measured signal-to-noise ratio due to less mismatch between template and data, and increases the detectability of lower mass BBH coalescences. We present the results of a performance study of *Chirplet Omega* in colored Gaussian noise at initial LIGO sensitivity.

1. Motivations

Gravitational wave burst searches [1] are designed to search for short duration transients. Burst search algorithms typically look for excess signal energy in the time-frequency plane after filtering the GW strain data with a particular template bank. Current burst algorithms, such as *Omega* [2] and *Coherent Waveburst* [3] use sine-Gaussian waveforms and a collection of Meyer wavelet bases respectively as templates. Matched filtering searches for the coalescence of compact binary systems with total mass 2-25 $M_{\odot}$ [4], 25-100 $M_{\odot}$ [5] and > 100 $M_{\odot}$ [6] have also been performed. These matched filtering algorithms use signal models such as the Post Newtonian approximation [7], the EOBNR analytical template family [8] and quasinormal ringdown modes, respectively.

Inspiralling binary black holes are characterized by chirping GW signals, where the frequency changes rapidly in the later inspiral and the merger phases. The duration of the signal in the most sensitive band of a ground based detector is dependent on the mass of the binary. For a system of total mass larger than 100 $M_{\odot}$, a ground based GW detector is sensitive to the merger

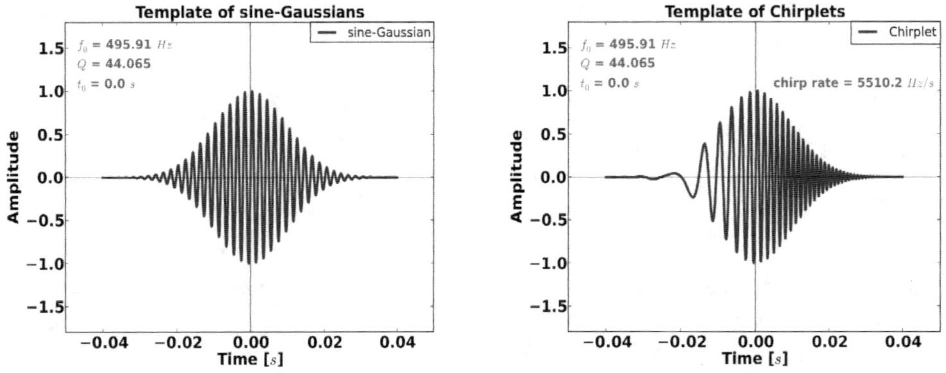

(a) A sample sine-Gaussian. (b) A sample Chirplet.

Figure 1: **Sine-Gaussian and Chirplet waveform:** (*left*) Example of a Sine-Gaussian waveform. (*right*) Example of a Chirplet waveform.

phase, and the in-band whitened waveform is a short-duration transient lasting 10-30 ms. For a symmetric binary system with a total mass between 10 and 100 $M_\odot$, the detector is sensitive instead to the inspiral phase and the in-band signal has a longer duration between 30 ms - 3 s.

An extension of the aforementioned pipeline Omega has been proposed in Ref. [9], referred here to as Chirplet-Omega. Instead of the monochromatic sine-Gaussian waveforms, it uses frequency-modulated wavelets i.e, chirplets, that better match GW signals which exhibits rapid variations of the frequency contents (as expected for orbiting or spinning astrophysical systems). In constrast to matched filtering searches, Chirplet-Omega does not rely on specific priors about the frequency evolution of GW signals.

BBH coalescences are an obvious target for Chirplet-Omega. In this paper we explore further the use of Chirplet-Omega and benchmark its detection performance for BBH coalescences, compared to the standard Omega at a constant false-alarm rate.

2. Chirplet Definition

The sine-Gaussian template bank used in the standard Omega burst search, are defined in the time domain as:

$$\psi(\tau) \equiv A \exp\left(-\frac{(2\pi f)^2}{Q^2}(\tau - t)^2\right) \exp\left(2\pi i \left[f(\tau - t)\right]\right), \tag{1}$$

Chirplets are a generalization of sine-Gaussian waveforms, defined in time domain as:

$$\psi(\tau) \equiv A \exp\left(-\frac{(2\pi f)^2}{Q^2}(\tau - t)^2\right) \exp\left(2\pi i \left[f(\tau - t) + d/2\,(\tau - t)^2\right]\right), \tag{2}$$

where $A = (8\pi f^2/Q^2)^{1/4}$ is a normalization factor ensuring that $\int |\psi|^2 d\tau = 1$; t and f are the central time and central frequency, respectively and Q is the dimensionless quality factor. See Fig. 1 for an example of sine-Gaussian and a chirplet.

The difference between a chirplet and a sine-Gaussian waveform is the *chirp rate*, an additional term in the phase, denoted as d, that linearly controls the slope of the chirplet frequency evolution as $f(\tau) = f + d(\tau - t)$. When $d = 0$, a chirplet becomes the standard sine-Gaussian. Chirplets are thus described with a four-dimensional parameter space of $\{t, f, Q, d\}$.

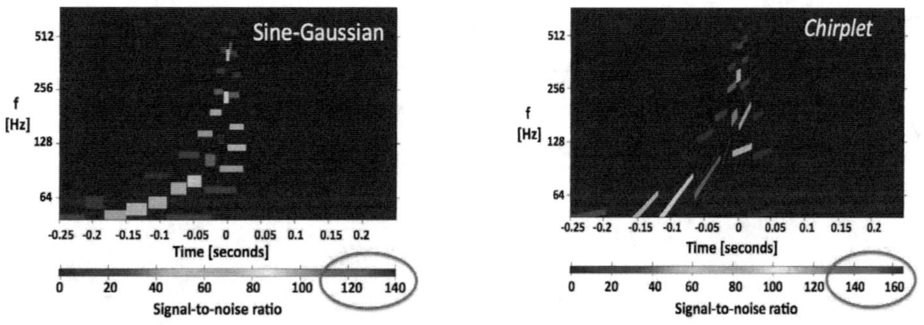

(a) Eventgram in sine-Gaussian bank. (b) Eventgram in chirplet bank.

Figure 2: **Eventgrams for sine-Gaussian and Chirplet banks:** The eventgrams are plotted for inspiralling black-hole binary (with $m_1 = 13\ M_\odot$ and $m_2 = 20\ M_\odot$) signal in simulated Gaussian LIGO noise. See the difference between the constant frequency tiles for the sine-Gaussian waveforms and the linear frequency variation for chirplets. An increase of SNR by $\sim$ 20% in the peak pixel (the most significant time-frequency pixel) is to be noted.

The *chirplet transform* T is obtained by correlating the data with chirplets. In the frequency domain, it reads:

$$T[x; \{t, f, Q, d\}] = \left| \int X(\xi)\Psi^*(\xi; \{t, f, Q, d\})d\xi \right|^2, \tag{3}$$

where $X(\xi)$ and $\Psi(\xi; \{t, f, Q, d\})$ denote the Fourier transforms of the (whitened) data stream $x(\tau)$ and chirplet $\psi(\tau)$ with the parameters $\{t, f, Q, d\}$ respectively.

3. Performances of Chirplet-Omega

3.1. Searching for binary coalescence with chirplets

The gravitational wave spectrum from the coalescence of a binary system is at lower frequency for system with larger total mass. In particular, the frequency associated with the inner-most stable circular orbit (ISCO), which corresponds to the transition between inspiral and merger phase of the coalescence, is below 80 Hz for systems with total mass $M \gtrsim 100\ M_\odot$. With this feature, the chirp phase of the waveform with a spectral content at frequencies below ISCO is outside the detector sensitive band and thus does not contribute significantly to the total detected SNR of the signal. Sine-Gaussian waveforms are a reasonable approximation for the few waveform cycles associated with the merger and ringdown portions of the coalescence waveform. For a symmetric system with lower mass (M $< 30\ M_\odot$), the coalescence time of BBH in a ground based detector's sensitive band is > 400 ms. There is loss of measured signal-to-noise ratio induced by the mismatch in the waveforms; between the chirp nature of the GW evolution and the constant central frequency sine-Gaussian template bank. The linear frequency variation of chirplets provide better match to the chirping signal of BBH in this mass range.

3.2. Increased signal-to-noise ratio

It has been shown [9] that Chirplet-Omega measures a higher signal-to-noise-ratio (SNR) compared to sine-Gaussian waveforms for a BBH signal with total mass $< 100\ M_\odot$ due to a better match. The SNR improvement is $\sim 20\%$ for total mass $M \lesssim 60\ M_\odot$ and the improvement may go up-to $\sim 40\%$ for $M \lesssim 20\ M_\odot$.

As an illustration, we show an eventgram (Fig. 2). An eventgram shows the significant time-frequency pixels for different Q values. In this particular eventgram a BBH signal is added to

Gaussian noise with the spectral characteristics of initial LIGO design. The signal is a non-spinning BBH system with mass components $m_1 = 13\ M_\odot$ and $m_2 = 20\ M_\odot$. In this example, chirplets with a positive slope are chosen to sine-Gaussian wavelets with constant frequency and the correlation of the most significative chirplet is, $\sim 20\%$ larger than the most significative sine-Gaussian.

3.3. Increased detectability

The detectability of BBH signals is limited by a background of noise fluctuations that trigger the search algorithms. Noise fluctuations can be reduced by eliminating non-GW triggers from the search with the application of a suitable coincident or coherent strategy for a network of GW detectors. The background in a chirplet search is in general different from the typical background in the standard Omega burst search. An additional 5% increase was reported [9] in the measured single detector SNR for background triggers in a chirplet burst search compared to standard Omega. Previous analysis [9] was conducted with different false alarm rates. In order to ensure an unbiased comparison, this paper addresses this issue and compares the detectability in the two search algorithms at the same false alarm rate (FAR). In the current study we added BBH signals to 2 months of simulated colored Gaussian noise mimicking the instrumental noise of the two Hanford LIGO detectors (H1 and H2), and the Livingston LIGO detector (L1). We split the range of BBH masses in two sets, total mass 4-35 $M_\odot$ and total mass 35-80 $M_\odot$. The distribution of mass was uniform in the component masses of the two black holes. We estimated the network's background by time-sliding the triggers from L1 detector with the coincident triggers from H1 and H2. Similarly foreground triggers were collected by coincidence of triggers from H1, H2 and L1. A time window of 10 ms was used to find the coincident triggers. The times at which the signals were added to noise are known. We used those times to identify significant events within 10 ms in the foreground triggers. We ranked the triggers from both the background and the simulated signals using the the geometric mean of the single detector signal-to-noise ratios as ranking statistics.

$$\rho_{GM} = (\rho_{H1} \times \rho_{H2} \times \rho_{L1})^{\frac{1}{3}} , \tag{4}$$

where $\rho_{detector}$ is the single detector SNR.

After all the triggers are ranked, the significance of each trigger was evaluated by computing its probability to belong to background (FAR). The performance comparison between chirplet and standard Omega searches was done with a common threshold on the FAR.

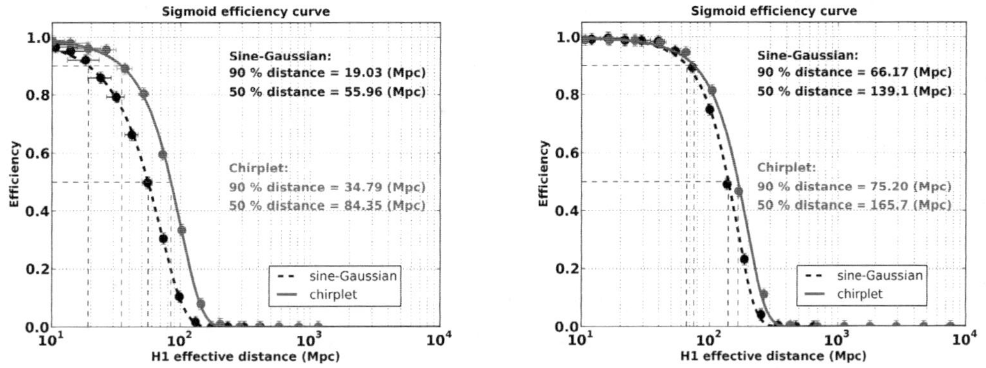

(a) Efficiency curves for set 4-35 $M_{\odot}$.

(b) Efficiency curves for set 35-80 $M_{\odot}$.

Figure 3: **Efficiency curve:** Efficiency sigmoids are plotted with a FAR of 10^{-8} Hz. Higher detection efficiency in a chirplet search is observed. The improvement in chirplet is more pronounced for the set with mass 4-35 $M_{\odot}$.

Efficiency curve — We compared the performance of the sine-Gaussian vs. the Chirplet-based algorithms by estimating the detection efficiency as a function of the effective distance of the BBH from the detector. We fit a sigmoid to the observed number of signals as a function of the distance, at a fixed false alarm rate of 10^{-8} Hz. The characteristic distance parameters, known as the 50% and 90% efficiency distances, which are the points on the sigmoid where 50% and 90% of the signals are recovered respectively, are both improved for chirplet compared to Omega as seen in Fig. 3. The numbers obtained from the sigmoid fit are shown below in Table 1. The overall improvement of > 50% in the distance parameters, is higher for the lower mass set (4-35 $M_{\odot}$) compared to, up-to 40% for the mass set with total mass ranging 35-80 $M_{\odot}$.

Set	Sine-Gaussians 50% (Mpc)	Sine-Gaussians 90% (Mpc)	Chirplets 50% (Mpc)	Chirplets 90% (Mpc)
4-35 $M_{\odot}$	56 ± 4	19 ± 1	84 ± 2	35 ± 2
35-80 $M_{\odot}$	139 ± 4	66 ± 2	166 ± 4	75 ± 3

Table 1: 50% and 90% efficiency distance at a FAR of 10^{-8} Hz.

Sensitivity distance — While efficiency sigmoids are good measures for the detection performance, as shown in Fig. 3, they don't provide details of how detectability changes across mass bins. For this reason, we compared the fiducial distance the detector network is sensitive at a false alarm rate of 10^{-8} Hz, or the *sensitivity distance*, as the function of the BBH mass parameters. Fig. 4a and Fig. 4b show sensitivity distance for chirplet and standard Omega search in the mass parameter space for the set with total mass 4-35 $M_{\odot}$. Fig. 4c shows the ratio of the sensitivity distance measured by chirplet to the sensitivity distance measured by standard Omega search. For this set, chirplet search improves detectability *more than* 30% over standard Omega search in all mass bins as seen in the ratio plot of Fig. 4c. Similarly, Fig. 4d and Fig. 4e show detectability for set with total mass 35-80 $M_{\odot}$. In this set the detectability in chirplet search also improves *up-to* 30% over standard Omega search for most of the mass bins and is shown in Fig. 4f.

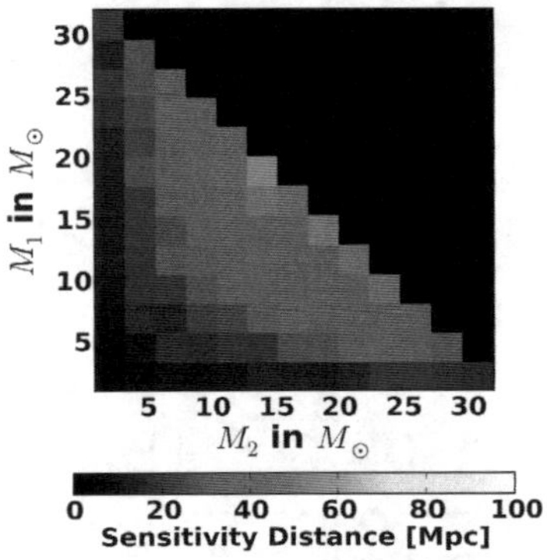

(a) Sine-Gaussians on 4-35 $M_\odot$.

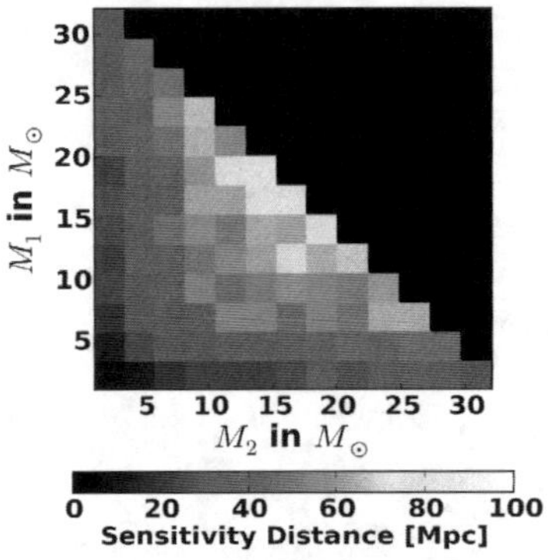

(b) Chirplets on 4-35 $M_\odot$.

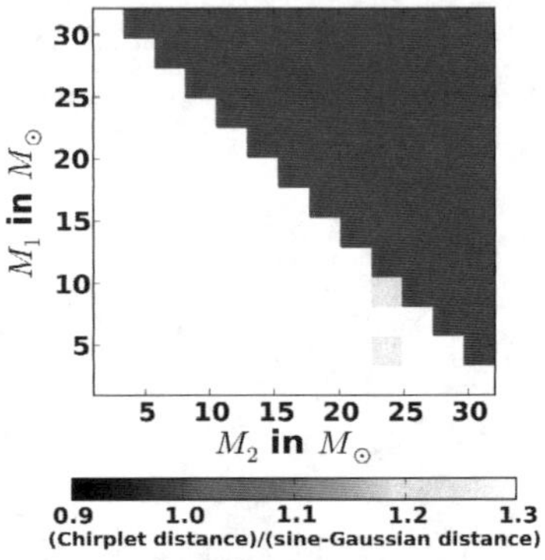

(c) Chirplet/sine-Gaussian:4-35 $M_\odot$.

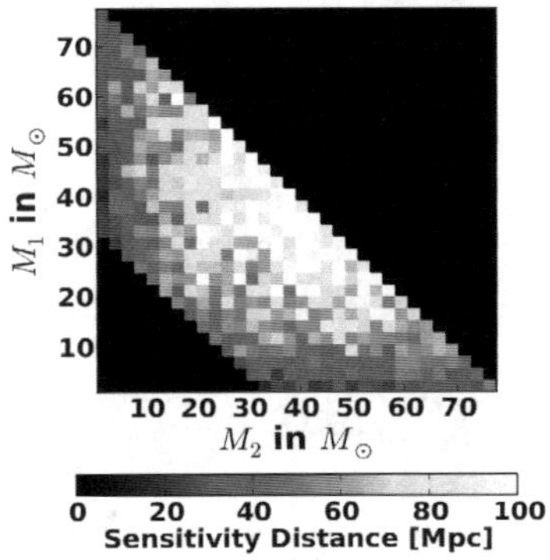

(d) Sine-Gaussians on 35-80 $M_\odot$.

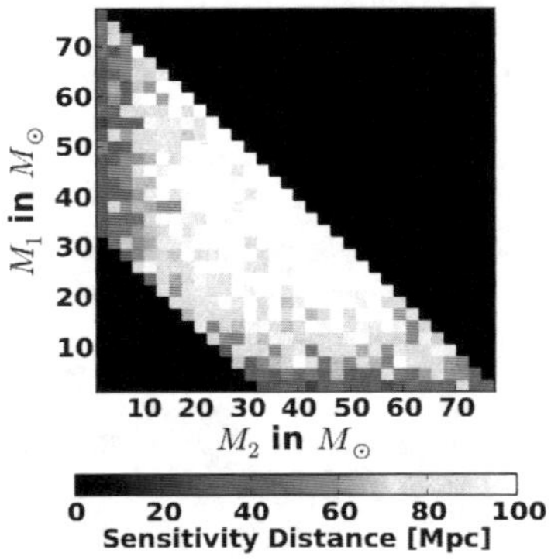

(e) Chirplets on 35-80 $M_\odot$.

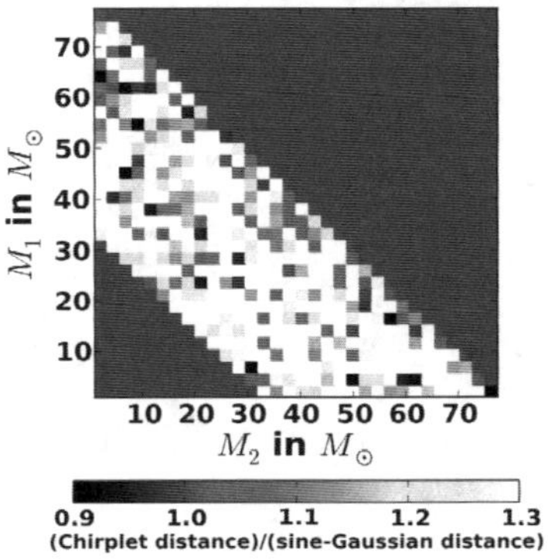

(f) Chirplet/sine-Gaussian:35-80 $M_\odot$.

Figure 4: **Sensitivity distance:** Plots of the fiducial distance that can be observed at a FAR of 10^{-8} Hz for both chirplet and sine-Gaussian in the mass parameter space are shown in Fig. 4a, 4b, 4d and 4e. The ratio plots of the detectable distance are shown in Fig. 4c and 4f. *More than* 30% of improvement in the sensitivity distance for the set 4-35 $M_\odot$ in all the mass bins and *up-to* 30% improvement in the sensitivity distance for most of the mass bins in set 35-80 $M_\odot$ should be noted.

The above comparisons of efficiency sigmoids and sensitivity distances at a false alarm rate of 10^{-8} Hz point very well to higher detectability of BBH signal in a chirplet burst search over standard Omega burst search, specially in the lower mass region (total mass $< 35\ M_\odot$). This is in accordance with the previous result of increased SNR [9].

4. Future plans

In this study, we have demonstrated the advantage of a chirplet burst search over the standard sine-Gaussian burst search using coincidence method as the multi-detector network strategy. There is room for additional modifications in the chirplet-based analysis, with the implementation of a coherent detection statistics and novel clustering strategies, which are currently under study. There have been studies [10, 11] based on matrix factorization techniques to construct orthonormal bases that span the space of high-mass merger waveforms. Using chirplets as templates, is a similar but distinct approach, with different advantages. In particular, a chirplet based search is not tied to the availability of an explicit representation of the family of waveforms to construct the basis, which gives it greater simplicity and robustness in the face of possible errors in the numerical relativity simulations, although, in exchange, it suffers some possible SNR loss.

In addition to an improved detectability, the chirplet-based analysis offers additional information on the frequency evolution of the observed signal, which will be useful in the *a posteriori* interpretation of an event.

The work presented in this paper does not address the question of the spin of the binary black hole system. Spins affect the detectability profoundly [12]. We are planning to extend our current studies to express the detectability measured in a chirplet burst search to spin parameter space.

References

[1] J. Abadie et al. All-sky search for gravitational-wave bursts in the first joint ligo-geo-virgo run. Phys. Rev. D, 81:102001, May 2010.
[2] S. K. Chatterji. The search for gravitational-wave bursts in data from the second LIGO science run. PhD thesis, MIT Dept. of Physics, 2005.
[3] S Klimenko, I Yakushin, A Mercer, and G Mitselmakher. A coherent method for detection of gravitational wave bursts. Classical and Quantum Gravity, 25(11):114029, 2008.
[4] J. Abadie et al. Search for gravitational waves from compact binary coalescence in ligo and virgo data from s5 and vsr1. Phys. Rev. D, 82:102001, Nov 2010.
[5] J. Abadie et al. Search for gravitational waves from binary black hole inspiral, merger, and ringdown. Phys. Rev. D, 83(12):122005, Jun 2011.
[6] B. P. Abbott et al. Search for gravitational wave ringdowns from perturbed black holes in LIGO S4 data. Phys. Rev., D80:062001, 2009.
[7] B. J. Owen and B. S. Sathyaprakash. Matched filtering of gravitational waves from inspiraling compact binaries: Computational cost and template placement. Phys. Rev., D60:022002, 1999. gr-qc/9808076.
[8] A. Buonanno et al. Approaching faithful templates for nonspinning binary black holes using the effective-one-body approach. Physical Review D (Particles, Fields, Gravitation, and Cosmology), 76(10):104049, 2007.
[9] Éric Chassande Mottin, Miriam Miele, Satya Mohapatra, and Laura Cadonati. Detection of gw bursts with chirplet-like template families. Classical and Quantum Gravity, 27(19):194017, 2010.
[10] Kipp Cannon, Adrian Chapman, Chad Hanna, Drew Keppel, Antony C. Searle, and Alan J. Weinstein. Singular value decomposition applied to compact binary coalescence gravitational-wave signals. Phys. Rev. D, 82:044025, Aug 2010.
[11] Kipp Cannon, Chad Hanna, and Drew Keppel. Efficiently enclosing the compact binary parameter space by singular-value decomposition. Phys. Rev. D, 84:084003, Oct 2011.
[12] P. Ajith, , et al. Inspiral-merger-ringdown waveforms for black-hole binaries with nonprecessing spins. Phys. Rev. Lett., 106(24):241101, Jun 2011.

Acknowledgments

We are grateful to the LIGO Scientific Collaboration for the use of its algorithm library available at `https://www.lsc-group.phys.uwm.edu/daswg/projects/lalapps.html`. This work is supported in part by NSF grant PHY-0653550.

This document has been assigned LIGO Laboratory document number P1100138.

Transient analysis with fast Wilson-Daubechies time-frequency transform

V Necula, S Klimenko and G Mitselmakher
University of Florida, P.O.Box 118440, Gainesville, Florida, 32611, USA

Abstract. The time-frequency transforms are important tools for identification of transient events in the output of the gravitational-wave detectors. Produced by the terrestrial and possibly by astrophysical sources, the transient events can be identified as patterns on the time-frequency plane with the excess power above stationary detector noise. In this paper we consider a particular case of the Wilson-Daubechies time-frequency transform for use in the gravitational-wave burst analysis. The presented Wilson-Daubechies basis shares some properties with the Gabor frames, but circumvents the Balian-Low theorem. It also shares similarity with the Meyer wavelet, which is actively used in the gravitational-wave burst analysis. The main advantages of the Wilson-Daubechies transform are the low computational cost, spectral leakage control, flexible structure of the frequency sub-bands, and the existence of the analytic time-delay filters, which are important for localization of the gravitational-wave sources in the sky. These properties of the Wilson-Daubechies transform may prove useful not only in the transient analysis, but also in other areas of the gravitational wave data analysis and detector characterization.

1. Introduction

Time-frequency (TF) analysis is a part of searches for gravitational wave burst signals expected to accompany collapses of massive stars, mergers of compact binary objects (consisting of neutron stars and black holes), gamma ray bursts and other energetic phenomena. Such searches [1, 2, 3, 4, 5, 6, 7, 8, 9, 10, 11] have been already conducted using the data collected with the first generation ground based laser interferometer detectors (LIGO [12], VIRGO [13], GEO [14]), but no detection has been reported. The second generation detectors will start operation after 2015 with the order of magnitude better sensitivities, and the hunt for burst signals will be resumed with increased chances of success. To maximize the discovery potential of the second generation detectors, the development of new more advanced data analysis techniques is also required [15]. This paper presents a time-frequency transform first considered by Wilson [16] and detailed by Daubechies et al. [17] which can significantly improve the performance of the burst search algorithms.

Optimal matched filter techniques are not always suitable for the burst searches because for some transient GW signals accurate waveforms are not available or difficult to obtain. Also, transient events due to environmental and instrumental disturbances need to be identified in the detector data streams for a better understanding of the non-stationary noise (detector characterization). Therefore, the existing burst algorithms exploit the localization of transient events in the TF domain. With an appropriate TF basis, it is anticipated that transient events (both GW bursts and environmental disturbances) can be identified as patterns (clusters) on the TF plane. The size, shape and the statistical properties of the clusters greatly depend on the TF transform used in the analysis.

One common approach to the burst TF analysis is to use Gabor bases (windowed Fourier) and their variations [18, 19, 20, 21, 22], where template waveforms densely cover a TF plane to produce an overcomplete representation of the data. Such TF transforms are convenient for the visualization of burst events, however their template waveforms do not form an orthogonal basis. This significantly complicates the reconstruction of burst signals, which are usually represented by multiple template waveforms. In this case, the inverse transform usually is not defined and the time domain GW signal can not be reconstructed from the TF data.

Another approach to the TF analysis is to project the data on a local orthonormal basis of wavelet functions [23]. In this case, the data in the TF domain is critically sampled (the same number of data samples before and after the transformation), the energy is conserved, the inverse transform is well defined, and the statistical independence of the TF data samples ensures a straightforward application of the clustering algorithms [24] and the coherent network reconstruction methods [25]. For example, one of the baseline LIGO-Virgo burst algorithms [26] uses Meyer's wavelet which has replaced older implementations of the TF transform based on the Symlet or Daubechies wavelets [24].

The TF transform presented in this paper combines advantages of the conventional Wilson-Daubechies transform and the Meyer wavelet. It has a better performance in several areas including: a) fast transformation by using the Fast Fourier Transform (FFT), b) shorter transformation filters (time localization), and c) better control over the spectral leakage (frequency localization). Also it offers a simple analytic expression for the time delay filters used to time-shift TF series. These filters are necessary for the sky-localization algorithms [26], which should synchronize the TF data streams from different detectors for an anticipated sky location of the GW source.

2. Wilson-Daubechies orthonormal basis

Conventional approaches to the time-frequency analysis include the windowed Fourier transform

$$\tilde{x}(\omega, t) = \int \mathrm{e}^{-\mathrm{i}\omega\tau} \phi(\tau - t) x(\tau) \mathrm{d}\tau \ , \tag{1}$$

with ϕ as the window function, and its discrete version in frequency ω and time t based on Gabor frames

$$g_{nm}(t) = \mathrm{e}^{im\omega\frac{t}{T}} \phi(t - nT) \quad , \quad n, m \in \mathbb{N} \ , \tag{2}$$

where the free parameter T defines the time resolution. In principle, with a careful choice of the window function an orthonormal basis can be constructed. However, it is a subject to the Balian-Low theorem stating that the basis functions are non-local in either time or frequency:

$$\int t^2 |\phi(t)|^2 \mathrm{d}t = \infty \qquad \text{or} \qquad \int \omega^2 |\tilde{\phi}(\omega)|^2 \mathrm{d}\omega = \infty \tag{3}$$

Wilson [16] circumvented the limitations of the Balian-Low theorem by introducing a general alternative to Gabor frames where each frequency band has its own window function. Daubechies and collaborators [17] showed that one can build a simple orthonormal Wilson basis by using just one window function $\phi(t)$ and construct the basis as:

$$g_{n0}(t) = \phi(t - nT) \tag{4}$$

$$g_{nm}(t) = \sqrt{2}\, (-1)^{mn} cos(2\pi m t/T) \phi(t - nT/2) \ , \quad m + n = 2k \ , \ m > 0 \tag{5}$$

$$g_{nm}(t) = \sqrt{2}\, sin(2\pi m t/T) \phi(t - nT/2) \ , \quad m + n = 2k + 1 \ , \ m > 0 \tag{6}$$

Also the window function should satisfy the condition

$$\sum_{l\in\mathbb{Z}} \tilde{\phi}(\omega+2\pi l)\tilde{\phi}(\omega+2\pi l+4\pi m) = (2\pi)^{-1}\delta_{m0}\,. \tag{7}$$

There are constructions of ϕ which guarantee exponential decay in both time and frequency [17]. But in this paper we will focus on a special case where ϕ is a generalization of the Meyer wavelet scaling function, which is compactly supported in the frequency domain and decays faster than any inverse polynomial in the time domain. This choice is particularly convenient in applications to band-limited (sampled) signals because a fast discrete version of the transform can be obtained.

3. Fast Wilson-Daubechies-Meyer transform

In this section we consider a discrete data $x[k]$ sampled at the frequency r_s with the maximum angular frequency $\Omega = \pi r_s$, and find the discrete Wilson-Daubechiers-Meyer (WDM) filters $f_{nm}[k]$ used to obtain the time-frequency representation of the signal:

$$w_{nm} = \sum_k f_{nm}[k]x[k] \tag{8}$$

The WDM basis functions in the Fourier domain are given by

$$\tilde{g}_{n0}(\omega) = \mathrm{e}^{-in\omega T}\tilde{\phi}(\omega) \tag{9}$$

$$\tilde{g}_{nm}(\omega) = \frac{1}{\sqrt{2}}\mathrm{e}^{-in\omega T/2}\tilde{\psi}_{nm}(\omega). \tag{10}$$

The exponential factors describe the time translation of the basis and the functions $\tilde{\psi}_{nm}(\omega)$ are constructed from the frequency-shifted $\tilde{\phi}(\omega)$ functions

$$\tilde{\psi}_{nm}(\omega) = C^*_{m+n}\tilde{\phi}(\omega+m\Delta\Omega) + C_{m+n}\tilde{\phi}(\omega-m\Delta\Omega)\ ,\quad m>0 \tag{11}$$

$$\Delta\Omega = 2\pi/T\quad,\quad C_{2k}=1\quad,\quad C_{2k+1}=\mathrm{i}\,. \tag{12}$$

The generalized Meyer scaling function $\tilde{\phi}(\omega)$ is defined as

$$\tilde{\phi}(\omega) = \frac{1}{\sqrt{\Delta\Omega}}\quad,\quad |\omega|<A \tag{13}$$

$$\tilde{\phi}(\omega) = \frac{1}{\sqrt{\Delta\Omega}}cos\left[\nu_n\left(\frac{|\omega|-A}{B}\right)\frac{\pi}{2}\right]\quad,\quad A\le|\omega|<A+B\,, \tag{14}$$

where A and B are two positive parameters

$$2A+B=\Delta\Omega\,, \tag{15}$$

and $\nu_n:[0,1]\to\mathbb{R}$ is a monotonically increasing function constructed from the incomplete Beta function $B(x;a,b)$

$$\nu_n(x) = \frac{B(x;n,n)}{B(1;n,n)}\quad,\quad B(x;a,b) = \int_0^x t^{a-1}(1-t)^{b-1}\mathrm{d}t\,. \tag{16}$$

The parameter n defines how sharp are the edges of the scaling function. For example, Figure 1 shows the two scaling functions with $n=2$ and $n=4$ (top plots) and their time domain counterpart (bottom plots).

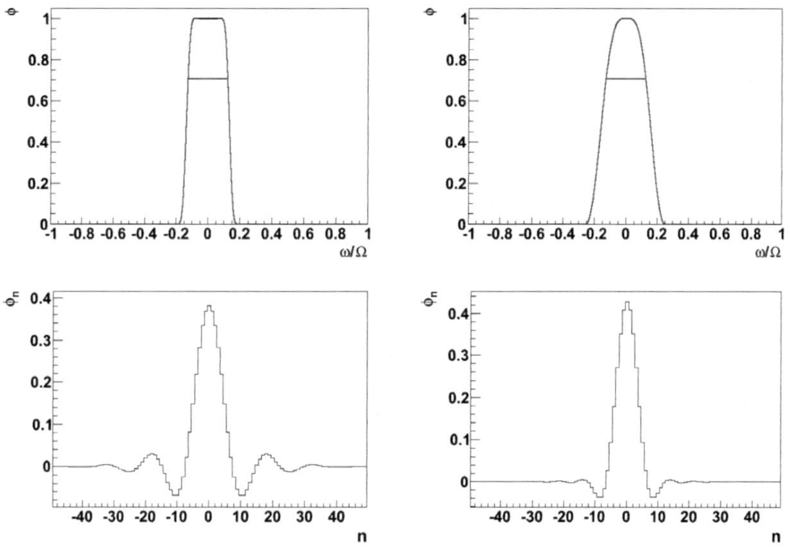

Figure 1. The top plots show $\phi(\omega)$ and highlight the defining parameters: the width of the top flat region (black) is $2A$, the transition regions (blue) are defined by B, and the red line spans the nominal band width $2A+B$. The shape of the transition region depends on the parameter n in ν_n; $n=4$ for the left plot and $n=2$ for the right plot. The bottom plots show the coefficients of their discrete time domain representation. In these examples $M=4$ (see Equation (17)).

Only those functions $\tilde{g}_{nm}$ which span the data frequency band $[-\Omega, \Omega]$ are relevant for the transformation. Therefore $m \leq M$, where the parameter M is given by

$$M = \frac{\Omega}{\Delta\Omega} = \frac{T}{2\tau} \tag{17}$$

and $\tau = 1/r_s$. Unless M is a natural number, the filters $f_{nm}[k]$ for $n \neq 0$ are not discrete translations of $f_{0m}[k]$, therefore $M \in \mathbb{N}$. In this case we can also redefine the functions $\tilde{g}_{nM}$ to be

$$\tilde{g}_{nM}(\omega) = e^{-\mathrm{i}(2n+q)M\omega\tau}\left[\tilde{\phi}(\omega+\Omega) + \tilde{\phi}(\omega-\Omega)\right] \quad , \quad |\omega| \leq \Omega \tag{18}$$

where $q=0$, if M is even, and $q=1$ otherwise. The new set $\tilde{g}_{nm}$ $(0 \leq m \leq M)$ forms an orthonormal basis on $[-\Omega, \Omega]$ and the explicit WDM expansion is

$$w_{n0} = \tau \sum_{k\in\mathbb{Z}} x[2nM+k]\phi[k] \tag{19}$$

$$w_{nm} = \tau\sqrt{2}\Re\, C_{m+n} \sum_{k\in\mathbb{Z}} \mathrm{e}^{\mathrm{i}\pi k m/M} x[nM+k]\phi[k] \quad , \quad 0 < m < M \tag{20}$$

$$w_{nM} = \tau \sum_{k\in\mathbb{Z}} (-1)^k x[2nM+qM+k]\phi[k] \tag{21}$$

where $\phi[k]$ are the sampled values of $\phi(t)$ - the time domain representation of $\tilde{\phi}(\omega)$. Note, the Fourier inverse of $\tilde{\phi}(\omega)$ does not have a compact support in the time domain, therefore, the filter

$\phi[k]$ should be truncated. As the result the transformation is approximately orthonormal, but for any practical application the error introduced by the truncation can be made insignificant.

The formula for $0 < m < M$ case also cover both $m = 0$ and $m = M$ cases up to the normalization factor $\sqrt{2}$ and picking only the odd or even coefficients as appropriate. Therefore, using the periodicity of the exponential factor we can rewrite

$$\sum_{j=0}^{j<2M} e^{i2\pi jm/2M} X_n[j] = \sum_{k\in\mathbb{Z}} e^{i\pi km/M} x[nM+k]\phi_k \tag{22}$$

$$X_n[j] = \sum_{k\in\mathbb{Z}} x[nM+2kM+j]\ \phi[2kM+j] \tag{23}$$

The data vectors X_n have the length of $2M$, and the summation on the left side of Equation (22) is the discrete Fourier transform of X_n. Therefore, the FFT algorithm can be used to speed up the calculations. The construction of the inverse transform is straightforward and also uses the FFT algorithm.

Only half of the Fourier components obtained from the vectors X_n represents the WDM expansion in Equations (19-21). The other half is the WDM expansion

$$\hat{w}_{n0} = \tau \sum_{k\in\mathbb{Z}} x[(2n+1)M+k]\phi[k]\ , \tag{24}$$

$$\hat{w}_{nm} = \tau\sqrt{2}\Im\, C_{m+n} \sum_{k\in\mathbb{Z}} e^{i\pi km/M} x[nM+k]\phi[k] \quad , \quad 0 < m < M\ , \tag{25}$$

$$\hat{w}_{nM} = \tau \sum_{k\in\mathbb{Z}} (-1)^k x[2nM+(1-q)M+k]\phi[k]\ , \tag{26}$$

in the complementary orthonormal basis constructed from the time-shifted (by $T/2$) WDM basis functions in Equations (9) and (10). Each WDM set (w_{nm} or $\hat{w}_{nm}$) gives a complete TF representation of the data. For $0 < m < M$ the coefficients $\hat{w}_{nm}$ represent the quadrature of the data w_{nm} in the WDM domain. The quadrature is often used in the transient analysis and obtained with the fast WDM transform at no additional computational cost.

4. WDM time-delay filters

In the coherent analysis of data from multiple detectors, the relative time delays between detectors need to be introduced in order to synchronize the expected signal with respect to a particular source location in the sky. Time-delay filters perform this operation in the time-frequency domain and in this section we present these filters for the Wilson-Daubechies-Meyer transform.

A time shifted data $w_{nm}(\delta t)$ in the TF domain can be calculated as

$$w_{nm}(\delta t) = \sum_{lk} w_{lk} \int_{-\infty}^{\infty} g_{nm}(t+\delta t) g_{lk}(t) \mathrm{d}t\ . \tag{27}$$

where w_{lk} is the original TF data. time-delay filter can be evaluated in the Fourier domain

$$\int_{-\infty}^{\infty} g_{nm}(t+\delta t) g_{lk}(t) \mathrm{d}t = \int_{-\Omega}^{\Omega} e^{-i\omega\delta t} \tilde{g}^*_{nm}(\omega) \tilde{g}_{lk}(\omega) \mathrm{d}\omega \tag{28}$$

Unless $|m-k| \le 1$, the integral vanishes because there is no overlap between the basis functions. The results take a simpler form, if the time shift δt is a multiple of the sampling interval τ. Evaluating the integral one finds

$$\int g_{nm}(t+\delta t)g_{n+l,m}(t)\mathrm{d}t = (-1)^{ln}\Re\, C_l^* \mathrm{e}^{\mathrm{i}m\Delta\Omega\delta t}\, T_l \;, \tag{29}$$

$$T_l = \int_{-\Omega}^{\Omega} \mathrm{e}^{\mathrm{i}\omega(Ml\tau+\delta t)} |\tilde{\phi}(\omega)|^2 \,\mathrm{d}\omega \;, \tag{30}$$

$$\int g_{nm}(t+\delta t)g_{n+l,m\pm1}(t)\mathrm{d}t = (-1)^{m+n}(-1)^{ln}\Re\, C_{l+1}^* \mathrm{e}^{\mathrm{i}(m\pm1/2)\Delta\Omega\delta t}(\pm \mathrm{i})^l T_l' \;, \tag{31}$$

$$T_l' = \int_{-\Omega}^{\Omega} \mathrm{e}^{\mathrm{i}\omega(Ml\tau+\delta t)} \tilde{\phi}^*(\omega-\Delta\Omega/2)\tilde{\phi}(\omega+\Delta\Omega/2)\,\mathrm{d}\omega \;. \tag{32}$$

Equation (31) has to be multiplied by $\sqrt{2}$ whenever the sub-band index (m or $m \pm 1$) is equal to 0 or M. For the WDM transform the calculation of the time-delay filters is essentially reduced to computing the integrals T_l and $T_l^{\pm}$. For comparison, for Meyer wavelet [26] the time-delay filters should be evaluated and stored for each frequency band.

5. Properties of the WDM transform

5.1. WDM as a band-pass filter

Similar to wavelets the WDM transform is representing a set of band-pass filters. Figure 2 (top plot) shows an example of the sub-band structure of the basis functions for $M = 16$. For the WDM transform presented in the paper, the first and the last WDM sub-bands cover half of the frequency resolution, however a transform with the uniform frequency resolution can be also constructed. For comparison, Fig. 2 (bottom plot) also shows the sub-band structure for the binary wavelet transform [24] which is less regular, each sub-band (frequency layer) having a different shape due to the iterative nature of the wavelet decomposition. Although, both transforms can be used for the TF representation of data, the WDM transform offers a more accurate separation of the data spectral components.

5.2. Time-frequency maps

Each data sample w_{nm} can be presented as a time-frequency pixel with dimensions $\Delta\Omega \times T/2$. Respectively, the w_{nm} can be presented as a pixel map showing the TF content of the data. Figure 3 shows the energy (w_{nm}^2) time-frequency maps for a GW signal from a merger of non-spinning binary black holes with $M_1 = M_2 = 10M_\odot$ and at distance of 10 Mpc. The WDM map (top left) shows the average energy of the signal and its quadrature, which produces a more clear signature of the signal and highlights the utility of the quadrature coefficients. The other three TF maps are obtained with different wavelet transforms. Except for the Daubechies wavelet (bottom left), which has poor TF localization due to its non-symmetric filter, the other wavelet representations are fairly similar to the WDM map. However, the wavelets are limited to $M = 2^i$ by their iterative decomposition procedure and cumbersome to use for large M.

5.3. Spectral density estimation

The WDM transform is an excellent tool for the power spectral density (PSD) estimation. Figure 4 shows the noise spectral density amplitude obtained with the PSD estimators based on the WDM ($M = 8192$, $K = 2M$, $P = 12$, $n = 7$) and windowed FFT (Hann) for one hour of data collected by the LIGO Livingston detector. In both cases the frequency resolution is 1 Hz. There is a very good agreement between the WDM and FFT in the quasi-white noise regions (around 130 Hz and 170 Hz), but the WDM transform offers a lower spectral leakage in regions dominated by the 60 Hz power line (and its harmonics) and the violin modes between 330 Hz - 350 Hz. Within the filter truncation errors, the PSD obtained with the WDM transform is

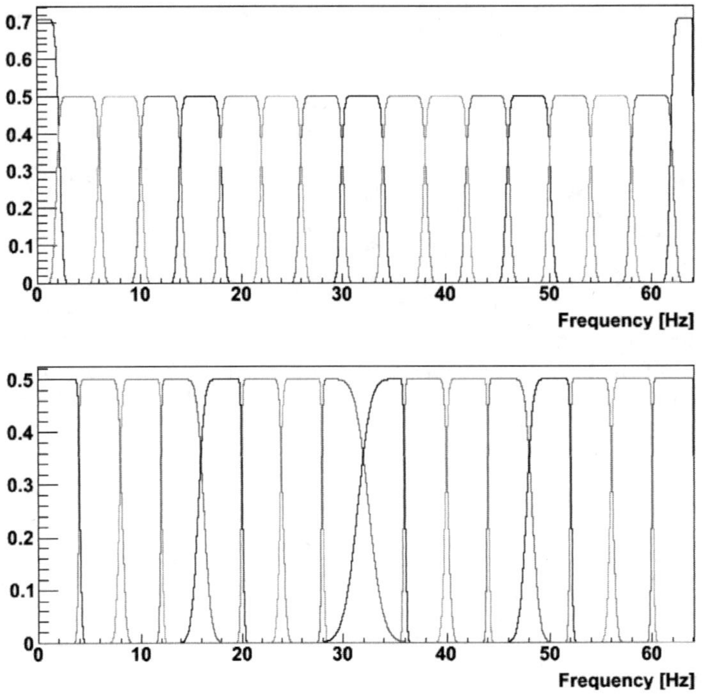

Figure 2. Spectral Fourier amplitudes of the basis functions for WDM transform (top) and binary Meyer wavelet (bottom).

an accurate representation of the data, while the windowed FFT is always altering the data by using a window in order to diminish spectral leakage.

5.4. Computational accuracy

The WDM transform is uniquely characterized by the parameters M, K

$$2A + B = \frac{\Omega}{M} \quad , \quad B = \frac{\Omega}{K} \quad , \quad K \geq M \tag{33}$$

and n from $\nu_n(x)$ defining the smooth edge of the window function. Since the time domain WDM basis functions are not compactly supported, in practice, they have to be truncated. Namely, the filter $\phi[k]$ is defined for $-L \leq k \leq L$, where $2L + 1$ is the filter length. The truncation introduces an error in the normalization of the basis functions, which can be characterized by the precision parameter

$$P = -log_{10} \left(1 - \sum_{k=-L}^{L} \phi^2[k] \right) . \tag{34}$$

For a given n, ratio K/M and the precision parameter P, the filter length is proportional to M. Figure 5 shows the ratio $(2L + 1)/M$ as a function of K/M and different values of the n and P parameters.

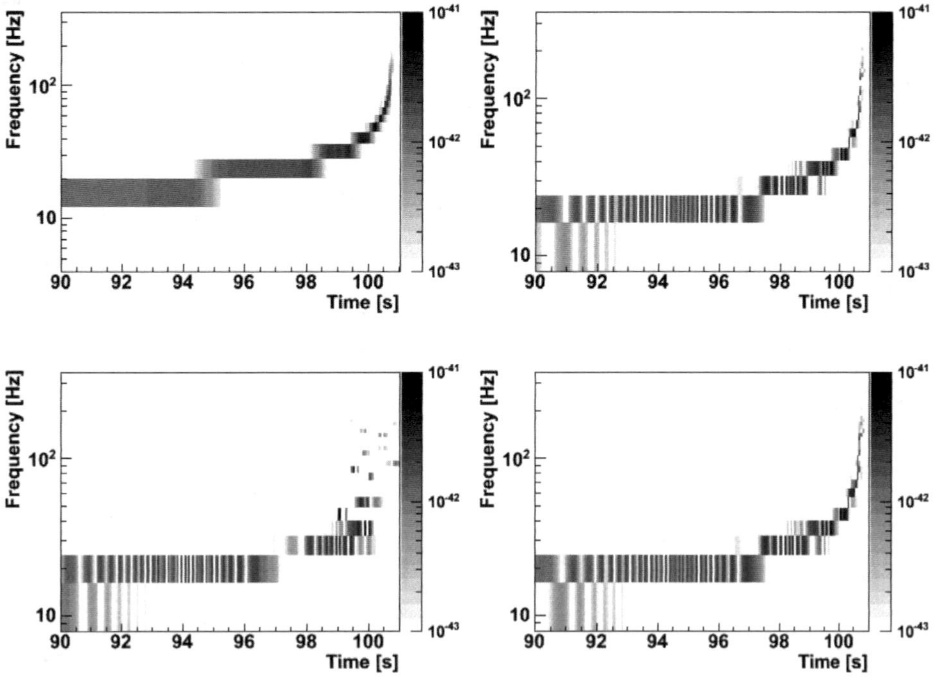

Figure 3. WDM transform (top left), Meyer (top right), Daubechies (bottom left) and Symlet (bottom right) wavelet time-frequency energy maps for a binary black hole inspiral signal.

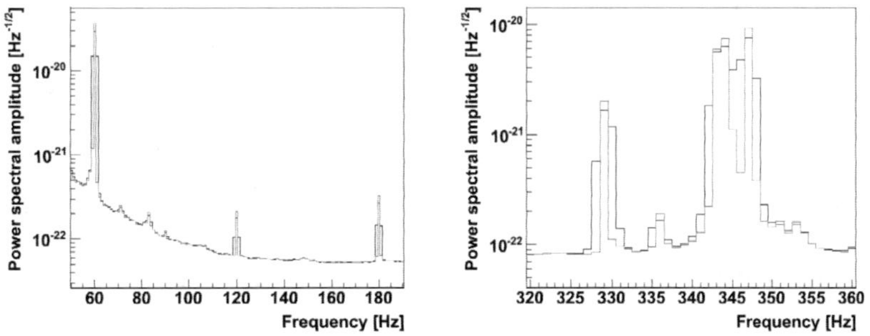

Figure 4. Spectral density amplitudes for LIGO Livingston data (GPS range 816260809 - 816264750) obtained with the WDM transform (red) and Hann windowed FFT (black).

When the n value increases, initially the filter length decreases. But for higher n the filter length starts to increase again, because in the limit $n \to \infty$ the $\phi(t)$ converges to the *sinc* function with a slowly time decay.

Another way to estimate the accuracy of the transform is to perform the forward and inverse transformations and subtract the resulting time series from the original one. Figure 6 shows the distribution of the residual data for $n = 4$, $M = 64$ and $K/M = 2$ when the input data is a

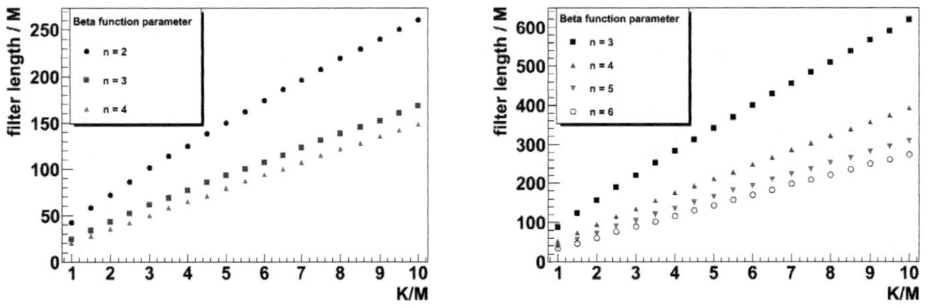

Figure 5. The ratio $(2L+1)/M$ as a function of K/M for different values of the Beta function parameter n and two precision requirements: $P = 8$ (left) and $P = 12$ (right)

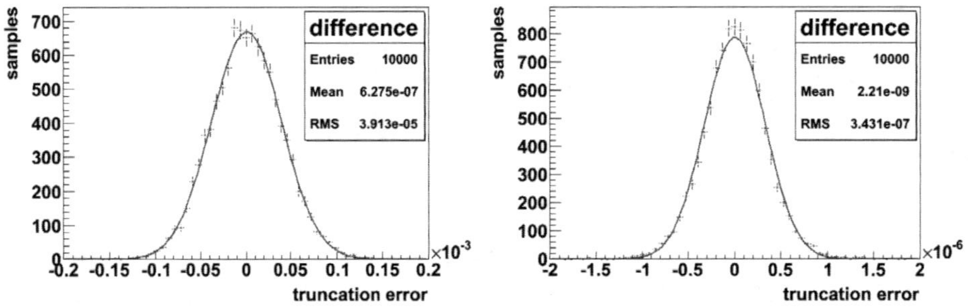

Figure 6. Truncation errors for the inverse/forward transformation ($n = 4$, $M = 64$, $K/M = 2$) of white Gaussian noise with the unity variance: $P = 8$ (left) and $P = 12$ (right).

white Gaussian noise with the unity variance.

5.5. Accuracy of the time delay filters

Truncation of $\phi[k]$ also affects the time-delay filters. By using the integrals T_l and T_l' (Eq. 30,32), they need to be constructed for every WDM frequency band m and the time-delay parameter δt. For a given m and δt, the achieved precision (calculated similar to Eq. 34 with the sum taken over the time-delay filter coefficients) of the time-delay transformation depends on the number of the filter coefficients (filter length L_{TD}). Figure 7 (left plot) shows the precision as a function of the sub-band index m and the delay index $\delta t/\tau$ for $L_{\text{TD}} = 123$. By selecting the central point ($m = \delta t/\tau = 32$) as a conservative representation of the precision of all filters, the filter length can be estimated as a function of K/M, n and P (see Fig. 8).

Acceptable accuracy of the time-delay transformation can be achieved for $L_{\text{TD}} > 60$. Figure 7 (right plot) shows the difference between the time-delay transformation ($L_{\text{TD}} = 123$ and $\delta t/\tau = 32$) implemented in the WDM domain and the correct values obtained with the WDM transform ($M = 64$, $K = 2M$, $n = 3$) of the delayed data in the time domain, which is a white Gaussian noise with the unity variance.

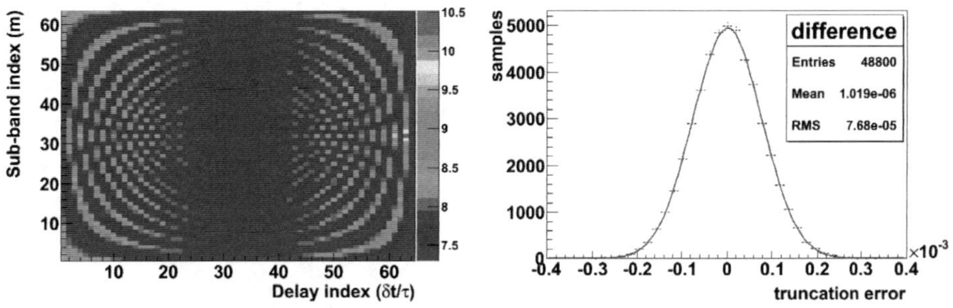

Figure 7. The left plot shows time delay filters precision with truncation at L_{TD}=123 for M=64, K=2M, n=3. The right plot shows the time-delay filter truncation errors for Gaussian noise with unity variance.

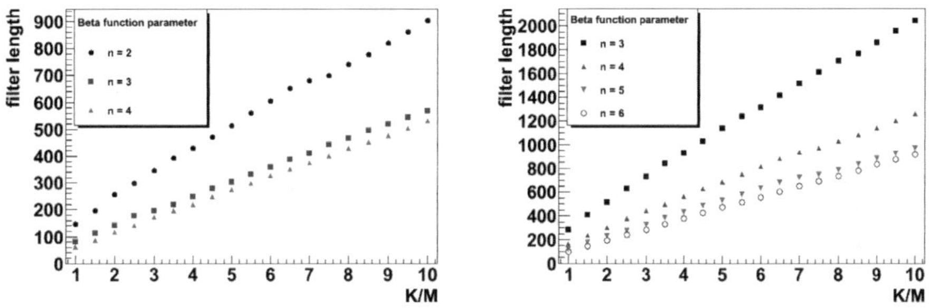

Figure 8. Time-delay filter length (L_{TD}) as a function of K/M and n for two different precision requirements, $P = 8$ (left) and $P = 12$ (right)

5.6. Complexity

Let N be the number of samples in the time series, and M, K, P the parameters defining the transformation. The number of multiplications needed to obtain the TF data is

$$C_{\mathrm{WDM}} = 2N\left[log_2\, 2M + \frac{L(K,M,P)}{M}\right] \tag{35}$$

where $2L(K,M,P)+1$ is the length of the WDM filter. The first term in square parenthesis represents the complexity of the FFT and the second term is the added complexity due to the multiplication by the WDM filter (per 1/2 data sample). For comparison, the windowed FFT performed with the 50% overlap between the adjacent windows has a complexity

$$C_{\mathrm{FFT}} = 2N\left[log_2 M + 1\right] \ . \tag{36}$$

The FFT is always faster, but for large M the computational performance of the WDM transform becomes comparable. The complexity of the binary wavelet transform is

$$C_{\mathrm{W}} = N\; L_{\mathrm{W}}\; log_2 M \ , \tag{37}$$

where L_{W} is the wavelet filter length. It is comparable with the FFT only for Haar wavelet ($L_{\mathrm{W}} = 2$). For longer wavelet filters ($L_{\mathrm{W}} \sim 1000$), required to reduce spectral leakage, the computational cost of the wavelet transform is significantly higher than for the FFT and the WDM.

6. Conclusion

The fast discrete Wilson-Daubechies-Meyer (WDM) transform is a promising tool for the gravitational-wave burst analysis. With the WDM transform the transient events can be efficiently localized on the TF plane. The transform is orthonormal and critically sampled, which significantly simplifies the statistical analysis in the TF domain. The conservation of energy is approximate, however, for any practical application the transformation errors can be made insignificant. Depending on the particular requirements of the analysis, the transform's parameters (M, K, n and P) can be customized to optimize the performance. Analytic time delay filters are readily available to perform time-shifts of data directly in the TF domain. The inverse transform is also available. These properties of the WDM transform make it a convenient tool for the coherent network analysis and the reconstruction of the GW waveforms and source coordinates. Due to a superior spectral leakage control, the WDM transform provides an accurate spectral density estimation which can be particularly useful for monitoring of a quasi-stationary noise typical for real detectors. With its unique properties, the WDM transform may prove useful not only for burst searches, but also in the other areas of the gravitational wave data analysis and detector characterization.

7. Acknowledgements

This work was supported by the US National Science Foundation grants PHY-0855044 and PHY-0855313 to the University of Florida, Gainesville, Florida.

References

[1] Abbott B *et al* 2004 *Phys. Rev.* D **69** 102001
[2] Abbott B *et al* 2005 *Phys. Rev.* D **72** 062001
[3] Abbott B *et al* 2005 *Phys. Rev.* D **72** 122004
[4] Abbott B *et al* 2006 *Class. Quantum Grav.* **23** S29
[5] Abbott B *et al* 2007 *Class. Quantum Grav.* **24** 5343
[6] Abbot B P *et al* 2009 *Phys. Rev.* D **80** 102001
[7] Abbot B P *et al* 2010 *Astrophys. J.* **715** 1438
[8] Abadie J *et al* 2010 *Phys. Rev.* D **81** 102001
[9] Abbott B *et al* 2008 *Phys. Rev. Lett.* **101** 211102
[10] Abbot B P *et al* 2009 *Astrophys.J.* **701** L68
[11] Abadie J *et al* 2011 *Astrophys.J.* **734** L35
[12] Abbott B P *et al* 2009 *Rep. Prog. Phys* **72** 076901
[13] Acernese F *et al* 2006 *Class. Quantum Grav.* **23** S635
[14] Luck H *et al* 2006 *Class. Quantum Grav.* **23** S71
[15] LIGO-Virgo Collaboration, LIGO Document T1100322
https://dcc.ligo.org/cgi-bin/DocDB/ShowDocument?docid=62893
[16] Wilson K G, preprint, Cornell University
[17] Daubechies I, Jaffard S and Journé J L 1991 *J. Math. Anal.* **22** pp 554-673
[18] Anderson W G, Brady P R, Creighton J D E and Flanagan É É 2001 *Phys. Rev.* D **63** 042003
[19] Searle A C, Sutton P J and Tinto M 2009 *Class. Quantum Grav.* **26** 155017
[20] Beauville F *et al* 2008 *Class. Quantum Grav.* **25** 045002
[21] Kalmus P, Khan R, Matone L and Marka S 2007 *Class. Quantum Grav.* **24** S659
[22] Chatterji S, Lazzarini A, Stein L, Sutton P, Searle A and Tinto M 2006 *Phys. Rev.* D **74** 082005
[23] Daubechies I 1992 *Ten Lectures on Wavelets* (Philadelphia, PA: SIAM)
[24] Klimenko S and Mitselmakher G, 2004 *Class. Quantum Grav.* **21** S1819
[25] Klimenko S, Mohanty S, Rakhmanov M and Mitselmakher G 2005 *Phys. Rev.* D **72** 122002
[26] Klimenko S, Yakushin I, Mercer A and Mitselmakher G 2008 *Class. Quantum Grav.* **25** 114029
[27] Meyer Y 1992 *Wavelets and Operators* (Cambridge: Cambridge University Press)

The Analysis of ROTSE Images of Potential Counterparts to Gravitational Wave Events

L. K. Nuttall[1] for the LIGO Scientific Collaboration and Virgo Collaboration, W. Zheng[2], C. Akerlof[2]

[1]Cardiff University, Cardiff, CF24 3AA, United Kingdom

[2]Randall Laboratory of Physics, University of Michigan, 450 Church Street, Ann Arbor, MI, 48109-1040, USA

E-mail: laura.nuttall@astro.cf.ac.uk

Abstract. During the most recent LIGO-GEO-Virgo science run a number of partner telescopes performed follow-up observations of gravitational wave (GW) candidates. One of these collaborators was the ROTSE project. Consisting of four optical telescopes, ROTSE responded to GW triggers and took over 700 follow-up images. Analysis of these images is currently under way using ROTSE's own image processing pipeline. We describe the analysis used to search for transients of astrophysical significance, and steps being taken to automate and optimise the analysis for rapid identification of electromagnetic (EM) counterparts to GW candidates.

1. Introduction

The most recent science run (July 2009 - October 2010) of LIGO [1] and Virgo [2] saw the first efforts to detect EM counterparts to candidate GW events. The detection of an EM counterpart to a GW signal would increase the confidence in the GW detection as well as provide complementary information about the progenitor. Several low latency pipelines were used to promptly analyse GW data, identify candidates and reconstruct their apparent sky positions. This effort is referred to as the 'LOOC UP' project [3]. Details on how this search was performed are documented in [4]. Several wide field-of-view optical telescopes took part in this search; including the four ROTSE-III telescopes.

2. The ROTSE-III telescope system

The Robotic Optical Transient Search Experiment (ROTSE) is dedicated to rapid follow up observations of gamma ray bursts (GRBs) as well as other fast optical transients on the time scale of seconds to days. ROTSE has consisted of two phases thus far, ROTSE-I and III. ROTSE-I consisted of a 2 x 2 array of telephoto camera lenses co-mounted on a rapid-slewing platform, located in northern New Mexico. The array was fully automated and started taking data in 1998. Observations made by ROTSE-I of GRB 990123 revealed the first detection of an optical burst occurring during the gamma-ray emission, demonstrating the value of autonomous robotic telescope systems [5]. The ROTSE-III telescope system [6] came online in 2003 and consists of four 0.45m robotic reflecting telescopes located in New South Wales, Australia (ROTSE-IIIa), Texas, USA (ROTSE-IIIb), Namibia (ROTSE-IIIc) and Turkey (ROTSE-IIId). The instruments are fully automated and make use of fast optics to give a 1.85 x 1.85 square degree field of view. ROTSE-III is capable of attaining 17th magnitude with a 5 second exposure and 18.5 magnitude with a 60 second exposure. If multiple images are stacked on top of one another or 'coadded' ROTSE-III can reach $\sim$19th magnitude [7].

The LOOC UP program ran in two phases. During the latter phase, between September 2 and October 20 2010, ROTSE-III took over 700 images in response to GW triggers. The images taken span from the first night after the event occurred to one month later. Images vary in exposure length (either 20 or 60 seconds) and all four telescopes were used to gather the data. With GW detectors at the sensitivities they were at the time of the search, most or all of the triggers are unlikely to represent true astrophysical events.

3. The ROTSE image processing pipeline

The ROTSE image processing pipeline was developed and is still used by the ROTSE collaboration to search for transient objects within images taken with the ROTSE-III telescopes. The pipeline makes use of cross convolution to perform image subtraction. This relaxes the need for high quality reference images and the computational efficiency

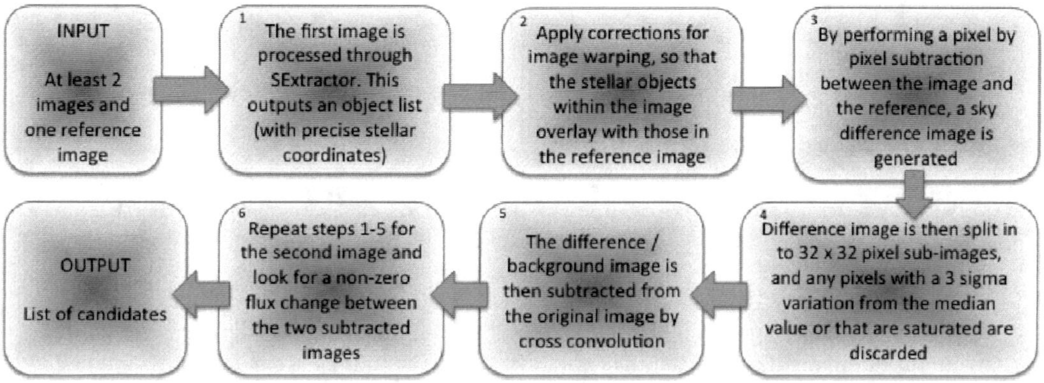

Figure 1. Flow chart describing how the ROTSE pipeline processes images to produce a list of potential candidates.

is similar to other procedures for image processing [8]. We present a brief summary of the main steps of the pipeline; for more details please see [8, 9].

When a LIGO/Virgo trigger was sent to the ROTSE telescopes, typically 30 images were taken on the first night with 8 images taken on subsequent follow-up nights, per telescope, for the first ten nights following the trigger, with additional observations around nights 15 and 30. These images were of the same part of the sky, so that images may be stacked or 'coadded' on top of one another. Coadding increases the limiting magnitude to which we are sensitive (i.e. to see fainter objects) without saturating the brightest objects within the image. This permits two different modes of analysis. In the coadded case, all images for a given night are combined in to two images (i.e. the first 15 (4) images are coadded to form the first image, and the second 15 (4) images are coadded to form a second image of the first night (follow-up night)), which are then compared. Image deficiency is used to pick out the transients, as described in Figure 1. This procedure is repeated separately for each night so that each night produces a separate list of candidates. The ROTSE pipeline can also perform a 'non-coadded' analysis, in which just the images taken from the first night are processed without coadding to see if there are any fast transients on the $\sim$ hour time scale. The non-coadded analysis does not stack images, meaning the images would have a shallower limiting magnitude than those images which have been coadded. In the non-coadded case each consecutive pair of images produces a list of candidates.

Both analyses (coadded or non-coadded) will produce several lists of candidates identified by the pipeline. The lists from within a single analysis therefore need to be compared to produce a single list of unique candidates. The vast majority of these potential candidates will be due to poor image subtraction, with a minority due to a variable, rather than transient, source (such as a star or asteroid). We identify and remove transients, which are not targeted, by comparing to catalogues such as SIMBAD‡ or the minor planet checker§ so that we only produce lightcurves of the most likely candidate EM counterparts.

4. Steps taken to automate the pipeline

The ROTSE image processing pipeline has been used to make some significant discoveries of optical transients. The drawbacks of the pipeline include being widget based, highly human intensive and therefore time consuming, and the limitation of only processing images from one GW candidate at a time. We have therefore taken steps to automate the pipeline and taken a slightly different approach to processing these images.

The detection of an EM counterpart to a GW event is particularly challenging in a number of respects. First, the error boxes associated with a GW candidate are large, often 100 square degrees or more. Second, the magnitude and decay timescale of an EM counterpart are unknown, necessitating observations both at early and late times, from seconds to weeks after the trigger. Both of these imply the need to process large numbers of optical images. In addition, the *a priori* unknown nature of lightcurve requires the ability to locate transients with lightcurves of different shapes. Finally, since there has not been a confirmed detection of a GW we will need to be able to assign a high statistical confidence in any putative EM counterpart, which likely will require repeating the analysis for many times not associated with GW candidates. All of these factors point to the need to automate the EM image analysis, to allow large-scale processing and quantitative characterisation of the pipeline.

The ROTSE image processing code originally took a series of commands from within IDL, as well as widget based commands to give a list of potential candidates. The automated pipeline is now able to run off one command with the end product not only being a list of unique candidates, but also all the lightcurve information for each potential candidate the pipeline identifies. In addition, multiple sets of images from multiple GW candidates are able to run simultaneously, and therefore can be submitted to condor‖. We have also removed the need for an IDL licence, as the pipeline can be run through IDL Virtual Machine¶. The essence of the pipeline has not been changed at all. The original pipeline and the new automated pipeline produce the same list of candidates.

5. Future Steps

Currently in development is a post processing code. This code will run after the automated code, and makes use of the already available lightcurve information for each

‡ http://simbad.u-strasbg.fr/simbad/
§ http://scully.cfa.harvard.edu/cgi-bin/checkmp.cgi
‖ http://research.cs.wisc.edu/condor/
¶ http://www.exelisvis.com/language/en-us/productsservices/idl.aspx

potential candidate the pipeline identifies. The code will apply a series of pass/fail tests to each potential candidate. At present these cuts are anticipated to test whether the candidate appears on more than one night, whether its coordinates overlap with a minor planet or known variable source (by querying the SIMBAD catalogue), as well as test to see if the lightcurve of the potential candidate has an overall negative gradient. We would also like to see if a potential candidate overlaps with a known galaxy (by making use of the gravitational wave galaxy catalogue [10]) and whether its lightcurve is similar to target theoretical lightcurves [11, 12, 13].

In addition, an injection code is being written which will inject decaying transients in to a sequence of ROTSE images. These injected transients will be modelled on objects which are already present in the image, and the code will alter its magnitude either in a random fashion or in a way which is consistent with theoretical lightcurves such as those previously suggested. With these altered images we will be able to test the efficiency of the image processing pipeline as well as obtain an independent measure of the limiting magnitude. For more details of this injection code please see the paper in these proceedings by White *et al.*.

Finally we will use images from the ROTSE archive as 'background' images as well as process the images the ROTSE telescopes took specifically in response to a potential GW event. This will allow us to estimate the 'false-alarm' rate (the rate of transients not associated with a GW candidate) to assess the statistical significance of any potential EM counterpart.

6. Concluding remarks

The automation of the ROTSE pipeline and further cuts which are being implemented will be essential to quantify the false alarm rate which in turn will be vital in adding confidence to a GW detection. In the next few months these steps will be implemented and the images analysed. These first efforts to detect EM counterparts of GW events will pave the way for the advanced detector era, where the detectors will have a ten-fold increase in sensitivity and GW detections are expected to be a regular occurrence [14].

This document has been assigned LIGO Laboratory document number LIGO-P1100188-v8.

7. References

[1] B. Abbott et al. LIGO: The Laser Interferometer Gravitational-Wave Observatory. *Rept. Prog. Phys.*, 72:076901, 2009.
[2] F. Acernese et al. Status of VIRGO. *Class. Quant. Grav.*, 25:114045, 2008.
[3] J. Kanner, T.L. Huard, S. Marka, D.C. Murphy, J. Piscionere, et al. LOOC UP: Locating and observing optical counterparts to gravitational wave bursts. *Class.Quant.Grav.*, 25:184034, 2008.

[4] The LIGO Scientific Collaboration, Virgo Collaboration, M. Boer, R. Fender, N. Gehrels, A. Klotz, E.O. Ofek, M. Smith, M. Sokolowski, B.W. Stappers, I. Steele, J. Swinbank, and R.A.M.J. Wijers. Implementation and testing of the first prompt search for gravitational wave transients with electromagnetic counterparts. *A&A In Press, arXiv:1109.3498*, 2011.

[5] R. Kehoe, C. Akerlof, R. Balsano, S. Barthelmy, J. Bloch, et al. Prompt Optical Emission from Gamma-Ray Bursts. In M. Livio, N. Panagia, and K. Sahu, editors, *Supernovae and Gamma-Ray Bursts: The Greatest Explosions Since the Big Bang*, Space Telescope Science Institute Symposium Series (no.13), pages 47–66. Cambridge University Press, 2002.

[6] C.W. Akerlof, M.C.B. Ashley, D.E. Casperson, H.W. Epps, R.L. Kehoe, et al. The ROTSE-III robotic telescope system. *Publ.Astron.Soc.Pac.*, 115:132–140, 2003.

[7] D.A. Smith, C. Akerlof, M.C.B. Ashley, D. Casperson, G. Gisler, et al. The ROTSE-IIIa Telescope System. *AIP Conf.Proc.*, 662:514–516, 2003.

[8] F. Yuan and C.W. Akerlof. Astronomical Image Subtraction by Cross-Convolution. *Astrophys.J.*, 677:808–812, 2008.

[9] F. Yuan. PhD thesis: Chasing the Brightest Cosmic Explosions with ROTSE-III. *http://hdl.handle.net/2027.42/75873*, 2010.

[10] D.J. White, E.J. Daw, and V.S. Dhillon. A List of Galaxies for Gravitational Wave Searches. *Class.Quant.Grav.*, 28:085016, 2011.

[11] B. D. Metzger et al. Electromagnetic Counterparts of Compact Object Mergers Powered by the Radioactive Decay of R-process Nuclei. *Monthly Notices of the Royal Astronomical Society*, 406:2650–2662, 2010.

[12] D.A. Kann, S. Klose, B. Zhang, D. Malesani, E. Nakar, et al. The Afterglows of Swift-era Gamma-Ray Bursts. I. Comparing pre-Swift and Swift era Long/Soft (Type II) GRB Optical Afterglows. *Astrophys.J.*, 720:1513–1558, 2010.

[13] D.A. Kann, S. Klose, B. Zhang, A.C. Wilson, N.R. Butler, et al. The Afterglows of Swift-era Gamma-Ray Bursts II.: Type I GRB versus Type II GRB Optical Afterglows. *Astrophys.J.*, 734:96, 2011.

[14] J. Abadie et al. Predictions for the rates of compact binary coalescences observable by ground-based gravitational-wave detectors. *Class. Quant. Grav.*, 27(17):173001, 2010.

Search for gravitational waves associated with the InterPlanetary Network short gamma ray bursts

V. Predoi for LIGO Scientific Collaboration and Virgo Collaboration and Kevin Hurley for the IPN

Cardiff University, Physics and Astronomy, The Parade, Cardiff CF24 3AA, UK

E-mail: valeriu.predoi@astro.cf.ac.uk

Abstract. We outline the scientific motivation behind a search for gravitational waves associated with short gamma ray bursts detected by the InterPlanetary Network (IPN) during LIGO's fifth science run and Virgo's first science run. The InterPlanetary Network localisation of short gamma ray bursts is limited to extended error boxes of different shapes and sizes and a search on these error boxes poses a series of challenges for data analysis. We will discuss these challenges and outline the methods to optimise the search over these error boxes.

1. Introduction

Short hard gamma ray bursts (short GRBs or SGRB) are believed to be produced by mergers of either two neutron stars or a neutron star and a stellar mass black hole [1, 2, 3]. These events are ideal sources for strong gravitational wave emission [4, 5]. If an observation of both gamma rays and gravitational waves (GW) originating from the same event could be achieved, it will increase confidence and allow for better science output. It is thus of great importance to constantly monitor and record GRBs to allow a GW search to be performed around the burst times. Systematic analyses of GW data around burst times of short GRBs have been done in the past and the most recent publications from the LIGO-Virgo group contain results from the Swift-observed GRBs during LIGO's fifth science run (S5) and Virgo's first science run (VSR1) [6, 7]. This paper reports the methodology and motivation for a proposed GW search around the burst times of 20 additional GRB during S5/VSR1. These bursts were observed by the InterPlanetary Network (IPN) [8, 9], a group of satellites orbiting the Earth, Mercury and Mars and operating, among other equipment on board, gamma ray detectors. These bursts were detected between 2006 and 2007 and have been localized, in both time and sky location, such that a targeted GW search is possible.

The InterPlanetary Network [8, 10] employs several space missions and synthesizes data obtained from the detection of the same burst by different spacecraft equipped with gamma ray detectors. The IPN has been operating for three successive generations; presently the third IPN (IPN3) began its operation in November 1990. Currently the spacecraft gathering data are Konus-WIND, Suzaku, INTEGRAL, RHESSI, Swift, Fermi/GBM (in Earth orbit), MESSENGER (in Mercury orbit) and Mars Odyssey (in Mars orbit) [9]. When the duty cycles and effective fields of view of all the missions in the network are considered, the IPN is an all-time, isotropic GRB monitor.

In this paper, we first review the motivation for performing a joint GW-GRB search as an overview and then provide details of the IPN short GRBs that were identified during the LIGO-Virgo S5-VSR1 science runs. We discuss the necessary changes to the current ongoing analysis (for the S6/VSR2-3 science runs Swift and Fermi-observed GRBs, [11, 12]) that are needed to implement a search on GRBs identified by the IPN network which may be less well localized in both sky position and time than the corresponding bursts identified by the Swift satellite and used in previous analyses [6, 7].

2. Search for GW associated with short GRB

Gamma ray bursts fall into two commonly accepted groups depending on their duration and spectral hardness [1, 13]. Characteristic duration is quantitatively expressed by the T_{90} parameter, defined as the time interval over which 90% of the total background-subtracted counts are observed (the total counts emitted by the source, found from the counts in the source region minus the contribution from the background), with the interval starting when 5% of the total counts have been observed [14]. The majority of bursts, with softer spectrum and duration $T_{90} > 2$s are called long GRBs. They are thought to be produced during a core-collapse supernova event. Those with a harder spectrum and $T_{90} \leq 2$s are called short GRBs.

Short GRBs ($T_{90} \leq 2$s) are widely believed to be mergers of either neutron star-neutron star or neutron star-black hole binaries [1]. Such compact binary coalescences are believed to generate strong gravitational waves in the sensitive frequency band of Earth-based gravitational-wave detectors [15, 16]. Although binary coalescence is the favoured progenitor model for short GRBs, this has not yet been confirmed by means of a direct observation. Consequently, the detection of gravitational waves associated with a short GRB would provide evidence that the progenitor is indeed a coalescing compact binary and also provide information on the parameters of the binary.

The coalescence of a compact stellar-mass binary (either two neutron stars or a neutron star and a black hole companion) is the endpoint of its ~Gyr life revolving around the common centre of mass while constantly losing energy and angular momentum through emission of gravitational radiation. The binary merger will lead to the formation of a black hole [2, 17]. A transient hyper massive highly magnetized neutron star with a lifetime of a few ms [18, 19] might precede the formation of the final black hole. In the favoured model of short GRBs [18, 5, 20, 21], the gamma-ray emission is contingent on the formation of a massive (and highly magnetized) torus around the final black hole. The matter in the torus is accelerated to relativistic velocities leading to the formation of a collimated jet of electromagnetic radiation. Differences in velocities between layers in the jet account for the prompt gamma-ray emission. Full relativistic and magnetohydrodynamic numerical simulations have shown that the time difference between the binary merger and the jet formation can be between a few milliseconds up to a few seconds [2, 17, 20, 21].

Supposing gravitational waves produced just before merger travel at the same speed as the gamma rays, the speed of light in vacuum, and suppose the observer is situated within the cone of the jet, one would expect to observe the GW within a few seconds prior to the arrival of the gamma rays. As the GWs emitted during the pre-merger inspiral dominate the signal-to-noise-ratio (SNR) observable in current GW detector data and they can be accurately modelled using post-Newtonian approximations [22, 23], our search will be aimed at this phase only.

3. The short GRB sample to be analysed

We gathered the sample of GRBs to be analysed using data provided in the IPN online table available publicly at [9] and by cross-checking this with NASA's HEASARC online table at [24]. Since these GRBs are not reported in any of NASA's Gamma Ray Burst Circulars (GCN), a manual check on each burst had to be performed in order to confirm its characteristics, involving checking several IPN satellite homepages e.g. Suzaku at [25], INTEGRAL at [26], Swift at [27] and Konus-WIND at [28].

The details of the sky position were obtained by manually processing raw data from the IPN satellites for every GRB. In order to perform a search of the GW data for a given GRB, the sky position and time of the GRB must be determined. This information is obtained from knowledge of: the IPN satellites that detected the burst together with their absolute and relative sky positions (information needed for constructing the GRB error boxes), locations of all the spacecraft (used to obtain the blocking constraints to reduce the size of the GRB error boxes), the burst time of arrival at the satellites and at Earth (used to determine the time interval on which we will perform the GW search) and its error. Constructing the GRB error boxes, determining their sizes and choosing the right GW data for analysis are entirely contingent on these pieces of information.

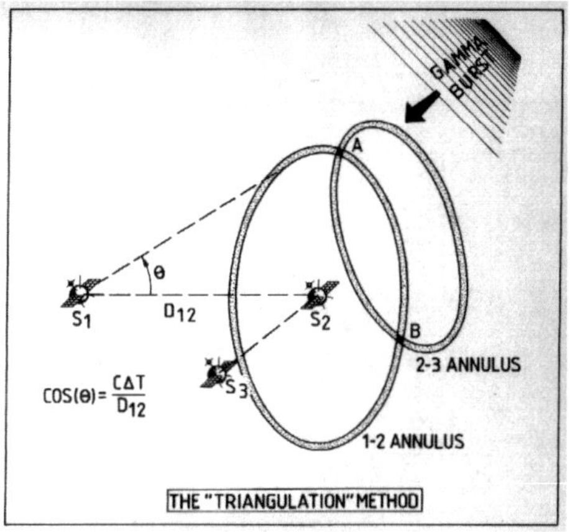

Figure 1. The IPN schematics: triangulating the position of a GRB using three IPN spacecraft (S1, S2 and S3). Using three satellites we obtain two IPN annuli that intersect to form two error boxes. Image reproduced from reference [9]

3.1. The IPN triangulation mechanism

The principle on which the IPN triangulation method is based uses the arrival time of the same burst at different spacecraft to determine the source sky location. Figure 1 illustrates how an IPN triangulation works. S1, S2 and S3 denote three spacecraft detecting the same GRB and θ is angle between the burst direction and the baseline between S1 and S2. Then, the burst will be detected by S2 at a time δT seconds earlier than S1

$$\cos(\theta) = c\delta T/D_{12} \tag{1}$$

where D_{12} is the distance between S1 and S2, and c the speed of light. Since D_{12} is known and δT is measured, θ is estimated. The solution to equation (1) is represented by a ring, or an *annulus*, whose width depends on the timing uncertainties ($\sigma(\delta T)$) and on the separation D_{12}. The farther apart the detectors, the more precise the localisation. The number of independent couples of detectors (and, therefore, the number of independent annuli) is two for the case of three spacecraft; thus, the burst direction must be inside one of the two intersection regions of the annuli. This intersection region is called the IPN error box of the GRB. Depending on the location of the IPN spacecraft and their timing errors this error box may vary in size from fractions of, to hundreds of square degrees. The annulus width is obtained by propagating the uncertainty on the time delay δT. Thus, from equation (1) it follows that

$$\sigma(\theta) = \frac{c\sigma(\delta T)}{D_{12}\sin(\theta)} \tag{2}$$

Equation (2) gives the uncertainty $\sigma(\theta)$ in the angle θ expressed in radians. The uncertainty on D_{12} has been neglected, as the main contribution comes from the timing uncertainties. One has to take into account not only the time resolution of each of the detectors, but also the difficulty of comparing different light curves, often derived from different energy bands. For example, when D_{12} spans the typical range: few 10^2 - few 10^3 light seconds, equation (2) shows that a minimum time resolution of the order of $10^{-2} - 10^{-3}$ s is required, to have $\sigma(\theta) \sim$ few arcminutes or less for a precise localisation.

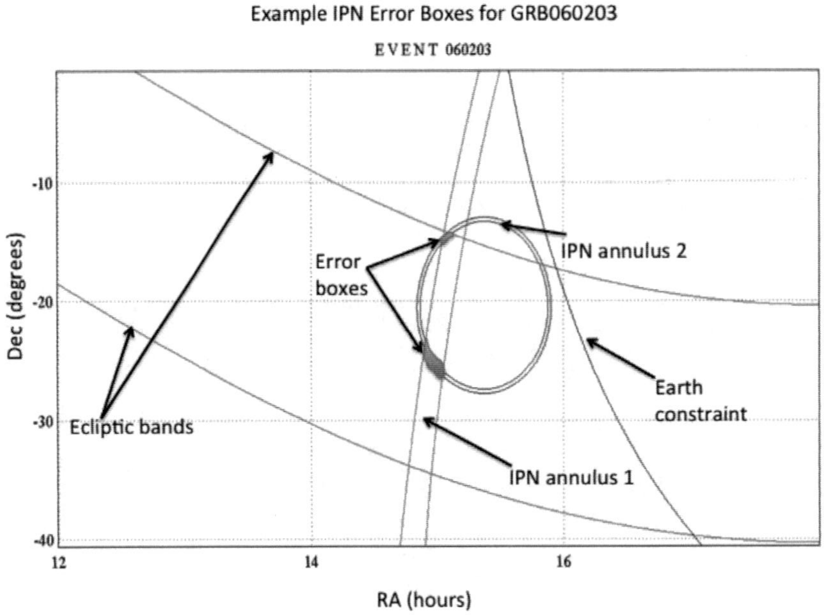

Figure 2. Error box construction: red lines correspond to the Konus-WIND ecliptic latitude bands, blue lines are the 3-σ IPN1-2 annulus and magenta lines are the 3-σ IPN1-3 annulus obtained by triangulation (in this case we have three IPN spacecraft observing the same burst). The error boxes are the solid regions bounded by the intersections of the IPN annuli and ecliptic bands.

3.2. The IPN GRB error box construction

The error boxes for the IPN GRBs are constructed from the intersection of the triangulated 3-σ IPN annuli and different field-of-view blocking constraints (if present). See figure 2 for an illustration example. When the constraints involve Konus-WIND ecliptic latitude bands, the error box that will be kept is located between the ecliptic bands. The reason for this is that the Konus-WIND spacecraft consists of two detectors, one facing the north ecliptic pole, and the other facing the south ecliptic pole. By comparing the count rates on these two detectors, the Konus team obtains an estimate of the ecliptic latitude of the burst. Typically, the band is 20-30 degrees wide, and it is good to a 90-95% confidence; systematic and statistical errors usually prevent it from being much more accurate than this value. In those cases where we have a single triangulation annulus, plus an ecliptic latitude band, the result is often two long, narrow error boxes, where the annulus intersects the band. Other intersections are possible, too, such as a single long, narrow error box if the IPN annulus grazes the ecliptic latitude band. Also, anything that overlaps a region where a planet blocks the view will be discarded.

3.3. Determining the GRB time of arrival

The Earth crossing time, also referred to as the time of arrival of the burst, is the time when the gamma ray signal would cross the centre of the Earth and is the reference time to be considered when constructing the time search window for gravitational waves. This time can only be estimated based on the burst arrival times at the different IPN satellites and based on their positions with respect to Earth. This way of choosing the burst time is prone to two types of uncertainties: the first is simply the fact that we may not have the time at Earth but at a satellite that is separated from Earth by a certain distance; the error is directly proportional to the distance to Earth where the closest IPN satellite is located at the time of the burst. The best estimate comes from any satellite that is not interplanetary (i.e. not on orbit around Mars or at a distant point from Earth). These "close" satellites are usually within a fraction of a light second distance from Earth. These uncertainties are small, less than 1 second for satellites orbiting Earth but may be of up to 5 seconds or more for interplanetary satellites. The second kind of uncertainty comes from the fact that the IPN consists of nine spacecraft with different energy ranges and different spectral sensitivities. So it is possible, in an extreme case, to have one spacecraft trigger on a GRB precursor, while the others trigger on the main burst, resulting in a time difference. This way, the trigger times can differ by tens of seconds or more. For short GRBs this effect is minimised due to the hard nature of their spectra and consequently reduces to under one second imprecisions. Altogether, time imprecisions are not more than 5s for the burst sample we will be analysing. The few GRBs that have a timing imprecision greater than 1 second are localised by distant satellites, usually MESSENGER/Mars Odyssey and/or Konus-WIND.

3.4. Gravitational wave detectors and data availability

Our aim is to perform a search for GW associated with the well or fairly well localised short GRBs detected by the IPN during LIGO's S5 run that lasted from 4 November 2005 to 30 September 2007, and Virgo's VSR1 that lasted from 18 May 2007 to 30 September 2007. The analysis will make use of data from four operational GW interferometers: the 4km and 2km co-located LIGO detectors at Hanford, WA (H1 and H2), the 4km detector at Livingston, LA (L1) and the 4km detector at Cascina, Italy (V1) [29, 30, 31]. The search will be using a method that coherently combines data from multiple operational GW detectors described in [12]; the method is being used in GW-GRB searches for S6/VSR2-3 [11] and is proven to be performing better than the method used for the S5/VSR1 search [12, 6, 7]; we require that all GRBs have data from at least two GW detectors. In order to perform the search, we require approximately forty minutes of data around the time of the GRB. The search pipeline identifies a foreground time representing the time interval around the actual burst when the signal is most likely to have been detected by the GW detectors. For the IPN GRBs that have a burst time of arrival error less than a second, based on the delay between the time of the arrival of the gamma ray and of the GW (described above) an interval of 5 seconds prior to the GRB time and 1 second following it will be used as foreground. This time interval will account for any timing imprecisions and has been previously used in the S5/VSR1 search for GW associated with the Swift GRBs [6].

For the IPN GRBs that have a time of arrival error larger than 1 second, the foreground will be extended on either side to account for this error. The data surrounding the time of the GRB are used for background estimation, in order to assess the data quality in the detectors around the time of the GRB.

The LIGO-Virgo detector network [29, 30, 31] is sensitive to a large fraction of the sky, albeit with relatively poor localization capability (on the order of tens or hundreds of square degrees) [32]. In much the same way as the IPN network, the GW network of detectors can reconstruct a sky position primarily through triangulation. A GW signal with an SNR large enough to be detected will have its sky location resolved by timing its arrival at the different GW detectors. The approximate timing resolution between two detectors in the network is ~0.5ms, giving a best angular resolution of around 2°. When performing a search for a GW signal associated with a GRB, the data from the various detectors in the network is coherently added with the appropriate time delays between detectors corresponding to a given sky location. Thus, if the IPN error box spans a large region of the sky, it is necessary to search several different sky positions for a GW signal.

The IPN GRB error boxes may differ in shape and size, with areas ranging from fractions to hundreds or even thousands of square degrees. The short GRBs we chose for analysis have either small or very narrow and elongated error boxes which will make the gravitational waves search much easier and less computing intensive. These error boxes are tiled by a set of points spaced by 1.8° in each direction. An empirical upper limit of 100 deg^2 square degrees was chosen for the maximum ΔA area of the error box, necessitating a maximum of about 30 independent sky points to search the error box. Searching over more than 30 points requires significant computational cost and much of the sensitivity improvement for the GRB triggered search over the all sky searches that have been performed would be lost.

The short GRBs for which GW data from only the H1 and H2 detectors is available are a special case. Since these two detectors are co-located and co-alligned, they would observe any gravitational wave signal at the same time, irrespective of the location of the signal. Hence, all the search points are degenerate, i.e. using just these two detectors would not allow us any spatial sensitivity since there is no time delay between these and triangulation is impossible. A limited size error box is not a requirement for these bursts any more and any short burst, no matter how extended the error box, as long as it has available data from H1 and H2, will be analysed. Furthermore, although all-time all-sky searches for GW during S5/VSR1 have been published [33, 34], these searches did not make use of the H1-H2 data in the way we will do. This was because, as the detectors are co-located, they share many common sources of noise. Consequently, the usual method of estimating background by introducing an artificial time shift between the detector data is not applicable. This renders an all time search difficult as we have no accurate way of measuring the noise background. For the previous searches only the few loudest H1-H2 coincident events were considered based on no background estimations and solely on the coincidence test. For a GRB search, we can make use of data away from the time of the GRB (without time shifting) to estimate the background, therefore giving us better confidence in assigning detection status to an event.

Depending on detector data availability and error box size we divide the GRB sample to be analysed into two groups: 14 short bursts with error boxes smaller than 100 deg^2 and available data from at least two sites and 6 short bursts with an arbitrarily sized error box area that have available data from H1H2 only. Six bursts have large error boxes and will be considered only for an archival look-up in the S5/VSR1 all-sky all-time data and three other bursts have already been analysed and published previously. The GRB sample is summarised in Table 1.

1. Short IPN GRBs with $\Delta A < 100$ deg^2 and data from two or more GW detector sites that will be analysed

GRB	IPN	GW	GRB Date	T_{90}(s)	ΔA(deg^2)	Δt(s)
060103	MO/I	H1H2L1	Jan 03 2006 08:42:17	2.00	4.60	<1
060107	K/MO/S	H1H2L1	Jan 07 2006 01:54:40	3.00	8.20	<1
060203	K/MO/H	H1H2L1	Feb 03 2006 07:28:58	0.40	0.80	<1
060415B	K/MO/S	H1H2L1	Apr 15 2006 18:14:44	0.44	0.20	<1
060522C	S/K/MO	H1H2L1	May 22 2006 10:10:19	1.10	0.40	<1
060708B	H/K/MO	H1H2L1	Jul 08 2006 04:30:38	1.00	0.06	<1
060930A	K/MO	H1H2L1	Sept 30 2006 02:30:11	1.00	2.40	<1
061006A	K/MO/S	H1H2L1	Oct 06 2006 08:43:38	1.60	3.20	<1
070321	I/K/MO/S	H1H2L1	Mar 21 2007 18:52:15	0.34	0.40	<1
070414	S/M	H1H2L1	Apr 14 2007 17:19:52	0.38	0.30	<1
070516	I/K/M/S	H1H2L1	May 16 2007 20:41:24	1.00	7.68	<1
070614	K/H	H1H2L1V1	Jun 14 2007 05:05:09	0.40	~20	<1
070915	Sw/I/M/K	H1H2L1V1	Sept 15 2007 08:34:48	0.50	0.10	<1
070927A	Sw/M/I	L1V1	Sept 27 2007 16:27:55	0.70	1.60	<1

2. Short IPN GRBs with data from H1H2-only that will be analysed

GRB	IPN	GW	GRB Date	T_{90}(s)	ΔA(deg^2)	Δt(s)
060317	K/S/I	H1H2	Mar 17 2006 11:17:39	0.70	9.24	<1
060601B	I/S	H1H2	Jun 01 2006 07:55:41	0.50	~600	<1
061001	I/Sw	H1H2	Oct 01 2006 21:14:28	1.00	~2000	<1
070129B	S/K	H1H2	Jan 29 2007 22:09:26	0.22	47.50	<1
070222	K/MO	H1H2	Feb 22 2007 07:31:55	1.00	0.45	5.0
070413	I/S	H1H2	Apr 13 2007 20:37:55	0.19	~350	<1

3. Short IPN GRBs with $\Delta A > 100$ deg^2 and data from two or more GW detector sites for archival data look-up only

GRB	IPN	GW	GRB Date	T_{90}(s)	ΔA(deg^2)	Δt(s)
060916	S/I	H1H2L1	Sept 16 2006 14:33:34	0.13	>3000	<1
061014	I/H	H1L1	Oct 14 2006 06:17:02	1.5	>3000	<1
061111B	K/Sw	H1H2L1	Nov 11 2006 10:54:27	0.6	~700	<1
070203	I/S	H1H2L1	Feb 03 2007 23:06:44	0.69	>2000	<1
070721C	K/I	H1H2V1	Jul 21 2007 14:24:09	1.00	495	<1
070910	K/S	H1H2L1V1	Sept 10 2007 17:33:29	0.38	>200	<1

4. Short IPN GRBs that have already been analysed and published

GRB	IPN	Reference
060427	K/MO/I/Sw	[6, 7]
060429A	S/K/MO	[6, 7]
070201	K/M/I	[35]

Table 1. The short S5/VSR1 IPN GRB sample - 14 with data from multiple non-H1H2-only GW detectors and well localised bursts (error box area ΔA <100 deg^2); 6 H1H2-only poorly localised bursts; 6 multiple GW detectors for poorly localised bursts; 3 bursts previously analysed and published. Δt represents the time of arrival error. The IPN satellites that observed the bursts: S - Suzaku, Sw - Swift, I - INTEGRAL, M - MESSENGER, MO - Mars Odyssey, K - Konus-WIND, H - HESSI (RHESSI).

4. The proposed search

For the IPN short GRBs, the data streams from the operational detectors will be combined coherently and searched using the methods described in [12]. The search for compact binary coalescing signals is done using match filtering [36] by correlating the detector data against theoretical waveforms that replicate the signal for a broad interval of binary parameters. In order to cope with the affects of non-stationary, transient noise "glitches" in the GW detectors' data, the pipeline uses a number of signal consistency tests, including the null stream, amplitude consistency and several χ^2 tests [37, 12, 38]. The pipeline can do the search on a single sky point as well as on multiple sky points. Simulated GW signals injected in the data streams are used to assess the sensitivity of the search. The two most important changes specific to the analysis of the IPN GRBs are the way we will perform the search over multiple sky points and the generation of simulations over the GRB error boxes.

We will generate a grid of search points to pave each GRB's error box in order to increase the chances of finding a signal over the entirety of the box. A fixed grid spacing (distance between adjacent grid points) of 1.8 degrees will be used for the search, motivated by the GW detectors' power of resolving the sky location, described above. Each error box is a 3-σ region but we will search and assign equal detection probability to each search point.

Aside from generating a search grid, we will also generate random positions in right ascension and declination within an IPN error box for simulations. To generate these positions, the error box is first paved with grid points with a denser 0.2 degree grid spacing. This is a finer spacing than the one which is used for the search to ensure efficient coverage of the error box by the simulated positions. It will also provide a verification that the 1.8 degree spacing of search points is adequate. Random positions are generated within a small square bin centred at each of these grid points and whose sides have lengths equal to the grid spacing. The relative number (or density) of positions generated for each grid point is weighted according to the estimated source position probability distribution. The probability distribution that will be used in the case of a single 3-σ IPN annulus is a one-dimensional Gaussian distribution centred at the central radius of the IPN annulus, and which has a sigma of 1/3 the width of the given annulus half-width. For error boxes which are formed by the intersection of two 3-σ IPN annuli, the probability distribution will be a two-dimensional Gaussian, with distances measured from the two central radii of the two annuli. This assures that there are proportionally more simulations for those positions with larger probabilities of having a signal. Then we will draw random locations for injections from this list of simulation points.

5. Discussion

We have presented the methodology for a search for gravitational waves around the times of short GRBs, detected by IPN during S5 and VSR1. This search has all the needed tools and will commence in a short time. The work in this paper was presented in the form of a poster at the 10th Amaldi Conference for Gravitational Waves organised in Cardiff, UK in July 2011. These types of searches are very promising for the future detection of gravitational waves and combined with the prospect of detecting other electromagnetic counterparts from GRBs, e.g. radio or optical pre- or afterglows (summarised in [39, 40]), may open the doors for true multi-messenger astronomy with gravitational waves.

Acknowledgements

The authors gratefully acknowledge the support of the United States National Science Foundation for the construction and operation of the LIGO Laboratory, the Science and Technology Facilities Council of the United Kingdom, the Max-Planck-Society, and the State of Niedersachsen/Germany for support of the construction and operation of the GEO600 detector, and the Italian Istituto Nazionale di Fisica Nucleare and the French Centre National de la Recherche Scientifique for the construction and operation of the Virgo detector. The authors also gratefully acknowledge the support of the research by these agencies and by the Australian Research Council, the International Science Linkages program of the Commonwealth of Australia, the Council of Scientific and Industrial Research of India, the Istituto Nazionale di Fisica Nucleare of Italy, the Spanish Ministerio de Educación y Ciencia, the Conselleria d'Economia Hisenda i Innovació of the Govern de les Illes Balears, the Foundation for Fundamental Research on Matter supported by the Netherlands Organisation for Scientific Research, the Polish Ministry of Science and Higher Education, the FOCUS Programme of Foundation for Polish Science, the Royal Society, the Scottish Funding Council, the Scottish Universities Physics Alliance, The National Aeronautics and Space Administration, the Carnegie Trust, the Leverhulme Trust, the David and Lucile Packard Foundation, the Research Corporation, and the Alfred P. Sloan Foundation. Kevin Hurley is grateful for IPN support from the following NASA grants: NNX10AR12G (Suzaku), NNX07AR71G (MESSENGER), NNX10AI23G (Swift), NNX09AR28G (INTEGRAL), and JPL 1282043 (Odyssey).

References

[1] Ehud Nakar. Short-hard gamma-ray bursts. *Phys. Rept.*, 442:166–236, 2007.

[2] Masaru Shibata and Keisuke Taniguchi. Merger of binary neutron stars to a black hole: Disk mass, short gamma-ray bursts, and quasinormal mode ringing. *Phys. Rev.*, D73:064027, 2006.

[3] Edo Berger. The Environments of Short-Duration Gamma-Ray Bursts and Implications for their Progenitors. *New Astron. Rev.*, 55:1–22, 2011.

[4] Theocharis A. Apostolatos, Curt Cutler, Gerald J. Sussman, and Kip S. Thorne. Spin induced orbital precession and its modulation of the gravitational wave forms from merging binaries. *Phys. Rev. D*, 49:6274–6297, 1994.

[5] Kenta Kiuchi, Yuichiro Sekiguchi, Masaru Shibata, and Keisuke Taniguchi. Exploring binary-neutron-star-merger scenario of short- gamma-ray bursts by gravitational-wave observation. *Phys. Rev. Lett.*, 104:141101, 2010.

[6] J. Abadie et al. Search for gravitational-wave inspiral signals associated with short Gamma-Ray Bursts during LIGO's fifth and Virgo's first science run. *Astrophys. J.*, 715:1453–1461, 2010.

[7] B. P. Abbott et al. Search for gravitational-wave bursts associated with gamma-ray bursts using data from LIGO Science Run 5 and Virgo Science Run 1. *Astrophys. J.*, 715:1438–1452, 2010.

[8] K. Hurley et al. The Current Performance of the Third Interplanetary Network. *AIP Conf. Proc.*, 662:473–476, 2003.

[9] http://www.ssl.berkeley.edu/ipn3/chronological.csv.

[10] K. Hurley et al. Precise Interplanetary Network Localization of the Bursting Pulsar GRO J1744-28. 1999.

[11] LIGO Scientific Collaboration and Virgo Collaboration. Search for gravitational waves associated with the Swift and Fermi detected gamma ray bursts during LIGO's S6 and Virgo's VSR2 and 3. *in preparation.*

[12] Ian W. Harry and Stephen Fairhurst. A targeted coherent search for gravitational waves from compact binary coalescences. *Phys. Rev.*, D83:084002, 2011.

[13] Y. P. Qinx, A. C. Gupta, J. H. Fan, C. Y. Su, and R. J. Lu. Duration distributions for different softness groups of gamma-ray bursts. *Sci. China Phys. Mech. Astron.*, 53:1375–1382, 2010.

[14] S. McBreen, F. Quilligan, B. McBreen, L. Hanlon, and D. Watson. Temporal properties of the short gamma-ray bursts. *Astronomy and Astrophysics*, 380:L31–L34, December 2001.

[15] Luc Blanchet, Bala R. Iyer, and Benoit Joguet. Gravitational waves from inspiralling compact binaries: Energy flux to third post-Newtonian order. *Phys. Rev.*, D65:064005, 2002.

[16] L. Blanchet and T. Damour. Post-newtonian generation of gravitational waves. *Annales Inst. H. Poincaré Phys. Théor.*, 50:377–408, 1989.

[17] Masaru Shibata and Keisuke Taniguchi. Merger of black hole and neutron star in general relativity: Tidal disruption, torus mass, and gravitational waves. *Phys. Rev.*, D77:084015, 2008.

[18] Masaru Shibata, Matthew D. Duez, Yuk Tung Liu, Stuart L. Shapiro, and Branson C. Stephens. Magnetized hypermassive neutron star collapse: a central engine for short gamma-ray bursts. *Phys. Rev. Lett.*, 96:031102, 2006.

[19] Matthew D. Duez, Yuk Tung Liu, Stuart L. Shapiro, Masaru Shibata, and Branson C. Stephens. Collapse of magnetized hypermassive neutron stars in general relativity. *Phys. Rev. Lett.*, 96:031101, 2006.

[20] Luciano Rezzolla et al. The missing link: Merging neutron stars naturally produce jet-like structures and can power short Gamma-Ray Bursts. *Astrophys. J.*, 732:L6, 2011.

[21] R. Oechslin and Hans-Thomas Janka. Torus Formation in Neutron Star Mergers and Well-Localized Short Gamma-Ray Bursts. *Mon. Not. Roy. Astron. Soc.*, 368:1489–1499, 2006.

[22] K G Arun, Luc Blanchet, Bala R. Iyer, and Moh'd S S. Qusailah. The 2.5pn gravitational wave polarisations from inspiralling compact binaries in circular orbits. *Class. Quantum Grav.*, 21:3771, 2004. Erratum-ibid. **22**, 3115 (2005).

[23] K G Arun, B R Iyer, M S S Qusailah, and B S Sathyaprakash. Testing post-newtonian theory with gravitational wave observations. *Class. Quantum Grav.*, 23:L37, 2006.

[24] http://heasarc.gsfc.nasa.gov/W3Browse/all/ipngrb.html.

[25] http://www.astro.isas.ac.jp/suzaku/HXD WAM/WAM-GRB/.

[26] http://www.isdc.unige.ch/integral/science/grb.

[27] http://swift.gsfc.nasa.gov/docs/swift/swiftsc.html.

[28] http://lheawww.gsfc.nasa.gov/docs/gamcosray/legr/bacodine/konus_grbs.html.

[29] B. Abbott et al. LIGO: The Laser Interferometer Gravitational-Wave Observatory. *Rept. Prog. Phys.*, 72:076901, 2009.

[30] J. Abadie et al. Calibration of the LIGO Gravitational Wave Detectors in the Fifth Science Run. *Nucl. Instrum. Meth.*, A624:223–240, 2010.

[31] Virgo Collaboration. Virgo status. *Classical and Quantum Gravity*, 25(18):184001, 2008.

[32] Stephen Fairhurst. Source localization with an advanced gravitational wave detector network. *Class. Quant. Grav.*, 28:105021, 2011.

[33] J. Abadie et al. Search for Gravitational Waves from Compact Binary Coalescence in LIGO and Virgo Data from S5 and VSR1. *Phys. Rev.*, D82:102001, 2010.

[34] B. P. Abbott et al. Search for Gravitational Waves from Low Mass Binary Coalescences in the First Year of LIGO's S5 Data. *Phys. Rev.*, D79:122001, 2009.

[35] B. Abbott et al. Implications for the Origin of GRB 070201 from LIGO Observations. *Astrophys. J.*, 681:1419–1428, 2008.

[36] Benjamin J. Owen and B. S. Sathyaprakash. Matched filtering of gravitational waves from inspiraling compact binaries: Computational cost and template placement. *Phys. Rev.*, D60:022002, 1999.

[37] Bruce Allen. A chi**2 time-frequency discriminator for gravitational wave detection. *Phys. Rev.*, D71:062001, 2005.

[38] Chad. Hanna. *Searching for gravitational waves from binary systems in non-stationary data.* PhD thesis, Louisiana State University, 2008.

[39] V. Predoi et al. Prospects for joint radio telescope and gravitational wave searches for astrophysical transients. *Class. Quant. Grav.*, 27:084018, 2010.

[40] D. M. Coward et al. Toward an optimal search strategy of optical and gravitational wave emissions from binary neutron star coalescence. *Mon. Not. Roy. Astron. Soc.*, 415:L26, 2011.

Origin and Implications of high eccentricities in massive black hole binaries at sub-pc scales

C. Roedig, A. Sesana

Max-Planck-Institut für Gravitationsphysik, Albert Einstein Institut, Am Mühlenberg 1, 14476 Golm, Germany

E-mail: croedig@aei.mpg.de

Abstract. We outline the eccentricity evolution of sub-parsec massive black hole binaries (MBHBs) forming in galaxy mergers. In both stellar and gaseous environments, MBHBs are expected to grow large orbital eccentricities before they enter the gravitational wave (GW) observational domain. We re–visit the predicted eccentricities detectable by space based laser interferometers (as the proposed ELISA/NGO) for both environments. Close to coalescence, many MBHBs will still maintain detectable eccentricities, spanning a broad range from $< 10^{-5}$ up to $\lesssim 0.5$. Stellar and gas driven dynamics lead to distinct distributions, with the latter favoring larger eccentricities. At larger binary separations, when emitted GWs will be observed by pulsar timing arrays (PTAs), the expected eccentricities are usually quite large, in the range $0.01 - 0.7$, which poses an important issue for signal modelling and detection algorithms. In this window, large eccentricities also have implications on proposed electromagnetic counterparts to the GW signal, which we briefly review.

1. Introduction

Multimessenger astronomy with massive black hole (MBH) binaries (MBHBs) has been discussed frequently in recent literature (e.g. [1–8]) as a unique way of fully understanding the mutual interaction of these systems with their environment. The general picture of MBHBs forming after a galaxy merger has been given in [9], but observational evidence for sub–parsec MBHBs is still lacking (see [10] for a review). Numerical studies down to parsec scales [11–14] have shown that the two MBHs continue to lose orbital energy and angular momentum under the large-scale action of gas/star-dynamical friction, and end up forming a Keplerian binary. At this point, the gaseous and stellar environment furthers the inspiral by either three-body scattering of individual stars and/or the interaction of the binary with a circumbinary gaseous disc (e.g. [15, 16]). At this late evolutionary stage, excitation of e was already noticed and studied in both environments (see, e.g., [2, 17]).

A non zero source eccentricity has important implications for gravitational wave (GW) observations: on one hand, eccentricity has an impact on source detection and parameter estimation; on the other hand, measured eccentricities carry valuable information about the dynamical evolution of the detected systems. In gas rich environments, the surviving detectable eccentricity in the LISA band ($10^{-4} - 10^{-1}$Hz) was addressed in [6], stating that $e_{\rm LISA} \sim 10^{-3} - 10^{-1}$ depending on the MBHB parameters. However, in view of the recent re–design of the LISA mission, estimates have to be re-adapted accordingly. The New Gravitational Observatory (ELISA/NGO)[1] will see a shorter portion of the binary inspiral, with inevitable consequences for parameter estimation, and for residual eccentricity detectability. Pulsar timing arrays (PTAs) [18] are expected to detect the first GWs in the low-frequency band by the end of this decade. Given their frequency window ($10^{-9} - 10^{-7}$Hz), they will be sensitive to very massive ($> 10^8\,M_\odot$) systems only , whose chirping is still marginal. This implies at the same time that observed MBHBs might still be coupled to their environment. We assess here the expected eccentricity distribution for these heavy sources in the PTA band. We then focus on the possibility of having observable PTA systems that are to some extent still coupled to either their stellar or gaseous environment, and assess the plausibility of detecting possible EM counterparts with current and up–coming missions, such as eROSITA [19], MAXI [20] and ATHENA[2].

The paper is structured as follows: First we describe the mechanisms of eccentricity growth in stellar environments in Sec. 2.1 and gaseous discs in Sec. 2.2. We then calculate the residual eccentricity distributions in the ELISA/NGO band in Sec. 3.1 and comment on the expected eccentricity population in the PTA band in Sec. 3.2. We review the implications this has on multimessenger signals in Sec. 3.2 and give our conclusions in Sec. 4.

2. Mechanisms of eccentricity growth:

In the following we discuss the main physical mechanisms driving the MBHB eccentricity evolution in the dense astrophysical environment of a post-merger galaxy. We consider a MBHB with total mass $M = M_1 + M_2$ ($M_2 < M_1$ are the individual MBH masses), semimajor axis a and eccentricity e. Throughout the manuscript, quantities without subscripts (such as M, a, e, E, L) always refer to the MBHB, whereas subscripts '1' and '2' refer to the individual MBHs.

2.1. Stellar environment

In dense stellar environments, the MBHB evolution is basically driven by three body interactions with individual stars [21–23]. To get a sense of the general picture, it is useful to consider the interactions between the stars and the binary one by one, as uncorrelated events, even though secular effects may also play a role (see, e.g., [24]).

At the moment of pairing at sub-pc separations, the MBHB is expected to be surrounded by a dense stellar distribution of stars. If a is the MBHB semimajor axis, three body interactions with stars approaching the binary center of mass within a result in superelastic scattering, usually followed by the ejection of the stars. These stars carry away energy and angular momentum, causing the binary to shrink. The efficiency of this mechanism depends on several environmental factors, such as the details of the stellar distribution and the effectiveness of relaxation mechanisms. We are here interested in the eccentricity evolution.

Assessing the eccentricity evolution is not straightforward, because it depends on a combination of energy (E) and angular momentum (L) exchange during the scattering: $\Delta e \propto c_i \Delta E + c_j \Delta L$, where the two c coefficients depend on the property of the MBHB. Let us consider, for simplicity, a bound star orbiting the MBHB at a given semimajor axis a_*. The orbital energy that can be extracted in the scattering is $\propto a_*^{-1}$, whereas the angular momentum of the orbit is proportional to $\sqrt{a_*}$. It follows that $\Delta e \propto -c_i a_*^{-1} + c_j \sqrt{a_*}$ [25]. If a_* is small enough, then the first term dominates and the binary circularizes; vice versa, if a_* is large enough, then the second term dominates, and the binary eccentricity grows. This behavior has to be expected, since stars with large a_* have large angular momentum but just little energy, and the opposite is true for stars with small a_*. It turns out that the transition point is at $a_* \approx a$, i.e., stars within the binary orbit on average promote circularization, whereas stars outside the binary orbit make

[1] `https://lisa-light.aei.mpg.de/bin/view/`
[2] `http://www.mpe.mpg.de/athena/home.php?lang=en`

the binary grow eccentric. Heuristically this can be interpreted as follows. Let us consider a MBHB with $M_2 \ll M_1$ and a given non-zero eccentricity. Such a binary spends most of its period near its apocenter, so in the case $a_* > a$ the probability of a close star-binary encounter (and subsequent star ejection) is maximal there. In the unequal mass case, the star is ejected following a series of close encounters with M_2. In such interactions the star gets a small 'kick' and the velocity of M_2 decreases. As this velocity is perpendicular to the MBH position vector close to the apocenter, the secondary is forced on a more radial orbit. On the contrary, a star with $a_* < a$ "feels" the secondary hole when this approaches the primary at a distance $< a$. At that point, the interaction with the star is unlikely to occur close to the pericenter of the MBHB orbit (because the time spent by the binary at the pericenter is very small), and typically also extracts a large radial component from the velocity of the secondary black hole, hence causing circularization. Considering the collective effect of the stellar distribution, the evolution of the orbital eccentricity is therefore governed by the relative weight of stars inside and outside a. The steeper is the stellar cusp, the less effective is the eccentricity growth. As a general trend, quasi circular-equal mass MBHBs experience just a mild eccentricity growth, while systems which are already eccentric at the moment of pairing, or with mass ratio significantly lower than 1, can evolve up to $e > 0.9$ [17]. Such a trend has been observed in several numerical studies [11, 26–28].

The above argument is strictly valid for isotropic stellar distributions. The evolution of the binary eccentricity can be extremely different for non isotropic systems. For example, [29] demonstrated that in a rotating stellar system, the eccentricity evolution of unequal MBHBs is dramatically affected by the level of co/counter rotation of the stellar distribution with respect to the binary, with corotating distributions promoting circularization rather than eccentricity growth. This is because, as a simple consequence of angular momentum conservation during the ejection process, stars counterrotating with the binary tend to extract a lot of angular momentum from the MBHB, causing the eccentricity growth, whereas corotating stars do not. Nonetheless, most of the simulations involving rotating bulges [30, 31], or merging systems [32, 33] find quite eccentric binaries at the moment of pairing (ranging from 0.4 to 0.8), and the subsequent evolution leads to a general eccentricity growth, in good agreement of what predicted for an isotropic stellar distribution. This may be because the binary evolution is mostly driven by loss cone refilling of unbound stars on almost radial orbits, with negligible initial angular momentum. On the other hand, [34] find milder eccentricity evolutions, along the lines of what is expected for a rotating stellar remnant. The eccentricity evolution of MBHBs in stellar environments is still a largely open issue, but a significant eccentricity seem to be a general outcome of the models.

2.2. Gaseous environments

The study of the secular evolution of a MBHB immersed in a gaseous disc is essentially similar to type-II planet migration [35]. However, when self-gravity of the disc and the accretion of matter onto the MBHs is included, the clear predictions of resonances and decay times coming from perturbation theory tend to be smeared out. In this paragraph, we will sketch the two effects that lead to the excitation and the saturation of the eccentricity evolution:

- excitation of eccentricity via asymmetric torques from the circumbinary disc
- saturation of eccentricity growth due to accumulated gas around the black holes

When dynamical friction becomes inefficient, the subsequent MBHB evolution relies on transfer and redistribution of energy and angular momentum from the binary to the so–called circumbinary disc. This usually occurs at sub-parsec binary orbital separations, and goes hand in hand with the appearance of a so–called *gap*. A *gap* is a hollow, low–density region which is found to have a radial extent δ of about twice the semi-major axis a of the binary (e.g. [6, 36, 37]) which is attributed to the excitation of outer Lindblad resonances (OLR) with the most prominent

one at $\delta_{m=1} = 2^{2/3}a \sim 1.59\,a$ [3]. However, gas will still continue to leak from the inner edge of the disc to the black holes and form *mini-discs* inside the respective Roche-lobes [6, 7, 38, 39]. As first shown by [2] and confirmed by [37], a binary inside a gap will not remain circular and neither will the inner edge of the disc (see also [36, 40, 41]).
As showed in [6] for a MBHB of total mass M and mass-ratio $q = M_2/M_1 = 1/3$, the eccentricity of the binary will continue to grow up to about $e \sim 0.6$ while the eccentricity of the disc stays small, $e_{\rm disc} < 0.15$, even for $e > 0.7$. The reason for this growth of e can be understood by comparing the angular velocity of the black holes to that of the fluid elements in the circumbinary disc. The movement of the MBHB inside the disc will induce overdensities close to the inner rim of the circumbinary disc, moving at lower angular speeds than the MBHs. If the MBHs are at apocenter, where they are slowest and thus remain the longest, they will experience a deceleration (negative torque) thus decreasing their associated angular momenta $L_{1,2}$ and increasing e. Since the disc is at the same time expanding radially to accommodate the larger maximum MBHB separation, the theoretical limit to this growth depends on the evolution of $\delta(e,t)$. Equating the average angular velocity of the inner rim of the disc to that of the binary at apocenter: $\langle\omega_{\rm fluid}\rangle = \omega_{\rm apo}$ yields the implicit relation for a limiting e:

$$\frac{(1+e)^3}{(1-e)} = \delta_a(e,t)^3, \qquad \delta_a = \delta/a \tag{1}$$

which shifts the question to determining the function $\delta(e,t)$. Assuming $\delta(e,t) = const = 2$, [6] found a limiting eccentricity of ≈ 0.6. However we expect $\delta(e,t)$ to generally be a growing function of e related to the strongest OLRs. A detailed discussion of the OLRs and their evolution will soon be found in [42]. In general, if δ grows with e, $L_{\rm fluid} < L_2$ is always true, and e continues to grow unless some other effect comes into play. When the binary is very eccentric, the Roche-lobes and the mini–discs attached to each individual MBH overlap at pericenter. Whenever this starts happening the secondary MBH will have to pass through the mini–disc of the primary and it will be slowed down at pericenter, decreasing e. This effect is thus strongly dependent on the thermodynamics of the gaseous disc and the mass contained within the mini–disc.

In the remainder of this section, we will elucidate the growth of e with numerical examples of Newtonian self-gravitating discs around a MBHB with $q = 1/3$, using a setup described in [6], the details of which will soon be found in [42]. The two examples use the same initial data and code, differing only in the thermodynamics treatment inside the gap. The *dry* run allows for an adiabatic evolution of the internal energy plus a β cooling inside the gap which effectively suppresses gas-inflows, whereas the *wet* run furthers the inflows by treating the gas isothermally when crossing the radius of $r_{\rm iso} = 1.73\,a$, thus allowing us to differentiate the effects of the gas in the gap. When plotting the torque density T exerted by the disc onto the MBHs in figure 1, we show the individual torques in the z-direction onto each MBH in the middle and right panels and the sum $T_z = M_1 T_{z_1} + M_2 T_{z_2}$ in the left panel. Since the binary is rotating counterclockwise, the negative portion of the torques always stems from the upstream region (yellow/red). Comparing the top and the bottom set of panels, we observe that altering the prescription for the thermodynamics inside the gap alters the torque density and the dynamical evolution of the entire disc. It is clear that the torque from the mini–discs are larger for the *wet* run (bottom panels), since the mass inside the mini–discs is higher and the gas is not allowed to cool (i.e. its internal energy remains constant) after crossing the threshold radius $r_{\rm iso}$. The torque density onto M_1 (middle panel) is more or less symmetric in the azimuthal direction, whereas M_2 clearly experiences larger torques from the part of the disc that is close to it. What

[3] The excitation of the $m = 1$ OLR is suppressed in equal mass binaries, nevertheless similar gap-sizes are also found for exactly equal mass MBHBs [36]

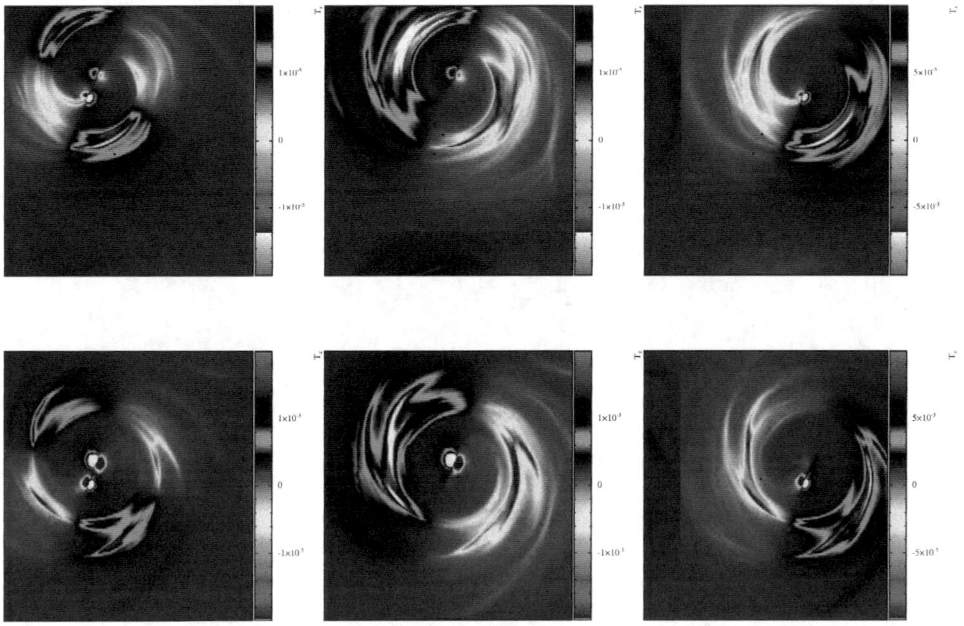

Figure 1. Torque density in the z-direction for two initially circular MBHB runs reaching $e \approx 0.1$ after 90 orbits. Top (bottom) panels: *dry* (*wet*) thermodynamic prescription for the cavity. In each line, the three panels are, from left to right: the total torque density on the binary $T_z = M_1 T_{z_1} + M_2 T_{z_2}$, torque onto the primary MBH T_{z_1}, torque onto the secondary MBH T_{z_2}. Figure made using SPLASH[43]

is moreover notable, is the fact that in the left panel, where the total torque is shown to have a quadrupolar structure, the contributions from the mini–discs do not have exactly the same orientation as the contributions from the disc. However, the growth of e is very similar for *wet* and *dry* runs, which leads us to conclude that the *excitation* of e stems mostly from the circumbinary disc and is robust under different numerical prescriptions. The precise value of $e_{\rm limit}$, however, depends on the density and angular momentum of the mini–discs, but is expected to be $e_{\rm limit} \in [0.6, 0.8]$. Further studies of the precise dependence on the thermodynamics are ongoing [42].

3. Implications for ELISA/NGO and PTA sources

Low-frequency GWs are expected to be rich in astrophysical information if the source parameter analysis is sufficiently accurate to differentiate between various proposed astrophysical scenarios. Great hopes lie in a coincident detection of EM counterparts to complement the GW information with an independent measurement of the source redshift and a connection with the host galaxy, opening new scenarios for precision cosmology (see, e.g, [1]). The proposed ELISA/NGO instrument is designed for $10^3 - 10^7\,M_\odot$ MBHBs chirping and merging in the $10^{-4} - 10^{-1}$ Hz band, whereas PTAs will be sensitive to GW frequencies in the range $10^{-9} - 10^{-6}$Hz with successful detections depending on the sensitivity of the radio instruments, the number of available stable millisecond pulsars and the observation time.

To guide the development of effective detection pipelines, it is reasonable to review the assumptions of GW sources emitting purely sinusoidal, monochromatic signals. To estimate the impact of the eccentricity growth via the mechanisms described above, we need to understand at which binary separation they are effective. We therefore compare the star/gas driven shrinking timescales to the GW energy loss timescale. The GW timescale is given by:

$$t_{\rm GW} = 7.84 \times 10^7\,{\rm yr}\, M_8 q_{\rm s}^{-1} a_3^4 F(e)^{-1}, \quad F(e) = (1-e^2)^{-7/2}\left(1 + \frac{73}{24}e^2 + \frac{37}{96}e^4\right) \tag{2}$$

where $q_{\rm s} = 4q/(1+q^2)$ is the symmetric binary mass ratio, $M_8 = M/10^8\,{\rm M}_\odot$ and a_3 is the binary separation in units of $10^3 R_S$ ($R_S = 2GM/c^2$).

In gas rich environments, assuming a β–disc model [44] for the circumbinary disc, the migration timescale is [7]

$$t_m = 2.09 \times 10^6\,{\rm yr}\,\alpha_{0.3}^{-1/2}\left(\frac{\dot m_{0.3}}{\epsilon_{0.1}}\right)^{-5/8} M_8^{3/4} q_{\rm s}^{3/8} \delta_a(e)^{7/8} a_3^{7/8}. \tag{3}$$

Here we defined the viscosity parameter $\alpha_{0.3} = \alpha/0.3$, the accretion efficiency $\epsilon_{0.1} = \epsilon/0.1$ and the mass accretion rate $\dot m_{0.3} = \dot m/0.3$, where $\dot m = \dot M/\dot M_{\rm Edd}$ is the accretion rate normalized to the Eddington rate.

In a star dominated system, assuming a full loss cone [17] the MBHB hardening timescale is instead given by

$$t_h = 2.89 \times 10^6\,{\rm yr}\,\sigma_{100} M_8^{-1} \rho_5^{-1} a_3^{-1} H_{15}^{-1}, \tag{4}$$

where we introduced the stellar velocity dispersion $\sigma_{100} = \sigma/100$ Km s^{-1}, the stellar density at the binary influence radius (see [17], for details) $\rho_5 = \rho/10^5\,{\rm M}_\odot\,{\rm pc}^{-3}$ and the dimensionless hardening rate $H_{15} = H/15$. Note that the combined M_8^{-1} and a_3^{-1} scaling implies that t_h is independent on the MBHB mass.

By equating t_m and t_h to $t_{\rm gw}$, one gets the characteristic a at which GW emission takes over in the binary evolution. It turns out that binaries are still coupled to their environment down to a separation of few hundred Schwarzschild radii, where the most massive systems fall in the PTA frequency band. This has two direct consequences: (i) PTA sources are generally eccentric, with an eccentricity distribution directly determined by the shrinking mechanism, and (ii) significant residual eccentricity might be present close to coalescence, in the ELISA/NGO band.

In the following we apply selfconsistent models for the evolution of the MBHBs in gaseous and stellar environments to the expected population detectable by LISA/NGO and PTAs. In gas rich systems, we assume MBHBs to maintain an eccentricity of $e_{\rm limit} = 0.6$ as long as they are coupled to their environment and then to circularize under the action of GW emission after decoupling. In star dominated systems, we use the self–contained model for the evolution of the MBHB under the effect of star ejection and GW radiation as outlined in [17].

3.1. Eccentricity population of ELISA/NGO sources

In view of the new sensitivity curve of the ELISA/NGO instrument, we re–visit here the question of the residual detectable eccentricity of a population of MBHBs evolving either in stellar or gaseous environments. We take the population of merging MBHBs predicted by a standard MBH cosmic evolution model where MBHs grow by merger and accretion starting from a population of light seed MBHs, thought to be remnants of the PopIII stars [45]. To be specific, we took the model "LE" from [46]. We stress here that in this model, MBHBs were assumed to instantly coalesce after a dynamical friction timescale from the pairing of the two galaxies hosting the two MBHs: none of the dynamical evolution models is implemented selfconsistently in deriving the MBHB population. We therefore have a single MBHB population to which we apply different dynamical evolution models to determine the residual eccentricity. For each

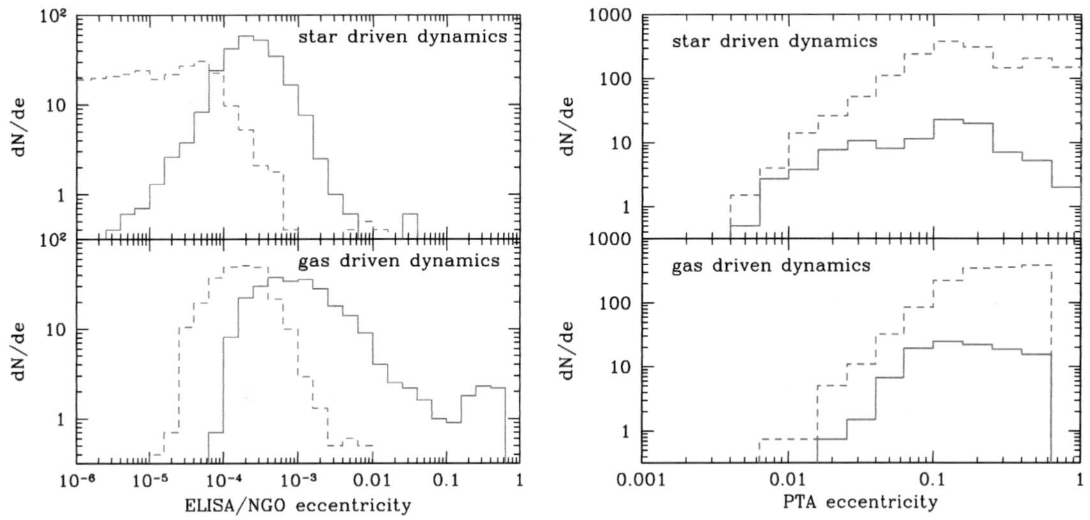

Figure 2. Eccentricity population of MBHBs detectable by ELISA/NGO and PTAs, expected in stellar and gaseous environments. Left panel: The *solid* histograms represent the efficient models whereas the *dashed* histograms are for the inefficient models. Right panel: *solid* histograms include all sources producing timing residuals above 3 ns, *dashed* histograms include all sources producing residual above 10 ns.

mechanism (gas/star) we consider two scenarios (efficient/inefficient), to give an idea of the expected eccentricity range. The models are the following

(i) gas-efficient: $\alpha = 0.3$, $\dot{m} = 1$. The migration timescale is maximized for this high values of the disc parameters, and the decoupling occurs in the very late stage of the MBHB evolution;

(ii) gas-inefficient: $\alpha = 0.1$, $\dot{m} = 0.1$. Decoupling occurs at much larger (factor 3-to-5) separations, and MBHBs have much more room to circularize;

(iii) stars-efficient: $p(e_i) \propto e$. Initial eccentricities are taken according to a thermal probability density function [33];

(iv) stars-inefficient: $e_i = 0$. Binaries are initially circular, a condition that minimizes the effectiveness of eccentricity growth.

Results are shown in the left panel of figure 2, where we plot the residual eccentricity of all sources detected with SNR>8, when they enter in the ELISA/NGO band[4]. In general, MBHBs evolving in gaseous environments are expected to retain a larger eccentricity, with a distribution tail possibly extending to $e \sim 0.5$. Residual eccentricities are mostly in the $10^{-4} - 10^{-2}$ range, and are generally lower for star driven systems. Note that the distribution predicted by the star-efficient model just overlaps to the one predicted by the gas-inefficient one. Therefore, apart from those two extreme cases, the distributions predicted by the two families of models are generally distinct, especially at the high e tail. A sufficient source detection statistics should allow discrimination between gas and star driven dynamics, providing valuable information of the MBHB environment.

[4] The band entrance is defined as the frequency where the strain of the source h_s squared equals the one-sided noise spectral density of the detector $S(f)$.

3.2. Eccentricity population of PTA sources

To investigate PTA sources, we instead use MBHB population models from [47]. Here we just want to highlight the fact that PTA sources might be very eccentric, and adequate signal processing algorithms need to be developed. We therefore use a single representative model for each mechanism. For the gas driven evolution we use $\alpha = 0.3$, $\dot{m} = 0.3$; whereas for the star driven model we use $e_i = 0.1$ for all the binaries. In the right panel of figure 2, we plot the eccentricity distribution of all sources contributing to the signal at timing residual level above 10ns or 3ns, which is expected to be in the reach of future PTA campaigns. The distributions are rather similar with more spread and higher extremal eccentricities for the stellar case. It is not clear, whether the two distributions are discernible; the figures, however, clearly suggest that a large portion of PTA signals would have fairly high eccentricities ($e > 0.1$).

Multimessenger signals. Concentrating on gaseous disc environments, [7] and [8] have, in a complementary way, outlined possible counterparts to resolvable PTA sources, which we will quickly review here. First, we will look at coupled systems, as done in [7], then conclude with a remark on how eccentricity impacts on decoupled systems as studied in [8].
If t_ν is the viscous time of the disc, then as long as $t_{GW} > t_\nu$, the disc can dissipate inwards fast enough to follow the MBHB and during this phase the MBHB is expected to be highly eccentric. In [6], it has been shown that the strength of a periodicity in the light curve related to the MBHB orbital frequency depends on the eccentricity of the orbit and the angular momentum distribution of the inspiralling gas. First, it is important to note that for circular or mildly eccentric orbits the periodicity in the gas inflow rate is much weaker than for high (e >0.4) eccentric systems. The number and quality of observations needed to identify the orbital periodicity is challenging for upcoming all-sky surveys, even assuming highly modulated inflow rates for $e \sim 0.6$ systems. This is simply because of the high number of pointings per orbit needed (which will be the main limitation of eROSITA) and the required flux limit of the instrument (which limits ongoing surveys like MAXI). Detecting a PTA counterpart in a gaseous environment will thus be likely biased towards eccentric systems that accrete the instreaming gas on time scales comparable to the binary period. If, however, in the above scenario the time scale for the accretion of the mini–disc is long compared to the binary period, the instreaming gas periodicity might be diluted in the process of angular momentum redistribution within the disk, and might not be reflected in the accretion luminosity onto the two MBHs. In this case, stable, broad double $K\alpha$ lines arising from the inner portion of the two mini–discs might be detectable by ATHENA [7], and could provide meaningful measurements of the two MBH spins. If, on the contrary, the gas is instantaneously accreted, there will be no detectable lines and the general contribution from of gap region to the emitted luminosity will likely be negligible; only if the spread in inflowing angular momentum is very large, colliding caustics can form small inviscid accretion discs [48, 49] making the MBHs variable, luminous X–Ray sources.

Looking at circular binaries, [8] have shown that gas can continue to leak towards the MBHB even after decoupling, finding surface densities as large as $\Sigma \sim 10^4\,\mathrm{g/cm^2}$ at $\sim 40\,GM/c^2$ for a total MBHB mass of $10^9\,\mathrm{M_\odot}$. While for eccentric systems the width of the gap is generally larger than for circular ones, their Green function approach is valid nonetheless, as long as the leaking gas is ripped out of the edge of the circumbinary disc on dominantly radial orbits, i.e. having a much lower angular momentum than the inner edge of the circumbinary disc. Their results (see, e.g. their figure 4) would not change much qualitatively, if they instead used a larger cavity to begin with and then adapted the inner boundary condition to a function $\delta_a(e(t))$ rather than using $2\,a(t)$. However, due to gas from the decoupled disc being required to diffuse over larger distances while the gas is also colder at the same time thus less diffusive, it is expected that the total amount of fossil gas is smaller the higher the eccentricity.

4. Conclusions

We have shown that the excitation of MBHB eccentricity in both star and gas dominated environments is a robust prediction of state of the art dynamical evolution models. In the gas dominated case, limiting eccentricity values $e_{\rm limit} \geq 0.6$ are expected following excitation by a cold dense circumbinary disk, whereas stellar dynamics can easily produce eccentricities higher than 0.9. As a consequence, PTA sources, many of which are at the evolutionary stage in which the environment still plays a role in their dynamical evolution, are expected to be significantly eccentric. Many binaries producing residuals at a 10ns level (in the reach of future PTA campaigns) have $e > 0.1$. The impact of such high eccentricities on the PTA signal should be appropriately taken into account when developing dedicated data analysis algorithms. For ELISA/NGO sources, which have long since decoupled from their environment, we have shown the distribution of the residual eccentricity expected for both environments. As the latter are disjoint at the high-eccentricity tail, enough GW observation statistics could distinguish among the different scenarios, providing valuable information on the environment of coalescing MBHBs. As far as multimessenger signals are concerned, we did not treat electromagnetic counterparts to ELISA/NGO events and focused on PTA observations. In this latter case, the electromagnetic signal is generally bright, due to the high mass and relatively small luminosity distance of the sources, making certain distinct features detectable by upcoming observatories. We have shown that counterpart identification is likely to be bias towards eccentric MBHB in periodicity searches, whereas, depending on the angular momentum of the instreaming material feeding the mini–discs, peculiar double $K\alpha$ lines might be identifiable by future hard X–ray deep spectroscopy.

Acknowledgments

The authors thank Takamitsu Tanaka for helpful conversations and Massimo Dotti for carefully reading the manuscript and giving detailed comments.

References

[1] Holz D E and Hughes S A 2005 *Astrophys. J.* **629** 15–22
[2] Armitage P J and Natarajan P 2005 *Astrophysical Journal* **634** 921–927
[3] Kocsis B, Frei Z, Haiman Z and Menou K 2006 *Astrophys. J.* **637** 27–37
[4] Phinney E S 2009 *Astronomy* **2010** 235
[5] Schnittman J D 2011 *Classical and Quantum Gravity* **28** 094021
[6] Roedig C, Dotti M, Sesana A, Cuadra J and Colpi M 2011 *Mon. Not. R. Astron. Soc* **415** 3033–3041
[7] Sesana A, Roedig C, Reynolds M T and Dotti M 2011 *Mon. Not. R. Astron. Soc.* (*Preprint* 1107.2927)
[8] Tanaka T, Menou K and Haiman Z 2011 *ArXiv e-prints 1107.2937* (*Preprint* 1107.2937)
[9] Begelman M C, Blandford R D and Rees M J 1980 *Nature* **287** 307–309
[10] Colpi M and Dotti M 2009 *(arXiv:0906.4339)* (*Preprint* 0906.4339)
[11] Milosavljević M and Merritt D 2001 *Astrophys. J.* **563** 34–62
[12] Escala A, Larson R B, Coppi P S and Mardones D 2005 *Astrophysical Journal* **630** 152–166
[13] Dotti M, Colpi M, Haardt F and Mayer L 2007 *Mon. Not. Roy. Astr. Soc.* **379** 956–962
[14] Dotti M, Ruszkowski M, Paredi L, Colpi M, Volonteri M and Haardt F 2009 *Mon. Not. Roy. Astr. Soc.* **396** 1640–1646
[15] Merritt D and Milosavljević M 2005 *Living Reviews in Relativity* **8** 8
[16] Armitage P J and Natarajan P 2002 *Astrophys. J. Lett.* **567** L9–L12

[17] Sesana A 2010 *Astrophys. J.* **719** 851–864

[18] Hobbs G e a 2010 *Classical and Quantum Gravity* **27** 084013 (*Preprint* 0911.5206)

[19] Predehl P e a 2010 eROSITA on SRG *Society of Photo-Optical Instrumentation Engineers (SPIE) Conference Series* (*Presented at the Society of Photo-Optical Instrumentation Engineers (SPIE) Conference* vol 7732) (*Preprint* 1001.2502)

[20] Matsuoka M e a 2009 *Publications of the ASJ* **61** 999– (*Preprint* 0906.0631)

[21] Mikkola S and Valtonen M J 1992 *Mon. Not. R. Astron. Soc.* **259** 115–120

[22] Quinlan G D 1996 *New Astron.* **1** 35–56

[23] Sesana A, Haardt F and Madau P 2006 *Astrophysical Journal* **651** 392–400

[24] Meiron Y and Laor A 2011 *ArXiv e-prints* (*Preprint* 1110.6491)

[25] Sesana A, Haardt F and Madau P 2008 *Astrophys. J.* **686** 432–447

[26] Merritt D, Mikkola S and Szell A 2007 *Astrophys. J.* **671** 53–72 (*Preprint* 0705.2745)

[27] Matsubayashi T, Makino J and Ebisuzaki T 2007 *Astrophysical Journal* **656** 879–896

[28] Amaro-Seoane P, Miller M C and Freitag M 2009 *Astrophys. J. Lett.* **692** L50–L53

[29] Sesana A, Gualandris A and Dotti M 2011 *Mon. Not. R. Astron. Soc.* **415** L35–L39

[30] Berczik P, Merritt D, Spurzem R and Bischof H P 2006 *Astrophys. J. Lett.* **642** L21–L24

[31] Berentzen I, Preto M, Berczik P, Merritt D and Spurzem R 2009 *Astrophys. J.* **695** 455–468

[32] Khan F M, Just A and Merritt D 2011 *Astrophys. J.* **732** 89

[33] Preto M, Berentzen I, Berczik P and Spurzem R 2011 *Astrophys. J. Lett.* **732** L26+

[34] Gualandris A and Merritt D 2011 *ArXiv e-prints* (*Preprint* 1107.4095)

[35] Goldreich P and Sari R 2003 *Astrophysical Journal* **585** 1024–1037

[36] MacFadyen A I and Milosavljević M 2008 *Astrophys. J.* **672** 83–93

[37] Cuadra J, Armitage P J, Alexander R D and Begelman M C 2009 *Mon. Not. Roy. Astr. Soc.* **393** 1423–1432

[38] Artymowicz P and Lubow S H 1996 *Astrophys. J. Letters* **467** 77

[39] Hayasaki K, Mineshige S and Sudou H 2007 *Publications of the Astronomical Society of Japan* **59** 427–441

[40] Lubow S H 1991 *Astrophysical Journal* **381** 259–267

[41] Papaloizou J C B, Nelson R P and Masset F 2001 *Astronomy and Astrophysics* **366** 263–275

[42] Roedig C, Sesana A, Dotti M and et al 2011 *In preparation*

[43] Price D J 2007 *Publ. Astron. Soc. Aust.* **24** 159–173

[44] Shakura N I and Sunyaev R A 1973 *Astronomy & Astrophysics* **24** 337–355

[45] Madau P and J R M 2001 *Astrophys.J* **551** L24

[46] Arun K G, Babak S, Berti E, Cornish N, Cutler C, Gair J, Hughes S A, Iyer B R, Lang R N, Mandel I, Porter E K, Sathyaprakash B S, Sinha S, Sintes A M, Trias M, Van Den Broeck C and Volonteri M 2009 *Classical and Quantum Gravity* **26** 094027 (*Preprint* 0811.1011)

[47] Kocsis B and Sesana A 2011 *Mon. Not. R. Astron. Soc.* **411** 1467–1479

[48] Beloborodov A M and Illarionov A F 2001 *Mon. Not. Roy. Astr. Soc.* **323** 167–176

[49] Zalamea I and Beloborodov A M 2009 *Mon. Not. Roy. Astr. Soc.* **398** 2005–2011

Estimating transient detection efficiency in electromagnetic follow up searches

Darren J. White[1] for the LIGO Scientific Collaboration and the Virgo Collaboration
[1]Department of Physics and Astronomy, University of Sheffield, Sheffield, UK S3 7RH
E-mail: d.white@sheffield.ac.uk

Abstract. During the most recent LIGO-Virgo science run (Dec 17 2009 to Jan 8 2010 and Sep 2 to Oct 20 2010) multi-messenger searches were performed using several partner telescopes. This resulted in large data sets with images covering several square degrees of the sky. Analysis of these images is currently underway using a variety of different tools. We present an overview of these efforts, in particular the development of new tools which enable us to establish the efficiency for transient images in the fields. This is critical in establishing the sensitivity of gravitational wave and electromagnetic multi-messenger searches to the astrophysical signals we expect to be associated with gravitational waves.

1. Introduction

During the summer of 2010, a joint science run between LIGO and Virgo allowed us to attempt an electromagnetic (EM) follow up campaign in order to image sky locations of possible gravitational wave (GW) sources in the hope of detecting any EM counterparts [1]. Low latency pipelines analysed GW data and produced triggers for human verification. Following data quality checks the most likely sky positions any viable triggers were sent to partner telescopes across the globe. Images taken for these triggers are currently being analysed using a variety of methods. In particular, images taken using the ROTSE telescopes and the Liverpool Telescope will be analysed using image subtraction [2], which allows us to detect down to faint magnitudes. In order to quantify the efficiency of our analysis method, it is necessary to inject transients into the images.

2. Follow-up images taken with ROTSE and the Liverpool Telescope

The ROTSE system [3] consists of 4 robotic telescopes situated at Los Alamos, New Mexico, USA, Coonabarabran, Australia, Mt. Gamsberg, Namibia and Bakirlitepe, Turkey, each with a 0.45 m aperture. This provides a 3.5 square degree Field of View (FoV). These telescopes imaged (using a 'clear' filter) areas of the sky relating to 5 triggers produced by the low latency pipelines. The strategy used was to image the region 30 times in one night as soon after the trigger as possible, followed by 8 images for as many nights as possible for up to approximately one month after the trigger. This produced a total of 782 images for analysis, with limiting magnitudes of around $17^{\rm th}$ magnitude. The Liverpool Telescope [4] is a 2 m robotic telescope with a 4.6' FoV situated at the Observatorio del Roque de los Muchachos, La Palma, Spain. Images were taken with the main telescope using the SDSS r' band filter. We followed up a

Journal of Physics: Conference Series **363** (2012) 012036 doi:10.1088/1742-6596/363/1/012036

single trigger, in which we took eleven images of the galaxy that is the most likely host of the GW source, each with a 300 second exposure time (with overheads). This allowed us to reach an image depth of $\sim 21^{\mathrm{st}}$ magnitude as soon after the trigger as possible, with a further 11 images taken approximately one month later. In addition, we were able to obtain images from two secondary finder scopes having 1 and 20 degree FoV centered on the same field which can reach 18th and 12th magnitude respectively. This produced a total of 22 narrow field images with a further 172 images from the secondary scopes.

3. Estimating efficiency

As described above, the follow-up campaign resulted in hundreds of images, all of which need to be analysed and characterised in order to provide a quantitative statement on any detection or limits on any non-detection. By injecting transients we can quantify several important parameters in our follow-up search, such as the efficiency with which we can detect specific afterglow models (see below), the distance ranges our searches are sensitive to given these models and the magnitude limit of the image subtraction method.

3.1. Afterglow models

In order to test the efficiency of our analysis, it is useful to test how well the pipelines detect the EM transients expected as counterparts. The most promising sources of GW signals are the mergers of compact objects (neutron stars and black holes) and core collapse supernovae. The EM sources believed to be associated with these events are long Gamma Ray Bursts (GRBs) [5], short GRBs [6] and the radioactive decay of neutron rich matter (known as the Kilonova model [7]). For the injections, we used the observed afterglows of on-axis GRBs and the decay model expected for kilonovae (see Fig. 1). The transient flux depends on both the distance to the source, and the difference in time between the original event and the time the image was taken.

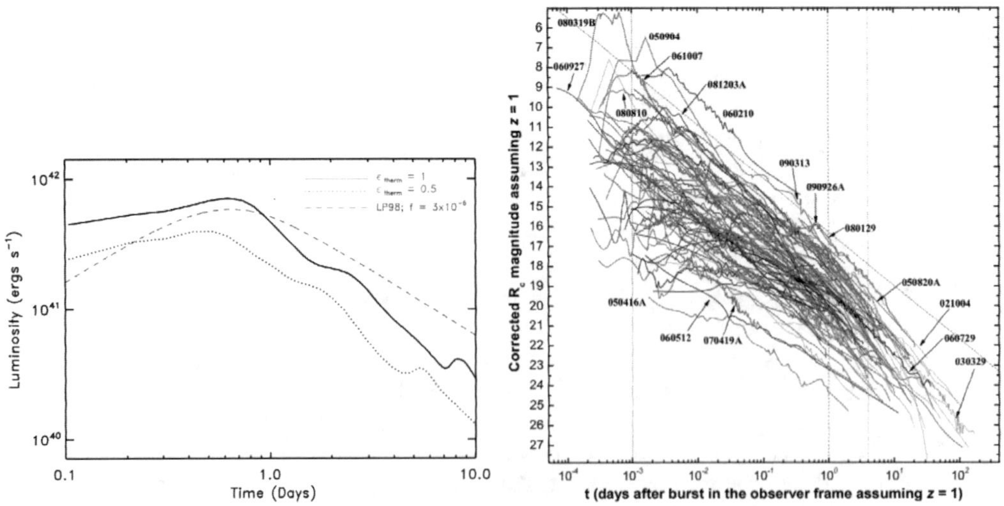

Figure 1. Examples of the light curves we used as models for transient injections. On the left is the radioactive afterglow model from [7]. On the right, an example of measured Long GRB light curves taken from [5].

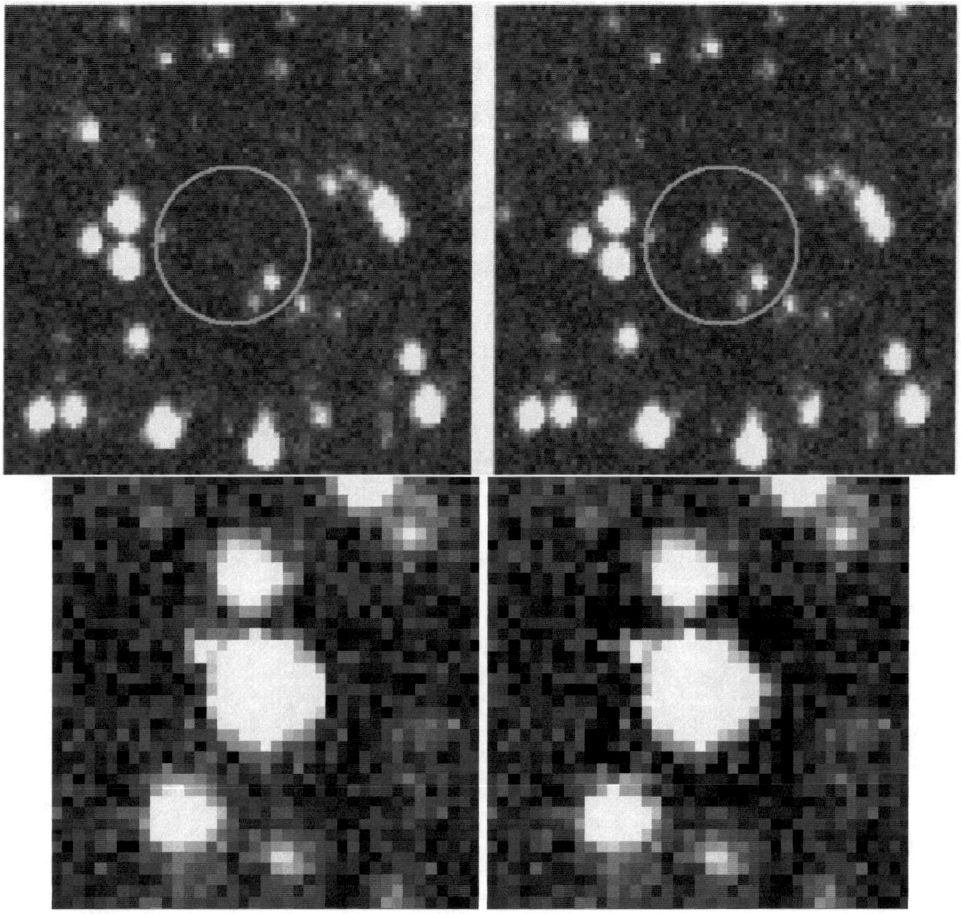

Figure 2. Examples of injected transients in a single image. The top images show a region of an image before (left) and after (right) injection of a bright object. The bottom image shows a different injection comparing the process with (left) and without (right) scaling in the manner described below.

3.2. Injection method

The method described here is originally only designed to work on the telescopes mentioned above, as other telescopes are being analysed with different methods and therefore have different requirements, but it has been created to be as robust as possible for future analyses. For our work in particular, the image subtraction method is sensitive to any variances in background around a transient, so a simple "cut-and-paste" of a model star into the image (either real or a 2 dimensional gaussian, with no care taken regarding the background noise or the varying PSF across the image) may not be sufficient.

The injection pipeline relies on the selection of bright, isolated stars in the field. This ensures that the object flux is dominated by Poissonian noise from the source, rather than sky or instrument noise, allowing a smooth, scaled injection which has a realistic point spread function (PSF), and with no nearby stars erroneously copied into the injection location. A position close to the source star in the field is chosen at random as the injection position, along with a randomly chosen model and associated distance to the source. In each image identified, the background subtracted flux of the original star is scaled according to the magnitude required for the image calculated using the time difference between the first image and the current image. The scaled flux, $F_{starscaled}$, is calculated using:

$$F_{\text{starscaled}} = F_{\text{star}} \Big/ 10^{\frac{mag_{\text{image}} - mag_{\text{inj.}}}{-2.5}}. \tag{1}$$

where F_{star} is the flux of the original star, mag_{image} is the magnitude of the original star in the image and $mag_{\text{inj.}}$ is the required magnitude of the injected transient. In order to obtain realistic backgrounds it is necessary to scale the injection region by a factor f. Without scaling the injection region, the background noise added in quadrature, post-injection, will be:

$$\sigma(postinj)^2 = \sigma(preinj)^2 + \sigma(star)^2 \tag{2}$$

where $\sigma(star)$ is the standard deviation of the background around the source star, and $\sigma(preinj)$ and $\sigma(postinj)$ are the standard deviations of the background of the injection region before and after injection respectively. The background noise, post injection, is therefore obviously larger than the original background noise. This can result in image subtractions in which this excess noise is clearly visible, possibly producing a higher than expected likelihood of detecting injections and increasing the error in photometry, depending on the methods used. In order to account for this, the injection region must be scaled so that the background before and after injection remains the same:

$$\sigma(preinj)^2 = f^2 \times \sigma(preinj)^2 + \sigma(star)^2 \tag{3}$$

which rearranges to:

$$f = \sqrt{1 - \frac{\sigma(star)^2}{\sigma(preinj)^2}}. \tag{4}$$

By scaling the injection region before injecting the transient we reduce the background noise in the injection region, but also the mean of the injection region, so we therefore add a constant value across all pixels such that the mean of the background and the noise in the background remains the same post-injection. This constant is calculated by comparing the mean before and after scaling, and the difference is simply added to each pixel in the injection region. After adding our transient object, we have a smooth injection in which both the mean and standard deviations of the background are the same as before the injection. Two example injections, both with and without scaling, can be found in Fig. 2.

4. Conclusion

The most recent joint science run allowed us to take hundreds of images of the sky. This presents a huge challenge to not only develop efficient methods of analysing these images, but also to be able to provide quantitative results. Calculating the efficiencies and limits of our analysis pipelines is extremely important for both detection and non-detection. Using injected transients across a series of images allows us to provide the efficiencies and limits of both the images and our analysis methods, and also allows us to fully understand the significance of the results we obtain.

References

[1] The LIGO Scientific Collaboration and Virgo Collaboration: J Abadie et al 2012 *A&A* **539** A124 (*Preprint* `1109.3498`)
[2] Alard, C 2000 *A&AS* **144** 363–370
[3] Akerlof, C W et al 2003 *PASP* **115** 132–140 (*Preprint* `arXiv:astro-ph/0210238`)
[4] Steele, I A 2001 *NAR* **45** 45–47

[5] Kann, D A et al 2010 *ApJ* **720** 1513–1558 (*Preprint* 0712.2186)
[6] Kann, D A et al 2011 *ApJ* **734** 96 (*Preprint* 0804.1959)
[7] Metzger, B D et al 2010 *MNRAS* **406** 2650–2662 (*Preprint* 1001.5029)

The NoEMi (Noise Frequency Event Miner) framework

T. Accadia[1], F Acernese[2ac], M Agathos[3a], P Astone[4a], G Ballardin[5], F Barone[2ac], M Barsuglia[6], A Basti[7ab], Th S Bauer[3a], M Bebronne[1], M Bejger[8c], M G Beker[3a], M Bitossi[7a], M A Bizouard[9a], M Blom[3a], F Bondu[10b], L Bonelli[7ab], R Bonnand[11], V Boschi[7a], L Bosi[12a], B Bouhou[6], S Braccini[7a], C Bradaschia[7a], M Branchesi[13ab], T Briant[14], A Brillet[10a], V Brisson[9a], T Bulik[8b], H J Bulten[3ab], D Buskulic[1], C Buy[6], E Calloni[2ab], B Canuel[5], F Carbognani[5], F Cavalier[9a], R Cavalieri[5], G Cella[7a], E Cesarini[13b], O Chaibi[10a], E Chassande-Mottin[6], A Chincarini[15], A Chiummo[5], F Cleva[10a], E Coccia[16ab], P-F Cohadon[14], C N Colacino[7ab], J Colas[5], A Colla[4ab], M Colombini[4b], A Conte[4ab], J-P Coulon[10a], E Cuoco[5], S D'Antonio[16a], V Dattilo[5], M Davier[9a], R Day[5], R De Rosa[2ab], G Debreczeni[17], W Del Pozzo[3a], L Di Fiore[2a], A Di Lieto[7ab], M Di Paolo Emilio[16ac], A Di Virgilio[7a], A Dietz[1], M Drago[18ab], G Endrőczi[17], V Fafone[16ab], I Ferrante[7ab], F Fidecaro[7ab], I Fiori[5], R Flaminio[11], L A Forte[2a], J-D Fournier[10a], J Franc[11], S Franco[9a], S Frasca[4ab], F Frasconi[7a], M Galimberti[11], L Gammaitoni[12ab], F Garufi[2ab], M E Gáspár[17], G Gemme[15], E Genin[5], A Gennai[7a], A Giazotto[7a], R Gouaty[1], M Granata[6], C Greverie[10a], G M Guidi[13ab], J-F Hayau[10b], A Heidmann[14], H Heitmann[10], P Hello[9a], G Hemming[5], P Jaranowski[8d], R J G Jonker[3a], M Kasprzack[5,9a], I Kowalska[8b], A Królak[8ae], N Leroy[9a], N Letendre[1], T G F Li[3a], N Liguori[18ab], M Lorenzini[13a], V Loriette[9b], G Losurdo[13a], E Majorana[4a], I Maksimovic[9b], V Malvezzi[16a], N Man[10a], M Mantovani[7ac], F Marchesoni[12a], F Marion[1], J Marque[5], F Martelli[13ab], A Masserot[1], C Michel[11], L Milano[2ab], Y Minenkov[16a], M Mohan[5], N Morgado[11], A Morgia[16ab], S Mosca[2ab], B Mours[1], L Naticchioni[4ab], F Nocera[5], L Palladino[16ac], C Palomba[4a], F Paoletti[7a,5], R Paoletti[7a], M Parisi[2ab], A Pasqualetti[5], R Passaquieti[7ab], D Passuello[7a], G Persichetti[2ab], F Piergiovanni[13ab], M Pietka[8d], L Pinard[11], R Poggiani[7ab], M Prato[15], G A Prodi[18ab], M Punturo[12a], P Puppo[4a], D S Rabeling[3ab], I Rácz[17], P Rapagnani[4ab], V Re[16ab], T Regimbau[10a], F Ricci[4ab], F Robinet[9a], A Rocchi[16a], L Rolland[1], R Romano[2ac], D Rosińska[8cf], P Ruggi[5], B Sassolas[11], D Sentenac[5], L Sperandio[16ab], R Sturani[13ab], B Swinkels[5], M Tacca[5], L Taffarello[18c], A P M ter Braack[3a], A Toncelli[7ab], M Tonelli[7ab], O Torre[7ac], E Tournefier[1], F Travasso[12ab], G Vajente[7ab], J F J van den Brand[3ab], C Van Den Broeck[3a], S van der Putten[3a], M Vasuth[17], M Vavoulidis[9a], G Vedovato[18c], D Verkindt[1], F Vetrano[13ab], A Viceré[13ab], J-Y Vinet[10a], S Vitale[3a], H Vocca[12a], R L Ward[6], M Was[9a], K Yamamoto[18bd], M Yvert[1],

A Zadrożny[8e], J-P Zendri[18c]

[1]Laboratoire d'Annecy-le-Vieux de Physique des Particules (LAPP), Université de Savoie, CNRS/IN2P3, F-74941 Annecy-Le-Vieux, France
[2]INFN, Sezione di Napoli [a]; Università di Napoli 'Federico II'[b] Complesso Universitario di Monte S.Angelo, I-80126 Napoli; Università di Salerno, Fisciano, I-84084 Salerno[c], Italy
[3]Nikhef, Science Park, Amsterdam, the Netherlands[a]; VU University Amsterdam, De Boelelaan 1081, 1081 HV Amsterdam, the Netherlands[b]
[4]INFN, Sezione di Roma[a]; Università 'La Sapienza'[b], I-00185 Roma, Italy
[5]European Gravitational Observatory (EGO), I-56021 Cascina (PI), Italy
[6]Laboratoire AstroParticule et Cosmologie (APC) Université Paris Diderot, CNRS: IN2P3, CEA: DSM/IRFU, Observatoire de Paris, 10 rue A.Domon et L.Duquet, 75013 Paris - France
[7]INFN, Sezione di Pisa[a]; Università di Pisa[b]; I-56127 Pisa; Università di Siena, I-53100 Siena[c], Italy
[8]IM-PAN 00-956 Warsaw[a]; Astronomical Observatory Warsaw University 00-478 Warsaw[b]; CAMK-PAN 00-716 Warsaw[c]; Białystok University 15-424 Białystok[d]; IPJ 05-400 Świerk-Otwock[e]; Institute of Astronomy 65-265 Zielona Góra[f], Poland
[9]LAL, Université Paris-Sud, IN2P3/CNRS, F-91898 Orsay[a]; ESPCI, CNRS, F-75005 Paris[b], France
[10]Université Nice-Sophia-Antipolis, CNRS, Observatoire de la Côte d'Azur, F-06304 Nice[a]; Institut de Physique de Rennes, CNRS, Université de Rennes 1, 35042 Rennes[b], France
[11]Laboratoire des Matériaux Avancés (LMA), IN2P3/CNRS, F-69622 Villeurbanne, Lyon, France
[12]INFN, Sezione di Perugia[a]; Università di Perugia[b], I-06123 Perugia,Italy
[13]INFN, Sezione di Firenze, I-50019 Sesto Fiorentino[a]; Università degli Studi di Urbino 'Carlo Bo', I-61029 Urbino[b], Italy
[14]Laboratoire Kastler Brossel, ENS, CNRS, UPMC, Université Pierre et Marie Curie, 4 Place Jussieu, F-75005 Paris, France
[15]INFN, Sezione di Genova; I-16146 Genova, Italy
[16]INFN, Sezione di Roma Tor Vergata[a]; Università di Roma Tor Vergata, I-00133 Roma[b]; Università dell'Aquila, I-67100 L'Aquila[c], Italy
[17]RMKI, H-1121 Budapest, Konkoly Thege Miklós út 29-33, Hungary
[18]INFN, Gruppo Collegato di Trento[a] and Università di Trento[b], I-38050 Povo, Trento, Italy; INFN, Sezione di Padova[c] and Università di Padova[d], I-35131 Padova, Italy

E-mail: `alberto.colla@roma1.infn.it`

Abstract.
The data collected by a gravitational wave interferometer are inevitably affected by instrumental artefacts and environmental disturbances. In particular, for continuous gravitational wave (CW) studies it is important to detect narrow-band disturbances (the so-called "noise lines") during science runs, and to help scientists to identify and possibly remove or mitigate their sources. The NoEMi (Noise Frequency Event Miner) framework exploits some of the algorithms implemented for the CW search to identify, on a daily basis, the frequency lines observed in the Virgo science data and in a subset of the environmental sensors, looking for lines that match in frequency. A line tracker algorithm reconstructs the lines over time, and stores them in a database, which is made accesible via a web interface. We describe the workflow of NoEMi, providing examples of its use for the investigation of noise lines in past Virgo runs (VSR2, VSR3) and in the most recent run (VSR4).

1. Introduction

Michelson interferometers (ITF) are presently the most promising instruments in the search for gravitational waves (GW) [1]. Gravitational waves traversing an ITF change the light paths in its arms, and the differential length variation is observed as a power fluctuation in the dark fringe

(DF[1]) of the ITF. The largest GW interferometers to have been built so far are the French-Italian Virgo [3] detector with 3 km-long arms, the two 4 km LIGO [4] detectors located in USA, and the German, 600 m-long GEO600 [5].

The signals expected at the detector are, however, extremely weak: current ITF data are still dominated by noise. Besides the intrinsic limitations to the detector sensitivity [1], interferometers are affected by instrumental artefacts and environmental disturbances of various origins. We usually distinguish between transient noise signals (lasting up to a few seconds) and continuous disturbances (lasting from a period of hours up to the full observation time). In this paper we will focus on the latter category, which is particularly harmful for continuous wave (CW) searches, where the analysis is based on the integration of the data over long periods of time.

CW signals are expected to be produced by rapidly rotating asymmetric neutron stars (e.g. pulsars) [6]. In fact, observations show that the rotation frequency of pulsars decreases in time (*Spin-Down*), and emission of GWs could explain part of the rotational energy loss. The frequency of the emitted GW is expected to be twice the spinning frequency of the star, and is Doppler-modulated because of the relative motion with respect to the detector. The search for signals is computationally cheap for "known" pulsars (whose location in the sky, frequency and frequency derivatives are known thanks to electro-magnetic observations), because it can be done by directly searching at the expected emission frequencies (Targeted search). The most famous examples of known pulsars are the Crab [7] (expected GW emission at 59.44 Hz) and Vela [8] (22.38 Hz). In this case it is important to constantly monitor the data quality at the target frequencies, ensuring that no disturbances degrade the detector sensitivity and thus the expected signal-to-noise ratio.

For the all-sky searches [9], in which as large a portion as possible of the source parameter space is explored (with the constraint given by the available computing power), the identification of continuous disturbances is even more important. Noise lines (narrow spectral features, so-called because they appear as persistent peaks in the DF power spectrum and lines in the power spectrograms) can mimic the signals produced by the pulsars and produce fake candidates in a given region of the sky. It is therefore crucial to locate and identify the noise sources ("line hunting"), remove or reduce them as soon as they appear whenever possible ("line mitigation"), or catalogue them as instrumental features in view of the off-line analysis.

GW interferometers are equipped with environmental sensors; seismometers, microphones and magnetometers are placed close to the main detector components and provide environment monitoring data. Relevant information about the noise coupling paths is provided by correlating the data of the channels[2] relating to these sensors with the GW channel data.

NoEMi (Noise Frequency Event Miner) is a Virgo software tool dedicated to noise line monitoring and identification. It was initially conceived as an online noise monitor for CW search purposes, aiming to guarantee fast feedback (with a maximum of 1 day of delay) whenever new lines appeared in the data. NoEMi analyses, on a daily basis, the GW channel and a subset of the environmental channels, identifying the noise lines in the various channels and looking for coincidences between the DF and the auxiliary channels (by *coincident lines* we mean lines found at the same frequency in different channels). A line tracker algorithm reconstructs the lines over time, and finally records them in a database.

NoEMi builds and publishes "daily summary pages", with plots and lists of the found noise lines, on the Virgo monitoring web pages. In particular, for the objectives of the targeted search, noise lines found at the same frequencies of the known pulsars are highlighted in the summary pages.

[1] We will refer to the DF as the uncalibrated GW detection data stream, and to $h(t)$ as the calibrated strain data [2].

[2] Channel refers to a time series obtained following conversion of an analague data stream to a digital data stream.

Particular care has been taken in the developement of a web interface to the NoEMi database, in order to make it easily accessible for detector characterisation and off-line noise studies. In this paper we present the NoEMi framework providing examples of its usage as a "noise monitor" and as an instrument for line identification in the most recent Virgo runs.

2. The NoEMi pipeline

NoEMi's main components are schematised in Fig. 1. They are the *Event finder*, the *Event analyser*, the *Coincidence finder* and the *Line tracker*.

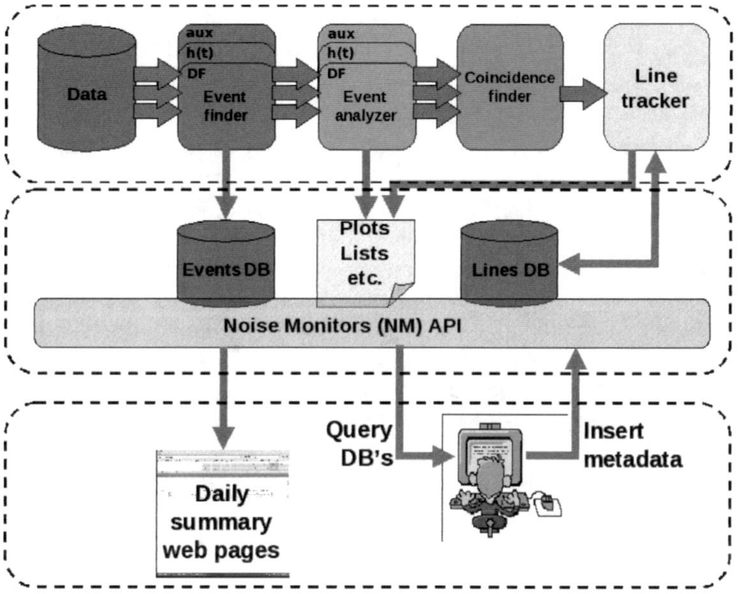

Figure 1. The NoEMi pipeline and the associated software architecture.

2.1. The Event finder

The Event finder runs Fast Fourier Transforms (FFTs) on the data, following a data-cleaning procedure, which removes the short transients [10]. For each FFT, the program, using an autoregressive (AR) technique, identifies the spectrum peaks, also called *frequency events* (EVF). The main Event finder configurable parameters are the length of the FFT, which determines the frequency resolution of the spectrum, and the threshold set on the peak Critical Ratio (CR), defined as the difference between the peak amplitude and the mean value of the spectrum, divided by the spectrum standard deviation. The peak parameters (frequency, CR, width, and amplitude) are recorded in a MySQL [11] database (Events DB).

2.2. The Event analyzer

The Event analyser makes a statistical analysis of the EVF parameters, collecting the events of a few tens of FFTs. In particular, two distributions are computed: persistency (frequency distribution of the events, normalised by the number of FFTs in the collection) and CR versus frequency (average CR of the events in each frequency bin). The program identifies the peaks in the two histograms, setting a threshold based on their average and standard deviation, recording their maximum value and width (defined as the full width over threshold). The peaks of the two histograms can be considered as a "projection" of the noise lines in the period covered by the collection: stationary lines generate peaks in the spectra which fall into the same frequency bin, thus forming a peak in the persistency distribution. Instead, the CR histogram helps to identify the strong non-stationary lines, the persistency of which may be too low to emerge in the first distribution. The algorithm also looks for peaks falling in or near the Doppler-shifted frequency band of the known pulsars, highlighting them in the summary web pages in order to alert the noise experts.

2.3. The Coincidence finder
The next step is the Coincidence finder: the set of peaks extracted in the DF or $h(t)$ are compared with those of the auxiliary channels, and marked as coincident in frequency if their peaks overlap.

2.4. The Line tracker
As a last step, the Line tracker algorithm reconstructs the time evolution of the noise lines, comparing the frequency of the latest peaks with those identified in the previous iterations. If more than one line is found near the peak frequency, the program chooses the best candidate, also comparing the other parameters (mean persistency, CR, amplitude). For each iteration NoEMi stores the line parameters and the list of coincident auxiliary channels in the Lines database (Lines DB), assigning them the ID that identifies the matched line. The set of parameters stored for each iteration in the Lines DB allows for the dynamical computation of the "global" line parameters, such as the mean frequency, persistency, CR, etc., or the number of iterations in which an auxiliary channel has been found in coincidence with the GW channel (thus making it possible to discard the "randomly" coincident peaks).

Fig. 2 shows examples of histograms and plots produced in each step of the analysis of the DF channel. The maximum "update frequency" of the lines depend on the length of the FFTs used in the first step of the analysis and on the number of FFTs collected to calculate the statistical distributions: low-frequency resolution requires shorter FFTs and allows more frequent iterations (leading to better line tracking) than high-resolution. In the last Virgo run (VSR4), NoEMi ran in two different configurations at the same time:

(i) "CW search-oriented" configuration: $h(t)$ channel sampled at 4096 Hz, 1 mHz frequency resolution, daily updates, "low" threshold on CR (3). This configuration enhances NoEMi's sensitivity at low frequency (up to 2 kHz), which is the CW analysis target. Moreover, the high frequency resolution increases the identification efficiency for the sharp and stationary lines, the most harmful for the CW all-sky searches;

(ii) "detector characterisation-oriented" configuration: DF channel sampled at 20 kHz, 10 mHz frequency resolution, updates every 2 hours, "high" threshold on CR (6). This configuration helps the identification and follow-up of the non-stationary lines up to 10 kHz, thanks to its higher time resolution.

3. NoEMi user interface and Lines DB
NoEMi is integrated in the Noise Monitors (NM) API framework [12], which provides a common interface to a web server for the publication of the daily results and for the access to the databases. The summary web pages are saved on disk and a calendar applet on the page allows for the results to be displayed for any day of interest. NoEMi produces various summary pages: for each channel the daily time-frequency diagrams of the peak maps (the set of frequency events) and their distributions, and the updated lists of noise lines.

The web interface to the NoEMi databases makes it possible to run queries using specific filtering criteria: for instance (concerning the Lines DB) a user can select lines found in a given

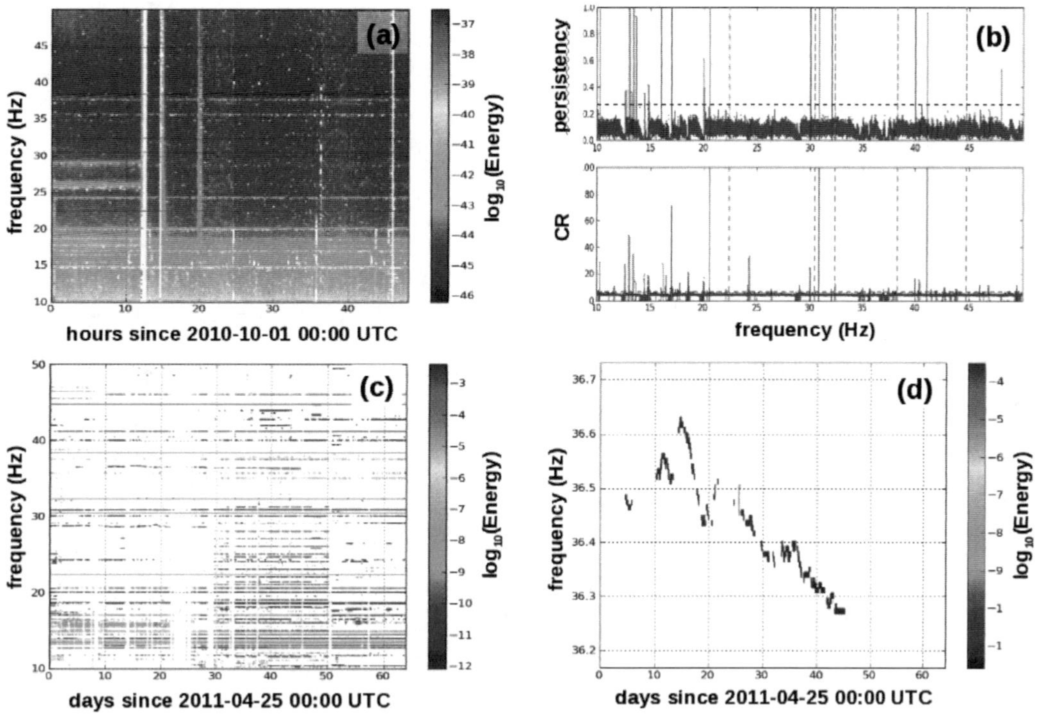

Figure 2. a) Time-Frequency diagram of the frequency events; the red lines show the frequencies of the known pulsars. b) Persistency and CR histograms built from the events distributions. The red dashed lines show the frequencies of the known pulsars, while the green solid lines show the identified peaks, i.e. those higher than the thresolds set on the histograms (the horizontal dashed lines). c) Time-Frequency plot of the reconstructed lines; note that the time range of this plot is 40 days, while the Event Time-Frequency diagram covers 48 hours. d) Example of a reconstructed non-stationary line in VSR4. The plots refer to the DF channel.

time or frequency interval, setting thresholds on their dynamic parameters (persistency, CR, line "duration", etc.) or even query the lines coincident with a subset of the auxiliary channels.

Each line is linked to the Metadata table [13], which is filled with user-defined information once the line is identified. Thanks to the Metadata table, the Lines DB can be considered a true "Line catalogue", dynamically updated by NoEMi and manually maintained by the noise experts. The NM API allows the users to easily associate metadata to the lines, and to use the metadata as additional filtering criteria: for instance, to distinguish between identified and not yet identified lines, or to select lines containing specific descriptions.

4. Lines identification and monitoring with NoEMi

During the Virgo runs, but also in commissioning periods, NoEMi was used as a line monitor and (with the Lines DB) as a tool for line identification.

4.1. Hardware injections

Hardware injections considered here are simulated pulsar GW signals injected in the detector by applying a calibrated force to displace the ITF end mirrors, in order to test the detection efficiency of the analysis pipelines. Fig. 3 shows the signal of an injected pulsar seen in the data from the Events and Lines DB. The simulation takes into account the pulsar spin-down and the Doppler modulation of the pulsar frequency, due to its relative motion with respect to the detector. This example helped to check the NoEMi framework and evidences its line extraction and tracking capabilities.

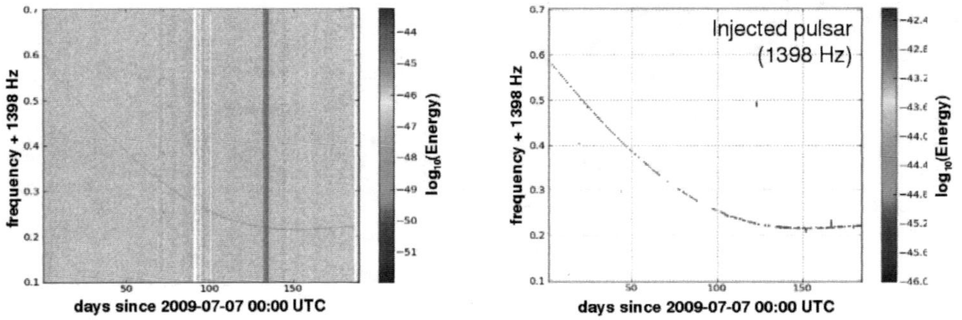

Figure 3. Signal of the 1398 Hz hardware simulated pulsar signal injection in the VSR2 data. Left: Time-Frequency plot of the frequency events detected by NoEMi in the $h(t)$ channel. Right: Time-Frequency plot of the reconstructed line. The mean CR of the signal is 6.

4.2. Detector resonance studies

The Lines DB, filled by NoEMi, was also used in an off-line detector characterisation analysis. ITF components (suspensions, mirrors, etc.) generate thermal noise charaterised by sharp resonances. Fig. 4 presents the time-frequency plots of two mirror resonances (bulk modes). Analysing and correlating the time evolution (and consequently the temperature behaviour) of the different resonances allowed us to associate them to the mirrors and suspensions to which they belong. This study was performed in the frame of the thermal noise investigations, in which the mirror and violin mechanical losses were characterised with the aim of predicting the effective thermal noise present in the detector [16]. The results of the analysis were used to optimise the mirror suspension design in view of the construction of the next generation Virgo detector [14].

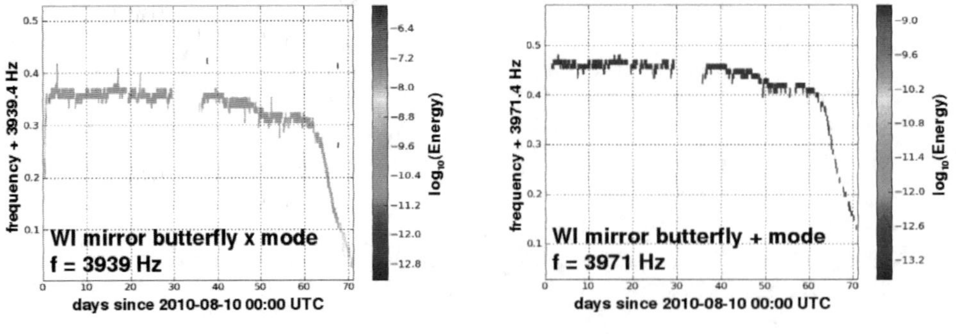

Figure 4. Time-frequency plots made with DF data extracted from the Lines database for VSR3. We clearly see the lines from west input mirror bulk modes.

4.3. The VSR2 "Vela killer line"

During VSR2 a non-stationary noise line affected the sensitivity of the Virgo ITF at the frequency of a GW possible signal from the Vela pulsar (22.38 Hz). The disturbance caused a loss of sensitivity at this frequency by about 20%. Unfortunately, NoEMi was not yet working at that time, therefore the noise line was only discovered during the off-line data analysis.

NoEMi found that the noise line (actually a "doublet", see Fig. 5a and 5b) was coincident with the data of an accelerometer monitoring the vibrations of the Thermal Compensation System (TCS) [15] optical benches (Fig. 5c). Further investigations identified the source of the disturbance as being in the two chiller devices, which refrigerate and circulate water to cool down the TCS lasers, located just out of the TCS room. The vibration source was identified in one mechanical component of the chillers, whose rotation frequency (22.4 Hz) matched the noise line. The vibration was probably transmitted to the TCS bench through the cooling pipes.

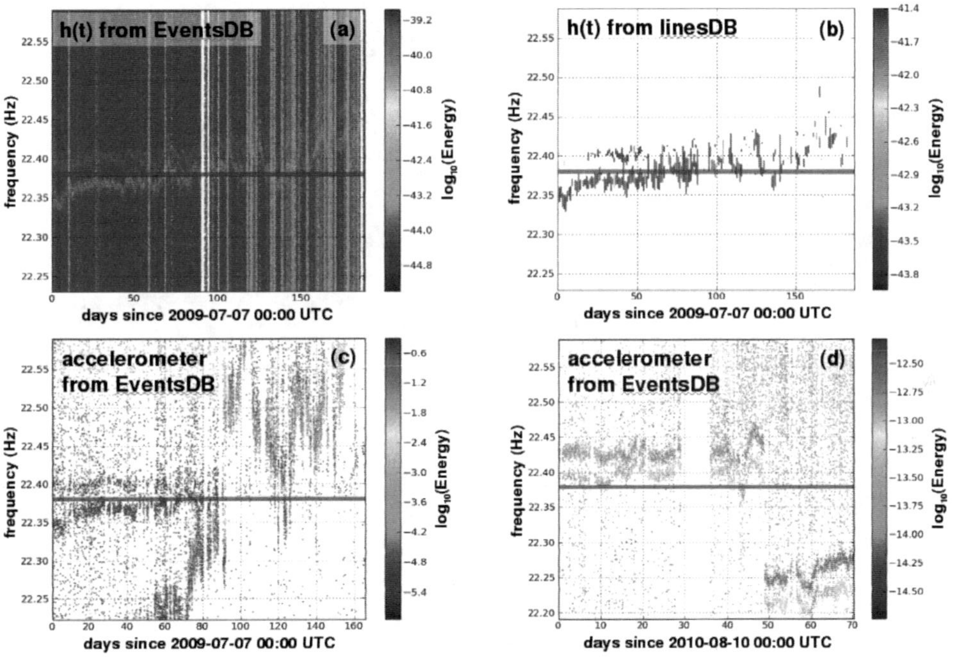

Figure 5. The disturbance at the Vela pulsar frequency seen in the Events DB (a), in the Lines DB (b) and in the Events DB of an accelerometer (c) (VSR2). Plot d) shows the effect (on the accelerometer data) of the shift away from the Vela region of the frequency of the chiller engine causing the disturbance, on the 49th day of VSR3. The blue band show the Doppler band of the Vela pulsar frequency.

During VSR3 the noise line near the Vela signal frequency was no longer visible in the DF, however, it was still in the accelerometer: therefore we suspected that the line was just buried under the ITF noise, which at the Vela frequency was 2 to 3 times worse with respect to VSR2. A small but indicative coherence was indeed found between the DF and the accelerometer data. The adopted solution, in order to move the disturbance away from the Vela region, was to install a variable frequency drive to change the rotation frequency of the chiller engines (Fig. 5d).

4.4. The VSR4 "Vela bump"

As a final example, we describe the case of another disturbance affecting the sensitivity at the Vela pulsar frequency, at the beginning of VSR4, which confirmed the efficiency of the NoEMi framework as a monitoring tool. Before the science run began, NoEMi detected a disturbance close to the Vela frequency and raised an alarm on the summary pages. In the peak map time-frequency plots the disturbance appeared as a region with a higher peak density (Fig. 6), over which a comb of lines emerged (which triggered the alarm).

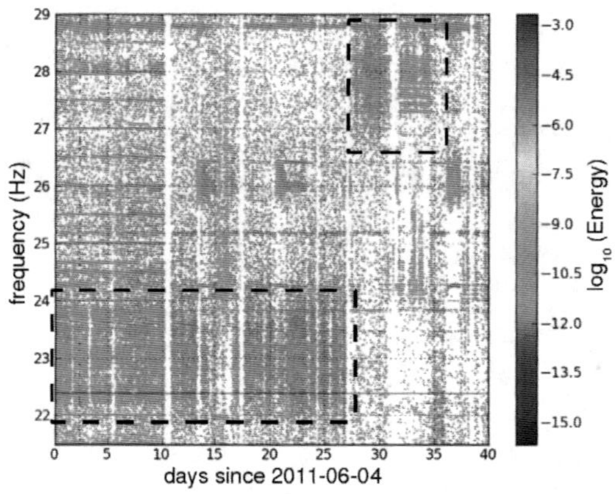

Figure 6. Time-frequency plot of the peak maps at the Vela frequency region, showing the noise bump (enclosed in the dashed rectangles) before and after the change of the frequencies of the calibration lines. The red line shows the frequency of the Vela pulsar.

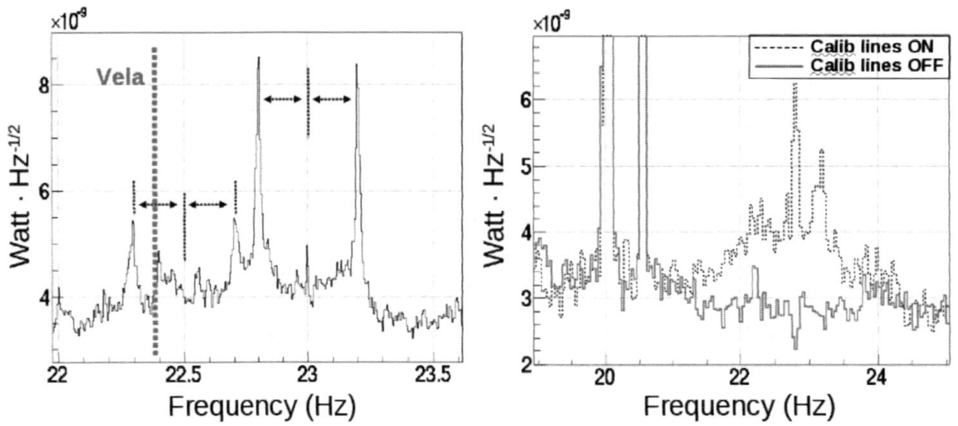

Figure 7. Left: DF spectrum at the Vela frequency region showing the comb of noise lines. Right: Effect of the calibration signal switch off on the spectrum.

The noise hunting began to observe that two pairs of lines were equally spaced by 0.2 Hz on each side of two "remarkable" frequencies: 22.5 and 23.0 Hz (Fig. 7 Left). Since 0.2 Hz is the fundamental resonance of the Virgo Super-Attenuator (SA) [17], it was assumed that the line side-bands were caused by some coupling mechanism between the SA mode and the two frequencies.

However, this was not yet the end of the story, as no "carrier" line was found at either 22.5 or 23.0 Hz! Following deeper investigations, the conclusion was that the two frequencies could be obtained as the difference between the frequencies of a set of calibration and control lines (sinusoidal signals injected into the interferometers for calibration and control purposes) at 379 and 356 (356.5) Hz. To confirm the hypothesis the calibration lines were switched off: the bump and line comb disappeared (Fig. 7 right). The disturbance was then assumed to be caused by a non-linear coupling mechanism in the Digital-to-Analogue converters used for control signals.

Since the calibration signals could not be switched off for the rest of the data-taking period, the solution adopted was to shift the 356 and 356.5 Hz lines down by 5 Hz: the disturbance then moved to $\approx$ 28 Hz, where no known pulsar signal is expected. Thanks to the intervention, the sensitivity at the Vela frequency was increased by $\approx$ 50%. This allowed us to obtain, for VSR4, an integrated sensitivity gain of about 1.7 with respect to VSR2, the data from which had been used to set the first direct upper limit on continuous GW emissions from the Vela pulsar [8].

5. Conclusions

In this paper we described the NoEMi framework, the Virgo tool for noise line monitoring and cataloguing. NoEMi exploits some of the algorithms implemented for the CW analysis as its main engine and is integrated into the Noise Monitor API framework [12], which provides a web application for the publication of daily analysis results and an interface for its databases. NoEMi monitored the third and fourth Virgo scientific runs, ensuring, in particular, the data quality at the frequencies of the known pulsars; it also proved to be a useful tool in off-line noise hunting and detector characterisation. The line catalogue filled by NoEMi will be used to veto the noisy frequency bands in the CW all-sky search analysis. The demonstrated potentialities of NoEMi promoted the plan to include it in the Noise Monitors of the Advanced generation of both the Virgo and LIGO detectors [14].

References

[1] Cella G and Giazotto A 2011 Interferometric gravity wave detectors *Rev. Sci. Instrum.* **82** 101101
[2] Accadia T *et al* 2011 Calibration and sensitivity of the Virgo detector during its second science run *Class. Quantum Grav.* **28** 025005
[3] Acernese F *et al* 2008 *Class. Quantum Grav.* **25** 114045
[4] Abbott B *et al* 2009 *Rep. Prog. Phys.* **72** 076901
[5] Grote H *et al* 2010 *Class. Quantum Grav.* **27** 084003
[6] Palomba C for the LIGO Scientific collaboration and the Virgo Collaboration 2011 Searches for continuous gravitational wave signals and stochastic backgrounds in ligo and virgo data *Proceedings of Moriond 2011* (to be published)
[7] Abbott P *et al* 2010 Searches for gravitational waves from known pulsars with science run 5 LIGO data *ApJ* **713** 671
[8] Abadie J *et al* 2011 Beating the spin-down limit on gravitational wave emission from the Vela pulsar *ApJ* **737** 93
[9] Antonucci F, Astone P, D'Antonio S, Frasca S and Palomba C 2008 Detection of periodic gravitational wave sources by Hough transform in the frequency and spin down plane *Class. Quantum Grav.* **25** 184015
[10] Acernese F *et al* 2009 Cleaning the Virgo sampled data for the search of periodic sources of gravitational waves *Class. Quantum Grav.* **26** 204002
[11] `http://www.mysql.com`
[12] Accadia T *et al* 2011 Noise monitor tools and their application to Virgo data *These proceedings*
[13] Carbognani F, Colla A, Cuoco E and Hemming G 2011 Lines database web interface software requirements *Virgo internal note* VIR-0227A-11
[14] Accadia T *et al* 2011 Advanced Virgo *Proceedings of Moriond 2011* (to be published)
[15] Accadia T *et al* 2011 Virgo+ Thermal Compensation System *These proceedings*
[16] Puppo P 2009 Virgo+MS sensitivity curve *Virgo internal note* VIR-0639E-09
[17] Accadia T *et al* 2011 The seismic Superattenuators of the Virgo gravitational waves interferometer *Journal of Low Frequency Noise, Vibration and Active Control* **30** 63

Coherent search of continuous gravitational wave signals: extension of the 5-vectors method to a network of detectors

P Astone[1], **A Colla**[2,1], **S D'Antonio**[3], **S Frasca**[2,1], **C Palomba**[1]

[1] INFN, Sezione di Roma, I-00185 Roma, Italy
[2] Universita' di Roma "Sapienza", I-00185 Roma, Italy
[3] INFN, Sezione di Roma 2, I-00133 Roma, Italy

E-mail: `cristiano.palomba@roma1.infn.it`

Abstract. We describe the extension to multiple datasets of a coherent method for the search of continuous gravitational wave signals, based on the computation of 5-vectors. In particular, we show how to coherently combine different datasets belonging to the same detector or to different detectors. In the latter case the coherent combination is the way to have the maximum increase in signal-to-noise ratio. If the datasets belong to the same detector the advantage comes mainly from the properties of a quantity called *coherence* which is helpful (in both cases, in fact) in rejecting false candidates. The method has been tested searching for simulated signals injected in Gaussian noise and the results of the simulations are discussed.

1. Introduction

The search for continuous gravitational wave signals, like those emitted by a deformed rotating neutron star, is an important target of the analysis of data produced by interferometric detectors, like GEO [1], LIGO [2] and Virgo [3]. In the so-called *targeted search* the source parameters (sky position, frequency and frequency derivatives) are assumed to be known. This is the case, .e.g., of the search of signals from known pulsars for which radio-astronomical observations can provide very accurate measures of the parameters themselves. For this kind of search coherent data analysis methods, which precisely correct the signal phase variations due to Doppler, spin-down, Einstein and, if necessary, Shapiro effects, allow in principle to approach the best sensitivity at a relatively small computing cost. Various coherent pipelines have been developed and have been used (and will be used) to analyze data from Virgo and LIGO scientific runs, see e.g. [4]. In particular, we consider here a method which goes through the computation of the so-called 5-vectors [5] which has been used in the analysis of Virgo VSR2 data for the search of continuous gravitational wave signals from the Vela pulsar [4].

Given the expected weakness of continuous signals, it is important to use the available data, which are typically affected by various kinds of disturbances, in the most effective way. This is done in the 5-vectors method by applying different techniques. For instance, a cleaning procedure is used both on the initial time domain data [6] and after Doppler, spin-down and Einstein effect corrections, in order to remove particularly disturbed periods. Moreover, a Wiener filter is applied for taking into account slow non-stationarities so to weight less the more noisy periods

[4]. In this paper we describe one more analysis step that can be applied in order to improve the sensitivity of the search: if more datasets are available, belonging to the same detectors (e.g. two different runs) or to different detectors, they can be properly combined in order to increase the output signal-to-noise ratio and to improve rejection of false candidates. Other methods of coherently combining data from different detectors have been developed in the past, like e.g. the network analysis using the $\mathcal{F}$-statistic [7] and a complex heterodyne method [8].

The plan of the paper is the following. In Sec.(2) the basic concepts of the 5-vectors method are summarized and the definition of signal-to-noise ratio is given. In Sec.(3) the coherent combination of more datasets is introduced and tests done with simulated data are described. Sec.(4) is devoted to discuss further future developments of the method.

2. The 5-vectors method

Here we briefly recall the 5-vectors method and define the signal-to-noise ratio (SNR). More details can be found in [5] and [4]. Let us indicate the data to be analyzed with

$$x'(t) = n'(t) + h'(t) \tag{1}$$

where $n'(t)$ is the noise and $h'(t)$ is a continuous gravitational wave signal. The data, after the extraction of a small band around the expected frequency of the signal, are corrected for Doppler, spin-down and Einstein effects by a re-sampling procedure described in detail in [4] and which effectiveness is demonstrated by the very good recovery of parameters of hardware injected pulsar signals in Virgo VSR2 data as discussed in [4]. Let us indicate the corrected data with $x(t) = n(t) + h(t)$. At this stage a signal present into the data is monochromatic apart from an amplitude and phase sidereal modulation. The gravitational wave strain at the detector can be then written as

$$h(t) = H_0(H_+A^+(t) + H_\times A^\times(t))e^{j(\omega_0 t+\Phi_0)} \tag{2}$$

(where the operation of taking the real part is understood in order to have the physical strain) In this equation the two amplitudes are given by

$$\begin{aligned} H_+ &= \frac{\cos(2\psi) - j\eta\ \sin(2\psi)}{\sqrt{1+\eta^2}} \\ H_\times &= \frac{\sin(2\psi) + j\eta\ \cos(2\psi)}{\sqrt{1+\eta^2}} \end{aligned} \tag{3}$$

where η is the ratio between the minor and the major semi-axis of the wave polarization ellipse ($\eta = 0$ for a linearly polarized signal, $\eta \pm 1$ for a circularly polarized signal) and ψ is the polarization angle defining the direction of the ellipse semi-major axis respect to the celestial parallel of the source (measured counterclockwise). The functions $A^+(t)$, $A^\times(t)$ are equal to the standard beam-pattern functions $F^+(t;\psi)$, $F^\times(t;\psi)$, defined e.g. in [9], computed at $\psi = 0$. The power of the signal in Eq.(2) is spread among the five angular frequencies ω_0, $\omega_0 \pm \Omega$, $\omega_0 \pm 2\Omega$, where Ω is the Earth sidereal angular frequency.

The noise $n(t)$ is assumed to be gaussian with mean $\mu = 0$ and standard deviation σ. The signal detection and parameter estimation procedure is based on the computation of data and signal 5-vectors, defined as their Fourier components at the 5 frequencies in which the signal is split by the sidereal modulation. More precisely, given a generic time series $f(t)$, the corresponding 5-vector is a complex vector

$$\mathbf{F} = \int_T f(t)e^{-j(\omega_0 - \mathbf{k}\Omega)t}dt \tag{4}$$

where $\mathbf{k} = [-2, -1, ..., 2]$ and T is the observation time. The signal 5-vector is defined as

$$\mathbf{A} = (H_+\mathbf{A}^+ + H_\times\mathbf{A}^\times) \tag{5}$$

Note that for a source with known position and frequency and for a given detector $\mathbf{A}^+$ and $\mathbf{A}^\times$ can be computed from Eq.(4) while H_0, η, ψ, Φ_0 are in general unknown (for a few pulsars, like Vela and Crab, astronomical observations provide estimations of η and ψ [10]). The data 5-vector can be expressed as

$$\mathbf{X} = H_0 e^{j\Phi_0} (H_+\mathbf{A}^+ + H_\times\mathbf{A}^\times) + \mathbf{N} = H_0 e^{j\Phi_0}\mathbf{A} + \mathbf{N} \tag{6}$$

where $\mathbf{N}$ is the noise 5-vector. It is important to stress that data or signal 5-vectors contains exactly the same information of the corresponding time series. Once the 5-vectors of data, $\mathbf{X}$, and signal, $\mathbf{A}^+$ and $\mathbf{A}^\times$, have been computed the following complex amplitude estimators are built:

$$\begin{aligned} \hat{H}_+ &= \frac{\mathbf{X}\cdot\mathbf{A}^+}{|\mathbf{A}^+|^2} \\ \hat{H}_\times &= \frac{\mathbf{X}\cdot\mathbf{A}^\times}{|\mathbf{A}^\times|^2} \end{aligned} \tag{7}$$

where the scalar product among two complex vectors $\mathbf{a}$ and $\mathbf{b}$, each with m components, is defined as $\mathbf{a}\cdot\mathbf{b} = \sum_{i=1}^{m} a_i b_i^*$. $\hat{H}_+$ and $\hat{H}_\times$ are the estimation of, respectively, $H_0 e^{j\Phi_0} H_+$, and $H_0 e^{j\Phi_0} H_\times$ and are used to compute the detection statistics

$$S = |\mathbf{A}^+|^4 \left|\hat{H}_+\right|^2 + |\mathbf{A}^\times|^4 \left|\hat{H}_\times\right|^2 \tag{8}$$

which is compared to its expected distribution in presence of noise alone to establish if a statistically significant candidate is present. In case of detection Eq.(7) is also used to estimate source parameters. In particular, the estimation of the signal amplitude is given by

$$\hat{H}_0 = \sqrt{|\hat{H}_+|^2 + |\hat{H}_\times|^2} \tag{9}$$

A preliminary comparison of our detection statistic with the $\mathcal{F}$-statistic is done in [5]. A more detailed comparison, also with other detection statistics like the $\mathcal{B}$-statistic introduced in [11], will be done in a forthcoming paper.

If a detection is claimed we can quantify its reliability by a parameter called *coherence* defined as [5]

$$c = \frac{\hat{H}\hat{\mathbf{A}}}{|\mathbf{X}|^2} \tag{10}$$

where $\hat{\mathbf{A}} = \hat{H}_+\mathbf{A}^+ + \hat{H}_\times\mathbf{A}^\times$ is the estimated total signal 5-vector and $\hat{H} = \frac{\mathbf{X}\cdot\hat{\mathbf{A}}}{|\hat{\mathbf{A}}|^2}$. The *coherence* is a number between 0 and 1 which measures the resemblance between the shape of the expected signal and the data. In fact it does not depend on scaling factors on the signal but just on its shape. We have numerically found that in the absence of any signal the *coherence* is described (assuming all the source polarization parameters are unknown) by a β distribution with parameters $\alpha_1 = 2,\ \alpha_2 = 3$:

$$p(c; \alpha_1 = 2, \alpha_2 = 3) = \frac{12c\cdot(1-c)^2}{B(\alpha_1 = 2, \alpha_2 = 3)} \tag{11}$$

where $B(\alpha_1, \alpha_2)$ is the β function which acts as a normalization constant. Large values of *coherence* have a small probability of being produced by noise alone. We will further understand the utility of the *coherence* when extending the analysis to more datasets.

If no significant candidate is found an upper limit on the signal amplitude can be computed through a Monte Carlo simulation [4].

The 5-vectors method presents various strong points. For instance, the computation of cross products among data and signal templates is very fast because it involves only 10 complex numbers instead of $o(10^7)$ as it would be working in the time domain and this, in conjunction with the re-sampling procedure used to correct the Doppler effect, is particularly useful when doing simulations. Moreover, the re-sampling procedure allows to easily extend the method to the so-called 'narrow band' searches in which a possible mismatch between the electromagnetic and gravitational wave signal can be accounted for. As we will see in Sec.3, another advantage is the straightforward extension to a network of detectors, which is the main topic of this paper.

2.1. *Signal-to-noise ratio*

In the absence of signals, assuming the noise has a Gaussian distribution with mean zero and variance σ^2, each component $\mathbf{X}_i$ ($i = 1, ..5$) of the data (noise) 5-vector is distributed as a complex gaussian variable, with $\mu_X = 0$ and $\sigma_X^2 = \sigma^2 T$, as it can be immediately seen from Eq.(4) (i.e. the real and imaginary parts are two independent gaussian variables with mean zero and variance $\sigma^2 T/2$). The expectation value of $|\mathbf{X}|^2$ is $E[|\mathbf{X}|^2] = \sum_{i=1}^{5} |\mathbf{X}_i|^2 = 5\sigma^2 T$. Using Eq.(7) it is easy to see that the two amplitude estimators also have a gaussian distribution with mean zero and variance

$$\begin{aligned} \sigma_+^2 &= E\left[\left|\hat{H}_+\right|^2\right] = \frac{\sigma^2 T}{|\mathbf{A}^+|^2} \\ \sigma_\times^2 &= E\left[\left|\hat{H}_\times\right|^2\right] = \frac{\sigma^2 T}{|\mathbf{A}^\times|^2} \end{aligned} \tag{12}$$

From Eq.(9)

$$\sigma_{\hat{H}_0}^2 = E\left[\left|\hat{H}_0\right|^2\right] = E\left[\left|\hat{H}_+\right|^2\right] + E\left[\left|\hat{H}_\times\right|^2\right] = \sigma_+^2 + \sigma_\times^2 = \sigma^2 T \frac{|\mathbf{A}^+|^2 + |\mathbf{A}^\times|^2}{|\mathbf{A}^+|^2 |\mathbf{A}^\times|^2} \tag{13}$$

Let us define the signal-to-noise ratio (SNR) at the filter output as the ratio between the estimated signal amplitude and the standard deviation of the estimation when only noise is present:

$$SNR = \frac{\hat{H}_0}{\sigma_{\hat{H}_0}} = \frac{\hat{H}_0 |\mathbf{A}^+||\mathbf{A}^\times|}{\sigma\sqrt{(|\mathbf{A}^+|^2 + |\mathbf{A}^\times|^2)T}} \tag{14}$$

From the definition of 5-vector, see Eq.(4), we see that each component of the signal 5-vectors $\mathbf{A}^+$ and $\mathbf{A}^\times$ increases linearly with the observation time T. Hence, from Eq.(14), we have $SNR \propto \sqrt{T}$, as expected for a coherent analysis.

3. Extension to more datasets

Let us consider n datasets. Let us assume for simplicity that they have all the same length, the same noise level and belong to the same detector. Let us compute the corresponding 5-vectors, $\mathbf{X}_i$, $i = 1, ...n$. Given a source target let us compute also the signal 5-vectors, $\mathbf{A}_i^+$, $\mathbf{A}_i^\times$, $i = 1, ...n$. We can now put together the n 5-vectors of each kind in one $5n$-vector:

$$\begin{aligned} \mathbf{X} &= [\mathbf{X}_1, \mathbf{X}_2, ...\mathbf{X}_n] \\ \mathbf{A}^+ &= \left[\mathbf{A}_1^+, \mathbf{A}_2^+, ...\mathbf{A}_n^+\right] \\ \mathbf{A}^\times &= \left[\mathbf{A}_1^\times, \mathbf{A}_2^\times, ...\mathbf{A}_n^\times\right] \end{aligned} \tag{15}$$

In this simple case $|\mathbf{A}_i^+| = |\mathbf{A}_j^+|$, $|\mathbf{A}_i^\times| = |\mathbf{A}_j^\times|$ $\forall i,j = 1,...n$ (because the datasets belong to the same detector and have the same length) so that

$$\begin{aligned} |\mathbf{A}^+|^2 &= n\left|\mathbf{A}_i^+\right|^2 \quad \forall i = 1,...n \\ |\mathbf{A}^\times|^2 &= n\left|\mathbf{A}_i^\times\right|^2 \quad \forall i = 1,...n \end{aligned} \tag{16}$$

Then from Eq.(14) we have

$$SNR = SNR_i \cdot \sqrt{n} \quad \forall i = 1,...n \tag{17}$$

This means that working with 5n-vectors the signal-to-noise ratio increases as in the case of a coherent analysis over the total observation time. This result can be easily generalized to the case in which datasets of different length, sensitivity or belonging to different detectors are considered. For instance, let us consider two datasets with different sensitivities, i.e. with two different values of variance σ_1^2, σ_2^2 for the components of the corresponding 5-vectors, and two different pairs of signal templates ($\mathbf{A}_1^+$, $\mathbf{A}_1^\times$) and ($\mathbf{A}_2^+$, $\mathbf{A}_2^\times$) (different because corresponding to different detectors and/or different observation periods). In this case the variance of the signal amplitude estimation is given by

$$\sigma^2_{\hat{H}_0} = \frac{\sigma_1^2|\mathbf{A}_1^+|^2 + \sigma_2^2|\mathbf{A}_2^+|^2}{\left(|\mathbf{A}_1^+|^2 + |\mathbf{A}_2^+|^2\right)^2} + \frac{\sigma_1^2|\mathbf{A}_1^\times|^2 + \sigma_2^2|\mathbf{A}_2^\times|^2}{\left(|\mathbf{A}_1^\times|^2 + |\mathbf{A}_2^\times|^2\right)^2} \tag{18}$$

It can be shown that also the probability distribution of the detection statistics is equal to that we would have in case of 5-vector analysis over the whole observation time. The estimators of the signal complex amplitudes are given by

$$\begin{aligned} \hat{H}_+ &= \frac{\mathbf{X}\cdot\mathbf{A}^+}{|\mathbf{A}^+|^2} = \frac{\sum_{i=1}^n \mathbf{X}_i \cdot \mathbf{A}_i^+}{\sum_{i=1}^n |\mathbf{A}_i^+|^2} \\ \hat{H}_\times &= \frac{\mathbf{X}\cdot\mathbf{A}^\times}{|\mathbf{A}^\times|^2} = \frac{\sum_{i=1}^n \mathbf{X}_i \cdot \mathbf{A}_i^\times}{\sum_{i=1}^n |\mathbf{A}_i^\times|^2} \end{aligned} \tag{19}$$

From these equations in case of detection we can estimate signal parameters exactly in the same way as in the $n = 1$ case. We also see that each data 5-vector 'interacts' only with the corresponding templates.

The practical usefulness of working with 5n-vectors is clear when we want to combine data from different detectors, but it can be appreciated also when we have two or more datasets of the same detector. In this case the advantage of using 5n-vectors comes mainly from the properties of the *coherence*, defined by Eq.(10). We have numerically found that the probability distribution of the *coherence* computed over 5n-vectors in presence of gaussian noise alone is a β distribution with parameters $\alpha_1 = 2$, $\alpha_2 = 5n - 2$:

$$p(c; \alpha_1 = 2, \alpha_2 = 5n-2) = \frac{(5n-2)(5n-1)c\cdot(1-c)^{5n-3}}{B(\alpha_1 = 2, \alpha_2 = 5n-2)} \tag{20}$$

which for $n = 1$ reduces to Eq.(11) . The probability of obtaining a value of *coherence* larger than a given threshold c_{thr} is

$$P(c > c_{thr}) = (5n-1)(1-c_{thr})^{5n-2} - (5n-2)(1-c_{thr})^{5n-1} \tag{21}$$

and then it is a decreasing function of the number n of datasets being considered. This is shown in Fig.(1) where Eq.(21) is plotted for $n = 1, 2, 3$. For instance, the probability of getting a value

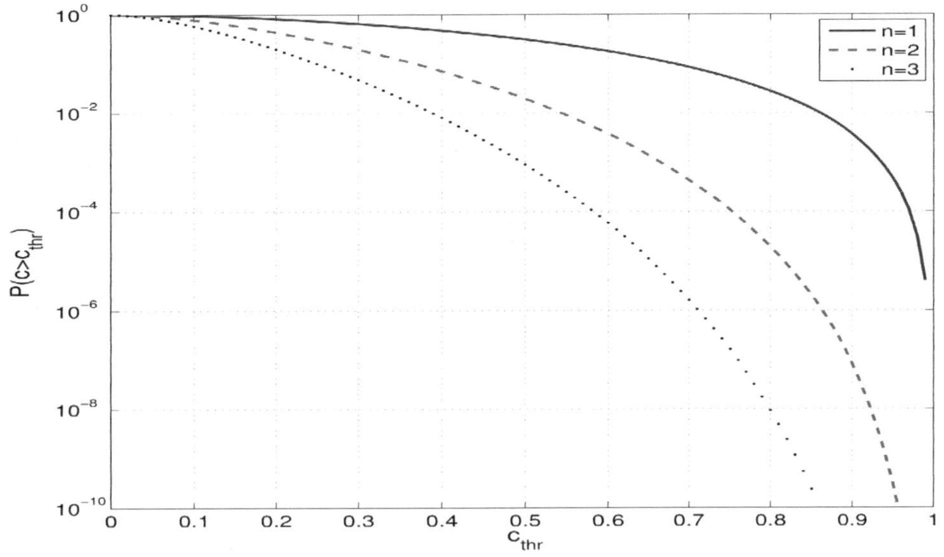

Figure 1. Probability to have a value of the *coherence* larger than a given threshold c_{thr} for a 5n-vector analysis with different values n of the number of datasets.

of the *coherence* larger than $c_{thr} = 0.8$ is ~ 0.03 for $n = 1$, and decreases to $\sim 10^{-5}$ for $n = 2$ and to $\sim 10^{-8}$ for $n = 3$. The computation of the *coherence* after a 5n-vector analysis is then a powerful tool to reject false candidates produced by noise fluctuations.

We have tested the 5n-vector method by analyzing data in which a simulated signal has been added to several realizations of white Gaussian noise. Tab.(1) shows the mean value and variance of the distribution of estimated signal parameters for a set of 500 realizations of noise (with output $SNR \simeq 15$). Columns 1 and 2 refer to the 5-vector analysis of each of the two datasets, which have the same length and same sensitivity in this example. Column 3 refers to the 5n-vector analysis over the two periods (n=2 in this case) while column 4 refers to the 5-vector analysis over the total dataset. We see that the results of the 5n-vector method are identical to those obtained coherently analyzing the whole observation period. On the other hand, while also in the case of analysis over the single periods the mean value of estimated parameters is very near to the injected ones, due to the high SNR we have in this example, the variance of the estimations is larger. In fact, it decreases as the total observation time and so, as expected, for the 5n-vector analysis and for the coherent analysis over the total dataset it reduces by a factor of about 2.

In Tab.(2) we show the results of the same kind of analysis for various values of SNR. The gain in jointly analyzing different datasets with the 5n-vector method is more evident for moderate or low signal-to-noise ratio, when the bias and variance of the estimations in the 5n-vector analysis are much smaller than that of the 5-vector analyses on the single datasets. For instance, for $SNR = 5$ the average relative error on the amplitude H_0 is of about 11.8% in the analysis of the single datasets, while it is of about 3.4% for the 5n-vector analysis. Similarly, for η we have a relative error of about 15% in the former case and 6.2% in the latter case. In Fig.(2) the distribution of estimated parameters for the 5-vector analysis over a single dataset and for the 5n-vector analysis over two datasets are shown for $SNR = 5$. The smaller bias and variance of the estimations with the 5n-vector method are clearly visible.

Table 1. Parameter estimation in the analysis of simulated data. The simulation consisted in the injection of a simulated signal into 500 realizations of white Gaussian noise. Injections were done into two datasets of length 100 days assumed to belong to the Virgo detector. The noise level was the same (output $SNR \simeq 15$). The injected parameters were: $H_0 = 1$, $\eta = 0.16$, $\psi = 24.439^o$. In columns 1 and 2 the mean value and variance of the estimated parameters obtained with the 5-vect analysis applied separately to the two datasets are given. In column 3 the results for the 5n-vect analysis over the two periods are given (with n=2 in this case). Column 4 refers to the 5-vector analysis of the whole period covered by the two datasets.

	dataset 1	dataset 2	5n-vect	total
H_0	1.0071	1.0138	1.0070	1.0070
$\sigma^2_{H_0}$	0.0045	0.0045	0.0025	0.0025
$\bar{\eta}$	0.1568	0.1601	0.1587	0.1587
σ^2_{η}	0.0048	0.0047	0.0023	0.0023
ψ	25.4491	25.4888	25.4746	25.4746
σ^2_{ψ}	4.3798	3.9501	2.0240	2.0240

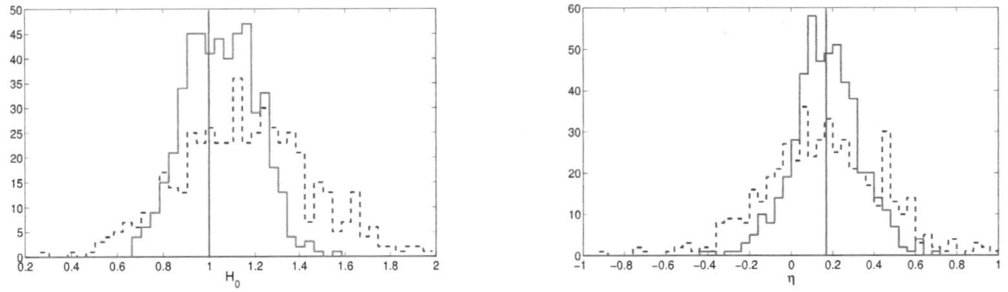

Figure 2. Distribution of estimated H_0 and η for single period analysis (blue, dashed line) and for the 5n-vector joint analysis (green, solid line) in the case $SNR = 5$, see Tab.(2). Vertical red lines indicate the injected parameters. Note that the estimation of parameters over a single period is clearly more biased and has a larger variance respect to the result of the joint analysis.

4. Further developments

The next step to test the method will be the analysis of real data from Virgo and LIGO detectors, searching for software and hardware injected simulated signals. After this validation it will be ready for the search of real signals. A priority will be the joint analysis of Virgo VSR2 and VSR4 and of LIGO S6 data targeted to the Crab pulsar, for which the sensitivities could allow to improve the best published upper limit [12].

Another relevant issue to be addressed is the possible presence of a phase discontinuity in the signal to be detected. This could be due, for instance, to a glitch of the emitting source, like those observed in several pulsars. Glitches are sudden spin-up of a neutron star followed by a long term relaxation period. The mechanism at the base of glitches is still poorly known but the most popular models invoke quakes in the neutron star crust [13] or the transfer of angular momentum from the superfluid to the inner crust [14]. If this happens we can always think to split the data into two datasets corresponding respectively to times before and after the glitch. In this case, however, the direct application of the 5n-vector analysis would not work due to the relative phase of the signal in the datasets. If the phase jump can be accurately estimated, e.g. from radio-astronomical observations, then it is easy to properly shift the signal templates in

Table 2. Parameter estimation in the analysis of simulated data. The simulation consisted in the injection of a simulated signal into 500 realizations of white Gaussian noise for different values of SNR. Injections were done into two datasets both of length 100 days: one assumed to belong to the Virgo detector, the other belonging to LIGO Livingston. The noise level was the same in the two cases. The values of injected parameters are the same as in Tab.(1). In the first two tables the mean value and variance of the estimated parameters obtained with the 5-vector analysis applied separately to the two datasets are given. In the third table the results for the 5n-vector analysis over the two periods are reported (with n=2 in this case). Values for the polarization angle are not shown to keep the table simpler.

Virgo

SNR	H_0	$\sigma^2_{H_0}$	$\bar{\eta}$	σ^2_{η}
5	1.1403	0.0417	0.1394	0.0451
15	1.0127	0.0044	0.1635	0.0049
30	1.0064	0.0012	0.1609	0.0013

LIGOL

SNR	H_0	$\sigma^2_{H_0}$	$\bar{\eta}$	σ^2_{η}
5	1.0962	0.0408	0.1344	0.0445
15	1.0172	0.0045	0.1619	0.0048
30	1.0082	0.0011	0.1603	0.0012

5n-vector

SNR	H_0	$\sigma^2_{H_0}$	$\bar{\eta}$	σ^2_{η}
5	1.0338	0.0226	0.1497	0.0240
15	1.0081	0.0023	0.1624	0.0025
30	1.0036	0.0006	0.1609	0.0007

order to correct it. If the phase shift is not known a simple 'brute force' approach could be to treat it as a free parameter and estimate it by numerically maximizing the total *coherence*. More sophisticated solutions need a detailed study. This issue will be discussed in detail elsewhere.

References

[1] Grote H *et al* 2010 *Classical & Quantum Gravity* **27** 084003
[2] Abbot B P *et al* 2009 *Reports on Progress in Physics* **72** 076901
[3] Accadia T *et al* 2011 *Classical & Quantum Gravity* **28** 114002
[4] Abadie J *et al* 2011 *The Astrophysical Journal* **737** 93
[5] Astone P *et al* 2010 *Classical & Quantum Gravity* **27** 194016
[6] Acernese *et al* 2009 *Classical & Quantum Gravity* **26** 204002
[7] Cutler C and Schutz B F 2005 *Physical Review D* **72** 063006
[8] Dupuis R J and Woan G 2005 *Physical Review D* **72** 102002
[9] Jaranowski P, Krolak A and Schutz B F 1998 *Physical Review D* **58** 063001
[10] Ng C -Y and Romani W 2008 *The Astrophysical Journal* **673** 411
[11] Prix R and Krishnan B 2009 *Classical & Quantum Gravity* **26** 204013
[12] Abbott B P *et al* 2010 *The Astrophysical Journal* **713** 671
[13] Ruderman M, Zhu T and Chen K 1998 *The Astrophysical Journal* **492** 267
[14] Anderson P and Itoh N 1975 *Nature* **256** 25

A Bayesian search for gravitational waves from the Vela Pulsar in Virgo VSR2 data.

C Gill[1], for the LIGO Scientific Collaboration and the Virgo Collaboration.

[1] SUPA, School of Physics & Astronomy, University of Glasgow, Glasgow, G12 8QQ, UK

E-mail: colingill.glasgow@gmail.com

Abstract.
The Virgo detector undertook its second science run (VSR2) from July 2009 to January 2010, providing unprecedented sensitivity to gravitational waves at frequencies below 40 Hz. The VSR2 dataset presented an ideal opportunity [1] to search for gravitational waves from the Vela pulsar (B0833-45, J0835-4510), for which gravitational wave emission is expected at $\sim$ 22 Hz assuming it is a non axi-symmetric rotator. We give a summary of the results obtained in [1], describing the Bayesian method more fully and presenting further details of the data used.

1. Introduction

A recent paper by the Virgo collaboration and the LIGO Scientific Collaboration [1] presented three searches for gravitational waves from the Vela pulsar (PSR B0833-45, PSR J0835-4510) in data from the Virgo interferometer during the Virgo Science Run 2 (VSR2). These searches assumed the rigid triaxial neutron star (NS) model [2], which predicts gravitational wave emission at twice the rotation frequency of the pulsar. The searches also assumed that the gravitational wave signal is phase locked to the observed radio pulses. Here we will give a summary of the results obtained, describing the Bayesian method more fully, the results from this method, and present further details of the data used.

PSR B0833-45 was discovered in 1968 [3], in the Vela supernova remnant, providing some of the first direct observational evidence that established pulsars as rotating NSs. Since the initial detection of radio pulses, weak visible and X-ray pulses have been detected, and strong γ-rays pulses seen.

Certain properties of Vela mark it out as an appealing target to search for gravitational waves. It is young (with a characteristic age of $\sim$ 11 000 y) and has a high spin-down ($\dot{f} \simeq -1.56 \times 10^{-11}\,\mathrm{Hz\,s^{-1}}$), meaning that there is a large amount of energy lost from the system which could be powering gravitational wave emission. Vela is a young pulsar and glitches regularly, with a glitch rate of $1/3\,\mathrm{y}^{-1}$. This would present a problem if Vela glitched during VSR2 as the gravitational wave phase evolution during a glitch is not known. Vela's relative proximity of $\sim 300\,\mathrm{pc}$ also marks it out as a prime candidate for gravitational wave detection.

A useful upper limit can be placed on the gravitational wave strain from a spinning star called the spin-down limit [4]. This limit is computed by assuming that all the rotational energy lost from a source is due to gravitational wave emission

$$h_0^{sd} = \left(\frac{5}{2}\frac{GI_{zz}|\dot{\nu}|}{c^3r^2\nu}\right)^{1/2}, \tag{1}$$

which for Vela is $h_0^{sd} = 3.59 \times 10^{-24}$. This scenario is unrealistic as we know that pulsar spin-down is due to a number of mechanisms, including magnetic dipole radiation. However, the spin-down limit does provide us with a useful benchmark. If we can place an upper limit on the gravitational wave emission below the spin-down limit, then we can constrain the fraction of spin-down energy due to gravitational waves. The spin-down limit for Vela is the highest for all known pulsars. However, Vela is at a disadvantage compared to many other pulsars as the rotation frequency of Vela ($\sim$ 11 Hz) [5], and hence the frequency of gravitational waves ($\sim$ 22 Hz), is near the low frequency limit of current detectors. This low rotation frequency, means that a search for gravitational waves from Vela gains no advantage from including data from the LIGO detectors, as their noise floors at low frequencies are significantly higher than that of the Virgo detector. Figure 1 shows the spin-down limit of Vela (3.29×10^{-24}) on the same axes as the noise curves for the current gravitational wave detector network. It is clear that our best chance of detecting a signal will be using data from the Virgo detector alone, where the advanced seismic isolation system results in a lower noise floor at $\sim$ 22 Hz [6].

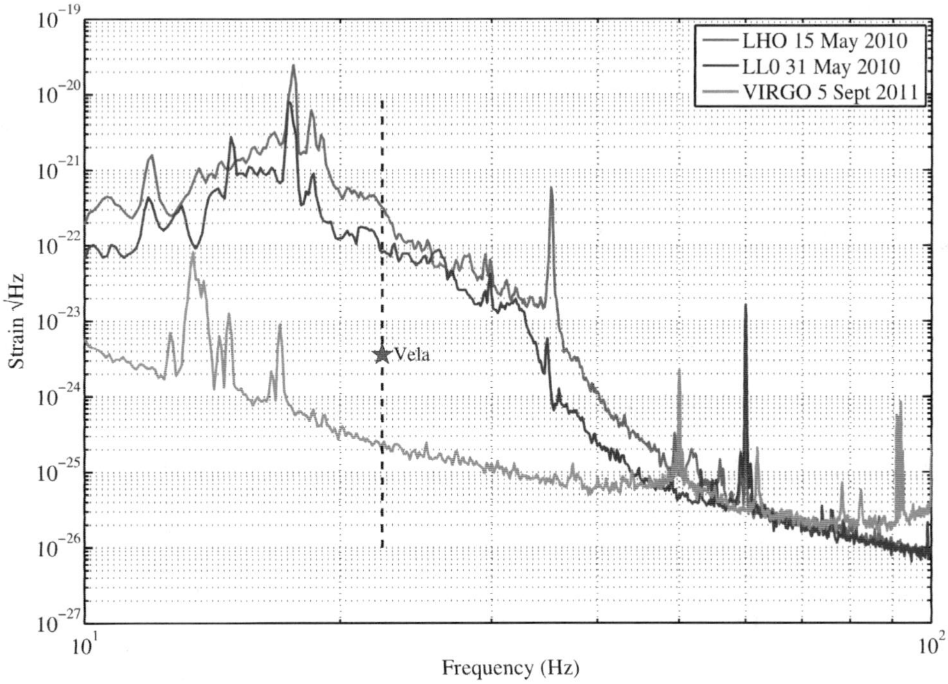

Figure 1. The RMS strain sensitivity noise curves for the LIGO and Virgo interferometers (scaled for the length of the VSR2 run, but not for antenna pattern attenuation), with the spin-down limit plotted for the Vela pulsar at twice its spin frequency. The points at which the dashed line intersects the noise curves shows the noise floor in Virgo data at twice the spin frequency of the Vela pulsar is more than 100 times lower than in LIGO data. [7, 8].

There has been only one previous targeted search for gravitational waves from Vela. That search used 16 days of data from the Cryogenic Laser Interferometer Observatory (CLIO) detector in Japan taken in 2007 and produced a strain upper limit of 5.3×10^{-20}, well above the spin-down limit of $\sim 3.3 \times 10^{-24}$ [9].

The method used for the current search is outlined in Section 2. Section 3 looks at the VSR2 data, highlighting a nuisance noise artefact present in the VSR2 data. The timing model used in the analysis is presented in Section 4. Section 5 briefly describes the recovery of hardware injections in the VSR2 data. The results from the analysis make up Section 6, and the discussion of the results finishes off the paper in Section 7.

2. The Bayesian search method

Three almost independent analysis methods were used to perform a search for gravitational waves from Vela during VSR2. The importance of using different and independent search methods lies not just in the obvious need for corroborating results and cross-checking the analyses' outputs, but also in the situation where different methods can have specific strengths making them more suitable than others in certain situations. The three methods are: a complex heterodyne method using Bayesian formalism and a Markov chain Monte Carlo [10]; a time-domain matched filter method using the $\mathcal{F}$-statistic and $\mathcal{G}$-statistic [11]; and a frequency domain matched filter method using the signal's sidereal modulation Fourier components [12]. A fuller description of the Bayesian method follows.

The Bayesian analysis method used was developed by [13], and has previously been used to search for gravitational waves from many other pulsars e.g. [10, 14]. It consists of two distinct stages. The first stage performs a precise complex heterodyne to remove the known spin phase evolution from the data and drastically down-samples the data in the process (from 16 384 Hz to 1/60 Hz). In order to remove the phase evolution from the signal we assume that the gravitational wave signal is phase-locked to its radio signal. The down-sampling stage is necessary in order to make the process computationally efficient. This Bayesian parameter estimation is the second stage of the search, and uses a Markov Chain Monte Carlo (MCMC) algorithm to explore the posterior probability volume and produce marginalised posteriors for the unknown signal parameters $h_0, \phi_0, \psi, \cos\iota$, defined below.

The heterodyned data undergoes a process aimed at removing any particularly large outliers, by simply removing data points whose absolute value is greater than five times the standard deviation of the data. This test is run on the heterodyned data twice, to account for any extreme outliers that might skew the standard deviation of the data and so degrade the effectiveness of the cleaning.

Once heterodyned, the expected gravitational wave signal has the form [1]

$$\begin{aligned} h'(t) &= h_0\left(\frac{1}{4}F_+(1+\cos^2\iota)\cos\phi_0 + \frac{1}{2}F_\times\cos\iota\sin\phi_0\right) + \\ & \quad ih_0\left(\frac{1}{4}F_+(1+\cos^2\iota)\sin\phi_0 - \frac{1}{2}F_\times\cos\iota\cos\phi_0\right), \end{aligned} \tag{2}$$

where F_+ and $F_\times$ are the antenna beam patterns of the interferometer to plus and cross polarisations, h_0 is the gravitational wave amplitude, ψ is the gravitational wave polarisation angle, ι is the inclination of the pulsar's rotation axis with respect to the line of sight, and ϕ_0 is the initial phase of the signal.

The parameter estimation assumes the data is stationary over 30-minute segments of contiguous data, and divides the data into as many 30-minute sections as possible. If there is a section of data of 5 minutes or longer on the end of a longer contiguous section of data already included, then these smaller segments are also included.

For this search two separate Bayesian parameter estimation runs are performed. The first run has uniform priors on all parameters; this assumes complete ignorance of the parameters prior to the analysis. The second run assumes uniform priors on the parameters h_0 and ϕ_0, but with Gaussian priors on $\cos\iota$ and ψ. The motivation for using Gaussian priors on $\cos\iota$ and ψ comes from observations of the pulsar wind tori as described in [15], and the premise that the orientation of the wind tori matches the orientation of the pulsar spin axis.

For a more detailed description of the gravitational wave signal and the complex heterodyne method see [1, 13].

3. The Data used in the Search

The Virgo science run 2 (VSR2) started on 7th July 2009 21:00:00 UTC and finished on 8th January 2010 22:00:01 UTC. A representative noise curve for this run is shown in Figure 1. Segments of data from the run have various flags assigned to them which indicate the state of the interferometer during those times. The flag that indicates the best possible quality data is known as "science mode"; all the data from "science mode" segments that pass the contiguity and outlier tests, described in Section 2, was used in the search. For VSR2 data, the outlier and contiguity tests removed a few percent of the data. The heterodyned data, with the contiguity and outlier tests applied is shown in Figure 2. A similar plot is shown in Figure 6 from [1]; however, the data from that figure are further processed, including an averaging of the data, to arrive at the data in Figure 2.

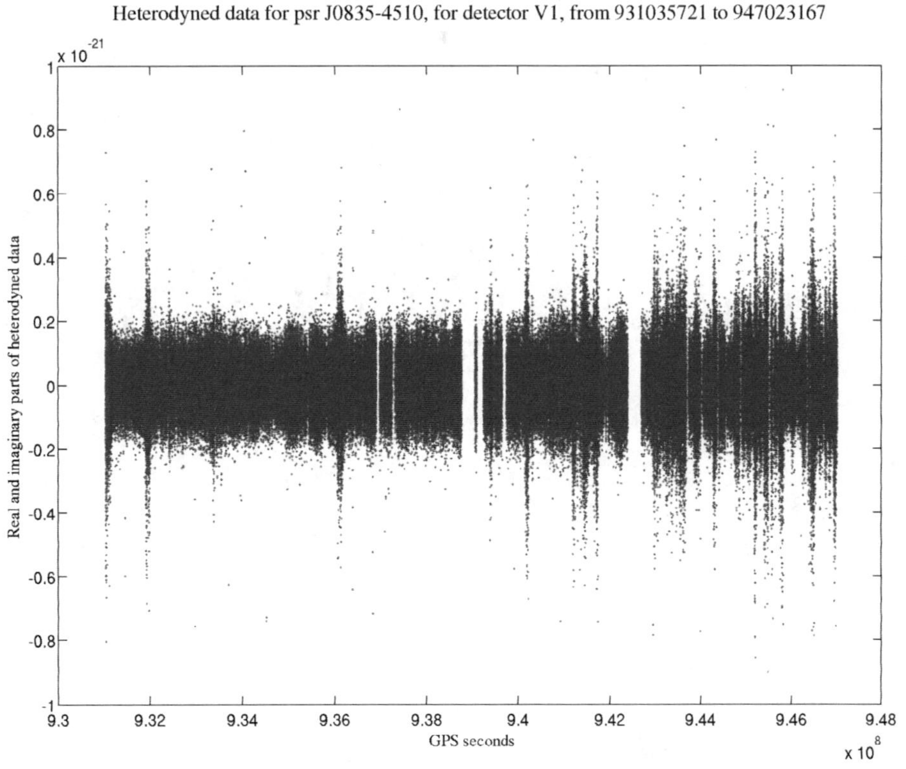

Figure 2. A plot of the heterodyned VSR2 data, with the GPS seconds along the x-axis, and the strain on the y-axis. The blue and red dots represent the real and the imaginary parts of the complex data respectively. The time period covered by this plot is the entire of VSR2, i.e. from 7th July 2009 21:00:00 UTC to 8th January 2010 22:00:01 UTC.

Post-run analysis of the VSR2 data found a previously unknown noise artefact in the frequency band at which we expect gravitational wave signals from Vela [1]. The noise presented itself in the form of a pair of wondering "lines" which can clearly be seen in Figure 3. The cause of the lines was found to be a chiller pump, and measures have been taken to remove the effects on the detector data of this pump in future science runs. It is estimated that the effect of the chiller pump noise lines degraded the sensitivity of the searches by 20% [1], for a more detailed investigation into this noise artefact see [16].

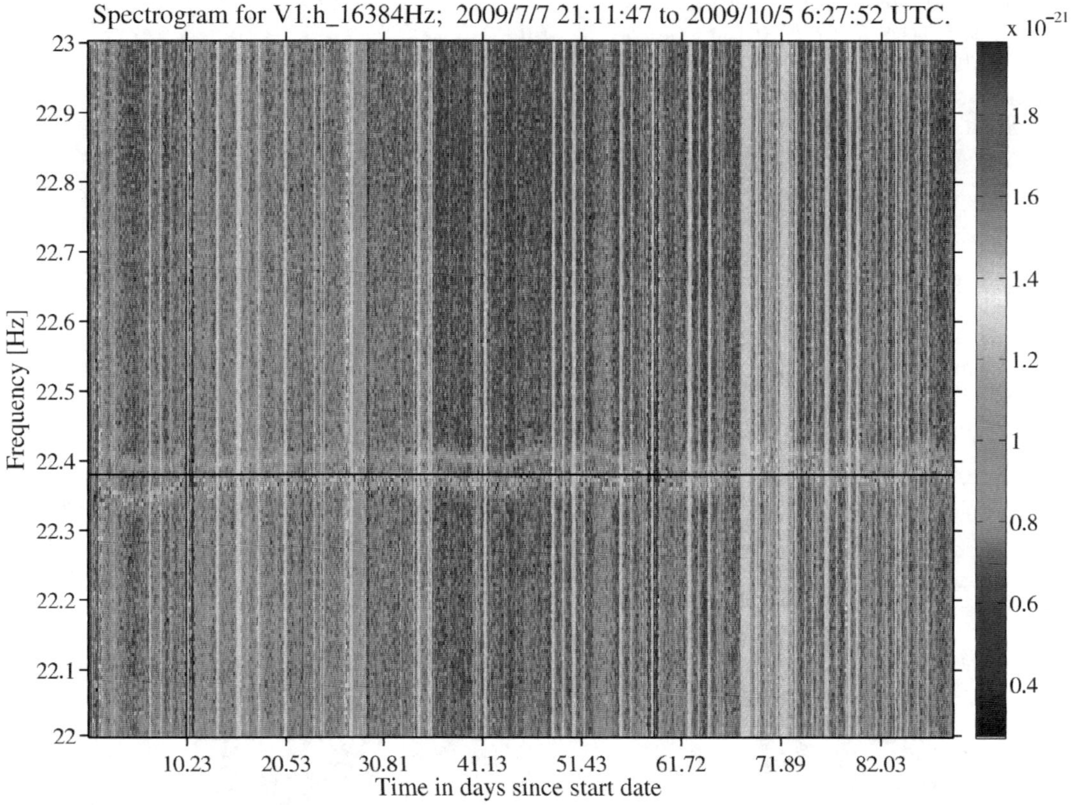

Figure 3. A spectrogram of the whole of VSR2. The expected frequency of gravitational waves from Vela is plotted in black at ~ 22.4 Hz. The chiller lines are clear to see at ~ 22.4 Hz and ~ 22.35 Hz. It is also clear to see the variation in the detector noise over time.

4. The timing model

The complex heterodyne stage of the search which removes the expected phase evolution of the gravitational wave signal from the data needs a precise model of the phase evolution of the expected signal from the pulsar in question. To determine this phase model we assume that the gravitational waves are phase locked with electromagnetic pulses emitted from the pulsar. Detailed timing of these pulses is carried out by several radio telescopes, and it is from this data the phase model is formed [1]. The time of arrival data (TOAs) of the radio pulses were supplied by two different observatories: the Mt. Pleasant Observatory near Hobart in Tasmania (which has two antennas, the 26-m Mt Pleasant antenna and the 14-m Vela antenna built specifically for observing the Vela pulsar), and the Hartebeesthoek Radio Astronomy Observatory (HartRAO) near Johannesburg in South Africa.

Tempo2 software [17] was used to fit a modelled phase evolution to the TOAs. Best fit values for the right ascension, declination, proper motion, rotation frequency, and first and second derivatives of the rotation frequency were produced over the epoch of VSR2. TOAs were used from 1st June 2009 to 31st March 2010. By ensuring that the TOAs span VSR2, we aim to produce an accurate fit to the data and therefore an accurate model of the gravitational wave phase evolution over the same period. The parameters of the timing model produced for VSR2 are shown in Table 1.

Right Acension	$08^{\mathrm{h}}\,35^{\mathrm{m}}\,20.^{\mathrm{s}}7543822 \pm 2.5 \times 10^{-05}$
Declination	$-45^{\circ}\,10'\,32.95068'' \pm 6.8 \times 10^{-4}$
f_{rot} [Hz]	$11.19105730233076 \pm 8.6 \times 10^{-11}$
$\dot{f}_{\mathrm{rot}}$ [Hz/s]	$-1.55838759952784 \times 10^{-11} \pm 3.9 \times 10^{-18}$
$\ddot{f}_{\mathrm{rot}}$ [Hz/s^2]	$4.90687976149349 \times 10^{-22} \pm 8.7 \times 10^{-26}$

Table 1. Table showing the parameters for the fitted timing model for Vela TOAs from 1st June 2009 to 31st March 2010.

Vela is known as a particularly glitchy pulsar, with a glitch rate of $\sim 1/3\,\mathrm{y}^{-1}$, and as the gravitational wave phase evolution is uncertain during a glitch, it was important to ensure that Vela did not glitch during VSR2. The observations of Vela by Hartesbessthoek and Mt. Pleasant radio antennae show no evidence for a glitch during VSR2.

5. Validation with hardware injections

As a test for each of the three analyses' pipelines presented in [1], 13 gravitational wave signals were injected into the VSR2 data by moving one of the interferometer mirrors. Each of the three methods were then used to search for these signals, labelled pulsar00 to pulsar12. All three search methods were able to detect the signal and recover its parameters accurately when the SNR of the injected signal was sufficiently large. The results from the searches for three of the pulsars (pulsars 03, 05, 08), for each of the three search methods are presented in [1]. We present the probability distribution functions (PDFs) for the four unknown signal parameters for pulsar03 in Figure 4. These plots show the PDFs in blue, and also show the injected value for each parameter plotted as black vertical dashed lines on the plots. The injected parameter values have been accurately recovered, with the PDF for the h_0 parameter showing largest discrepancy of $\sim 3\%$ between injected value and the peak in the PDF, which is well within the calibration uncertainty for the Virgo interferometer.

6. Results

The headline result of the three different search methods are upper limits on the gravitational wave signal amplitude h_0. Although the Bayesian and frequentist methods are not asking directly equivalent questions, we find that the upper limits from the three methods, presented in table 2, are very similar.

The output of the Bayesian method's parameter estimation runs are posterior probability distribution functions (PDFs) for each of the signal parameters. Each PDF is calculated by marginalising over the other signal parameters, meaning that at each value of the parameter for which the PDF has been calculated, the posterior probability is calculated considering all plausible combinations of values for the other signal parameters. The PDF of most interest is for the amplitude parameter h_0, as this is the strength of the gravitational wave signal. Where the h_0 PDF indicates there is no signal present, e.g. the PDF peaks at a h_0 value of zero, we

Journal of Physics: Conference Series **363** (2012) 012039 doi:10.1088/1742-6596/363/1/012039

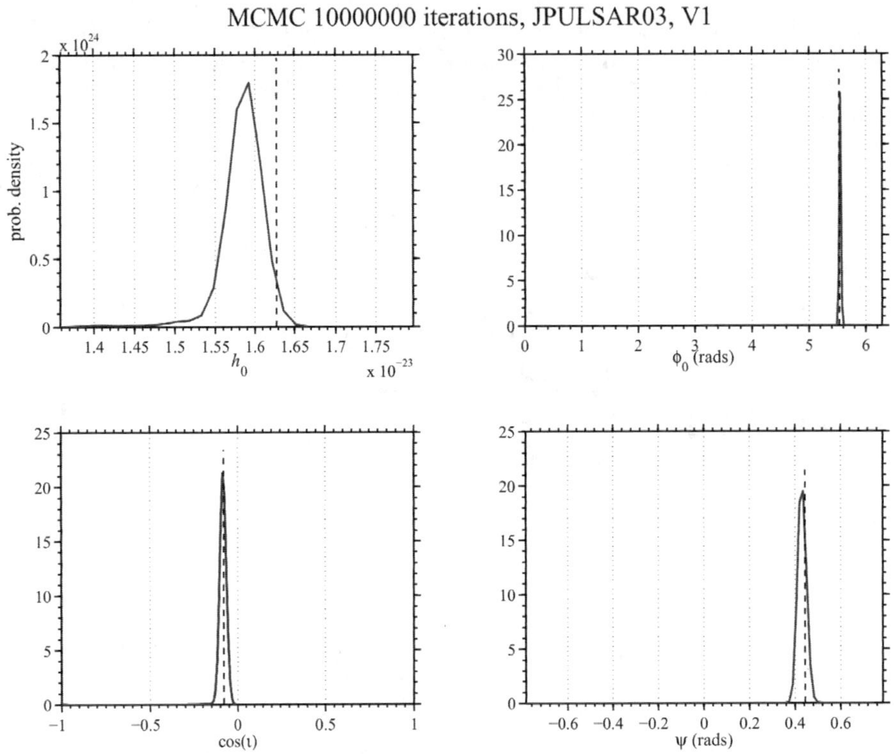

Figure 4. The PDFs for the hardware injection pulsar03. The black vertical dashed lines on each plot shows the injected value for that parameter, it can be seen that the injected parameter values have been accurately recovered. The PDF for the h_0 parameter appears as the worst match to the injected value, however the discrepancy is only $\sim 3\%$.

Analysis method	95% upper limit for h_0
Bayesian heterodyne, restricted priors	$(2.1 \pm 0.1) \times 10^{-24}$
Bayesian heterodyne, unrestricted priors	$(2.4 \pm 0.1) \times 10^{-24}$
$\mathcal{G}$-statistic	$(2.2 \pm 0.1) \times 10^{-24}$
$\mathcal{F}$-statistic	$(2.4 \pm 0.1) \times 10^{-24}$
MF on signal Fourier components, 2 d.o.f.	$(1.9 \pm 0.1) \times 10^{-24}$
MF on signal Fourier components, 4 d.o.f.	$(2.2 \pm 0.1) \times 10^{-24}$

Table 2. Table showing the upper limits obtained by the three different search methods [1].

find the h_0 value that bounds 95% of the probability density. This value is a 95% credible upper limit on h_0. The upper limit on h_0 can also be used to place an inferred upper limit on the equatorial ellipticity of the NS.

The PDFs using uniform priors on the parameters can be seen in Figure 5. The important PDF from Figure 5 is the h_0 PDF, as this tells us how likely different signal strengths are; the most likely value for h_0 is at zero for this PDF, i.e. there is no evidence to suggest there is a signal present. The 95 % credible upper limit on h_0 can be seen as the vertical dashed line, it is $\sim 2.4 \times 10^{-24}$.

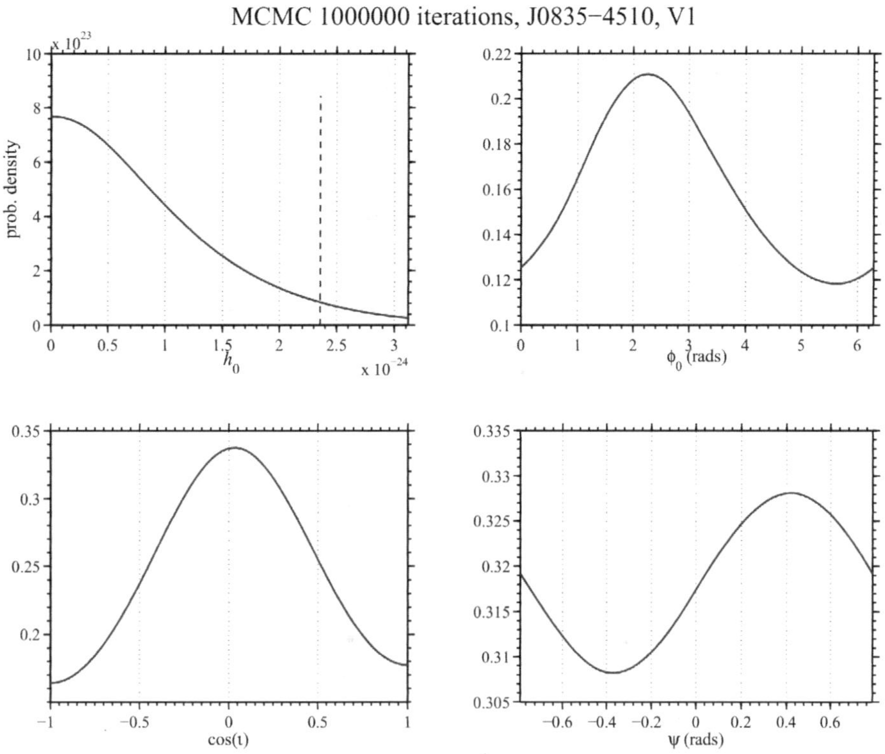

Figure 5. The PDF outputs for Vela from VSR2, produced by MCMC with 1 000 000 iterations [1]. The MCMC was run with uniform priors on all parameters. The 95 % upper limit on h_0 derived from this MCMC is $\sim 2.4 \times 10^{-24}$.

The PDFs obtained using Gaussian priors on ψ and $\cos(\iota)$ can be seen in Figure 6. Again, the h_0 PDF indicates that there is no signal present. It should also be clear that the PDFs for ψ and $\cos(\iota)$ are strongly peaked compared to those in Figure 5, because of the priors placed on these parameters. The 95 % credible upper limit is again plotted as a vertical dashed line at $\sim 2.1 \times 10^{-24}$.

7. Discussion

With the analysis of VSR2 data we have been able to surpass the spin-down limit for Vela with three almost independent search methods [1]. Each search method was used to perform two searches making different assumptions about our knowledge of the signal parameters ψ and ι. Vela is only the second pulsar for which this benchmark has been achieved. We now discuss the results obtained with the Bayesian method in more detail.

For the analysis where we assume no prior knowledge of the unknown signal parameters we have placed a 95 % credible upper limit of 2.4×10^{-24} on h_0, corresponding to ~ 73 % of the spin-down limit of 3.3×10^{-24}. In the analysis where we assume some prior knowledge of the

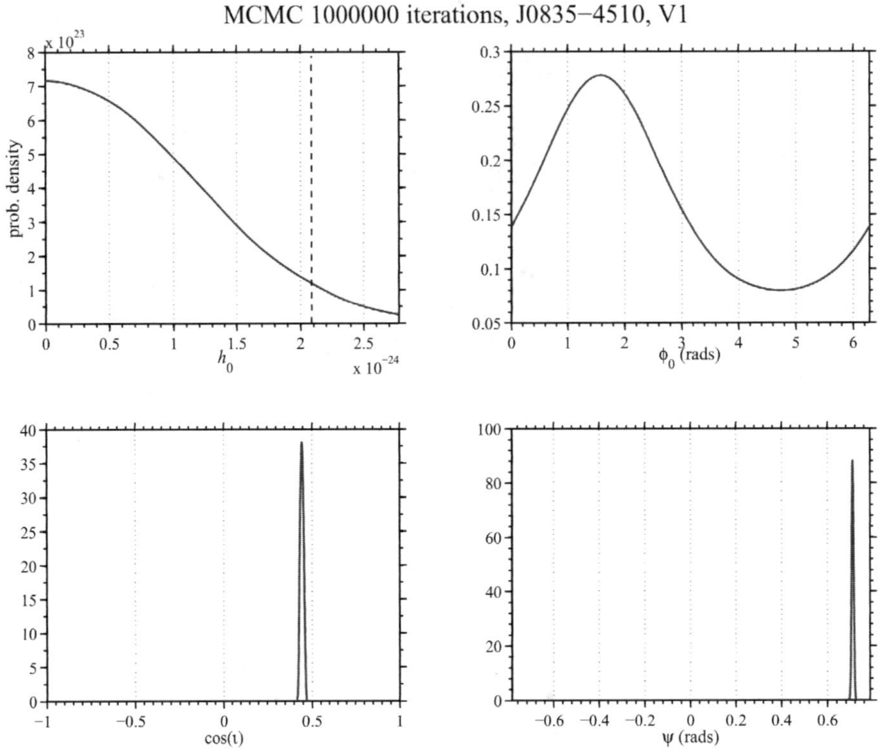

Figure 6. The PDF outputs for Vela from VSR2, produced by MCMC with 1 000 000 iterations [1]. The MCMC was run with Gaussian priors on the orientation parameters ψ and $cos(\iota)$. The 95 % upper limit on h_0 derived from this MCMC is $\sim 2.1 \times 10^{-24}$.

parameters ψ and ι we are able to place a slightly lower 95 % credible upper limit on h_0 of 2.1×10^{-24}. This slightly lower upper limit corresponds to $\sim$ 64% of the spin-down limit.

The upper limits placed on h_0 can be used to infer an upper limit on the equatorial ellipticity of the NS using the canonical value for a NS's moment of inertia of $I = 10^{38}\,\mathrm{kg\,m^2}$, this yields corresponding upper limits on the ellipticity for the two different analyses of $\sim 1.2 \times 10^{-3}$ and $\sim 1.3 \times 10^{-3}$, both of which are greater by several orders of magnitude than the maximum allowable equatorial ellipticity predicted by standard NS equations of state. However, these upper limits represent conservative estimates, as the theoretical predictions for I range from $\sim 1 - 3 \times 10^{38}\,\mathrm{kg\,m^2}$ [10]. It is also useful to represent the upper limits as a fraction of the total spin-down energy, by inferring an upper limit on the power emitted as gravitational waves by $\dot{E}_{GW} = -\frac{32\pi^6 G}{5c^5} I_{zz}^2 \epsilon^2 f^6$; this gives values of 52 % and 45 %. Although we have surpassed the spin-down limit, these results remain far from those for the Crab pulsar [10], where upper limits on the fraction of spin-down energy lost as gravitational waves are $\sim$ 2 %; however, tighter upper limits for Vela are expected to be obtained with the analysis of Virgo VSR4 data, which was collected from 3rd June 2011 to 3rd September 2011.

Acknowledgements

The authors would like to acknowledge S. Buchner of the School of Physics, University of the Witwatersrand, South Africa, A. Hotan of CSIRO Astronomy and Space Science, Australia, and J. Palfreyman of the School of Maths and Physics University of Tasmania, Australia, for providing the TOAs used to construct the timing model for Vela in [1].

The authors gratefully acknowledge the support of the United States National Science Foundation for the construction and operation of the LIGO Laboratory, the Science and Technology Facilities Council of the United Kingdom, the Max-Planck-Society, and the State of Niedersachsen/Germany for support of the construction and operation of the GEO600 detector, and the Italian Istituto Nazionale di Fisica Nucleare and the French Centre National de la Recherche Scientifique for the construction and operation of the Virgo detector. The authors also gratefully acknowledge the support of the research by these agencies and by the Australian Research Council, the Council of Scientific and Industrial Research of India, the Istituto Nazionale di Fisica Nucleare of Italy, the Spanish Ministerio de Educación y Ciencia, the Conselleria d'Economia Hisenda i Innovació of the Govern de les Illes Balears, the Foundation for Fundamental Research on Matter supported by the Netherlands Organisation for Scientific Research, the Polish Ministry of Science and Higher Education, the FOCUS Programme of Foundation for Polish Science, the Royal Society, the Scottish Funding Council, the Scottish Universities Physics Alliance, The National Aeronautics and Space Administration, the Carnegie Trust, the Leverhulme Trust, the David and Lucile Packard Foundation, the Research Corporation, and the Alfred P. Sloan Foundation.

References

[1] Abadie J *et al.* 2011 *ApJ* **737** 93 (*Preprint* 1104.2712)

[2] Jaranowski P, Królak A and Schutz B F 1998 *Phys. Rev. D* **58** 063001–+ (*Preprint* arXiv:gr-qc/9804014)

[3] Large M I *et al.* 1968 *Nature* **220** 340–341

[4] Abbott B *et al.* 2008 *ApJ* **683** L45–L49 (*Preprint* 0805.4758)

[5] Radhakrishnan V *et al.* 1969 *Nature* **221** 443–446

[6] Acernese F *et al.* 2008 *Journal of Physics Conference Series* **120** 032007–+

[7] 2010 http://wwwcascina.virgo.infn.it/DataAnalysis/Calibration/Sensitivity/

[8] 2010 http://www.ligo.caltech.edu/~jzweizig/distribution/LSC_Data/

[9] Akutsu T *et al.* 2008 *Classical and Quantum Gravity* **25** 184013–+

[10] Abbott B P *et al.* 2010 *ApJ* **713** 671–685

[11] Jaranowski P and Królak A 2010 *Classical and Quantum Gravity* **27** 194015 (*Preprint* 1004.0324)

[12] Astone P *et al.* 2010 *Classical and Quantum Gravity* **27** 194016

[13] Dupuis R J and Woan G 2005 *Phys. Rev. D* **72** 102002–+

[14] Abbott B P *et al.* 2007 *Phys. Rev. D* **76** 042001–+

[15] Ng C and Romani R W 2008 *ApJ* **673** 411–417 (*Preprint* 0710.4168)

[16] Accadia T *et al.* 2012 **The NoEMi (Noise Frequency Event Miner) framework** *Journal of Physics Conference Series* [In these proceedings]

[17] Hobbs G B, Edwards R T and Manchester R N 2006 *MNRAS* **369** 655–672

Journal of Physics: Conference Series **363** (2012) 012040 doi:10.1088/1742-6596/363/1/012040

The cross-correlation search for a hot spot of gravitational waves : Numerical study for point spread function

Yuta Okada[1], Nobuyuki Kanda[1*], Sanjeev Dhurandhar[2], Hideyuki Tagoshi[3] and Hirotaka Takahashi[4,5]

[1] Graduate School of Science, Osaka City University, Sumiyoshi-ku, Osaka 558-8585, Japan
[2] Inter-University Centre for Astronomy and Astrophysics, Post Bag 4, Ganeshkhind, Pune 411007, India
[3] Department of Earth and Space Science, Graduate School of Science, Osaka University, Toyonaka, Osaka 560-0043, Japan
[4] Department of Humanities, Yamanashi Eiwa College, 888, Yokone, Kofu, Yamanashi 400-8555, Japan
[5] Earthquake Research Institute, University of Tokyo, Bunkyo-Ku, Tokyo 113-0032, Japan
[*] Paper presented at the conference by N. Kanda

E-mail: `kanda@sci.osaka-cu.ac.jp`

Abstract. The cross-correlation search for gravitational waves, also known as 'radiometry', has been previously applied to map the gravitational wave stochastic background in the sky and also to target gravitational waves from rotating neutron stars/pulsars. We consider the Virgo cluster which may appear as a 'hot spot' spanning few pixels in the sky in a radiometry analysis. Our results show that sufficient signal to noise ratio can be accumulated with integration times of the order of a year. We also present a numerical simulation of radiometric analysis, assuming ground-based detectors which are currently under construction or being upgraded. The point spread function of the injected sources is confirmed by numerical tests. The typical resolution of radiometry analysis is a few square degrees, as compared to the several thousand pixels for the full sky in an all-sky map.

1. Introduction

Ground-based gravitational wave detectors currently being constructed or upgraded may reach the strain sensitivity of the order of $10^{-24}[1/\sqrt{\mathrm{Hz}}]$ in the near future. The detector sites are widely distributed around the globe, which makes independent observation of gravitational waves possible. Since the distributed detectors may not have correlated noise but will have a correlation in the gravitational signal, the cross-correlation of data from two or more detectors makes is a possible way to search for unknown gravitational waveforms. This is a useful property for gravitational wave searches esecially for short duration bursts, or for long duration stochastic gravitational waves. The measurements from spatially separated detectors allow triangulation of the source. The time delay between detectors can be used to determine the incident direction of the gravitational wave in short duration signal analysis. On the other hand, employment of the time delay term in searches for long duration signals or stochastic gravitational waves which come from an arbitrary direction is not simple.

However, we can extract the signal strength if we deconvolve the Earth's rotation from the signal for any point in the sky. Using multiple detectors and long integration times, the deconvolution is possible. This type of directional cross-correlation search for gravitational waves is known as 'radiometry'.

Radiometry and any other triangulation analysis requires widely separated detectors - long baselines, and the variation of orientation (zenith direction, and azimuthal rotation of interferometers) to cover the sky and resolve the polarizations. The recent beginning of the construction of LCGT[1] in addition to previously constructed advanced detector sites of LIGO[2] and Virgo[3] encourage the development of radiometric analysis.

1.1. Radiometry Filter

Gravitational wave radiometry is a type of stochastic gravitational wave search, but it is also directed or targeted. Radiometric analysis assumes that gravitational waves come from a particular direction such as a point like source or a 'hot spot' in the sky. The radiometry filter uses data from at least two detectors and does not require the knowledge of the waveform. Gravitational waves coming from a particular direction will have a time of arrival delay at two detectors as shown in Fig.1. Since gravitational wave signals at both detectors are coherent in the same polarization component, we can extract gravitational wave signals in the cross-correlation product from two detectors observations with appropriate time delay. Moreover, the signal-to-noise ratio for the gravitational wave will be increased with long integration time, because the two detector noises are independent. By changing the source direction sequentially after every step, we can construct a gravitational wave 'sky map' by radiometric analysis.

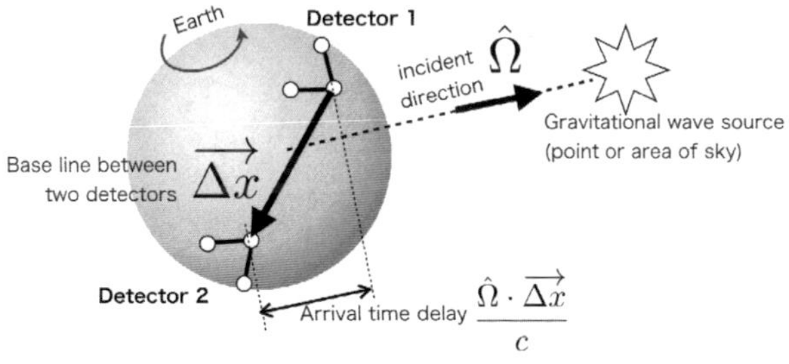

Figure 1. Ideal setup of a radiometry analysis.

Radiometry analysis has been proposed and studied in previous works[4][5]. We explain the radiometry filter briefly here. The source direction or trial direction of the search can be represented by a unit vector $\hat{\Omega}$. The two detectors are labelled as 1 and 2, and the base-line from detector 1 to detector 2 is denoted by the vector $\overrightarrow{\Delta x}$. The arrival time delay for the gravitational wave between the two detectors is $\frac{\hat{\Omega}\cdot\overrightarrow{\Delta x}}{c}$, where c is the speed of light. The output from each detector $s_1(t), s_2(t - \frac{\hat{\Omega}\cdot\overrightarrow{\Delta x}}{c})$ in the time domain has the same gravitational wave timing. In the Fourier domain, this time delay represents a phase difference of the detectors output $\tilde{s}_1(f), \tilde{s}_2(f)$, the Fourier transform of $s_1(t)$ and $s_2(t)$. Therefore, we can define the radiometry filter as follows. A directed overlap reduction function $\gamma(f,\hat{\Omega})$ can be defined to correct the phase difference as

$$\gamma(f,\hat{\Omega}) = \sum_{A=+,\times} F_1^A F_2^A e^{i2\pi f \frac{\hat{\Omega}\cdot\overrightarrow{\Delta x}}{c}} , \quad (1)$$

Journal of Physics: Conference Series **363** (2012) 012040 doi:10.1088/1742-6596/363/1/012040

where $F^A_{1,2}$ are the antenna pattern[6] of detector 1 or 2 for each polarization $A = +$ or $\times$ of the gravitational wave. For optimal filtering, we employ the detectors noise power spectrum $P_{1,2}(f)$ respectively, and assume that gravitational wave spectrum is $H(f)$. The radiometry filter can defined as

$$Q(f,\hat{\Omega}) = \lambda \frac{\gamma^*(f,\hat{\Omega})H(f)}{P_1(f)P_2(f)} , \tag{2}$$

and the filter output is

$$\Delta S(\hat{\Omega}) = \int_{-\infty}^{\infty} df\ \tilde{s}_1^*(f)\tilde{s}_2(f)Q(f,\hat{\Omega}). \tag{3}$$

The base-line $\overrightarrow{\Delta x}$ will change in celestial coordinates according to Earth's rotation in the case of ground-based detectors. Therefore, we calculate ΔS over a short time slice ('chunk') $t - \Delta t/2$ and $t + \Delta t/2$. Therefore the quantities defined in the above equations also become functions of t. Thus the Fourier transforms of detector data can be represented using the short Fourier transforms of $\tilde{s}^*_{1,2}(t;f)$, and $Q(f)$ becomes $Q(t,f)$ and so on. Then

$$\Delta S(t,\hat{\Omega}) = \int_{-\infty}^{\infty} df\ \tilde{s}_1^*(t;f)\tilde{s}_2(t;f)Q(t,f,\hat{\Omega}). \tag{4}$$

Processing many chunks for long duration observational data, we can determine the statistics of ΔS. Since the gravitational wave component will appear in the real part of ΔS, we employ the mean $\mu(\hat{\Omega})$

$$\mu_{\Delta S}(\hat{\Omega}) = \langle \Re[\Delta S] \rangle , \tag{5}$$

and the standard deviation $\sigma_{\Delta S}(\hat{\Omega})$ for the real part of ΔS.

We now need to sum up over the contributions to the SNR, which we denote by $\rho(\hat{\Omega})$, for all the chunks making up the total observation period. The statistic for the full observation time is denoted by $S(\hat{\Omega})$ and its mean and standard deviation by $\mu(\hat{\Omega})$ and $\sigma(\hat{\Omega})$ respectively. The signal-to-noise ratio of particular direction $\hat{\Omega}$ is then given by:

$$\rho(\hat{\Omega}) = \mu(\hat{\Omega})/\sigma(\hat{\Omega}). \tag{6}$$

2. Hot Spot of Gravitational Wave Radiometry

A radiometry analysis might find a hot spot in the sky map. A hot spot may be formed by a concentration of a large number of unresolved astronomical sources. One of the promising targets is the Virgo cluster, which could contain $\sim 10^{11}$ neutron stars. Our results show that sufficient signal to noise can be accumulated with integration times of the order of a year. We published this study in the reference [7]. If the average ellipticity of neutron stars is 10^{-6}, Virgo cluster might appear as a hot spot with observation of order of year, using 2nd generation ground-based detectors. The numerical estimation of the expected GW spectrum and signal-to-noise ratio of a Virgo cluster hot spot is explained in this reference in detail.

Here we remark on the average detector response as a function of source declination. In eq.(1), the magnitude of two detector's combined detector response $\Gamma(\hat{\Omega},t)$ is given as

$$\Gamma(\hat{\Omega},t) = F_{+1}(\hat{\Omega},t)F_{+2}(\hat{\Omega},t) + F_{\times 1}(\hat{\Omega},t)F_{\times 2}(\hat{\Omega},t). \tag{7}$$

Since $\Gamma(\hat{\Omega},t)$ will change according to Earth's rotation, we wish to use the average over day:

$$\langle \Gamma^2 \rangle_{1\text{day}}(\hat{\Omega}) = \frac{1}{T_{1\text{day}}} \int_0^{1\text{day}} \Gamma^2(\hat{\Omega},t)\, dt. \tag{8}$$

$\langle \Gamma^2 \rangle$ will not depend on the right ascension. Fig.2 displays the square root of $\langle \Gamma^2 \rangle$. For the Virgo cluster whose center is at declination of about $12°4'59''$, LCGT and the India site (IndIGO[8]) is a good combination. Using several detector combinations will cover the whole sky in radiometry analysis.

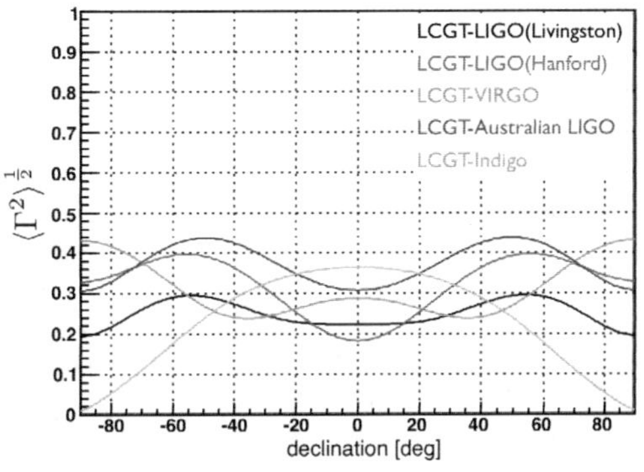

Figure 2. Average of detector response Γ^2 as a function of celestial declination angle.

3. Numerical Simulation: Source Injection Test

To confirm the behavior of a hot spot in the radiometry sky map, we construct a numerical test of the radiometry filter. It consists of the following steps.

1) **segment the data stream into short duration *'chunks'*:** In real observational data processing, steps 2) 5) are performed time series data split into short duration 'chunks'.
2) **generate gravitational wave detector noise:** Assuming the design noise spectrum [9] [10] [11] [12] of each detector and injecting random gaussian noise, we generate time series of noise $n(t)$. In this paper, the data sampling frequency is 4000 Hz, and the chunk length is 65.536 sec. In the study of the point spread function in next subsection, we skip noise generation and use the gravitational wave signal only.
3) **generate gravitational wave signal:** The gravitational wave signal at the source is generated. We test with a sinusoidal wave, and random phase waveform which has a flat power spectrum in a certain frequency band, e.g. flat power spectrum in 300±30 Hz.
4) **inject gravitational wave signal into detector noise in the time domain:** The gravitational wave signal in the detectors will be affected by the Doppler effect, antenna patterns and time delay between the detectors. The Doppler effect is appears as a frequency modulation of the gravitational wave in the detector according to the orbital and rotational motion of the Earth. The antenna patterns $F_+, F_\times$ are calculated for the incident direction. Also the arrival time difference of the signal on two (or more) detectors are given in this step. These quantities will change as a function of time due to the Earth's motion.
5) **calculate radiometry filter output $\Delta S(t, \hat{\Omega})$:** We calculate radiometry filter output of this chunk in eq.(4) using two detectors for the particular direction $\hat{\Omega}$ of a *'pixel'*. Changing $\hat{\Omega}$ step by step, sweeping over all the sky, we calculate $\Delta S(t, \hat{\Omega})$ for all pixels.
6) **repeat step 1) – 5) for next chunk:** We iterate steps above for all chunks.

Finally,

7) **sum up ΔS and calculate the signal-to-noise ratio $\rho(\hat{\Omega})$ of each pixel:** We sum up each chunk's output $\Delta S(t, \hat{\Omega})$ with respect to each pixel. Then we get $\mu(\hat{\Omega}), \sigma(\hat{\Omega})$ and calculate the signal-to-noise ratio $\rho(\hat{\Omega})$ in eq.(6). The $\rho(\hat{\Omega})$ are displayed as pixels of a sky map with color according to its magnitude as in Fig.3.

Fig.4 displays a schematic data flow of our numerical test of the radiometry filter. We tried various combinations of current detectors, and Fig.4 displays LCGT and advanced LIGO as an example. For studying the point spread behavior in the following sections of this paper, we use the case of LCGT and advanced LIGO at Livingston.

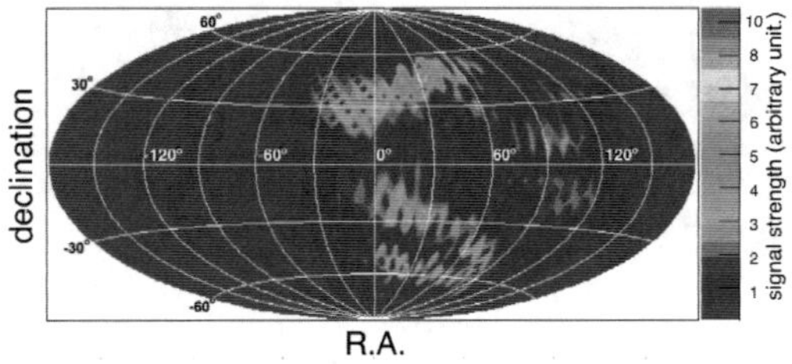

Figure 3. Example radiometry sky map. This map is displayed in celestial coordinates. Red small area (R.A. 30 degree, declination 45 degree) is corresponding to the injected gravitational wave source.

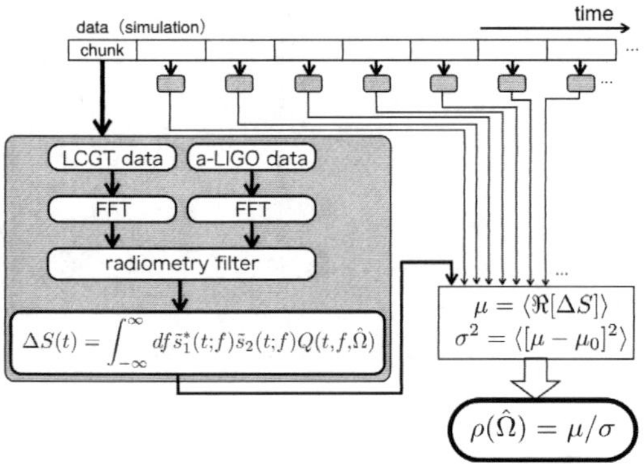

Figure 4. Data flow of numerical test of the radiometry filter.

3.1. Point Spread Function

How spread out a point source appears on the sky is important for evaluating the feasibility of searches. This is called the *'point spread function'*. The point spread function depends on the gravitational wave frequency or spectrum, the position on the sky (right ascension, celestial declination), and on the detector orientation on the Earth. In the point spread function study in this subsection, we used noiseless detector output, but the other filter settings are the same as above. Thus, the magnitude of the gravitational wave in this section is arbitrary. Our test gravitational wave signal has a flat spectrum around f_0 [Hz] given by

$$H(f) = \begin{cases} 1 & (0.9 f_0 \leq f \leq 1.1 f_0) \\ 0 & (\text{otherwise}). \end{cases} \tag{9}$$

The injected signal for each chunk fluctuates as a stochastic signal whose phase is random.

Fig.5 shows the point spreading for some frequency samples in a single chunk process. In a single chunk, we can find wide and multiple spreading images from the injection of a point source. These multiple images occur for sky positions where the time delay along the two-detector baseline differs from the correct delay by an integer number of periods of the central frequency of the narrowband signal. The spatial interval is wide for low frequency band width, and narrow for higher frequency bandwidth. But in general, it is hard to identify the exact position of the point source in a single chunk map, even if the gravitational wave is strong.

As the Earth rotates, the multiple (aliased) images will change position as the detector orientation relative to the source changes. Therefore, during a long integration time, aliases will be reduced and the true signal position will build up with stacking filter output coherently. Fig.6 displays a radiometry map with one day integration for a narrow region around the

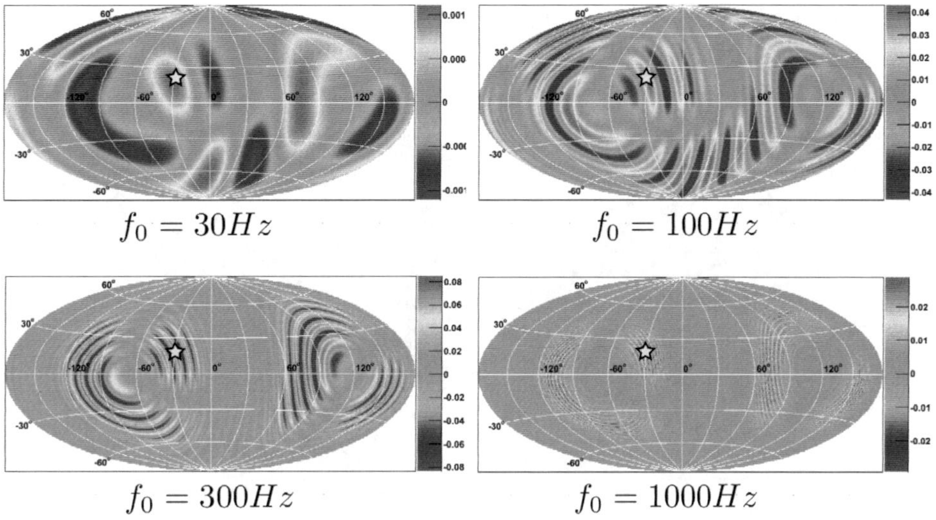

Figure 5. Radiometry filter output of a single chunk with various frequency signals with the LCGT and LIGO Livingston. The star symbol shows the true injection position. The gravitational wave has a flat spectrum around f_0 [Hz] with $\pm$ 10% of f_0 [Hz] width; e.g. f_0=300Hz means from 270Hz to 330Hz. The chunk duration is 65.536 sec.

injection sources. The figure displays variation in declination angle of $\delta = 15°, 45°$and $75°$. The point spread function is not dependent on right ascension with more than one day integration. According to the combined detector response as eq.(7) and eq.(8) with the LCGT and LIGO Livingston, mid declination angles have a circular image spread. On the other hand, in low and high declination areas, the point spread function is an ellipsoid.

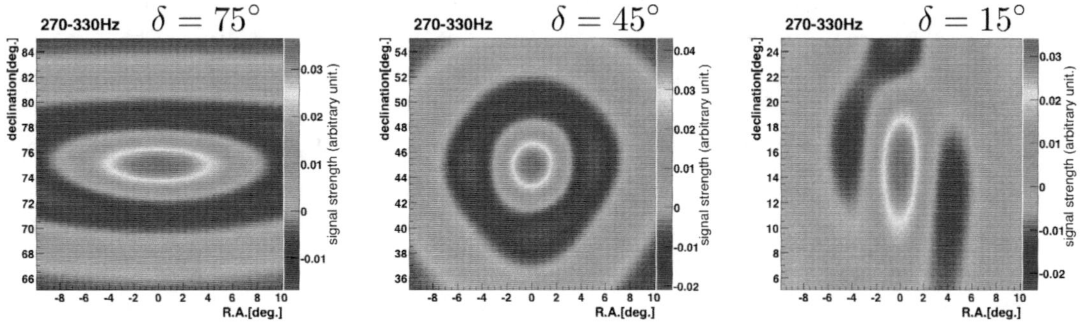

Figure 6. Point spread image in radiometry map (zoomed in) with one day integration for various declination angles with the LCGT and LIGO Livingston. The frequency f_0 of the injected signal is 300 Hz. The declination angles are $\delta = 15°, 45°$and $75°$ respectively.

The spread of a point source is strongly dependent on the gravitational wave frequency. Its order is determined by the diffraction limit $\frac{\lambda}{|\overrightarrow{\Delta x}|}$, where λ is the wave length of the gravitational wave and the distance between the detectors is $|\overrightarrow{\Delta x}|$. Fig.7 displays some examples of the image spread for a point injection at $\delta = 60°$. The signal at 30 Hz spreads widely. The 1kHz signal has aliases as concentric circles, but the true injected point appears as in intense peak.

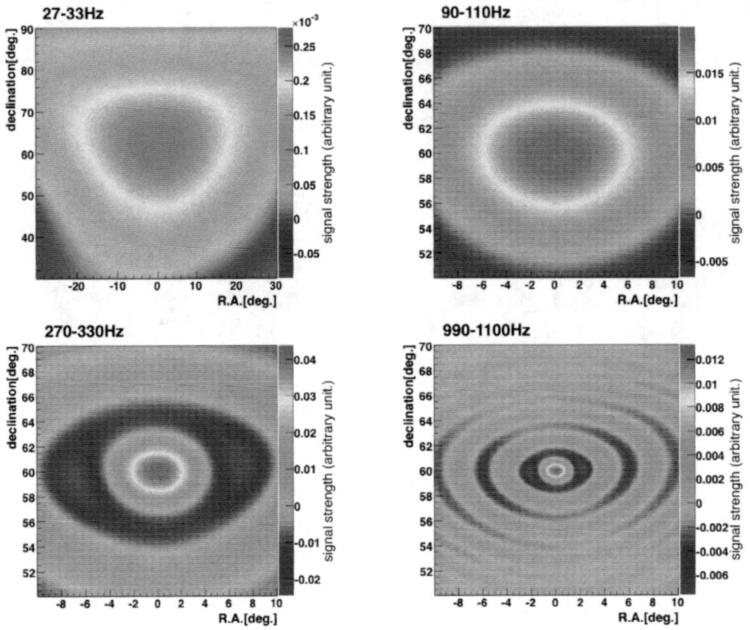

Figure 7. Point spread image in a radiometry map with one day integration for variation of gravitational wave frequencies with the LCGT and LIGO Livingston. The injection point is at $\delta = 60°$. The central frequency of each figure is 30Hz, 100Hz, 300Hz and 1kHz respectively.

To qualify this spreading, we define the spread size as the solid angle $\Delta\Omega$ [sr] of the area which is larger than half the height of the local peak. Fig.8 displays the relation between $\Delta\Omega$ and the frequency of the gravitational wave. The solid line is $\frac{\lambda}{|\overrightarrow{\Delta x}|}$ and the dashed line is an empirical approximation for the high δ region :

$$\sim \pi \left(\frac{1}{2} \frac{c}{f\ |\overrightarrow{\Delta x}|} \right)^2 \times \left(\frac{80 - \delta\ [\mathrm{deg.}]}{100} \right) [\mathrm{sr}]. \tag{10}$$

In the Fig.8, we also display the corresponding apparent diameter for the y-axis in the case of a circular image.

3.2. Optimal Pixel Size

Point spreading suggests an optimal size of pixels. It is possible to employ pixels of a similar size to the point spread function, because this is the angular resolution of the radiometry analysis. However, there is no need to divide the celestial sphere finely that one pixel is far smaller than the point spread function. Fig.9 displays the optimal number of pixels in the sky when we use the as point spread size (solid angle) as the pixel size. In this study with the LCGT and LIGO Livingston base line can employ up to several thousand pixels for the good sensitivity frequency band of the detectors, around $100 \sim 300$ Hz.

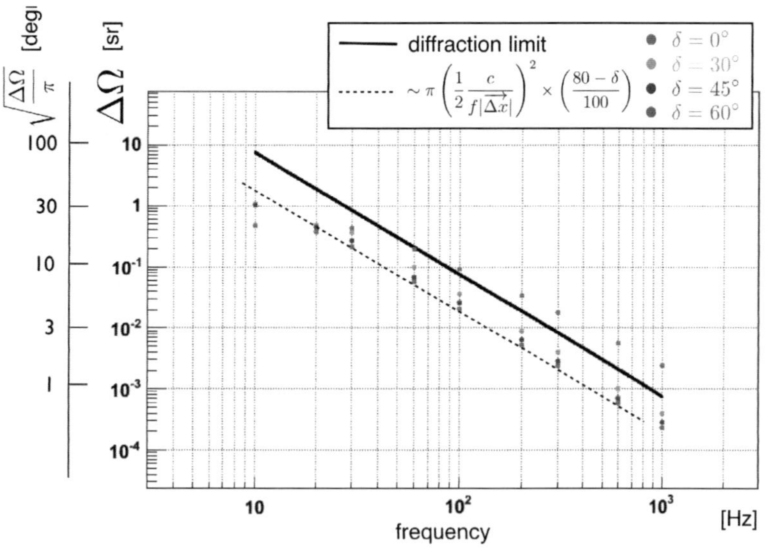

Figure 8. Point spread size of injected gravitational waves. Points are injected gravitational wave signals with different central frequency f_0 and variation of declination angle δ. The x-axis is the gravitational wave frequency, and the y-axis is the solid angle of point spreading. The y-axis also displays the corresponding apparent diameter for the case of a circular image.

The Earth rotation during one chunk length ($\sim$ 65 sec.) in this analysis is approximately 0.25 degree. This is closer to the diffraction limit in high frequency bandwidth $\sim$ 1 KHz. Therefore, we have to use shorter chunks to get the fine resolution for high frequency bandwidth.

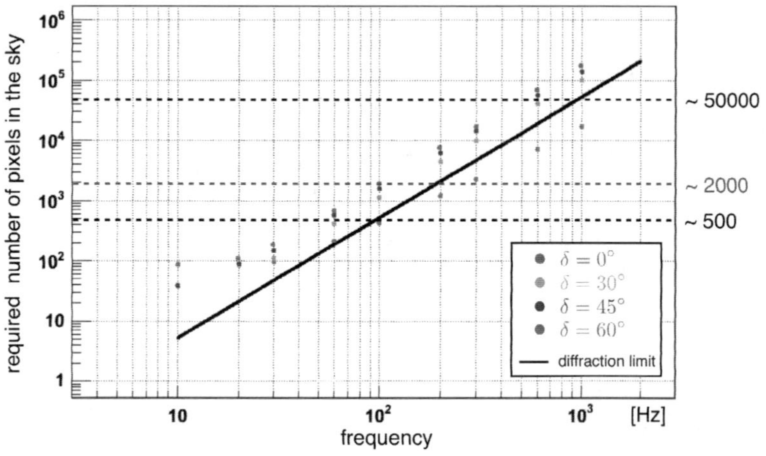

Figure 9. Optimal number of pixels to cover the sky when we use the as point spread size (solid angle) as the pixel size. The solid line corresponds to the diffraction limit size.

4. Summary

We have studied radiometry analysis for stochastic gravitational waves of astronomical origin, which might have an anisotropy or local excess. The Virgo cluster may appear as a 'hot spot' on the radiometry map of the sky with advanced and third generation gravitational wave detectors. We have performed numerical simulations with a radiometry filter, and displayed the image spread for point like gravitational wave sources, with the LCGT and advanced LIGO combination. Other combinations of the global network detectors will be useful to determine the signal properties. We are planning to proceed with injection tests for multiple sources and sources spread over areas larger than the point spread function.

Acknowledgments

S. Dhurandhar acknowledges the DST and JSPS Indo-Japan international cooperative programme for scientists and engineers for supporting visits to Osaka City University, Japan and Osaka University, Japan. H. Tagoshi, N. Kanda and H. Takahashi thank JSPS and DST under the same Indo-Japan programme for their visit to IUCAA, Pune, India. N.Kanda's work was supported in part by a Monbu Kagakusho Grant-in-aid for Scientific Research of Japan (No. 23540346). H.Tagoshi's work was also supported in part by a Monbu Kagakusho Grant-in-aid for Scientific Research of Japan (Nos. 20540271 and 23540309). H.Takahashi's work was also supported in part by a Monbu Kagakusho Grant-in-aid for Scientific Research of Japan (No. 23740207).

References

[1] K. Kuroda and the LCGT Collaboration, Classical Quantum Gravity 23, S215 (2006).
[2] B. P. Abbott et al., Rep. Prog. Phys. 72 (2009) 076901
[3] T. Accadia et al., Class. Quantum Grav. 28 114002
[4] S. Mitra, S. V. Dhurandhar, T. Souradeep, A. Lazzarini, V. Mandic, S. Bose, and S. Ballmer, Phys. Rev. D 77, 042002 (2008).
[5] D. Talukder, S. Mitra, and S. Bose, Phys. Rev. D 83, 063002 (2011).
[6] P. Jaranowski, A. Krolak, and B. F. Schutz, Phys. Rev. D 58, 063001 (1998).
[7] S. V. Dhurandhar, H. Tagoshi, Y. Okada, N. Kanda and H. Takahashi, Phys. Rev. D 84, 083007 (2011)
[8] http://www.gw-indigo.org/
[9] http://gwcenter.icrr.u-tokyo.ac.jp/researcher/parameters
[10] https://dcc.ligo.org/cgi-bin/DocDB/ShowDocument?docid=2974
[11] https://wwwcascina.virgo.infn.it/advirgo/
[12] http://www.et-gw.eu/etsensitivities

Journal of Physics: Conference Series **363** (2012) 012041 doi:10.1088/1742-6596/363/1/012041

A new code for parameter estimation in searches for gravitational waves from known pulsars

M Pitkin[1], C Gill[1], J Veitch[2,3], E Macdonald[1] and G Woan[1]

[1] SUPA, School of Physics & Astronomy, University of Glasgow, Glasgow, G12 8QQ, UK

E-mail: matthew.pitkin@glasgow.ac.uk, colingill.glasgow@gmail.com, e.p.macdonald.glasgow@gmail.com, graham.woan@glasgow.ac.uk

[2] School of Physics & Astronomy, Cardiff University, Queens Buildings, The Parade, Cardiff, CF24 3AA, UK

[3] Nikhef – National Institute for Subatomic Physics, Science Park 105, 1098 XG Amsterdam, The Netherlands

E-mail: johnv@nikhef.nl

Abstract. We describe the consistency testing of a new code for gravitational wave signal parameter estimation in known pulsar searches. The code uses an implementation of nested sampling to explore the likelihood volume. Using fake signals and simulated noise we compare this to a previous code that calculated the signal parameter posterior distributions on both a grid and using a crude Markov chain Monte Carlo (MCMC) method. We define a new parameterisation of two orientation angles of neutron stars used in the signal model (the initial phase and polarisation angle), which breaks a degeneracy between them and allows more efficient exploration of those parameters. Finally, we briefly describe potential areas for further study and the uses of this code in the future.

1. Introduction

Over the last decade several searches have looked for gravitational waves from a large selection of known pulsars using data from the LIGO, GEO 600 and Virgo gravitational wave detectors [1, 2, 3, 4, 5, 6]. No evidence for signals has been observed, but these searches have produced upper limits on the gravitational wave amplitude for many pulsars by using Bayesian inference to calculate posterior probability distributions on the unknown signal parameters [7]. These searches have all assumed a single signal model in which the pulsars are triaxial and emit at precisely twice their rotational frequency, with the complex heterodyned and narrow-banded time domain signal described by

$$\begin{aligned} h(t) =& h_0 \left(\frac{1}{4} F_+(t,\psi)(1+\cos^2\iota)\cos\phi_0 + \frac{1}{2} F_\times(t,\psi)\cos\iota\sin\phi_0 \right) + \\ & ih_0 \left(\frac{1}{4} F_+(t,\psi)(1+\cos^2\iota)\sin\phi_0 - \frac{1}{2} F_\times(t,\psi)\cos\iota\cos\phi_0 \right), \end{aligned} \quad (1)$$

where the four unknown parameters are the gravitational wave amplitude, h_0, polarisation angle, ψ, cosine of the inclination angle, $\cos\iota$ and the initial phase, ϕ_0, and F_+ and $F_\times$ are the detector antenna patterns for the '+' and '×' polarisations for the pulsar's known sky position.

In early searches the posterior probability distribution was calculated on a 4-dimensional grid and the marginalisation integral required to produce posteriors on individual parameters was performed numerically using the trapezium rule. Later searches, which began allowing for inclusion of narrow uncertainties on extra signal parameters (such as frequency and sky position), performed the posterior exploration using a very basic Markov chain Monte Carlo (MCMC) method [5]. Both these methods, as applied in the search code, present practical implementation issues that mean they are not very flexible to all situations we may encounter: the former grid-based method quickly becomes computationally unworkable for larger parameters spaces, wastes time on posterior volumes with tiny probabilities and requires tuning of the grid for different strength signals; the MCMC method implemented was crude, required extensive tuning for certain signals and was not able to produce a Bayesian evidence value for model comparison[1] (which could in the future be used as a detection statistic as in [9]).

Recently a new package of codes for Bayesian inference and evidence calculation (`lalinference`) have been created within the gravitational wave software repository `lalsuite` [10]. This suite contains more sophisticated MCMC methods (e.g. [11, 12]) and a method of Bayesian evidence calculation called nested sampling [13] that are able to more effectively explore posteriors without detailed tuning, can handle many parameters and provide values of the Bayesian evidence. We have used this new suite, and in particular its implementation of nested sampling (which can provide posterior probability distribution samples as well as an evidence value) to create a new code for the known pulsar search. Both new and old codes use the same likelihood as defined in [7], which assumes Gaussian noise with an its unknown variance analytically marginalised over.

This paper has two objectives: i) to show consistency between the new code and the previous code that has been used in astrophysical searches, and ii) to introduce a new parameterisation of two of the expected signal's angular parameters, which breaks a degeneracy that leads to a bimodal posterior distribution. Finally, we briefly discuss how this new code will be used in future known pulsar searches.

2. The algorithm

Bayesian inference is underpinned by Bayes theorem, which for a problem defined by a set of variables $\vec{\theta}$, a dataset d and some prior assumptions I, gives the posterior probability density

$$p(\vec{\theta}|d, I) = \frac{p(d|\vec{\theta}, I)p(\vec{\theta})}{p(d|I)}, \tag{2}$$

where $p(d|\vec{\theta}, I)$ is the likelihood function, $p(\vec{\theta})$ is the prior probability distribution of the variables, and $p(d|I)$ is a normalising factor, often called the Bayesian evidence or the *marginalised likelihood*, which makes the total probability unity. The Bayesian evidence value can be very useful in testing different hypotheses on the same dataset i.e. models with different variables. It can be found by integrating, or *marginalising*, the numerator of (2) over the prior ranges of all variables

$$Z = p(d, I) = \int_{\vec{\theta}} p(d|\vec{\theta}, I)p(\vec{\theta})\mathrm{d}\vec{\theta}. \tag{3}$$

For many problems this integral will not be analytical, so numerical integration is required, which will become computationally unfeasible if calculating the integral on a grid for a large parameter space.

[1] Note that the problems with the MCMC method that was implemented are not intrinsic issues of all MCMC methods, but were a problem with the particular basic approach used. Indeed MCMC outputs can even be used to compute Bayesian evidences (e.g. [8]).

Nested sampling was developed [14] as an efficient method of performing the integral given by (3) even for large parameter spaces. For good descriptions of the algorithm see [14, 13, 15]. As well as producing the evidence the routine will also output points sampled from the prior parameter volume and their associated likelihoods. These collected samples can be used to reconstruct the posterior probability distributions of the parameters.

3. Change of parameters

Other than the major change in the underlying algorithm used, a notable change that has been made in the new code is how it samples two of the signal's angle parameters: the initial phase ϕ_0 and the polarisation angle ψ. It is well known that these two parameters are degenerate to a $\pi/2$ radians shift in ψ and π radians shift in ϕ_0 (see Figure 1). This issue can cause problems when trying to efficiently sample this parameter space. Within the nested sampling algorithm new samples are drawn by generating an MCMC with a proposal distribution calculated from the covariance of a set of previously sampled points. For a parameter that is circular about a particular prior range (e.g. an angle parameter wrapping round from 0 to 2π) circularity can be easily taken into account when calculating the covariance, but the degeneracy in the two mentioned parameters can lead to a bimodal distribution in ϕ_0. This can be seen in Figure 1, for example with the signal represented by the blue circles. This bimodality will give a vastly distorted covariance matrix and lead to very low acceptance rates of new points within the MCMC sampling.

A way to avoid this is to find a new parameterisation in which both parameters are circular without causing any discontinuity and subsequent bimodality. Such a parameterisation is given by

$$\begin{aligned} {\phi'}_0 &= \phi_0 \sin\theta + \psi\cos\theta, \\ \psi' &= -\phi_0 \sin\theta + \psi\cos\theta, \end{aligned} \tag{4}$$

where $\theta = \arctan(1/2)$. These new axes are shown by the dashed black lines in Figure 1. It can be seen that signals lying at the edges of this parameter space wrap around cleanly (the blue and red circles) at the edge of the ranges of $-(\pi/2)\cos\theta \leq {\phi'}_0, \psi' \leq (\pi/2)\cos\theta$.

Given flat priors on ϕ_0 and ψ, the above change in variables will mean that the priors on ${\phi'}_0$ and ψ' remain flat (the Jacobian is constant). The new parameters can be converted back to ϕ_0 and ψ via inversion of the transformation matrix giving

$$\begin{aligned} \phi_0 &= {\phi'}_0/(2\sin\theta) - \psi'/(2\sin\theta), \\ \psi &= {\phi'}_0/(2\cos\theta) + \psi'/(2\cos\theta). \end{aligned} \tag{5}$$

However, it should be noted that this will return values in the range $-\pi \leq \phi_0 \leq \pi$ and $-\pi/2 \leq \psi \leq \pi/2$, so these have to be appropriately wrapped to cover the standard range.

An example of this new parameterisation on the calculation of a signal's parameter posteriors can be seen in Figure 2, where a strong fake signal was created at the ψ boundary, with $\phi_0 = \pi/2$ rads and $\psi = -\pi/4$ rads. It was recovered using the new parameterisation and then converted back into the original parameters correctly, where the bimodality in ϕ_0 can clearly be seen. The MCMC acceptance rate improved from a fraction of a percent to of order 50%. The same result could also be achieved by using this new parameterisation to define a prior volume for ϕ_0 and ψ.

4. Code testing

The old code used in the analyses up to now has been sufficient, but as stated previously has some disadvantages. However, as it is well understood it provides a good check against which to test

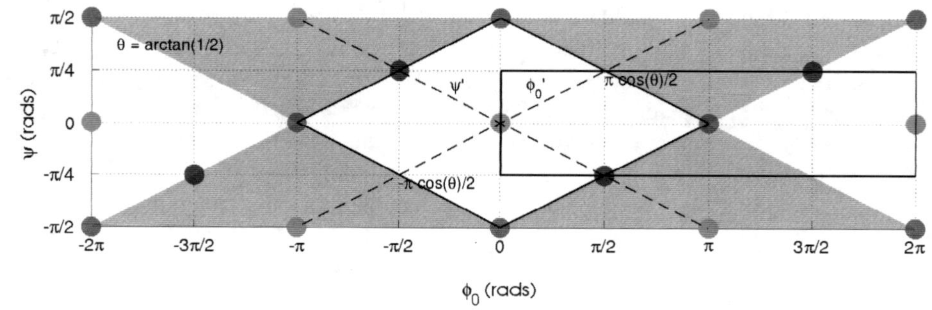

Figure 1. A diagram of the ϕ_0–ψ parameter space demonstrating the new ${\phi'}_0$–ψ' parameterisation. The filled circles represent an area of probability for a signal, with the same colours representing the same signal and showing the degeneracy between parameters. The rectangular box shows the parameters space used in the original code and the diamond box shows the new parameter space.

that the new code is working correctly. We have tested the posterior probability distributions extracted with simulated signals and simulated noise using the new and old codes.

An example is shown in Figure 3, which shows three sets of posterior probability distributions for the four unknown signal parameters of a simulated pulsar signal injected into fake noise from the LIGO Hanford (H1) detector. Two were calculated with the old code in a grid-based mode (with a 50^4 grid over the parameter ranges) and an MCMC mode (with a burn-in of 50 000 points and 100 000 points sampling the posterior) and the third was produced using the new code with the nested sampling algorithm (using a total of 1 250 000 likelihood evaluations). Here we make no assessment of the relative speed/efficiencies of the methods, which mainly depends on the number of likelihood evaluations they each perform, but just test their consistency. As with other injections we have studied all three methods show very good agreement in the posteriors they produce. It should be noted that in setting up the codes the nested sampling code required no tuning, whereas with the old code for the grid-based search and the MCMC the parameter ranges and proposal distributions had to be specified to quite closely match the posterior ranges. All three methods also recover the true signal values within the posteriors.

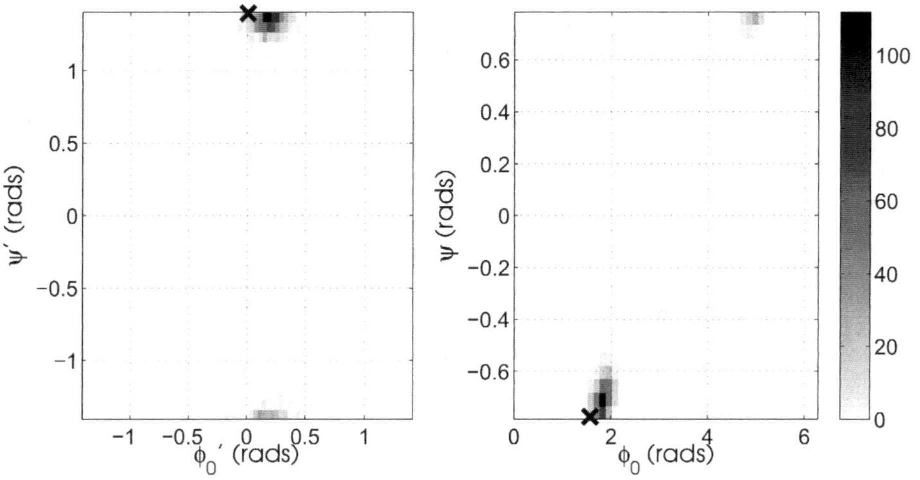

Figure 2. The 2D posterior probability distributions of a fake signal (true parameters marked with the $\times$) extracted with the new ϕ'_0–ψ' parameterisation (left) and converted back into the ϕ_0–ψ parameters (right).

4.1. Upper limit testing

Previous astrophysical searches have not seen any signals, so have instead set upper limits on the gravitational wave amplitude, h_0. Monte Carlo tests have shown the distribution of upper limits (in our case limits at 95% degree-of-belief) that would be expected given many datasets consisting of Gaussian noise of known power spectral density for pulsars randomly distributed on the sky [7]. So, again a good way to test the new code is to check that it can produce the same distribution of upper limits as that found with the old code. Figure 4 shows the distribution of 95% upper limits normalised by the noise power spectral density (S_n in Hz^{-1}) and data length (T in seconds) using the nested sampling code and the old grid-based code. These two independent methods are very consistent with each other. The agreement between these distributions and that shown in [7] is good, but not prefect. Without knowing the exact set-up of that earlier analysis we cannot track down the root cause of this discrepancy, but are confident that our current analysis with these two independent methods is correct.

5. Additional models and parameters

The simple 4-dimensional parameter space of the signal given in (1) can easily be covered on a grid (although as stated still requires some tuning of grid spacing and parameter ranges). However, there are many cases where the number of parameters in known pulsar searches can greatly increase and model comparisons are required. Here we briefly discuss a couple of cases of this and plans for the future.

5.1. Inclusion of phase parameter uncertainties

As in [5] radio pulsar observations will often come with uncertainties on the parameters, which although in general are very small can mean that the single phase model used in (1) is not sufficient to described the signal. These extra parameters describing the phase change need to be incorporated into the gravitational wave search. The new code can incorporate these observed

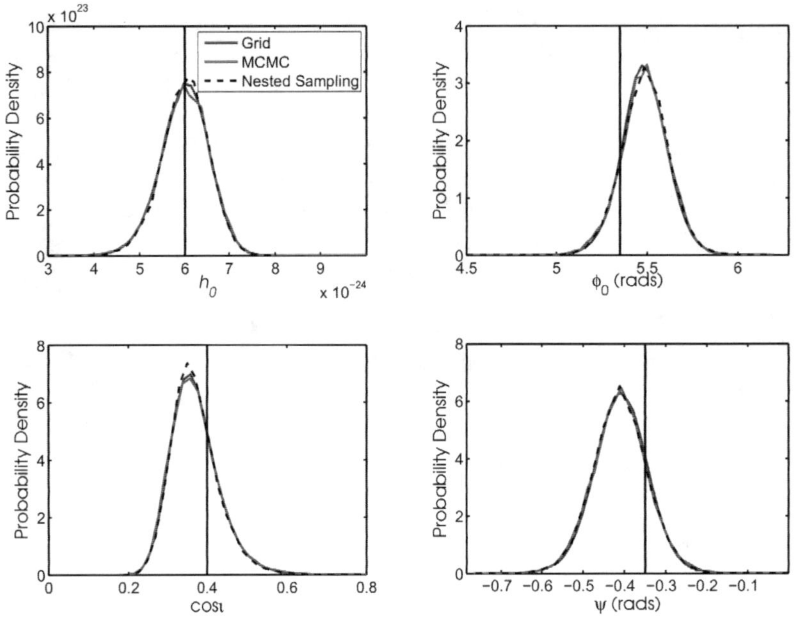

Figure 3. Marginalised posterior probability densities for a fake pulsar signal injected into simulated LIGO Hanford detector noise. The injection parameters are shown as the thin vertical lines. The solid blue lines are the posteriors from the old code with a grid-based approach, the solid red lines are the posteriors from the old code with the basic MCMC approach, and the dashed black lines are from the new code using nested sampling.

	RA	Dec	Frequency
RA	1.0	0.5	-0.2
Dec	0.5	1.0	-0.7
Frequency	-0.2	-0.7	1.0

	RA	Dec	Frequency
RA	1.0	0.467	-0.173
Dec	0.467	1.0	-0.694
Frequency	-0.173	-0.694	1.0

Table 1. The correlation coefficients of the priors (left) on frequency, right ascension and declination used to produce a posterior (right) for data consisting of Gaussian noise.

uncertainties, via the phase parameter covariance matrix, to be used as priors. Figure 5 shows the marginalised posteriors for three parameters (frequency, right ascension and declination) that have been estimation from data just containing Gaussian noise, using priors similar to uncertainties from radio observations, with other parameters held fixed. The priors used are also plotted and show that for these phase parameters the posteriors are completely defined by the priors as would be expected for data containing only noise. These priors were also correlated, with the prior correlation coefficients used and correlation coefficients calculated from the posterior given in Table 1. Again these show that the expected prior distribution is being obtained.

There are also potential sources and candidates from all sky searches for unknown pulsars/neutron stars (e.g. [16, 17]) that may have parameters e.g. frequency, frequency derivatives and sky position, with relatively large uncertainties spanning many phase models.

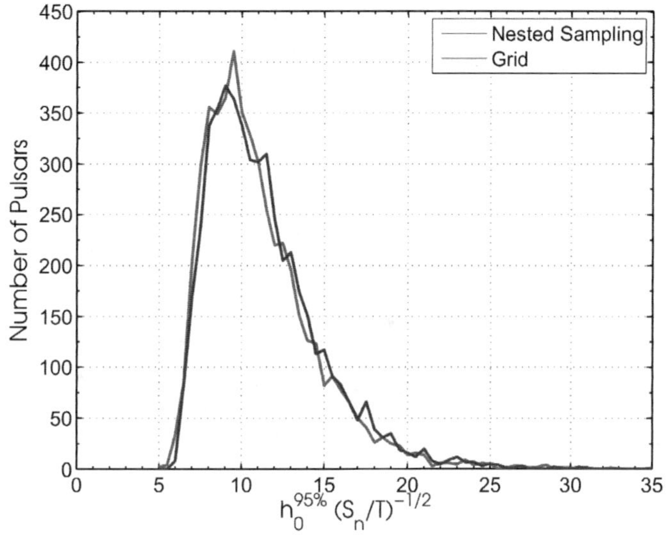

Figure 4. The distribution of 95% degree-of-belief upper limits on h_0 for 5000 realisations of noise for sources uniformly distributed on the sky using the nested sampling code (blue line) and grid-based code (red line).

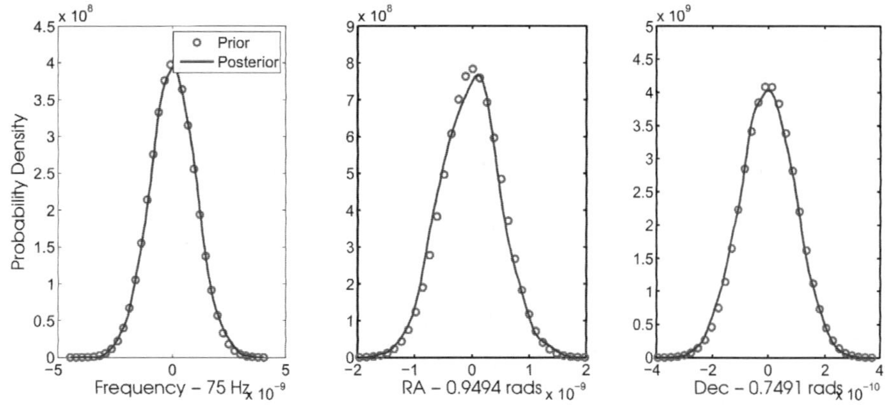

Figure 5. The marginalised posteriors for frequency, right ascension and declination calculated from data consisting of Gaussian noise, using the priors shown.

The new code has preliminarily been tested in a search search containing a fake signal, with a prior range on the frequency band of 2×10^{-3} Hz, with the recovered h_0–frequency posterior shown in Figure 6. The code recovers the injected signal well, but in a future paper we will fully characterise the code for searches over these wider frequency bands, and compare it to similar searches using the $\mathcal{F}$-statistic as applied in [4], or the $\mathcal{B}$-statistic of [9].

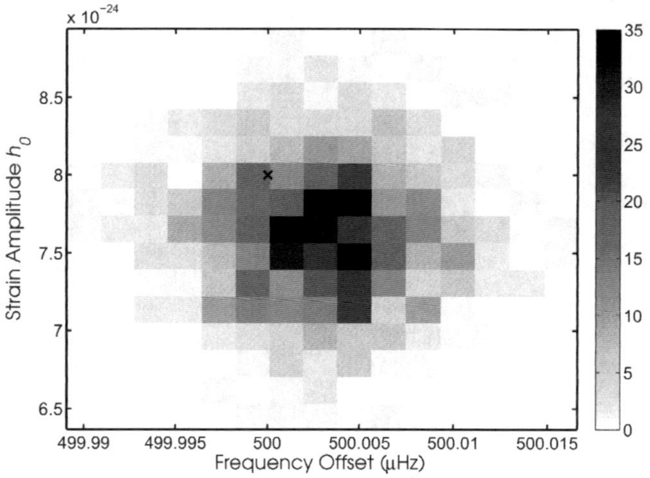

Figure 6. The 2D posterior of signal frequency offset (from the central search frequency) against h_0 for a fake pulsar signal with a signal-to-noise ratio of $\sim$ 19. The true injection parameters are marked with a black cross. The prior volume covered $\pm 10^{-3}$ Hz in frequency offset and 0 to 10^{-22} in h_0.

5.2. Model selection

Nested sampling produces a value of the evidence which can be used for model comparisons. In a future paper the code will be fully characterised for comparing the evidence for the signal model described in Section 1 against a model that the data contains just Gaussian noise. The ratio of these evidences can be used as a detection statistic to assess a candidate signal's significance. The code is also flexible enough to allow additional models including extra parameters. We plan to assess a new signal model, which can include emission at once and twice the rotation frequency [18] and includes additional angular parameters, and compare it to the standard model given by (1). Further signal and noise models (for example models including interference lines that are incoherent between multiple detectors [19]) could easily be added in the future in the current framework.

6. Conclusions

A new code for parameter estimation and model comparison, based on the nested sampling algorithm, has been developed for continuous-wave searches in gravitational wave data. It has been shown to be consistent with a previous code, used successfully in many searches, in extracting signal parameters and producing upper limits on gravitational wave amplitude. The new code has advantages over the old code in that it requires less tuning for specific searches and allows comparison between different models.

The new code is flexible enough to be expanded to larger numbers of parameters and parameter ranges. This will allow it to be used in searches for pulsars with less well defined parameters, or as a method to follow up candidates from all-sky neutron star searches. These will be characterised in future papers. Additional MCMC proposal distributions for drawing new samples will also be investigated to study improvements in the algorithm's efficiency. These will including proposals that only update one parameter at a time when drawing from a multivariate Gaussian, using a k-D tree of the samples to form a proposal (e.g. [20]) and potentially using MultiNest [21], which has been integrated into the `lalinference` package of `lalsuite` [10].

In the near future this code will be used in searches for a large number of known pulsars in data from the recent LIGO S6 and the Virgo VSR2, 3 and 4 data runs.

Acknowledgments

The authors would like to thank the LIGO Scientific Collaboration and Virgo Collaboration continuous wave and CBC Bayesian parameter estimation working groups, and the developers of the `lalinference` software suite, for many useful discussions. We also thank the referee for their many useful comments on the draft. The document has LIGO DCC number LIGO-P1100171.

References

[1] Abbott B, Abbott R, Adhikari R, Ageev A, Allen B, Amin R, Anderson S B, Anderson W G, Araya M, Armandula H and et al 2004 *Phys. Rev. D* **69** 082004 (*Preprint* arXiv:gr-qc/0308050)
[2] Abbott B, Abbott R, Adhikari R, Ageev A, Allen B, Amin R, Anderson S B, Anderson W G, Araya M, Armandula H and et al 2005 *Physical Review Letters* **94** 181103 (*Preprint* arXiv:gr-qc/0410007)
[3] Abbott B, Abbott R, Adhikari R, Agresti J, Ajith P, Allen B, Amin R, Anderson S B, Anderson W G, Arain M and et al 2007 *Phys. Rev. D* **76** 042001 (*Preprint* arXiv:gr-qc/0702039)
[4] Abbott B, Abbott R, Adhikari R, Ajith P, Allen B, Allen G, Amin R, Anderson S B, Anderson W G, Arain M A and et al 2008 *ApJ* **683** L45–L49 (*Preprint* 0805.4758)
[5] Abbott B P, Abbott R, Acernese F, Adhikari R, Ajith P, Allen B, Allen G, Alshourbagy M, Amin R S, Anderson S B and et al 2010 *ApJ* **713** 671–685 (*Preprint* 0909.3583)
[6] Abadie J, Abbott B P, Abbott R, Abernathy M, Accadia T, Acernese F, Adams C, Adhikari R, Affeldt C, Allen B and et al 2011 *ApJ* **737** 93 (*Preprint* 1104.2712)
[7] Dupuis R J and Woan G 2005 *Phys. Rev. D* **72** 102002 (*Preprint* arXiv:gr-qc/0508096)
[8] Weinberg M D 2009 *ArXiv e-prints* (*Preprint* 0911.1777)
[9] Prix R and Krishnan B 2009 *Classical and Quantum Gravity* **26** 204013 (*Preprint* 0907.2569)
[10] URL https://www.lsc-group.phys.uwm.edu/daswg/projects/lalsuite.html
[11] van der Sluys M, Raymond V, Mandel I, Röver C, Christensen N, Kalogera V, Meyer R and Vecchio A 2008 *Classical and Quantum Gravity* **25** 184011 (*Preprint* 0805.1689)
[12] Röver C, Meyer R and Christensen N 2007 *Phys. Rev. D* **75** 062004 (*Preprint* arXiv:gr-qc/0609131)
[13] Veitch J and Vecchio A 2010 *Phys. Rev. D* **81** 062003 (*Preprint* 0911.3820)
[14] Skilling J 2006 *Bayesian Analysis* **1** 833–860
[15] Sivia D S and Skilling J 2006 *Data Analysis: A Bayesian Tutorial* (Oxford University Press) chap 9
[16] Abbott B P, Abbott R, Adhikari R, Ajith P, Allen B, Allen G, Amin R S, Anderson S B, Anderson W G, Arain M A and et al 2009 *Phys. Rev. D* **80** 042003 (*Preprint* 0905.1705)
[17] Abadie J, Abbott B P, Abbott R, Abbott T D, Abernathy M, Accadia T, Acernese F, Adams C, Adhikari R, Affeldt C and et al 2011 *ArXiv e-prints* (*Preprint* 1110.0208)
[18] Jones D I 2010 *MNRAS* **402** 2503–2519 (*Preprint* 0909.4035)
[19] Keitel D, Prix R, Alessandra Papa M and Siddiqi M 2012 *ArXiv e-prints* (*Preprint* 1201.5244)
[20] Farr W M and Mandel I 2011 *ArXiv e-prints* (*Preprint* 1104.0984)
[21] Feroz F, Hobson M P and Bridges M 2009 *MNRAS* **398** 1601–1614 (*Preprint* 0809.3437)

Prospects for transient gravitational waves at r-mode frequencies associated with pulsar glitches

I Santiago-Prieto[1], I S Heng [1], D I Jones[2] and J Clark[3,4]

[1] School of Physics and Astronomy, University of Glasgow, Glasgow G12 8QQ, UK
[2] Department of Mathematics, University of Southampton, Southampton SO17 1BJ, UK
[3] School of Physics and Astronomy, Cardiff University, 5 The Parade, Cardiff CF24 3AA, UK
[4] Physics Department, University of Massachusetts, Amherst, Massachusetts 01003, USA

E-mail: ignacio.santiago-prieto@ligo.org

Abstract. t Glitches in pulsars are likely to trigger oscillation modes in the fluid interior of neutron stars. We examined these oscillations specifically at r-mode frequencies. The excited r-modes will emit gravitational waves and can have long damping time scales $\mathcal{O}$(minutes - days). We use simple estimates of how much energy the glitch might put into the r-mode and assess the detectability of the emitted gravitational waves with future interferometers.

1. Introduction

One possible source of burst-like gravitational waves is a *pulsar glitch*. These are anomalies in the observed spin frequency of a neutron star (NS) wherein there is a sudden increase in its rotational frequency followed by an exponential recovery to nearly the pre-glitch rotation rate on timescales of a few days–weeks. In most cases, these glitches are produced by young pulsars. It is believed that these glitches are a consequence of global events within the star's fluid interior and its solid crust [1]. These glitches could excite a variety of quasi-normal modes [2]. A previous search for gravitational waves associated with a glitch in the Vela pulsar in 2005 focussed on f-modes whose frequencies are believed to lie in the range 1–3 kHz [3] and last $\mathcal{O}(100\,\text{ms})$.

A possibility to be considered in gravitational wave searches of this type is the excitation of r-modes oscillations during a pulsar glitch [4]. These r-modes, which are a type of inertial modes (oscillation modes restored by the Coriolis force), have frequencies proportional to the star's angular velocity and they are considered relatively more efficient GW emitters compared to other inertial modes [5]. Also, these modes evolve in time with an exponential decrement that depends on the mode's frequency and dissipation processes [6]. Reference [7] details that when a rotational lag between the fluid component and the crust reach a critical value, an instability that excite r-mode oscillations within the NS interior sets in. Because of this lag, the fluid component (already 'pinned' to the crust by a series of vortices) will be unpinned and will transfer angular momentum to the crust which will consequently spin up. As a result of this process a pulsar glitch will be produced. In that work, the authors considered short-wavelengths modes at which the quadrupole moment $l = m = 2$ would not be excited.

In section 2 we describe the form and duration of the GW signal one might expect from r-mode oscillations. In section 3, we explore the energetics of the r-mode signals associated with pulsar glitches and the plausibility of detecting GWs from r-modes. Finally, in section 4, we summarise our findings.

2. Gravitational waves signals from r-modes in neutron stars

We parameterise the expected GW signal from an r-mode oscillation in terms of harmonic ringdown gravitational waves signals [8],

$$h_+(t) = h_0 \cos(\omega_r t)e^{-t/\tau_r} \qquad h_\times(t) = h_0 \sin(\omega_r t)e^{-t/\tau_r}, \tag{1}$$

where h_+ and $h_\times$ are the gravitational wave signal polarizations 'plus' $(+)$ and 'cross' $(\times)$, $\omega_r = 2\pi f_r$ is the frequency of the r-mode oscillation, τ_r is the damping time of the signal and h_0 is the amplitude of the signal at the detector given by equation (23) in [9]

$$h_0 = \sqrt{\frac{8\pi}{5}\frac{G}{c^5}\frac{1}{r}}\alpha\omega_r^3 M R^3 \tilde{J}. \tag{2}$$

This expresses the amplitude of the signal in terms of the NS mass M, radius R, distance from the observer r, a dimensionless parameter $\tilde{J}$ ($1.635\times\ 10^{-2}$) that depends on the stellar mass distribution and the amplitude of the mode oscillation α. We consider the amplitude α in more detail in section 3.

According to [10], for sufficiently slowly rotating stars, the r-mode oscillation frequency f_r is proportional to the spin frequency of the star $f_{\rm spin}$. It is expected that the $\ell = m = 2$ mode dominates the GW emission [11]. Under these assumptions the r-mode frequency is given by,

$$\omega_r \approx \frac{4}{3}\Omega, \tag{3}$$

where Ω is the angular spin frequency of the star ($= 2\pi f_{spin}$) [10]. There are several sources of uncertainty in this relation. These include rotational effects [12], which modify the mode frequency by an amount that depends upon the average density, and relativistic effects [11], which introduce fractional corrections of the order of the stellar compactness M/R, so that equation (3) is accurate only to the $\sim 10\%$ level.

The damping time of the r-mode is given by,

$$\frac{1}{\tau_r} = \frac{1}{\tau_{\rm viscosity}} + \frac{1}{\tau_{\rm GRR}}, \tag{4}$$

where $\tau_{viscosity}$ is the dissipation time-scale due to the viscosity in the fluid component of the NS and $\tau_{\rm GRR}$ is the gravitational radiation reaction time-scale. Following [6], the radiation reaction time-scale is a function of the mode frequency,

$$\frac{1}{\tau_{\rm GRR}} = -\frac{32\pi}{15^2}\frac{G}{c^7}\omega_r^6 \tilde{J} M R^4, \tag{5}$$

Note that the damping time of the signal τ_r is a balance between dissipative processes with time-scale $\tau_{\rm viscosity} > 0$ and the radiation reaction time-scale $\tau_{\rm GRR} < 0$; while $\tau_{\rm viscosity}$ acts to dissipate energy and reduce the oscillation amplitude of the mode, $\tau_{\rm GRR}$ forces the mode's amplitude to grow due to the emission of gravitational radiation [13]. In this work we assume that the mode is in the stable regime so that $\tau_{\rm viscosity} << |\tau_{\rm GRR}|$, and so the viscosity time-scale dominates in the duration of the signal.

Levin and Ushomirsky proposed in [14] the existence of a viscous boundary layer in the crust-core interface of the NS which yields the following estimate of the dissipative time-scale,

$$\frac{1}{\tau_{\rm viscosity}} \simeq 0.01 s^{-1} \frac{R_6^2 F^{1/2}}{M_{1.4} T_8}\frac{\rho_b}{\rho}\left(\frac{f_{\rm spin}}{\rm kHz}\right)^{1/2}\left(\frac{\delta u}{u}\right)^2, \tag{6}$$

where R_{10}, $M_{1.4}$, ρ and ρ_b are the radius (10 Km), mass (1.4 $M_\odot$), the density at the crust-core interface, and an estimate of this density ($\rho_b = 1.5\times10^{14}$ g cm^{-3}), respectively. T_8 is an assumed internal temperature of 10^8 K in the NS. The value of the fitting parameter F depends on the dominant scattering processes in the fluid interior [15]:

$$F \sim \begin{cases} 1/15 & \text{for electron–proton scattering,} \\ (\rho/\rho_b)^{5/4} & \text{for neutron scattering,} \\ 5\rho/\rho_b & \text{for electron scattering.} \end{cases} \tag{7}$$

Here, we assume that neutron-scattering dominates and we have set $\rho = \rho_b$.

Finally, the ratio $\delta u/u$ measures the fractional velocity mismatch between the crust and the core. As demonstrated in [14], for a sufficiently slowly rotating star, the crust does not significantly participate in the r-mode oscillation, so that $\delta u/u \approx 1$. Just how slow the rotation needs to be for this to be the case depends upon the thickness of the crust, but Figure 1 of [14] indicate that all young glitching pulsars are likely to have $\delta u/u \approx 1$. Note, however, that in more rapidly rotating stars $\delta u/u \approx 0.1$, lengthening the viscous decay timescale by a factor of ~ 100.

3. Detectability

The detectability of the GW signal from r-mode oscillations excited by a pulsar glitch is determined by the total amount of energy that the glitch deposits in the r-mode, and by the fraction of that energy which is radiated as GWs by the r-mode oscillations.

Following [9], the instantaneous gravitational wave luminosity from a source at distance r, undergoing damped oscillations with initial amplitude h_0, angular frequency ω and damping time τ is,

$$\dot{E}_{\rm GW} = \frac{c^3}{G}\frac{1}{10}r^2\omega^2(h_0 e^{-t/\tau})^2. \tag{8}$$

The total time-integrated energy emitted in GWs is, therefore,

$$\Delta E_{GW} = \frac{c^3}{G}\frac{1}{20}r^2\omega^2 h_0^2\tau. \tag{9}$$

The mode energy is

$$\tilde{E} = \alpha^2\Omega^2 M R^2 \tilde{J}. \tag{10}$$

A rough estimate of the total energy associated with a pulsar glitch is (see e.g. [3]),

$$E_{\rm glitch} \approx I\Omega^2\left(\frac{\Delta\Omega}{\Omega}\right), \tag{11}$$

where I the moment of inertia of the NS, Ω is the spin frequency and $\Delta\Omega/\Omega$ is the size of the glitch relative to the spin-frequency. An upper limit on the mode amplitude α can be obtained by assuming all of the energy associated with the glitch is channelled into r-mode excitation and we find,

$$\alpha = \left(\frac{\tilde{I}}{\tilde{J}}\right)^{1/2}\left(\frac{\Delta\Omega}{\Omega}\right)^{1/2}, \tag{12}$$

where $\tilde{I}$ (typically $\approx$ 0.261) is a dimensionless parameter dependant on the stellar mass distribution [6]. To evaluate the total power contained in short-duration, narrow-band signals, it is convenient to use the root-sum-squared amplitude:

$$h_{\rm rss} = \left[\int_0^\infty h_+^2(t) + h_\times^2(t)\,{\rm d}t\right]^{\frac{1}{2}} \approx h_0\tau^{1/2}, \tag{13}$$

where the right hand side is the approximate value for the damped sinusoid we consider for our GW signal. Combining equations (2), (11) and (12), the root-sum-squared GW amplitude from r-mode excitation in terms of the stellar parameters and the size of the glitch is,

$$h_{\mathrm{rss}} = \frac{128}{27}\sqrt{\frac{2\pi}{5}\frac{G}{c^5}}\frac{M}{r}(\Omega R)^3\left(\tilde{I}\tilde{J}\frac{\Delta\Omega}{\Omega}\tau\right)^{1/2}. \tag{14}$$

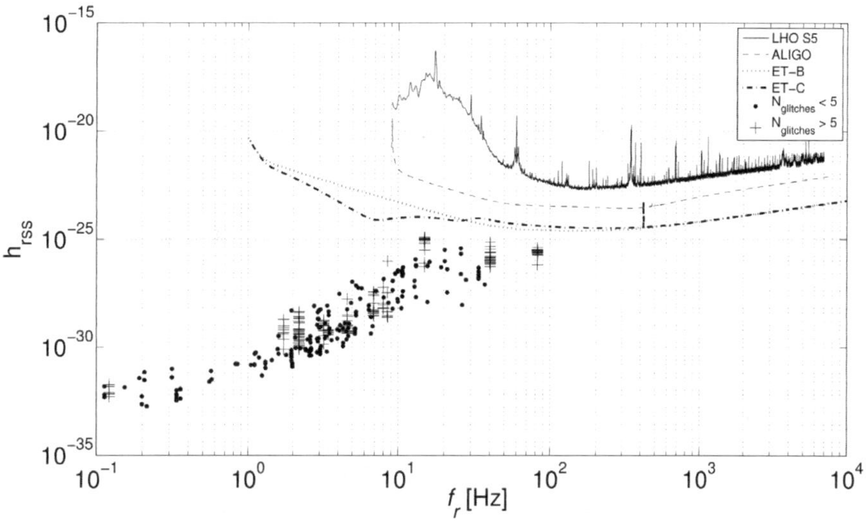

Figure 1. Sensitivity curves of gravitational waves detectors and upper limit estimates h_{rss} of the gravitational wave signal as a function of the r-mode frequency $f_r = \omega_r/(2\pi)$.

In figure 1, we compare the estimated GW upper limits from glitch-induced r-mode excitations in various known pulsars with the noise spectral densities of the initial LIGO instrument during its fifth science run [16], the advanced LIGO detector [17],[18] (currently under construction) and the proposed 3rd generation Einstein Telescope (ET) [19]. We compare the estimates for frequently glitching pulsars (observed to have glitched more than five times, shown in crosses) and all other pulsars, shown in dots. The glitch observations are compiled from [20, 21, 22] and [23]. We find that, in this optimistic case, the estimated GW amplitude is a factor ~ 6 below the noise floor of even ET in its most favorable configuration

4. Discussion

We have presented some estimates of the energy associated with r-modes excited by pulsar glitches and the corresponding estimated upper limit GW amplitudes. We see that, even in the most optimistic scenario of energy transfer from the glitch to the mode, the GW amplitude is well below the sensitivity curves of both existing and planned gravitational wave detectors. Specifically, even for the most rapidly spinning pulsar we consider, where the r-mode frequency lies near the most sensitive part of the detectors, the expected amplitude is $\sim 4 \times 10^{-26}$.

However, the strong frequency dependence of the r-mode oscillation implies that rapidly rotating NS have a significantly greater chance of detection in 3rd generation instruments such as ET. In addition to this, it is possible that these faster rotators possess a significantly smaller slippage parameter $\delta u/u$ (see section 2), leading to longer duration and more detectable GW signals. There is clearly some interest in exploring r-mode excitation in more rapidly rotating stars.

Related to this, we end by noting that, despite the greater frequency of glitches occurring in young pulsars, there is a report of a millisecond pulsar (MSP) glitch in [24]. As well as pushing the expected GW signal closer to the sensitive regions of interferometric GW detectors, higher spin frequencies yield intrinsically stronger signals. Accreting millisecond pulsars present a particularly tantalising opportunity for future analyses and the GW r-mode detectability from such sources is to be considered in future work.

References

[1] Lyne A G and Graham-Smith F 2006 *Pulsar Astronomy* (Cambridge University Press)
[2] Sidery T, Passamonti A and Andersson N 2010 *MNRAS* **405** 1061–1074
[3] Abadie J *et al.* (The LIGO Scientific Collaboration) 2011 *Phys. Rev. D* **83** 042001
[4] Rezania V and Jahan-Miri M 2000 *MNRAS* **315** 263–268
[5] Lockitch K H and Friedman J L 1999 *The Astrophysical Journal* **521** 764
[6] Owen B J, Lindblom L, Cutler C, Schutz B F, Vecchio A and Andersson N 1998 *Phys. Rev. D* **58** 084020
[7] Glampedakis K and Andersson N 2009 *Phys. Rev. Lett.* **102**(14) 141101
[8] Abadie J *et al.* (The LIGO Scientific Collaboration and The Virgo Collaboration) 2010 *Phys. Rev. D* **81** 102001
[9] Owen B J 2010 *Phys. Rev. D* **82** 104002
[10] Provost J, Berthomieu G and Rocca A 1981 *Astron. & Astrophys.* **94** 126
[11] Lockitch K H, Friedman J L and Andersson N 2003 *Phys. Rev. D* **68** 124010
[12] Lindblom L, Mendell G and Owen B J 1999 *Phys. Rev. D* **60** 064006
[13] Lindblom L, Owen B J and Morsink S M 1998 *Phys. Rev. Lett.* **80**(22) 4843–4846
[14] Levin Y and Ushomirsky G 2001 *Monthly Notices of the Royal Astronomical Society* **324** 917–922 ISSN 1365-2966
[15] Bildsten L and Ushomirsky G 2000 *The Astrophysical Journal Letters* **529** L33
[16] Abbott B *et al.* 2009 *Rep. Prog. Phys.* **72** 076901
[17] Harry G M (for the LIGO Scientific Collaboration) 2010 *Classical and Quantum Gravity* **27** 084006
[18] Public sensitivity curves URL `https://dcc.ligo.org/cgi-bin/DocDB/ShowDocument?docid=T0900288`
[19] Hild S *et al.* 2011 *Classical and Quantum Gravity* **28** 094013
[20] Manchester R N, Hobbs G B, Teoh A and Hobbs M 2005 *The Astronomical Journal* **129** 1993 http://www.atnf.csiro.au/research/pulsar/psrcat
[21] Melatos A, Peralta C and Wyithe J S B 2008 *The Astrophysical Journal* **672** 1103
[22] Yuan J P, Wang N, Manchester R N and Liu Z Y 2010 *Monthly Notices of the Royal Astronomical Society* **404** 289–304 ISSN 1365-2966
[23] Espinoza C M, Lyne A G, Stappers B W and Kramer M 2011 *Monthly Notices of the Royal Astronomical Society* **414** 1679–1704 ISSN 1365-2966
[24] Mandal R D R, Konar S, Dey M and Dey J 2009 *Monthly Notices of the Royal Astronomical Society* **399** 822–826 ISSN 1365-2966

Coherent follow-up of Continuous Gravitational-Wave candidates: minimal required observation time

Miroslav Shaltev

E-mail: miroslav.shaltev@aei.mpg.de
Albert-Einstein-Institut, Callinstr. 38, 30167 Hannover, Germany
LIGO-P1100172-v3 Fri Mar 2 16:52:05 2012 +0100

Abstract. We derive two different methods to compute the minimal required integration time of a fully coherent follow-up of candidates produced in wide parameter space semi-coherent searches, such as global correlation StackSlide searches using Einstein@Home. We numerically compare these methods in terms of integration duration and computing cost. In a Monte Carlo study we confirm that we can achieve the required detection probability.

1. Introduction

Isolated neutron stars as potential sources of continuous gravitational waves are optimally studied with fully coherent matched filtering methods. These methods are not directly applicable to previously unknown objects due to the large parameter space that needs to be covered in all-sky wide parameter space searches and the related enormous computing cost [1]. Advanced semi-coherent techniques, e.g. StackSlide searches on the distributed computing environment Einstein@Home [2], produce candidates that require follow-up in greatly reduced parameter space regions. A follow-up scheme consists of two basic stages. In the first *refinement* stage, we find the maximum-likelihood estimator and associated optimal search volume V_0. In the second *zoom* stage, we zoom in on the optimal search volume by semi-coherent or fully-coherent integration. In this paper we focus on a fully-coherent zoom for which we derive and discuss two different methods to compute the minimal required coherent integration time in order to distinguish real signals from noise.

2. Properties of $\mathcal{F}$-statistic searches

The $\mathcal{F}$-statistic was first derived in [3] for the single detector case and generalized to multi-detector searches in [4]. Continuous gravitational-wave signals are monochromatic and sinusoidal in the frame of the gravitational-wave source and undergo phase- and amplitude modulation due to the rotation and orbital motion of the detector. The $\mathcal{F}$-statistic is analytically amplitude-maximized, thus the parameter space to search for signals is spanned by the remaining "Doppler parameters" λ, namely sky position (α - right ascension, δ - declination) and intrinsic frequency and frequency derivatives ($f, \dot{f}, \ddot{f}$...), further referred to as spindowns. Searching for previously unknown objects with matched filtering implies computing matched filters for different points in parameter space, also referred to as templates. As realized in [5, 6] in the context of searches for gravitational waves from inspiraling binaries, a geometrical approach is best suited for optimal template placement and template counting. This is made possible by the introduction of a metric tensor g_{ij} on the parameter space and mismatch m

$$m = g_{ij}\Delta\lambda^i\Delta\lambda^j + \mathcal{O}(\Delta\lambda^3) \,, \tag{1}$$

where the mismatch m measures the fractional loss of (squared) signal to noise ratio (SNR) ρ^2 due to the usage of a nearby template λ_c with offset $\Delta\lambda = \lambda_c - \lambda_s$ from the true parameters of a putative signal λ_s

$$m = \frac{\rho_s^2 - \rho_c^2}{\rho_s^2} \,, \tag{2}$$

with the squared SNR ρ_s^2 and ρ_c^2 obtained at point λ_s and λ_c, respectively. Given the metric, the problem of efficient lattice and alternative random and stochastic template-bank construction is studied in [7, 8, 9].

2.1. *Fully-coherent search*

A fully-coherent search is the classical and most sensitive $\mathcal{F}$-statistic-based search in the case of unlimited available computing power or a sufficiently cheap computing cost requirement. The squared SNR ρ^2 scales linearly with the observation time T, according to the following formula:

$$\rho^2 = h_0^2 R N_{\rm d} T S^{-1}(f) \,, \tag{3}$$

where h_0 is the intrinsic signal amplitude, R represents the geometrical "detector response" , S is the one-sided noise spectral density, which is assumed constant in a narrow frequency band around f, and $N_{\rm d}$ is the number of detectors [10]. In the presence of a signal, the $\mathcal{F}$-statistic follows a non-central χ^2 distribution with four degrees of freedom and non-centrality parameter ρ^2. Thus the expectation value is

$$E[2\mathcal{F}_S] = 4 + \rho^2 \,, \tag{4}$$

with standard deviation

$$\sigma(2\mathcal{F}_S) = \sqrt{2(4 + 2\rho^2)} \,. \tag{5}$$

2.2. *Semi-coherent search*

At fixed and limited computing cost a more sensitive detection statistic can be constructed from the incoherent combination of results obtained by coherent integration of shorter data segments. In particular we consider a Stack-Slide search [11, 12, 13], where the statistic is the sum of the $\mathcal{F}$-statistic over the segments:

$$\Sigma = \sum_{k=1}^{N} 2\mathcal{F}_k(\lambda). \tag{6}$$

This new statistic Σ follows a non-central χ^2 distribution with $4N$ degrees of freedom, thus the expectation value is

$$E[\Sigma] = 4N + \rho_\Sigma^2 \,, \tag{7}$$

where the non-centrality parameter is the sum of the squared SNRs over different segments

$$\rho_\Sigma^2 = \sum_{k=1}^{N} \rho_k^2 \,. \tag{8}$$

A trivial but useful reformulation of Eq. (7) is in terms of average $2\bar{\mathcal{F}} = \frac{1}{N}\sum_k 2\mathcal{F}_k$ and $\bar{\rho^2} = \frac{1}{N}\sum_k \rho_k^2$, namely

$$E[2\bar{\mathcal{F}}] = 4 + \bar{\rho^2} \,. \tag{9}$$

2.3. Template counting

The number of templates sufficient to cover the search volume V_0 is given by [7]

$$\mathcal{N}_n = \theta m^{-n/2} \mathcal{V}_n \; , \tag{10}$$

where θ is the normalized thickness characterizing the geometric structure of covering, m is the maximum allowed mismatch , n the number of dimensions and

$$\mathcal{V}_n = \int d^n\lambda \; \sqrt{\det g} \; , \tag{11}$$

is the metric template-bank volume with g_{ij} the parameter space metric. This is the general form of the template counting formula, which is valid for arbitrary lattices and curved parameter spaces. In practice, using the flat metric approximation, where the metric coefficients are constant, we can take the determinant out of the integral. Moreover, if the parameter space is a n-dimensional "box", we can replace the integral over infinitesimal displacement $d\lambda$ by a product of n "search bands" $\Delta\lambda$, namely

$$\mathcal{V}_n = \sqrt{\det g} \prod_{i=1}^{n} \Delta\lambda_i \; . \tag{12}$$

Follow-up of candidates from semi-coherent searches involves a semi-coherent metric, shown in [11, 14] to be the average of the metric computed for every segment. The semi-coherent metric allows us to estimate the search band $\Delta\lambda_i$ around the follow-up candidate using the diagonal elements of the inverse Fisher matrix [15, 16], i.e.

$$\Delta\lambda_i \equiv \kappa\sqrt{\bar{\Gamma}^{ii}} \; , \tag{13}$$

with

$$\bar{\Gamma}^{ii} = \bar{g}^{ii}/\rho^2 \; , \tag{14}$$

where κ defines the confidence level and g^{ij} is the inverse matrix to g_{ij}. In the present work we use an analytical semi-coherent metric first derived by Pletsch [14]. For coherent integration time longer than a day, but much shorter than a year, the number of sky templates at fixed frequency f converges to

$$\mathcal{N}_{sky} = \frac{2\pi^3 \tau_{\rm E}^2 f^2}{m} \; , \tag{15}$$

where $\tau_{\rm E} \approx 21 \times 10^{-3} s$ is the light travel time from the Earth's center to the detector [14]. The semi-coherent parameter space is finer than the coherent one by a refinement factor γ. Using the notion of *refinement per direction* γ_n we can also obtain the search bands from the extents of the fully coherent metric, namely

$$\Delta\lambda_i = \kappa\sqrt{\frac{g^{ii}}{\gamma_i^2 \rho^2}} \; . \tag{16}$$

For uniformly distributed segments of data without gaps, based on [14] the refinement factors can be obtained as

$$\gamma_f = 1 \; , \tag{17}$$

$$\gamma_{\dot{f}} = \sqrt{5N^2 - 4} \; , \tag{18}$$

$$\gamma_{\ddot{f}} = \sqrt{(35N^4 - 140N^2 + 108)/3} \; , \tag{19}$$

$$\gamma_{\dddot{f}} = \sqrt{(105N^8 - 1260N^6 + 5012N^4 - 6160N^2 + 2304)/(5N^2 - 4)} \; . \tag{20}$$

Finally, for simplicity of the template-bank construction, we use a hyper-cubic lattice to place templates, though hyper-cubic lattices are in general suboptimal, compared to better solutions, e.g. A_n^* lattice. The normalized thickness for an n-dimensional hyper-cubic grid is [7]

$$\theta_n = n^{n/2}\, 2^{-n} \; . \tag{21}$$

The proper choice of the number of dimensions that maximizes the number of templates [1, 11, 12, 13] $\mathcal{N}$ is:

$$\mathcal{N} = \max_n \mathcal{N}_n \; . \tag{22}$$

2.4. Computing cost

In the follow-up of real candidates, especially weak signal candidates, along with the constraint of the total amount of available data, the computing cost constraint may limit significantly the feasibility of the search. Thus the computing-cost requirement is of particular interest. There are currently two different strategies to implement an $\mathcal{F}$-statistic search code in LIGO's reference software suite `lalsuite`[17], namely the SFT-method based on short Fourier transforms of the data with duration T_{SFT} [10] and the FFT-method based on barycentric resampling [18]. Regarding the computational cost, the FFT method is preferable, as the computational requirement to calculate the $\mathcal{F}$-statistic, for a single point in the parameter space, scales only with $\log T$, while the cost of the SFT algorithm scales with T. However, for historical reasons the SFT method is currently still more often used by LIGO/LSC [19, 20, 21], is well tested and we can use recent timing information. The computing cost of a SFT-based $\mathcal{F}$-statistic search is

$$C = \mathcal{N} c_0 N_{\mathrm{SFT}} \; , \tag{23}$$

where N_{SFT} is the number of used SFTs, namely

$$N_{\mathrm{SFT}} = N_{\mathrm{d}} T / T_{\mathrm{SFT}} \tag{24}$$

and c_0 is the fundamental implementation- and hardware-specific computing constant per SFT and template.

3. Minimal required observation time

The main scope of the present work is to find the minimal required observation time that guarantees a certain detection probability of a putative signal buried deep in the detector noise at a certain confidence level by using the fully-coherent $\mathcal{F}$-statistic search technique. We consider two different methods to compute the required integration duration. In method 1, which is closely related to hypothesis testing, we use the concept of false-alarm and false-dismissal probability to achieve certain detection probability. This is the natural way to compute the required integration time. In method 2 we alternatively use the more intuitive notion of expectation value to find the observation duration that guarantees the required detection probability.

3.1. Method 1

In absence of a signal, the probability density function of the $\mathcal{F}$-statistic reduces to a central χ^2-distribution, and the false-alarm probability is given by

$$p_{\mathrm{fA}}^1 = \int_{2\mathcal{F}_{\mathrm{th}}}^{\infty} d(2\mathcal{F}) \chi_4^2(2\mathcal{F}; 0) \; , \tag{25}$$

where $p^1_{\rm fA}$ denotes single trial false-alarm probability and $\chi^2_4(2\mathcal{F};0)$ is the central χ^2-distribution with 4 degrees of freedom. The integration of $\chi^2_4(2\mathcal{F},0)=\frac{1}{2}\mathcal{F}e^{-\mathcal{F}}$ yields

$$p^1_{\rm fA}=(1+\mathcal{F}_{\rm th})e^{-\mathcal{F}_{\rm th}}\ . \tag{26}$$

The overall false-alarm probability of crossing the threshold $2\mathcal{F}_{\rm th}$ in $\mathcal{N}$ trials is

$$p_{\rm fA}=1-(1-p^1_{\rm fA})^{\mathcal{N}}\approx p^1_{\rm fA}\mathcal{N}, \tag{27}$$

when $p^1_{\rm fA}\mathcal{N}\ll 1$ [3, 22], thus

$$p^1_{\rm fA}=p_{\rm fA}/\mathcal{N}\ . \tag{28}$$

We cannot solve Eq. (26) analytically, but numerical solution gives a threshold $2\mathcal{F}_{\rm th}$ value. This allows us to numerically integrate the false-dismissal probability

$$p_{\rm fD}(2\mathcal{F}_{\rm th},\rho^2)=\int_{-\infty}^{2\mathcal{F}_{\rm th}}(d2\mathcal{F})\chi^2_4(2\mathcal{F},\rho^2)\ , \tag{29}$$

where $p_{\rm fD}(2\mathcal{F}_{\rm th})=1-p_{\rm det}$, with the desired detection probability $p_{\rm det}$ and $\chi^2_4(2\mathcal{F},\rho^2)$ is the non-central χ^2-distribution with 4 degrees of freedom and non-centrality parameter ρ^2. At fixed $p^*_{\rm fA}$ and $p^*_{\rm fD}$, using the above equation, we can compute a threshold SNR $\rho_{\rm th}(p^*_{\rm fA},p^*_{\rm fD})$. The required T is such that the inequality

$$\rho^2_{\rm ac}(T)\geq\rho^2_{\rm th}(p^*_{\rm fA},p^*_{\rm fD}) \tag{30}$$

holds, where $\rho^2_{\rm ac}(T)$ is the accumulated SNR due to the presence of signal in the analyzed data. Assuming that the follow-up search will use data of similar constant noise floor, we can rewrite Eq. (3) as

$$\rho^2_{\rm ac}(T)=\rho^2_{\rm c}\frac{N_{\rm d}T}{N^{\rm c}_{\rm d}\Delta T}\ , \tag{31}$$

where ΔT is the length of one segment in the semi-coherent search using data from $N^{\rm c}_{\rm d}$ number of detectors. With the average $2\bar{\mathcal{F}}_{\rm c}$ value of the candidate, we can compute its SNR $\rho_{\rm c}$ from Eq. (9), namely

$$\rho^2_{\rm c}=E[2\bar{\mathcal{F}}_{\rm c}]-4\ . \tag{32}$$

Substitution in the equations above yields the accumulated SNR in presence of signal

$$\rho^2_{\rm ac}=\left(E[2\bar{\mathcal{F}}_{\rm c}]-4\right)\frac{N_{\rm d}T}{N^{\rm c}_{\rm d}\Delta T}\ , \tag{33}$$

which gives the required minimal T.

3.2. Method 2

Computation of the $\mathcal{F}$-statistic on data with no signal, has a certain expectation value, therefore we ask what is the expected maximal $2\mathcal{F}$ value $E[2\mathcal{F}_{\mathcal{N}}]$ in $\mathcal{N}$ trials in Gaussian noise, where $\mathcal{F}_{\mathcal{N}}\equiv\max\{\mathcal{F}\}^{\mathcal{N}}_{i=1}$. The probability to get $(\mathcal{N}-1)$ values of $2\mathcal{F}$ less than $2\mathcal{F}_{\mathcal{N}}$ follows a binomial distribution, namely

$$\begin{aligned}p_{\mathcal{N}}(2\mathcal{F}_{\mathcal{N}}) &= \binom{\mathcal{N}}{1}\chi^2_4(2\mathcal{F},0)(1-\alpha_1)^{\mathcal{N}-1} && (34)\\ &= \frac{1}{2}\mathcal{N}\mathcal{F}_{\mathcal{N}}e^{-\mathcal{F}_{\mathcal{N}}}\left(1-(1+\mathcal{F}_{\mathcal{N}})e^{-\mathcal{F}_{\mathcal{N}}}\right)^{\mathcal{N}-1}\ . && (35)\end{aligned}$$

With this we can numerically integrate the expectation value

$$E[2\mathcal{F}_{\mathcal{N}}] = \int_0^\infty d(2\mathcal{F}_{\mathcal{N}})\, 2\mathcal{F}_{\mathcal{N}}\, p_{\mathcal{N}}(2\mathcal{F}_{\mathcal{N}}) \;, \tag{36}$$

and standard deviation

$$\sigma_{\mathcal{N}}(2\mathcal{F}_{\mathcal{N}}) = \left(\int_0^\infty d(2\mathcal{F}_{\mathcal{N}})\, (2\mathcal{F}_{\mathcal{N}} - E[2\mathcal{F}_{\mathcal{N}}])^2\, p_{\mathcal{N}}(2\mathcal{F}_{\mathcal{N}}) \right)^{1/2} \;. \tag{37}$$

To safely distinguish a real signal from pure noise, we can require the following inequality to hold:

$$E[2\mathcal{F}_S] - h\sigma_S(2\mathcal{F}_S) > E[2\mathcal{F}_{\mathcal{N}}] + h\sigma_{\mathcal{N}}(2\mathcal{F}_{\mathcal{N}}) \;, \tag{38}$$

where the expectation value $E[2\mathcal{F}_S]$ of a real signal and its standard deviation $\sigma_S(2\mathcal{F}_S)$ are computed using Eqs. (4) and (5). As all terms in inequality (38) are function of the observation time, this gives an alternative method to compute the minimal required integration time. Fine-tuning of Eq. (38) is possible through the safety parameter h, which we quantify by using Chebyshev's inequality. For a random variable X, with expected value $E[X]$ and standard deviation σ,

$$P(|X - E[X]| \geq h\sigma) \leq 1/h^2 \;, \tag{39}$$

which means that at least a fraction

$$p = 1 - 1/h^2 \tag{40}$$

of the data is within h standard deviations on either side of the mean [23]. Rearranging the above equation yields

$$h = 1/\sqrt{1-p} \;. \tag{41}$$

Having two independent random variables, $2\mathcal{F}_S$ and $2\mathcal{F}_{\mathcal{N}}$, we can label the fraction of data around each mean as p_S and $p_{\mathcal{N}}$ and introduce the joint probability $p_J = p_S p_{\mathcal{N}}$. We see, that the same joint probability can be achieved for different combinations of p_S and $p_{\mathcal{N}}$. However, a natural choice is $p_S = p_{\mathcal{N}}$, thus

$$h = 1/\sqrt{1 - \sqrt{p_J}}. \tag{42}$$

We give a set of p_J values and related h in Table 1.

p_J	0.75	0.90	0.95	0.99
h	2.73	4.41	6.28	14.12

Table 1. Joint probability p_J and corresponding required h standard deviations.

Fixing p_J to some value and with this h in inequality (38), we can compute the minimal required coherent observation time T, such that (38) holds. For this integration time, the joint probability p_J becomes the separation probability $p_{\mathrm{sep}} = p_J$. This is the probability, that a candidate due to the presence of a signal is consistent with the signal hypothesis and a candidate due to the noise is consistent with the noise hypothesis. Taking into account that $p_S = 1 - p_{\mathrm{fD}}$ and $p_{\mathcal{N}} = 1 - p_{\mathrm{fA}}$, we find the relation of the separation probability to the detection probability, namely $p_{\mathrm{sep}} = p_{\mathrm{det}}(1 - p_{\mathrm{fA}})$, or for negligible false-alarm $p_{\mathrm{det}} \approx p_{\mathrm{sep}}$.

4. Method comparison

4.1. Numerical predictions

In the following we compare the two methods to find the minimal required integration time described in the previous section in terms of observation duration and computing cost. We consider a StackSlide search with $N = 205$ segments of duration $\Delta T = 25$ hours, each using data from $N_{\rm d}^{\rm c} = 2$ detectors. For a hypothetical candidate with fixed Doppler parameters $\alpha = 1.45$ rad, $\delta = 0$ rad $f = 185$ Hz, $\dot{f} = -1 \times 10^{-9}$ Hz/s, we pick an average strength in the range $2\bar{\mathcal{F}}_c \in [5, 13]$. Then using Eq. (13) with $\kappa = 1$ and the semi-coherent metric we compute the search bands associated with such a candidate. Having that, for mismatch $m = 0.01$ and a hyper-cubic lattice, we can compute the number of templates using Eq. (10) and the fully-coherent metric. Using method 1, requiring detection probability $p_{\rm det}^* = 0.9$ at overall false-alarm probability $p_{\rm fA}^* = 0.01$ using Eq. (29) we compute $\rho_{\rm th}^2(p_{\rm fA}^*, p_{\rm fD}^*)$ and the minimal required observation time T_1, which substituted in Eq. (33) with $N_{\rm d} = N_{\rm d}^{\rm c}$ satisfies Eq. (30). For method 2 a separation probability equal to $p_{\rm det}^*$ yields safety factor $h = 4.41$, see Table 1. We label the integration time that satisfies Eq. (38) as T_2 and plot both integration times $T_1(2\bar{\mathcal{F}}_c)$ and $T_2(2\bar{\mathcal{F}}_c)$ in Figure 1 (a) as function of $2\bar{\mathcal{F}}_c$. With the number of templates for T_1 and T_2 we estimate the computing cost C_1 and C_2 using the fundamental computing cost constant $c_0 = 7 \times 10^{-8} s$ in Eq. (24) and assuming SFTs of duration $T_{\rm SFT} = 1800$ s in Eq. (25). $C_1(2\bar{\mathcal{F}}_c)$ and $C_2(2\bar{\mathcal{F}}_c)$ are plotted in Figure 1 (b). In Figure 1 (c) we plot how the expectation value from a real signal grows with increasing T compared to loudest candidate from Gaussian noise. In this plot the candidate strength is fixed to $2\bar{\mathcal{F}}_c = 8.5$.

We see that method 2 yields much longer observation time, at same candidate strength compared to method 1. Due to the resulting much larger number of templates, the computing cost, especially for weak candidates, is much higher. The inferiority of method 2 compared to method 1 in terms of required integration duration and computing power can be explained by the *ad hoc* construction of method 2 and the use of Chebyshev's inequality, which is only a lower bound. In this sense method 2 is a more conservative approach, though the important information about false-alarm and false-dismissal probability gets lost in this framework. The computing cost of method 1 looks very promising even for weak candidates, however we should keep in mind that this is lower limit and the cost of a search with real data would most likely be much higher. The reason for this is that gaps in the data are direct penalty for the growth of $\rho_{\rm ac}^2$, while $\rho_{\rm th}^2$ remains unaffected. Furthermore, for very weak signals, the required integration duration may violate the assumption of constant sky resolution, thus we would underestimate the number of templates, resulting in a higher false-dismissal.

4.2. Monte Carlo results

To confirm the numerical predictions of method 1 we perform the following Monte Carlo studies. We create a set of 205 segments with duration 25 hours of Gaussian noise and draw a set of pulsar parameters $\alpha \in (0, 2\pi)$, $\delta \in (-\pi/2, \pi/2)$, $\cos\iota \in (-1, 1)$, $\psi \in (0, 2\pi)$, $\phi_0 \in (0, 2\pi)$ at fixed frequency of $f = 185$ Hz and spindown value in the range $\dot{f} \in (-f/\tau, 0)$, where $\tau = 2220$ yr is the minimal spindown age of the source [1]. We inject a signal with the above parameters and intrinsic signal amplitude h_0 high enough to produce a candidate with expected average strength $E[2\bar{\mathcal{F}}_S] \in [12, 13]$. To find the actual injected value we first do a targeted StackSlide search at the point of the injection. With this measured injected $2\bar{\mathcal{F}}_S$ value, using Eq. (13) we compute Fisher extents, from which we draw a random parameter point λ_c satisfying

$$\bar{\Gamma}_{ij} \Delta\lambda^i \Delta\lambda^j < 1 \; . \tag{43}$$

The point λ_c is within the 1-σ Fisher ellipsoid of the true signal location and becomes the candidate to follow up. Following the scheme for method 1 as described above, we compute

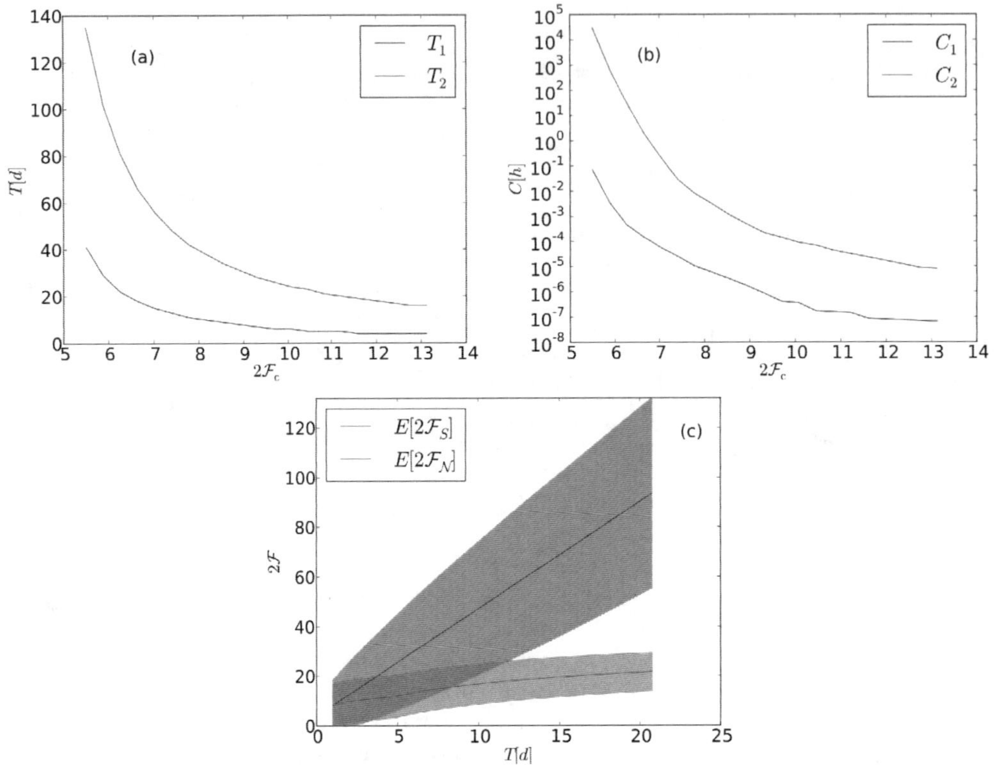

Figure 1. Numerical comparison between method 1 and method 2 (quantities labeled with 1 and 2, respectively). Figure (a) shows the required coherent integration time as function of the strength of the candidate, (b) shows the computing cost depending on the strength of the candidate, (c) shows expected value of signal, noise and related $h = 4.41$ standard deviations for detection probability $p_{\rm det} = 0.9$ of a candidate with $2\bar{\mathcal{F}}_{\rm c} = 8.5$.

the minimal required coherent observation time targeting detection probability $p^*_{\rm det} = 0.9$ and search for the signal. After computation of $2\mathcal{F}_S$ using the data with the injected signal, we compute $2\mathcal{F}_\mathcal{N}$ with the same grid and integration duration using the noise only data. We claim "detection" whenever the loudest measured $2\mathcal{F}_S$ value in the data with injected signal is higher than the loudest measured $2\mathcal{F}_\mathcal{N}$ of the noise. The result of the Monte Carlo simulations is as follows: in 897, out of 1000 trials, the measured $2\mathcal{F}_S$ value in the data containing injected signal exceeds the measured $2\mathcal{F}_\mathcal{N}$ value of the noise only data. With this the achieved detection probability $p_{\rm det} = 0.897 \pm 0.023$ is in accordance with the targeted detection probability $p^*_{\rm det}$.

5. Discussion

We derived two different methods to compute the minimal required coherent integration time in a fully-coherent $\mathcal{F}$-statistic search in the *zoom* stage of follow-up of candidates from a semi-coherent StackSlide search. By numerical comparison we showed that method 1 is superior to method 2 in terms of required integration duration and computing cost. We confirmed in a Monte Carlo study that the predicted coherent integration time is sufficient to achieve the desired detection probability. The results of this paper have been derived for Gaussian data without gaps and two detectors of equal noise floor. Further extension of this work is closely related to the data selection problem.

Acknowledgments

This work significantly benefited from numerous suggestions of Reinhard Prix. I also thank Holger Pletsch, Karl Wette and Paola Leaci for useful discussions. Finally I would like to acknowledge the support of Bruce Allen and the IMPRS on Gravitational Wave Astronomy of the Max-Planck-Society.

References

[1] Brady P R, Creighton T, Cutler C and Schutz B F 1998 *Phys. Rev.* **D57** 2101–2116 (*Preprint* gr-qc/9702050)
[2] 2011 Einstein@home http://www.einsteinathome.org
[3] Jaranowski P, Krolak A and Schutz B F 1998 *Phys. Rev.* **D58** 063001 (*Preprint* gr-qc/9804014)
[4] Cutler C and Schutz B F 2005 *Phys. Rev. D.* **72** 063006
[5] Balasubramanian R, Sathyaprakash B S and Dhurandhar S V 1996 *Phys. Rev. D.* **53** 3033–3055
[6] Owen B J 1996 *Phys. Rev. D.* **53** 6749–6761
[7] Prix R 2007 *Class. Quant. Grav.* **24** S481 (*Preprint* 0707.0428)
[8] Messenger C, Prix R and Papa M A 2009 *Phys. Rev. D.* **79** 104017–+ (*Preprint* 0809.5223)
[9] Harry I, Allen B and Sathyaprakash B 2009 *Phys.Rev.D* **80** 104014 URL http://arxiv.org/abs/0908.2090
[10] Prix R 2010 The F-statistic and its implementation in ComputeFStatistic_v2 Tech. rep. LIGO Scientific Collaboration (LIGO-T0900149-v2) URL https://dcc.ligo.org/cgi-bin/DocDB/ShowDocument?docid=1665
[11] Brady P R and Creighton T 2000 *Phys. Rev.* **D61** 082001 (*Preprint* gr-qc/9812014)
[12] Cutler C, Gholami I and Krishnan B 2005 *Phys. Rev. D* **72** 042004
[13] Prix R and Shaltev M 2011 (*Preprint* LIGO-P1100156-v1) URL http://arxiv.org/abs/1201.4321
[14] Pletsch H J 2010 *Phys. Rev. D.* **82** 042002–+ (*Preprint* 1005.0395)
[15] Prix R 2007 *Phys. Rev. D.* **75** 023004 (*Preprint* gr-qc/0606088)
[16] Vallisneri M 2008 *Phys. Rev. D* **77**(4) 042001 URL http://link.aps.org/doi/10.1103/PhysRevD.77.042001
[17] 2011 lalsuite https://www.lsc-group.phys.uwm.edu/daswg/projects/lalsuite.html
[18] Patel P, Siemens X, Dupuis R and Betzwieser J 2010 *Phys. Rev. D.* **81** 084032–+ (*Preprint* 0912.4255)
[19] Abbott B *et al.* (LIGO Scientific Collaboration) 2008 *Phys. Rev.* **D77** 022001 (*Preprint* 0708.3818)
[20] Abbott B P *et al.* (LIGO Scientific Collaboration) 2009 *Phys. Rev.* **D80** 042003 (*Preprint* 0905.1705)
[21] Abadie J, Abbott B P, Abbott R, Abernathy M, Adams C, Adhikari R, Ajith P, Allen B, Allen G, Amador Ceron E *et al.* (LIGO Scientific Collaboration) 2010 *ApJ* **722** 1504–1513 (*Preprint* 1006.2535)
[22] Watts A, Krishnan B, Bildsten L and Schutz B F 2008 *Mon. Not. Roy. Astron. Soc.* **389** 839–868 (*Preprint* 0803.4097)
[23] Ghahramani S 2000 *Fundamentals of Probability* (Prentice Hall, Inc.)
[24] Prix R 2007 (*Preprint* gr-qc/0702068) URL http://arxiv.org/abs/gr-qc/0702068
[25] Prix R and Krishnan B 2009 Targeted search for continuous gravitational waves: Bayesian versus maximum-likelihood statistics Tech. Rep. LIGO-P0900066-v2 URL http://arxiv.org/abs/0907.2569

State-space modelling for heater induced thermal effects on LISA Pathfinder's Test Masses

F Gibert[1,2], M Nofrarias[1,2], M Diaz-Aguiló[1,3], A Lobo[1,2], N Karnesis[1,2], I Mateos[1,2], J Sanjuán[4], I Lloro[1,2], L Gesa[1,2] and V Martín[1,2]

[1] Institut d'Estudis Espacials de Catalunya (IEEC), Barcelona, Spain
[2] Institut de Ciències de l'Espai (CSIC-IEEC), Barcelona, Spain
[3] Escola Politècnica Superior de Castelldefels (EPSC), Universitat Politècnica de Catalunya (UPC), Castelldefels, Spain
[4] Department of Physics, University of Florida, Gainesville, FL, United States

E-mail: gibert@ieec.cat

Abstract. The OSE (Offline Simulations Environment) simulator of the LPF (LISA Pathfinder) mission is intended to simulate the different experiments to be carried out in flight. Amongst these, the thermal diagnostics experiments are intended to relate thermal disturbances and interferometer readouts, thereby allowing the subtraction of thermally induced interferences from the interferometer channels. In this paper we report on the modelling of these simulated experiments, including the parametrisation of different thermal effects (radiation pressure effect, radiometer effect) that will appear in the Inertial Sensor environment of the LTP (LISA Technology Package). We report as well how these experiments are going to be implemented in the LTPDA toolbox, which is a dedicated tool for LPF data analysis that will allow full traceability and reproducibility of the analysis thanks to complete recording of the processes.

1. Introduction

The Thermal Diagnostic Subsystem of the LISA Technology Package (LTP) onboard LISA Pathfinder (LPF) [1, 2] includes the generation of thermal disturbances in some thermal sensitive locations of the LTP to induce perturbing signals in the interferometer (IFO) readouts. The objective of such controlled perturbations is to obtain information of the effects that potential temperature gradients in the LTP may induce in the final IFO readout during the main operational phase, enabling the possibility of removing thermally induced noise from the signal.

The LISA Pathfinder main noise requirements in the LTP state for the test masses residual differential acceleration:

$$S^{1/2}_{\Delta a,\,\mathrm{LPF}}(\omega) \leq 3\times 10^{-14}\left[1+\left(\frac{\omega/2\pi}{3\,\mathrm{mHz}}\right)^2\right]\ \mathrm{m\,s^{-2}\,Hz^{-1/2}} \qquad (1)$$

in the bandwidth between 1 mHz and 30 mHz [3] .

Therefore, a set of Thermal Diagnostics experiments consisting of the injection of controlled loads of heat at specific spots must be carefully designed, producing in-band perturbing signals with high Signal-to-Noise Ratio ($SNR \approx 50$). Moreover, a model linking both the heat propagation and temperature distributions to final IFO readouts must be designed, and it is intended to be introduced into two LPF key simulators:

- The OSE (*Offline Simulation Environment*) simulator: Located at ESAC (*European Space Astronomy Centre*), this simulator is intended to validate all LPF mission operations considering the different interfaces and data transfer processes.
- The LTPDA simulator: Developed by the LTP Data Analysis (LTPDA) team, within the LTPDA software [4]. It is a linear simulator based on State-Space Modelling (SSM) intended to analyse the data delivered by the mission.

Part of this model is presented here in this paper in Section 3, and some first results are reported in Section 4. Nevertheless, it is important to first introduce the different Thermal Diagnostics experiments and the architecture of the Data Analysis associated to them.

1.1. Thermal Diagnostics Experiments

The LTP is equipped with a set of 14 heaters and 24 temperature sensors (see Figures 1 and 2). Different experiments are intended to apply heat signals through the heaters and record the temperature readouts. These experiments are gathered in the Experiment Master Plan (EMP) [5] which aims to characterise the thermal effects in Table 1 by switching on different heaters at each case, and consequently thermally disturbing different parts of the LTP.

Table 1. Types of Thermal Diagnostics Experiments. Heaters, associated thermal effects and expected main consequences.

Heaters activated	Thermal effects	Main consequence
Electrode Housing (EH)	Radiometer effect, radiation pressure	Force and torques on Test Masses (TMs)
Optical Window (OW)	Glass and clamp thermal expansion	Optical Window glass refractive index change
Suspension Struts (SS)	Optical Bench asymmetric thermal expansion	Optical Bench (OB) thermoelastic stress

1.2. Thermal Diagnostics Data Analysis

A complete model simulating the different consequences of the experiments (temperature responses, dynamics on the TM, and phase shift in the IFO) on one hand, and the telemetry data of the IFO and the temperature sensor readouts, on the other, will be used in the subsequent data analysis, when all the data will be collected and processed so more accurate values can be obtained for the model coefficients.

The global scheme is presented in Figure 3, where blue blocks refer to the thermal model-obtained temperature, while red ones refer to the blocks representing conversion from temperature distributions to the different perturbations that can disturb the readouts of the interferometer.

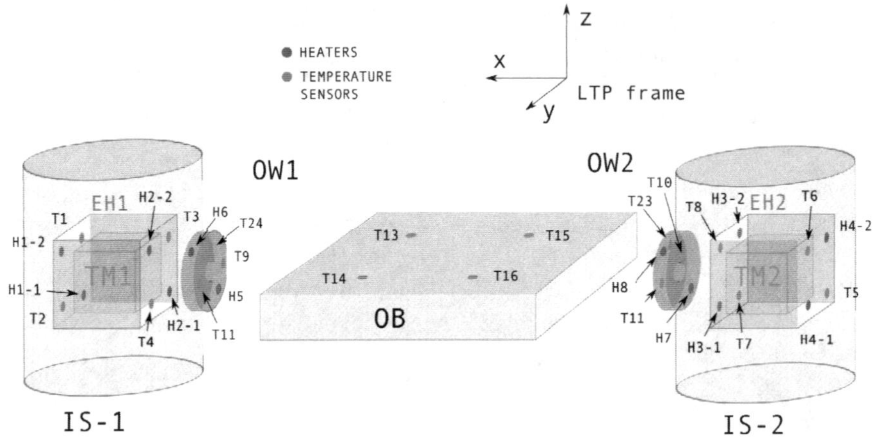

Figure 1. Scheme of the Thermal DS (Diagnostics Subsystem) items locations in the Electrode Housings, on the Optical Windows and on the Optical Bench, with their enumeration.

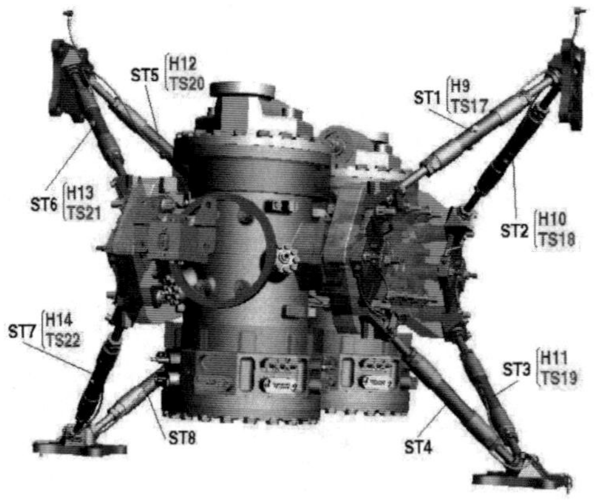

Figure 2. Scheme of the Thermal DS items location on the Suspension Struts. CAD rendering of the LTP courtesy of EADS Astrium GmbH.

The Data Analysis work must handle two different types of simulations: (1) a simulation providing data that will be observed during the mission, as temperature sensor readouts, and (2) a simulation providing temperatures at non-observable points whose temperatures are inputs to the blocks converting temperature fluctuations into IFO perturbations. Block (1) will be used to validate the thermal model used in (2).

The modelling of the different blocks in Figure 3 varies depending on the feasibility of carrying out experimental on-ground campaigns able to reproduce the disturbing thermal effects of each experiment in Table 1. Consequently, experiments for thermal qualification purposes on the Optical Window and on the Optical Bench were planned, the first carried out in AEI Hannover Laboratory (2007, [6]) and the latter taking place at the time of writing in the frame of the LPF Thermo-Optical Qualification Model (TOQM) campaign in IABG facilities in Ottobrunn, near Munich, being run by EADS Astrium-Germany and EADS Astrium-UK (November 2011, [7]). As thermal effects on the free falling TMs can not be directly tested, an accurate model for

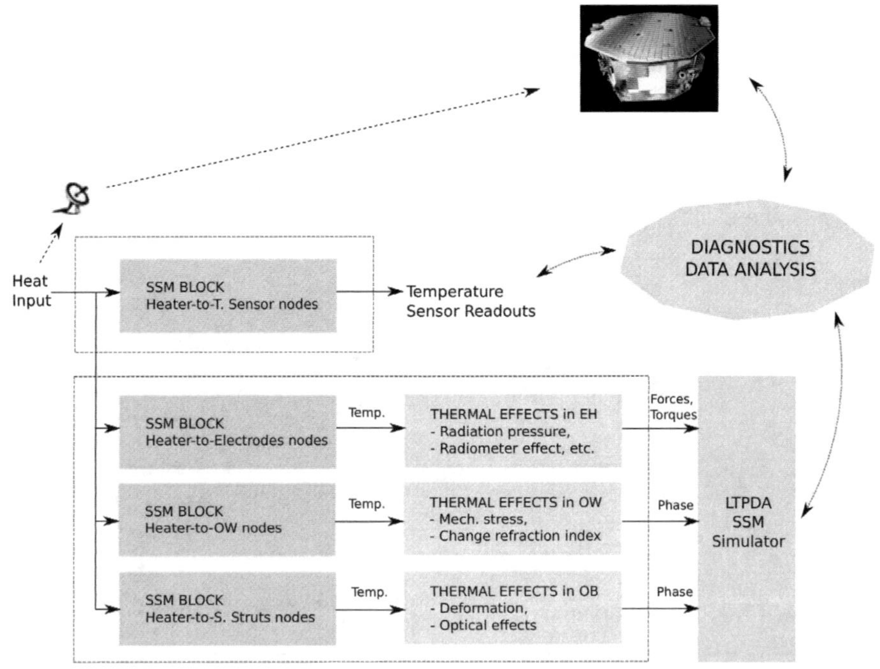

Figure 3. Scheme of the Thermal Diagnostics model. LPF picture courtesy of ESA.

the EH thermal effects is being developed, the results of which are going to be compared with similar experiments carried out at Trento University with a torsion pendulum [8].

2. Thermal effects in the EH

Temperature differences across the TMs environment can produce differential pressures that turn into net forces and torques on the TMs. Three different thermal effects have been identified [9] as possible mechanisms that will create noticeable dynamic effects on the TMs:

2.1. Radiation pressure effect

The radiation pressure effect is based on the temperature-dependence of the radiation emitted by a surface, following the Stefan-Boltzmann Law. The force applied on a surface j by the presence of another surface i at a different temperature is expressed as:

$$F_{ij}^{RP} = \frac{8}{3}\frac{S_{ij}^{RP}\epsilon_i A_j \sigma T_0^3}{c}\Delta T_{ij} \quad [N] \tag{2}$$

where σ is the Stefan-Boltzmann constant, c is the speed of light, S_{ij}^{RP} is a system geometric factor from surface i to surface j which contains information of both the system geometry and potential multi-reflections, ϵ_i is the emissivity of surface i, A_j is the area of the section where the force is applied, T_0 is the absolute temperature and ΔT_{ij} is the temperature difference between two surfaces, written as $\Delta T_{ij} = T_i - T_j$.

2.2. Radiometer effect

The radiometer effect appears in rarefied atmospheres where the particles have a mean free path much longer than the distance between the surfaces of the enclosing volume, i.e in systems with large Knudsen number. The consequent force is represented here as:

$$F_{ij}^{RM} = \frac{1}{4}\frac{S_{ij}^{RM} A_j p}{T_0}\Delta T_{ij} \quad [N] \tag{3}$$

where p stands for the remaining gas static pressure in the enclosure and S_{ij}^{RM} is a system geometry coefficient for the radiometer effect, which includes multi-reflections as well. A_j, T_0 and ΔT_{ij} have the same meanings as in Equation 2.

2.3. Outgassing effect

The outgassing effect is caused by the detachment of gas molecules from the surface, in a low pressure environment. Its behaviour is strongly related with the surface roughness and the material properties; although its consequences could be similar to the ones from the other thermal effects, it is not yet included in the model because of the uncertainty of some of its parameters.

More details of all these effects can be found in [9] and the values considered for their main parameters are presented in Table 2, where the radiative emissivities presented there refer to the TM and EH surfaces [10].

Table 2. Default values considered for the thermal effects expressions.

Symbol	Parameter	Value
p	Environment static pressure	10^{-5} Pa
T_0	System temperature	293 K
ϵ_{TM}	TM surface radiative emissivity	0.02
ϵ_{EH}	EH surface radiative emissivity	0.05

3. Thermal effects modelling

In this Section, we focus on how the thermal effects inside the Electrode Housing are being modelled. Such a model can be divided into two main blocks, a thermal block that provides temperatures at specific spots when heaters are activated and another block that implements the different thermal effect expressions presented in Section 2 taking into account the geometry of the system and other effects as the potential multi-reflections between surfaces.

3.1. ESATAN model block

A thermal model of the LTP in ESATAN software (developed by *Carlo Gavazzi Space*) together with a general model of the whole spacecraft is maintained at ESTEC (*European Space Research and Technology Centre*). It provides node-to-node transfer function samples, where the input is the heat power applied to a node and the output is the temperature response at another requested node. Temperature responses are calculated as

$$T_s(s) = H_{h\to s}(s) \cdot q_h(s) \tag{4}$$

where q_h is the heat power applied to heater h, $H_{h\to s}$ the transfer function and T_s the temperature response at sensor s, in Laplace Domain.

Different representative nodes in the thermal model were selected, and series of transfer functions relating all the DS heaters to these interesting nodes were obtained and implemented into a single state-space model.

3.2. Geometry and multi-reflections considerations

Since the real system is not composed of infinite parallel surfaces, it is necessary to consider finite size effects when modelling the EH.

First of all, the surfaces of the system composed of the TM and the inner walls of the EH need to be discretised properly so potential temperature gradients across a surface can be represented. The discretisation adopted is shown in Figure 4, where each TM/EH surface is divided into 4 new facets, thereby enabling the representation of both forces and torques in the three space axis.

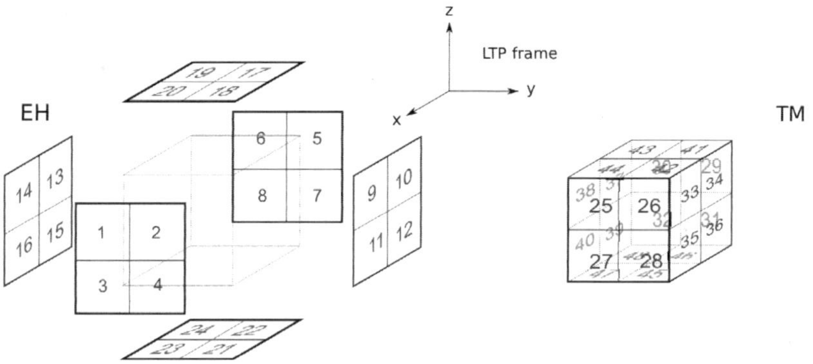

Figure 4. Discretisation considered for the temperature distribution in the inner faces of the EH (exploded view) and on the TM with numeration of the facets. Each facet is considered isothermal.

On the other hand, in order to model radiation reflection effects in the EH it is necessary to develop a multi-reflection model for our specific system. Additionally, the model must include the possibility of generating not only pressures but also shear forces on the TM surfaces, as they are also important [8].

Such conditions have required the calculation of all facet-to-facet force contributions coming from the different surfaces in a vectorial way (through the computation of the Radiative Exchange Factors [11]), so the different resulting components can be disentangled on each surface. The main trade-off of these considerations is the assumption that all reflections are diffuse [12].

This multi-reflection model is then extrapolated and applied to the Radiometer effect, where equivalent Geometric Exchange Factors have been considered in replacement of the Radiative Exchange Factors, which only apply to the Radiation Pressure Effect. The parameters used in such extrapolations are still under study, but first results are quite satisfactory (see Section 4). Table 3 presents the current values considered for the different effects and surfaces.

Finally, as its implementation is linear and it has no dynamics associated, this block can be considered a post-processing part of the temperature state-space model block.

4. Results

Under the assumption of constant temperature at the TM, we report here first results obtained from the implementation of the model. Specifically, we present forces and torques obtained from the simulation of one planned run from the EMP [13]. In this case, the exercise is applied to EH1 and the heaters activated are H1 and H2. The parameters of the input signal are given in Table 4 and Figure 5 shows graphically the input signal shape.

Table 3. Parameters considered in the multi-reflection model. Radiative emissivities have already been presented in Table 2.

Symbol	Parameter	Value
$\rho_{TM,RP}$	TM radiative reflectivity	0.98
$\rho_{EH,RP}$	EH radiative reflectivity	0.95
$\epsilon_{TM,RM}$	TM equivalent emissivity for radiometer effect	0.02
$\epsilon_{EH,RM}$	EH equivalent emissivity for radiometer effect	0.05
$\rho_{TM,RM}$	TM equivalent reflectivity for radiometer effect	0.8
$\rho_{EH,RM}$	EH equivalent reflectivity for radiometer effect	0.8

Table 4. Input signal parameters.

Heaters activated	H1 and H2 alternatively
Period	2000 s
Duty cycle	50.00%
Power	10 mW (5 mW to each physical heater)
Duration	4000 s
Total signal duration	20000 s

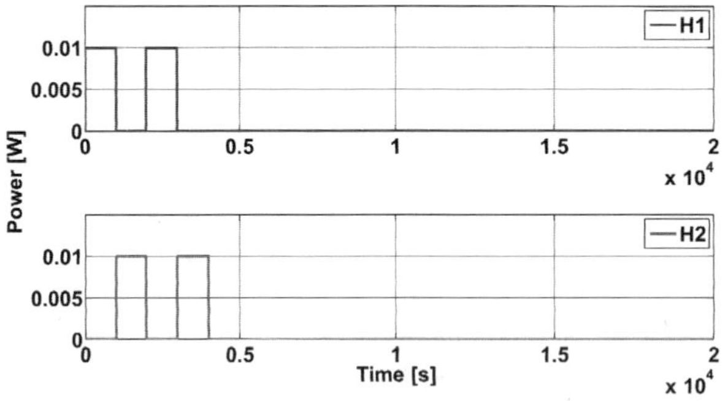

Figure 5. Input signals. H1 and H2 are activated alternatively.

Figure 6 shows the evolution of the EH temperature and the temperature differences, as measured by the sensors in the indicated faces. Temperature sensor readouts in the same face of the EH are not identical, but their differences are negligible in this context.

Note the following temperature evolution features:

(i) When the heater H1 on the $+X$ face is activated, the temperature of the latter starts to rise until the heat input is discontinued, during the second half of the heating signal period. At the same time, the -X face also feels the heating, and its temperature increases, though significantly less.

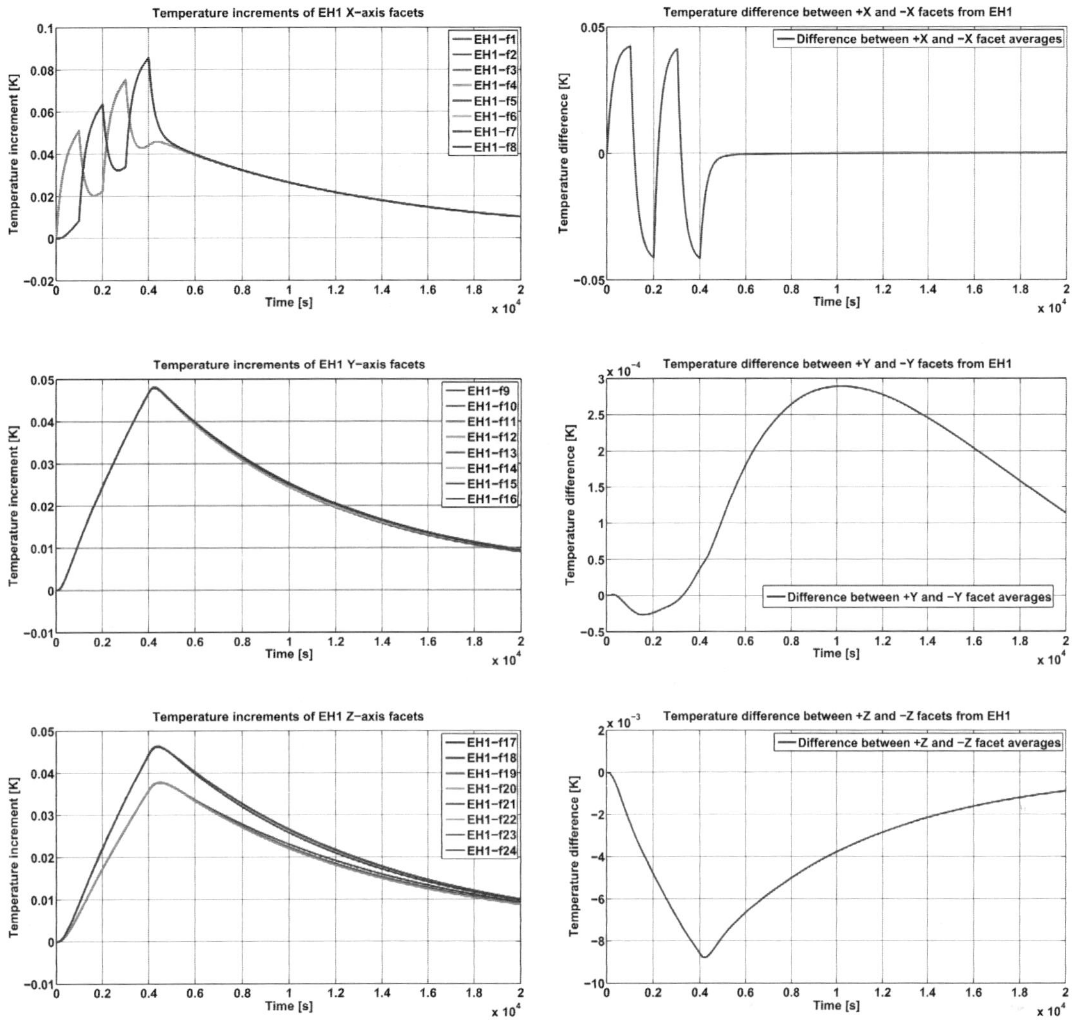

Figure 6. Plots of temperature increments and differences on EH1 facets. The temperature differences have been obtained by previously averaging the temperature at all the facets from each wall. The facet position follows the notation in Figure 4.

(ii) During the second half-period, when H1 is off, the $+X$ face cools down, but does not end up at the original temperature. The activation of H2 in the $-X$ face actually causes the temperature of the $+X$ face to slightly start rising towards the end of the first H2 cycle.

(iii) As the process continues, there is a consistent drift upwards in the temperature of both faces. When the heaters are completely switched off, the temperatures of the faces match each other -approximately from 5000 seconds-, then they continue in parallel a smooth decline which asymptotically approaches the initial temperature.

In Figure 6 we see as well plots of the temperature differences between the different faces. As can be observed, a very symmetric pattern shows up in the case of $+X$ and $-X$ faces.

Finally, the forces and torques obtained on TM1 considering the contribution of each different effect are presented in Figure 7. At each plot the contributions of each thermal effect are presented separately, and the parameters applied are the default ones used in the algorithm.

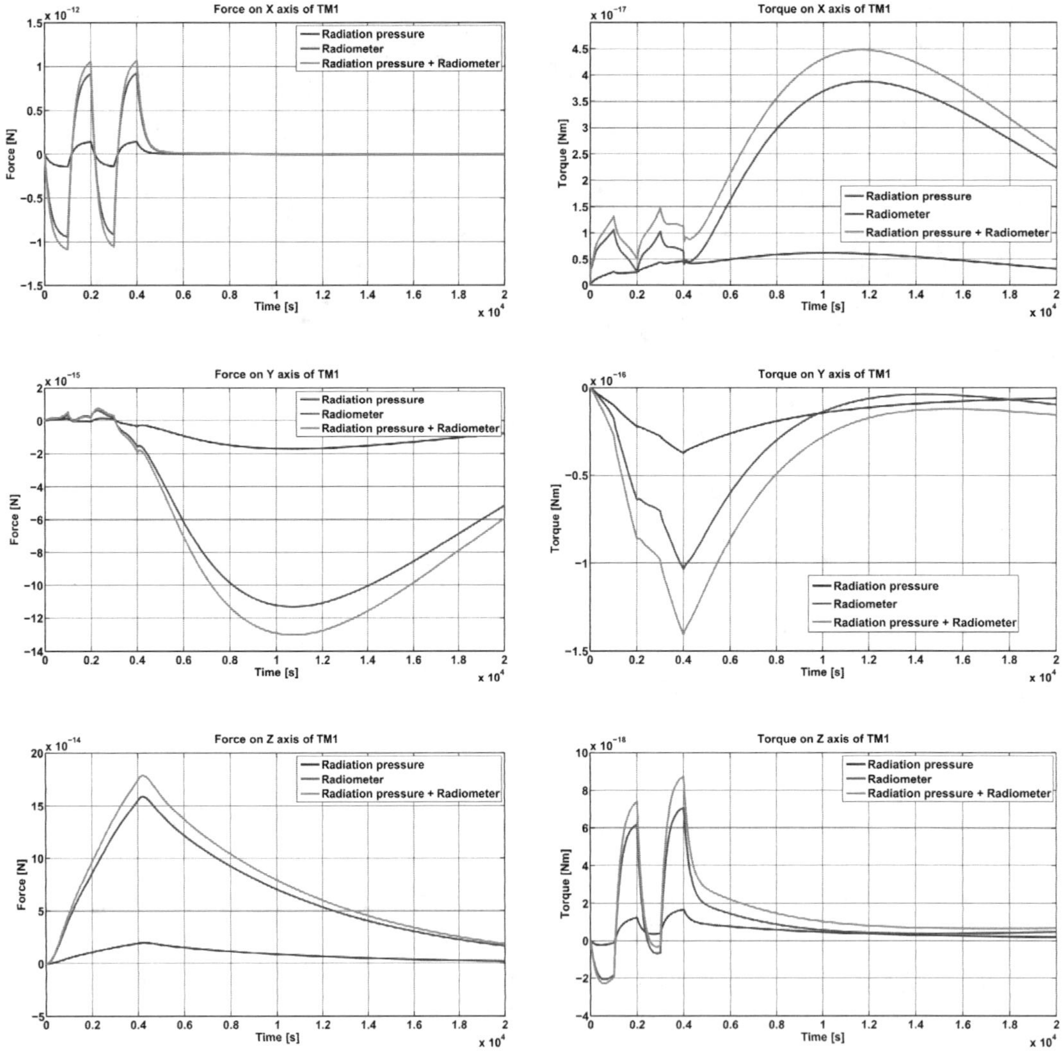

Figure 7. Plots of forces and torques on TM1, with the disentangled contributions of the different thermal effects under the default parameters of the algorithm.

5. Conclusions

A model for the thermal Diagnostics Experiments in the Electrode Housing to be carried out in the LPF has been presented. Improved from older versions, this model can now represent multi-reflections for the different thermal effects in a vectorial way, and it includes representations of the radiation pressure and the radiometer effect.

First results show that noticeable thermally-created forces can be obtained in the IFO-sensitive axis -X axis- and that the rest of perturbations, i.e. forces on the Y and Z axis and the different torques, present much smaller effects. On the other hand, all torques observed are negligible as they are far below the DC 1.1×10^{-11} Nm specified in [14]. Additionally, an interesting temperature gradient is observed on the Z axis, though it is out of the bandwidth of interest.

However, some of the parameters considered for the model are still under on-going study so improved values are expected to be obtained in the future. These parameters refer mostly to the multi-reflections block.

Future work also includes the integration of all the Thermal Diagnostics Experiments into a single Module able to simulate the thermal disturbances in the LTP caused by the different DS heaters, i.e. adding to the current model the OW phase shift and potentially thermoelastic effects on the OB caused by the OW and the Suspension Strut heaters respectively. The implementation of the part involving the EH and the OW in the LTPDA simulator so as to obtain IFO readouts from the thermal experiments is already under way.

Acknowledgments
We acknowledge support from Project ESP2007-61712 of Plan Nacional del Espacio of the Spanish Ministry of Science and Innovation (MICINN).

References

[1] M Armano *et al* 2009, *LISA Pathfinder: the experiment and the route to LISA*, *Class. Quantum Grav.* **26** 094001
[2] F Antonucci *et al* 2011, *From laboratory experiments to LISA Pathfinder: achieving LISA geodesic motion*, *Class. Quantum Grav.* **28** 094002
[3] G Kahl 2010, *DDS Subsystem Specification*, Requirements Specification S2-ASD-RS-3004 v4.6
[4] M Hewitson *et al* 2009, *Data analysis for the LISA Technology Package*, *Class. Quantum Grav.* **26** 094003
[5] N Brandt *et al* 2007 *Experiment Performance Budget*, Report S2-ASD-RP-3036 v2
[6] M Nofrarias *et al* 2007 *Thermal diagnostic of the optical window on board LPF*, *Class. Quantum Grav.* **24** 5103
[7] A Lobo *et al* 2011 *Diagnostics thermal experiments in the LTP TOQM*, Technical Note S2-IEC-TN-3086 v3
[8] L Carbonne *et al* 2007 *Thermal gradient-induced forces on geodetic reference masses for LISA*, *Phys. Rev. D* **76** 102003
[9] M Nofrarias 2007 *Thermal diagnostic in the LISA technology package experiment*, Chapter 2. PhD Thesis
[10] A Franzoso 2007 *LISA Core Assembly Thermal Model Description*, Technical Note S2-CGS-TN-3031-v2
[11] ALSTOM Power Technology Centre 2004 *ESARAD Kernel Algorithms Document* v7.3
[12] F Gibert *et al* 2011 *Algorithm Specification for the Thermal Diagnostics on board LPF*, Technical Note S2-IEC-TN-3088 v1
[13] M Nofrarias *et al* 2009 *Thermal experiments on board LTP*, Technical Note S2-IEC-TN-3042 v1.1
[14] S Vitale *et al* 2005 *Science Requirements and Top-Level Architecture Definition for the LISA Technology Package (LTP) on Board LISA Pathfinder*, Science Requirement Document LTPA-UTN-ScRD Iss3-Rev1

Amaldi 9 and NRDA 2011 IOP Publishing
Journal of Physics: Conference Series **363** (2012) 012045 doi:10.1088/1742-6596/363/1/012045

Scientific goals achievable with radiation monitor measurements on board gravitational wave interferometers in space

C. Grimani[1,2], C. Boatella[3], M. Chmeissani[4], M. Fabi[1], N. Finetti[2,5], A. Lobo[6], I. Mateos[6]

[1]DiSBeF, Università degli Studi di Urbino "Carlo Bo", Urbino (PU), Italy
[2]Istituto Nazionale di Fisica Nucleare, Florence, Italy
[3]Centre National d'Etudes Spatiales, Toulouse, France
[4]Institut de Física d'Altes Energies (IFAE), Barcelona, Spain
[5]Dipartimento di Fisica, Università degli Studi dell'Aquila, L'Aquila, Italy
[6]Institut d'Estudis Espacials de Catalunya (IEEC), Barcelona, Spain

E-mail: catia.grimani@uniurb.it

Abstract. Cosmic rays and energetic solar particles constitute one of the most important sources of noise for future gravitational wave detectors in space. Radiation monitors were designed for the LISA Pathfinder (LISA-PF) mission. Similar devices were proposed to be placed on board LISA and ASTROD. These detectors are needed to monitor the flux of energetic particles penetrating mission spacecraft and inertial sensors. However, in addition to this primary use, radiation monitors on board space interferometers will carry out the first multipoint observation of solar energetic particles (SEPs) at small and large heliolongitude intervals and at very different distances from Earth with minor normalization errors. We illustrate the scientific goals that can be achieved in solar physics and space weather studies with these detectors. A comparison with present and future missions devoted to solar physics is presented.

1. Introduction

When the gravitational wave hunting moved from Earth to space, the problem of the control of the noise induced by the environment on the experiments changed completely. Free-floating metal test masses constitute the heart of the inertial sensors carried by interferometers in space [1, 2, 3]. Galactic and solar particles with energies larger than 100 MeV per nucleon (MeV/n) charge the test masses [4, 5, 6] and induce spurious forces that might mimic genuine gravitational wave signals [7].

ESA found as necessary to fly radiation monitors on LISA-PF [8], the technology test mission for LISA [2], the first interferometer devoted to the detection of low frequency gravitational waves in space. LISA-PF particle detectors will allow us to monitor the integral flux in energy of protons and helium nuclei (98% in composition of both galactic and solar particles) incident on the spacecraft. These in-situ measurements will help in evaluating the test-mass charging process. It was proposed to place particle detectors on the LISA and ASTROD [9] missions as well.

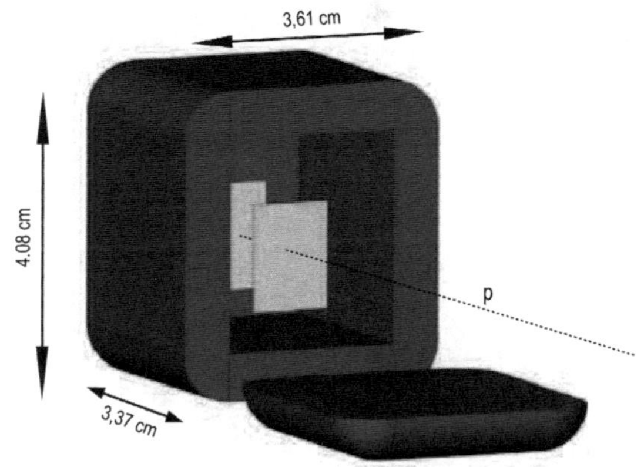

Figure 1. LISA-PF radiation monitor set-up. Silicon wafers are sketched inside the shielding copper box. An incident proton trajectory is shown. See text for details.

We show that radiation monitors on board space interferometers may provide important clues on solar physics and space weather investigations. We recall that by *space weather* we mean the conditions on the Sun and in the solar wind, magnetosphere, ionosphere, and thermosphere that can influence the performance and reliability of space-borne and ground-based technological systems and endanger human life or health.

Measurements carried out with particle detectors on LISA and ASTROD would allow for the mapping of solar particles above 100 MeV/n at different steps in heliolongitude and at various distances from Sun and Earth. Identical instruments on board all spacecraft would carry out SEP observations with no systematic errors. Solar Sentinels [10] is the only proposed mission devoted to solar physics providing similar measurements but the Sentinel constellation will cover different distances from Earth and Sun and different intervals in heliolongitude compared to space interferometers as reported in Section 6. It would be possible to obtain a more complete representation of intense SEP event evolution if solar missions and space interferometers will be in orbit at the same time. We point out that a particular interest is devoted to the study of high energy solar particles before men return to the Moon and before the Mars exploration.

2. Radiation monitors on board space interferometers

The LISA-PF radiation monitors consist of two, 300 μm thick silicon wafers of 1.4 x 1.05 cm^2 area placed in a telescopic arrangement at a distance of 2 cm [11]. The silicon detectors are located inside a shielding, copper box (see figure 1). The box thickness of 6.4 mm prevents from detection protons and helium nuclei with energies smaller than 75 MeV/n for particle normal incidence [12]. This energy threshold was set, conservatively, slightly below the minimum energy needed to the most abundant components of cosmic rays to penetrate the spacecraft (100 MeV/n) in order not to underestimate the particle flux reaching the test masses.

Ionization energy losses for particles traversing both silicon wafers (coincidence mode) and counts on each silicon layer will be measured by on board LISA-PF radiation monitors [11]. Below 100 MeV/n the radiation monitor geometrical acceptance decreases rapidly with increasing incidence angle. Above 100 MeV/n the geometrical factor is energy independent and equal to 9 cm^2 sr for particle isotropic incidence on each silicon layer. In coincidence mode the geometrical factor is about one tenth of this value. The radiation monitor data will be stored over periods of 614.4 seconds and then sent to the on board computer in the form of histograms. The maximum detectable countrate is 6500 counts/s, corresponding to a SEP event integrated fluence of the order of 10^8 protons/cm^2. In coincidence mode up to 5000 energy deposits per second can be stored in the histogram. Only extreme SEP events might lead to bin saturation.

Neither electron nor heavy ion monitoring will be carried out on LISA-PF. The possibility to modify the LISA-PF radiation monitors for LISA in order to include solar electron detection for SEP event forecasting was presented in [13]. It was also recently proposed to place radiation monitors on board the ASTROD missions. SEP detection on board ASTROD would add an extra topic of solar physics to the scientific goals of this experiment already including detection of solar g-modes and the measurement of the solar Lense-Thirring effect to 10 parts per million [9].

3. Interplanetary medium characteristics

The Sun flings 10^6 tons of fully ionized plasma from the corona out into the space every second. This material constitutes the solar wind. Solar wind particles (mainly electrons and protons) travel at supersonic speeds of 200-800 km/s. Near the Earth the solar wind plasma has a density ranging between 0.4 and 100 particles/cm^3 with an average value of 6 particles/cm^3. Typical particle energies range between 0.5 and 3 keV [14]. The expanding solar wind drags also the solar magnetic field outward, forming what is called the interplanetary magnetic field. Although the solar wind moves out almost radially from the Sun, both solar wind and magnetic field are twisted into a spiral by the Sun rotation. At the orbit of the Earth the angle between the field lines and the radial is about 45°. The value of the interplanetary magnetic field ranges between 0.2 and 80 nT with an average value of 6 nT. Sectors (typically four) with alternating inward and outward directed magnetic fields can be identified. In addition to the solar wind which blows continuously, much more energetic electrons and protons (solar energetic particles with energies larger than 1 MeV) generated by impulsive or gradual solar flares can reach the Earth. Short duration events are more properly indicated as solar flares while long duration events are thought to be associated with coronal mass ejections (CMEs). CMEs are believed to be caused by sudden disruptions in the Sun's magnetic field. Magnetic field lines are supposed to stretch until they break. Large CMEs may contain 1000 tons of matter presenting speeds at the Earth of up to 2000 km/s. It has to be stressed that only 1-2% of CMEs (fast CMEs driven shocks) produce SEPs. Finally, the interplanetary medium is filled by interplanetary and galactic cosmic rays (GCR) consisting of 90% of protons, 8% of helium nuclei, 1% of heavy nuclei and 1% of electrons. The energy range of GCR varies between 100 MeV/n and beyond 10^{20} eV. Both GCRs and solar particles show latitude, longitude and distance from the Sun dependence. The magnitude of galactic cosmic-ray gradients varies from 10% per AU at one AU to 4% per AU at five AU.

An accurate study of cosmic-ray variations and fluctuations and solar event occurrence is mandatory to evaluate the performance of missions like LISA and ASTROD. In [15] we have discussed the capability of the LISA-PF radiation detectors to monitor the solar event evolution above the minimum and maximum GCR background expected in 2014.

Cosmic rays of planetary origin and solar particles below 100 MeV/n play no role on board space interferometers being stopped by the spacecraft material.

4. SEP energy spectra

Pure impulsive or gradual events are rare. On short time scales of the order of the duration of the event at one AU, the particle fluxes rise over a period of 1/2-1 day with a slow decay of about a few days. A strong increase after a first decay phase is found in the largest events and is due to a series of CMEs and shocks. The majority of solar proton events occurs during years 5-8 of the solar cycle [16]. It has to be underscored that in most cases SEPs present energies smaller than 10 GeV even if observations of particles with energies as high as 50 GeV and more have been reported. In particular, in Karpov, Miroshnichenko and Vashenyuk [17] integral proton intensities of about 10^{-5} cm^{-2} s^{-1} sr^{-1} at energies larger than 500 GeV were found for the 29 September 1989, 15 June 1991 and 12 October 1981 ground level events. Often impulsive-flare events cluster in or behind a CME that magnetic field lines connect to an observer. In these cases, a particle velocity dispersion is observed since particles arrive in inverse order to their velocity. The energy content in impulsive events appears to lie below 1 MeV [18]. Particles from a large, magnetically well connected gradual event associated with a CME-driven shock also show velocity dispersion early in the event but present identical time profiles that last for several days late in the event. The most energetic particles (energies larger than 1 GeV) arrive with a delay of only a few minutes after the visual recognition of the flare. Lower energy particles appear in increasing numbers while the high energy particles fade away. The dynamics of SEP events magnetically well connected to the observer is also characterized by a strong anisotropy at the onset while late in the event solar energetic particles show a smaller anisotropy. The proton flux dynamics associated with the solar medium-strong gradual event dated May 7th 1978 was reported, for example, by Grieder [19]. Recently, a satellite experiment for cosmic-ray antimatter detection, PAMELA, measured the proton and helium differential fluxes associated with the evolution of the solar event dated December 13th 2006 [20].

Attempts to parameterize the rigidity (R=pc/Ze; particle momentum per unit charge) spectra of solar energetic protons were reported by Grieder ([19] and references therein):

$$J(R,t) = I_o(t)e^{-\frac{R}{G(t)}} \qquad (1)$$

where G ranges between 40 MV and 400 MV and I_o between 2 and 1000 protons cm^{-2} sr^{-1} s^{-1}. Energy spectra with an exponential trend were reported also by Vashenyuk, Balabin and Miroshinchenko [21] for the strongest event ever recorded dated February 23rd 1956. In order to show how intense the solar proton flux can be at a few hundreds of MeV, in figure 2 we have compared the proton peak flux associated with the February 23rd 1956 event with the galactic proton flux expected in 2014 at the time of LISA-PF [15].

SEP production and acceleration processes are still matter of discussion and investigation and the relative role of flaring phase and shock acceleration in large, strong SEP events remains, in general, to be understood due to the different characteristics of individual events.

5. Space interferometer radiation monitor contribution to solar physics and space weather forecasting

In addition to the main role of noise monitoring, particle detectors placed on board missions for gravitational wave detection in space may provide precious clues on solar physics for space weather applications. In order to better study and model solar energetic proton fluxes that affect manned and unmanned space missions and life on Earth, multipoint, inner heliospheric SEP observations are needed. We recall that cosmic rays and solar particles with energies larger than 100 MeV/n penetrate spacecraft and astronaut suits. A crew on the surface of the Moon exposed to the August 1972 flare during the Apollo program would have gotten really sick or worse as in the case of the giant flare dated September 1859. Unfortunately, because of the difficulty to separate galactic cosmic rays from solar particles above 100 MeV/n and since only the strongest solar events generate particles at these energies, the majority of experiments intended for solar physics do not allow for high energy particle discrimination from lower energy ones. Consequently, very few data are available in the literature on very energetic solar particles. These data are naturally provided by cosmic-ray experiments carrying magnetic spectrometers in space. The PAMELA experiment, in orbit since June 15th 2006 during the very long lasting solar minimum, did not detect many events. This mission is supposed to end before the next solar

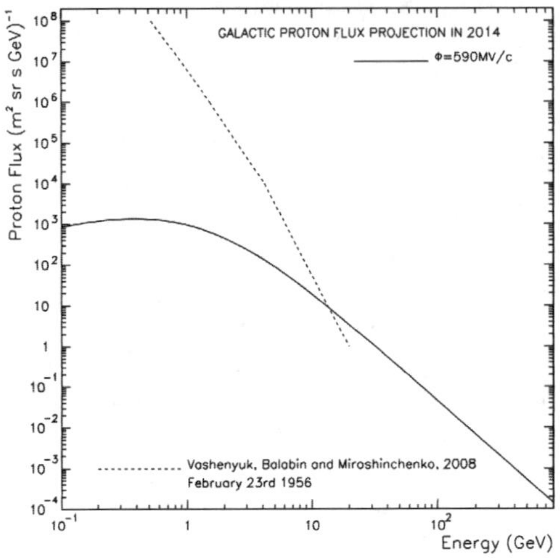

Figure 2. Comparison of the galactic proton spectrum expected in 2014 (solid line) with the proton peak flux associated with a 'worst case' SEP event such as that dated February 23rd 1956 (dashed line).

maximum. The Alpha Magnetic Spectrometer [22], launched in the summer 2011 is supposed to take data for the next ten years.

LISA and ASTROD experiments are natural multipoint observatories for cosmic and solar particles. The LISA mission [2] consists of three spacecraft located at the vertices of an equilateral triangle inclined at 60 degrees on the ecliptic. The details of this mission are still under discussion, however the arm of the interferometer might range between 1 and 5 million km. This means that the three LISA spacecraft will cover 0.4-2 degrees in heliolongitude. The orbit will be heliocentric. The LISA spacecraft constellation will slowly drift away from Earth. The distance from Earth will not exceed 50 million km in 4 years and 65 million km in 6 years. The mission lifetime is supposed to be 6 years.

In other words LISA would allow for a contemporary observation of SEP events at small (0.4-2 degrees in heliolongitude between LISA spacecraft) and increasing longitude interval ranging between 0 and 26 degrees with respect to experiments devoted to solar physics taking data near Earth during the mission whole lifetime. LISA would play the role of a sentinel for the detection of SEPs magnetically well connected to this mission and associated with CMEs emitted in the Earth direction.

The european-chinese ASTROD program consists of three missions: ASTROD I [9], ASTROD II (or ASTROD) [23] and ASTROD III (or Super-ASTROD) [24]. ASTROD I is one-spacecraft, interplanetary mission in solar orbit. The maximum distance from Earth that will be reached by the spacecraft is 1.7 million km. The most optimistic launch date is 2017. ASTROD II will consist of three spacecraft allowing for probing primordial gravitational waves below the LISA bandwidth. For the third phase, the mission ASTROD III is supposed to be characterized by larger orbits to map the outer solar system and to probe the primordial gravitational waves at frequencies below the ASTROD II bandwidth. The ASTROD spacecraft would orbit at different distances and heliolongitude intervals with respect to LISA.

LISA-PF consists of one spacecraft stationed in a Lissajous orbit around Lagrangian point

L1 at 1.5 million km from Earth. The satellite is supposed to spin with a period of six months. The same period of time is needed to the spacecraft to move on its orbit. Data will be taken for six months during the second half of 2014.

Space interferometers are supposed to be in orbit for years detecting solar events at very different intervals in heliolongitude and distances from Earth for at least two solar cycles.

Tens of medium-strong SEP events are estimated to be detected by LISA on the basis of the predictions we have carried out for LISA-PF [15].

The role of solar physics with and for LISA was described also in [25].

6. Future missions devoted to solar physics allowing for SEP measurements

STEREO [26] is the first multispacecraft observatory aiming to study the Sun and the nature of its CMEs. Launched on October 25th 2006 from Cape Canaveral (Florida), STEREO consists of two nearly identical observatories in heliocentric orbit (one moving ahead of the Earth and one moving behind). Electron and proton observations will be carried out on board STEREO but only below 4 MeV and 100 MeV, respectively.

On October 4th 2011 ESA selected Solar Orbiter [27] as a class M mission for implementation and launch in 2017 within the Cosmic Vision program (2015-2025). The cruise phase to the Sun should last approximately three years. The mission lifetime is supposed to be seven years. Solar Orbiter will have an elliptical heliocentric orbit with perihelion as low as 0.28 AU and with increasing inclination up to more than 25° with respect to the solar equator. This mission aims to answer fundamental questions such as how the Sun creates and controls the inner heliosphere. The Sun's magnetised atmosphere, the properties and dynamics of solar wind and solar energetic particle fluxes will be monitored by the onboard instrumentation. In particular, the Energetic Particle Detector (EPD) is devoted to the observation of particles in the energy range 2 keV/n-200 MeV/n.

Solar Sentinels [10] is a mission devoted to solar physics that will carry out multipoint, inner heliospheric, in-situ observations meant to discover, understand and model initiation, propagation and solar connections of those energetic phenomena dangerous to humans for space explorations and life on Earth. Solar Sentinels were designed to investigate high-energy solar particle generation and propagation processes but also to study CME origin, evolution and interactions, to characterize the interplanetary environment and finally to allow for the development of forecasting capabilities for Earth, Mars and for spacecraft transit. Solar Sentinels consist of a constellation of four identical spacecraft stationed between Venus and Mercury allowing for solar wind ion and electron composition measurements and solar energetic electron, proton and ion observations up to 10 MeV, 500 MeV/n and 100 MeV/n, respectively, associated with strong SEP events. A near-Earth sentinel, a single probe orbiting near Earth will carry a coronagraph for observing the Sun's faint corona where CMEs get started. The farside sentinel is another single probe meant to watch the farside of the Sun. Together with the other spacecraft, this sentinel would provide a complete picture of the Sun including the area not visible from Earth. Solar Sentinels might be launched within the solar cycle 24. Solar Orbiter and Solar Sentinels constitute the Heliophysical Explorers (HELEX): a joint science and technological effort of NASA and ESA (NASA/TM-2008-214159).

Finally, Solar Probe Plus [28] is the NASA mission supposed to reach the solar corona appoaching the Sun as close as 9.5 solar radii (8.5 solar radii above the Sun's surface). Solar Probe Plus will sample the near-Sun environment, improving our knowledge and understanding of coronal heating and the origin and evolution of the solar wind. Solar energetic particles will be studied in-situ where they are energized. Measurements of composition and energy spectra of ions extending from energies through about 100 MeV/n will be carried out. Solar Probe Plus will provide complementary observations with respect to the missions described above and with respect to radiation monitors on board space interferometers to complete the scenario of the most hazardous solar events and solar particle acceleration and propagation in the interplanetary medium. The Solar Probe Plus launch date is expected in July 2018.

7. Conclusions

Radiation monitors placed on interferometers for gravitational wave detection in space are primarily used to estimate test-mass charging and experiment overall noise. However, they may also provide precious clues on solar physics and space weather investigations before manned missions will return to the Moon and before the beginning of the Mars exploration. We have shown that particle detectors on board the LISA and ASTROD missions would allow for contemporary observations of solar particles above 100 MeV/n at small and large steps in heliolongitude (between a fraction of a degree and tens of degrees) and at various distances from the Sun with identical detectors avoiding systematic errors. SEP observations gathered by radiation monitors on space interferometers and by future solar experiments like Solar Orbiter, Solar Sentinels and Solar Probe Plus are expected to clarify the role of flares and CMEs in accelerating solar particles up to very high energies.

Acknowledgements

This work was financially supported by the Italian Space Agency within the program "Study Exploration of the Solar System" .

References

[1] Antonucci F *et al* 2011 *Class. Quantum Grav.* **28** 094001
[2] Amaro-Seoane P *et al* 2012 *Preprint* gr-qc/1202.0839 to appear on *Class. Quantum Grav.*
[3] Ni W.-T. 2008 *Int. J. Mod. Phys.* **D17** 921-940
[4] Araújo H M *et al* 2005 *Astropart. Phys.* **22** 451-469
[5] Grimani C *et al* 2004 *Class. Quantum Grav.* **21** S629-S633
[6] Grimani C *et al* 2005 *Class. Quantum Grav.* **22** S327-S332
[7] Shaul D Sumner T and Lobo A 2008 *Imperial College Technical Note, S2-ICL-TN-3019*
[8] Cañizares P *et al* 2009 *Class. Quantum Grav.* **26** 094005
[9] Braxmaier C *et al Preprint* gr-qc/1104.0060 to appear on *Exper. Astron.*
[10] `http://science.nasa.gov/science-news/science-at-nasa/2006/01sep_sentinels/`
[11] Mateos I *et al* 2010 *J. Phys. Conf. Ser.* **228** 012039
[12] Wass P 2007 *PHD Thesis* Imperial College of London University of London
[13] Grimani C *et al* 2009 *Class. Quantum Grav.* **26** 215004
[14] Reames D V 2003 *Adv. Sp. Res.* **34** 381-390
[15] Grimani C *et al* 2012 to appear on *Class. Quantum Grav.*
[16] Shea M A and Smart D F 2001 *Proc. 27th Int. Cosmic Ray Conf. (Hamburg)* **8** 3401-3404
[17] Karpov S N Miroshnichenko L I and Vashenyuk E V 1997 *Proc. 25th Int. Cosmic Ray Conf. (Durban)* **1** 205-208
[18] Reames D V Kahler S W and Ng C K 1997 *Ap. J.* **491** 414-420
[19] Grieder P 2001 *Cosmic Rays at Earth: Researcher's reference manual and data book, Elsevier*
[20] Casolino M *et al Preprint* astro-ph/0904.4692
[21] Vashenyuk E V Balabin Y V and Miroshnichenko L I 2008 *Adv. Sp. Res.* **41** 926-935
[22] Alpat B 2003 *Nucl. Sc. and Tech.* **14** 182-194
[23] Burston R *et al* 2008 *J. Phys. Conf. Ser.* **118** 012043
[24] Ni W.-T. 2009 *Class. Quantum Grav.* **26** 075021
[25] Grimani C and Vocca H 2005 *Class. Quant. Grav.* **22** S333-S338
[26] `http://stereo.gsfc.nasa.gov`
[27] `http://sci.esa.int/science-e/www/area/index.cfm?fareaid=45`
[28] `http://solarprobe.gsfc.nasa.gov`

Gravitational ringdown signal from coalescences of SMBH binaries - detection rates for LISA and ET

Ch. Filloux[1], J.A. de Freitas Pacheco[1], F. Durier[2] and J.C.N. de Araujo[3]

[1]Observatoire de la Côte d'Azur, Université de Nice-Sophia Antipolis - France
[2]Max Planck Institute for Extraterrestrial Physics - Germany
[3]Instituto Nacional de Pesquisas Espaciais, Brazil

E-mail: jcarlos.dearaujo@inpe.br

Abstract. The coalescence history of massive black holes is derived from cosmological simulations, in which their evolution and that of the host galaxies are followed in a consistent way. With the coalescence rate per comoving volume and per mass interval derived from the simulations we estimate the expected detection rate distribution of "ring-down" gravitational wave signals along frequencies accessible by LISA and Einstein Telescope (ET). For LISA, a total detection rate of about 15 yr^{-1} is predicted for events having a signal-to-noise ratio equal to 10. For ET, one event each 14 months down to one event each 4 years is expected with a signal-to-noise ratio of 5. The detection of these gravitational signals and their distribution in frequency would be in the future an important tool able to discriminate among different scenarios for the origin of supermassive black holes.

1. Introduction

It seems to be a well established fact that the center of massive galaxies host SMBHs (see, e.g., Ref. [1]). Presently, the general belief is that these objects have been formed by the growth of primordial "seeds" via matter accretion and coalescences resulting from merger episodes involving their hosts. However, the detection of bright SLOAN quasars associated to SMBHs of $10^9\ M_\odot$ at $z \sim 6.5$ [2], led some authors to propose an alternative evolutionary scenario, in which these objects would have been formed directly via the gravitational collapse. Processes leading to the formation of SMBHs release, in particular when coalescences occur, an important amount of gravitational radiation.

The evolution of primordial massive stars [3] may lead to the formation of BH "seeds" having masses in the range $100 - 500\ M_\odot$. The newly formed BHs are likely to be "distorted" with respect to the quiescent err geometry, emitting gravitational waves (GWs), the so-called "ring-down" radiation. More massive BHs in a process of merging emit GWs during the inspiral and the plunge phases and lately, when the two horizons merge. In this case, the GW signal has frequencies that are accessible by LISA.

Coalescences of SMBHs are certainly one of the primary sources of GWs for LISA and many investigations have been devoted to this subject aiming, in particular, to describe in detail the inspiral phase as well as the plunge in order to obtain adequate templates for the waveform [4] and to estimate the expected rate of events. These estimates require not only a previous knowledge of the merging rate at different redshifts, but also of the evolution of the BH mass distribution. Early studies considered that the coalescence rate of BH binaries is related to the overall merger rate of galaxies, displaying a broad maximum around $z = 3$. Crude estimates based on this picture indicate a total coalescence rate ranging from one up to a hundred events per year [5, 6].

In the present study, cosmological simulations were performed in which the growth of $100M_\odot$ BHs seeds as well as the evolution of the host galaxy were followed in a self-consistent way. Seeds grow mainly by accretion of baryonic matter and by coalescences occurring after merger events, forming as a result SMBHs of up $\sim 10^9\ M_\odot$. We then focus essentially on the "ring-down" emission and estimate the expected detection rate for LISA and Einstein.

2. The simulations

General aspects of the code and results concerning the main correlations between the BH mass and the properties of the host galaxy were already reported in Ref. [7] (see, also Ref. [8]). Here we only give the main characteristics of the code.

Simulations were performed by using the parallel TreePM-SPM code GADGET-2 in a formulation, despite the use of fully adaptive smoothed particle hydrodynamics (SPM), that conserves energy and entropy [9]. Different physical mechanisms affecting the gas dynamics were introduced, such as cooling (free-free transitions, radiative recombinations, H_2 molecular transitions, atomic fine-structure level excitation, Compton interactions with CMB photons), local heating by the UV-radiation of newly formed stars, mechanical energy injected by type II and type Ia supernovae as well as by AGNs and diffusion of heavy elements.

All simulations were performed in a cube with a size of $50h^{-1}\ Mpc$ with two different mass resolutions for the gas/stellar particles respectively equal to $5.35 \times 10^8\ M_\odot$ and $3.09 \times 10^8\ M_\odot$. In all runs a flat ΛCDM cosmology was adopted, characterized by a Hubble parameter $h = 0.7$ in units of $H_0 = 100\ km\ s^{-1}\ Mpc^{-1}$, by a "vacuum" energy density parameter $_\Lambda = 0.7$ and by a total matter energy density parameter $_m = 0.3$. The fraction of baryonic matter was taken as $h^2\ _b = 0.0224$ and the normalization of the matter density fluctuation spectrum was taken to be $\sigma_8 = 0.9$. Initial conditions were fixed according to the algorithm COSMICS and all simulations were performed in the redshift interval $60 \geq z \geq 0$.

3. Coalescence rate

This paper is essentially focused on the coalescence process involving black holes and, in particular, on the rate at which objects with a given mass M_{bh} are being formed at a given redshift. As is well known, such a process is intimately related to mergers suffered by the host galaxies. However, other mechanisms may affect the resulting BH mass distribution and, consequently, the expected frequency of the "ring-down" signal.

In Fig. 1, the coalescence rate per unit of volume and per mass interval $\Psi(M,z)$ (in $Mpc^{-3}yr^{-1}M_\odot^{-1}$) is shown as a function of the resulting BH mass and for different redshifts. This function is defined such that the local rate of coalescences R in a given redshift interval is given by

$$\frac{dR}{dz} = \int_{M_1}^{M_{max}} \Psi(M,z)\frac{dV}{dz}dM \tag{1}$$

where $M_1 = 200\ M_\odot$ is the minimum BH mass resulting from a fusion of two primordial seeds and M_{max} is the maximum BH mass resulting from the coalescences process, that in our simulations is of the order of $5 \times 10^9\ M_\odot$. Fig. 1 clearly shows the appearance of very massive BHs for decreasing redshifts, consequence of the accretion process and coalescences. This is the main result of the present investigation since it permits to predict the expected event rate for ground based and space interferometers.

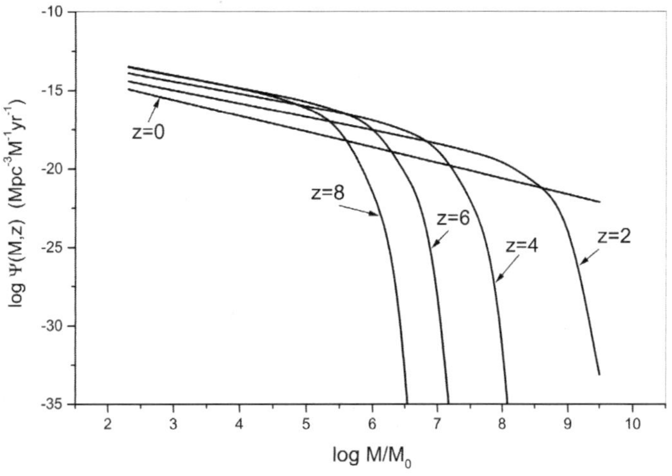

Figure 1. Coalescence rate per unit of volume and per mass interval $\Psi(M, z)$ as a function of the resulting BH mass. Curves are labeled for different redshifts. For a better visibility in the plot, curves were smoothed. (From Ref. [8].)

4. The expected detection rates for LISA and Einstein

The coalescence rate per unit of volume and per mass interval $\Psi(M, z)$ derived from our simulations permits to compute the expected event rates for LISA and Einstein Telescope. The coalescence rate in a given interval of mass and redshift seen by an observer at $z = 0$ is

$$dR = \frac{\Psi(M,z)}{(1+z)} \frac{dV}{dz} dM dz\,, \tag{2}$$

where dV is the comoving volume element.

Note that the factor $(1 + z)$ in the denominator takes into account the time dilation. The comoving volume element dV for a flat cosmology is given by

$$dV = 4\pi \left(\frac{c}{H_0}\right) \frac{r^2(z)dz}{\sqrt{\ {}_\Lambda + \ {}_m(1+z)^3}} \tag{3}$$

and the comoving distance by

$$r(z) = \frac{c}{H_0} \int_0^z \frac{dx}{\sqrt{\ {}_\Lambda + \ {}_m(1+x)^3}}. \tag{4}$$

With a change of variable in Eq. (2), i.e., expressing the BH mass in terms of the observed "ring-down" frequency, we obtain the expected coalescence rate per logarithm interval of frequency, namely

$$\frac{dR}{d\ln(\nu)} = \int_{z_{min}(\nu)}^{z_{max}(\nu)} \frac{\Psi(\nu,z)}{(1+z)} \left| \frac{dM}{d\ln(\nu)} \right| \frac{dV}{dz} dz\,, \tag{5}$$

where the relation between the black hole mass and the observed characteristic "ring-down" frequency is given by [10]

$$\nu = \frac{\nu_m}{(1+z)} = \frac{1.2 \times 10^4}{(1+z)} \left(\frac{M_\odot}{M}\right) F(a)\ \ Hz\,, \tag{6}$$

where $a = Jc/GM^2$ is the spin parameter and the function $F(a)$ is given approximately by[10]

$$F(a) = \left[\frac{100}{37} - \frac{63}{37}(1-a)^{0.30}\right]. \tag{7}$$

The lower limit $z_{min}(\nu)$ appearing in the integral defined by Eq. (5) is fixed by the maximum BH mass $M_{bh}(max)$ present in our simulations. For a given frequency, using Eq. (6), the minimum redshift is given by the condition

$$z_{min}(\nu) = Max\left[0, \frac{1.2\times 10^4 F(a)}{\nu M_{bh}(max)}\right]. \tag{8}$$

On the other hand, the upper limit $z_{max}(\nu)$ represents the redshift below which a gravitational signal can be detected at a given signal-to-noise ratio.

Recall that the optimal signal-to-noise ratio derived from a matched filtering technique is

$$\left(\frac{S}{N}\right)^2 = 4\int_0^\infty \frac{|\tilde{h}(\nu)|^2}{S_n(\nu)} d\nu\,, \tag{9}$$

where $|\tilde{h}(\nu)|^2$ is the spectral density of the signal averaged over both polarizations states and $S_n(\nu)$ is the effective one-sided spectral density of the noise in the detector [11, 12].

We model the "ring-down" waveform by a simple damped sinusoidal, namely

$$h(t) = h_0 e^{-t/\tau_m}\cos(\omega_m t)\,, \tag{10}$$

where the amplitude h_0 of the signal is related to the total energy E carried out from the source under the form of gravitational waves by

$$h_0^2 = \frac{16GE}{c^3 d_L^2}\frac{\tau_m}{1+4Q_m^2}\,, \tag{11}$$

where G is the gravitational constant, d_L is the luminosity distance and $Q_m = \pi\nu_m\tau_m = \omega_m\tau_m/2$ is the quality factor of the oscillation, given approximately by [10]

$$Q_m = K(a) \simeq 2(1-a)^{9/20}. \tag{12}$$

It is worth noting that more recent and more accurate fits to the quasinormal modes of a err black hole can be found in Ref. [13]. However, our simplifying modeling of the quasinormal mode signal is not invalidated by this more recent study.

The Fourier transform of the signal gives

$$\tilde{h}(\omega) = \frac{h_0\tau_m}{[1+(\omega-\omega_m)^2\tau_m^2]} = h_0\tau_m f(\omega_m, \tau_m). \tag{13}$$

Note that $\tilde{h}(\nu)$ has a Lorentz profile, indicating that the spectral density is peaked around the characteristic frequency ω_m.

Using the above equations and assuming that the total energy released under the form of GWs is given by $E = \varepsilon M_{bh}c^2$, one obtains for the S/N ratio

$$\left(\frac{S}{N}\right) = \frac{2.83\times 10^{-17}}{d_L}\left(\frac{\varepsilon M}{M_\odot}\right)^{1/2}\frac{G(a)}{\sqrt{\nu^2 S_n(\nu)}}\,, \tag{14}$$

where

$$G(a) = \frac{Q_m(a)}{\sqrt{1 + 4Q_m^2(a)}}. \tag{15}$$

The above estimate for SNR considers in fact a delta-function approximation, which was originally introduced in Ref. [14] (see also Ref. [13]).

In Eq. (14) the luminosity distance is given in Mpc and we have assumed that in the evaluation of the integral in Eq. (9), the noise spectral density does not vary considerably near the frequency defining the maximum of the spectral density of the signal. The "ring-down" efficiency ε was estimated in Ref. [15] to be about 7.8×10^{-4} from fully relativistic calculations of head-on collisions between a black hole and a neutron star. In our numerical estimates, a more conservative value equal to $\varepsilon = 10^{-4}$ was also considered.

In the next two subsections we apply the above prescriptions to evaluate $z_{max}(\nu)$, the expected coalescence rates per logarithm interval of frequency $dR/d\ln(\nu)$ and the total expected rates of events to LISA and Einstein.

4.1. LISA

In order to evaluate $z_{max}(\nu)$ from Eq. (14), we have computed the spectral noise density using an online simulator [16], where the confusing noise introduced by white dwarf binaries in our Galaxy is also taking into account. In Fig. 2 it is shown the maximum redshift as a function of the observed frequency of the signal for S/N=10 and for different values of the spin parameter a and the "ring-down" efficiency. Note that for the most favorable situation, the "ring-down" signal can be seen up to $z \sim 22$ around $\nu \sim 4$ mHz. In practice, the situation is a little bit different. The thin line in Fig. 2 indicates, for a fixed frequency, the effective redshift above which coalescences do not contribute significantly to the signal. The reason is essentially due to the evolution of the BH mass spectrum. Once the frequency is fixed, the mass contributing to the signal is related to the redshift via Eq. (6) and the absence of objects with the required mass produces the aforementioned effect. This critical redshift increases with frequencies probing lower and lower mass BHs which are more numerous than supermassive objects.

With $z_{max}(\nu)$ at hand, the integral in Eq. (5) can be computed numerically using the results of our simulations. Fig. 3 shows the expected distribution of coalescence rates in the frequency domain of LISA for S/N=10. This plot shows that most of the coalescences will produce "ring-down" signals mainly the frequency range 4−9 mHz, probing BHs in the mass range $2.2\times10^5\ M_\odot$ up to $2.6 \times 10^6\ M_\odot$. The event rate distributions shown in Fig. 3 reflect the formation history of these SMBHs resulting from the present simulations. Had most of the SMBHs been formed early in the universe, say around $z \sim 6$, most of the coalescences would occur at frequencies of only few μHz and not in the mHz domain as found in the present investigation. This is an important aspect indicating that LISA in its original configuration would be able to contribute significantly to astrophysics, discriminating the different evolutionary scenarios leading to the formation of SMBHs.

Integration of the coalescence rate gives the total expected rate of events. Table 1 shows the results for different values of the signal-to-noise ratio, the spin parameter and the "ring-down" efficiency. Inspection of Table 1 indicates that the predicted rates are in a very narrow range of values despite the different values of the spin parameter, GW emission efficiency and S/N ratios. In fact, increasing the efficiency or decreasing the S/N rate permits to probe effectively a higher volume of the universe. However, since the integral "saturates" above a certain redshift, the predicted total event rate does not change considerably, excepting that the rate at the high frequency side of the maximum increases slightly.

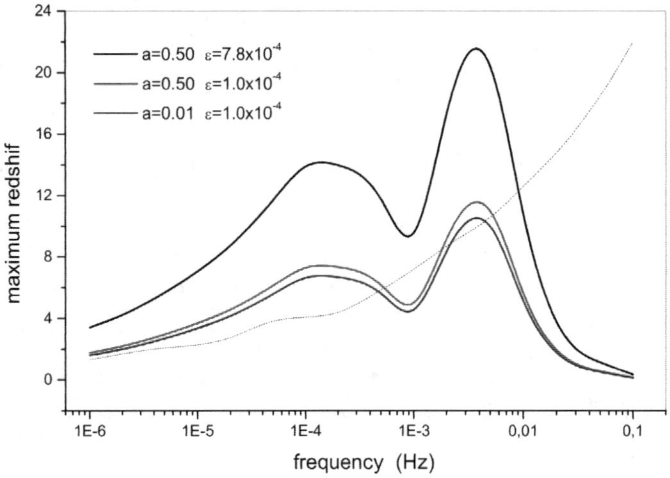

Figure 2. Maximum redshift seen by LISA at a given frequency for S/N=10. Different curves correspond to the different values of the spin parameter and the gravitational wave emission efficiency. The thin curve indicates the critical redshift value above which the contribution of coalescences to the signal becomes negligible. (From Ref. [8].)

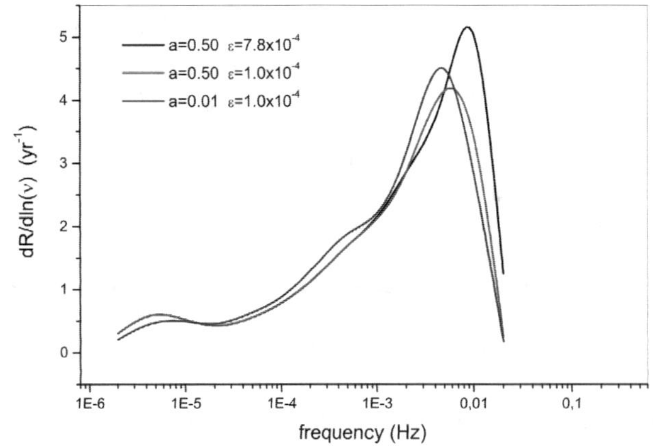

Figure 3. Expected detection rate per logarithm interval of frequency for LISA. The adopted signal-to-noise ratio is S/N=10. (From Ref. [8].)

4.2. Einstein

The third generation of ground based interferometers like ET [17] is expected to be much more sensitive in the frequency range 1-60 Hz than present antennas. In this range, these interferometers would be able to detect the "ring-down" signal originated from coalescences giving origin to BHs with masses below 500 $M_\odot$. These masses represent the very early stage of growth of "seeds". The detection of a gravitational signal emanating from coalescences involving these low mass objects could shed some light on the initial mass of "seeds" as well as on the environment conditions in which they evolve.

In order to estimate the event rate, the same procedure described above was adopted. Here, the sensitivity curve for the planned ET, version B, was taken from Ref. [18]. In Fig. 4, the

Table 1 - Coalescence Rates: columns give respectively: (1) the S/N ratio, (2) the spin parameter **a**, (3) the gravitational wave radiation efficiency ε and (4) the coalescence rate R.

S/N	a	ε	R (yr^{-1})
5	0.50	1.0×10^{-4}	15.9
5	0.50	7.8×10^{-4}	18.0
5	0.01	1.0×10^{-4}	16.7
10	0.50	1.0×10^{-4}	14.4
10	0.01	1.0×10^{-4}	14.8
10	0.50	7.8×10^{-4}	16.3

value of $z_{max}(\nu)$ is shown as a function of the frequency for different values of the spin and the efficiency parameters. Within an optimistic perspective, ET could be able to detect coalescences up to $z \sim 4$ at frequencies of few Hz. At these redshifts, low mass BHs will be probed and, according to our simulations, they are not located in the center of galaxies but are wandering in their halos. They constitute a population of seeds having suffered a small number of coalescences, since they live in an environment in which the probability to merger or to accrete mass is quite small.

In Fig. 5 the expected event rate per logarithm interval of frequency for ET is plotted as a function of the frequency. Most of the events are expected to occur in the range 10-20 Hz and total rates for $S/N = 5$ are about one event each 14 months for $\varepsilon = 7.8 \times 10^{-4}$ and one event each 4 years for $\varepsilon = 1.0 \times 10^{-4}$. A previous investigation [20], using the procedure developed in Ref. [19] to obtain the merger history of seeds, derived an event rate of about one per year for S/N=5. This is comparable to the present results if BHs have an average spin parameter $a = 0.50$ and the efficiency for emitting GWs is that derived from head-on collisions. The coalescence rate derived in Ref. [20] corresponds to a scenario in which seeds have the same mass, as assumed in our own investigation, but with a slight higher value, i.e., 150 $M_\odot$ instead of 100 $M_\odot$. If seeds have a mass distribution (log-normal in the range $10 - 600$ $M_\odot$) a higher event rate is obtained for the same S/N ratio, i.e., about $2 - 3$ events per year [20].

Recall that Ref.[21] also considers a detailed and interesting discussion of ET event rates related to different models of formation and coalescence of intermediate massive BHs.

5. Conclusions

The coalescence history of massive black holes is derived from cosmological simulations, in which their evolution and that of the host galaxies are followed in a consistent way. With the coalescence rate per comoving volume and per mass interval we estimate the expected detection rate distribution of "ring-down" gravitational wave signals along frequencies accessible by LISA and Einstein Telescope. For LISA in its original configuration, our study predicts a total event rate of about 15 coalescences per year with S/N=10 and with most of the GW bursts occurring in the frequency range 4-9 mHz. For ET, the expected rates are in the range of one event per year down to one event every 4 years, occurring in the frequency range 10-20 Hz. These characteristics reflect the growth history of SMBHs resulting from our simulations and can be used, in the future, to discriminate among different scenarios proposed to explain the origin of these objects.

Acknowledgments

JCNA thanks CNPq and Fapesp for financial support.

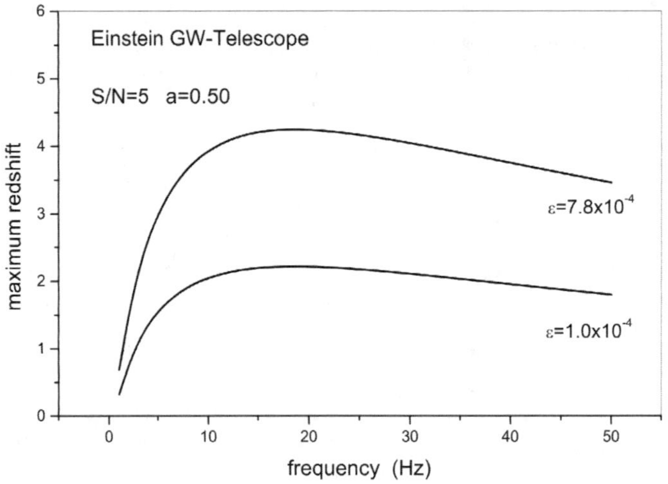

Figure 4. Maximum redshift probed by the gravitational wave telescope Einstein for S/N=5. (From Ref. [8].)

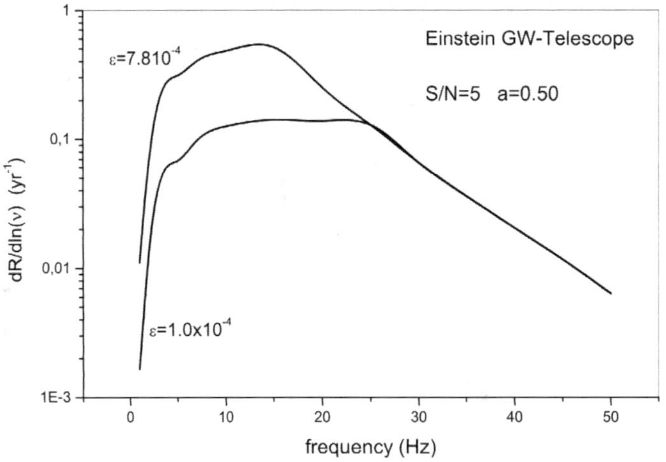

Figure 5. Expected detection rate per logarithm interval of frequency for Einstein-B and for S/N=5. (From Ref. [8].)

References

[1] J. Kormendy and D. Richstone, *ARA&A* **33** (1995) 581.
[2] X. Fan, V.K. Narayanan, R.H. Lupton et al., *AJ* **122** (2001) 2833.
[3] V. Bromm, P.S. Coppi and R.B. Larson, *ApJ* **527** (1999) L5.
[4] M. Preto, I. Berentzen, P. Berczik, D. Merrit and R. Spurzen, *JPhCS* **154** (2009) 2049.
[5] M.G. Haehnelt, AIP Conference Proceedings **456** (1998) 45.
[6] D. Merritt and R.D. Ekers, *Science* **297** (2002) 1310.
[7] Ch. Filloux, F. Durier, J.A. de Freitas Pacheco and J. Silk, *IJMPD* **19** (2010) 1233.
[8] Ch. Filloux, J.A. de Freitas Pacheco, F. Durier and J.C.N. de Araujo, *IJMPD* **20** (2011) 2399; ArXiv:1108.2638.
[9] V. Springel, *MNRAS* **364** (2005) 1105.
[10] F. Echeverria, *Phys.Rev.* **D40** (1989) 3194.

[11] N. J. Cornish, *Phys.Rev.* **D65** (2001) 022004.
[12] J. D. E. Creighton, *Phys.Rev.* **D60** (1999) 022001.
[13] E. Berti, V. Cardoso and C.M. Will, *Phys.Rev.* **D73** (2006) 064030; ArXiv: gr-qc/0512160.
[14] E.E. Flanagan and S.A. Hughes, **57**, 4535 (1998); gr-qc/9701039
[15] F. Loffler, L. Rezzolla and M. Ansorg, *Phys.Rev.* **D74** (2006) 104018.
[16] S. L. Larson, W. A. Hiscock and R. W. Hellings, *Phys.Rev.* **D62** (2000) 062001; *Online Sensitivity Curve Generator, "http://www.srl.caltech.edu/ shane/sensitivity:".*
[17] S. Hild, S. Chelkowski and A. Freise, [ArXiv:0810.0604v2].
[18] Gravitational Wave Einstein Telescope - "http://www.et.gw.eu".
[19] M. Volonteri, F. Haardt and P. Madau, *ApJ* **582** (2003) 599.
[20] A. Sesana, J. Gair, I. Mandel and A. Vechio, *ApJ* **698** (2009) L129.
[21] J.R. Gair, I. Mandel, M. Coleman Miller and M. Volonteri, *Gen. Rel. Grav.* **43** (2011) 485; arXiv:0907.5450

From black holes to their progenitors: A full population study in measuring black hole binary parameters from ringdown signals

Ioannis Kamaretsos

School of Physics and Astronomy, Cardiff University, Queens Buildings, CF24 3AA, Cardiff, United Kingdom

E-mail: ioannis.kamaretsos@astro.cf.ac.uk

Abstract. A perturbed black hole emits gravitational radiation, usually termed the ringdown signal, whose frequency and time-constant depends on the mass and spin of the black hole. I investigate the case of a binary black hole merger resulting from two initially non-spinning black holes of various mass ratios, in quasi-circular orbits. The observed ringdown signal will be determined, among other things, by the black hole's spin-axis orientation with respect to Earth, its sky position and polarization angle - parameters which can take any values in a particular observation. I have carried out a statistical analysis of the effect of these variables, focusing on detection and measurement of the multimode ringdown signals using the reformulated European LISA mission, Next Gravitational-Wave Observatory, NGO, the third generation ground-based observatory, Einstein Telescope and the advanced era detector, aLIGO. To the extent possible I have discussed the effect of these results on plausible event rates, as well as astrophysical implications concerning the formation and growth of supermassive and intermediate mass black holes.

1. Introduction

Astrophysical observations to date have provided sturdy evidence that black holes (BHs) may exist and play an important role in many physical processes [1, 2, 3]. With direct evidence still lacking, it is expected that observation of gravitational waves (GWs) from merged BHs will not only provide indisputable evidence for the existence of BHs, but also the ability to extract accurate information about the progenitor system and the BH.

The ringdown radiation consists of a superposition of, in principle, an infinite number of essentially damped sinusoids, termed quasi-normal modes. Their frequencies and time-constants depend only on the mass and spin of the BH – a consequence of the *no-hair* theorem [4]. In a recent work [5] we have argued that the amplitude terms of the various quasi-normal modes encode important information about the origin of the perturbation that caused them, such as the component masses of the progenitor binary. This allows performing parameter estimation on the system from the strong-field regime, as opposed to using the inspiral phase. However, in that study, as well as in previous studies of parameter estimation from ringdown signals [4, 6], only a small region of the parameter space was explored. In a realistic scenario, a BH ringdown signal can have any sky location and polarization. All the while, it could have originated from a BH with any spin-axis orientation with respect to Earth. These variables have a significant effect on the observed ringdown signal and a direct impact on the science we can achieve by observing GWs from merged BHs.

In the present study, I have investigated in detail how these angular parameters affect the detection and measurement of the ringdown signals. To this end, I have varied the angular parameters over their full range, thereby considering a large population of BH mergers. I have considered supermassive BHs (SMBHs of mass $\geq 10^6 M_\odot$) visible in NGO and intermediate mass BHs (IMBHs of mass $\sim 10^3 M_\odot$) observable in ET and aLIGO. I have computed the probability distribution functions of signal to noise ratios (SNR), as well as measurement errors of a chosen set of parameters, for a wide range of the BH mass and for mass ratios between 2 and 20. Finally, I translate these probabilities to proportions of observed events in NGO and ET that will yield parameter errors below certain thresholds and discuss how observations of ringdown signals could help in dealing with open questions on the existence and history of SMBHs and IMBHs.

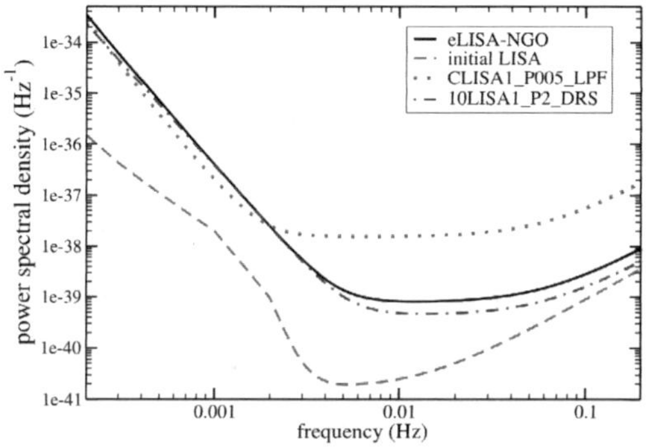

Figure 1: Noise power spectral densities for various proposed configurations of LISA-like space detectors. The dashed red line corresponds to the original LISA mission, while the rest of them refer to European designs of LISA. In all cases, the galactic binary white dwarf confusion noise [7, 8] is included, which has a negligible visible effect on the newer LISA curves though, due to their, almost two orders of magnitude worse sensitivity. Additionally, a low frequency cut-off - not shown - was induced at $5 \times 10^{-5} Hz$. In this study, I am using the latest arrangement for the European mission of LISA, labeled eLISA-NGO in this graph.

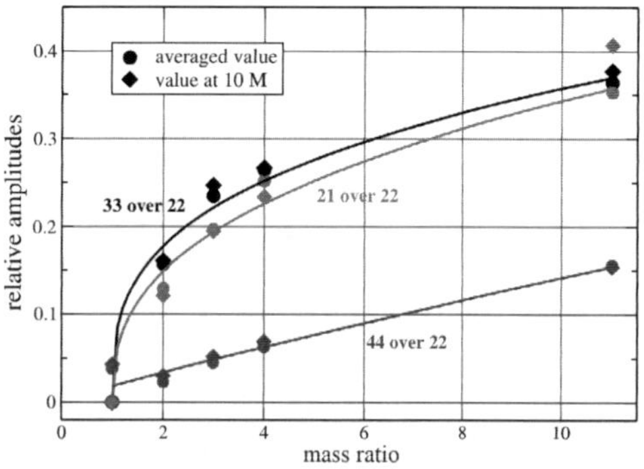

Figure 2: This plot shows the relative amplitudes of modes (3, 3), (2, 1) and (4, 4) over (2, 2) as a function of the mass ratio. The diamond points show the values that correspond to a time $10M$ after the peak luminosity of the 22 mode in the equal mass case. (see also Fig.3 in [5]) Fits to these points were used in this previous work. On the other hand, the circle points, which are used in the present study, were computed by taking into account all the points in the waveform, in a time region starting at the peak luminosity of 22 and ending $30M$ later. The solid lines shown, constitute fits to these circle points, given by expressions (6)-(8).

2. Full population Analysis

I have closely followed the procedure of Ref. [5] for estimating the signal-to-noise ratios and measurement errors. In this Section I will discuss the signal model and the parameter space covered in this study.

In the generic case where we have a network of detectors, we write the response to a ringdown signal as:

$$h^a(t) = \sum_{\ell,m,n\geq 0} B^a_{\ell mn} e^{-t/\tau_{\ell mn}} \cos\left(\omega_{\ell mn} t + \gamma^a_{\ell mn}\right), \tag{1}$$

where the superscript a is an index denoting the detector in question and $\omega_{\ell mn}$, $\tau_{\ell mn}$ are the frequencies and time-constants of each mode, which are functions of the mass and spin magnitude of the BH. For further reference, see [9, 10]. In this study I neglect modes with overtone index $n \geq 1$, thereby considering only the least damped modes. From now on, the n index (n=0) is omitted. The terms $B_{\ell m}$ and $\gamma_{\ell m}$ are the following combinations of the antenna pattern functions F^a_+, $F^a_\times$, amplitude factors $\alpha_{\ell m}$ and the angular functions $Y^{\ell m}_+(\iota)$, $Y^{\ell m}_\times(\iota)$:

$$B^a_{\ell m} = \frac{M\alpha_{\ell m}}{D_{\rm L}}\sqrt{(F^a_+ Y^{\ell m}_+)^2 + (F^a_\times Y^{\ell m}_\times)^2}, \tag{2}$$

$$\gamma^a_{\ell m} = \phi_{\ell m} + m\,\phi + \tan^{-1}\left[\frac{F^a_\times Y^{\ell m}_\times}{F^a_+ Y^{\ell m}_+}\right]. \tag{3}$$

Here, $\phi_{\ell m}$ are arbitrary constant phases of each mode. The effective amplitudes $B_{\ell m}$ vary inversely with the luminosity distance and proportionally to the intrinsic amplitudes $\alpha_{\ell m}$ of the modes, which are determined by the numerical simulations. The angular functions $Y^{\ell m}_{+,\times}(\iota)$ are the following combinations of the spin-weighted spherical harmonics [6]:

$$\begin{aligned} Y^{\ell m}_+(\iota) &\equiv {}_{-2}Y^{\ell m}(\iota,0) + (-1)^\ell\, {}_{-2}Y^{\ell -m}(\iota,0), \\ Y^{\ell m}_\times(\iota) &\equiv {}_{-2}Y^{\ell m}(\iota,0) - (-1)^\ell\, {}_{-2}Y^{\ell -m}(\iota,0). \end{aligned} \tag{4}$$

The antenna pattern functions are functions of the sky location coordinates, θ and ϕ and the polarization angle ψ, that is, $F^a_+(\theta,\varphi,\psi)$, $F^a_\times(\theta,\varphi,\psi)$. The spheroidal harmonics are angular functions of the inclination angle, ι and the azimuth angle ϕ. The first refers to the angle formed by the BH's spin angular momentum and the line-of-sight, while the latter is the azimuth angle defined in a non-rotating frame fixed to the BH.

2.1. Chosen waveform

The ringdown waveform used is of the form described by Eqs. (1)-(3). It is a signal comprised of four modes, with mode indices $(\ell, m, n) = (2,2,0), (3,3,0), (2,1,0), (4,4,0)$. Our choice is based on the ordering of the various modes according to power output, as determined from numerical simulations of non-spinning unequal mass binaries [5, 11, 4].

Numerical simulations of merging black-hole binaries were performed using the BAM code [12, 13], so as to obtain the amplitudes, $\alpha_{\ell m}$ of the various modes in Eq.(2), as well as their frequencies and time-constants (see also Table I of [5]). These simulations involve the case of *initially non-spinning* BHs in quasi-circular orbits and for different mass ratios of the binary. For the mass ratios, $q = \{1, 2, 3, 4\}$ the simulation results were first presented in [14, 15, 16], while an additional simulation of a $q = 11$ binary was carried out in [5].

The amplitude terms $\alpha_{\ell m}$ in Eq. 2, are given by the expressions:

$$\alpha_{22}(q) = 0.25\, e^{-q/7.5}, \tag{5}$$

$$\alpha_{33}(q) = 0.18\,\alpha_{22}(q)\,(q-1)^{0.32}, \tag{6}$$

$$\alpha_{21}(q) = 0.15\,\alpha_{22}(q)\,(q-1)^{0.38}, \tag{7}$$

$$\alpha_{44}(q) = 0.018\,\alpha_{22}(q)\,q^{0.89}. \tag{8}$$

These constitute fits, that were produced by fitting the merger-ringdown part of the numerical simulations data, taking into account all the different mass ratios for which these were performed. All points in a time region beginning at the peak luminosity of 22 and ending $30M$ later were considered, for each mass ratio. As opposed to the method that was applied in our previous work [5], where the relative amplitude values at $10M$ were used, this approach is expected to be more robust and to average out any numerical noise that might be present in this part of

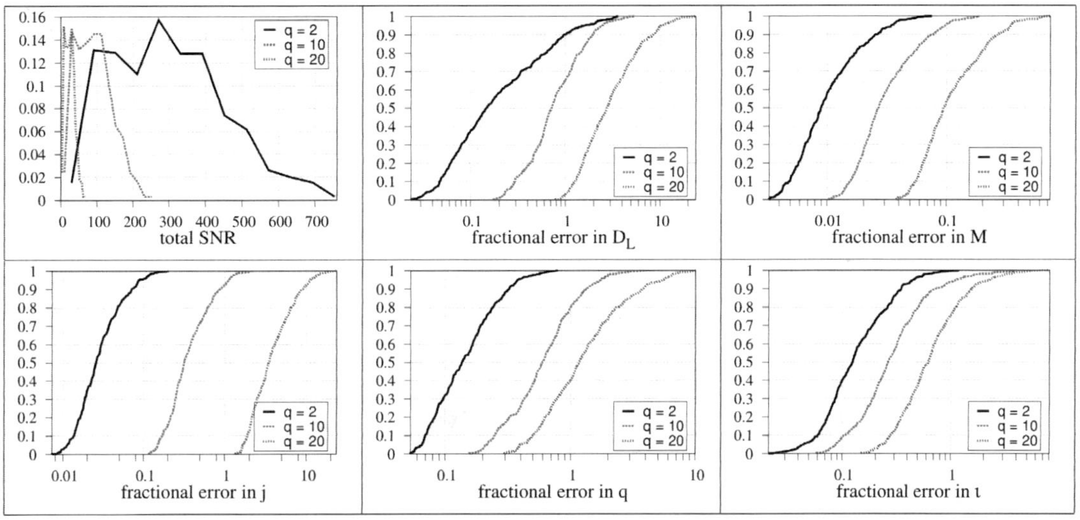

Figure 3: Frequency distributions for the total SNR and the measurement errors in NGO, for a $5\times10^6 M_\odot$ BH situated at a luminosity distance of 6.73 Gpc ($z \simeq 1$). The top left plot shows the probability to have a detector-BH configuration which will yield the SNR shown, while the rest of the plots concern cumulative frequency distributions for the measurement errors. In each graph, the comparison is shown among different mass ratios, q, specifically taking the values of 2, 10 and 20. For q=2 - black solid lines - the parameter estimation for the BH mass and spin is outstanding in all configurations. This effectiveness degrades considerably with increasing q. For q of around 10 it is still acceptable, while at q≃20, all parameters except the mass are very likely to have huge errors, of the order of 200%.

the waveform. Nevertheless, as can be seen from Fig. 2, these two methods do not give very different results. Note also that the above fitting functions, as well as the mode frequencies and time-constants, may be less accurate in the higher mass ratio values of around 20, where extrapolation has been performed.

2.2. The simulations

In Ref. [5] we had ignored the effect of the various angles $\{\theta,\ \varphi,\ \psi,\ \iota\}$ on the quasi-normal mode spectrum and their impact on the detection and measurement of ringdown signals. To assess this effect, the aforementioned analysis was repeated by varying the angular parameters $\{\theta,\ \varphi,\ \psi,\ \iota\}$. Specifically, six uniformly spaced values were chosen for these angles. This results in $6^4 = 1296$ distinct relative orientations between the detector and the BH and its spin axis. Additionally, a couple of simulations with eight uniformly spaced values were performed, that is $8^4 = 4096$ configurations, to allow the comparison. Hence, it was decided that six values in each parameter was acceptable in capturing the behaviour of the observable quantities.

The values in the polarization angle, ψ and the azimuth sky location angle, φ were linearly sampled in the ranges $[0, \pi]$ and $[0, 2\pi]$ respectively. Whereas, the values in the inclination angle ι and sky position, θ - which range from 0 to π - are deduced from the uniformly spaced values of $\cos(\iota)$ and $\cos(\theta)$ in the range $[-1, +1]$. Note that this excludes configurations of optimally oriented binaries, that is of $\iota = 0$ and $\iota = \pi$.

2.3. The parameter space

The parameter set of the ringdown signal in the case of a non-spinning black hole binary, consists of the following nine parameters: $\{M,\ j,\ q,\ D_{\rm L},\ \theta,\ \varphi,\ \psi,\ \iota,\ \phi\}$. Namely, the mass M, the dimensionless Kerr parameter or spin magnitude, j of the BH, the mass ratio q of the

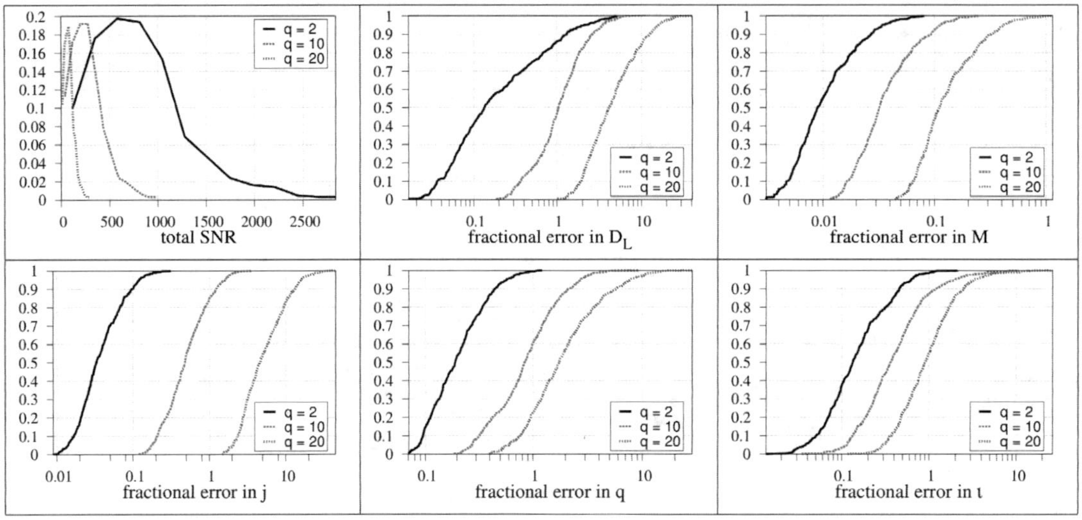

Figure 4: Same as in Fig. 3, but concerning a $25 \times 10^6 M_\odot$ BH. This figure concerns a BH 5 times more massive than that of Fig. 3, but some of the results are actually slightly worse, as the multimodal signal's power spectrum is shifted away from the lowest part of NGO's sensitivity curve.

progenitor binary, the sky location vector (D_L, θ, φ) of the BH with respect to Earth, the polarization angle ψ, the inclination angle ι and the BH azimuth angle ϕ. Note however, that in this case, the final spin of the BH is directly mapped to the mass ratio of the progenitor binary, therefore q and j are not treated as independent.

The above parameters are the standard ones, as they pertain to all of the modes. Additional parameters can be introduced, that are characteristic to each mode, such as an initial phase factor $\phi_{\ell m}$, see Eq. (3). Therefore, the total number of parameters can increase with the number of modes. We are considering a four mode signal, therefore we have a total of 13 parameters. One thing to note is that the $\ell = m$ modes have a nearly consistent rotational phasing, while the $\ell \neq m$ modes seem to have somewhat distinct associated dynamics, with differentiated amplitude and phasing during the merger process [11].

By virtue of the large number of parameters involved, it was unmanageable to treat the effect of all of them in this analysis. Thus, some of the above mentioned parameters had to be fixed. Specifically, the initial phase angles, $\phi_{\ell m 0}$ in all the four modes considered, were plainly chosen as zero. In addition, the luminosity distance is chosen to be 6.73 Gpc for NGO, and 1 Gpc for ET and aLIGO. Lastly, the BH azimuth angle, ϕ was given the value $\pi/3$. Note that this angle does not have an effect on the SNR, but of course needs to be considered in the waveform.

2.4. Choice of masses and mass ratios

The BH mass and the binary mass ratio constitute key parameters and the results depend quite strongly on them. The reasoning behind the choice of mass values is the following. First of all, the low and high mass end is limited by the sensitivity band of the detectors. The existence of BHs in the mass range $10^5 - 10^6 M_\odot$, is highly predicted by the mass - velocity dispersion in the galactic bulge of low-mass, low-luminosity galaxies, as well as in a number of galaxies which contain active galactic nuclei [17, 18]. The evidence for SMBHs ranging from $10^6 M_\odot$ to $10^9 M_\odot$, is quite abundant. They are thought to dwell at the centers of most galaxies. For various recent SMBHs mass estimation results and methods see for instance [19, 20, 21, 22, 23]. The SNR in NGO is quite low for BH masses of less than about $5 \times 10^5 M_\odot$. Therefore, I take the lower end

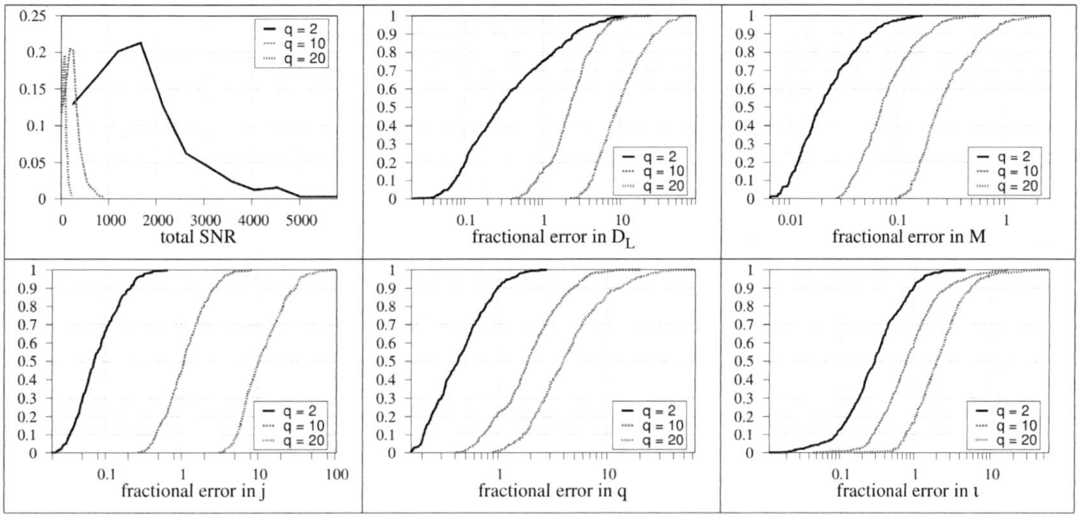

Figure 5: Same as in Figs. 3 and 4, but for a $10^8 M_\odot$ BH. The results slightly deteriorate with respect to the lower mass BH of Fig. 4.

of the mass range to be $5 \times 10^6 M_\odot$ and consider two other values of $25 \times 10^6 M_\odot$ and a $10^8 M_\odot$ BH, to cover the interesting range of masses potentially observable in NGO.

In the case of ET and aLIGO I have considered three IMBHs of mass $200 M_\odot$, $600 M_\odot$ and $1000 M_\odot$. It is believed that BHs in this range are situated in the centers of many globular clusters. However, their existence is being questioned, the evidence is thought to be strong but circumstantial [24, 18, 25, 26, 27].

Concerning the mass ratios, only unequal mass binaries are presented, except for one case in aLIGO. We do not consider the equal mass case as such systems are not as likely to occur in nature as slightly asymmetric ones. For NGO we examine the mass ratios of 2, 10 and 20, whilst lower mass ratios of 2, 5 and 10 are considered for ET. We, therefore, have in total, 9 different sets of simulations for each detector. Let me emphasize here that there is a possibility that massive BH binaries in the early universe ($z \leq 10$) will have a mass ratio significantly larger than one. For instance, in [28] it is suggested that low-redshift massive BH mergers occur predominantly with a mass ratio of 10 or higher.

3. Signal detectability and Parameter estimation

I will discuss the total signal to noise ratio, as well as the fractional errors in estimating the following 5 parameters: $\{M, j, q, D_L, \iota\}$. These errors are actually the quantities, $\sigma_\lambda = \sqrt{C_{\lambda\lambda}}$, which are computed from the covariance matrix, $C_{\ell m}$ [29, 30, 31, 32].

The results from all the distinct arrangements of the system (see Section 2.2) are presented via cumulative frequency distribution plots. That is to say, the different system configurations are classified according to the value they render in the error of the observable quantity in question. The proportion of a number of occurrences in a small width of values should, to a good approximation, equal the probability that a randomly placed observer will measure that quantity to take this range of values.

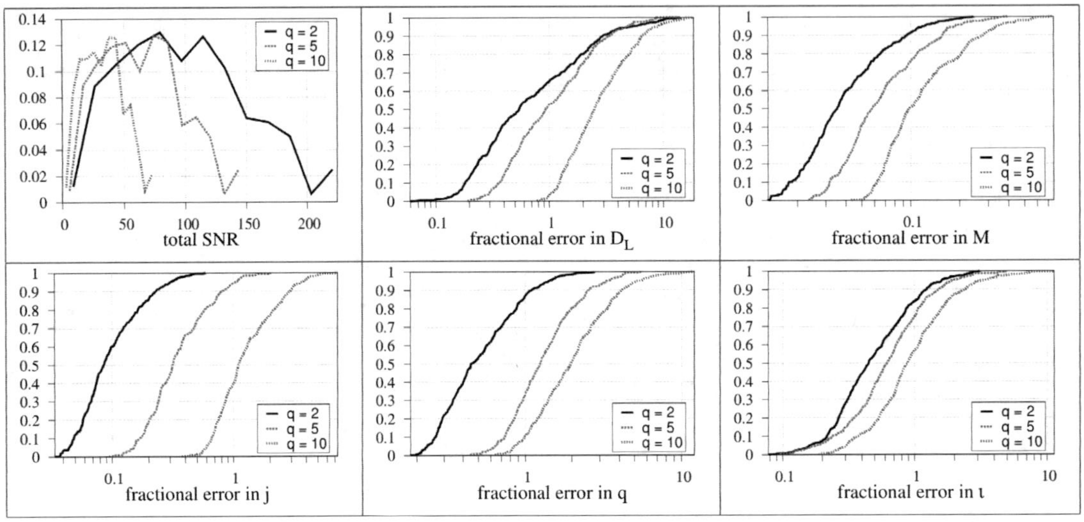

Figure 6: Similar graphs as those in Figs. 3 - 5, although for the ET detector, concerning a IMBH of $200 M_\odot$. The luminosity distance of the BH is now picked closer, at 1 Gpc ($z \simeq 0.2$). Also, note that the progenitor mass ratios now take lower values, namely 2, 5 and 10.

3.1. NGO

The sensitivity curve[1] that we use is what is thought to be possible for NGO and it is contrasted with other LISA-like sensitivity curves in Fig. 1. It refers to a 4-link interferometer, comprised of one mother and two daughter spacecraft, having armlengths of $10^9\,m$ and trailing a few degrees behind the Earth, in heliocentric orbit.

The main noise contributions are the acceleration noise, the shot noise, as well as some other measurement noise. These are respectively:

$$\begin{aligned} S_{acc,m}(f) &= 1.37 \times 10^{-32}(1 + \frac{10^{-4}}{f})f^{-4}\ m^2 Hz^{-1}, \\ S_{SN,m} &= 5.25 \times 10^{-23}\ m^2 Hz^{-1}, \\ S_{OMN,m} &= 6.28 \times 10^{-23}\ m^2 Hz^{-1}. \end{aligned}$$

The formula for the amplitude sensitivity curve is,

$$\sqrt{S_h(f)} = \sqrt{5}\frac{2}{\sqrt{3}}T(f)\frac{\sqrt{4S_{acc} + S_{SN} + S_{OMN}}}{L}\ Hz^{-1/2}, \tag{9}$$

while the transfer function is

$$T(f) = \sqrt{1 + \left(\frac{f}{0.41\,(\frac{c}{2L})}\right)^2}. \tag{10}$$

with $L = 10^9\,m$ and $c = 299,792,458$ metres per second.

3.2. Results for NGO

Our results for NGO are plotted in Figs. 3-5. For the $5 \times 10^6 M_\odot$, $25 \times 10^6 M_\odot$ and $10^8 M_\odot$ BHs the SNR curves (top left subplots of Figs. 3-5) peak at around 300, 600 and 1700 respectively.

[1] see https://lisa-light.aei.mpg.de/bin/view/DetectorConfigurations

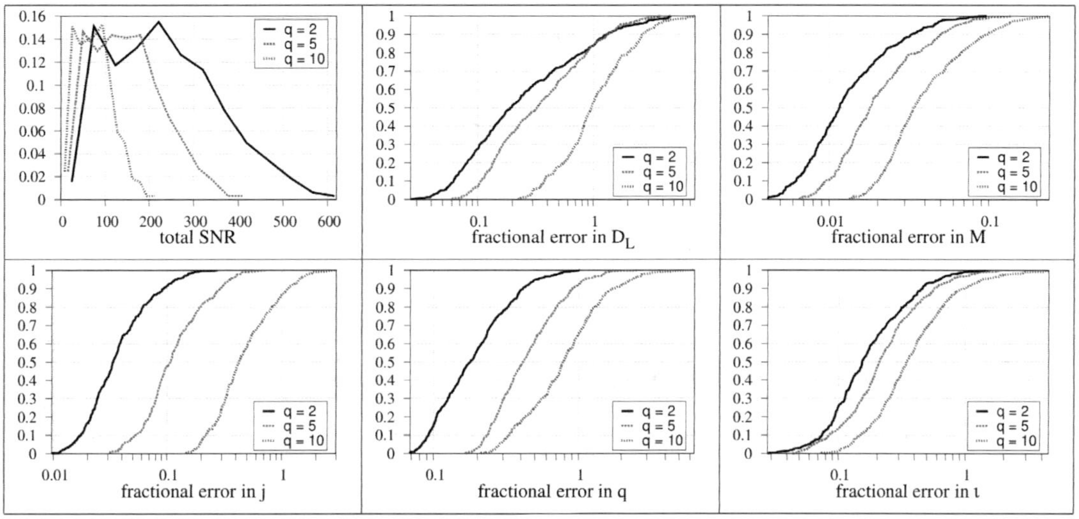

Figure 7: As in Fig. 6, but for a $600M_\odot$ black hole.

Relative frequency distributions were chosen for their plotting, as they portray clearly where the maximum occurs, as well as how they fall off. The higher mass ratio SNR curves have a similar outline. The fact that all of the curves resemble log-normal distributions, with steep risings and long tails, is mostly attributed to the fact that only a small fraction of the configurations, those close to the optimal orientation of the binary, will yield the highest SNRs.

Regarding the rest of the plots in Figs. 3-5, the general trend is that the mass and spin have comparable, low fractional errors, whilst the other group of parameters, namely the luminosity distance, mass ratio and inclination angle, yield an order of magnitude higher error values. This is not surprising, considering that the mass and spin have a direct effect on the modes' frequencies, $\omega_{\ell mn}$ and time-constants, $\tau_{\ell mn}$, quantities that determine to first order the shape of the ringdown waveform.

Quoting a few numbers for a progenitor of mass ratio 2, the probability to get a binary merger event that will yield a mass error less than 1% is correspondingly 58%, 55% and 10%, as we go from the lower mass to the higher mass value. The lowest mass value actually gives the best results, whereas the signal's main power content takes place near the lowest sensitivity area of the NGO detector. The spin magnitude exhibits a similar trend with the probabilities to fall below a 10% fractional error being 96%, 90% and 67% respectively.

For the second group of parameters, I again quote how likely it is to do better than 10%. For the luminosity distance, the values are 38%, 41% and 20%, corresponding to the BH masses $5 \times 10^6 M_\odot$, $25 \times 10^6 M_\odot$ and $10^8 M_\odot$. The mass ratio is harder to determine accurately, with the likelihoods being 32%, 15% and only 2% respectively. Lastly, there is a 40%, 37% and 13% likelihood of achieving an accuracy better than 10% in the inclination angle.

3.3. Einstein Telescope and Advanced LIGO

ET's very low sensitivity curve accounts for impressive results in the mass range $\sim 500M_\odot$ to $\sim 1000M_\odot$ and for mass ratios between 1 and 5. I consider the sensitivity curve designated ET-B [33], whose noise power spectral density is given by $S_h(f) = 10^{-50} h_n(f)^2\,\mathrm{Hz}^{-1}$, with:

$$\begin{aligned} h_n(f) &= 2.39 \times 10^{-27}\, x^{-15.64} + 0.349\, x^{-2.145} \\ &+ 1.76\, x^{-0.12} + 0.409\, x^{1.10}, \end{aligned} \tag{11}$$

where $x = f/100\,\text{Hz}$.

As for advanced LIGO, the noise spectral density is [2]

$$S_h(f) = 10^{-49}\left[10^{16-4(f-7.9)^2} + 0.08\,x^{-4.69} + 123.35\,\frac{1-0.23\,x^2+0.0764\,x^4}{1+0.17\,x^2}\right]\text{Hz}^{-1}, \tag{12}$$

where $x = f/215\,\text{Hz}$.

3.4. Results for ET and aLIGO

The results obtained for ET and advanced LIGO are plotted in Figs. 6-9. We fix the luminosity distance of the BH to be 1 Gpc. For the lowest mass considered (a $200M_\odot$ BH) although the SNR could be pretty high (in the range 30-200), errors in the estimation of parameters are poor (see Fig. 6). $\{D_L, q, \iota\}$ have 50% probability to be measured to an accuracy of $\sim$ 50%, while errors on $\{M, j\}$ are 90% and 60% likely to be below 10%. For the higher mass ratios the results, as expected, are worse.

The results for $600M_\odot$ and $1000M_\odot$ BHs are shown in Figs. 7 and 8. The parameter estimation accuracies for these systems observed with ET is almost as good as that for a SMBH with NGO. Referring to the heaviest BH and mass ratio in the range 2-5, it is 100%-99% and 96%-53% likely to acquire errors below 10% for the BH mass and spin. For the $\{D_L, q, \iota\}$, the efficiencies are correspondingly 50%-15%, 23%-0% and 60%-17%. The results are similar for the $600M_\odot$ case (see Fig. 7).

Fig. 9 shows two examples for aLIGO: a BH resulting from an equal mass binary and one for which the mass ratio is 5. Most configurations give ringdown SNR values in the range 10 to 30, not large enough for a good estimation of parameters. This translates to a 30% likelihood for a fair measurement of 10% accuracy in the mass and mass ratio, while the luminosity distance, spin and inclination angles are all measured to accuracies far worse than 10%.

4. Astrophysical implications

4.1. Supermassive black holes

The presence of SMBHs in the centers of massive galaxies seems to be a well established fact. Detecting their gravitational wave signals can give additional clues on their spatial[3] and mass distribution, as well as help discriminate among the different scenarios of their formation and growth. This could be done, for instance, by measuring the BH mass function as a function of red-shift. Additionally, determining the mass ratios [5] of these early universe merger events will be an important piece of information in selecting out the current models on SMBH formation. Note that, in NGO, BHs of mass higher than $10^{7-8}M_\odot$ are visible almost entirely due to the ringdown signal that they emit rather than the inspiral signal.

Several studies have been realized in the field of predicting the coalescence rates of SMBHs [34, 35, 36, 37, 38, 39, 40, 41]. Let us admit an event rate of $\sim 10\ yr^{-1}$ at $z \simeq 1$, which ascertains a scenario of most efficient BH coalescence [42]. Then, assuming that the BH masses are both around $5\times10^6 M_\odot$, Figs. 3 and 4, there is a good chance that in 6 of the events, the BH mass will be measured more accurately than 1% and that in 9 of the events the BH spin magnitude will have an accuracy better than 10%. As for the parameters $\{D_L, q, \iota\}$, in approximately 3-4 of the events they will feature errors lower than 10%, while in 7 of the events, the errors will lie below 30% for D_L and below 20% for q and ι.

[2] This fit was provided by C. Capano, Syracuse University and is tuned for detecting binary neutron stars.

[3] Unless the inspiral phase is inside NGO's band, it will not be possible to determine the sky position of the signal. The only hope in this case would be the existence of an electromagnetic counterpart to the merger event.

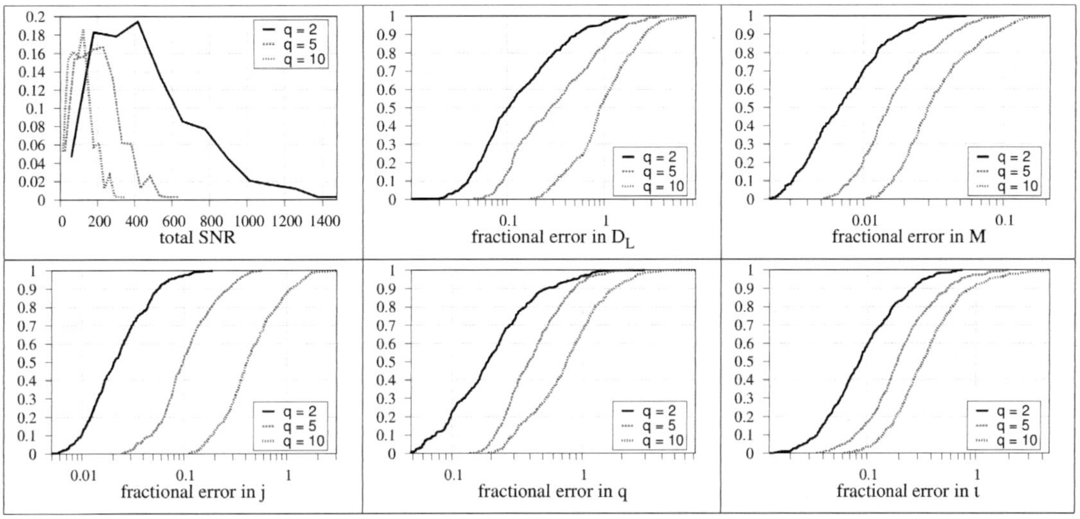

Figure 8: As in Figs. 6 and 7, but for a $1000M_\odot$ BH, a sweet spot in ET. The results are very encouraging for this size of BH and distance, with a very high probability to get errors lower than 10% in most of the parameters at mass ratios 2-5.

Let me clarify that ringdown signals in NGO, will be detectable out to about $z \simeq 5$, but the distance of $z = 1$ was chosen to make the quote on the errors. If we take the distance at $z = 3$, an optimistic rate of merger events will be of the order of 100 [42]. If we assume that they all involve BHs of masses $\simeq$ 2-3 $\times 10^6 M_\odot$, then in around 90 of them the error[4] in the mass will be below 10% and in 60 of them the spin magnitude error will fall below the 10% threshold. For the rest of the parameters, about 50 of the observed events should yield measurements better than 50% accuracy. Although this looks poor at the outset, it should suffice for a statistical test of different models of BH formation and growth.

4.2. Intermediate-mass black holes

The existence of intermediate-mass black holes remains uncertain, as is their mass distribution. Colliding globular clusters in interacting galaxies could be a mechanism to obtain a compact binary IMBH system [43, 3]. Another possibility could be the formation of a binary IMBH inside a young dense stellar cluster, especially when the fraction of binary stars is adequately high [26].

Estimates of IMBH-IMBH coalescence rates can be found in [44, 26, 3, 45]. A relatively optimistic rate is $R_{opt} = 0.007\ GC^{-1}Gyr^{-1}$, where it has been assumed that 10% of star clusters are sufficiently massive and have a sufficient stellar binary fraction to form an IMBH-IMBH binary once in their lifetime, taken at 13.8 Gyr. The maximum possible rate would come from assuming that all of the star clusters satisfied the above conditions. The corresponding rate value is then $R_{max} = 0.07\ GC^{-1}Gyr^{-1}$.

If the number of relatively luminous galaxies within a distance of 1 Gpc is approximately 5.3×10^7, [46] and the number of young dense stellar clusters per such galaxy is of the order of 100, then the optimistic estimate gives 0.037 events per year, while the maximum rate would be 0.37 events per year. An event within this distance will in some likelihood, involve IMBHs with masses between $\sim 6 \times 10^2 M_\odot$ and $\sim 10^3 M_\odot$ which means a relatively fair chance for errors to be low for several of the parameters using Einstein Telescope (see Figs. 7, 8).

[4] An additional simulation at $z = 3$, of a $5 \times 10^6 M_\odot$ BH, with a progenitor mass ratio of 2 was performed.

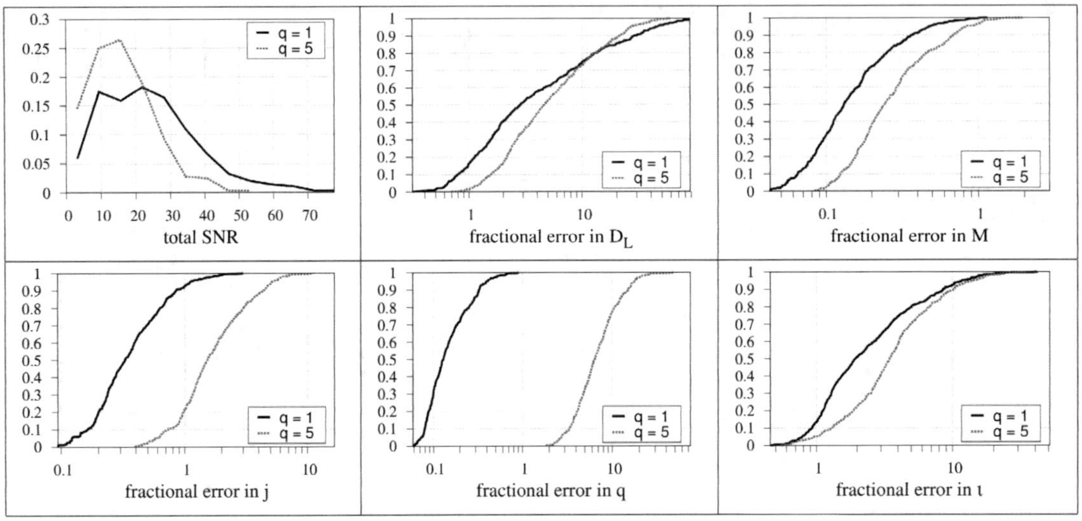

Figure 9: Frequency distributions involving the most optimistic scenario for advanced LIGO: a $1000M_\odot$ BH which is the merger of an equal mass binary, that is $q = 1$. The BH is again situated at 1 Gpc. A mass ratio of 5 is shown as well for contrasting.

5. Conclusions

The present study constitutes a sensible and realistic approach to the subject of parameter estimation from a multimodal ringdown signal, inasmuch as it is supposed to be emitted from a merged binary in a generic configuration. Parameters such as the inclination angle ι, along with the sky location θ and φ and the signal polarization ψ have an effect on the observed quasi-normal mode spectrum. Their impact on the detectability and parameter estimation has been assessed, by performing a large sample of Fisher-matrix analysis simulations, allowing for a simple statistical analysis of the results.

I am quoting frequency distributions for the errors at the representative distances of $z \simeq 1$ (6.73 Gpc) for supermassive BHs, in NGO, and of $z \simeq 0.2$ (1 Gpc) for intermediate-mass BHs, in the Einstein Telescope and advanced LIGO. An additional simulation was performed for a $5 \times 10^6 M_\odot$ BH at $z \simeq 3$. The results are quite satisfactory in determining the mass and spin, especially for the low mass ratios from 1 to ~ 5, where in typically 90% of the cases their errors fall below 10%. The effectiveness in measuring the luminosity distance, mass ratio and inclination angle is almost an order of magnitude worse.

The effects of these results on, as much as possible, realistic event rates in NGO and ET are discussed. The likelihood to have a waveform parameter measured to an accuracy of a certain threshold translates to the same proportion of observed events featuring error values below that threshold. As an example, if supermassive BHs within a luminosity distance of $z = 1$ coalesce at a rate of $\simeq 10$, then NGO could act as a SMBH dynamics probe, as almost all of these events will yield very low errors $1\% - 10\%$ in the BH mass and spin, while in half of the events the errors in the luminosity distance, mass ratio and inclination angle will be of the order of 20%.

6. Acknowledgements

I am grateful to B.S. Sathyaprakash for his supervision and exceptional advice throughout this study. I would like to thank Mark Hannam for discussions and him and Sascha Husa for making available numerical relativity simulations used in this study. I am thankful to Mr Antoine Petiteau for clarifications on the sensitivity response of the European mission of LISA.

References

[1] Nandra K, Aird J, Alexander D, Ballantyne D, Barcons X *et al.* 2009 * Brief entry * (*Preprint* 0903.0547)
[2] Somerville R S 2008 * Brief entry * (*Preprint* 0808.1254)
[3] Amaro-Seoane P and Santamaria L 2010 *Astrophys.J.* **722** 1197–1206 (*Preprint* 0910.0254)
[4] Berti E, Cardoso V and Will C M 2006 *Phys. Rev. D* **73** 064030–+ (*Preprint* gr-qc/0512160)
[5] Kamaretsos I, Hannam M, Husa S and Sathyaprakash B S 2012 *Phys. Rev. D* **85**(2) 024018 URL http://link.aps.org/doi/10.1103/PhysRevD.85.024018
[6] Berti E, Cardoso J, Cardoso V and Cavagliá M 2007 *Phys. Rev.* **D76** 104044 (*Preprint* 0707.1202)
[7] Nelemans G, Yungelson L and Portegies Zwart S 2004 *Mon.Not.Roy.Astron.Soc.* **349** 181
[8] Van Den Broeck C, Trias M, Sathyaprakash B and Sintes A 2010 *Phys.Rev.* **D81** 124031 (*Preprint* 1001.3099)
[9] Berti E, Cardoso V and Starinets A O 2009 *Class. Quant. Grav.* **26** 163001 (*Preprint* 0905.2975)
[10] Kokkotas K D and Schmidt B G 1999 *Living Rev.Rel.* **2** 2 (*Preprint* gr-qc/9909058)
[11] Baker J G, Boggs W D, Centrella J, Kelly B J, McWilliams S T and van Meter J R 2008 *Phys.Rev.* **D78** 044046 (*Preprint* 0805.1428)
[12] Brügmann B *et al.* 2008 *Phys. Rev.* **D77** 024027 (*Preprint* gr-qc/0610128)
[13] Husa S, González J A, Hannam M, Brügmann B and Sperhake U 2008 *Class. Quant. Grav.* **25** 105006 (*Preprint* 0706.0740)
[14] Berti E, Cardoso V, Gonzalez J, Sperhake U, Hannam M, Husa S and Brugmann B 2007 *Phys. Rev.* **D76** 064034 (*Preprint* gr-qc/0703053v2)
[15] Hannam M, Husa S, Sperhake U, Bruegmann B and Gonzalez J A 2008 *Phys.Rev.* **D77** 044020 (*Preprint* 0706.1305)
[16] Hannam M, Husa S, Ohme F, Muller D and Bruegmann B 2010 *Phys.Rev.* **D82** 124008 (*Preprint* 1007.4789)
[17] Valluri M, Ferrarese L, Merritt D and Joseph C L 2005 *Astrophys. J.* **628** 137–152 (*Preprint* astro-ph/0502493)
[18] Safonova M and Shastri P 2010 *Astrophys. Space Sci.* **325** 47–58 (*Preprint* 0910.2551)
[19] Beifiori A, Courteau S, Corsini E M and Zhu Y 2011 (*Preprint* 1109.6265)
[20] Li Y R, Ho L C and Wang J M 2011 *Astrophys. J.* **742** 33 (*Preprint* 1109.0089)
[21] Rafiee A and Hall P B 2011 *Astrophys. J. Suppl.* **194** 42 (*Preprint* 1104.1828)
[22] Burkert A and Tremaine S 2010 *Astrophys. J.* **720** 516–521 (*Preprint* 1004.0137)
[23] Harris G L H and Harris W E 2010 (*Preprint* 1008.4748)
[24] Miller M C 2009 *Class. Quant. Grav.* **26** 094031 (*Preprint* 0812.3028)
[25] Noyola E and Baumgardt H 2011 (*Preprint* 1108.4425)
[26] Fregeau J M, Larson S L, Miller M C, O'Shaughnessy R W and Rasio F A 2006 *Astrophys. J.* **646** L135–L138 (*Preprint* astro-ph/0605732)
[27] Bash F N, Gebhardt K, Goss W M and Bout P A V 2008 *The Astronomical Journal* **135** 182 URL http://stacks.iop.org/1538-3881/135/i=1/a=182
[28] Volonteri M, Madau P, Quataert E and Rees M J 2005 *Astrophys. J.* **620** 69–77 (*Preprint* astro-ph/0410342)
[29] Wainstein L A and Zubakov V D 1962 *Extraction of Signals from Noise* (Englewood Cliffs: Prentice-Hall)
[30] Finn L S 1992 *Phys. Rev. D* **46** 5236
[31] Balasubramanian R, Sathyaprakash B S and Dhurandhar S V 1996 *Phys. Rev. D* **53** 3033 erratum-ibid. D **54**, 1860 (1996) (*Preprint* gr-qc/9508011)
[32] Sathyaprakash B S and Schutz B F 2009 *Living Rev. Rel.* **12** 2 (*Preprint* arXiv:0903.0338)
[33] Hild S, Chelkowski S, Freise A, Franc J, Morgado N *et al.* 2010 *Class.Quant.Grav.* **27** 015003 (*Preprint* 0906.2655)
[34] Filloux C, Pacheco J A d F, Durier F and de Araujo J C N 2011 (*Preprint* 1108.2638)
[35] Berti E 2006 *Classical Quantum Gravity* **23** S785 (*Preprint* astro-ph/0602470)
[36] Haehnelt M G 2003 *Classical and Quantum Gravity* **20** S31 URL http://stacks.iop.org/0264-9381/20/i=10/a=304
[37] Rhook K J and Wyithe J S B 2005 *Mon. Not. Roy. Astron. Soc.* **361** 1145–1152 (*Preprint* astro-ph/0503210)
[38] Erickcek A L, Kamionkowski M and Benson A J 2006 *Mon. Not. Roy. Astron. Soc.* **371** 1992–2000
[39] Micic M, Holley-Bockelmann K, Sigurdsson S and Abel T 2007 *Mon. Not. Roy. Astron. Soc.* **380** 1533
[40] Wen Z L, Jenet F A, Yardley D, Hobbs G B and Manchester R N 2011 *Astrophys. J.* **730** 29 (*Preprint* 1103.2808)
[41] de Freitas Pacheco J A, Filloux C and Regimbau T 2006 *Phys. Rev.* **D74** 023001 (*Preprint* astro-ph/0606427)
[42] Menou K, Haiman Z and Narayanan V K 2001 *Astrophys.J.* **558** 535–542 (*Preprint* astro-ph/0101196)
[43] Amaro-Seoane P and Freitag M 2006 *Astrophys. J.* **653** L53–L56 (*Preprint* astro-ph/0610478)
[44] Abadie J *et al.* (LIGO Scientific Collaboration) 2010 (*Preprint* arXiv:1003.2480)
[45] Mandel I, Gair J R and Miller M C 2009 (*Preprint* 0912.4925)
[46] White D J, Daw E and Dhillon V 2011 *Class.Quant.Grav.* **28** 085016 (*Preprint* 1103.0695)

Amaldi 9 and NRDA 2011 IOP Publishing
Journal of Physics: Conference Series **363** (2012) 012048 doi:10.1088/1742-6596/363/1/012048

Improving Markov Chain Monte Carlo algorithms in LISA Pathfinder Data Analysis

N Karnesis, M Nofrarias, C F Sopuerta and A Lobo

Institut de Ciències de l'Espai (CSIC-IEEC), Facultat de Ciències, Campus UAB, Torre C5 parells, Bellaterra, 08193 Barcelona, Spain

E-mail: karnesis@ieec.uab.es

Abstract. The LISA Pathfinder mission (LPF) aims to test key technologies for the future LISA mission. The LISA Technology Package (LTP) on-board LPF will consist of an exhaustive suite of experiments and its outcome will be crucial for the future detection of gravitational waves. In order to achieve maximum sensitivity, we need to have an understanding of every instrument on-board and parametrize the properties of the underlying noise models. The Data Analysis team has developed algorithms for parameter estimation of the system. A very promising one implemented for LISA Pathfinder data analysis is the Markov Chain Monte Carlo. A series of experiments are going to take place during flight operations and each experiment is going to provide us with essential information for the next in the sequence. Therefore, it is a priority to optimize and improve our tools available for data analysis during the mission. Using a Bayesian framework analysis allows us to apply prior knowledge for each experiment, which means that we can efficiently use our prior estimates for the parameters, making the method more accurate and significantly faster. This, together with other algorithm improvements, will lead us to our main goal, which is no other than creating a robust and reliable tool for parameter estimation during the LPF mission.

1. Introduction

The European Space Agency (ESA) is working on a new space-based Gravitational-Wave (GW) observatory that follows the concept of the NASA-ESA LISA mission [1]. LISA is a constellation of three spacecrafts that will access GW signals at frequencies of around 1 mHz. The scientific working principle of LISA is the detection of picometric relative displacements between pairs of proof masses in nominally geodesic motion, or free fall, induced by passing GWs. Due to the technological complexity of LISA, ESA approved a precursor mission, LISA PathFinder [2, 3], to test the readiness of the main LISA technology. LPF consists in a single spacecraft hosting two proof masses in nominal free fall, whose motions are monitored by means of Mach-Zender laser interferometry. The top level requirement on LPF is usually expressed in terms of relative acceleration between the Test Masses spectral density of noise, and is given by

$$S_{\Delta a}^{1/2}(\omega) = 3 \times 10^{-14} \left[1 + \left(\frac{\omega/2\pi}{3\,\mathrm{mHz}}\right)^2\right] \mathrm{ms}^{-2}\mathrm{Hz}^{-1/2}, \qquad 1\,\mathrm{mHz} \leq \frac{\omega}{2\pi} \leq 30\,\mathrm{mHz}$$

The LPF mission is expected to be launched in 2014 and its ultimate objective is to demonstrate the purity of the test mass free fall up to the sensitivity expressed above, and to understand the physical origin of the different noise contributions. There are many sources of noise identified (thermal, magnetic, particles of cosmic origin, etc), and properly modeling them requires a careful planning of the measurement sequence, plus the use of suitable analysis tools to process the various data channels. During flight operations the LPF is going to perform a series of experiments with the main goal to fully characterize the physical model of the instrument. Each experiment will determine a set of parameters that are going to be used as additional information for the next experiment in the sequence. Using a Bayesian framework for LPF Data Analysis [7], we are flexible enough to combine knowledge obtained from different experiments. Moreover we are able to use this knowledge as *a priori* information for the next in line parameter estimation experiment, making the algorithm significantly faster. In this work we take advantage of exactly this feature and we present the benefits of using prior information in combination with the MCMC algorithm developed for LPF Data Analysis. First we applied the enhanced algorithm to a simple toy model and then to a more complex LTP model widely used in LTP Data Analysis Operational Exercises [8].

For the LPF Data Analysis, a MATLAB Toolbox has been created [12] by the Data Analysis Team. This package is object-oriented code, based on so called Analysis Objects. The philosophy is to create a highly capable and flexible tool, with strong history tracking capacity to keep memory of the work done on any data segment, or derived object by anybody using the tool. The present work has been implemented as a part of the LTP Data Analysis toolbox (LTPDA).

2. A Bayesian Framework

Markov Chain Monte Carlo (MCMC) is a subset of Monte Carlo methods that involve a Markov process in which a sequence of samples $\{\theta_n\}$ is generated, each sample θ_n having a probability distribution that depends on the previous one, θ_{n-1} [5, 6]. This means that new samples in the parameter space are generated using a Markov Chain mechanism and in each jump (or step) in the parameter space, the likelihood is calculated. After a sufficient number of samples, one can investigate the shape of the likelihood surface. More precisely, one can be based to some prior information about the system and update this information based on the observed data $\vec{\mathbf{x}}$. This procedure is based on the well-known Bayes Theorem:

$$\pi(\vec{\theta}|\vec{\mathbf{x}}) = \frac{f(\vec{\mathbf{x}}|\vec{\theta})p(\vec{\theta})}{f(\vec{\mathbf{x}})} \quad (1)$$

where $\vec{\theta}$ is a parameter vector, $f(\vec{\mathbf{x}}|\vec{\theta})$ is the *Likelihood* of the dataset $\vec{\mathbf{x}}$, given the parameters $\vec{\theta}$ and $\pi(\vec{\theta}|\vec{\mathbf{x}})$ and $p(\vec{\theta})$ are the *posterior* and the *prior* distributions respectively. Note that the *evidence* $f(\vec{\mathbf{x}})$ is independent of the parameters $\vec{\theta}$ and it is often neglected in parameter estimation algorithms, as it serves only as a normalising constant:

$$\pi(\vec{\theta}|\vec{\mathbf{x}}) \propto f(\vec{\mathbf{x}}|\vec{\theta})p(\vec{\theta}) \quad (2)$$

As already mentioned we need to successfully integrate the posterior distribution. A most used MCMC algorithm is the *Metropolis algorithm.* The Metropolis algorithm can be summarized in the following steps. First, an initial set of parameters $\vec{\theta}_0$ is chosen. Then a new point in the parameter space is drawn from a multivariate Gaussian distribution. The covariance matrix Σ of the multivariate Gaussian is extracted from the calculation of the inverse of the Fisher Information Matrix [7]. The FIM is calculated once in a point on the parameter space that is our first "guess" of the parameters and the proposal distribution is constant during the MCMC run. In the newly created point, the likelihood is calculated and then the *acceptance probability* α is evaluated. The acceptance probability,

$$\alpha = \min\left(1, \frac{f(\vec{\mathbf{x}}|\vec{\theta})p(\vec{\theta})q(\vec{\theta}_0|\vec{\theta})}{f(\vec{\mathbf{x}}|\vec{\theta}_0)p(\vec{\theta}_0)q(\vec{\theta}|\vec{\theta}_0)}\right) \quad (3)$$

will determine whether if the new point is accepted or rejected by the algorithm. If the point is accepted, we set $\vec{\theta}_0 = \vec{\theta}$ and propose another point from the parameter space. The proposal densities q are vanished from equation 3 since they are symmetrical. This procedure is repeated until we converge to a set of parameters $\vec{\theta}_{\mathrm{MAP}}$ that maximize the posterior distribution.

2.1. MCMC and the LTPDA

From Eq. (2) one can see that prior Probability Density Functions (PDFs) can play an important role to the evolution of the algorithm. Until now, in the MCMC developed for the LTPDA toolbox [7], the prior information about the parameters was limited to flat PDFs. This translates to a "closed box" in the parameter space, where we assume that the $\vec{\theta}_{\mathrm{MAP}}$ is located. The nature of LPF mission allows us to use any previous estimates for the parameters for the next experiment in the sequence. This makes the introduction of priors into the algorithm an essential improvement. The result of the first experiment is a set of PDFs for the parameters. These posterior PDFs can be used as prior information for the next experiment and what we expect theoretically is faster convergence to $\vec{\theta}_{\mathrm{MAP}}$. The MCMC LTPDA algorithm was improved taking into account the prior PDFs, as in Eq. (3). The next step was to study what exactly is our gain in terms of time and accuracy and apply our findings to a more complete and complex LTP model.

3. Numerical Investigations

In order to discover and verify the benefits of using results from previous experiments as prior knowledge for the following ones, we designed a series of studies using a toy model. A simple one-dimensional low pass filter was chosen, since the results from MCMC runs were obtained faster. In section 3.2 we applied the same to a LTP model.

3.1. Special case: Toy model

A simple toy model was used to investigate the behavior of the algorithm. The model is a first-order low-pass filter with cutoff frequency $f_0 = 0.01\,\mathrm{Hz}$. A data set and noise were created after filtering a single frequency sinusoidal input. Then we used the simple model and inputs to estimate the supposedly unknown cutoff frequency f_0 of the low pass filter toy model. The posterior PDF of f_0 estimated is then imported to the next experiment, of the same type as the first one, as prior PDF. The idea behind this is to illustrate any effect to the speed of convergence of the algorithm. In order to properly estimate the time of convergence, we estimate the unknown f_0 fifty times for each configuration. For the first runs we used flat prior PDFs and we were able to fit a Gaussian curve to the posterior PDFs computed and use it as a prior PDF for the following runs. Following, we estimated f_0 using the Gaussian priors with their variance σ multiplied by a constant in order to see the exact effects of the prior on the convergence of the algorithm (the σ is in the order of 10^{-4}). In particular we did:

- normal runs with flat priors.
- Runs with priors with original σ.
- Runs with priors $50 \times \sigma$
- Runs with priors $500 \times \sigma$
- Runs with priors $1000 \times \sigma$

In figure 1, the complete progress of the chains is presented. In these chains the rejected points are also included. During the first burn-in phase, new points are proposed with the aim to explore aggressively the parameter space. If a point is far away from the area defined from the σ of our prior probability, then the value of the logarithm of the acceptance ratio is dominated

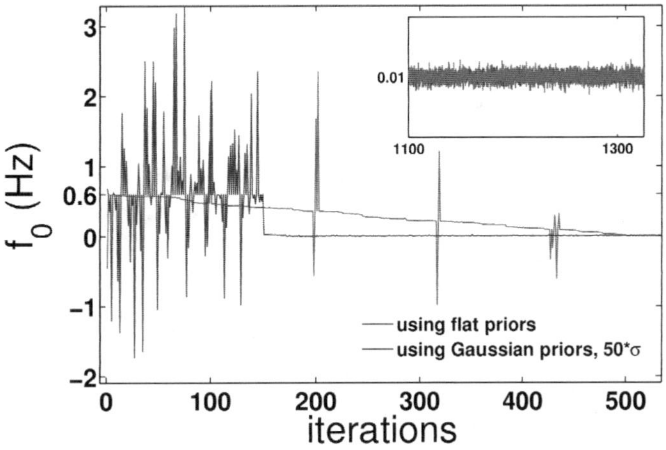

Figure 1. Graph of the chains *including the rejected points* from one of the fifty runs for two σ configurations. In the embedded figure the chains after convergence are displayed.

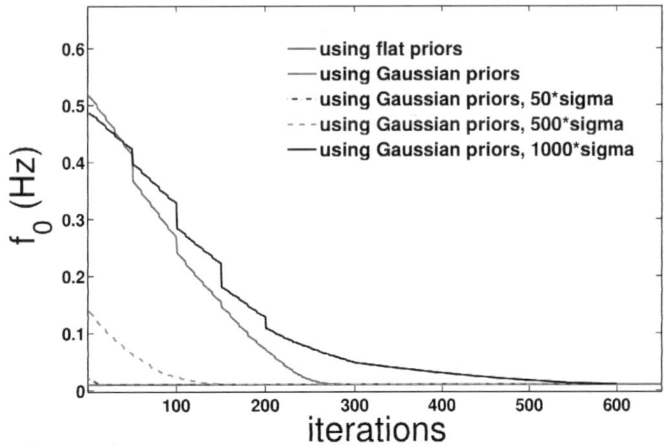

Figure 2. For fifty runs for each σ we can see the mean path of the chains of the accepted points.

by the very large terms of $log(p(\vec{\theta}))$ and the log-likelihood values are negligible compared to them. This results to a rejection of the new point proposed and while the acceptance ratio is computed, the costly calculation of multiplication of large matrices can be skipped. Now, this first searching phase becomes much more rapid. In figure 2, we can clearly see that as the σ gets narrower, the path of the chains becomes shorter in their way to the value of f_0 that maximizes the posterior.

3.2. Application to a LTP model

In order to understand the future LPF mission, a number of models of the LTP [10] have been implemented by the LTP DA team. One widely used for operational exercises [8, 9] is depicted in figure 3. It represents the experiment as a close-loop system, where noise contributions and cross-couplings are expresed as a set of transfer functions in frequency domain. This particular

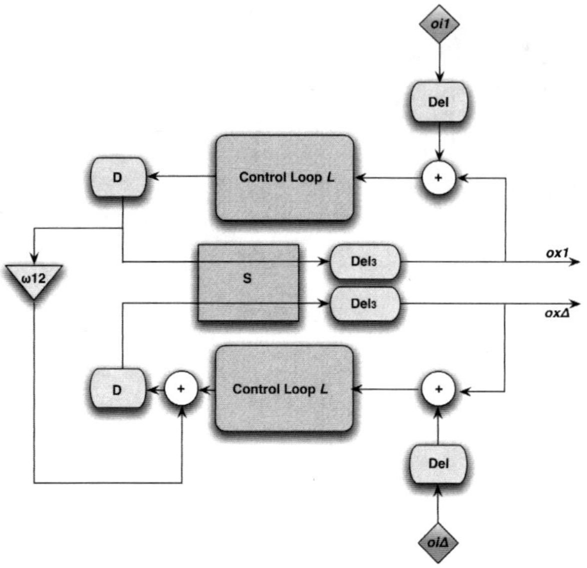

Figure 3. Illustration describing the simplified LTP model. This model describes the LTP operating in the Main Science Mode [11]. See text for a description of terms appearing in the picture.

model is a simplified model for the LTP, thus, only two heterodyne interferometers are taken into account: one measuring the distance between the two test masses ($o_{x\Delta}$) and one measuring the distance between test mass one (TM1) and the spacecraft (represented here as o_{x1}). The complete system, after the assumption that all noise terms are independent from the unknown parameters, can be expressed by:

$$\vec{o} = \mathbf{H}_s(\vec{\theta})\vec{o}_i + \vec{n} \tag{4}$$

This is justified as we are working in the high signal-to-noise ratio regime and consequently the dependance of the noise from the parameters is negligible. The terms appearing in Eq. (4) are: $\vec{o}$ is the vector of the interferometer readouts, $\vec{o}_i$ is vector of the injected signals, $\vec{n}$ is the overall instrument noise and $\mathbf{H}_s(\vec{\theta})$ is the complete transfer function describing the LTP. Each of the elements of Eq. (4) read as:

$$\vec{o} = \begin{pmatrix} o_{x1} \\ o_{x\Delta} \end{pmatrix}, \qquad \vec{o}_i = \begin{pmatrix} o_{i1} \\ o_{i\Delta} \end{pmatrix}, \qquad \vec{n} = \begin{pmatrix} n_{x1} \\ n_{x\Delta} \end{pmatrix}$$

$$\mathbf{H}_s(\vec{\theta}) = (\mathbf{D} \times (\mathbf{Del_3} \times \mathbf{S}) + \mathbf{L})^{-1} \times \mathbf{L} \times \mathbf{Del}, \tag{5}$$

where $\mathbf{H}_s(\vec{\theta})$ is the 4×4 matrix transfer function of the complete system. $\mathbf{H}_s(\vec{\theta})$ is composed of $\mathbf{D}$; the matrix describing the dynamics of the system; the $\mathbf{S}$ matrix, which describes the interferometric read-out and possible construction imperfections; the **Del$_3$** and **Del**, describing time delays; and finally the **L**, which describes the *Control Loop* of the LTP. The Control Loop contains three main subsystems: the inertia decoupling matrix, a matrix with the Drag-Free Attitude and Control System transfer functions and the electrostatic actuator gain and delays. A more detailed description can be found in [9].

The steps followed here are the same as in the Operational Exercise 6 [9]. A frequency sweep of sinusoidal signals was injected to the system forming two independent investigations. For the first investigation we injected a signal o_{i1} into the drag-free loop and for the second we injected a signal $o_{i\Delta}$ into the low-frequency suspension loop. The Operational Exercise 6 was designed with the aim of fully characterizing all the possible dynamical responses of the system-model of the LTP using parameter estimation methods developed by the LTPDA team. These particular investigations and model were chosen in order to perform the same complex analysis for LTP as in operational exercises and prove that the use of prior information applies to up-to-date parameter estimation methods.

As in the toy model, we used previous estimates of the parameters as prior information for our next experiments. The PDFs of the parameters of the first estimation were fitted with Gaussian curves and multiplied by a factor before we input them into the algorithm. The parameters to estimate are: ω_1, the stiffness coupling of TM1 to S/C, ω_{12}, the difference between stiffness coupling TM1 to S/C and TM2 to S/C, A1, the thrusters actuation gain, A2, the electrostatic actuation gain, DT1, the delay in the application of the thrusters actuation, DT2, the delay in the application of the electrostatic actuation and δ_{21} - the cross coupling between the interferometer channels.

In Figure 4 we can clearly see the difference of evolution of the MCMC chains for four out of seven total parameters of the LTP model using flat and Gaussian prior PDFs. While flat prior PDFs give little information about the parameter space that maximize the posterior, the Gaussian prior PDFs rapidly force the chains to explore the small area of interest. But in order to study the benefits of this practice we measured the time the algorithm needed until it achieved convergence to $\vec{\theta}_{\mathrm{MAP}}$. The algorithm was executed five times for each configuration and the mean time of execution was calculated. Basically, following this practice we are able to reduce the first burn-in period in terms of samples. Depending on the problem, the burn-in period could range between 3000 to 5000 samples. In this particular investigation it is reduced by 2000 samples, thus using a total of 5000 and 7000 for each case respectively.

Table 1. Mean computation time for both methods.

Method	mean time (s)
using flat priors	4602
using normal priors	2528

4. Results and future work

In table 1 the results of this investigation are summarized. The numbers presented are strongly correlated to the machine operating the toolbox, but it is clear that we gain almost a factor of two in terms of time. For the MCMC algorithm, faster convergence means that we can have efficient estimation of the parameters with fewer sample points. This is exactly what happens when we use Gaussian prior PDFs. Without *a priori* information we have to explore a relatively large

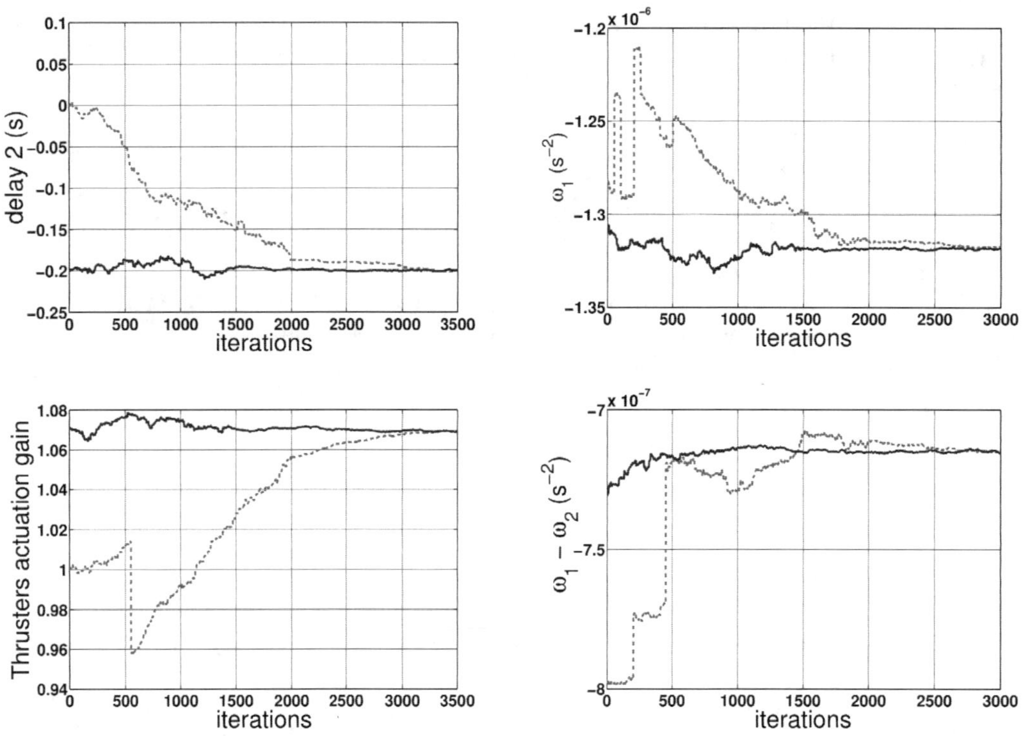

Figure 4. MCMC chains evolution for the estimation of the thrusters actuation gain A1, the delay2, the ω_1 and the $\omega_1 - \omega_2$ of the LTP model. The chains of the rest of the parameters of the model follow the same pattern. Here only the burn-in and cooling parts of the MCMC chains are presented. The dashed lines represent the chains when we use flat PDFs as prior knowledge, while the solid lines represent the chains when we use Gaussian prior information.

area of the parameter space. This means that again, a large number of samples has to be spent in the burn-in and cooling periods of the MCMC chains. These samples are then discarded and the remaining parts of the chains are used for the estimation of the parameters. The convergence of the chains is empirical, since, it is considered that given sufficient time, the MCMC will almost always converge to the $\vec{\theta}_{MAP}$. Also in the majority of system identification experiments of LPF, the Signal to Noise Ratio is huge and the MCMC reaches the MAP area quickly. In our case, the burn-in and cooling periods required more than half of the total samples gathered by the algorithm and usually the final 2000 to 3000 samples are used to make an estimation of the parameters. In our study the burn-in and cooling periods become significantly smaller, since the area to explore in the parameter space reduces to a small fraction of the original one. Although this practice seems to be consistent with the expected results we did not take in to account the correlation between the parameters in the calculation of the prior PDFs. That means that we used only the information contained in the diagonal of the covariance matrix extracted from the first estimation of the parameters. A next step is to include all the elements of the covariance matrix for a more complete calculation of the prior probability. Furthermore, a more complete version of the algorithm will take into account the fact that the narrow Gaussian priors affect the computation of the Fisher Information matrix and as a result the covariance matrix used in the proposal distribution. After completing this first step of enhancing the LTPDA

MCMC algorithm we plan further improvements and adaptations, like tuning the proposal distribution [13] and implementing various types of annealing and parallelizing.

We have shown how prior information can be used for parameter estimation and how it affects the time consumed by the algorithm. A simple model was implemented and a series of experiments were carried out. The results showed that *a priori* information provide accurate estimates using fewer samples. As expected, the same results were obtained for a more complex model describing the LTP onboard LPF. The time consumed by a statistical tool like MCMC is significantly reduced, making the method more suitable for Data Analysis between the experiments during the mission. This is a first step towards a more complete parameter estimation tool for LPF.

Acknowledgments

The authors acknowledge support from contract AYA2010-15709 of the Spanish Ministry of Science and Innovation. NK acknowledges support from the FI PhD program of the AGAUR (Generalitat de Catalunya). MN acknowledges support from a JAEDoc contract of the National Spanish Research Council (CSIC). CFS acknowledges support from the Ramón y Cajal Programme of the Ministry of Education and Science of Spain and by a Marie Curie International Reintegration Grant (MIRG-CT-2007-205005/PHY) within the 7th European Community Framework Programme.

5. References

[1] Danzmann K, and LISA Science Team, 2003 *Adv. in Space Res.*, **32**, Issue 7, 1233-1242
[2] Armano M *et al* 2009, *Class. Quantum Grav.* **26**, 09400.
[3] Dolesi R *et al* 2003, *Class. Quantum Grav.* **20** S99
[4] Kendall W S, Liang F, Wang J S, *Markov Chain Monte Carlo Innovations and Applications*, (Singapore:World Scientific Publishing)
[5] MacKay D J C, 2003, *Information Theory, Inference, and Learning Algorithms*, (Cambridge: C.U.P.)
[6] Andrieu C, Doucet A and Jordan M I, 2003, *Mach. Learn.*, **50**, 543
[7] Nofrarias M, Röver C, M Hewitson, A Monsky, G Heinzel and K Danzmann 2010 *Phys. Rev.* D **82** 122002
[8] F Antonucci *et al*, 2010, *Class. Quantum Grav.* **28**, 094006
[9] Nofrarias M, Ferraioli L, Congedo G, 2011, *Tech. Rep.* S2-AEI-TN-3070, Issue 1
[10] Hewitson M, Grynagier A, Diaz-Aguilo M, *Tech. Rep.* S2-AEI-TN-3069, Issue 1
[11] Fichter W, Gath P, Vitale S and Bortoluzzi D, 2005 *Class. Quantum Grav.* **22** S139
[12] Hewitson M *et al* 2009, *Class. Quantum Grav.* **26** 094003
[13] Röver C 2010 *J. Phys.: Conf. Series* **228** 012008

A double torsion pendulum with two cascade soft degrees of freedom

L Marconi[1], R Stanga[1] and M Bassan[2]
1 Università degli Studi di Firenze and INFN Firenze
2 Università di Roma Tor Vergata and INFN Roma 2
E-mail: marconi@fi.infn.it

Abstract. We report on a double torsion pendulum, where motion along two degrees of freedom (DoFs) is almost free. The Test Mass (TM) is enclosed in a replica of the LISA-Pathfinder electrostatic readout and actuation system. This apparatus is designed to perform extensive ground testing of undesired effects such as leakage of the readout noise from one DoF to another, or actuation cross talks with closed feedback loop. Such investigation is relevant to the noise budget of LISA and LISA-Pathfinder missions, as the TM will be sensitive to weak forces along all 6 degrees of freedom (DoFs). The instrument being developed in Firenze is capable of measuring the forces and stiffnesses acting simultaneously along the 2 soft DoFs. We have completed an upgrade of the apparatus to a definitive configuration and we report on both advances in the commissioning tests and on measurements of residual charge, with the first DoF released.

1. Introduction

One of the main requirements of the LISA mission is the drag-free operation mode, that means isolation of its test masses (TM) from all kind of spurious, i.e. non gravitational, disturbances.

The LISA requirement for the residual acceleration noise in the free-falling frame of the test mass is [1]:

$$S_a^{1/2} \leq 3 \cdot 10^{-15} \left[1 + \left(\frac{f}{0.003Hz} \right)^2 \right] \frac{\mathrm{m}}{\mathrm{s}^2} \frac{1}{\sqrt{\mathrm{Hz}}}, \tag{1}$$

in the frequency band $0.1\mathrm{mHz} \leq f \leq 0.1\mathrm{Hz}$.

Among the perturbing causes, special attention is to be given to the coupling with the electrostatic sensing and actuation system, that could introduce a coupling between test mass and spacecraft. The capacitive sensing and actuation device, called the Gravitational Reference System (GRS) [2], is used to provide the relative position input to the contol loop that is closed on micro-thrusters, used to force the spacecraft to follow the test-mass. Extensive ground tests are required to study the residual weak forces that may couple the test mass to the GRS. The challenge for these tests is to cancel gravity, in order to simulate free fall conditions, along as many DoFs as possible. Up to now, excellent results in the measurement and reduction of residual acceleration have been obtained with torsion pendulums, that only allows to study one rotational single DoF [3], [4] or one transtational single DoF [5]. The purpose of our experiment is to work with a double torsion pendulum that allows the test mass to be "almost free" on two DoFs. With our apparatus we aim to extend the above referenced work and to measure and analyze cross talks between two soft DoFs. A crucial goal of this experiment consists in evaluating the effects on one DoF, when feedback is applied to the second DoF. A similar experimental scheme is now pursued also by other groups [6], [7].

2. The PETER apparatus in Firenze

In the last few years we have built and begun to operate a double pendulum, nicknamed PETER (PEndulum Translational and Rotational). We use two Tungsten torsion fibers to suspend a hollow, cubic test mass (Figure 1): the lower, θ fiber allows the test mass to be *quasi free* in rotation around its symmetry vertical axis. This fiber hangs from the the tip of the arm of a crossbar that is, in turn, suspended to an upper, φ fiber: torsion of this upper fiber allows *quasi free* motion of the test mass along an arc of a 30cm diameter circumference: for all practical purposes, this can be considered a translational motion. Dummy loads hang from the other three arms of the crossbar. The sensitivity goal for this apparatus, limited by the mechanical setup and the electronics noise is $\leq 10^{-13}\mathrm{ms}^{-2}/\mathrm{Hz}^{1/2}$ around 1mHz (on each DoF), namely 1 order of magnitude worse than the LISA PF goal along the sensitivity axis. For further details we refer to [8].

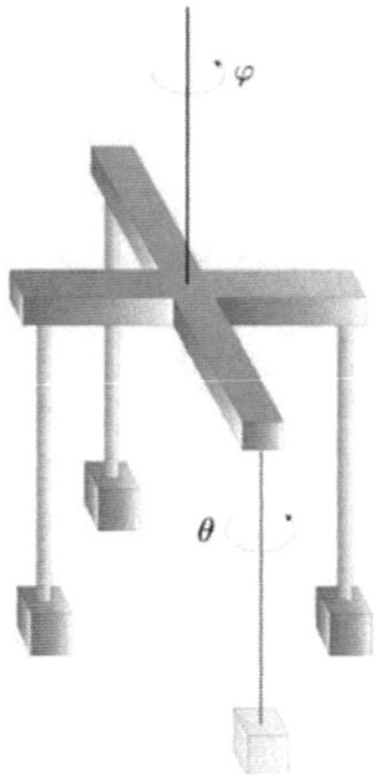

Figure 1. Scheme of the roto-translational pendulum.

During the 2010 operation stop, we installed a new GRS (on loan from Trento University) with capacitive sensors in Mo and SHAPAL (Figure 2) and a new, purposely made, hollow Al TM with 46mm side (Figure 3) . This set-up is a closer replica of the LISA-Pathfinder flight model geometry, and, very important for our set-up, leaves a 4mm gap between the sensors and the test mass, allowing larger amplitudes of initial oscillations without the TM hitting the GRS electrodes. At the same time we substituted our home-developed readout and sensing electronics with a replica of the flight model capacitive read out electronics produced by ETH Zurich. Each channel is read with a capacitive-inductive resonant bridge at 100kHz. The electronics also generates the waveform for the actuation. The apparatus was also equipped with an Optical Read Out (ORO) [9] that will provide an independent measurement of the test mass along the 2 soft DoFs. This is useful both for diagnostic purposes and because cross-correlation of two independent outputs can significantly reduce the readout noise.

Journal of Physics: Conference Series **363** (2012) 012049 doi:10.1088/1742-6596/363/1/012049

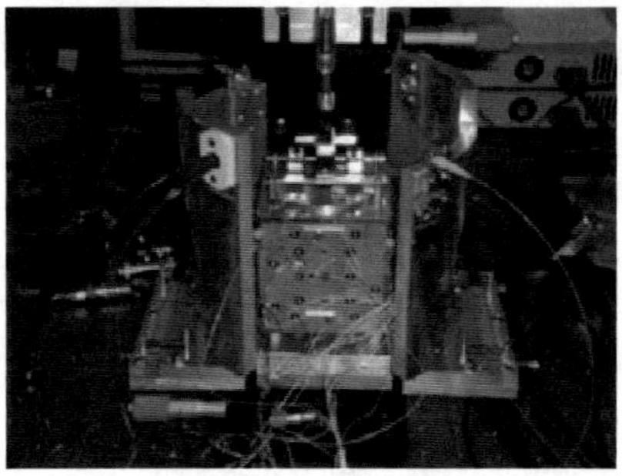

Figure 2. Picture of the GRS and the ORO.

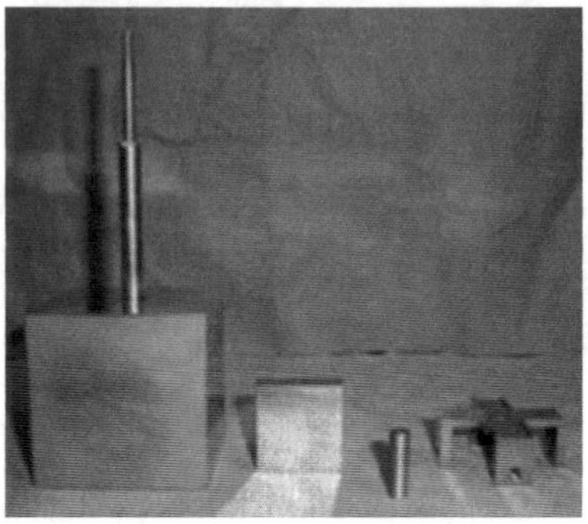

Figure 3. Picture of the 46mm test mass.

2.1. Calibration

The first step in preparing the apparatus consists of fine tuning the input cables capacitance, so that all capacitors resonate at the resonance frequency of the read-out bridge. We then mounted the GRS on a traditional micropositioner and rigidly hanged the TM inside it. By moving the sensor assembly of a known amount of both translation and rotation, we obtained, for both channels, a calibration $51\mu\mathrm{V/nm}$ with a $V_{pump} = 10\mathrm{V}pp$. Given this calibration factor, we verified that the electronics noise is $\approx 1\mathrm{nm/Hz}^{1/2}$, as specified by ETH Zurich.

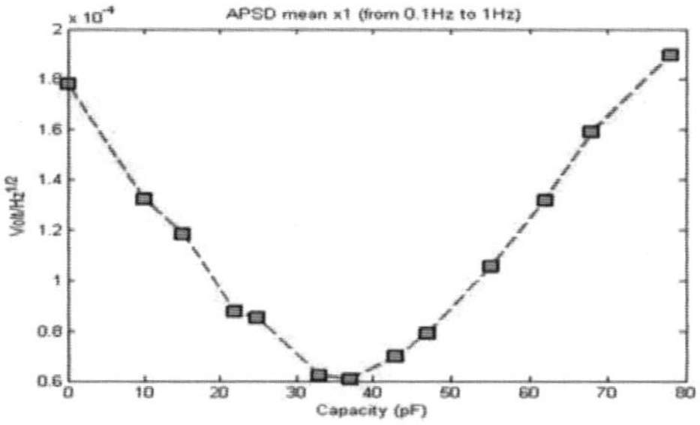

Figure 4. Fine tuning of the input cables capacitance to match the resonance frequency of the read-out bridge: the plot shows how the readout noise is minimized by a proper choice of total capacitance.

3. Present sensitivity

We have so far operated the double pendulum with just the rotation degree of freedom. First, we implemented a feedback control [10], using the electrostatic actuators, to damp the rotational mouvement of the test mass. In the control loop we have also integrated the movement of the motorized stages, needed to center the test mass in the GRS. In Figure 5 we show the amplitude spectral density of the angular displacement. This can be converted in acceleration (amplitude) spectral density (Figure 6) by considering an arm length of 0.02m, i.e. the sensing electrodes separation, and assuming a TM of 2kg, i.e. the mass of the TM of LISA PF. It can be seen that, in the high frequency band, we are limited by the electronics noise while in the low frequencies we are a factor ~ 50 above the intrinsic limit given by thermal noise of the suspension fiber. We are presently investigating the origin of the excess noise.

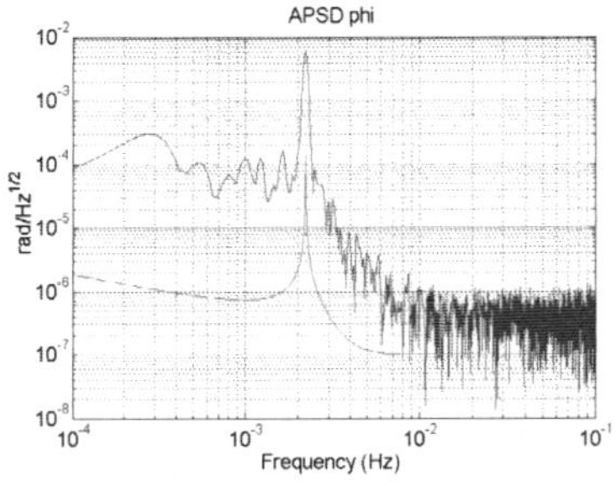

Figure 5. Angular noise spectrum, the instrument limit in red.

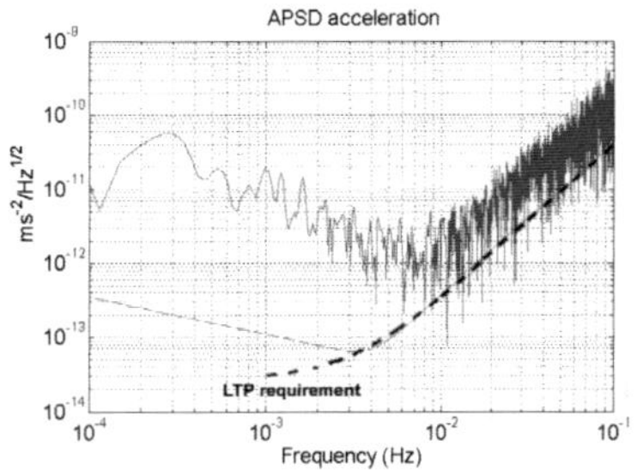

Figure 6. Acceleration noise spectrum, the instrument limit in red. The acceleration has been obtained for an arm length 0.02m and for an equivalent TM of 2kg

4. DC bias measurement

Using our apparatus with only the rotational DoF free, we have performed preliminary measurements for the electrostatics characterization of the GRS. The procedure used here follows closely that outlined in [11]. We measured the electrostatis patch fields on the sensor surfaces that might arise due to contamination from the assembly and from material out-gassing over a long mission [12]. This spurious DC potentials couple to the test mass charge fluctuations and can be a relevant noise source on force for LISA and LISA PF:

$$S_F^{1/2} = \frac{q}{C_{tot}} \left| \frac{\partial C_x}{\partial x} \right| S_{\Delta_x}^{1/2} \approx 1.6\text{fN/Hz}^{1/2} \left(\frac{q}{10^7 e} \right) \left(\frac{S_{\Delta_x}^{1/2}}{100\mu\text{V/Hz}^{1/2}} \right), \tag{2}$$

where q is the accumulated test mass charge, C_{tot} is the total capacitance of the test mass to all sensor surfaces and C_x is the the capacitance of x-electrode to the test mass, Figure 8. Δ_x is a combination of the DC bias distribution inside to the sensor and can be represented as:

$$\Delta_x = \left| \frac{\partial C_x}{\partial x} \right|^{-1} \sum_i \frac{\partial C_i}{\partial x} V_i, \tag{3}$$

where V_i are the spurious potentials of the surfaces inside to the sensor and C_i are the capacitance of each electrode to the test mass [11].

As our apparatus is sensitive to torque, we will consider the derivative with respect to the rotational angle φ rather than the displacement x.

Here we want to show how it is possible to compensate this effect by applying DC voltages to the sensing electrodes, so that the resulting average DC potential is zero. In the electrostatic model considered here, we assume each electrode as having a single uniform potential, neglecting spatial variation [11]: in this case the istantaneous torque N_φ and the test mass potential V_M can be expressed as:

$$N_\varphi = \frac{1}{2} \sum_i \frac{\partial C_i}{\partial \varphi} (V_i - V_M)^2, \tag{4}$$

$$V_M = \frac{q}{C_{tot}} + \frac{1}{C_{tot}} \sum_i C_i V_i. \tag{5}$$

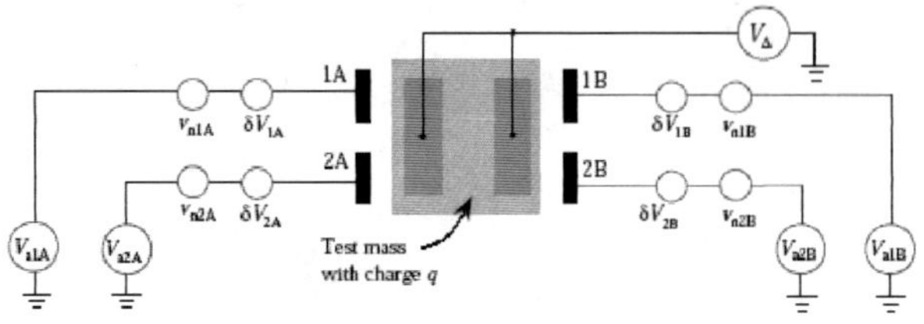

Figure 7. Schematic of the simplified, four electrode electrostatic model and measurement tecnique analyzed here. Stray DC biases, dielectric noise and test mass charge are denoted δV_i, v_{ni} and q. In the main measurement described in the text, a modulated bias V_Δ is applied to the test mass via the z electrodes. Actuation voltages V_{ai} can be applied to compensate δV_i [11].

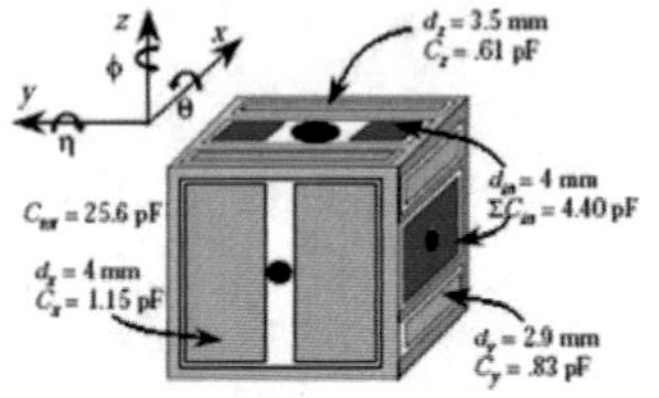

Figure 8. Schematic of proposed capacitive sensor, with capacitance and gap values, adapted from [12]. The injection electrodes used to provide the sensing bias are red, the sensing electrodes are green and grounded guard ring surfaces are light grey [11].

The DC bias effect can be measured by applying a sinusoidal signal to z electrodes, Figure 9: this simulates a modulated test mass charge. The torque produced at the voltage modulation frequency ω_z can be expressed as:

$$(N_\varphi)_{1\omega_z} = -\frac{4C_z}{C_{tot}} V_z \sin(\omega_z t) \left(\frac{\partial C_x}{\partial \varphi}\right) (\Delta_\varphi + 4V_{comp}), \tag{6}$$

where C_z, C_{tot} and C_x are indicate in Figure 8 and Δ_φ is the sum, with suitable signs, of the four sensing electrodes potential:

$$\Delta_\varphi = \delta V_{1B} + \delta V_{2B} - \delta V_{1A} - \delta V_{2A}. \tag{7}$$

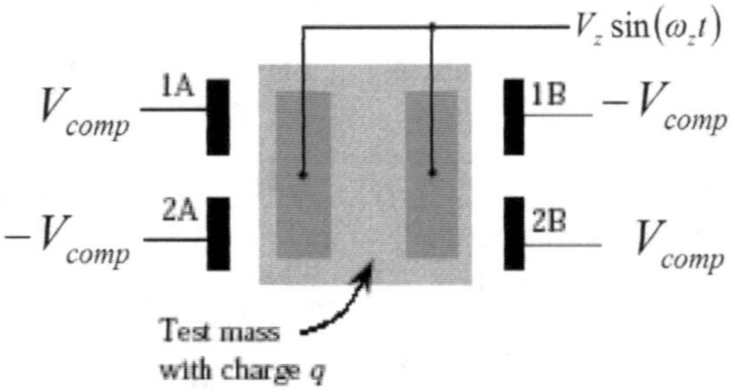

Figure 9. Schematic of the measurement.

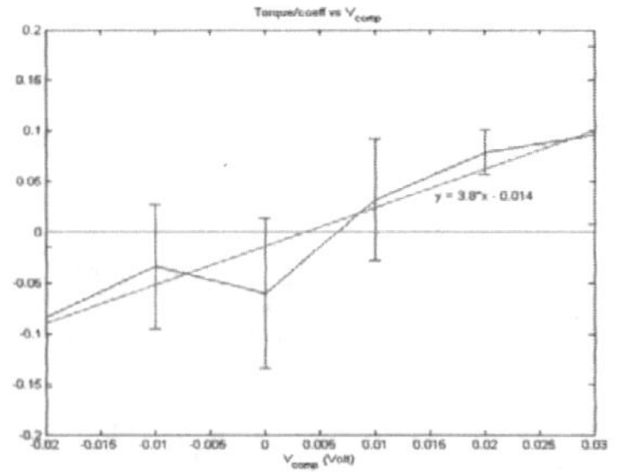

Figure 10. Applied torque/coeff (demodulated at ω_z) vs. V_{comp}, where: $coeff = -\frac{4C_z}{C_{tot}} V_z \sin(\omega_z t) \left(\frac{\partial C_x}{\partial \varphi}\right)$.

We can now scan (Figure 10) the V_{comp} amplitude of the demodulating signal at ω_z and, extracting the coherent torque, measure Δ_φ and the best V_{comp} that compensates it. Using

$V_z = 3\text{V}$ and $\omega_z = 3\text{mHz}$ for the modulation parameters we obtained a $\Delta_\varphi \approx -14\text{mV}$ and consequently:

$$(V_{comp})_{BEST} = -\frac{\Delta_\varphi}{4} \approx 3.5\text{mV}. \tag{8}$$

With the measurement of the V_{comp} needed to cancel the stray DC bias due to spatially varying DC surface potentials, we can also measure the residual charge on the TM. By applying a further modulation signal on the x-electrodes, as indicated in Figure 11, the torque produced at the voltage modulation frequency ω_{MOD} can be written as:

$$(N_\varphi)_{1\omega_{MOD}} = -4V_{MOD}\sin(\omega_{MOD}t)\left(\frac{\partial C_x}{\partial \varphi}\right)\frac{q}{C_{tot}}. \tag{9}$$

Finally, using the modulation parameter $V_{MOD} = 3\text{V}$, $\omega_{MOD} = 3\text{mHz}$ and $V_{comp} = 3.5\text{mV}$, we

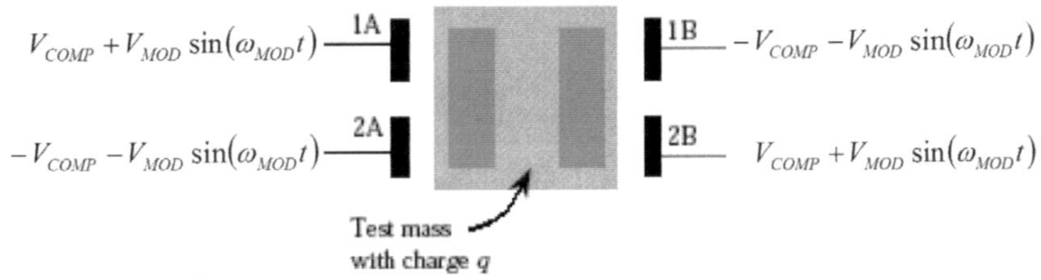

Figure 11. Schematic of the measurement of the residual test mass charge.

obtained a residual test mass charge $q \approx 3.7 \cdot 10^7 e$. This value is comparable to those derived in similar experiments, and shows that the test mass is properly isolated [13].

5. Next steps

We have satisfactorily tested the capabilities of our apparatus, when operated as a single DoF torsion pendulum. In the immediate future we shall release the upper fiber in order to enable the second, translational DoF and will implement a 2 DoF feedback control to keep the TM in a controlled motion. When that is achieved, we will have a test bench to simulate free fall in the laboratory along two degrees of freedom: this will enable us to carry on the desired measurements of cross talks between the 2 soft DoFs.

6. References

[1] Vitale S et al 2002 *Physics Letters* **B 110** 210
[2] Dolesi R et al 2003 *Class. and Quant. Grav.* **20** S99
[3] Hueller M et al 2002 *Class. and Quant. Grav.* **19** 1757
[4] Hoyle C D et al 2003 *Proc. Marcel Grossman Meeting* (Rio de Janeiro, Brazil)
[5] Cavalleri A et al 2009 *Class. and Quant. Grav.* **26** 094012
[6] Zhou Z B et al 2010 *Class. Quantum Grav.* **27** 175012
[7] Tu H B et al 2010 *Class. Quantum Grav.* **27** 205016
[8] Stanga R et al 2009 *J. Phys.: Conf. Series* **154** 012032
[9] Acernese F et al 2005 *Class. Quantum Grav.* **22** S279S285
[10] Marconi L et al 2003 *Proc. NSS/MIC IEEE* (Roma, Italy)
[11] Weber W Jet al 2007 *Advances in Space Research* **39** 213
[12] Weber W J et al 2002 *Proc. of the SPIE Astronomical Telescopes and Instrumentation Conference* **4856** 31
[13] Antonucci F et al 2011 *Class. Quantum Grav.* **28** 094002

Amaldi 9 and NRDA 2011 IOP Publishing
Journal of Physics: Conference Series **363** (2012) 012050 doi:10.1088/1742-6596/363/1/012050

LISA PathFinder radiation monitor proton irradiation test results

I. Mateos[1], M. Diaz-Aguiló[3], F. Gibert[1], C. Grimani[5], D. Hollington[4], I. Lloro[1], A. Lobo[1], M. Nofrarias[1] and J. Ramos-Castro[2]

[1] Institut de Ciències de l'Espai (CSIC-IEEC),Barcelona, Spain
[2] Departament d'Enginyeria Electrònica, Universitat Politècnica de Catalunya (UPC), Barcelona, Spain
[3]Departament de Física Aplicada, Universitat Politècnica de Catalunya (UPC), Castelldefels, Spain
[4] Department of Physics, Imperial College London, London, UK
[5] Università degli Studi di Urbino and Istituto Nazionale di Fisica Nucleare, Florence, Italy

E-mail: mateos@ice.csic.es

Abstract. The design of the Radiation Monitor in the LISA Technology Package on board LISA Pathnder is based on two silicon PIN diodes, placed parallel to each other in a telescopic configuration. One of them is able to record spectral information of the particle hitting the diode. A test campaign for the flight model Radiation Monitor was done in the Paul Scherrer Institute Proton Irradiation Facility in September 2010. Its purpose was to check correct functionality of the Radiation Monitor under real high energy proton fluxes. Here we present the results of the experiments done and their assessment by means of a simulated flight model geometry using GEANT4 toolkit. No deviation from nominal RM performance was detected, which means the instrument is fully ready for flight.

1. Introduction

LISA Pathfinder (LPF) [1] is a technologically sophisticated mission to demonstrate that two proof masses (TMs) can be put into free-fall to a certain level of accuracy. This idea is reflected in the differential acceleration noise requirements for the LTP

$$S^{1/2}_{\delta a,\mathrm{LPF}}(\omega) \leq 3\times 10^{-14}\left[1+\left(\frac{\omega/2\pi}{3\ \mathrm{mHz}}\right)^2\right]\ \mathrm{ms^{-2}Hz^{-1/2}}$$

in the frequency band between 1 mHz and 30 mHz. This noise is the result of various disturbances which limit the performance of the instrumentation on-board.

LPF will be stationed for operations in a Lissajous orbit around the Earth-Sun Lagrange point L1. This is about 1.5 million kilometers from Earth, i.e. well beyond its radiation belts. There, the spacecraft will be exposed to various ionising radiations coming from the Galaxy and from the Sun, so that, solar and galactic proton and helium particles will charge the onboard test masses (TMs) generating a significant source of noise for the experiments [3] [4]. The charge management device (CMD) will be used to discharge the TMs by means of a system of ultraviolet lamps to ensure they remain in operational modes. However, the CMD can only provide charging average information over long periods of time (several hours), which is insufficient to address TM charging noise in the LTP MBW. In order to better estimate the latter, LISA PathFinder will contain a Radiation Monitor (RM), which will allow the relationship between the incident radiation and test mass charging to be studied in an environment that will be similar to that of LISA, a space-based gravitational wave detector with the main goal of observing astrophysical and cosmological sources of low frequency gravitational waves. The experience gained during LISA Pathfinder mission will provide important keys to optimize particle detectors and TM monitoring for LISA, as well as will be of interest to both the solar physics and the cosmic-ray communities. It should be noted that the charging disturbances tend to increase with decreasing frequency and hence are more significant for LISA than for LPF. However, the aim of LPF on this topic is to investigate charging and charge management for LISA. For detailed information of the RM aims during the mission, the reader is referred to [5].

The RM is based on two silicon PIN diodes placed parallel to each other in a telescopic configuration, housed within a thick copper shield. The signal of the diodes is processed by a suitable electronics, which is able to identify single-particles and determine the energy deposition spectrum of an incoming particle flux when events are detected in coincidence in both PIN diodes [6]. The expected deposited energy in the diodes (with added Gaussian noise to mimic the electronic noise) is known by simulation.

Before it can be flown on-board LISA Pathfinder, it is necessary to test the flight model (FM) RM so as to verify its performance [7]. Here we describe the verification tests under real high energy proton fluxes that were performed in September 2010 at the Paul Scherrer Institute (PSI) in Switzerland. Along with the tests themselves, a GEANT4 simulation was also created to further understand the RM and its properties. Part of this work was carried out earlier [8], however, an update was needed due to the redesign of the proton beam facility in the PSI.

The experiments were split into three parts. The first (sections 4 and 5) involved irradiating the RM at various proton energies and at various incident angles in order to check the effectiveness of the copper shield and the angular acceptance of the diodes telescopic configuration. The second part (section 6) involved irradiating the RM at various incident proton energies in order to test that the RM was recording the deposited energy spectrum as expected, and in the third part (section 7) the RM was irradiated with a fixed proton energy but with increasing flux to determine the maximum count rate. The test was carried out to verify the performance of the FM RM, whose results should assess whether or not:

(i) The shield thickness surrounding the RM diodes is sufficient to stop soft protons with energy less than 70 MeV.

(ii) The dynamic range of the FEE+ADC is set properly to accommodate the smallest (50 keV) and the largest (5 MeV) energy deposition in the Si PIN diodes.

(iii) Angular acceptance for coincidence detection complies with the expected value ($\leq 10\,\%$ of 4π solid angle).

(iv) Each Si PIN diode, with its FEE, is able to handle 5 kHz triggers. When the two PIN diodes are triggered in coincidence, the ADC can handle a rate of 500 Hz.

2. Test Setup

The initial proton beam for proton irradiation facility (PIF) is delivered from the PROSCAN accelerator (590 MeV). The beam passes through a set of exchangeable copper-graphite blocks (primary degrader), which allows to set the initial beam energy to 250 MeV and an approximately Gaussian energy profile. The beam is subsequently guided to the experimental area where the PIF is located. Figures 1 and 2 show different parts of the setup configuration in the facility. The PIF experimental set-up consists of the local energy degrader, beam collimating and monitoring devices. The FM RM was placed in the target area, mounted on a rotary table used to set the orientation relative to the beam. After the RM had been positioned and carefully aligned the testing room was sealed and the beam settings and rotary table were controlled remotely from a separate control room.

Figure 1. Beam test setup.

Figure 2. PIF view with ionization chamber, energy degrader and wire chamber (Left). FM RM placed on the rotary table (Right).

3. Simulations

The RM and beam line were simulated using the GEANT4 toolkit. The model geometry simulates the beam degrader plates, wire and ionisation chambers, collimator, the aluminium box housing of the FM RM, the shield, a detailed model of the two silicon diodes, the two PCBs where the electronic circuitry is placed and the base plate of the rotary table. Figures 3 and 4 show the model geometry simulated with GEANT4.

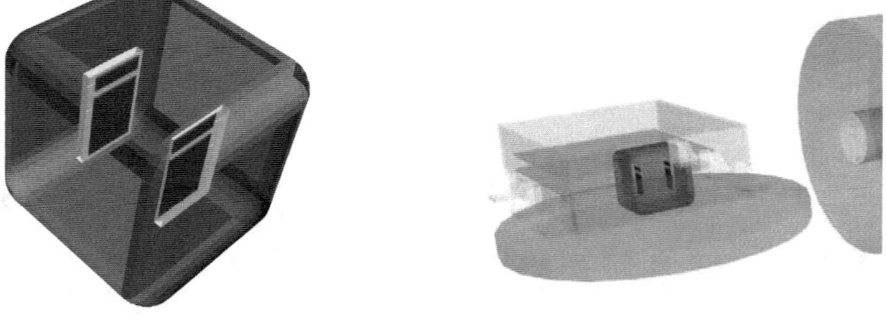

Figure 3. Left: GEANT4 geometry showing the PIN diodes inside the FM RM shield. Right: GEANT4 clear volume of the FM RM showing shield, PIN diodes and PCBs.

The diodes were modeled in greater detail than the rest of the simulation geometry in order to achieve as accurate results as possible —see figure 3. The full package is modeled including silicon regions, aluminium light shields, silicone resin windows and ceramic package. For the PIN diodes geometry, it was noticed during the data analysis that the sensitive silicon regions do not lie exactly at the centre of the ceramic substrate and are in fact offset by 0.5 mm [9]. As it is the ceramic substrates that are aligned inside the copper shield, and both diodes face each other, the problem is enhanced so that the sensitive regions are actually offset by 1.0 mm with respect to each other. Considering the sensitive regions are only 10.5 mm by 14.0 mm this leads to a subtle but measurable effect, which was included in the simulation.

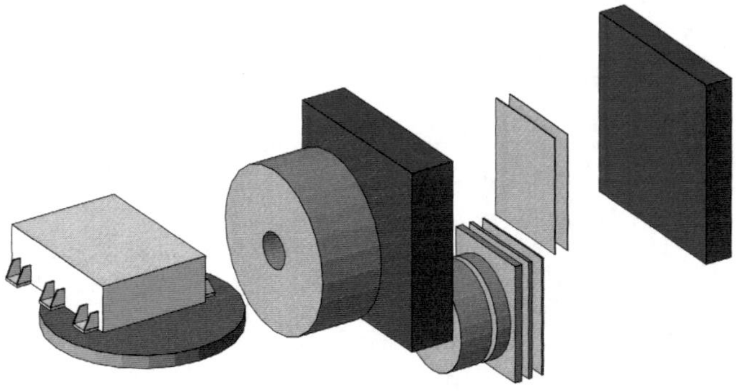

Figure 4. Beam test simulation setup in GEANT4 showing the FM RM aluminium box. Exchangeable degraders to select beam energies, ionisation/wire chamber, collimator and RT's base are also shown.

Several values of the experimental setup considered in the simulation are shown in the list below:

- Initial mean energy of the Beam: 250 MeV with an energy spread of 0.25 MeV.
- Thickness of the diode: 320 μm [9].

- Nominal area of the PIN Diode: 147 mm^2
- A FWHM Gaussian profile of 40 keV to mimic the effect of the electronic noise.

3.1. Beam simulation

The beam is simulated as a parallel beam with a Gaussian energy profile and a flat radial profile with a diameter of 42 mm (wide enough to irradiate the diode uniformly).

The simulation results of the beam energy profiles after it passes through the set of movable degraders, are shown in figure 5, whose degradation was simulated in an attempt to make the calculation as representative as possible. Based on these simulations, the properties of the energy spread of the beam for each energy setting ($\overline{E}_{\text{simulated}}$) are given in table 1, as well as the different degraders used and their thickness. The maximum error between the measured value of the beam and its simulated value is $\simeq 2\%$.

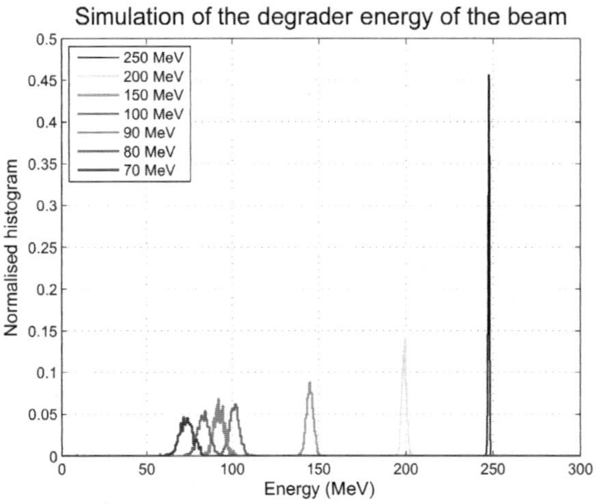

Figure 5. Simulation of the energy profiles of the final beam between 70 MeV and 250 MeV.

Table 1. Beam energy properties for each degrader settings used in the simulation. Ticks indicate the degraders necessary to obtain the corresponding energies.

E_{nominal} MeV	$\overline{E}_{\text{simulated}}$ MeV	σ MeV	Plate 1 0.5 mm	Plate 2 1 mm	Plate 3 2 mm	Plate 4 4 mm	Plate 5 8 mm	Plate 6 16 mm	Plate 7 38 mm
70	72.38	4.04			✓			✓	✓
80	82.54	3.54						✓	✓
90	92.07	3.25			✓	✓	✓		✓
100	100.8	2.81				✓	✓		✓
150	144.5	2.1							✓
200	198.8	1.39	✓	✓	✓			✓	
250	247	0.38							

4. Low energy cut-off of the RM shield

The shield design is intended to stop soft-energy charged nuclei before they reach the PIN diodes in the RM. The performance of the shield is assessed as a function of the energy of the proton beam as well as its angle of incidence on the shield.

Results considering the ratio obtained from the number of coincident detections over the total number of detections recorded in the RM show that, as expected, the energy cutoff below which protons cannot penetrate the RM shield is approximately 70 MeV. The Count ratios measured by the RM as a function of the proton energy for angles sweeping from 0 to 40 degrees are shown in figure 6. The simulated results show good agreement with the measured data, particularly around 90 MeV and higher energies.

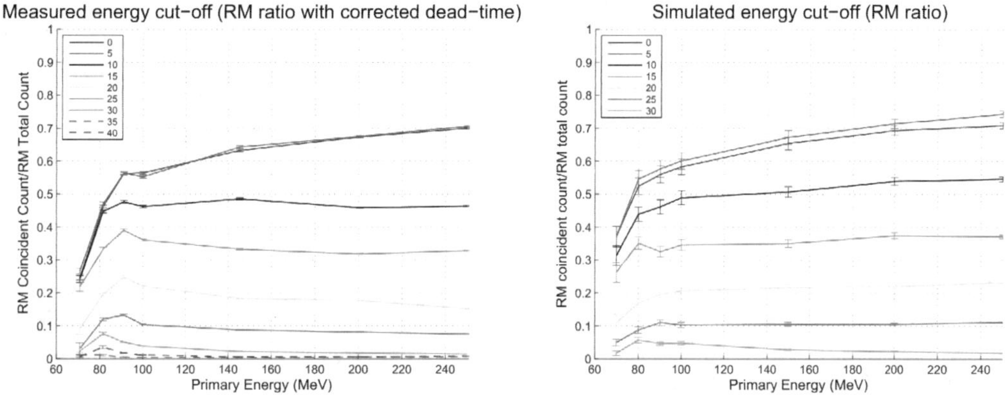

Figure 6. Measured (left) and simulated (right) low energy cut-off of the RM shield for different angles of incidence. A dead time of $\tau = 35\,\mu s$ is considered for the FM RM (non-paralyzable dead time). Note that the red curve (5 degrees) slightly overshoots the blue one (perpendicular incidence) at 150 MeV, it is caused by the small effect due to the linear offset of the PIN diodes, as previously mentioned in section 3.

5. Angular acceptance of the RM

Simulations show that the charging rate of the TMs depends on whether the primary particles are galactic cosmic rays (GCR) or correspond to flares in the Sun (SEP events, solar energetic particles). These two types of flux have different energy spectra, and the way to distinguish between them is therefore to do spectroscopy [2]. Although GCR fluxes are isotropic, SEP events may be directional, therefore, it is important to know the angular acceptance of the RM, particularly for coincidence detection which will be used for spectral measurements. The geometrical angular acceptance of the FM RM was calculated by a directional flux of protons. By plotting the count rate against RM angle we can determine the efficiency of the telescopic configuration of the diodes.

From purely geometrical considerations, the angular cut-off for the diodes telescopic configuration should be $\theta_{CE} = \arctan\frac{10.5}{20} = 27.7$ degrees. Figure 7 shows good agreement between measured and simulated data, particularly at higher energies, considering the ratio obtained from the number of coincident detections over the total number of detections recorded in the RM. The measured results suggest an angular acceptance of $\simeq 30°$.

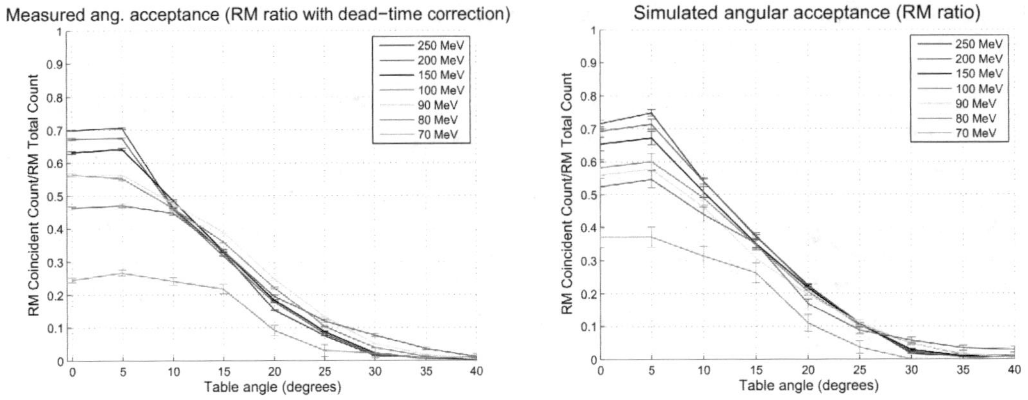

Figure 7. Measured and simulated angle acceptance for coincident to total counts ratio as a function of FM RM orientation (Y axis rotation) considering the FM RM dead time.

6. Measured energy deposited in PIN diodes

In order to ensure the FM RM measures the deposited energy spectra correctly, different runs were performed with nominal beam energies between 70 MeV and 250 MeV at perpendicular incidence. Using the parameters described in section 3, good agreement between the measured data and simulation was found, as can be seen in figure 8.

Experimental measurements show the energy deposited by the proton beam in the PIN diodes: low energy protons deposit higher energies than high energy protons, while the latter have a smaller spread in deposited energy. At 60 MeV almost no protons penetrate the copper shield and hence do not reach the diodes. Also, oblique beam incidence on the FM RM causes the distribution peaks to shift towards higher energies, as this increases the path-length of the protons inside the PIN diodes.

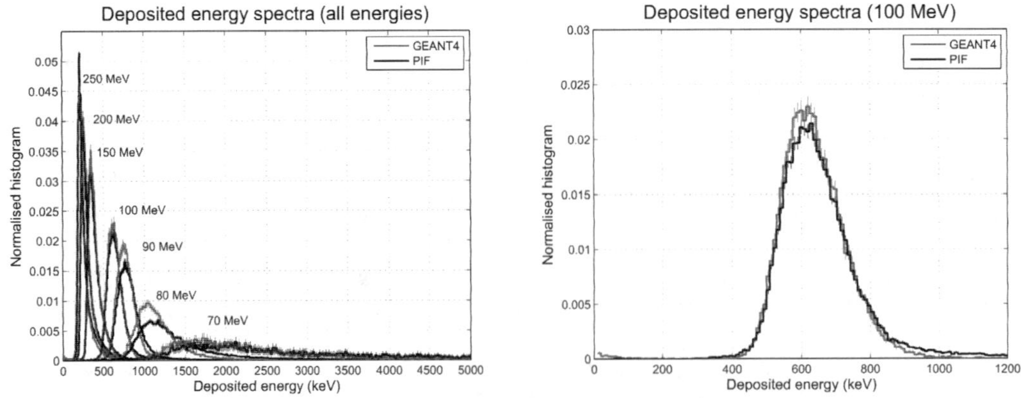

Figure 8. Comparison between simulation and real data in the FM RM for all incident energies (left) and a more detailed plot for incident protons with 100 MeV (right).

Journal of Physics: Conference Series **363** (2012) 012050 doi:10.1088/1742-6596/363/1/012050

7. Count Rate

In order to verify the maximum count rate at which the RM can operate, a series of measurements at a nominal energy of 100 MeV and with perpendicular incident beam were taken while increasing the beam flux from $\simeq$ 500 protons/s/cm^2 to $\simeq$ 5700 protons/s/cm^2 (according to the plastic scintillator data). By plotting the count rates measured by the RM against the flux measured by the plastic scintillator, we can set an upper limit for the maximum count rate based on when the linear relationship between the two breaks down. Figure 9 shows this relationship and suggests a good linear agreement above 5000 protons/s/cm^2 in the plastic scintillator equating to around 6300 protons/s total counts in the RM. In space, even the most violent SEP events are expected to produce count rates less that $\simeq$ 2000 protons/s/cm^2 in the RM while the GCR background less than $\simeq$ 20 protons/s/cm^2.

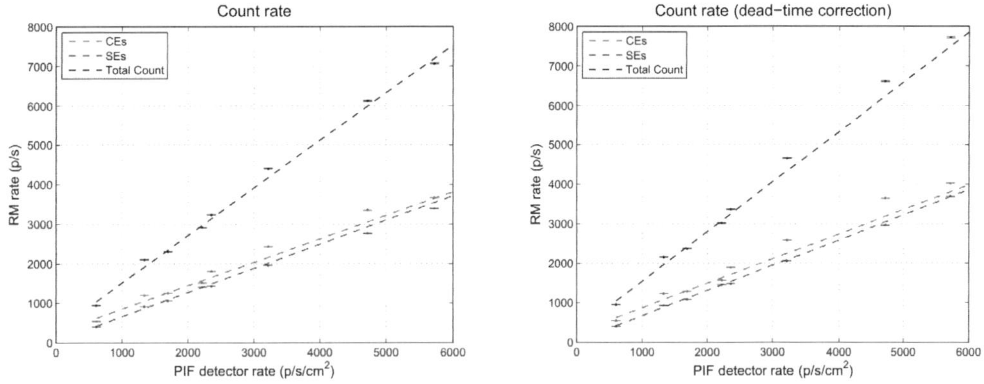

Figure 9. Demonstrates the linear relationship between the PIF scintillator flux measurements and the RM count rate at fluxes less than about 5000 protons/s/cm^2. Above this, dead time caused by the RM electronics begins to have a significant effect. Right figure shows the corrected count rates if one assumes a non-paralyzable dead time $\tau = 35\,\mu$s. By correcting for this effect the count rate continues to be linear up to 6000 protons/s/cm^2 and above .

A possible systematic error in the beam flux measurements by the plastic scintillator could also introduce a discrepancy between results.

8. Conclusions

The test results of the FM RM can be summarised as follows:

- Simulations and experimental data from the FM RM are in good agreement across all incident energies and angles.
- The shield surrounding the PIN diodes prevents the detection of particles with energy below $\sim$ 70 MeV, complying with the requirement.
- According to expectations, the acceptance angle of the coincident mode of the FM RM is $\sim$ 30 degrees of the rotary table for the diodes telescopic configuration.
- The ADC dynamic range of the deposited energy measurements (0 to 5 MeV) is sufficient to capture the entire spectra for incident protons down to 70 MeV.
- The maximum count rate before dead time in the electronics becomes significant is found to be $\simeq$ 6300 protons/s total counts in the RM.
- Measurements under real high energy proton fluxes have confirmed the correct functionality of the FM RM.

Acknowledgments
We acknowledge support from Project ESP2007-61712 of Plan Nacional del Espacio of the Spanish Ministry of Science and Innovation (MICINN).

References

[1] Anza S *et al* 2005 *Class. Quantum Grav.* **22** 125
[2] Grimani C *et al* 2010 *Class. Quantum Grav.* **28** 10
[3] Araújo H M, Wass P, Shaul D, Rochester G and Summer T J 2005 *Astropart. Phys.* **22** 451-69
[4] Sumner T J *et al* 2009 *Class. Quantum Grav.* **26** 9
[5] Shaul D, Sumner T and Lobo A 2008 *Analysis of Data from the Radiation Monitor on LISA Pathfinder*, report S2-ICL-TN-3019
[6] Cañizares P *et al* 2009 *Class. Quantum Grav.* **26** 11
[7] Mateos I and Lobo A 2011 *FM Radiation Monitor Test Report at PSI*, report S2-IEC-TR-3118
[8] Mateos I *et al* 2010 *J. Phys.: Conf. Series* **228** 012039
[9] Aude M 2002 *Hamamatsu photodiodes (DPD-S8576) evaluation report (GLAST Project)*, report E01P1435

Temperature coefficient improvement for low noise magnetic measurements in LISA

I. Mateos[1], M. Diaz-Aguiló[1], F. Gibert[1], I. Lloro[1], A. Lobo[1], M. Nofrarias[1] and J. Ramos-Castro[1,2]

[1] Institut de Ciències de l'Espai (CSIC-IEEC), Barcelona, Spain
[2] Departament d'Enginyeria Electrònica, Universitat Politècnica de Catalunya, Barcelona, Spain

E-mail: mateos@ice.csic.es

Abstract. Previous research with Anisotropic Magnetoresistive sensors (AMR) have shown significant improvements for weak magnetic field applications using dedicated noise reduction techniques in the signal conditioning circuit. However, an important source of error that must be addressed is the thermal dependence of the sensor system, more significant in the AMR sensitivity. The external temperature fluctuations affect the output of the sensors due to the temperature coefficient of the magnetoresistors, which may cause an increase of the estimation of the noise spectral density at low frequencies. Ongoing research using a low noise/low temperature coefficient current source to supply the sensor's bridge enhances the thermal performance of the sensors at the lower end of the LISA bandwidth. Preliminary results are shown in this paper.

1. Introduction

LISA (Laser Interferometer Space Antenna) is planned to be the first space-based gravitational wave detector with the main goal of detecting and observing gravitational waves within the frequency range between 0.1 mHz to 0.1 Hz. The required acceleration noise for the LISA mission is set at:

$$S^{1/2}_{\delta a,\mathrm{LISA}}(\omega) \leq 3\times 10^{-15} \left\{ \left[1 + \left(\frac{\omega/2\pi}{8\ \mathrm{mHz}} \right)^4 \right] \left[1 + \left(\frac{0.1\ \mathrm{mHz}}{\omega/2\pi} \right) \right] \right\}^{\frac{1}{2}} \frac{\mathrm{ms}^{-2}}{\sqrt{\mathrm{Hz}}} \tag{1}$$

in the frequency band between $0.1\,\mathrm{mHz} \leq \omega/2\pi \leq 100\,\mathrm{mHz}$. For LISA *Pathfinder*, a technology precursor of LISA, the acceleration noise budget has been relaxed ($30\,\mathrm{fms}^{-2}\,\mathrm{Hz}^{-1/2}$) in the frequency band between $1\,\mathrm{mHz} \leq \omega/2\pi \leq 30\,\mathrm{mHz}$ [1]. These noises are the result of various disturbances which limit the performance of the instrumentation onboard. Magnetic disturbances are foreseen to contribute a significant fraction of the total acceleration budget, which takes place as a consequence of the magnetic properties of the test masses (TMs).

The onboard LISA *Pathfinder* magnetometers are fluxgate technology, which were chosen on grounds of their excellent sensitivity, low noise and availability for space applications [2]. Nevertheless, a number of drawbacks have been later identified which point to the need of a different technology more suitable for LISA [5].

We are currently investigating alternative magnetic sensing techniques for LISA, based on magnetoresistive sensors [4] where previous research has shown significant improvements for weak magnetic field applications using dedicated noise reduction techniques in the signal conditioning circuit [6].

An important issue to take into account with respect to the magnetic sensor is its temperature coefficient (TC). The main concern comes from the thermal dependence of the sensor's bridge sensitivity, which may cause an increase of the noise spectral density in the bandwidth of interest. This paper describes a preliminary research and results to improve the thermal performance of the sensor at submilli-Hertz frequencies. In sections 2, 3 and 4 a brief description of the electronic chain is shown. In section 5 the noise results and TC obtained for different configurations are presented, and section 6 summarises the main conclusions and further work.

2. Influence of the temperature coefficient of the magnetoresistor

Temperature fluctuations of the resistors forming the Wheatstone bridge show up as an error in the measurements; assuming the worst-case condition, the thermal sensitivity of the bridge output voltage (in $\mathrm{V\,K^{-1}}$) is

$$\alpha_{\mathrm{b}}(T) = 2V_{\mathrm{b}}\left[\frac{R_1R_2}{(R_1+R_2)^2} + \frac{R_3R_4}{(R_3+R_4)^2}\right]\cdot\alpha_{\mathrm{R}} \tag{2}$$

where V_{b} is the bridge voltage, $R_{i=1,2,3,4}$ are the magnetoresistances shown in Figure 3, and α_{R} its TC, whose maximum permitted value in the desired bandwidth can be calculated by

$$\frac{\alpha_{\mathrm{b}}(T)}{s_{\mathrm{b,\ AMR}}(B)}\cdot S^{1/2}_{T,\ \mathrm{FEE}}(\omega) \le S^{1/2}_{B,\ \mathrm{system}}(\omega) \tag{3}$$

The TC of the magnetoresistor needed for the magnetic stability requirements[1] depends on the temperature spectral density of the Front-End Electronics. Assuming $S^{1/2}_{T,\ \mathrm{FEE}} \le 0.1\,\mathrm{K\,Hz^{-1/2}}$, comfortably met in the satellite, this leads to maximum values of $\alpha_{\mathrm{b}} \le 14\,\mu\mathrm{V/K}\,(100\,\mathrm{nT/K})$ and $\alpha_{\mathrm{R}} \le 1.6\,\mathrm{ppm/K}$. In view of these demanding values, research on dedicated techniques in the signal conditioning circuit to reduce the thermal dependence of the magnetometer is necessary. A common method is to supply the sensors' bridge by means of a low noise/low TC current source, as explained in the next sections.

3. Low noise/Low temperature coefficient current source

The floating load current source topology has been the preferred option because of its lower noise in the very low frequency range [5]. The theoretically expected total noise for the current source is calculated as

$$S_{I_{\mathrm{o}}}(\omega) = S_{I_{\mathrm{o}},noise}(\omega) + \alpha^2_{\mathrm{theoretical}}S_T(\omega) \tag{4}$$

where $S_{I_{\mathrm{o}},noise}(\omega)$ is the current noise density excluding random temperature fluctuations and $\alpha_{\mathrm{theoretical}}$ is the global TC of the current source.

Figure 1 shows the equivalent circuit for noise analysis purposes. Noise contributions due to the operational amplifier in the model are:

- Equivalent noise sources: e_n, i_{n1}, i_{n2}
- Thermal drift caused by the *input bias current* I_{b}, *input offset current* I_{os} and *input offset voltage* V_{os}

[1] The magnetic stability required for LISA is still not formally defined, but it will likely be one order of magnitude more demanding in frequency than in LISA Pathfinder, i.e. $S^{1/2}_{B,\mathrm{system}} \le 10\,\mathrm{nT\,Hz^{-1/2}}$ at 0.1 mHz.

The latter introduce a negligible effect, which is not so critical in terms of temperature dependence, specially if the temperature of the environment is stable in the MBW ($\leq 1\,\mathrm{K\,Hz^{-1/2}}$). Getting into numbers, the TC due to I_b, I_{os}, V_{os} and their thermal dependences is $0.1\,\mathrm{nA\,K^{-1}}$.

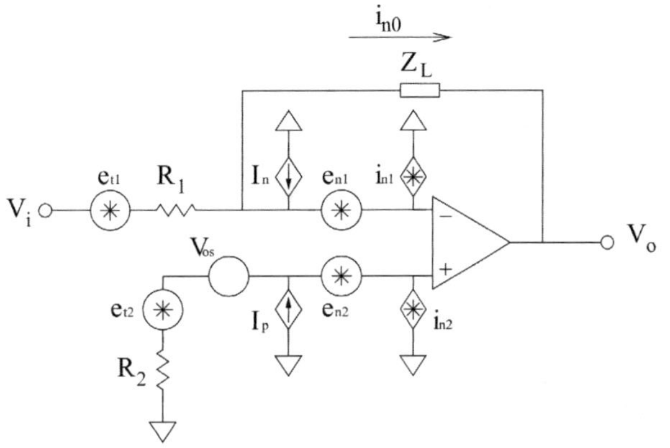

Figure 1: Noise model of the floating current source, which uses the load itself as the feedback element.

To sum up, using high stability resistors ($0.6\,\mathrm{ppm\,K^{-1}}$), the current source has a theoretical $\alpha_{\mathrm{global}} = 2.4\,\mathrm{nA\,K^{-1}}$ (assuming a worst-case condition). Combining thermal and noise model terms, equation 4 gives the overall equivalent current noise density of the bridge, which yields $\sim 25\,\mathrm{nA\,Hz^{-1/2}}$ at 0.1 mHz. Putting this number in context, the current noise of the designed *floating load* source is more than one order of magnitude lower than the *Howland* current source used in LISA Pathfinder to generate a controlled magnetic field by onboard induction coils [3]. Figure 2 shows theoretical and measured amplitude spectral noise densities of the current source.

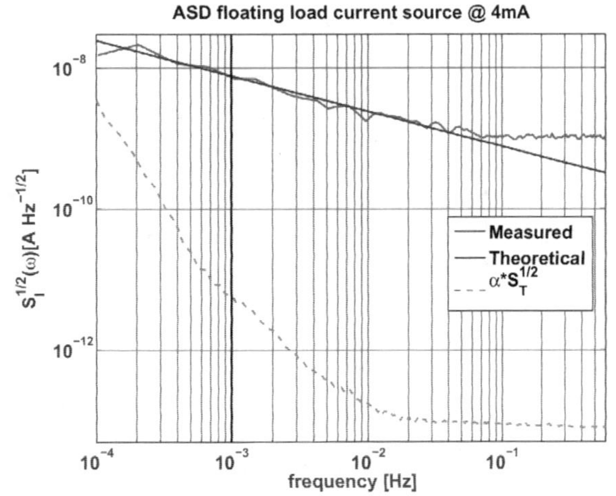

Figure 2: Current noise measurement for *floating load* source is compared with the theoretical prediction considering the measured temperature fluctuations, showing good agreement between both. Discrepancy between prediction and measurements above 100mHz is due to the noise spectral density of the ADC. The green line shows the small thermal contribution to the noise.

4. Front-End Electronics for AMR sensor

The analog signal conditioning for the magnetic field sensing is shown in Figure 3. The signal output from the bridge is modulated using a flipping technique [6], amplified and low-pass filtered. Once the signal has been sampled by the ADC, the demodulating process of the signal $d[n]$ is done digitally.

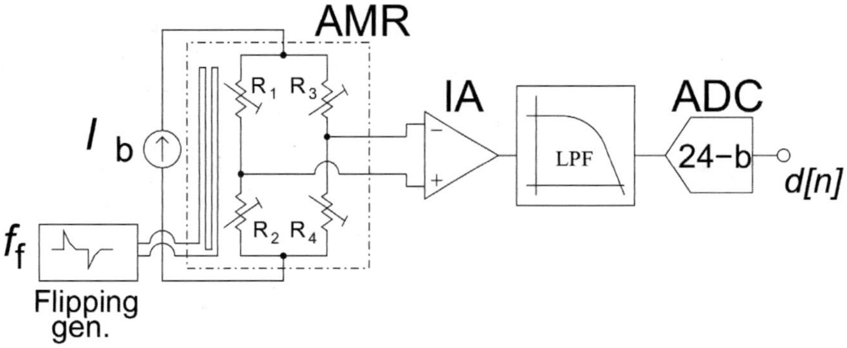

Figure 3: Analog signal processing scheme with flipping technique. For further details see [6].

In order to reduce the temperature dependence of the magnetometer response, the Wheatstone bridge circuit is powered with the *floating load* current source as explained in the previous section. The input voltage of the current source (V_i in Figure 1) is the analog-to-digital converter (ADC) voltage reference, permitting a voltage reference-independent conversion (*ratiometric conversion*) [7].

5. Experimental test results

5.1. Equivalent Magnetic field noise spectral density

Magnetic measurements have been done in suitable conditions of magnetic shield, placing the sensor inside a small Mu-metal chamber. Constant current source and voltage source for bridge supply have been compared. The results in Figure 4 show a significant decrease of the noise in the LISA Bandwidth when the AMR's bridge is powered with the designed current source.

Ambient temperature noise is about $2\,\mathrm{K\,Hz^{-1/2}}$ at 0.1 mHz for both topologies (voltage and current source), and the voltage noise is reduced from $\sim 1.3\,\mu\mathrm{V\,Hz^{-1/2}}$ to $\sim 0.5\,\mu\mathrm{V\,Hz^{-1/2}}$. Therefore, we can infer that using a voltage source, the excess noise is due to the greater thermal dependence of the AMR, which appears as error voltage fluctuations in the measurement. The value obtained in terms of equivalent magnetic field is about $4\,\mathrm{nT\,Hz^{-1/2}}$ at 0.1 mHz with a current source.

5.2. Noise and temperature coefficient in the presence of a static magnetic field

The test consisted in determining the experimental noise model and TC of the system under a constant magnetic field ($\sim 21\,\mu$T), which is applied by a coil inside the Mu-metal shielding. The purpose is to unbalance the Wheatstone bridge, where the error due to the thermal dependence of the AMR's sensitivity is greater. Test setup is shown in Figure 5.

Magnetic fluctuations created by the signal injection to the coil must be at least one order of magnitude less noisy than the required magnetic noise of the system, i.e $S_B^{1/2} \leq 1\,\mathrm{nT\,Hz^{-1/2}}$ within the MBW, which is mostly a requirement on the coil's current. Considering a coil at a certain distance x, the coil's magnetic field generated by a current is estimated with a simple

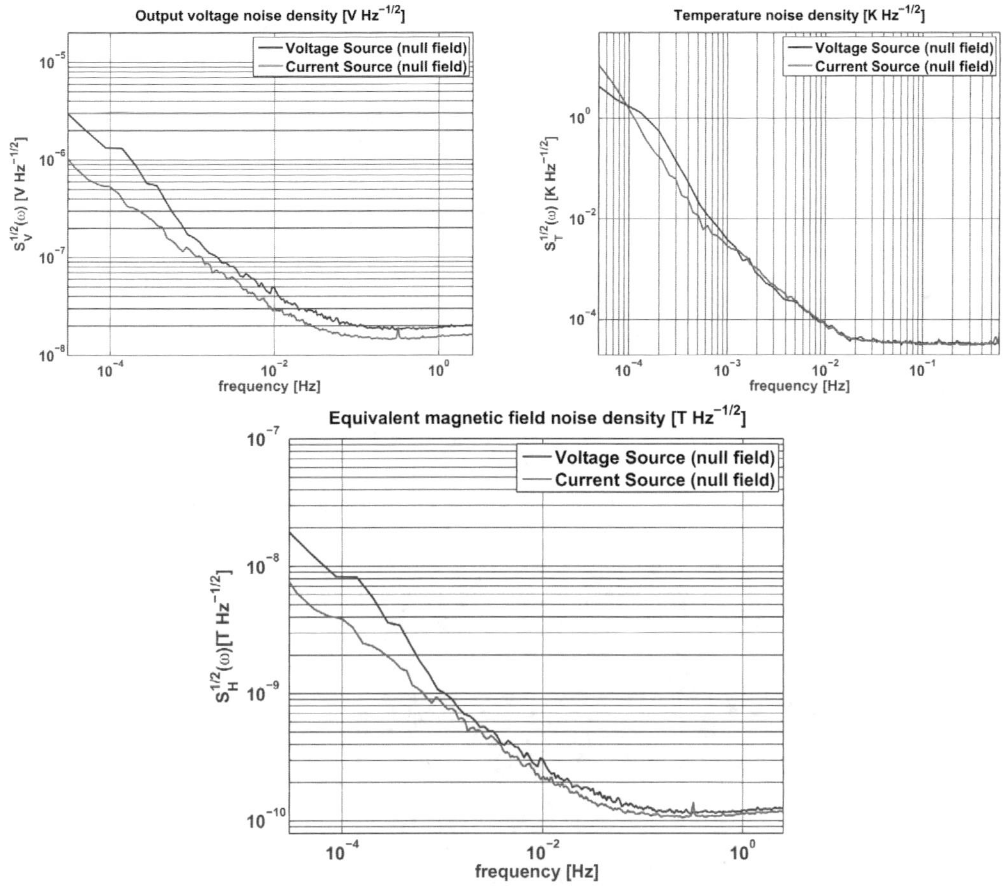

Figure 4: Spectral density in terms of voltage, temperature and magnetic field using voltage and current sources to supply the AMR's Wheatstone bridge.

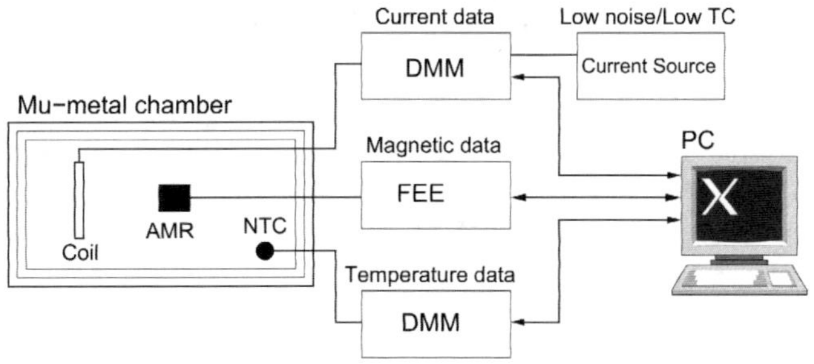

Figure 5: Setup for magnetic noise test with low fluctuation external magnetic field.

calculation based on Ampere's Law. Thus the stability of the current injected by the source is given by:

$$S_I^{1/2} \leq S_{B_x}^{1/2} \frac{2}{\mu_0 N} \frac{(r^2 + x^2)^{3/2}}{r^2} \tag{5}$$

Table 1: Comparison of the results at 0.1 mHz using voltage and current sources. A significant improvement of the TC is achieved with the current source topology.

Bridge Source	$S_B^{1/2}$ [nT/$\sqrt{\text{Hz}}$]	$S_T^{1/2}$ [K/$\sqrt{\text{Hz}}$]	TC [nT/K]	TC [%/K]
Voltage	150	2.5	60	0.3
Current	33	2	16.5	0.08

Current stability obtained with the *floating load* source of section 3 at 0.1 mHz is perfectly compatible with the needed requirements for the test. Figure 6 shows the map of the estimated magnetic field noise ,which has axial symmetry about the x-axis. From this Figure, we can infer the distance needed between the coil and the sensor, so that the applied magnetic field does not exceed the required noise level.

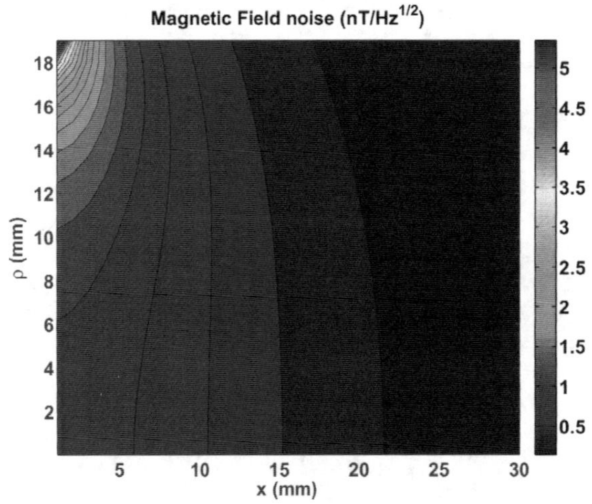

Figure 6: Magnetic noise applied by the coil at 0.1 mHz using the current source explained in the text. The sensor is placed at $\sim$ 15 mm from the coil along the axis. x and ρ are the distances to the centre and to the longitudinal axis of the coil, respectively.

Test results are shown in Figure 7 and Table 1. In summary, due to the thermal sensitivity of the bridge output voltage α_b, the greater the measured magnetic field, the greater the equivalent magnetic field noise at the frequency of the milli-Hertz. Therefore, techniques to reduce the thermal dependence of the system are totally necessary.

6. Conclusions

This paper shows how the temperature coefficient of the AMR using a current source for bridge excitation is improved by a factor of $\simeq 4$ compared with a voltage source. Hence, it helps to reduce the noise caused by thermal dependence of the AMR at the submilli-Hertz LISA band. Under an applied magnetic field, it has been observed that, across the MBW, the temperature drift of sensor sensitivity produces an extra noise, so the further from zero is the AMR bridge output, the larger the noise is.

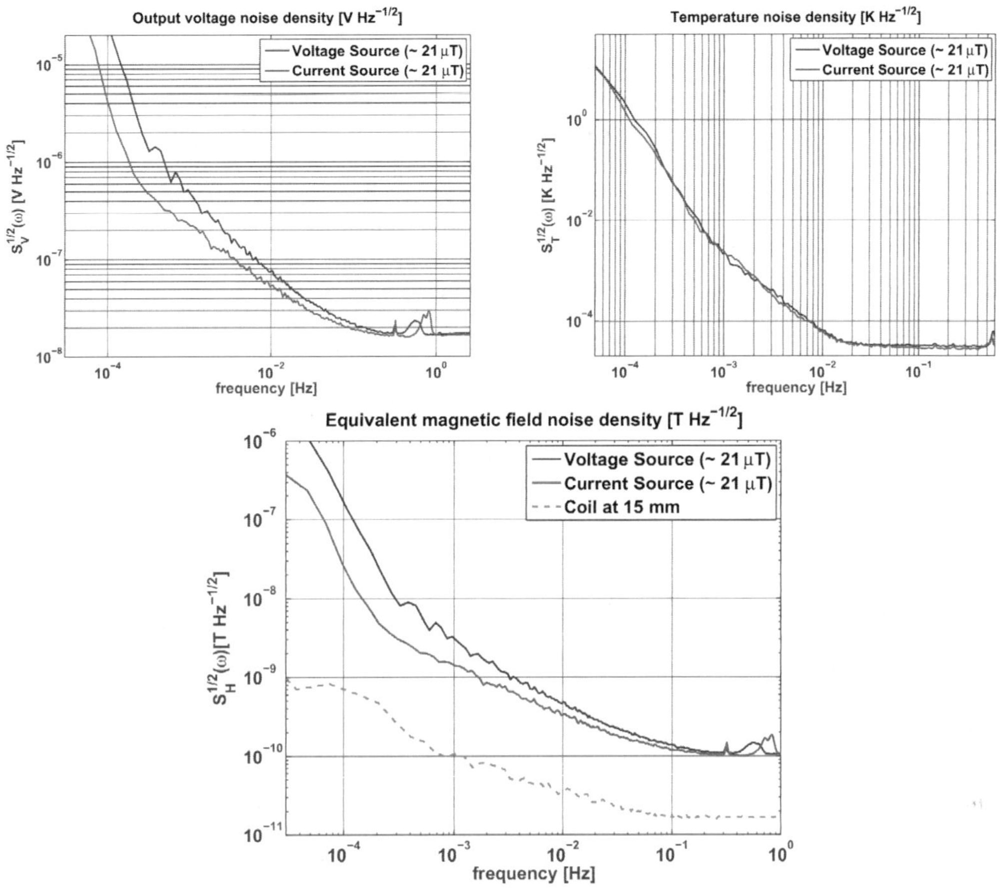

Figure 7: Spectral density in terms of voltage, temperature and magnetic field using voltage and current sources to supply the AMR's Wheatstone bridge. Green trace shows magnetic field noise generated by the coil at 15 mm from the sensor.

Further improvements involve using the AMR's *Offset* strap as a feedback element in a closed loop circuit to operate the bridge in the balanced resistance mode, and thus minimise the effect of the gain temperature coefficient of the sensor. This method could produce desirable results for measuring magnetic fields at the LISA frequencies and can be useful in other applications beyond the scope of LISA.

Acknowledgments
We acknowledge support from Project ESP2007-61712 of Plan Nacional del Espacio of the Spanish Ministry of Science and Innovation (MICINN).

References
[1] Vitale S 2005 Science Requirements and Top-level Architecture Definition for the *LISA* Technology Package (*LTP*) on Board *LISA* Pathfinder (*SMART-2*), report LTPA-UTN-ScRD-Iss003-Rev1
[2] Díaz-Michelena M 2009 Small magnetic sensors for space applications *Sensors* 2009 **9** 2271-2288

[3] Diaz-Aguiló M *et al* 2012 Design of the magnetic diagnostic unit onboard LISA Pathfinder *Aerospace Science and Technology*
[4] Hauser H, Stangl G, Fallmann W, Chabicovsky R and Riedling K 2000 *Proc. Workshop Preparation, Properties, and Applications of Thin Ferromagnetic Films (Viena)* pp 15-27
[5] Mateos I *et al* 2009 Towards an improved magnetic diagnostic system for LISA *J. Phys.: Conf. Series* **154** 012005
[6] Mateos I *et al* 2011 Low-noise magnetic field measurement for LISA *J. Phys.: Conf. Series* (In progress)
[7] Anderson R 2004 Understanding Ratio Metric Conversions, Texas Instruments Application Report SBAA110

Gravitational-wave standard siren without redshift identification

Atsushi Nishizawa[1], Kent Yagi[2], Atsushi Taruya[3,4], and Takahiro Tanaka[1]

[1] Yukawa Institute for Theoretical Physics, Kyoto University, Kyoto 606-8502, Japan
[2] Department of Physics, Kyoto University, Kyoto 606-8502, Japan
[3] Research Center for the Early Universe, Graduate School of Science, The University of Tokyo, Tokyo 113-0033, Japan
[4] Institute for the Physics and Mathematics of the Universe, The University of Tokyo, Kashiwa, Chiba 277-8568, Japan

E-mail: anishi@yukawa.kyoto-u.ac.jp

Abstract. Proposed space-based gravitational-wave (GW) detectors such as DECIGO and BBO will detect $\sim 10^6$ neutron-star (NS) binaries and determine the luminosity distances to the binaries with high precision. Combining the luminosity distances with cosmologically-induced phase corrections on the GWs, cosmological expansion out to high redshift can be measured without the redshift determinations of host galaxies by electromagnetic observation and can be a unique probe for dark energy. This article is based on the results obtained in [1] where we investigated constraining power of the GW standard siren without redshift information on the equation of state of dark energy with future space-based GW detectors. We also compare the results with those obtained with other instruments and methods.

1. Introduction

Future space-based gravitational-wave (GW) detectors such as DECI-hertz Interferometer Gravitational-wave Observatory (DECIGO) [2,3] and Big-Bang Observer (BBO) [4] (see also [5] for updated information) are the most sensitive to GWs in $0.1 - 1\,\mathrm{Hz}$ band and will aim at detecting the primordial GW background, the mergers of intermediate-mass black holes (BH), and a large number of neutron-star (NS) binaries in an inspiraling phase.

It is known that the continuous GW signal from a compact-binary object provides a unique way to measure the luminosity distance to the source with high precision. Such binary sources are often referred to as the standard siren. With the redshift information determined by an electromagnetic (EM) follow-up observation, the standard siren can be an accurate tracer of the cosmic expansion [6]. The potential power of this method as a dark-energy probe has been investigated with ground- and space-based detector configurations for GWs [1, 5, 7–17]. The most of the preceding works assume that the redshifts of all GW sources are known by EM observations. However, this assumption is rather strong and is too optimistic to be justified because the spectroscopic follow-up observation of galaxies is, in general, time-consuming, particularly at high redshifts [18]. In addition, not all galaxies can be observed due to intrinsic faintness, the absence of spectral features, limited sky coverage, and a limited redshift range (redshift desert). As a result, in a practical follow-up observation, the fraction of the binary sources whose redshifts are spectroscopically obtained is significantly reduced. From a rough estimate based on the number density of galaxies potentially observable and the number of galaxies to be observed in the future galaxy redshift surveys such as JDEM/WFIRST [19] and Euclid [20], it turns out that the fraction of redshift identification is $\sim 10^{-4}$ with uncertainty of about one order of magnitude [1]. This means that the GW sources with redshift information are quite rare unless we perform a large-scale follow-up campaign dedicated for GW events.

Another approach is to measure the cosmic-expansion rate from GW observations alone without electromagnetically-estimated redshifts. As suggested in [2], the cosmic expansion affects not only the amplitude (luminosity distance) of GWs but also the phase, and it is possible to directly obtain information about the cosmic acceleration by accurately measuring the GW phase shift of a binary source at a certain redshift. Although the redshifts of the binaries are assumed to be determined by the EM follow-up observations in the previous works [2, 21], if we combine the independent information from the luminosity distance and the cosmological phase shift, we can measure the expansion history of the universe without any reference to the EM counterpart or host-galaxy identification. Furthermore, this method based on a purely GW observation enables us to compare the observational data with those obtained in other EM observations. Since proposed space-based GW detectors such as DECIGO and BBO would detect $\sim 10^6$ NS binaries, they provide a novel opportunity to measure the property of dark energy without using the cosmic ladders.

Our primary interest here is in the potential of the standard siren without any help of redshift information of the binary sources and how the sensitivity is improved by partially adding the redshift information of a fraction of NS binaries. We show the constraining power of the GW standard siren with/without redshift information on the equation of state (EOS) of dark energy. As GW detectors, we use future space-based detector, DECIGO and BBO, allowing their noise curves to be scaled appropriately and including confusion noises produced by astrophysical sources. Throughout the paper, we adopt units $c = G = 1$.

2. Detector noise and astrophysical foregrounds

In this section, we summarize the detector-noise curves of DECIGO and BBO and confusion noises from astrophysical GW foregrounds.

DECIGO and BBO are the most sensitive to GWs in $0.1 - 1\,\mathrm{Hz}$ band [2, 4]. Using the currently-proposed design parameters [3], we obtained fitting formula for the sky-averaged noise curve of DECIGO single interferometer:

$$\hat{S}^{\mathrm{inst}}_{\mathrm{h,\,D}}(f) = 3.30 \times 10^{-50}\left(\frac{f}{1\mathrm{Hz}}\right)^{-4} + 3.09 \times 10^{-47}\left[1 + \left(\frac{f}{f_{\mathrm{c}}}\right)^2\right]\ \mathrm{Hz}^{-1}\,, \tag{1}$$

where $f_{\mathrm{c}} = 7.69\,\mathrm{Hz}$. BBO has a slightly different noise shape from DECIGO, because of the different interferometer type and optical parameters. In the presence of WD confusion noise below $\sim 0.1\,\mathrm{Hz}$, the BBO sensitivity to a NS binary is nearly three times better in amplitude than that of DECIGO.

DECIGO and BBO are sensitive to a large number of astrophysical GW sources, in particular NS, BH, and WD binaries. It is expected that the low-frequency side of the noise curve would be dominated by the GWs from extra-galactic population of WD binaries, which remains after subtracting individually identified signals and are stochastic in nature. According to the estimation by [22], the fitting formula for the power spectrum of WD confusion noise [23] (residual contribution after the subtraction) is given by

$$S_h^{\mathrm{WD}}(f) = 4.2 \times 10^{-47}\left(\frac{f}{1\,\mathrm{Hz}}\right)^{-7/3} \exp\left[-2\left(\frac{f}{5\times 10^{-2}\,\mathrm{Hz}}\right)^2\right]\ \mathrm{Hz}^{-1}\,.$$

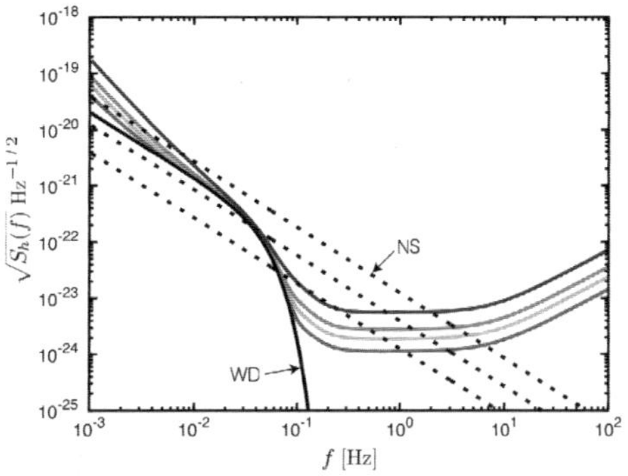

Figure 1. Noise curves of default and scaled DECIGO: $r_n = 1$ (blue), 1/2 (green), 1/3 (orange), and 1/5 (red) from the top to the bottom. These noise curves include the confusion noise from a number of WD binaries. The WD-binary foreground $\sqrt{S_h^{\rm WD}}$ is shown with solid, black curve on the left side. The three diagonal dashed lines represent the NS-binary foreground $\sqrt{S_h^{\rm NS}}$ before subtraction (or $\mathcal{R}_{\rm NS} = 1$) with $\dot{n}_0 = 10^{-5}$, 10^{-6}, $10^{-7}\,{\rm Mpc}^{-3}\,{\rm yr}^{-1}$ from the top to the bottom.

There is also contribution from galactic WD binaries [23]. However, we verified that the galactic contribution is negligible above the frequency 5×10^{-3} Hz.

Another astrophysical source that we have to take into account is the NS-binary foreground. According to [24], the energy density of GWs from NS binaries per logarithmic frequency bin normalized by the critical energy density of the universe at present is written as

$$\Omega_{\rm gw}^{\rm NS}(f) = \frac{8\pi^{5/3}}{9H_0^2} M_c^{5/3} f^{2/3} n_0 \,, \qquad n_0 = \int_0^\infty \frac{\dot{n}(z)}{(1+z)^{4/3} H(z)} dz \,. \tag{2}$$

H_0 is the Hubble constant and M_c is the chirp mass defined as $M_c \equiv \eta^{3/5} M_t$, together with the total mass $M_t = m_1 + m_2$ and the symmetric mass ratio $\eta = m_1 m_2 / M_t^2$. $\dot{n}(z)$ is the NS merger rate per unit comoving volume per unit proper time at a redshift z. We adopt the following fitting form of the NS-NS merger rate given in [25]: $\dot{n}(z) = \dot{n}_0\, s(z)$, $s(z) = 1 + 2z$ for $z \le 1$, $\frac{3}{4}(5 - z)$ for $1 < z \le 5$, 0 for $5 < z$. This function $s(z)$ is estimated based on the star formation history inferred from the UV luminosity [26]. The quantity $\dot{n}_0$ represents the merger rate at present. By assuming the flat ΛCDM universe with $\Omega_{\rm m} = 0.3$ ($\Omega_\Lambda = 0.7$) and the Hubble constant $H_0 = h_{72} \times 72\,{\rm km\,s^{-1}\,Mpc^{-1}}$, we obtain the NS confusion-noise power spectrum

$$S_h^{\rm NS}(f) = 1.55 \times 10^{-47} h_{72}^{-1} \left(\frac{M_c}{1.22 M_\odot}\right)^{5/3} \left(\frac{f}{1\,{\rm Hz}}\right)^{-7/3} \left(\frac{\dot{n}_0}{10^{-6}\,{\rm Mpc}^{-3}\,{\rm yr}^{-1}}\right) ,$$

where we used the relation $S_{\rm h}^{\rm NS}(f) = (3H_0^2/4\pi^2 f^3)\Omega_{\rm gw}^{\rm NS}(f)$ [27].

Including astrophysical contributions, a total-noise power spectrum is given by

$$S_h(f) = S_h^{\rm inst}(f) + S_h^{\rm WD}(f) + S_h^{\rm NS}(f)\mathcal{R}_{\rm NS} \,. \tag{3}$$

Since our purpose here is to assess cosmology achieved by space-based detectors such as DECIGO and BBO and to make clear the requirement for the experimental design, we consider only DECIGO, but allowing the noise curve in Eq. (1) to vary by an overall factor $r_n = 1$, 1/2, 1/3, 1/5 in amplitude. Namely, $S_h^{\rm inst}(f) = r_n^2 \times \hat{S}_{\rm h,D}^{\rm inst}(f)$. Note that BBO noise power spectrum approximately corresponds to that of DECIGO with $r_n = 1/3$. The factor $\mathcal{R}_{\rm NS}$ denotes a suppression factor due to the subtraction of individually identified NS binaries, whose value strongly depends on the detector noise curve. According to [1] in which the authors have fitted the numerical result obtained by Yagi and Seto [28], the suppression factor is $\mathcal{R}_{\rm NS} = 0$ except for $r_n = 1$ case. When $r_n = 1$, some amount of the foreground residuals is left and gives $\mathcal{R}_{\rm NS} \approx 4.6 \times 10^{-3}$. The contribution of each term in Eq. (3) is shown in Fig. 1.

3. Standard siren as a probe for dark energy

3.1. Standard siren

The continuous GW signal from a compact-binary object is often referred to as standard sirens [6]. For a single binary system, the Fourier transform of the GW waveform is expressed as a function of frequency f [27, 29],

$$\tilde{h}(f) = \frac{A}{d_L(z)} M_z^{5/6} f^{-7/6} e^{i\Psi(f)} , \tag{4}$$

where d_L is the luminosity distance, and the quantity $M_z = (1+z)M_c$ is the redshifted chirp mass. M_c is the proper chirp mass defined in the source rest frame. The constant A is given by $A = (\sqrt{6}\,\pi^{2/3})^{-1}$, which includes the factor $\sqrt{4/5}$ for a geometrical average over the inclination angle of a binary [30]. The function $\Psi(f)$ represents the frequency-dependent phase arising from the orbital evolution, and at the order of the *restricted* 1.5 post-Newtonian (PN) approximation, it is given by [27, 29]

$$\begin{aligned}\Psi(f) &= 2\pi f\, t_c - \phi_c - \frac{\pi}{4} + \frac{3}{128}(\pi M_z f)^{-5/3} \\ &\times \left[1 + \frac{20}{9}\left(\frac{743}{336} + \frac{11}{4}\eta\right)\eta^{-2/5}(\pi M_z f)^{2/3} - 16\pi\eta^{-3/5}(\pi M_z f) - \frac{25}{768}X(z)M_z(\pi f M_z)^{-8/3}\right] ,\end{aligned} \tag{5}$$

where t_c and ϕ_c are the time and phase at coalescence, respectively. The last term in the bracket is phase correction due to cosmic expansion (-4 PN order) [2, 21], where $X(z)$ is defined as $X(z) \equiv \frac{1}{2}[H_0 - H(z)/(1+z)]$ or equivalently expressed as $[\dot{a}(0) - \dot{a}(z)]/2$. Thus, $X(z) > 0$ ($X(z) < 0$) corresponds to the accelerating (decelerating) universe.

There are various methods of the standard siren. The simplest minimal one is to measure the amplitude of GWs, or the luminosity distance as a function of redshifts. However, seen from Eq. (4), GW observation itself cannot provide a source redshift, since the mass parameter we can determine is $M_z = (1+z)M_c$ and the source redshift degenerates with the proper source mass M_c. So the redshift has to be determined from EM observation of host galaxies. This method as a tool to measure the cosmic expansion has been investigated by many authors [5, 7–14], assuming that the source redshifts are known by spectroscopic follow-up observations. Even in the situation that the redshift of a source is not uniquely determined due to multiple candidates of a host galaxy in a detector error cube, one can infer the most likely cosmological model by statistically analyzing large samples of observed events [6, 31]. However, to do that, one needs a complete catalog of galaxy redshifts. This assumption is rather strong and is too optimistic to be justified, because of the smallness of the fraction of redshift identification ($\sim 10^{-4}$) [1].

To break the redshift degeneracy and measure the cosmic expansion by GW observation alone, recently some elaborate methods have been proposed. The redshift degeneracy could be broken if we know the EOS of a NS *a priori* [32] or if the observed chirp mass distribution of NS is sufficiently narrow [33]. We will discuss the sensitivities of several methods later in Sec. 4.

In this article, we investigate another method breaking the redshift degeneracy by the phase modulation of GWs due to the cosmic expansion, which is the last term including $X(z)$ in

Eq. (5) [2, 21]. Although in the preceding works the redshifts of the binaries are assumed to be determined by the EM follow-ups [2, 21], the redshift information in principle is not mandatory if independent information from both the luminosity distance and the phase shift is available. In fact, as indicated by our results shown later, we can measure a cosmic-expansion rate only with GW observations without electromagnetically-estimated redshifts. This is a great advantage of this method, because a pure GW observation enables us to compare its observational data with those obtained in other EM observations.

In the subsequent sections, we investigate the power of utilizing the phase correction without assuming any electromagnetically-estimated redshift, and give the figure of merits. Then we compare this method with the previous one, in which source redshifts are given, and combine both methods to maximize the accuracy of the estimation.

3.2. Numerical estimation of measurement accuracy

We consider a spatially flat universe with dark energy whose EOS is parameterized as $w(z) = w_0 + w_a z/(1+z)$. The parameters to be determined from GW observations are w_0, w_a, Ω_m, and H_0. For the analysis of error estimation, we adopt a fiducial set of cosmological parameters: $w_0 = -1$, $w_a = 0$, $\Omega_m = 0.3$, $H_0 = 72\,\mathrm{km\,s^{-1}\,Mpc^{-1}}$.

In what follows, we calculate the estimation errors of binary parameters θ_a: M_z, η, t_c, ϕ_c, d_L, and X in the waveform of Eq. (4) with the Fisher information matrix. It is well known that the Fisher matrix analysis underestimates the parameter errors if SNR is not sufficiently large. Even default DECIGO (8 identical interferometers) is able to observe a NS binary at $z = 1$, 3, and 5 with SNR about 24, 31, and 67, respectively. Therefore, it is justified to use the Fisher matrix for the parameter estimation. The Fisher matrix for a single binary is given by [29, 34]

$$\Gamma_{ab}^{(\mathrm{single})} = 4\sum_{i=1}^{8} \mathrm{Re} \int_{f_{\mathrm{min}}}^{f_{\mathrm{max}}} \frac{\partial_a \tilde{h}^*(f)\, \partial_b \tilde{h}(f)}{S_{\mathrm{h}}(f)} df \,, \tag{6}$$

where ∂_a denotes a derivative with respect to a parameter θ_a. Since the 8 interferometers of DECIGO and BBO are assumed to be identical, the summation is reduced to just multiplying a factor 8. The noise power spectrum and GW signal are given by Eqs. (3) and (4), respectively. The lower cutoff frequency f_{min} is given by, $f_{\mathrm{min}} = 0.233\,(1M_\odot/M_z)^{5/8}\,(1\,\mathrm{yr}/T_{\mathrm{obs}})^{3/8}$ Hz, and the upper cutoff of the frequency is set to $f_{\mathrm{max}} = 100\,\mathrm{Hz}$ which naturally arises from the upper cutoff of the noise curve. Given the numerically evaluated Fisher matrix, the marginalized 1-σ error of a parameter, $\Delta\theta_a$, is estimated from the inverse Fisher matrix, $\Delta\theta_a = \sqrt{\{\mathbf{\Gamma}^{-1}\}_{aa}}$.

Once the estimation error of X is obtained, it is straightforward to calculate the estimation errors of the cosmological parameters. Since the error of d_L is much smaller than that of X (see Appendix of [1]), we can replace observed d_L with the corresponding redshift in the fiducial cosmological model when we derive the measurement accuracies of cosmological parameters from X. Thus, for the simplicity of the analysis, we use $X(z)$ instead of $X(d_L)$. Furthermore, we assume that the Hubble constant H_0 is known *a priori*, and fix it to $H_0 = 72\,\mathrm{km\,s^{-1}\,Mpc^{-1}}$, because the Hubble constant has been determined at a-few-percent level from the observation of nearby Cepheids and supernovae [35]. Thus, the free parameters of the Fisher matrix are w_0, w_a, and Ω_m. The Fisher matrix is given by

$$\Gamma_{ab} = \int_0^\infty \frac{\partial_a X(z) \partial_b X(z)}{\sigma_X^2(z)} \frac{dN(z)}{dz} dz \,, \quad \frac{dN(z)}{dz} = T_{\mathrm{obs}} \frac{4\pi r^2(z)}{H(z)} \frac{\dot{n}(z)}{1+z} \,, \tag{7}$$

where $dN(z)/dz$ is the number of NS binaries in the redshift interval $[z, z+dz]$ observed during T_{obs} [25]. $r(z)$ is the comoving radial distance defined as $r(z) = d_L(z)/(1+z)$ and $\dot{n}$ is given in Sec. 2. Since the normalization of $\dot{n}$ is still uncertain, we adopt the most recent estimate [36], $\dot{n}_0 = 10^{-6}\,\mathrm{Mpc^{-3}\,yr^{-1}}$. Since the observation time T_{obs} is a crucial parameter, we suppose 3-yr, 5-yr, and 10-yr observation.

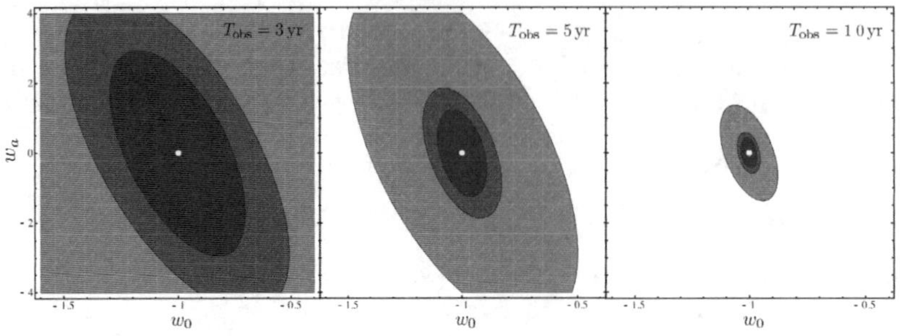

Figure 2. w_0 - w_a error ellipses marginalized over Ω_m. The merger rate is fixed to $\dot{n}_0 = 10^{-6}\,{\rm Mpc}^{-3}\,{\rm yr}^{-1}$. From the left to the right panels, the observation time is 3 yr, 5 yr, and 10 yr. In each panel, the larger to smaller ellipses denote those with $r_n = 1$, 1/3, and 1/5, respectively. The dot at the center is our fiducial value: $w_0 = -1$ and $w_a = 0$.

3.3. Results

The measurement accuracies of w_0 and w_a marginalized over the other remaining parameters are shown in Fig. 2 for the fixed $\dot{n}_0$ varying the values of r_n and $T_{\rm obs}$. The size of error ellipses change significantly depending on the parameters $T_{\rm obs}$ and r_n. For example, if we choose $T_{\rm obs} = 5\,{\rm yr}$, the errors Δw_0 and Δw_a are 0.330 and 3.254 for $r_n = 1$, and 0.113 and 1.231 for $r_n = 1/3$. For 10-yr observation, Δw_0 and Δw_a reach at the levels of several $\times 10^{-2}$ and a few $\times 10^{-1}$, respectively. The more comprehensive parameter surveys are found in [1].

To show the results more clearly, let us define the figure of merit (FoM) for dark energy, FoM $\equiv \sqrt{\det \gamma_{ab}}$, where γ_{ab} is the inverse matrix of $(\Gamma^{-1})_{ab}$ with $a, b = w_0, w_a$. This FoM is inversely proportional to the area of an error ellipse on the w_0 - w_a plane and ranges from $\sim 10^{-1}$ to $\sim 10^3$. Since DECIGO and BBO will be launched in the late 2020s, the FoM should be compared with other future projects of an EM observation probing dark energy at that time: type-Ia supernovae, baryon acoustic oscillation, or weak lensing surveys. So typical criteria in FoM would be from 10 to 100, which correspond to future projects of stage III and IV in the dark energy task force [37], respectively. To achieve these criteria, observation time longer than 5 yr is preferable. Since the FoM rapidly improves being roughly proportional to $T_{\rm obs}^{31/8}$ [1], the observation time is a crucial factor. Given 10-yr observation and the typical rate of binary mergers, the FoM is $\sim$ 100 and is comparable to a stage-IV project in the dark energy task force. It should be emphasized that this method requires no redshift information of the GW sources and is completely independent of any EM observation.

3.4. Adding redshift information of binaries

In the previous subsections, we did not assume any redshift information to measure the dark-energy parameters. Although it would be too ideal to assume that all binary redshifts are well determined by EM follow-up observations, some fraction of them will be determined by the follow-up observations. Then, we can optimize the constraints on the dark-energy parameters by using thus-determined redshifts. In this subsection, we combine z - d_L and d_L - X information and evaluate what fraction is needed to achieve FoM = 100, which is a typical value expected in future dark energy surveys [37].

The relevant Fisher matrix is defined with the redshift-identification fraction α by

$$\Gamma_{ab} = \alpha \int_0^\infty \frac{\partial_a d_L(z) \partial_b d_L(z)}{\sigma_{d_L}^2(z)} \frac{dN(z)}{dz} dz + \int_0^\infty \frac{\partial_a X(z) \partial_b X(z)}{\sigma_X^2(z)} \frac{dN(z)}{dz} dz \,, \tag{8}$$

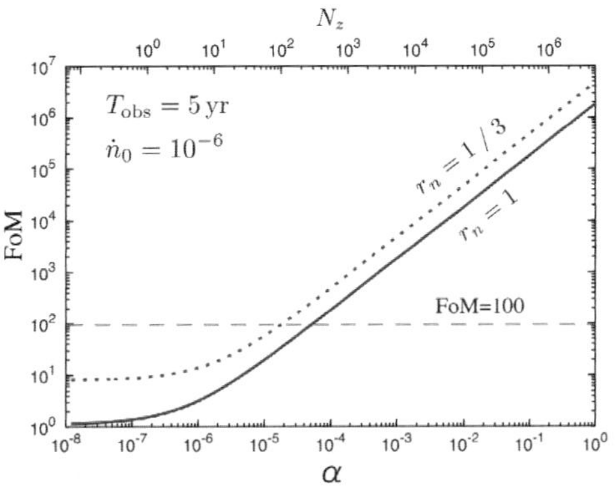

Figure 3. FoM as a function of α defined in Eq. (8) or N_z. The observation time and the merger rate are fixed to $T_{\rm obs} = 5\,{\rm yr}$ and $\dot{n}_0 = 10^{-6}\,{\rm Mpc}^{-3}\,{\rm yr}^{-1}$. The horizontal dashed line represents FoM $= 100$. The solid and dotted curves are the case with $r_n = 1$ and $1/3$, respectively.

where σ_{d_L} is an error in the luminosity distance. Strictly speaking, the variables d_L and X cannot be treated independently as in Eq. (8). However, the correlation coefficient between d_L and X is of the order of 10^{-6}. Hence, we can safely use the expression in Eq. (8). The fraction factor α would be a function of a redshift in a real galaxy survey, but we take it as a constant for simplicity of the analysis.

We estimate the error size of the luminosity distance via the same procedure as in Sec. 3.2, but including possible systematic errors: weak-lensing magnification due to the matter inhomogeneities along the line of sight [38] and the Doppler effect due to the random peculiar velocities of binary sources [39]. These systematic errors to the luminosity distance are summarized as

$$\sigma^2_{d_L}(z) \equiv \left[\frac{\Delta d_L(z)}{d_L(z)}\right]^2 = \sigma^2_{\rm inst}(z) + \sigma^2_{\rm lens}(z) + \sigma^2_{\rm pv}(z)\,, \tag{9}$$

$$\sigma_{\rm lens}(z) = 0.066\left[\frac{1-(1+z)^{-0.25}}{0.25}\right]^{1.8}\,, \qquad \sigma_{\rm pv}(z) = \left|1 - \frac{(1+z)^2}{H(z)d_L(z)}\right|\sigma_{\rm v,gal}\,,$$

where $\sigma_{\rm inst}$, $\sigma_{\rm lens}$, and $\sigma_{\rm pv}$ are induced by the instrumental noise, the lensing magnification, and the peculiar velocity of binaries, respectively. $\sigma_{\rm v,gal}$ is the one-dimensional velocity dispersion of the galaxy and is set to be $\sigma_{\rm v,gal} = 300{\rm km\,s}^{-1}$. For the plot of the redshift dependence of each error term and the details, see [16]. The important thing is that the lensing error dominates at almost all redshift range.

Then using Eq. (8), we estimate the errors of w_0 and w_a and corresponding FoM as a function of α. We also calculate the corresponding number of NS binaries N_z whose redshifts are identified. The result is shown in Fig. 3 for typical values of parameters, $T_{\rm obs} = 5\,{\rm yr}$ and $\dot{n}_0 = 10^{-6}\,{\rm Mpc}^{-3}\,{\rm yr}^{-1}$. At low α, of course, the z - d_L information plays no significant role in improving the FoM. But at $\alpha \approx$ a few $\times 10^{-6}$, z - d_L information begins to contribute to FoM, and enables the FoM to reach 100 at around $\alpha \approx$ a few $\times 10^{-5}$. As anticipated, z - d_L measurement is much more powerful than the d_L - X measurement because ${\rm FoM}|_{\alpha=0}$ is much smaller than ${\rm FoM}|_{\alpha=1}$. Therefore, identifying the source redshifts strongly assists the d_L - X measurement.

One may concern how much fraction α is feasible in the future galaxy redshift surveys. From our rough estimate [1] based on the number density of galaxies potentially observable and the number of galaxies to be observed in the future galaxy redshift surveys such as JDEM/WFIRST [19] and Euclid [20], it turns out that the fraction of redshift identification is $\sim 10^{-4}$ with uncertainty of about one order of magnitude. This means that the GW sources with redshifts are quite rare unless we perform a large-scale follow-up campaign dedicated for GW events. So it turns out that d_L - X measurement may be more sensitive than z - d_L measurement and that the standard sirens without redshifts guarantee minimally achievable FoM.

4. Comparison with previous results with other detectors and methods

Preceding works for other detectors are listed in Table 1, in which only the detectors that can probe at cosmological distance are shown. As for DEICGO/BBO, our results should be compared with the work by Cutler and Holz [5]. They assume that one can identify a single host galaxy due to the good angular resolution of DECIGO/BBO and that the redshift of the host galaxy is always obtained by an EM follow-up observation. However, as estimated in [1], the assumption is too optimistic to be justified, because the redshift identification of host galaxies is time-consuming and inefficient and whose fraction would be at worst $\sim 10^{-4}$ even for the future galaxy surveys such as JDEM/WFIRST and Euclid. Thus, our results are conservative because we do not require any redshift identification and any prior from other observations. We can conclude that the standard siren without source redshifts guarantee minimally achievable FoM, which can be improved by increasing the fraction of the galaxy redshift identification and combining other EM observations.

One may also compare the results in this paper with those obtained from Einstein Telescope (ET). The total number of NS binaries observed by ET is comparable to DECIGO. However, ET has to rely on short gamma-ray bursts (GRB) to determine the source redshift so that the number of sources with redshifts is limited by short GRB directed toward us and is reduced to $\sim 10^3$ or smaller. Consequently, the sensitivity of DECIGO/BBO without redshift information is twice better than that of ET by Sathyaprakash et al. [12]. If the Planck prior of CMB is added as done by Zhao et al. [14], the size of the error is significantly reduced. On the other hand, the method proposed by Messenger and Read [32] does not assume source redshifts. However, to break the parameter degeneracy between the chirp mass and redshift of a NS binary, one needs to assume the EOS of a NS to be known. Furthermore, a tiny phase signal due to tidal deformation of NS (5 PN order) has to be detected. ET would be able to detect this signal, but the measurement accuracy of the redshift is not so good, $\Delta z/z \sim 0.1$, and that of the luminosity distance also would be of the same order. To clarify its feasibility as a cosmological probe, further studies are needed.

LISA has marginal angular resolution so as to identify a single host galaxy. Van Den Broeck et al. [13] considered the observation of a SMBH binary that fortunately accompanies a single host galaxy in its sky position error area. If only w_0 is a free parameter, the sensitivity is ~ 0.2. Although multiple sources improve this sensitivity, such fortunate events would be rare. To avoid this difficulty, a statistical analysis method has been developed [15]. This method utilizes all redshifts, which obtained a priori by EM follow-up observations, of possible host galaxies in a LISA error cube and determines the best fitting cosmological model for all GW sources. Petiteau et al. [15] performed the analysis with LISA, including the CMB prior, and obtained the accuracy of a few percents for w_0, though other parameters are fixed in their analysis. The applications of this statistical method to ET and DECIGO/BBO are not straightforward. The problem is that the spectroscopic catalog of galaxies is far from a complete one for a large number of sources observed by these detectors. As mentioned above, the fraction of the redshift identification of host galaxies would be $\sim 10^{-4}$. Therefore, it is interesting to ask which methods are more sensitive, performing the statistical analysis with an incomplete galaxy catalog or using all GW sources without redshift information as we have done in this paper. We should address

Table 1. List of cosmological constraints obtained with other detectors and methods. Here we list only the detectors that can probe at cosmological distance and works that take into account weak lensing error. SMBH and CMB refer to supermassive black hole and cosmic microwave background, respectively.

reference	detector	GW source	number of events	source redshift	method to determine z
Van Den Broeck et al. [13]	LISA	SMBH	1	host galaxy	identifying a single source
Petiteau et al. [15]	LISA	SMBH	~ 40	host galaxy	statistical
Sathyaprakash et al. [12]	ET	NS	10^3	EM counterpart (short GRB)	identifying a single source
Zhao et al. [14]	ET	NS	10^3	EM counterpart (short GRB)	identifying a single source
Messenger & Read [32]	ET	NS	N/A	unnecessary	tidal effect on GW phase
Cutler & Holz [5]	DECIGO or BBO	NS	$\sim 10^5$	host galaxy	identifying a single source
Nishizawa et al. [1] (this work)	DECIGO or BBO	NS	$\sim 10^6$	unnecessary	cosmological phase drift

reference	free parameters	fixed parameters	prior	sensitivity
Van Den Broeck et al. [13]	w_0	w_a, Ω_m, Ω_k, H_0	none	$\Delta w_0 \sim 0.2$
Petiteau et al. [15]	w_0	w_a, Ω_m, Ω_k, H_0	CMB (WMAP)	$\Delta w_0 \approx 0.036$
Sathyaprakash et al. [12]	w_0, Ω_m, Ω_k, H_0	w_a	none	$\Delta w_0 \approx 0.23$
Zhao et al. [14]	w_0, w_a, Ω_m, Ω_k, H_0	none	CMB (Planck)	$\Delta w_0 \approx 0.099$ $\Delta w_a \approx 0.302$
Messenger & Read [32]	N/A	N/A	NS EOS	$\Delta z/z \sim 0.1$ (for an event)
Cutler & Holz [5]	w_0, w_a, Ω_m, H_0	Ω_k	CMB (Planck)	$\Delta w_0 \sim 0.01$ $\Delta w_a \approx 0.1$
Nishizawa et al. [1] (this work)	w_0 w_a, Ω_m	Ω_k, H_0	none	$\Delta w_0 \approx 0.113$ $\Delta w_a \approx 1.23$

this issue in a future work.

5. Conclusion

Proposed space-based GW detectors, DECIGO/BBO, can be a unique probe for dark energy. Using millions of NS binaries detected during the observation, we have estimated the sensitivity to the EOS parameters of the dark energy with/without identifying the redshifts of host galaxies. As a result, we found that the detectors without the redshift information have constraining power competitive to the future EM observations. This is a great advantage since the GW detector alone can probe for the cosmological expansion and enables us to compare the data of purely GW observation with those in other EM observations. With the help of the redshift information, FoM ≈ 100 corresponding to the stage III or IV of the dark energy task force is easily achieved with at most a few hundreds of sources. The standard sirens without redshifts guarantee minimally achievable FoM and strongly supports the feasibility of the space-based GW detectors as a dark energy probe.

Acknowledgments

A. N. and K. Y. are supported by a Grant-in-Aid through JSPS. K. Y. is also supported in part by the Grant-in-Aid for the Global COE Program from the MEXT of Japan. A. T. is supported in part by a Grants-in-Aid for Scientific Research from the JSPS No. 21740168.

References

[1] Nishizawa A, Yagi K, Taruya A and Tanaka T 2012 *Phys. Rev.* **D 85** 044047
[2] Seto N, Kawamura S and Nakamura T 2001 *Phys. Rev. Lett.* **87** 221103
[3] Kawamura S *et al.* 2011 *Classical Quantum Gravity* **28** 094011
[4] E. S. Phinney *et al.*, The Big Bang Observer, NASA Mission Concept Study (2003)
[5] Cutler C and Holz D E 2009 *Phys. Rev.* **D 80** 104009
[6] Schutz B F 1986 *Nature (London)* **323** 310
[7] Holz D E and Hughes S A 2005 *Astrophys. J.* **629** 15
[8] Dalal N, Holz D E, Hughes S A and Jain B 2006 *Phys. Rev.* **D 74** 063006
[9] Arun K G, Iyer B R, Sathyaprakash B S, Sinha S and Broeck C V D 2007 *Phys. Rev.* **D 76** 104016
[10] Deffayet C and Menou K 2007 *Astrophys. J.* **668** L143
[11] Nissanke S, Holz D E, Hughes S A, Dalal N and Sievers J L 2010 *Astrophys. J.* **725** 496
[12] Sathyaprakash B S, Shutz B F and Broeck C V D 2010 *Classical Quantum Gravity* **27** 215006
[13] Broeck C V D, Trias M, Sathyaprakash B S and Sintes A M 2010 *Phys. Rev.* **D 81** 124031
[14] Zhao W, Broeck C V D, Baskaran D and Li T G F 2011 *Phys. Rev.* **D 83** 023005
[15] Petiteau A, Babak S and Sesana A 2011 *Astrophys. J.* **732** 82
[16] Nishizawa A, Taruya A and Saito S 2011 *Phys. Rev.* **D 83** 084045
[17] K. Yagi, A. Nishizawa, and C. -M. Yoo, arXiv:1112.6040 (2011)
[18] Wang Y *et al.* 2010 *Mon. Not. R. Astron. Soc.* **409** 737
[19] JDEM: http://jdem.gsfc.nasa.gov/.
[20] Eucrid: http://sci.esa.int/euclid.
[21] Takahashi R and Nakamura T 2005 *Prog. Theor. Phys.* **113** 63
[22] Farmer A J and Phinney E S 2003 *Mon. Not. R. Astron. Soc.* **346** 1197
[23] Barack L and Cutler C 2004 *Phys. Rev.* **D 69** 082005
[24] E. S. Phinney, arXiv:astro-ph/0108028 (2001)
[25] Cutler C and Harms J 2006 *Phys. Rev.* **D 73** 042001
[26] Schneider R, Ferrari V, Matarrese S and Zwart S F P 2001 *Mon. Not. R. Astron. Soc.* **324** 797
[27] M. Maggiore, *Gravitational Waves* (Oxford university press, 2008)
[28] Yagi K and Seto N 2011 *Phys. Rev.* **D 83** 044011
[29] Cutler C and Flanagan E E 1994 *Phys. Rev.* **D 49** 2658
[30] The geometric factors arising from the nonorthogonal detector arms (60°) and the sky-averaged detector response functions are incorporated into the detector noise curve $S_{\rm h}^{\rm inst}(f)$.
[31] W. Del Pozzo, arXiv:1108.1317 (2011)
[32] Messenger C and Read J 2012 *Phys. Rev. Lett.* **108** 091101
[33] S. R. Taylor, J. R. Gair, and I. Mandel, arXiv:1108.5161 (2011)
[34] Finn L S 1992 *Phys. Rev.* **D 46** 5236
[35] Riess A G *et al.* 2011 *Astrophys. J.* **730** 119
[36] Abadie J *et al.* 2010 *Classical Quantum Gravity* **27** 173001
[37] A. Albrecht et al., arXiv:astro-ph/0609591 (2006)
[38] Hirata C M, Holz D E and Cutler C 2010 *Phys. Rev.* **D 81** 124046
[39] Gordon C, Land K and Slosar A 2007 *Phys. Rev. Lett.* **99** 081301

Amaldi 9 and NRDA 2011 IOP Publishing
Journal of Physics: Conference Series **363** (2012) 012053 doi:10.1088/1742-6596/363/1/012053

Parameter estimation in LISA Pathfinder operational exercises

M Nofrarias[1], L Ferraioli[2,3] , G Congedo[2], M Hueller[2], M Armano[4], M Diaz-Aguiló[5], A Grynagier[6], M Hewitson[7] and S Vitale[2]

[1] Institut de Ciències de l'Espai (CSIC-IEEC), Facultat de Ciències, 08193 Bellaterra, Spain
[2] Dipartimento di Fisica, Università di Trento and INFN, Gruppo Collegato di Trento, 38050 Povo, Trento, Italy
[4] European Space Astronomy Centre, European Space Agency, Villanueva de la Cañada, 28692 Madrid, Spain
[5] Universitat Politècnica de Catalunya (UPC), EPSC, Esteve Terrades 5, 08860 Castelldefels, Spain
[6] Institut für Flugmechanik und Flugregelung (IFR), 70569 Stuttgart, Germany
[7] Albert-Einstein-Institut (AEI), Max-Planck-Institut für Gravitationsphysik und Universität Hannover, 30167 Hannover, Germany

E-mail: `nofrarias@ice.cat`

Abstract. The LISA Pathfinder data analysis team has been developing in the last years the infrastructure and methods required to run the mission during flight operations. These are gathered in the LTPDA toolbox, an object oriented MATLAB toolbox that allows all the data analysis functionalities for the mission, while storing the history of all operations performed to the data, thus easing traceability and reproducibility of the analysis. The parameter estimation methods in the toolbox have been applied recently to data sets generated with the OSE (Off-line Simulations Environment), a detailed LISA Pathfinder non-linear simulator that will serve as a reference simulator during mission operations. These simulations, so called operational exercises, are the last verification step before translating these experiments into tele-command sequences for the spacecraft, producing therefore very relevant datasets to test our data analysis methods. In this contribution we report the results obtained with three different parameter estimation methods during one of these operational exercises.

1. Introduction

The main objective of the LISA Pathfinder (LPF) mission [1, 2] is to characterise the purity of the free fall of two test masses in nominal geodesic motion in its orbit around the Lagrange point L1. The main scientific objective of the mission is specified in a spectral density of relative acceleration fluctuations between test masses of $S_{\Delta a} \leq 3 \times 10^{-14} \mathrm{m\,s^{-2}\,Hz^{-1/2}}$ at 1 mHz. Such an unprecedented metrology sensitivity in a space mission would be considered as a technology achievement sound enough to guarantee a successful development of a future gravitational wave observatory in space. The most mature concept for this space gravitational wave mission is LISA [3], which was recently abandoned as a joint ESA/NASA effort, although the concept is still being actively pursued independently by both space agencies.

[3] Present address: APC, UMR 7164, Université Paris 7 Denis Diderot 10, rue Alice Domon et Léonie Duquet, 75025 Paris Cedex 13, France

An important aspect of the LISA Pathfinder mission is the need of a flexible data analysis infrastructure able to help the taking of decisions since, during operations, the LPF must be understood as an in-flight *experiment* with hundreds of channels being sampled and a tight schedule of experiments to be executed sequentially.

The LISA Pathfinder data analysis team has been developing in the last years the infrastructure and methods required to run the mission during flight operations. These are gathered in the LTPDA (LISA Technology Package Data Analysis) toolbox [4], an object oriented MATLAB toolbox that allows all the data analysis functionalities for the mission, while storing the history of all operations performed to the data, thus easing traceability and reproducibility of the analysis.

In this contribution we focus on three parameter estimation methods —time-domain linear estimation, time-domain non-linear estimation and Markov Chain Monte Carlo in frequency domain— implemented in the toolbox and the comparison of their results when applied with simulated data. The aim of the team when performing this exercise was twofold: to compare the consistency between methods and to test themselves against the most realistic data available.

2. The OSE simulator and the sixth operational exercise

The OSE (Off-line Simulations Environment) is a detailed LISA Pathfinder non-linear simulator that will serve as a reference simulator during mission operations. This simulator is being used as a black box during a series of runs being regularly executed, the so called *operational exercises*. These exercises with simulated data aim at testing the on-orbit experiments in a realistic environment in terms of software and time constraints. The operational exercises are the last verification step before translating experiments into tele-command sequences for the spacecraft, producing therefore very relevant datasets to test our data analysis methods.

The data being analysed here correspond to the sixth operational exercise, which was run on April 2011. The simulated data correspond to two runs —called *investigations*— of around 8 hours duration. In each of these a sequence of sinusoids of different frequency is injected in the first and the second channel, for the first and the second investigation respectively. These two runs excite the system making use of the two only input channels that we consider in the current model, however the real LPF will count with many other excitation mechanism, like the application of direct forces to the test mass, and hence will require a longer sequence of investigations to fully characterise its dynamics.

It is important to notice that the investigations analysed here are not fully representative in terms of the instrument performance. As shown below in Fig. 2 and 3, the noise level for the differential channel at 1 mHz was $S_x(\omega) \simeq 2.5\,\mathrm{nm}/\sqrt{\mathrm{Hz}}$ whilst it should be $S_x \approx S_{\Delta \mathrm{a}}/(2\pi\,\mathrm{f})^2 \simeq 0.76\,\mathrm{nm}/\sqrt{\mathrm{Hz}}$ in order to meet the acceleration noise requirement at 1 mHz. This must be assigned to a particular choice of the simulator parameters during these exercises, but it will not have a major impact in our results on parameter estimation, which will have a slightly lower signal-to-noise ratio that the one expected during operations.

3. Model description

The model of the LPF considered here for the analysis is an improved version of the one previously used in previous works [5], which includes actuators and delays. This same model has been considered for similar studies in [6, 7]. In the current analysis, we describe the measurement of an experiment in the frequency domain as

$$\vec{o} = \mathbf{G} \times \vec{o}_{\rm in} + \vec{n}, \qquad (1)$$

where $\vec{o}$ stands for the measured interferometer displacement; $\vec{o}_{\rm in}$ are the injected test signals and $\vec{n}$ the noise vector. The spectral shape of the latter can be seen in the lower panels of Fig.

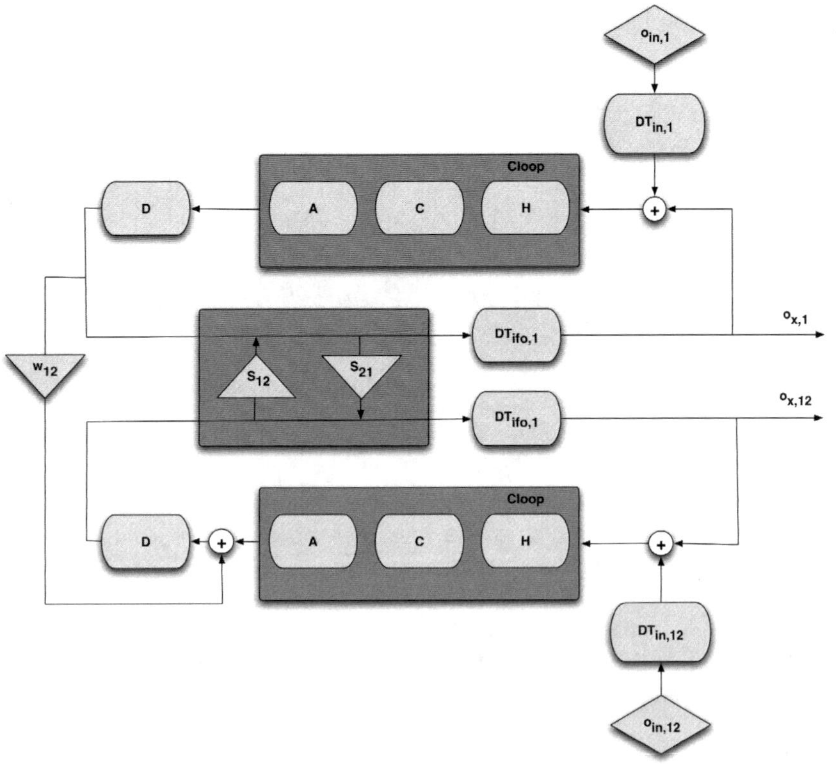

Figure 1. Box diagram of the model considered in the current analysis. Boxes are systems, rhombus represent injection points and triangles stand for cross-couplings. Notation and symbols are explained in the text.

2 and 3. Noise projections of the different contributions to this noise shape can be found in [8]. All of the previous are two components vectors, one component for each of the measurement channels considered here, the first one measuring the distance between the spacecraft and the first test mass, $o_{x,1}$, and the second one measuring the distance between the two test masses, $o_{x,12}$:

$$\vec{o} = \begin{pmatrix} o_{x,1} \\ o_{x,12} \end{pmatrix}, \quad \vec{o}_{\text{in}} = \begin{pmatrix} o_{in,1} \\ o_{in,12} \end{pmatrix}, \quad \vec{n} = \begin{pmatrix} n_1 \\ n_{12} \end{pmatrix}. \tag{2}$$

The transfer functions describing input to output relation in Eq.(1) are contained in the matrix $\mathbf{G}$,

$$\mathbf{G} = \left(\mathbf{D} \times (\mathbf{DT}_{\text{ifo}} \times \mathbf{S})^{-1} + \mathbf{C_{loop}}\right)^{-1} \times \mathbf{C_{loop}} \times \mathbf{DT}_{\text{in}} \tag{3}$$

with

$$\mathbf{C}_{\text{loop}} = \mathbf{C} \times \mathbf{A} \times \mathbf{H} \tag{4}$$

where we have split the control box, $\mathbf{C}_{\text{loop}}$, in three subsystems: inertia decoupling matrix, $\mathbf{C}$; a matrix with the Drag-Free and Attitude Control Subsystem (DFACS) controllers expressed as transfer functions, $\mathbf{H}$; and the matrix $\mathbf{A}$ with actuator gain and delays. These matrices can be expressed as

$$\mathbf{C} = \begin{pmatrix} -\frac{1}{m_{SC}} & \frac{1}{m_{SC}} \\ 0 & \frac{1}{m_2} \end{pmatrix}, \tag{5}$$

$$\mathbf{H} = \begin{pmatrix} \mathrm{H_{df}}(\omega) & 0 \\ 0 & \mathrm{H_{sus}}(\omega) \end{pmatrix} \tag{6}$$

where $\mathrm{H_{df}}(\omega)$ and $\mathrm{H_{sus}}(\omega)$ are (known) transfer function in frequency domain. And,

$$\mathbf{A} = \begin{pmatrix} A_1\, e^{-s\cdot\tau_1} & 0 \\ 0 & A_2\, e^{-s\cdot\tau_2} \end{pmatrix} \tag{7}$$

where A_1 and A_2 are actuator gains for the first —the Field-emission electric propulsion (FEEP)— and the second —the electrostatic capacitive actuators— channel. τ_1 and τ_2 are delays simulating actuators responses. Default values are $\tau_1 = 0.2\,\mathrm{s}$ and $\tau_2 = 0$ s.

The dynamics of the test mass are contained in the following matrix

$$\mathbf{D} = \begin{pmatrix} s^2 + \omega_1\left(1 + \frac{m_1}{m_{SC}} + \frac{m_2}{m_{SC}}\right) + \frac{m_2}{m_{SC}}\,\omega_{12} & \frac{m_2}{m_{SC}}\,(\omega_1 + \omega_{12}) + \Gamma_x \\ \omega_{12} & s^2 + \omega_1 + \omega_{12} - 2\,\Gamma_x \end{pmatrix} \tag{8}$$

where we define the first and second test mass masses, m_1 and m_2; the spacecraft mass, m_{SC}; first and differential stiffness, ω_1 and $\omega_{12}(=\omega_2^2 - \omega_1^2)$, and gravitational cross-coupling between both test masses, Γ_x. In this model, the interferometer is represented by a sensing matrix, $\mathbf{S}$, translating physical test mass displacements into interferometric read-out,

$$\mathbf{S} = \begin{pmatrix} S_{11} & S_{12} \\ S_{21} & S_{22} \end{pmatrix} \tag{9}$$

The four elements are interferometric calibration factors. The non-diagonal are the possible cross-couplings mixing both channels. Processing of interferometer data introduces a delay of 0.4 s on both channels which is indicated by $\mathbf{DT}_{\mathrm{ifo}}$ in the model —see figure 1. We consider as well a second delay coming from the guidance inputs, this could be useful to compensate for a lack of information on the true time that guidance command is applied inside the DFACS software. Guidance delays are indicated by $\mathbf{DT}_{\mathrm{in}}$ block in the model. Mathematically, both blocks are equivalent:

$$\mathbf{DT}_{\mathrm{ifo}} = \begin{pmatrix} e^{-s\cdot\Delta T_{\mathrm{ifo},1}} & 0 \\ 0 & e^{-s\cdot\Delta T_{\mathrm{ifo},2}} \end{pmatrix}, \quad \mathbf{DT}_{\mathrm{in}} = \begin{pmatrix} e^{-s\cdot\Delta T_{in,1}} & 0 \\ 0 & e^{-s\cdot\Delta T_{in,2}} \end{pmatrix}. \tag{10}$$

The model is implemented in the LTPDA toolbox as an analytical expression for each of the components. In future version, this model will be implemented as an state-space model in order to enhance its integration with the rest of the models of the toolbox.

4. Description of the parameter estimation methods

In the following we provide a summary of the three methods used in the analysis of the sixth operational exercise. More details for each method can be found in the interleaved references.

Linear Fit The procedure for this method is based on a model linearisation to first order. The method performs a time domain parameter estimation. The different steps of the fit process are summarized as follow [8]:

(i) Data for all investigations and channels are whitened. The whitening filter is obtained through the inverse of the power spectrum for a data segment where no signal is injected, i.e. where we are truly evaluating the noise performance of the instrument.

Table 1. Results of the parameter estimation methods for the sixth operational exercise.

Parameter	Linear $\hat{x} \pm \sigma$	Non-linear $\hat{x} \pm \sigma$	MCMC $\hat{x} \pm \sigma$
A_1	1.0699 ± 0.0005	1.0705 ± 0.0006	1.0694 ± 0.0003
A_2	0.99998 ± 0.00003	0.99998 ± 0.00003	0.99996 ± 0.00002
S_{21}	$(1.2 \pm 0.4) \times 10^{-6}$	$(1.2 \pm 0.4) \times 10^{-6}$	$(1.9 \pm 0.3) \times 10^{-6}$
$\Delta T_{\text{in},1}$	-0.1982 ± 0.0005	-0.1985 ± 0.0005	-0.1998 ± 0.0002
$\Delta T_{\text{in},2}$	-0.199 ± 0.001	-0.199 ± 0.001	-0.199 ± 0.001
ω_1^2	$(-1.319 \pm 0.002) \times 10^{-6}$	$(-1.319 \pm 0.002) \times 10^{-6}$	$(-1.319 \pm 0.002) \times 10^{-6}$
$\omega_2^2 - \omega_1^2$	$(-7.160 \pm 0.006) \times 10^{-7}$	$(-7.160 \pm 0.006) \times 10^{-7}$	$(-7.150 \pm 0.005) \times 10^{-7}$

(ii) Model response to current parameters values is calculated and whitened. This is then subtracted from the data.

(iii) A first order expansion (in terms of the parameters) of the model is calculated and whitened. The set of different series corresponding to different parameters represent the fit basis.

(iv) Each experiment can provide only a limited amount of information, therefore some parameters will be undetermined causing an impair of fit capability. Fit basis is then changed for each experiment by singular value decomposition (SVD) which provides a new fit basis in terms of a linear combination of the elements of the old (physical) basis. The corresponding parameters are linear combination of the original fit parameters.

(v) SVD parameters are estimated for each experiment separately by the solution of the corresponding system of normal equations.

(vi) Once the values of all SVD parameters is known with the corresponding uncertainties. The results for the different experiments are collected together and the system is inverted in order to come back to the desired fit parameters.

(vii) New values for fit parameters are then updated and the procedure can start again in a loop until convergence is reached. If the ratio between squared difference of two consecutive parameters estimation and parameters variance is less than 1, then convergence is assumed and the loop is stopped. Parameters which have demonstrated minimum value for mean squared error during the loop are selected for the output.

Non-linear fit In this case, the method implements a non-linear parameter estimation in time domain. The key features of the algorithm are [6]:

(i) The joint test statistic, in χ^2 sense, is computed on all available investigations and channels.

(ii) Data are decorrelated by applying whitening filters previously estimated on noisy data stretches.

(iii) To ease the search, parameters are bounded with lower and upper limits.

(iv) The optimization starts from an initial guess by using a derivative-free algorithm (simplex). Eventually, if the minimum is too far away from the optimum, a preliminary large-scale gradient-based algorithm (BFGS Quasi-Newton) is also applied.

(v) Estimated 1-σ error are obtained by inverting the Fisher matrix of analytical first-order derivatives.

Markov Chain Monte Carlo (MCMC) The MCMC method implemented in LTPDA performs a Monte Carlo integration of the log-likelihood function computed in frequency domain. In summary, the algorithn can be split in following steps [5]:

(i) The Fast Fourier transform is applied to segments where signals are injected and the power spectrum is computed to evaluate the noise level in a signal free segment.

(ii) The expected covariance matrix of the parameters is computed as the inverse of the Fisher matrix.

(iii) The method can optionally look for the maximum likelihood parameters using the simplex algorithm.

(iv) Sample the log-likehood surface with the Metropolis algorithm. Initially, the method applies an *annealing* to guarantee a correct exploration of the parameter space. There is a cooling down phase before starting the integration of the likelihood surface around the likelihood maxima. During this phase, the covariance matrix which defines the proposal distribution is rescaled in some samples to improve convergence. The user sets all parameters that define these phases.

(v) Sample the log-likehood surface around the maxima obtained in the previous phase until the chain reaches the number of samples set by the user.

5. Parameter estimation methods comparison

The previous methods were applied to the two investigations of the sixth operational exercise. Our Fisher matrix analysis showed that, given the two investigations available, we are only able to estimate 7 parameters independently. We selected the subset shown in Table 1, which produces a full rank Fisher matrix. As we show below, estimating the numerical value for these parameters allows to completely subtract the injected signal in most cases, implying that the selected parameters are the most relevant ones for the current investigation.

In the table, the results for the three parameter estimation methods show a good agreement for the numerical value as well as for the estimated error on the parameter. Although not shown here, this uncertainty agrees with the one computed from the Fisher matrix analysis —more details on the Fisher matrix computation can be found in [5].

Although the methods seem in good agreement we developed some tools to quantify the agreement and check if the different parameter estimation methods provide us different information about the data being analysed. We discuss this in the next section.

6. Statistical goodness-of-fit tests

We test accuracy of our parameter estimation methods by statistical tests on residuals. Fit residuals are obtained subtracting model response from the measured signal. If the model and the fit process are accurate, then we expect to remove the deterministic signal from our measured signal. Therefore the residuals are expected to be equivalent to the noise output of the system —figures 2 and 3 show the response predicted for each model and the residuals for the two investigations under analysis, respectively.

The main idea behind the test is that subtracting the estimated from the measured response should retrieve a noise curve which is statistically equivalent to the noise level of the system evaluated previously to the signal injection. There are two main limitation to such kind of test: i) The noise out of the system should be stationary and ii) A noise only time series must be available for comparison. The test we construct is based on the comparison of residuals and noise power spectral densities. In order to build a test for power spectral densities we should in general face the problem of the statistical properties of the spectrum, which, in the case of Welch overlap windowed averaged periodogram, can be highly complex [9] and annoyingly dependent on data underlying distribution. We overcome the problem with the Kolmogorv-Smirnov test (KS-test)

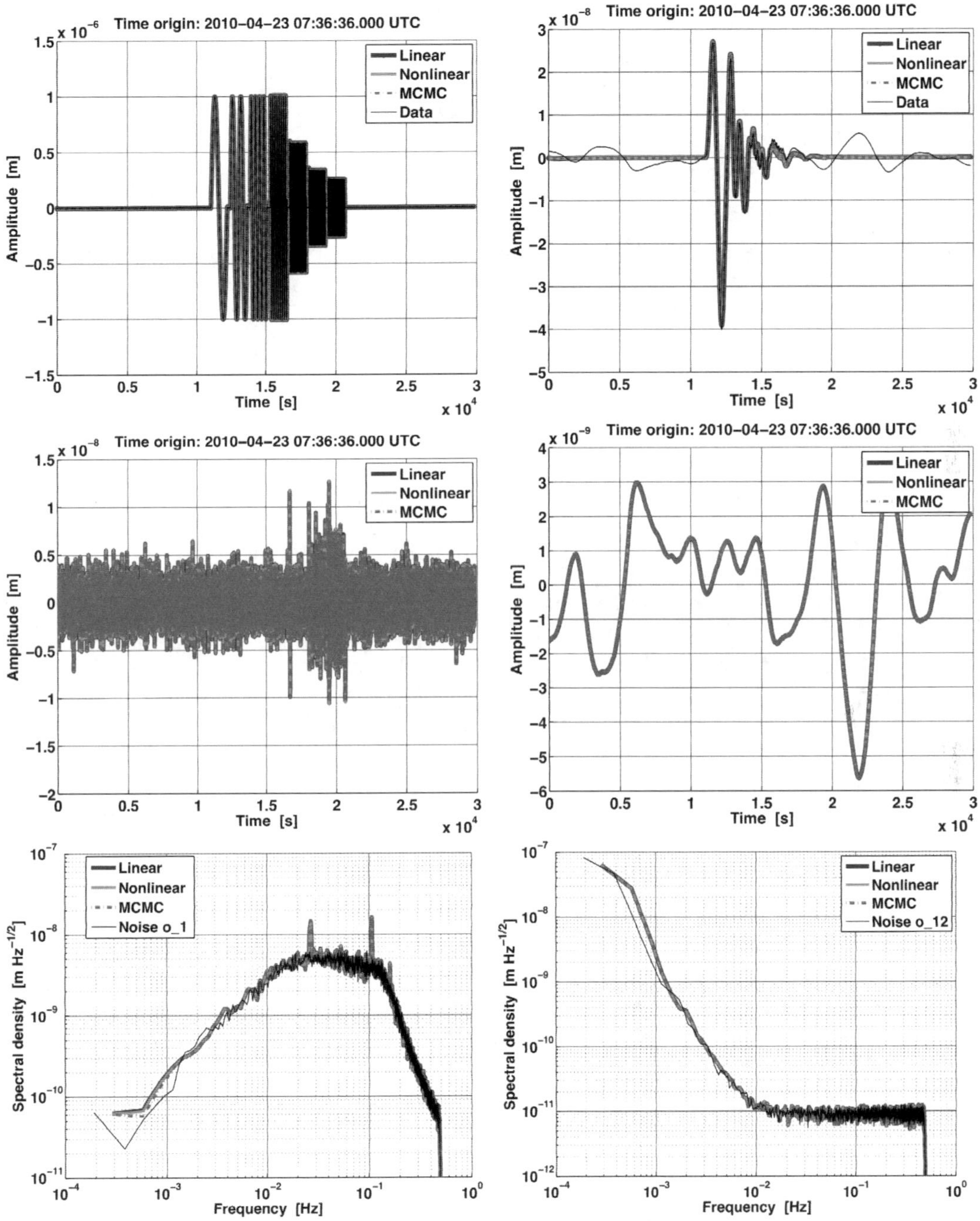

Figure 2. Investigation 1 in the 6th operational exercise. Top to bottom for channel $o_{x,1}$ (left) and $o_{x,12}$ (right): Comparison in time domain of the measured output against output predicted by the models resulting from each method (top), residuals in time domain (middle) and residuals in frequency domain compared to independent noise level estimate (bottom).

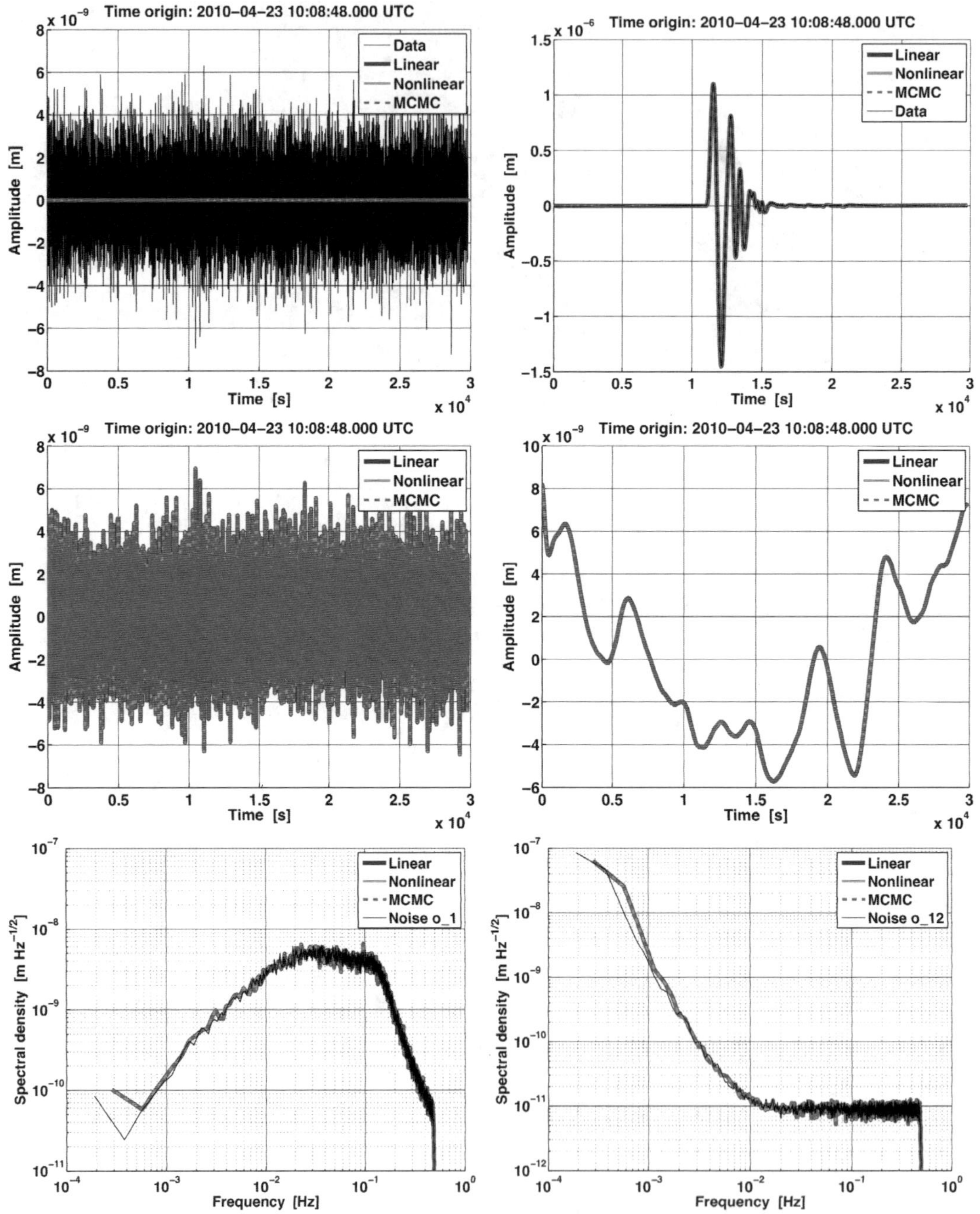

Figure 3. Investigation 2 in the 6th operational exercise. Top to bottom for channel $o_{x,1}$ (left) and $o_{x,12}$ (right): Comparison in time domain of the measured output against output predicted by the models resulting from each method (top), residuals in time domain (middle) and residuals in frequency domain compared to independent noise level estimate (bottom).

Table 2. Results of the Kolmogorov - Smirnov test for the sixth operational exercise.

Investigation	Method	Test Result	P-Value	Test Statistic	Critical Value @95%
Inv. 1 $o_{x,1}$	Linear	Reject	0.0049	0.052	0.041
Inv. 1 $o_{x,1}$	Non-Linear	Reject	0.0046	0.052	0.041
Inv. 1 $o_{x,1}$	MCMC	Reject	0.0048	0.052	0.041
Inv. 1 $o_{x,12}$	Linear	Not Reject	0.6480	0.022	0.041
Inv. 1 $o_{x,12}$	Non-Linear	Not Reject	0.6480	0.022	0.041
Inv. 1 $o_{x,12}$	MCMC	Not Reject	0.6543	0.022	0.041
Inv. 1 $o_{x,1}$	Linear	Not Reject	0.9801	0.014	0.041
Inv. 2 $o_{x,1}$	Non-Linear	Not Reject	0.9801	0.014	0.041
Inv. 2 $o_{x,1}$	MCMC	Not Reject	0.9801	0.014	0.041
Inv. 2 $o_{x,12}$	Linear	Not Reject	0.4631	0.026	0.041
Inv. 2 $o_{x,12}$	Non-Linear	Not Reject	0.4478	0.026	0.041
Inv. 2 $o_{x,12}$	MCMC	Not Reject	0.4478	0.026	0.041

whose statistic is not dependent form the statistical properties of the data under test —a more detailed description of this method can be found in [7].

The null hypothesis for the test is to consider that residuals and noise are two random processes with the same underlying distribution, i.e. residuals and noise are two realizations of the same physical process. The null hypothesis is rejected or not on the basis of a given significance level which is fixed to 5% in the present case. Null hypothesis is rejected if the probability associated with the KS-statistic of the two data series is less than the required significance level. Results of the KS-test for our methods are reported in table 2. The test provides the same results for the three methods. Spectra of fit residuals for the first channel on the first investigation are not compatible with expected noise spectra, which shows that our methods are not able to completely remove the injected signal in the first channel for this investigation. This effect, that could show a mismatch between the model being used for the analysis and the one used to generate the data, is currently being investigated. In the other 3 experiments the subtraction works effectively and the residuals spectra are compatible with the expected noise spectra.

7. Summary and conclusions

In this contribution we have presented the comparison between three different parameter estimation methods implemented in LTPDA applied to the the analysis of a set of data generated by a complex LPF simulator, the OSE. Results show that the three methods are in agreement when we perform the analysis of the two investigations in the sixth LPF operational exercise.

We also presented a statistical analysis of the residuals obtained by each method when subtracting the expected response of the system to the measured one. All methods agree as well under this framework, however the Kolmogorov-Smirnov analysis show that the methods are not able to completely subtract the injected signal from the measured output in the first channel of the first investigation, i.e. the experiment where we inject a sequence of sinusoids in the first channel. This effect, enhanced by the high signal-to-noise ratio of the signal in this channel, may be due to a mismatch between the model used for the data analysis and the one used for data generation. It is worth reminding here that the OSE is a black-box simulator developed by Astrium GmbH and delivered to ESA under industrial contract.

Next steps of the data analysis team include the improvement of the data analysis methods and focus on the analysis of the different experiments planned during flight operations. In parallel, the models implemented in the LTPDA toolbox to describe the LPF mission are being implemented as state-space models, which will allow a more modular design and ease the code maintenance and testing.

References

[1] Armano M, Benedetti M, Bogenstahl J, Bortoluzzi D, Bosetti P, Brandt N, Cavalleri A, Ciani G, Cristofolini I, Cruise A M, Danzmann K, Diepholz I, Dixon G, Dolesi R, Fauste J, Ferraioli L, Fertin D, Fichter W, Freschi M, García A, García C, Grynagier A, Guzmán F, Fitzsimons E, Heinzel G, Hewitson M, Hollington D, Hough J, Hueller M, Hoyland D, Jennrich O, Johlander B, Killow C, Lobo A, Mance D, Mateos I, McNamara P W, Monsky A, Nicolini D, Nicolodi D, Nofrarias M, Perreur-Lloyd M, Plagnol E, Racca G D, Ramos-Castro J, Robertson D, Sanjuan J, Schulte M O, Shaul D N A, Smit M, Stagnaro L, Steier F, Sumner T J, Tateo N, Tombolato D, Vischer G, Vitale S, Wanner G, Ward H, Waschke S, Wand V, Wass P, Weber W J, Ziegler T and Zweifel P 2009 *Classical and Quantum Gravity* **26** 094001 URL `http://stacks.iop.org/0264-9381/26/i=9/a=094001`

[2] Antonucci F, Armano M, Audley H, Auger G, Benedetti M, Binetruy P, Boatella C, Bogenstahl J, Bortoluzzi D, Bosetti P, Brandt N, Caleno M, Cavalleri A, Cesa M, Chmeissani M, Ciani G, Conchillo A, Congedo G, Cristofolini I, Cruise M, Danzmann K, Marchi F D, Diaz-Aguilo M, Diepholz I, Dixon G, Dolesi R, Dunbar N, Fauste J, Ferraioli L, Fertin D, Fichter W, Fitzsimons E, Freschi M, Marin A G, Marirrodriga C G, Gerndt R, Gesa L, Giardini D, Gibert F, Grimani C, Grynagier A, Guillaume B, Guzmán F, Harrison I, Heinzel G, Hewitson M, Hollington D, Hough J, Hoyland D, Hueller M, Huesler J, Jeannin O, Jennrich O, Jetzer P, Johlander B, Killow C, Llamas X, Lloro I, Lobo A, Maarschalkerweerd R, Madden S, Mance D, Mateos I, McNamara P W, Mendes J, Mitchell E, Monsky A, Nicolini D, Nicolodi D, Nofrarias M, Pedersen F, Perreur-Lloyd M, Perreca A, Plagnol E, Prat P, Racca G D, Rais B, Ramos-Castro J, Reiche J, Perez J A R, Robertson D, Rozemeijer H, Sanjuan J, Schleicher A, Schulte M, Shaul D, Stagnaro L, Strandmoe S, Steier F, Sumner T J, Taylor A, Texier D, Trenkel C, Tombolato D, Vitale S, Wanner G, Ward H, Waschke S, Wass P, Weber W J and Zweifel P 2011 *Classical and Quantum Gravity* **28** 094002 URL `http://stacks.iop.org/0264-9381/28/i=9/a=094002`

[3] Bender P *et al.* 2000 Laser interferometer space antenna: a cornerstone mission for the observation of gravitational waves Tech. Rep. ESA-SCI(2000)11 ESA

[4] Hewitson M, Armano M, Benedetti M, Bogenstahl J, Bortoluzzi D, Bosetti P, Brandt N, Cavalleri A, Ciani G, Cristofolini I, Cruise M, Danzmann K, Diepholz I, Dolesi R, Fauste J, Ferraioli L, Fertin D, Fichter W, García A, García C, Grynagier A, Guzmán F, Fitzsimons E, Heinzel G, Hollington D, Hough J, Hueller M, Hoyland D, Jennrich O, Johlander B, Killow C, Lobo A, Mance D, Mateos I, McNamara P W, Monsky A, Nicolini D, Nicolodi D, Nofrarias M, Perreur-Lloyd M, Plagnol E, Racca G D, Ramos-Castro J, Robertson D, Sanjuan J, Schulte M O, Shaul D N A, Smit M, Stagnaro L, Steier F, Sumner T J, Tateo N, Tombolato D, Vischer G, Vitale S, Wanner G, Ward H, Waschke S, Wand V, Wass P, Weber W J, Ziegler T and Zweifel P 2009 *Classical and Quantum Gravity* **26** 094003 URL `http://stacks.iop.org/0264-9381/26/i=9/a=094003`

[5] Nofrarias M, Röver C, Hewitson M, Monsky A, Heinzel G, Danzmann K, Ferraioli L, Hueller M and Vitale S 2010 *Phys. Rev.* D **82** 122002

[6] Congedo G, Ferraioli L, Hueller M, De Marchi F, Vitale S, Armano M, Hewitson M and Nofrarias M 2011 *ArXiv e-prints* (*Preprint* `1108.0862`)

[7] Ferraioli L, Congedo G, Hueller M, Vitale S, Hewitson M, Nofrarias M and Armano M 2011 *Phys. Rev. D* **84**(12) 122003 URL `http://link.aps.org/doi/10.1103/PhysRevD.84.122003`

[8] Antonucci F, Armano M, Audley H, Auger G, Benedetti M, Binetruy P, Boatella C, Bogenstahl J, Bortoluzzi D, Bosetti P, Caleno M, Cavalleri A, Cesa M, Chmeissani M, Ciani G, Conchillo A, Congedo G, Cristofolini I, Cruise M, Danzmann K, Marchi F D, Diaz-Aguilo M, Diepholz I, Dixon G, Dolesi R, Fauste J, Ferraioli L, Fertin D, Fichter W, Fitzsimons E, Freschi M, Marin A G, Marirrodriga C G, Gesa L, Giardini D, Grimani C, Grynagier A, Guillaume B, Guzmán F, Harrison I, Heinzel G, Hewitson M, Hollington D, Hough J, Hoyland D, Hueller M, Huesler J, Jeannin O, Jennrich O, Jetzer P, Johlander B, Killow C, Llamas X, Lloro I, Lobo A, Maarschalkerweerd R, Madden S, Mance D, Mateos I, McNamara P W, Mendes J, Mitchell E, Monsky A, Nicolini D, Nicolodi D, Nofrarias M, Pedersen F, Perreur-Lloyd M, Perreca A, Plagnol E, Prat P, Racca G D, Rais B, Ramos-Castro J, Reiche J, Perez J A R, Robertson D, Rozemeijer H, Sanjuan J, Schulte M, Shaul D, Stagnaro L, Strandmoe S, Steier F, Sumner T J, Taylor A, Texier D, Trenkel C, Tombolato D, Vitale S, Wanner G, Ward H, Waschke S, Wass P, Weber W J and Zweifel P 2011 *Classical and Quantum Gravity* **28** 094006 URL `http://stacks.iop.org/0264-9381/28/i=9/a=094006`

[9] Percival D B and Walden A T 1993 *Spectral Analysis for Physical Applications* (Cambridge University Press)

Precision laser development for interferometric space missions NGO, SGO, and GRACE Follow-On

K Numata[1,2], J Camp[2]
[1]University of Maryland, Department of Astronomy/CRESST, College Park, Maryland 20741 USA
[2]NASA Goddard Space Flight Center, Greenbelt, Maryland 20771, USA
E-mail: kenji.numata@nasa.gov

Abstract. Optical fiber and semiconductor laser technologies have evolved dramatically over the last decade due to the increased demands from optical communications. We are developing a laser (master oscillator) and optical amplifier based on those technologies for interferometric space missions, including the gravitational-wave missions NGO/SGO (formerly LISA) and the climate monitoring mission GRACE Follow-On, by fully utilizing the matured wave-guided optics technologies. In space, where simpler and more reliable system is preferred, the wave-guided components are advantageous over bulk, crystal-based, free-space laser, such as NPRO (Non-planar Ring Oscillator) and bulk-crystal amplifier.

1. Introduction

Three space missions are now being planned that will use precision interferometry to achieve very high displacement sensitivity. In Europe, the gravitational wave mission NGO, aiming for launch in 2022, will use laser interferometry to monitor the motion of freely falling test masses in spacecraft separated by $\sim 10^9$ m. In the US, a number of concepts are under study for the SGO mission, including both laser and atom interferometry, requiring the use of low-noise lasers to enable the measurement of gravitational waves. SGO may launch by the end of the next decade. (While the designs of NGO and SGO are under discussion, they are both closely related to LISA [1], the former joint US-European collaboration, and to be conservative we refer to the more stringent LISA laser requirements in this article.) Finally, the GRACE Follow-On mission [2], planned for launch in 2016, will use laser interferometry between spacecraft separated by ~200 km to monitor the time-varying gravity fields of the earth.

Laser technology for the telecom industry has undergone dramatic advances in the past decade, replacing lasers and optical amplifiers based on bulk optics with technology including fiber laser/amplifier, waveguide devices, and semiconductor lasers. These components naturally fit into the precision laser systems needed for the interferometric missions, because of their high mechanical robustness, high reliability, compact form factor, and high wall-plug efficiency. We are pursuing all-fiber/waveguide space laser solution based on the MOFA (master oscillator fiber amplifier) configuration, namely, a waveguide-based oscillator followed by a pre-amplifier and a power amplifier. We are developing both 1.06 μm and 1.55 μm systems for NGO, SGO, GRACE Follow-On, and potentially other lidar missions.

2. Master oscillator

We have developed fiber ring laser and fiber DBR laser, both of which are based on wide-bandwidth fiber Bragg grating (FBG), Fabry-Perot FBG, and Yb-doped gain fiber for 1.06 μm [3]. Although these lasers have comparable performance to the non-planar ring oscillator (NPRO) at low (<10 kHz) frequency, they have larger relaxation oscillation peak around 1 MHz, which could affect the heterodyne interferometry typically operated near this frequency. These fiber lasers also tend to have a large number of optical components, resulting in questions about their reliability. In this paper, we introduce the planar-waveguide external-cavity diode laser (PW-ECL), which we have identified to be a simple and robust design and thus promising for the space applications. It also has the proven reliability by the Telcordia GR-468 standard and the best low-frequency performance among the lasers we have tested.

2.1. PW-ECL design

Fig.1. (*Left*) schematic of the PW-ECL. HR: high-reflective coating, AR: anti-reflective coating, PLC: planar lightwave circuit. (*Right*) Photo of core optics of NPRO and the PW-ECL package. In the PW-ECL butterfly package, the output is PM fiber-coupled after the lens and an optical isolator.

The PW-ECL evaluated here is commercial product (named PLANEX), built by Redfern Integrated Optics (RIO). Figure 1 (*left*) shows schematic of the cavity of the PW-ECL. The laser cavity is formed by two reflectors: a high-reflective (HR) coated facet on an InP multiple quantum well gain chip, and an anti-reflection (AR) coated waveguide grating formed in a silica-on-silicon planar lightwave circuit (PLC). All components are integrated into a standard 14-pin butterfly package on top of a thermoelectric cooler. As shown in Fig.1 (*right*), the package is much more compact than that of NPRO, in which a strong magnet fundamentally limits the size. The narrow reflection peak of the Bragg reflector in the PLC enables stable, low-noise, single-mode lasing at a selected wavelength within the telecom C-band (1528~1565 nm). Typical output power is ~15 mW.

2.2. PW-ECL noise performance

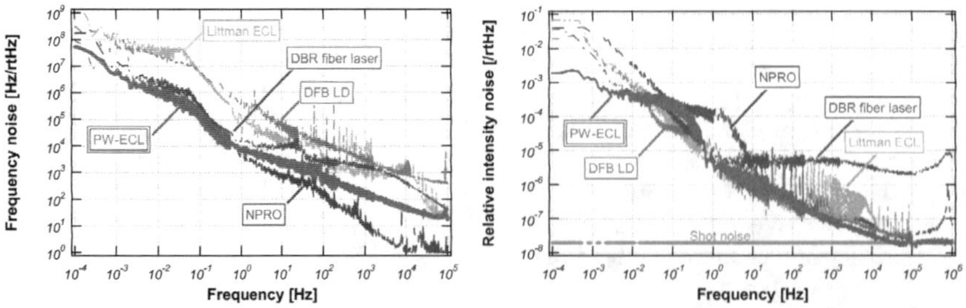

Fig.2. Free-running frequency (*left*) and relative intensity (*right*) noise spectra of various single-frequency lasers. PW-ECL: PLANEX from Redfern Integrated Optics (1542 nm), NPRO: Model 125

from Lightwave Electronics (1064 nm), DBR fiber laser: ROCK from NP Photonics (1542 nm), Littman ECL: Lion from Sacher Lasertechnik (1064 nm), DFB-LD: FRL15DCWD from Fitel (1578 nm).

As shown in Fig. 2, the PW-ECL has a sufficiently low level of frequency and intensity noise to be suitable for precision measurement applications. The noise level was comparable or better than the NPRO and fiber laser between 0.1 mHz to 100 kHz. Controllability of the PW-ECL was also demonstrated by stabilizing its frequency to hyperfine line of acetylene at 10^{-13} level of Allan deviation [4], by phase locking between two PW-ECLs, and by Pound-Drever-Hall frequency locking to high finesse (~20,000) optical cavity.

2.3. Space qualification tests

We have performed various tests on the PW-ECL to investigate its suitability for a space qualification program. All performance and reliability testing done to date has indicated that the PW-ECL is robust for spaceflight. Some of the test results are introduced below.

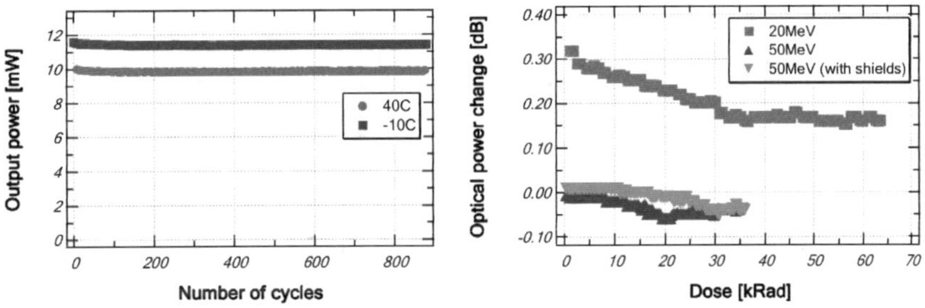

Fig.3. (*Left*) Vacuum thermal cycling test result of PW-ECL. Output powers of the two temperature extremes are plotted. (*Right*) Proton radiation test result of PW-ECL.

- Active vacuum thermal cycling was carried out between – 10 to + 40°C, over 1000 cycles. The power levels at the temperature extremes had differences in the range of 0.5 to 1.8 mW at the nominal 10 mW output level (Fig.3, (*left*)). This was explained as due to a slight temperature difference between the gain chip and PLC at the two temperature extremes. All module tested show very little or no degradation. Additional tests over a temperature cycling range of -25 to + 80°C showed similar steady performance of the PW-ECL.

- Hermeticity testing commenced with a 100°C bake for 16 hours, followed by lid sealing with a nitrogen-helium fill gas, fine-leak testing and thermal cycling (– 43 to + 93°C, 20 cycles). The final fine leak, gross leak and residual gas analysis showed that the modules remained hermetic, which greatly reduces possibility of material degradation over long time.

- The PW-ECL module was subjected to a total dose of 200 krad of 1 MeV gamma rays from a Co^{60} source behind an aluminum plate with 2.5-mm thickness. Small changes were noted in the wavelengths (≈ +10 pm) and the output power (≈ – 0.30 dB), both in situ and 2 days after the termination of the irradiation. Under proton beam at 20 MeV and 50 MeV and total dose of few kRad, the maximum change in wavelengths was ≈ + 50 pm. There was recovery several days after that stabilized at ≈ + 25 pm. The maximum change in output powers was ≈ – 0.15 dB (Fig.3, (*right*)). There was no mode hop during those irradiations and no significant changes in the phase noise performance after the irradiation. These tests were performed at a highly accelerated rate to fluence values on the order of 100 times those expected over 5 years on the orbit.

2.4. Future activities

Future activities planned for the PW-ECL for use in space include: 1) Development of 1064-nm version by using different gain chip material and PLC design while keeping other mechanical design, 2) Introduction of small phase modulation section within the gain chip in order to expand the frequency tuning bandwidth from ~10 kHz to ~10 MHz, 3) Irradiation testing with low-energy proton, 4) Frequency locking to optical cavity to achieve thermal noise level frequency noise performance (at Univ. of Texas Brownsville), and 5) 1/f frequency noise level reduction by using different gain-chip design.

3. Pre-amplifier

The ~10 mW output of PW-ECL itself may be sufficient for the GRACE-FO mission, and as a seed laser for the power amplifier in the LISA mission where ~2 W level of power is required. However, in order to have power margin and to be flexible for changing mission requirements, we are developing fiber pre-amplifiers with output power of ~200 mW both at 1.06 μm and at 1.55 μm. Both are common wavelengths for laser-communication and lidar applications.

3.1. Design

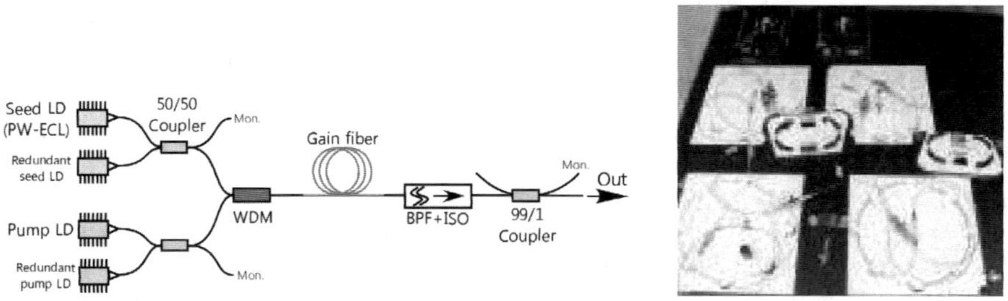

Fig.4. (*Left*) Conceptual design of seed laser + pre-amplifier system. WDM: wavelength-division multiplexing coupler, BPF: band-pass filter, ISO: optical isolator. (*Right*) Photograph of preamplifier system under test.

Figure 4 (*left*) shows the design of the seed laser and the pre-amplifier system. The laser unit will house two seed lasers (PW-ECLs) and two pump laser diodes for redundancy, as well as their control circuits. All fiber components are single-mode, polarization-maintaining. The gain fiber is ~2-m long and core-pumped. The pump LD is at ~976 nm, which is more reliable than the 808 nm LD for the Nd:YAG lasers in the space.

3.2. Preamplifier noise performance

Using 1542-nm PW-ECL seed and Er-doped gain fiber, we built test pre-amplifier (without redundant LDs). For 1064-nm system, we are using Yb-doped gain fiber and seed NPRO, which will be eventually replaced by 1064 nm PW-ECL, currently under development by RIO. Figure 5 shows noise performance of the 1542-nm preamplifier, where 9-mW PW-ECL input is amplified to 180 mW. The preamplifier adds negligible frequency noise (Fig.5 (*left*)). The preamplifier adds intensity noise especially at low frequency (Fig.5 (*right*)). This excess noise can be suppressed by controlling the pump power.

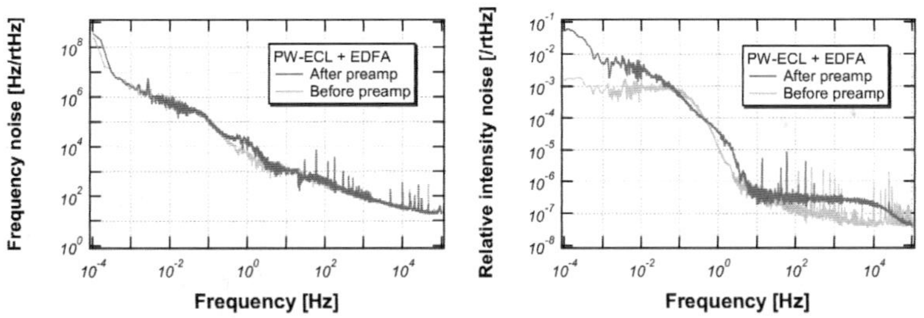

Fig.5. (*Left*) Frequency noise and (*Right*) relative intensity noise before and after the Er-doped preamplifier seeded by 1542-nm PW-ECL.

3.3. Space qualification tests

The fiber-based preamplifier power level is low and thus the system components are not expected to present risk for flight; however, radiation hardness is an important remaining factor in the qualification of pre-amplifier components. We have performed gamma irradiation tests on 1.06-μm Yb amplifier components, which are less matured than the 1.55-μm components. The tested components include Yb-doped gain fiber, optical band-pass filter with isolator, and coupler.

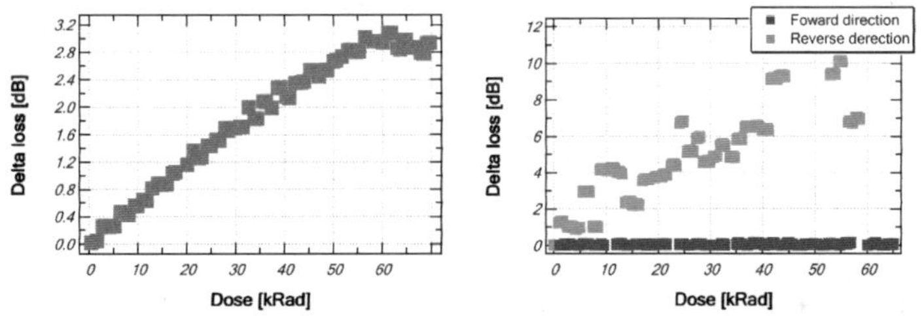

Fig.6. Gamma irradiation testing results of preamplifier components. Vertical axes represent change in the insertion loss. (*Left*) Yb-doped gain fiber (5-m length). Irradiation was terminated at 60 kRad. (*Right*) 1.064-μm band-pass filter isolator.

Figure 6 shows results example. Yb-doped gain fiber (1200-dB/m small signal absorption at 976 nm) showed typical increase in loss of ~0.6 dB/m with 60 kRad dose (Fig.6, (*left*)). This value is typical for doped fibers under the high dose rate. Such damage will be much lower at lower dose rate expected during the mission and will be mitigated by the photo-annealing effect. The filter-isolator showed ~0.15 dB and ~10 dB increase in loss for the forward and reverse directions, respectively (Fig.6, (*right*)). The performance and functionality of the isolator were not affected by the irradiation.

3.4. Current activities

We are identifying components to be used in the preamplifier package and evaluating test preamplifiers. We are also designing space-qualifiable low-noise current driver based on the design shown in [5]. The goal is to have a compact oscillator-preamplifier module with integrated electronics at a total electrical power consumption of 10 W.

4. Power amplifier

We are working with Lucent Government Solutions to develop Yb fiber amplifier for the NGO/SGO missions. The wavelength is 1.06 μm.

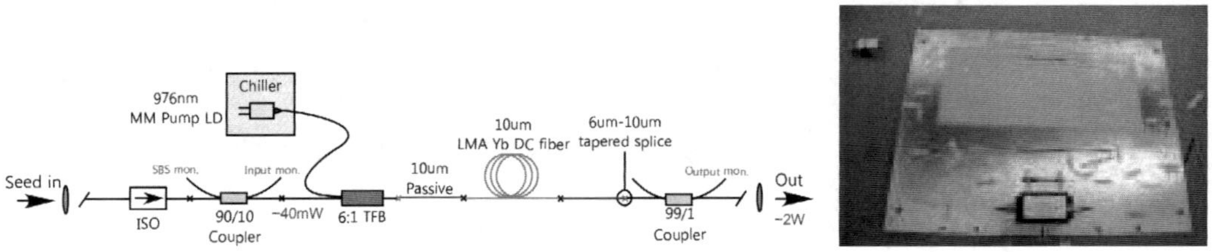

Fig.7. (*Left*) fiber amplifier configuration. ISO: optical isolator, TFB: tapered fiber bundle. (*Right*) photograph of the amplifier breadboard.

4.1. Design

After various trade studies and simulations, the fiber amplifier shown in Fig. 7 was constructed. A 976-nm pump LD was selected over 915-nm, since it was expected to have better performance (efficiency, noise figure, and temperature insensitivity) in our simulation. A 10-μm core, double-clad (DC), large mode area (LMA) fiber was selected as a gain media to elevate the SBS (stimulated Brillouin scattering) threshold. The length of the gain fiber is about 5 m. It is forward, clad-pumped through a tapered fiber bundle (TFB, 6:1 pump combiner). The forward-pumping design minimizes potential sources of feedback (e.g. from the TFB). Such feedback could result in lasing and in catastrophic damage, which we found to occur in our prototype backward-pumped amplifiers. The simple single-stage amplifier design gives ~2 W output (~6 W maximum) with 40-mW seed input and 5-A pump current on the fiber-coupled 10-W pump LD. Redundant pump LDs can be attached to unused ports of the TFB. The TFB was identified as the most risky component within the amplifier, and was pre-screened by a thermal imager and re-packaged using suitable compounds for higher reliability in space. This design was characterized for optical efficiency (~68%), polarization extinction ratio (PER, 17~32 dB, with 20 dB typical), and noise figure of 5 dB.

4.2. Amplifier performance

We are evaluating frequency (phase) noise performance of the power amplifier and the results are shown in Fig. 8. In these figures, commercial NPRO was used as a seed source.

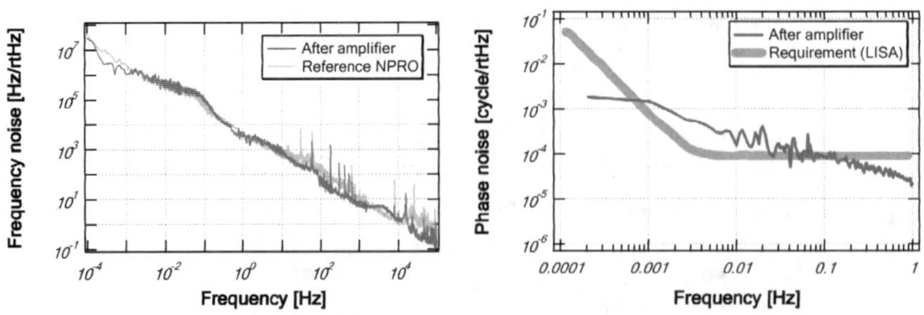

Fig.8. (*Left*) Free-running frequency noise of the NPRO and the amplifier output at 2 W. (*Right*) Differential phase noise of the amplifier at 2 W.

The power amplifier adds negligible frequency noise to the carrier light of the seed as depicted in Fig. 8 (*left*). Any frequency noise added by the amplifier would be suppressed by the frequency stabilization control loop, whose light is taken after the amplifier. The differential phase noise (phase noise between the carrier and the clock transfer sideband at 2 GHz) [6] is not suppressed by the control loop and has to be kept low. As shown in Fig. 8 (*right*), the noise level is factor ~4 higher than the LISA requirement level. The excess differential phase noise had some correlations to temperature variations (which are not stabilized in our setup) and to the increase in the backscattered power (which indicates the existence of SBS). Our goal is to improve this performance by implementing a stable temperature (which will be present in the spacecraft) and also by modifying the amplifier optics. We have confirmed that the possible SBS component is not affected by the seed laser's linewidth by using an iodine-stabilized NPRO as seed.

The intensity noise of the amplifier was also measured and stabilized. The bandwidth of the intensity control servo was 10 kHz. Figure 9 (*left*) shows the relative intensity noise (RIN) of the amplifier at 2 W output at low frequency. The stabilized level is below the LISA requirement above 1 mHz. We think the excess noise below 1 mHz is not inherent to the amplifier, but due to the temperature variation in the stabilization system. At the higher frequency region (Fig. 9, (*right*)), the RIN is at the shot noise level of ~1.5-mW detected power above ~5 MHz with and without the stabilization, as in the seed NPRO only. This indicates shot-noise-limited heterodyne interferometry is possible with the amplifier at this detection power level and with a proper choice of the RF heterodyne frequency.

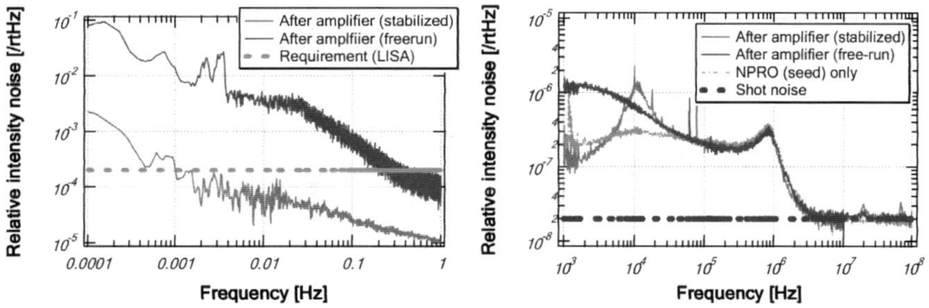

Fig.9. (*Left*) Free-running and stabilized RIN of the amplifier output at 2 W at low frequency. (*Right*) amplifier RIN at high frequency together with seed NPRO's RIN. The stabilized RIN was measured with out-of-loop photo-detector and the band-pass filter.

4.3. Space qualification tests

Vacuum thermal cycling test of the amplifier was conducted by splicing the amplifier breadboard into a thermally controlled vacuum chamber. The temperature of the breadboard was set using temperature-controlled fluid circulated through the base-plate into the vacuum chamber.

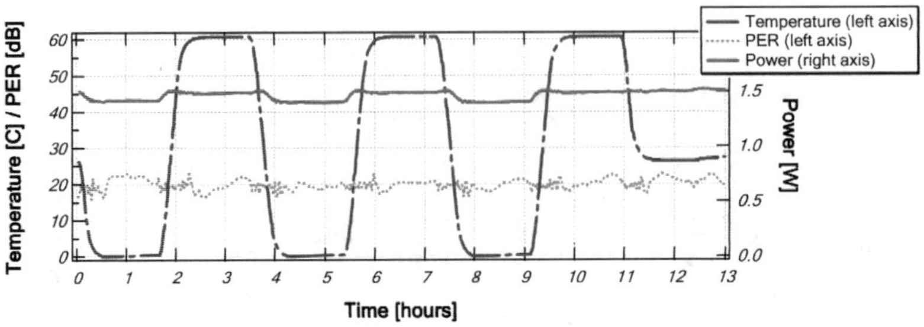

Fig.10. Vacuum thermal cycling result of the power amplifier.

The amplifier showed negligible performance change after 100 hours of vacuum thermal cycling between 0 °C and 60 °C. As an example, Fig. 10 shows data collected during the temperature cycles, including power, PER, and the baseplate temperature.

4.4. Current activities

We are evaluating radiation hardness of a set of gain fibers, including the one used in the prototype amplifier and two other commercial brands. We are further investigating the noise issues, especially its relationship to the SBS (threshold), by adopting slightly different amplifier configurations. The 5-m gain fiber, which generates 6 W maximum, is too long for the NGO/SGO missions that require ~1-W level output. Reducing the lengths of the gain and delivery fibers should improve the noise performance through increased SBS threshold. Selection of an output isolator, a filter to cut out amplified spontaneous emission, and a high-power fiber collimator are also underway.

5. Summary

We presented our efforts at the NASA Goddard Space Flight Center on precision laser systems for the interferometric missions NGO, SGO, and GRACE Follow-On. Our approach is based on fibers and waveguides, adapting the matured wave-guided optics technologies of the telecom industry. The fiber approach has numbers of advantages over traditional bulk optics, and will be used as possible in space laser missions in the future. We are characterizing the laser and the amplifier both at the component level and at the system level. We will continue to pursue the development of low-noise space laser systems, bringing in the latest technical advances.

Acknowledgement

The authors would like to thank Alfonso Piccirilli, Robert Ahrens, Anthony Fields, Gregory Johnson, Leslie Reith, and Gunther Schwarz at Lucent Government Solutions for their technical support on this project. The authors also thank Lew Stolpner and Mazin Alalusi at Redfern Integrated Optics for their development work on the PW-ECL.

References

[1] P. Bender, K. Danzmann, and the LISA Study Team, "Laser interferometer space antenna for the detection of gravitational waves, pre-Phase A report," Tech. Rep. MPQ233, Max-Planck-Institut fur Quantenoptik, Garching (1998). 2nd ed.

[2] M. Stephens, et al., "Demonstration of an Interferometric Laser Ranging System for a Follow-On Gravity Mission to GRACE," in Proceedings of IEEE International Conference on Geoscience and Remote Sensing Symposium, (IEEE, 2006), pp. 1115~1118.

[3] K. Numata, J. Chen, and J. Camp, J. Phys.: Conf. Ser., 228 (2010) 012043.

[4] K. Numata, J. Camp, M. A. Krainak, and L. Stolpner, Optics Express, 18 (2010) 22781.

[5] C. J. Erickson, M. V. Zijll, G. Doermann, and D. S. Durfee, Rev. Sci. Instrum. 79 (2008) 073107.

[6] M. Tröbs, S. Barke, Th. Theeg, D. Kracht, G. Heinzel, and K. Danzmann, Optics Letters, 35 (2010) 435.

Tackling gravity wave confusion noise with template optimizers

Alexander Stroeer & Matthew Benacquista

Center for Gravitational Wave Astronomy, University of Texas at Brownsville, 80 Ft. Brown, Brownsville, TX 78521, USA

E-mail: astroeer@phys.utb.edu

Abstract. The Mock LISA Data Challenge 4.0 simulated the joint two-year recording of gravitational wave signals from mergers of spinning black holes, extreme mass ratio inspirals, Galactic white dwarf binaries, bursts from cosmic strings, and a stochastic background—all over LISA instrument noise. We analysed this data using a global multi-start box and bound optimization scheme, incorporating multi-dimensional Nelder Mead simplex 2 optimization. Our scheme identified 2658 binaries. Of these, 2246 were found to systematically decompose the power in a strong spinning black hole merger into a "white dwarf binary transform". The remaining 416 binaries were identified with a false alarm rate of $\sim 23\%$.

1. Introduction

The Laser Interferometer Space Antenna (LISA) was a proposed joint venture of ESA and NASA to observe gravitational waves in the low frequency range from $10^{-4} - 10^{-1}$ Hz [1]. The proposed configuration was a constellation of 3 spacecraft in heliocentric orbit. The constellation forms an interferometer from a nearly equilateral triangle with 5 million km arms. The likely sources of gravitational radiation in this range are supermassive black hole mergers at cosmological distances, extreme mass ratio inspirals of compact objects into supermassive black holes, and the Galactic population of close white dwarf binaries. A stochastic background of gravitational waves and bursts from cosmic strings are less likely, but nevertheless possible sources. All of these sources will be recorded by LISA simultaneously, and this forms the main challenge of LISA data analysis—how to disentangle all of these sources.

The Mock LISA Data Challenges were created to encourage the data analysis community to develop solutions to the problem of identifying all of these sources [2]. There have been 5 challenges associated with the MLDC. Challenge 1 introduced several source types, each with its own data stream [2]. The Challenge 2 included a data stream containing a Galactic population of $\sim 26 \times 10^6$ white dwarf binaries, although all the systems were monochromatic (i.e. $\dot{f} = 0$) [3]. The most successful search algorithm to date was able to extract $\sim 20,000$ binaries, many of which had cross-correlations greater than 0.9 [3]. Unfortunately, this code [4, 5] has been lost [6], and is no longer in use. Following a reprise of Challenge 1 that saw the introduction of new algorithms [7], Challenge 3 incorporated chirping and mass-transferring systems. The three submissions to Challenge 3 had varying degrees of success, recovering 494 [8], 1940 [9], and 14,838 [10], but with somewhat low cross-correlations with the true population [11]. In Challenge 4, a single two year data stream was simulated containing Galactic binaries, supermassive black hole mergers, extreme mass ratio inspirals, bursts from cosmic strings, and a stochastic background. We have developed an analysis algorithm to identify Galactic binaries from within the LISA data stream [11]. We tested our algorithm against Challenge 3 of the MLDC, which had only the Galactic population of white dwarf binaries and recovered 7131 binaries with cross-correlations greater than 0.9. When applied to the Challenge 4 data, the additional signals from the supermassive black hole mergers have proven to be problematic and significantly reduced the efficacy of our algorithm. These signals must be removed for the algorithm to be fully successful.

2. The Algorithm

The gravitational wave signal from a close white dwarf binary can be described by 8 parameters. The biggest challenge with identifying individual close white dwarf binaries from within the combined Galactic signal of 30 million sources is that a global fit requires searching through an $8N$-dimensional space for the optimal fit with the data, but N is an unknown number to be determined by the analysis. Previous solutions to this problem have included fixing N, or through single signal subtraction schemes where individual binaries are identified and subtracted from the signal [10, 12, 13]. We use an adaptive multi-dimensional/multi-signal layered reduction approach. We work in the frequency domain using the rigid adiabatic approximation for our template models and process only isolated data snippets covering the frequency spread of a single signal. We adaptively add signals to the single-signal solution parameter space if the signals are found to be overlapping.

A match between template and signal is determined by minimizing the value of χ^2. The minimization technique that we use is the Nelder-Mead simplex 2 (NM2) algorithm. The Nelder-Mead optimization technique was proposed in 1965 by John Nelder and Roger Mead [14] in order to minimize an objective function in a many-dimensional space without the use of derivatives. As first proposed, the algorithm scaled with the dimension n of the space as $O(n^2)$. Recent advances in computing have allowed the implementation of the simplex 2 algorithm, an improvement over the initial one, with an $O(n)$ scaling.

This improvement has allowed us to more efficiently search over larger parameter spaces without relying on more costly MCMC techniques. We use a multi-start global optimization scheme in which we invoke the NM2 at multiple starting points in parameter space and record the many local minima that are obtained. The global minimum is then found by comparing the local minima. We guide and tune the multi-start by using a box-and-bound method in which we restrict the search over parameter space by resampling from an initial random distribution of sampling points. The χ^2 surface is sampled with a coarse grid for an initial distribution and the probability of resampling points is weighted by the χ^2 values from the initial sample. In this way we focus our multi-start distribution on regions of parameter space where the solution is most likely to be found. This can be seen in Figure 1.

3. Tuning on MLDC 3.1.0

We tuned our algorithm on the data set from MLDC Challenge 3 that contained only Galactic binaries. The MLDC 3.1.0 data set contained 30 million close white dwarf binaries and included systems with a drifting frequency. The change in frequency could be in either direction and was due to either gravitational wave inspiral ($\dot{f} > 0$) or stable mass transfer ($\dot{f} < 0$). With a run time of 15840 CPU-hours on the UTB cluster *Futuro*, the algorithm identified 7131. We determined the efficiency of the algorithm by using the cross-correlation tool from the MLDC assessment suite and identified any source that was found with a cross-correlation greater than 0.9 as a successful recovery. This allowed us to impose thresholds in amplitude and frequency in order to eliminate many false alarms. The thresholds are shown in Figure 2. With these thresholds, we determined a 98.1% efficiency of detection and had 6996 binaries with a cross-correlation greater than 0.9.

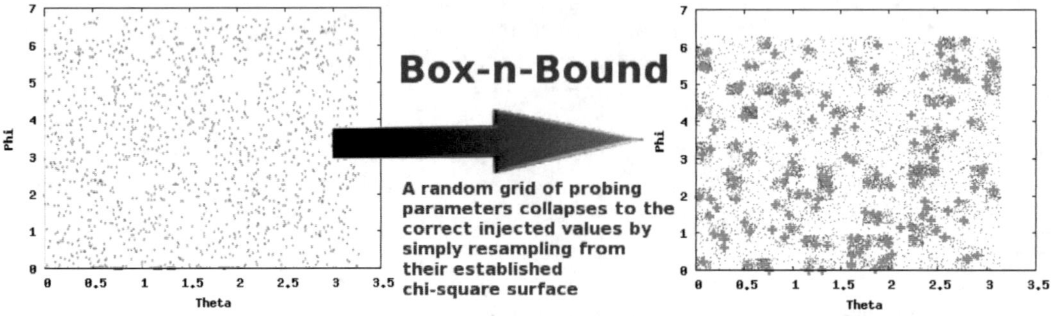

Figure 1. An example of the box-and-bound method in the two-dimensional slice of parameter space for sky location. The initial distribution of multi-start points is randomly distributed throughout the parameter space, but after a few resamplings, the distribution has collapsed about the true values shown by crosses.

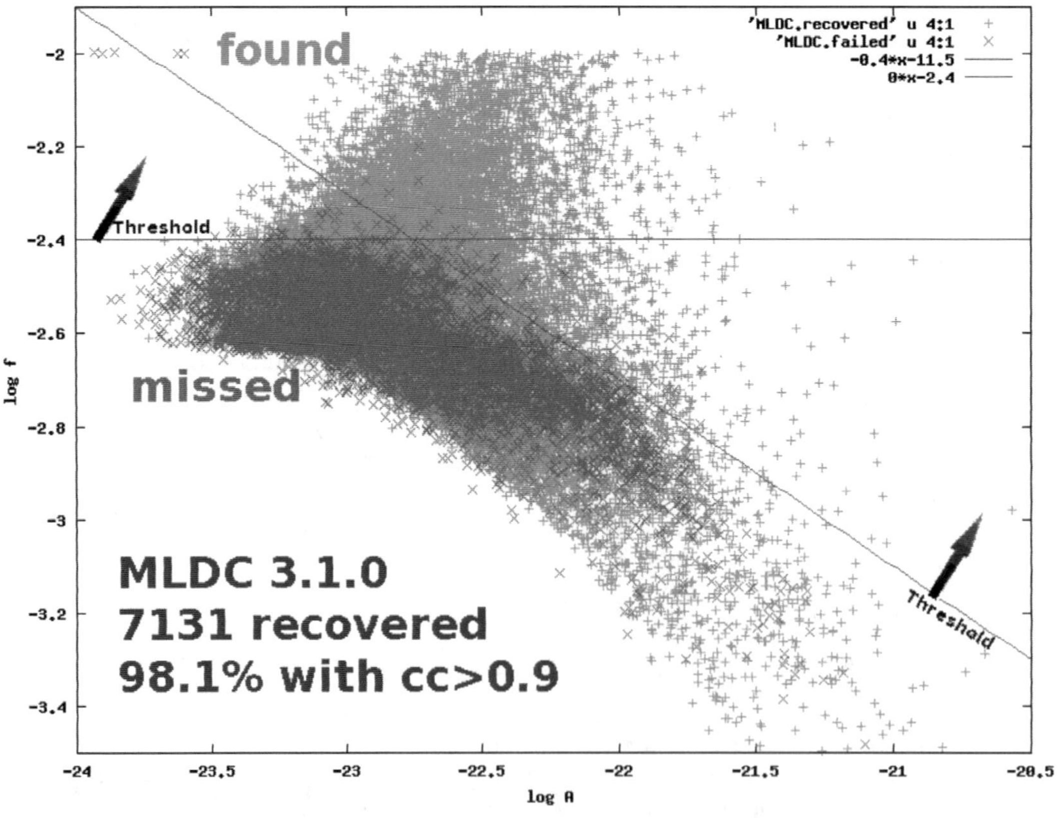

Figure 2. Recovered binaries in the frequency-amplitude plane of parameter space from the tuning runs on MLDC 3.1.0. The found binaries are shown in light grey (green in the online version), while missed binaries are shown in dark grey (red in the online version). The frequency and amplitude thresholds are also shown.

4. Application to MLDC 4

The data set that was provided with the fourth round of the Mock LISA Data Challenge was known as the "whole enchilada." This is because it was a single set containing all sources of the separate MLDC 3 challenges combined together. These sources include supermassive black hole mergers, extreme mass ratio inspirals, a cosmological stochastic background, cosmic string bursts, and the full Galactic population of interacting and detached white dwarf binaries. Although our code was intended solely to operate on a cleaned data set, we applied it to the whole enchilada without removing any of the other signals. With a runtime on *Futuro* of 16810 CPU-hours, our code identified 8192 binaries. After thresholding based on tuning with known binaries from the training set, 2658 signals remained. However, 2246 of these in the very low frequency regime had a strange feature of nearly matching characteristics. We assumed that these binaries were the result of a flaw in the analysis and discarded them as well. This left us with 412 recovered binaries that we submitted to the MLDC task group—a rather poor performance. After comparison with the key-file for challenge 4 and applying the cross-correlation assessment, we found a detection efficiency of 77.1%. The missed-found plot for these binaries is shown in Figure 3.

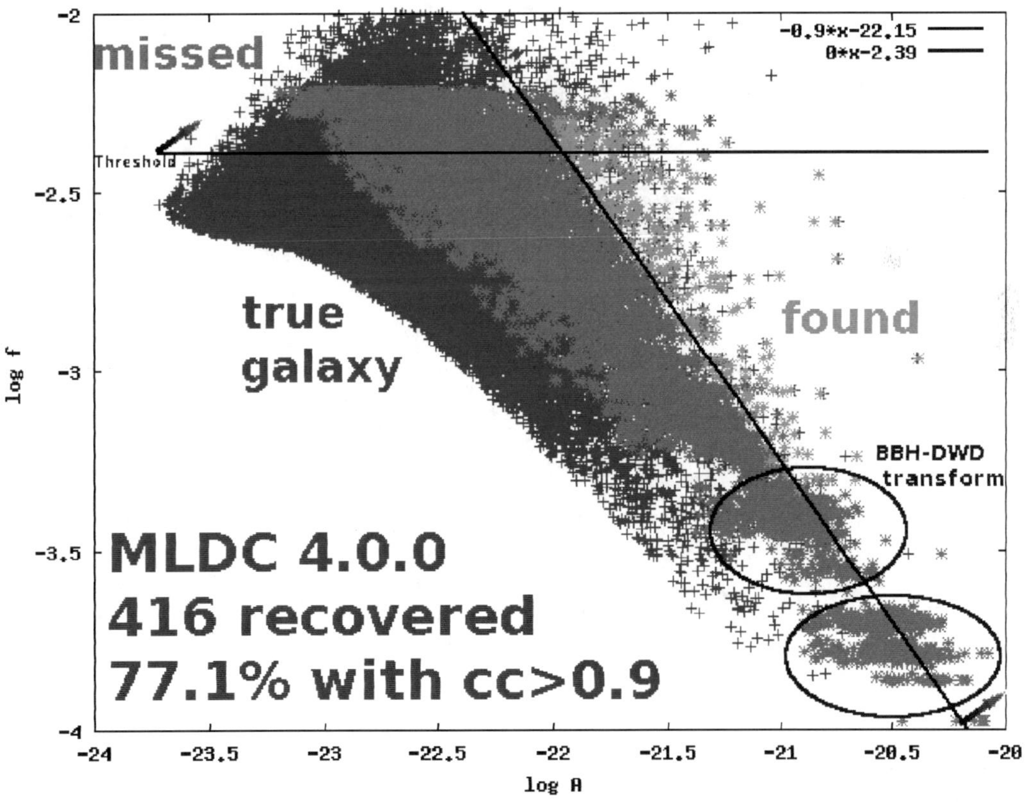

Figure 3. Recovered binaries in the frequency-amplitude plane of parameter space from the application of our code to the MLDC 4 data set. The true Galaxy is shown in dark grey (blue in the online version), the missed or false binaries are shown in grey (red in the online version), and the found binaries are shown in light grey (green in the online version). The 2246 discarded binaries are circled in the lower right-hand corner.

Journal of Physics: Conference Series **363** (2012) 012055 doi:10.1088/1742-6596/363/1/012055

5. The White Dwarf Binary Transform

After the key to the data was released, post-processing showed that a very strong supermassive black hole merger was placed just at the frequency range where the unusual population of binaries was found. We therefore hypothesized that these matching binaries were actually a white dwarf binary wavelet transform of the supermassive black hole binary. In order to check this result, we superposed the frequency-domain of the time-delay interferometry signal of the supermassive black hole binary with the unusual, discarded white dwarf binaries. We found that 1319 of the white dwarf binaries corresponded to a match with the supermassive black hole binary. This is shown in Figure 4 We then compared the time domain response of the black hole merger and

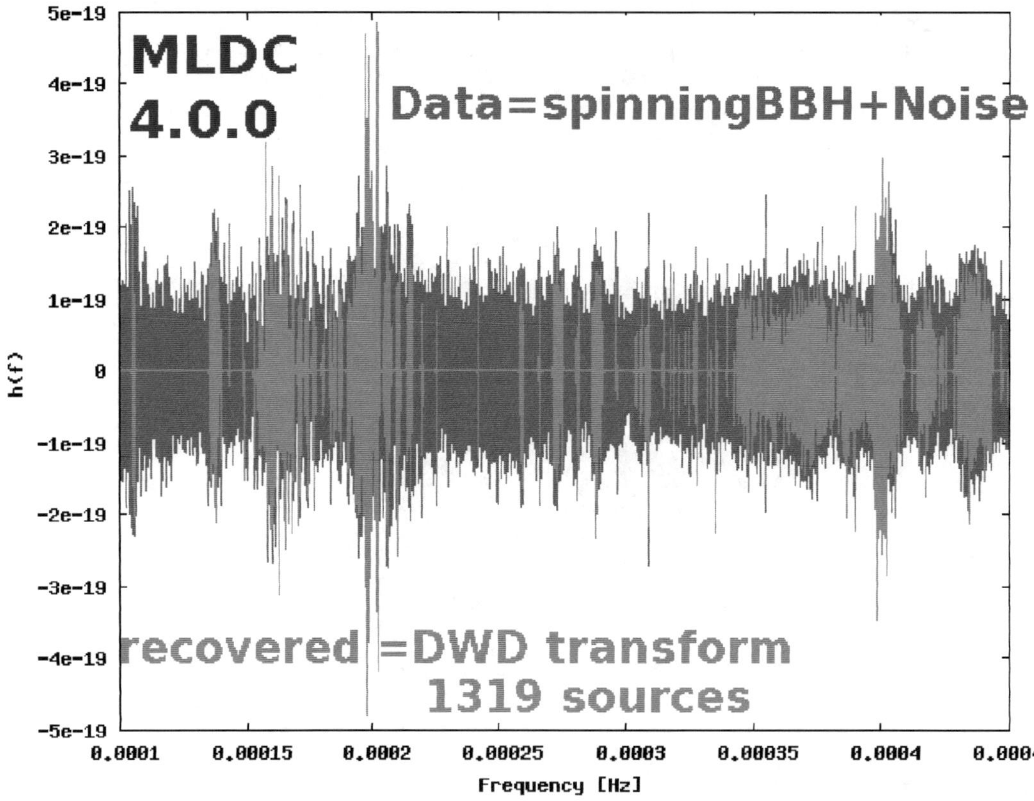

Figure 4. The TDI frequency response of the injected black hole merger and the recovered compact object binary wavelet transform. The frequency response shows obvious gaps towards lower power, which translates to a poor match in time.

the superposition of these false alarms in Figure 5. We notice a good visual match of the white dwarf binary wavelet transform with the generic waveform for a black hole merger. However, the match is qualitative at best and lacks the necessary fidelity to give a quantitative match to the exact waveform.

6. Discussion

We have demonstrated that our code is CPU-efficient, yet powerful enough to untangle source confusion in high numbers at low false alarm rates. However, this success is predicated on a clean

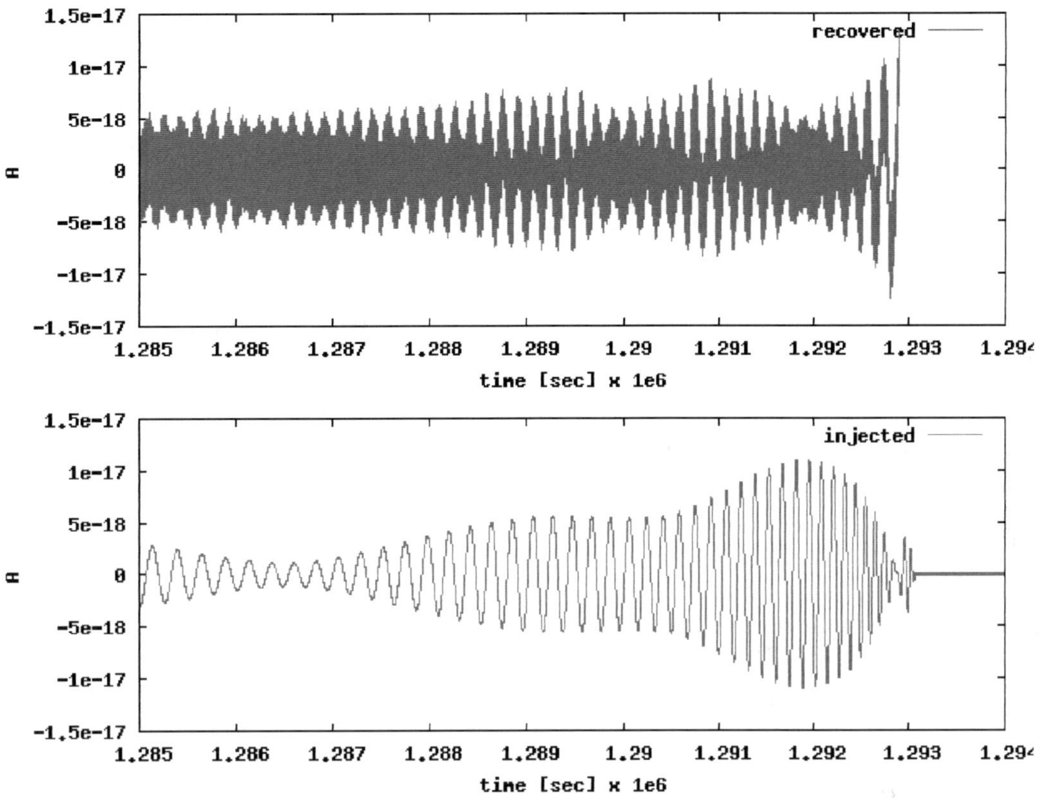

Figure 5. The barycentric time response of the white dwarf binary transform (above) compared with the true supermassive black hole binary merger signal (below). There is a qualitative match, but not a quantitative one.

data stream, and so a subtraction of loud supermassive binary black hole mergers is required in order to fully realize the potential of our technique. Despite this, our technique has yielded 412 signals at an efficiency of 77.1% in the presence of very strong source confusion from other source classes. We further presented qualitative evidence that our data analysis can also identify and decompose a strong binary black hole merger at low frequencies by means of a white dwarf binary transform. We acknowledge the accidental character of the white dwarf binary transform and therefore do not claim conclusive quantitative proof. It is up to future work to concentrate on this property and fine-tune the approach to completely and quantitatively decompose a black hole merger signal beyond doubt. We highlight the prospect of such an approach, as it may be able to render the frequency and time domain response of a spinning black hole merger without ever applying any kind of numerical template. If successful, this work would show that it may be possible to recover strong waveforms even if the most sophisticated template calculated today may mismatch the reality of a gravitational waveform. [15]

Acknowledgments

We acknowledge the support of the Center for Gravitational Wave Astronomy which is funded by CREST (NSF HRD-0734800) and URC (NASA grant NNX09AV06A). Extensive use was made of *Futuro* which was supported by MRI (NSF CNS-0923456).

References

[1] Bender P, Ciufolini I, Danzmann K, Folkner W, Hough J, Robertson D, Rüdiger A, Sandford M, Schilling R, Schutz B, Stebbins R, Sumner T, Touboul P, Vitale S, Ward H, Winkler W, Cornelisse J, Hechler F, Jafry Y and Reinhard R 1996 Lisa. laser interferometer space antenna for the detection and observation of gravitational waves. a cornerstone project in esa's long term space science programme "horizon 2000 plus". pre-phase a report, december 1995 Tech. rep. Garching

[2] Arnaud K A, Babak S, Baker J G, Benacquista M J, Cornish N J, Cutler C, Larson S L, Sathyaprakash B S, Vallisneri M, Vecchio A and Vinet J Y 2006 *Laser Interferometer Space Antenna: 6th International LISA Symposium* (*American Institute of Physics Conference Series* vol 873) ed S M Merkovitz & J C Livas pp 619–624 (*Preprint* `arXiv:gr-qc/0609105`)

[3] Babak S, Baker J G, Benacquista M J, Cornish N J, Crowder J, Cutler C, Larson S L, Littenberg T B, Porter E K, Vallisneri M, Vecchio A, data challenge task force t M L, Auger G, Barack L, Błaut A, Bloomer E, Brown D A, Christensen N, Clark J, Fairhurst S, Gair J R, Halloin H, Hendry M, Jimenez A, Królak A, Mandel I, Messenger C, Meyer R, Mohanty S, Nayak R, Petiteau A, Pitkin M, Plagnol E, Prix R, Robinson E L, Roever C, Savov P, Stroeer A, Toher J, Veitch J, Vinet J, Wen L, Whelan J T, Woan G and Challenge-2 participants t 2008 *Classical and Quantum Gravity* **25** 114037 (*Preprint* `0711.2667`)

[4]

[5] Crowder J and Cornish N J 2007 *Classical and Quantum Gravity* **24** 575 (*Preprint* `0704.2917`)

[6] Littenberg T B 2011 *Phys. Rev. D* **84** 063009 (*Preprint* `1106.6355`)

[7] Trias M, Vecchio A and Veitch J 2008 *Classical and Quantum Gravity* **25** 184028 (*Preprint* `0804.4029`)

[8] Trias M, Vecchio A and Veitch J 2009 *ArXiv e-prints* (*Preprint* `0904.2207`)

[9] Whelan J T, Prix R and Khurana D 2010 *Classical and Quantum Gravity* **27** 055010 (*Preprint* `0908.3766`)

[10] Błaut A, Babak S and Królak A 2010 *Phys. Rev. D* **81** 063008 (*Preprint* `0911.3020`)

[11] Babak S, Baker J G, Benacquista M J, Cornish N J, Larson S L, Mandel I, McWilliams S T, Petiteau A, Porter E K, Robinson E L, Vallisneri M, Vecchio A, Data Challenge Task Force t M L, Adams M, Arnaud K A, Błaut A, Bridges M, Cohen M, Cutler C, Feroz F, Gair J R, Graff P, Hobson M, Shapiro Key J, Królak A, Lasenby A, Prix R, Shang Y, Trias M, Veitch J, Whelan J T and participants t C 2010 *Classical and Quantum Gravity* **27** 084009 (*Preprint* `0912.0548`)

[12] Błaut A, Królak A and Babak S 2009 *Classical and Quantum Gravity* **26** 204023

[13] Stroeer A and Veitch J 2009 *Phys. Rev. D* **80** 064032 (*Preprint* `0907.2198`)

[14] Nelder J A and Mead R 1965 *Computer Journal* 308–313

[15] Peters P C and Mathews J 1963 *Physical Review* **131** 435–440

Probing the Inhomogeneous Universe with Gravitational Wave Cosmology

Kent Yagi
Department of Physics, Kyoto University, Kyoto, 606–8502, Japan
E-mail: kent@tap.scphys.kyoto-u.ac.jp

Atsushi Nishizawa and Chul-Moon Yoo
Yukawa Institute for Theoretical Physics, Kyoto University, Kyoto 606–8502, Japan

Abstract.
If we assume that we live in the center of a spherical inhomogeneous universe, we can explain the apparent accelerating expansion of the universe without introducing the unknown dark energy or modifying gravitational theory. Direct measurement of the cosmic acceleration can be a powerful tool in distinguishing ΛCDM and the inhomogeneous models. If ΛCDM is the correct model, we have shown that DECIGO/BBO has sufficient ability to detect the positive redshift drift of the source by observing gravitational waves from neutron star binaries for 5-10 years. This enables us to rule out any Lemaître-Tolman-Bondi (LTB) void model with monotonically increasing density profile. Furthermore, by detecting the positive redshift drift at $z \sim 0$, we can even rule out generic LTB models unless we allow unrealistically steep density gradient at $z \sim 0$. We also show that the measurement accuracy is slightly improved when we consider the joint search of DECIGO/BBO and the third generation Einstein Telescope. This test can be performed with GW observations alone without any reference to electromagnetic observations.

1. Introduction

When we assume that our universe is homogeneous and isotropic, current cosmological observations (e.g. type Ia supernovae (SNe) [1]) indicate that the cosmic expansion is accelerating. Once we allow the possibility that our universe has cosmological scale spherical inhomogeneity with the observer at the center, the observations can be explained without introducing the unknown dark energy or alternative theories of gravity. In such models, the Copernican Principle is apparently violated and the cosmic expansion is not necessarily accelerating. Therefore, future direct detection of the cosmic acceleration is very useful in distinguishing them. The direct detection of the acceleration of the universe provides not only a key to solve the dark energy problem, but also a critical test of the Copernican Principle.

The simplest example of the inhomogeneous model is the Lemaître-Tolman-Bondi (LTB) spacetime which is a spherically symmetric dust solution of the Einstein Equations. The metric is given as

$$ds^2 = -dt^2 + \frac{\partial_r R(t,r)^2}{1-k(r)r^2}dr^2 + R^2(t,r)d\Omega^2, \tag{1}$$

where $R(t, r)$ and $k(r)$ are arbitrary functions. These functions are related to the density of the dust $\rho(r)$ via Einstein Equations. If we live at the center of the LTB spacetime with a Gpc-scale void, the apparent cosmic acceleration can be explained. The LTB model has been partially tested with the cosmological observations like, the cosmic microwave background, baryon acoustic oscillations, the kinetic Sunyaev-Zeldovich effect etc., [2–14], but it has not been completely ruled out yet. For example, Ref. [13] claimed that LTB void models are in conflict with current observations with nearly scale invariant primordial spectrum, but Nadathur and Sarkar [14] claim that that observational results can be explained by assuming different primordial spectrum ansatz. Therefore, the observations which are unaffected by the primordial information are very crucial to test a wide class of LTB models. One of such observations is the redshift-distance relation of type-Ia SNe. However, it has been shown that one can construct the LTB void model that exactly reproduces the redshift-distance relation in ΛCDM [15]. Therefore, we need other observations that do not depend on the primordial information.

Redshift-drift measurement is the one that meets our demands [16] (see also Refs. [15, 17, 18]). Redshift drift is the time evolution of the redshift due to the cosmic acceleration, hence its detection means the direct measurement of the acceleration of the cosmic expansion.

In the Friedmann-Lemaître-Robertson-Walker (FLRW) spacetime, the redshift drift is given as

$$\Delta_t z = H_0 \Delta t_o \left(1 + z - \frac{H(z)}{H_0}\right), \tag{2}$$

where Δt_o denotes the observation period, and H_0 and $H(z)$ are the Hubble parameter at present and at redshift z [19], respectively. In the ΛCDM universe, $\Delta_t z$ is *positive* in the range $z = 0 - 2$ [17]. On the other hand, $\Delta_t z$ in LTB spacetime obeys the following differential equation [18];

$$\frac{d}{dz}\left(\frac{\Delta_t z}{1+z}\right) = \frac{1}{(1+z)^2} \frac{\partial_t^2 \partial_r R}{\partial_t \partial_r R} \Delta t_o, \tag{3}$$

where $R(t, r)$ is an arbitrary function. Recently, Yoo *et al.* [18] have shown that when the matter density is monotonically increasing, the right hand side of the above equation is negative. By combining this with $\Delta_t z|_{z=0} = 0$, we can show that $\Delta_t z$ must be negative at any z in this model. Furthermore, they have shown that for any LTB density profile, $d\Delta_t z/dz < 0$ for $z \ll 1$ unless we allow unrealistically steep density profile at $z \sim 0$. Therefore, it is crucial to measure the sign of $\Delta_t z$ at $z < 2$ in order to distinguish the ΛCDM and LTB models.

The order of magnitude of $\Delta_t z$ is roughly given as the cosmic age divided by the observation time, hence $\Delta_t z \sim 10^{-10}$ for 1yr observation. It is this tiny value that makes it difficult to measure $\Delta_t z$ with current technology. Recently, Quartin and Amendola [17] have shown that by measuring the shift of the Lyman α forest of quasar spectrum at $z = 2 - 5$ with the proposed E-ELT instrument CODEX [20] for 10 yrs, it will be possible to distinguish ΛCDM and typical LTB void models. However, CODEX would not be able to measure $\Delta_t z$ at low z since Lyα forest can be measured from ground only at $z \geq 1.7$ [20]. Hence, they can only test typical LTB models but not generic ones.

In this paper, we estimate how accurately we can measure the redshift drift with future gravitational wave (GW) interferometers. It seems that DECIGO [21, 22] and BBO [23] are the only proposed detectors that can measure $\Delta_t z$ at $z \lesssim 2$. We consider neutron star (NS) binaries as GW sources, which are often called as the standard sirens and can be unique tools to probe the cosmic expansion [24–26]. When the expansion is accelerating, we may find an additional phase shift in gravitational waveforms [21]. We assume that ΛCDM is the correct model and estimate whether we can tell the positivity of the redshift drift at low z with GW observations.

Throughout this paper, we use the unit $G = c = 1$.

2. Correction in gravitational waveform due to the redshift drift

Let us first derive the correction in the gravitational waveform phase due to the accelerating expansion of the universe. We here consider a binary consisting of two bodies with masses m_1 and m_2. We define the time to coalescence measured in the observer frame as $\Delta t \equiv t_c - t$ with t_c representing the coalescence time. This Δt includes the effect of cosmic acceleration. On the other hand, we denote the time to coalescence measured in the source frame as Δt_e and define $\Delta T \equiv (1+z_c)\Delta t_e$ where z_c is the source redshift at coalescence. Then, the relation between Δt and ΔT is [21, 27]

$$\Delta t = \Delta T + X(z_c)\Delta T^2 , \tag{4}$$

where $X(z)$ is the acceleration parameter defined as

$$X(z) \equiv \frac{H_0}{2}\left(1 - \frac{H(z)}{(1+z)H_0}\right) . \tag{5}$$

Notice that $X(z)$ is related to the redshift drift $\Delta_t z$ as

$$\Delta_t z = 2(1+z)\Delta t_o X(z). \tag{6}$$

By using the stationary phase approximation [28], the waveform in the Fourier domain can be expressed as

$$\tilde{h}(f) = e^{i\Psi_{\rm acc}(f)}\tilde{h}(f)|_{\rm no\ accel} , \tag{7}$$

where

$$\Psi_{\rm acc}(f) \equiv -2\pi f X(z_c)\Delta T(f)^2 \tag{8}$$

and $\tilde{h}(f)|_{\rm no\ accel}$ corresponds to the gravitational waveform in the Fourier domain without cosmic acceleration explained in the Appendix. The leading $\Delta T(f)$ is given as [28]

$$\Delta T(f) = 5(8\pi \mathcal{M}_z f)^{-8/3}\mathcal{M}_z, \tag{9}$$

where $\mathcal{M}_z \equiv M(1+z_c)\eta^{3/5}$ denotes the redshifted chirp mass with $M \equiv (m_1+m_2)$ and $\eta \equiv m_1 m_2/M^2$ representing the total mass and the symmetric mass ratio, respectively. With this at hand, $\Psi_{\rm acc}(f)$ is given as [21, 27]

$$\Psi_{\rm acc}(f) \equiv -\Psi_N(f)\frac{25}{768}X(z_c)\mathcal{M}_z x^{-4} , \tag{10}$$

where $x \equiv (\pi\mathcal{M}_z f)^{2/3}$. A term proportional to $\Psi_N(f)x^n$ represents the n-th post-Newtonian (PN) order relative to the leading $\Psi_N(f) \equiv \frac{3}{128}(\pi\mathcal{M}_z f)^{-5/3}$, hence this is "-4PN" correction. We used the restricted-2PN waveform including spin-orbit coupling at 1.5PN order , where "restricted" means that we only take the leading Newtonian quadrupole contribution for the amplitude and neglect the ones from higher harmonics.

3. Numerical Setups

The measurement accuracy on $\Delta_t z$ has been estimated mainly using ultimate DECIGO (which is three orders of magnitude more sensitive than DECIGO) [21, 27]. Especially, Ref. [27] used this measurement accuracy on the acceleration parameter $X(z)$ to estimate the ones on cosmological parameters. We improve their analyses in the following way and apply the result to probe the inhomogeneity of the universe: (I) We use DECIGO/BBO (and not ultimate DECIGO), including the confusion noises from white dwarf (WD) binaries. (II) Rather than sky-averaged analysis, we perform Monte Carlo simulations by randomly distributing the directions and

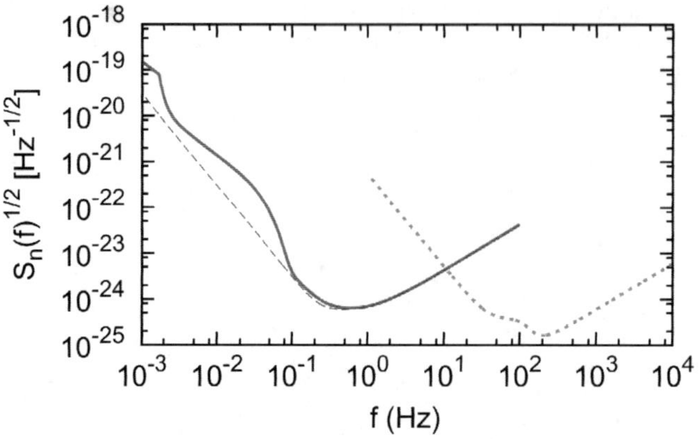

Figure 1. The root noise spectral density of DECIGO/BBO with (solid) and without (dashed) WD confusion noises, and the one of ET (dotted).

orientations of sources. (III) We include the spin-orbit coupling into binary parameters. (IV) We use the merger rate that reflects the star formation history.

We take the binary parameters as

$$\theta^i = (\ln \mathcal{M}_z, \ln \eta, \beta, t_c, \phi_c, D_L, \bar{\theta}_{\rm S}, \bar{\phi}_{\rm S}, \bar{\theta}_{\rm L}, \bar{\phi}_{\rm L}, X_H)\,. \tag{11}$$

Here, β, ϕ_c and D_L represent the spin-orbit coupling parameter, the coalescence phase and the luminosity distance, respectively. $(\bar{\theta}_{\rm S}, \bar{\phi}_{\rm S})$ are the direction of the source in barycentric frame which is tied to the ecliptic and centered in the solar system barycenter (see Fig. 1 of Ref. [29]), and $(\bar{\theta}_{\rm L}, \bar{\phi}_{\rm L})$ are the orientation of the source orbital axis in the same frame. We have introduced a new parameter X_H which is defined as $X_H \equiv X(z)/H_0$. X_H is related to $\Delta_t z$ through Eq. (6).

We estimate how accurately we can measure binary parameters θ^i using Fisher analysis. Assuming that the detector noise is stationary and Gaussian, the measurement accuracy is given as $\Delta\theta^i \equiv \left(\tilde{\Gamma}^{-1}\right)_{ii}^{1/2}$ [28], where, $\tilde{\Gamma}_{ij}$ is defined as

$$\exp\left[-\frac{1}{2}\tilde{\Gamma}_{ij}\delta\theta^i\delta\theta^j\right] \equiv p^{(0)}(\boldsymbol{\theta})\exp\left[-\frac{1}{2}\Gamma_{ij}\delta\theta^i\delta\theta^j\right]. \tag{12}$$

Here, $\delta\theta^i \equiv \theta^i - \theta^i_{\rm true}$ with $\theta^i_{\rm true}$ representing the true values of binary parameters and $p^{(0)}(\boldsymbol{\theta})$ is the prior information. Γ_{ij} is the Fisher matrix which is defined as

$$\Gamma_{ij} \equiv 4{\rm Re}\int_{f_{\rm in}}^{f_{\rm fin}} df\, \frac{\tilde{h}^*(f)\tilde{h}(f)}{S_n(f)}, \tag{13}$$

where $S_n(f)$ denotes the noise spectrum of the detector. In this paper, we use the instrumental noise spectrum of BBO when we perform Fisher analyses and assume that DECIGO also has the same sensitivity as BBO. The noise curves are shown in Fig. 1. The actual expression of the total noise spectrum is given in Eq. (36) of Ref. [30].

$f_{\rm in}$ and $f_{\rm fin}$ in Eq. (13) are given as [30]

$$f_{\rm in} = (256/5)^{-3/8}\pi^{-1}\mathcal{M}_z^{-5/8}\Delta t_o^{-3/8}, \qquad f_{\rm fin} = 100{\rm Hz}, \tag{14}$$

respectively. $f_{\rm in}$ represents the frequency at Δt_o before coalescence and $f_{\rm fin}$ denotes the higher cutoff frequency of DECIGO/BBO. For fiducial values, we set $m_1 = m_2 = 1.4 M_\odot$ and take $t_c = \phi_c = \beta = 0$. Since the dimensionless spin parameter is less than 1, we adopt the prior as $|\beta| < 9.4$ [31]. We assume that a flat ΛCDM is the correct model. In the following computations, we set the fiducial values as $H_0 = 70$km/s/Mpc and the cosmological parameters as $\Omega_m = 0.3$ and $\Omega_\Lambda = 0.7$.

The number of binaries $\Delta N(z)$ that exists in each bin with size $\delta z = 0.1$ is estimated as [32]

$$\Delta N(z) = 4\pi \left[a_0 r(z)\right]^2 \dot{n}(z)(d\tau/dz)\delta z \Delta t_o, \tag{15}$$

where

$$a_0 r(z) = \int_0^z dz'/H(z'), \qquad d\tau/dz = \{(1+z)H(z)\}^{-1} \tag{16}$$

with the Hubble parameter at redshift z given as

$$H(z) \equiv H_0 \sqrt{\Omega_m (1+z)^3 + \Omega_\Lambda}. \tag{17}$$

Here, a_0, $r(z)$, and τ each represents current scale factor, comoving distance and proper look back time, respectively. $\dot{n}(z) \equiv \dot{n}_0 R(z)$ shows the NS/NS merger rate per unit comoving volume per unit proper time, where we assume the merger rate today as $\dot{n}_0 = 10^{-6}$ Mpc^{-3} yr^{-1} [33] and the merger rate evolution against z as

$$R(z) = \begin{cases} 1+2z & (z \le 1) \\ \frac{3}{4}(5-z) & (1 \le z \le 5) \\ 0 & (z \ge 5), \end{cases} \tag{18}$$

which is based on the current observation of star formation history [34].

Following Refs. [29–31], we randomly generate 10^4 sets of $(\bar{\theta}_{\rm S}, \bar{\phi}_{\rm S}, \bar{\theta}_{\rm L}, \bar{\phi}_{\rm L})$ for each fiducial redshift $z_k \equiv 0.1k + 0.05$ $(k = 0, 1, \cdots)$. Then, we calculate $\left(\tilde{\Gamma}^{-1}\right)_{ii}^{1/2}$ for each set and take the average to yield $\left[\left(\tilde{\Gamma}^{-1}\right)_{ii}^{1/2}\right]_{\rm ave}$. The measurement accuracy at each z_k is estimated as

$$\Delta\theta^i = N_{\rm int}^{-1/2} \Delta N(z_k)^{-1/2} \left[\left(\tilde{\Gamma}^{-1}\right)_{ii}^{1/2}\right]_{\rm ave}, \tag{19}$$

where $N_{\rm int} = 8$ shows the number of effective interferometers. DECIGO/BBO consists of four clusters of triangular detectors (see e.g. Fig. 2 of Ref. [30] for the proposed configurations of DECIGO/BBO) and for simplicity, we assume that all the clusters are placed on the same site. We also use the sky-averaged analysis to see how the new effects that we considered in this paper affect our results.

4. Results

In Fig. 2, we show the measurement accuracies of $\Delta_t z$ using DECIGO/BBO for 5 yr observations. We see that the positivity of $\Delta_t z$ can be marginally detected at redshift around $z \sim 0.5$. Figure 3 is same as Fig. 2 but for 10 yr observations. If this observational period is realized, $\Delta_t z > 0$ can be detected with 3-σ confidence level.

Next, we show the effects of including β and WD confusion noises. Figure 4 shows $\Delta(\Delta_t z)/\Delta_t z$ for the sky-averaged analyses with 5 yr observations where the sky-averaged waveform is given in Eq. (A.12) [1] We see that the spin-orbit coupling β affects the result

[1] We have assumed multiple detectors, but in principle, sky-averaged analysis can be performed with a single detector.

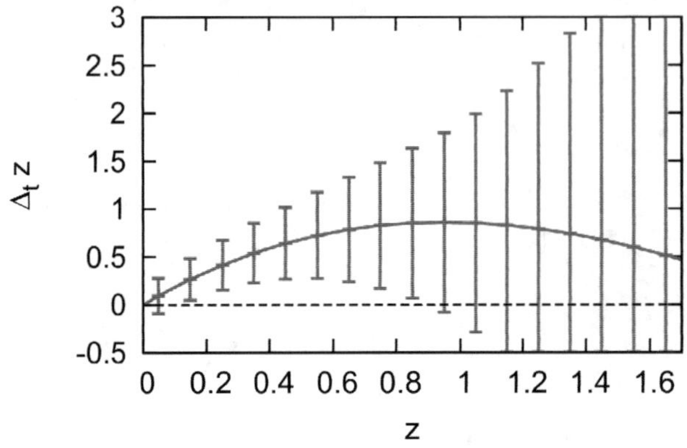

Figure 2. The measurement accuracies of $\Delta_t z$ using DECIGO/BBO for 5 yr observations.

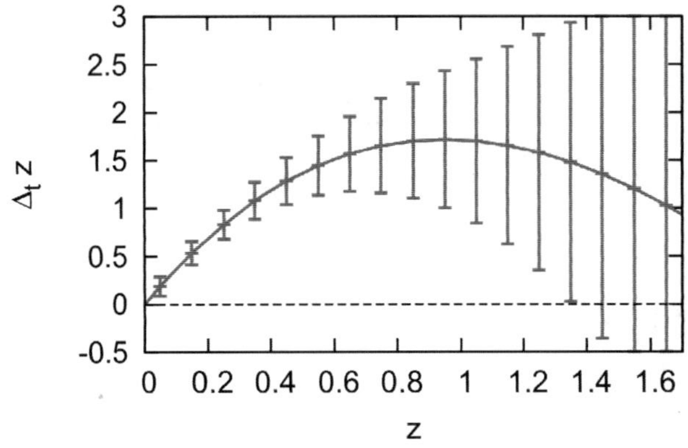

Figure 3. The same as Fig. 2 but the observation period changed to 10 years.

on lower z side. This is because the redshifted mass M_z is lower for lower z source, which leads to smaller orbital velocity v at a given frequency f and larger degeneracies between β and other binary parameters. On the other hand, WD confusion noise affect higher z side. This is because higher z source leads to GW signal with lower frequency where the effect of the confusion noise is larger.

Similarly, let us compare the results using sky-averaged analysis and Monte Carlo simulation. Figure 5 shows the former in solid and the latter in dashed lines. Also, sky-averaged analysis without taking β nor WD confusion is shown in dotted line. By comparing the dashed curve with this dotted one, we see that taking β, WD confusion noise, directions and orientations of sources into account deteriorate the determination accuracies of $\Delta_t z$ by a factor of a few. Since 5 yr observation can only marginally detect the positivity of $\Delta_t z$, this difference is crucial in detecting $\Delta_t z > 0$.

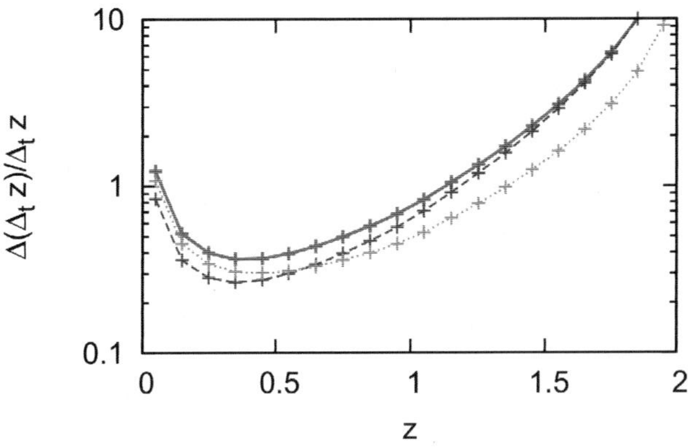

Figure 4. $\Delta(\Delta_t z)/\Delta_t z$ using sky-averaged analysis for 5 yr observations, taking both β and WD confusion noise into account (solid), without β (dashed) or without WD confusion noise (dotted).

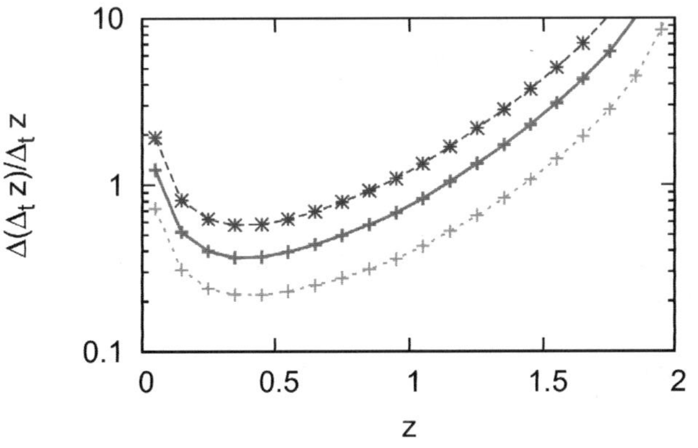

Figure 5. Solid curve is same as the one in Fig. 4 while dotted curve does not include β nor WD confusion noise. Dashed one shows the result with the Monte Carlo simulation.

5. Conclusions and Discussions

In this paper, we have estimated how accurately we can directly measure the cosmic acceleration using future space-borne GW interferometers such as DECIGO and BBO. If we assume that the ΛCDM model is correct, we have shown that we will be able to measure the positivity of the redshift drift with 5-10 yr observations, which enables us to rule out *any* LTB void model with monotonically increasing density profile. Furthermore, since 10 yr observation allows us to detect the positive redshift drift at $z \sim 0.05$, it seems that we can even rule out generic LTB void models unless we allow unrealistic density profile at $z \sim 0$. In order to measure the positivity of $\Delta_t z$, it is more useful to consider the accumulated $\Delta_t z$ [35].

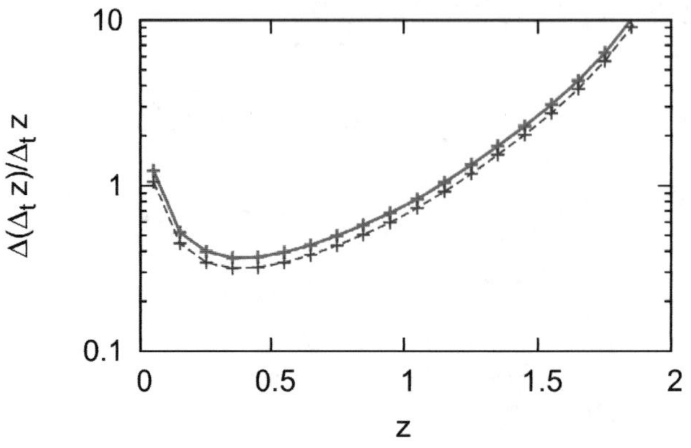

Figure 6. $\Delta(\Delta_t z)/\Delta_t z$ using sky-averaged analysis for 5 yr observations using DECIGO/BBO only (solid), and with joint searches of DECIGO/BBO and ET (dashed).

In this paper, we have assumed that the correct model is ΛCDM but it can be other dark energy model. However, it has been shown that a variety of dark energy models predict the similar values of $\Delta_t z$ to ΛCDM one (see Quartin and Amendola [17] and references therein). Therefore, we emphasize that our results are not restricted to ΛCDM only.

Unfortunately, it seems very difficult to measure $\Delta_t z$ with the ground-based detectors, even with the third-generation ones like ET [36]. (The noise curve is shown as dotted curve in Fig. 1.) The main reasons for this is because it has less number of GW cycles, the event rate is smaller, and the frequency is higher which leads to smaller contribution of "-4PN" effect. However, ET can help improving the measurement accuracy of $\Delta_t z$ when combined with DECIGO/BBO. Figure 6 shows the results comparing the measurement accuracy of $\Delta_t z$ with DECIGO/BBO observation only (solid), and the one obtained by the joint search of DECIGO/BBO *and* ET (dashed.) We see that the errors are slightly reduced for the latter case.

There can be peculiar acceleration of each binary which acts as an additional "noise" when measuring the cosmic acceleration. However, Amendola *et al* [37] have estimated the peculiar accelerations for typical clusters and galaxies, and found that they are almost the same magnitude as the cosmological acceleration. While the peculiar acceleration is different for each binary source, the cosmic acceleration is a universal effect. As such, we can safely neglect the effect of peculiar accelerations as a noise source since the one from the detector overwhelms it. (See also Uzan *et al* [38] for the discussion about the peculiar acceleration.)

In this paper, we have assumed that the binary orbits are circular. Once we include the eccentricity into parameters, this may have large degeneracy with $\Delta_t z$. We need to estimate how the measurement accuracy of $\Delta_t z$ is reduced for the eccentric binaries in future. Also, we note here that we have used the LTB metric as a simple effective model. To be more realistic, we need to use more sophisticated models such as Swiss-Cheese model [39] (see a recent discussion by Célérier [40]). We leave these issues for future work.

Acknowledgments

We thank Takahiro Tanaka, Naoki Seto, Takashi Nakamura and Daniel Holz for having discussions and giving us valuable comments. KY is supported by the Japan Society for the Promotion of Science (JSPS) grant No. 22 · 900. AN and CY is supported by a Grant-in-Aid through JSPS. This work is also supported in part by the Grant-in-Aid for the Global COE Program "The Next Generation of Physics, Spun from Universality and Emergence" from the Ministry of Education, Culture, Sports, Science and Technology (MEXT) of Japan.

Appendix A. Binary gravitational waveform without the effect of cosmic acceleration

The restricted binary gravitational waveform in Fourier space without cosmic acceleration is given as [31, 41]

$$\tilde{h}(f)|_{\text{no accel}} = \frac{\sqrt{3}}{2}\mathcal{A}f^{-7/6}e^{i\Psi(f)}\left[\frac{5}{4}A_{\text{pol},\alpha}(t(f))\right]e^{-i(\varphi_{\text{pol},\alpha}+\varphi_D)}, \tag{A.1}$$

where the amplitude $\mathcal{A}$ is given as $\mathcal{A} = \frac{1}{\sqrt{30}\pi^{2/3}}\frac{\mathcal{M}_z^{5/6}}{D_L}$ and the phase up to second Post-Newtonian (PN) order can be expressed as [21, 31]

$$\begin{aligned}\Psi(f) &= 2\pi f t_c - \phi_c - \frac{\pi}{4} + \Psi_N(f)\times\left[1+\left(\frac{3715}{756}+\frac{55}{9}\eta\right)x - 4(4\pi-\beta)x^{3/2}\right.\\ &\left.+\left(\frac{15293365}{508032}+\frac{27145}{504}\eta+\frac{3085}{72}\eta^2\right)x^2\right].\end{aligned} \tag{A.2}$$

Here, ϕ_c is the coalescence phase and $A_{\text{pol},\alpha}$, $\varphi_{\text{pol},\alpha}$ and φ_D are given as [29, 31]

$$A_{\text{pol},\alpha}(t) = \sqrt{(1+(\hat{\boldsymbol{L}}\cdot\hat{\boldsymbol{N}})^2)^2F_\alpha^+(t)^2+4(\hat{\boldsymbol{L}}\cdot\hat{\boldsymbol{N}})^2F_\alpha^\times(t)^2}, \tag{A.3}$$

$$\cos(\varphi_{\text{pol},\alpha}(t)) = \frac{(1+(\hat{\boldsymbol{L}}\cdot\hat{\boldsymbol{N}})^2)F_\alpha^+(t)}{A_{\text{pol},\alpha}(t)}, \tag{A.4}$$

$$\sin(\varphi_{\text{pol},\alpha}(t)) = \frac{2(\hat{\boldsymbol{L}}\cdot\hat{\boldsymbol{N}})F_\alpha^\times(t)}{A_{\text{pol},\alpha}(t)}, \tag{A.5}$$

$$\varphi_D(t) = 2\pi f(t)R\sin\bar{\theta}_S\cos[2\pi t/T-\bar{\phi}_S], \tag{A.6}$$

where $\hat{\boldsymbol{L}}$ is the unit vector parallel to the orbital angular momentum, $\hat{\boldsymbol{N}}$ is the unit vector pointing towards the center of mass of the binary, T=1yr and R=1AU. F_α^+ and $F_\alpha^\times$ are the beam pattern functions defined as

$$F_{\text{I}}^+(\theta_S,\phi_S,\psi_S) = \frac{1}{2}(1+\cos^2\theta_S)\cos(2\phi_S)\cos(2\psi_S)-\cos(\theta_S)\sin(2\phi_S)\sin(2\psi_S), \tag{A.7}$$

$$F_{\text{I}}^\times(\theta_S,\phi_S,\psi_S) = \frac{1}{2}(1+\cos^2\theta_S)\cos(2\phi_S)\sin(2\psi_S)+\cos(\theta_S)\sin(2\phi_S)\cos(2\psi_S), \tag{A.8}$$

$$F_{\text{II}}^+(\theta_S,\phi_S,\psi_S) = F_{\text{I}}^+(\theta_S,\phi_S-\pi/4,\psi_S), \tag{A.9}$$

$$F_{\text{II}}^\times(\theta_S,\phi_S,\psi_S) = F_{\text{I}}^\times(\theta_S,\phi_S-\pi/4,\psi_S), \tag{A.10}$$

where the direction (θ_S,ϕ_S) is measured in the detector's frame and ψ_S is the polarization angle. These angles and $\hat{\boldsymbol{L}}\cdot\hat{\boldsymbol{N}}$ are related to the angles $(\bar{\theta}_S,\bar{\phi}_S,\bar{\theta}_L,\bar{\phi}_L)$ measured from the center of mass of the solar system as explained in Appendix A of Ref. [29]. We have neglected the spin-spin coupling at 2PN order since it has been shown that this effect is negligible for NS/NS binaries [32]. $t(f)$ in Eq. (A.1) up to 2PN order is given as

$$\begin{aligned}t(f) &= t_c - t_N(f)\left[1+\frac{4}{3}\left(\frac{743}{336}+\frac{11}{4}\eta\right)x-\frac{8}{5}(4\pi-\beta)x^{3/2}\right.\\ &\left.+2\left(\frac{3058673}{1016064}+\frac{5429}{1008}\eta+\frac{617}{144}\eta^2\right)x^2\right],\end{aligned} \tag{A.11}$$

with $t_N(f) \equiv (5/256)\mathcal{M}_z(\pi\mathcal{M}_z f)^{-8/3}$ [2]. When we take the average over the direction and orientation of sources, Eq. (A.1) just reduces to

$$\tilde{h}(f)|^{\rm (sky-averaged)}_{\rm no\ accel} = \frac{\sqrt{3}}{2}\mathcal{A}f^{-7/6}e^{i\Psi(f)}. \tag{A.12}$$

References

[1] Riess A G *et al.* (Supernova Search Team) 2004 *Astrophys. J.* **607** 665–687
[2] Alnes H, Amarzguioui M and Gron O 2006 *Phys. Rev.* **D73** 083519
[3] Yoo C M, Nakao K and Sasaki M 2010 *JCAP* **1007** 012
[4] Yoo C M, Nakao K and Sasaki M 2010 *JCAP* **1010** 011
[5] Garcia-Bellido J and Haugboelle T 2008 *JCAP* **0804** 003
[6] Garcia-Bellido J and Haugboelle T 2008 *JCAP* **0809** 016
[7] Garcia-Bellido J and Haugboelle T 2009 *JCAP* **0909** 028
[8] Bolejko K and Wyithe J S B 2009 *JCAP* **0902** 020
[9] Caldwell R R and Stebbins A 2008 *Phys. Rev. Lett.* **100** 191302
[10] Clarkson C and Regis M 2011 *JCAP* **1102** 013
[11] Zhang P and Stebbins A 2011 *Phys. Rev. Lett.* **107** 041301
[12] Zibin J P, Moss A and Scott D 2008 *Phys. Rev. Lett.* **101** 251303 (*Preprint* 0809.3761)
[13] Moss A, Zibin J P and Scott D 2011 *Phys. Rev.* **D83** 103515
[14] Nadathur S and Sarkar S 2011 *Phys. Rev.* **D83** 063506
[15] Yoo C M, Kai T and Nakao K 2008 *Prog. Theor. Phys.* **120** 937–960
[16] Uzan J P, Clarkson C and Ellis G F R 2008 *Phys. Rev. Lett.* **100** 191303
[17] Quartin M and Amendola L 2010 *Phys. Rev.* **D81** 043522
[18] Yoo C M, Kai T and Nakao K 2011 *Phys. Rev.* **D83** 043527
[19] Loeb A 1998 *Astrophys. J.* **499** L111–L114
[20] Liske J *et al.* 2008 *Mon. Not. Roy. Astron. Soc.* **386** 1192–1218
[21] Seto N, Kawamura S and Nakamura T 2001 *Phys. Rev. Lett.* **87** 221103
[22] Kawamura S *et al.* 2011 *Class. Quant. Grav.* **28** 094011
[23] Phinney E S *et al.* 2003 *Big Bang Obsever Mission Concept Study* (NASA)
[24] Schutz B F 1986 *Nature* **323** 310–311
[25] Nishizawa A, Taruya A and Kawamura S 2010 *Phys. Rev.* **D81** 104043 (*Preprint* 0911.0525)
[26] Nishizawa A, Yagi K, Taruya A and Tanaka T 2012 *Phys. Rev.* **D85** 044047 (*Preprint* 1110.2865)
[27] Takahashi R and Nakamura T 2005 *Prog. Theor. Phys.* **113** 63–71
[28] Cutler C and Flanagan E E 1994 *Phys. Rev.* **D49** 2658–2697 (*Preprint* gr-qc/9402014)
[29] Yagi K and Tanaka T 2010 *Phys. Rev.* **D81** 064008
[30] Yagi K, Tanahashi N and Tanaka T 2011 *Phys. Rev.* **D83** 084036
[31] Berti E, Buonanno A and Will C M 2005 *Class. Quant. Grav.* **22** S943–S954
[32] Cutler C and Harms J 2006 *Phys. Rev.* **D73** 042001
[33] Abadie J *et al.* (LIGO Scientific) 2010 *Class. Quant. Grav.* **27** 173001
[34] Schneider R, Ferrari V, Matarrese S and Portegies Zwart S F 2001 *Mon. Not. Roy. Astron. Soc.* **324** 797
[35] Yagi K, Nishizawa A and Yoo C M 2011 (*Preprint* 1112.6040)
[36] Einstein Telescope www.et-gw.eu
[37] Amendola L, Balbi A and Quercellini C 2008 *Phys. Lett.* **B660** 81–86
[38] Uzan J P, Bernardeau F and Mellier Y 2008 *Phys. Rev.* **D77** 021301
[39] Brouzakis N, Tetradis N and Tzavara E 2007 *JCAP* **0702** 013
[40] Celerier M N 2011 (*Preprint* arXiv:1108.1373)
[41] Cutler C 1998 *Phys. Rev.* **D57** 7089–7102

[2] When we consider the effect of cosmic acceleration, there appears a leading "-4PN" correction in $t(f)$, but we have checked that it does not affect our results.

Computer-games for gravitational wave science outreach: *Black Hole Pong* and *Space Time Quest*

L. Carbone, C. Bond, D. Brown, F. Brückner, K. Grover, D. Lodhia, C.M.F. Mingarelli, P. Fulda, R.J.E. Smith, R. Unwin, A. Vecchio, M. Wang, L. Whalley and A. Freise

School of Physics and Astronomy, University of Birmingham
Edgbaston - Birmingham, B15 2TT, United Kingdom

E-mail: lc@star.sr.bham.ac.uk

Abstract. We have established a program aimed at developing computer applications and web applets to be used for educational purposes as well as gravitational wave outreach activities. These applications and applets teach gravitational wave physics and technology. The computer programs are generated in collaboration with undergraduates and summer students as part of our teaching activities, and are freely distributed on a dedicated website. As part of this program, we have developed two computer-games related to gravitational wave science: 'Black Hole Pong' and 'Space Time Quest'. In this article we present an overview of our computer related outreach activities and discuss the games and their educational aspects, and report on some positive feedback received.

1. Introduction

The next generation of gravitational wave (GW) detectors - most notably Advanced LIGO [1], GEO-HF [2] and Advanced VIRGO [3] - are expected to achieve the first direct detection of GWs during 2015, when Advanced LIGO begins operation or soon thereafter, giving life to the new field of the GW astronomy. The impact on cosmology, fundamental physics, and of course on conventional astronomy itself will be enormous; the interest in GW observation is therefore growing rapidly in the scientific community. But while the public interest in cosmology and relativity is high, public knowledge of the principles behind GW detection is still rather limited, and even the existence of the large GW observatories is little known by the more general public, compared to larger experimental facilities such as, for instance, the Large Hadron Collider [4]. There is a clear need to better inform and inspire the general public and prospective students in GW astronomy and related sciences. Within the LIGO Scientific Collaboration (LSC) the 'Education and Public Outreach' (EPO) group aims to combine ideas and approaches across the collaboration to successfully communicate the vision and benefits of GW observation throughout the world.

To contribute to these international efforts in the promotion of GW astronomy, at the University of Birmingham we have established a unique program aimed to the development of small educational computer applications that can be used to illustrate the basics of GW science and GW detector technology in a playful but informative way. The aim is to present GW science to younger generations within one of the environments they are more familiar with, i.e.

computer-games, and to exploit the communication channels that the *new technologies* offer to possibly get an even larger international audience in contact with GW science.

This activity led us to the successful development of a number of interactive computer applets describing a variety of concepts connected to GW science and, eventually, to the realisation of two full-scale computer-games based on the subjects of gravity and GWs. In this article we overview our computer related outreach activity and present and discuss our GW related games, Black Hole Pong and Space Time Quest.

2. 'Processing' programs for science outreach

The idea of developing small computer applications for educational purposes builds upon the need to introduce new students to the world of computer programming and modelling of physical systems in a manner which is enjoyable. This will not only help them learn successfully, but also engages them with GW research. During the initial 'induction' phase, undergraduate or summer-students involved in research projects with our group are encouraged to generate a small computer program on a GW subject of their interest, which is then developed, as a conventional student-project, by the student with the supervision of more senior members of the group.

The small computer programs, called 'sketches', are developed using the open-source programming environment Processing [5, 6], originally developed at the MIT Media Lab in 2001 as a software prototyping environment and to teach fundamentals of computer programming within a graphical context. Processing has eventually reached a wide audience and is now largely used in many professional communities, such as designers, artists, and architects to create graphical applications, animations, interactive tools and for visual arts in general. Processing offers an intuitive approach to programming for the beginner or an efficient sketchbook for rapid prototyping by experienced programmers. Thus Processing allows students with very different computing-backgrounds to collaborate and to successfully produce graphically impressive sketches in a relatively short period of time.

The successful sketches are published online on our outreach website gwoptics.org [7], on individual webpages where the student-author can provide instructions and a short description of the physics illustrated in the sketch. The program is either embedded within the HTML code to run as an applet in the webpage itself or, where more appropriate, distributed for download as an application to install and run on the computer of the interested person.

The collection of sketches developed so far covers a wide spectrum of subjects related to GW science and technologies. The programs range from illustrating the most fundamental properties of GWs, for example the deformation of space-time produced by a propagating GW or the characteristic 'sound' of the GW signal of colliding black holes, to the illustration of the vital technologies and phenomena used in GW detectors such as lasers, vibration isolation systems or interference of light. Different combinations of these sketches have been successfully used as interactive tools during seminars about GW detection and during more general lectures in schools and universities, and the sketches webpages are regularly consulted online by people interested in learning about the specific subject.

3. The gravitational wave games

The positive feedback we received about the different interactive sketches has encouraged us to realise two properly defined computer-games based on subjects related to gravity and GW science. The motivation behind the development of the two particular games was as follows: in one case, the aim was to produce an intuitive and graphically attractive computer-game that could engage and entertain children and teenagers during science exhibitions; in the other, the goal was vice-versa to develop an interactive element that could function as an engaging and playful supplement for illustrating and explaining the secrets of GW detectors and their

Figure 1. Screen-shots of the Black Hole Pong game. The image on the left shows the start-up screen of the game and the right snapshot has been taken mid-game, showing one of the astronomical background images, the two 'black holes' controlled by the players and several bright disks as the stars currently in play.

technology to a more educated audience, such as high-school students onwards. The two games, Black Hole Pong and Space Time Quest, are presented and discussed in this section.

3.1. Black Hole Pong

Black Hole Pong (BHP) is a new arcade-style game with a reference to one of the very first computer-games, *Pong* [8]. Pong involved two players, each one controlling a paddle which they would move vertically up and down in order to bounce a ball back towards their opponent: each time the ball touched the opponent's far edge of the screen, the player would score a point. In BHP the idea of the split-screen has been kept. However each player controls a black hole which can move horizontally as well as vertically, and the objective is to make use of the gravitational potential of the black hole to fling free roaming stars towards the other player.

BHP has been designed as a simple, fun game for people of all ages. At the same time, BHP features several educational elements that make it a fascinating tool for teaching, learning and discovering new physical concepts. For example, by learning how to manoeuvre the incoming stars only using the black hole's gravitational potential, the player develops an intuitive

Figure 2. Children, students and adults enjoying the BHP game at the University of Birmingham Community Day 2011 (left), the British Science Festival 2010 (center) and the 9th Amaldi Conference (right).

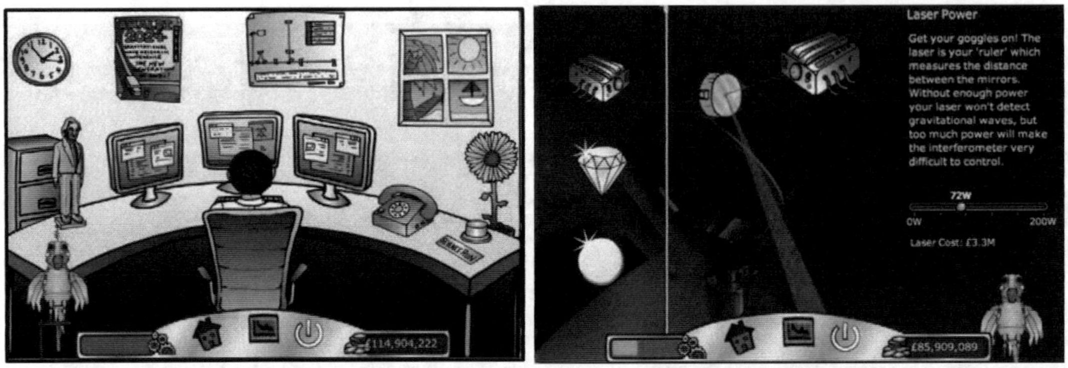

Figure 3. Screenshots of STQ. On the left, the 'PI's office' is the game's hub from which the player can access all the detector's subsystems. On the right an image of the 'Optics' subsystem screen.

understanding of concepts such as gravitational attraction, orbital mechanics and gravitational slingshot effect. The background graphics of the game is a slideshow containing images of the night sky as well as astronomical objects taken by both amateur astronomers and large ground/space based telescopes. Furthermore, several astrophysical phenomena are graphically featured in the game, such as 'worm holes', 'star capture' and 'gravitational lensing', adding to the overall simple but attractive graphics, see Fig. 1. When used as part of an exhibition, science fair or similar activity we have supported the arcade-style attraction by state-of-the art hardware: the game is designed to be controlled with the well-known Microsoft Xbox controllers and whenever possible we use large Apple iMac computers to run the game, as shown in Fig. 2.

3.2. Space Time Quest

Space Time Quest (STQ), here shown with screenshots in Fig. 3 and Fig. 4, is a manager-simulation type game: the player is the 'Principal Investigator' (PI) of a future ground-based GW interferometer who has the goal to design the most sensitive GW detector. The PI is assigned a limited budget that he has to wisely distribute between the different detector's subsystems to tweak the instrument parameters and to achieve the best sensitivity. Once the player is satisfied with their design, they can operate the detector in 'Science-Run' mode and see how well it performs: the final score is determined by how far in the universe the detector itself can explore, based on the achieved sensitivity curve, and it is recorded in the web-based STQ 'hall-of-fame'.

STQ is complemented by the 'Gravitational Waves E-Book' [9], effectively a collection of webpages with short introductions to a number of topics relevant to GW science. The E-Book offers a description of the main instrument subsystems that comprise the detector and illustrates how each noise source relates to the different subsystem parameters that the player chooses. The E-Book text is purposely aimed to the general person who has interested in GW science, and as such it is written with a simple style to make it accessible to a broad public of all ages and backgrounds. The E-Book is also independently available online as more general reading material on GW science, and is now also offered in multiple languages.

STQ contains many educational merits for the public. First of all, the game showcases the physics behind a real GW detector and it presents all the main subsystems that comprise it. Furthermore, it illustrates all the most important noise sources which can limit the detector's sensitivity. By reading the E-Book and by looking at the changes in the sensitivity curve, the player can see how each subsystem is affected by the different noise sources, and discover

Figure 4. Left: a screen-shot of the 'sensitivity curve' in the STQ game. Right: school students playing STQ with a demonstrator during an exhibition at University of Birmingham.

some of the ways in which physicists try to reduce the noise sources in the detector. The game also presents some of the typical challenges that physicists face when designing a real physics experiment, such as making trade-offs between performances of different interlinked subsystem parameters and managing the available resources wisely. Finally, STQ features images of astronomical objects in the background, similar to BHP. Furthermore the STQ graphics are based on photographs of components from real GW detectors complemented with realistic cartoons of the detector parts, offering a realistic picture of how a GW interferometer looks like to the user.

STQ is mainly targeted for science teachers, A-level, Higher and Advanced Higher science students and is mainly suited for use in science fairs and exhibitions and initially played with the help of demonstrators. However, STQ has also proved to be an entertaining tool to teach the basics of GW science also to beginners in GW research-projects and PhD schools, or to attract prospective research students towards the GW field.

4. Use in exhibitions and distribution

Early prototypes of the games have been used for the first time during the exhibition about GW science 'Looking for Black Holes with Lasers', organised by the Birmingham GW Group within the 'British Science Festival', held in Birmingham in September 2010 [11]. The very positive feedback collected with the games during this first exhibition gained us the attention of our university and of local schools and associations. This led to our displays being routinely used in University Open/Admission days and in our university's outreach events, e.g. the University

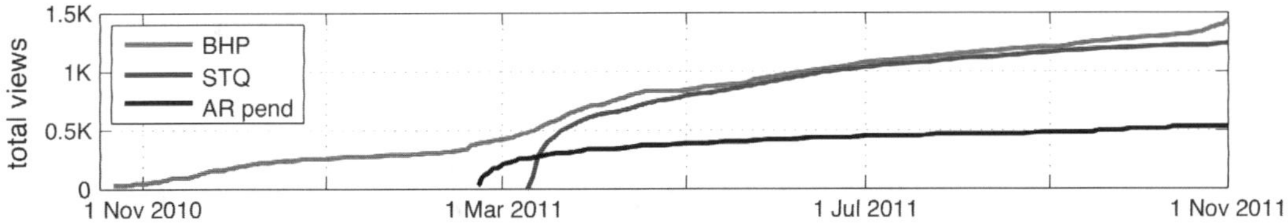

Figure 5. Total number of unique views of the online video tutorials for the Black Hole Pong and Space Time Quest games and for one of our processing sketches, the 'Augmented Reality Pendulum'. Data from youtube.com.

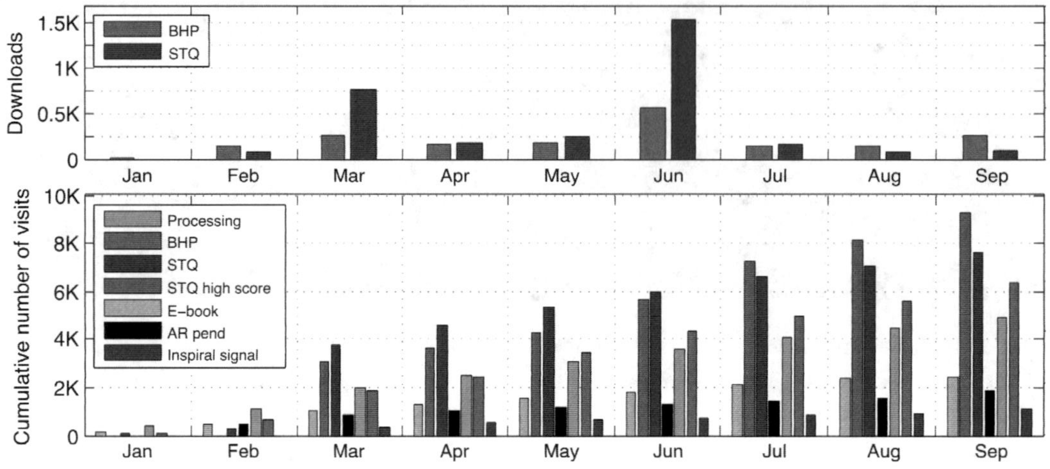

Figure 6. Top panel: number of download of the Black Hole Pong and Space Time Quest games over the year 2011. Bottom panel, cumulative number of visits during 2011 to some of the gwoptics.org outreach pages, such as the main page collecting all Processing programs, the games BHP and STQ, the STQ high-score 'hall of fame', the E-Book and the pages of two other Processing sketches, the 'Augmented Reality Pendulum' and the 'Inspiral signal'.

Community Day 2011 [12], and gained members of our group invitations to visit schools and give public seminars, where the games were used to complement the seminar. At the same time, positive feedback has been received from school teachers concerning our other Processing sketches which have been used as support material in physics lectures and during science activities.

Since their official release, BHP and STQ are freely distributed on our website gwoptics.org and on the outreach pages of the ligo.org website [13], which hosts links and multimedia material of interest for the EPO group. The launch of the two games was also announced online via social-media networks and with online videos, with the main aim of raising both the profile of the games and its visibility. Indicative figures of merit on how positive this campaign has been can be inferred from the total number of download of the two games, the entries in the high score 'hall of fame', the unique views of the online video-tutorials, and more generally from the number of visits to the webpages presenting our online material. Examples of such data are presented in Fig. 5 and Fig. 6. The response is so far very positive and seem to indicate a slow but constant growth of new contacts and an increasing interest in the products themselves. In particular, sensible increments in the number of contacts can be successfully correlated with our contribution at science events and exhibitions, with release of new outreach material and with our communication campaign via online social networks.

STQ and BHP are also accompanied by short questionnaires, handed-out at exhibitions and during visits to schools as well as online, which are distributed to collect anonymous feedback amongst the users, targeting in particular teacher's and student's categories. The aim of the questionnaires is primarily to evaluate the success of the games among the users and in particular about the effectiveness of their educational aspects. As a further step in the future, the goal is to develop a proper analysis of the feedback provided in the questionnaires that will allow us to better link the games to specific elements of education, such as formal education, and to improve their integration within the syllabus.

5. Conclusion and future activities

Our program aimed at the development of small computer applications for educational purposes. This has been successful, with the realisation of several interactive sketches and two full scale games related to GW science and technology. Thanks to our participation at popular science events, and to our online presence and communication campaign, the sketches and the games are now becoming popular within schools and science associations and, as shown by the response gathered from the online audience, the prospective feedback looks promising for the future. In particular, BHP and STQ have allowed us to significantly increase the visibility of our online presence and as such also of GW science within the general public, as well as within the local scientific community.

Encouraged by these positive results, we will continue our computer related outreach activity in the next years, and we plan to realise new Processing sketches for outreach in the near future. In parallel, BHP and STQ will be treated as running projects. We will take advantage of the feedback and advice from teachers, students and other users to make further modifications and improvements to both games. Furthermore, we will continue presenting BHP, STQ and our other interactive sketches during visits to local schools and in popular science events. We hope to increase and improve our online communication campaign on GW subjects to make GW detectors more and more popular among the general public and to gain GW science the largest audience possible.

Acknowledgments

We are very grateful to the Processing community for all the online examples, code libraries and online forums that helped us in the development of our sketches and games and without whom most of this work would not have been possible. We thank the astronomers who allowed us to use their impressive photographs of the night sky as background images for the BHP and STQ games (see the credits page in the games for the details) and we are thankful to the GEO600 collaboration, for providing images of components of the detector used in the STQ game and for extensive beta-testing of the initial game prototype. This document has been assigned the LIGO Laboratory document number LIGO-P1100145.

References

[1] G.M. Harry and the LIGO Scientific Collaboration, 2010 Class. Quantum Grav. 27 084006
[2] B. Willke et al, Class. Quantum Grav., 23, S207, 2006
[3] FAcernese et al, Class. Quantum Grav. 25, 184001, 2008
[4] `http://lhc.web.cern.ch/`
[5] C. Reas and B. Fry, *Processing: a programming handbook for visual designers and artists* (MIT Press Ltd, Boston, MA - USA , 2007), ISBN 10: 0262182629
[6] `http://www.processing.org`
[7] `http://www.gwoptics.org/processing`
[8] `http://en.wikipedia.org/wiki/Pong` and `http://www.ponggame.org`
[9] `http://www.gwoptics.org/ebook`
[10] `http://www.youtube.com/gwoptics`
[11] `http://www.britishscienceassociation.org/web/BritishScienceFestival/`
[12] `http://www.birmingham.ac.uk/community/communityday`
[13] `http://www.ligo.org`